Push your Career Publish your Thesis

Wissenschaft sollte allen zugänglich sein. Teilen Sie Ihr
Wissen, Ihre Ideen und Ihre Leidenschaft für die Forschung.
Veröffentlichen Sie Ihre Bachelor- und Masterarbeit sowie Ihre
Dissertation und Habilitationsschrift bei Infinite Science.

www.publishing.infinite-science.de

Dissertationsreihen der Universität zu Lübeck

Institut für
Biomedizinische Optik

Institut für
Medizintechnik

Institut für
Medizinische Informatik

Klinik für Orthopädie
und Unfallchirurgie

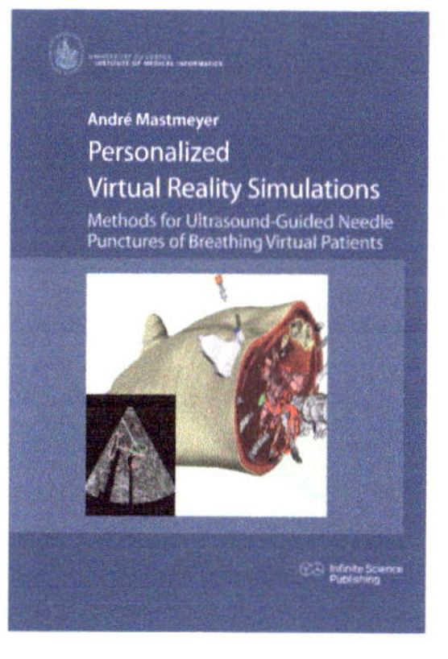

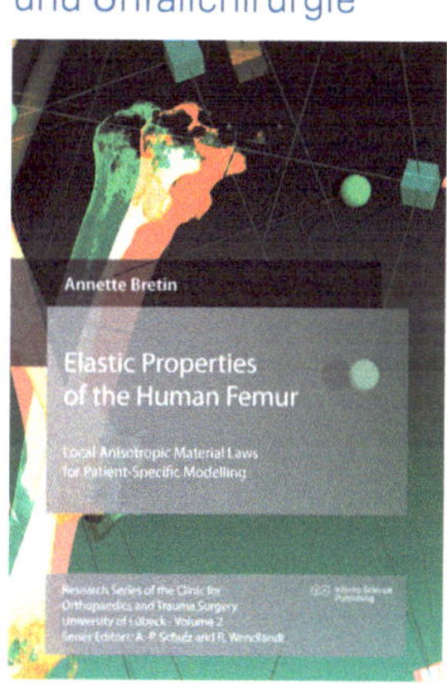

Imprint of Infinite Science GmbH,
Technikzentrum | MFC 1
Maria-Goeppert-Straße 1
23562 Lübeck, Germany

Cover Design and Illustration: Uli Schmidts, metonym
Editorial and Copy Editing: University of Lübeck

Publisher: Infinite Science GmbH, Lübeck, www.infinite-science.de
Printed in Germany, BoD, Norderstedt

ISBN: 978-3-945954-62-1

Bibliografische Information der Deutschen Nationalbibliothek:
Die Deutsche Nationalbibliothek verzeichnet diese Publikation in der Deutschen Nationalbibliografie; detaillierte bibliografische Daten sind im Internet über http://dnb.d-nb.de abrufbar.

Student Conference Proceedings 2020

9th Conference on Medical Engineering Science
5th Conference on Medical Informatics
3rd Conference on Biomedical Engineering
2nd Conference on Auditory Technology

Lübeck, March 3-5, 2020

Editors in Chief

T. M. Buzug, H. Handels, S. Klein, A. Mertins

Associate Editors

C. Debbeler, K. Gräfe, J.-H. Wrage, S. Venker, A. Möller

Editors

E. Barth, R. Birngruber, M. Brandenburger, G. Buntrock, N. Bunzeck, H. Busch, T. M. Buzug, C. Damiani, F. Ernst, H. Gehring, M. Grzegorzek, T. Gutsmann, H. Handels, M. Heinrich, H. Hellbrück, R. Huber, G. Hüttmann, J. Ingenerf, T. Jürgens, S. Karpf, S. Klein, D. Lühmann, A. Mastmeyer, S. Meier, A. Mertins, S. Müller, J. Obleser, Ö. Özçep, H. Paulsen, M. Rafecas, F. Reinholz, P. Rostalski, E. Rückert, G. Schildbach, A.-P. Schulz, J. Schröter, Y.-H. Song, F. Spitzenberger, J. Tchorz, R. Wendlandt

Exhibitors

Studenten und Absolventen gesucht!

VisiConsult ist führend als Entwickler und Hersteller von Röntgenlösungen und ist ein Inhabergeführtes Familienunternehmen mit flachen Hierarchien, daher ist unser Arbeitsalltag sehr dynamisch. Unser engagiertes, offenes und internationales Team ist ein Grund für das starke Wachstum unseres Unternehmens.

Was wir bieten

- Berufserfahrung in einem internationalen und innovativen Unternehmen
- Praxisrelevanten Thema
- Flexible und familienfreundliche Arbeitszeiten
- Aufstiegs- und Entwicklungsmöglichkeiten
- Großzügige und moderne Arbeitsplätze und Aufenthaltsräume

Studierende bieten wir Nebentätigkeiten und Abschlussarbeiten in den Bereichen

- Applikation (m/w/d)
- Informatik (m/w/d)
- **Software Support** (m/w/d)

Absolventen bieten wir folgende Stellenangebote

- Projektleiter (m/w/d)
- Technischen Redakteur (m/w/d)
- HR-Assistenz (m/w/d)
- Elektroingenieur
- Abteilungsleitung Mechanische Konstruktion (m/w/d)
- Fachinformatiker Anwendungsentwicklung (m/w/d)
- Abteilungsleiter Automatisierung (m/w/d)
- Informatiker / Softwareentwickler (m/w/d)
- Roboterprogrammierer (m/w/d)

Bewerbung (inkl. Motivation, Lebenslauf und Arbeitszeugnissen) und Gehaltsvorstellung an communication@visiconsult.de. Wir freuen uns darauf, Sie kennen zu lernen!

PHILIPS
Careers
Don't just
make a living
Make a
difference
philips.to/campus_de

Dräger

Warum
Praktikanten bei
Dräger Kickboxen lernen?

Lingxu Lin
Praktikant
Forschung und Entwicklung

Weil wir im unternehmensweiten Kickbox-Innovationsprogramm viel Freiraum zum Ausprobieren bekommen und unsere Ideen wirklich zählen. Mitarbeiter aus verschiedenen Abteilungen entwickeln mit diesem Programm innovative Ideen. Wir können richtig kreativ sein, teamübergreifend zusammenarbeiten und uns gegenseitig inspirieren. Ich konnte mit der Kickbox eine Maske entwickeln, die Beatmung angenehmer für Patienten macht. Ein tolles Projekt und eben typisch Dräger: Denn Leben schützen, unterstützen und retten sind die Ziele, die uns hier alle miteinander verbinden. Findet heraus, wie gut das zu Euren persönlichen Zielen passt. www.draeger.com/karriere

Dräger. Technik für das Leben®

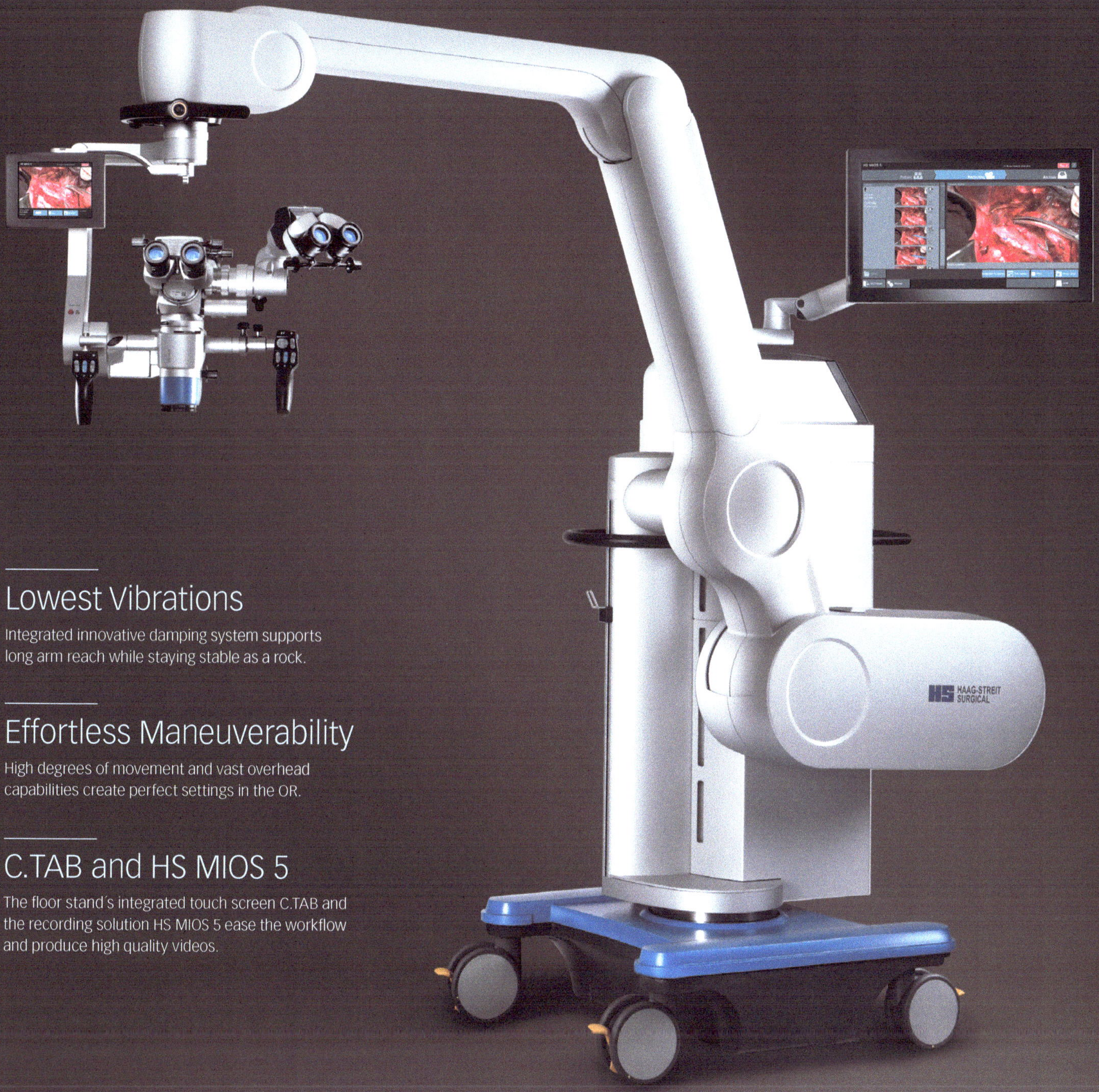

Lowest Vibrations

Integrated innovative damping system supports
long arm reach while staying stable as a rock.

Effortless Maneuverability

High degrees of movement and vast overhead
capabilities create perfect settings in the OR.

C.TAB and HS MIOS 5

The floor stand's integrated touch screen C.TAB and
the recording solution HS MIOS 5 ease the workflow
and produce high quality videos.

HS 5-1000
Maneuverability. Stability. Reach.
The Ultimate Surgical Experience.

Sponsor

Angebote der IHK zu Lübeck für Studierende

Erst das Studium und dann... Über den besten Studienplatz, interessante Praktikumsplätze, Finanzierung und Förderung sowie die Möglichkeiten, sich selbstständig zu machen und von gestandenen Unternehmerinnen und Unternehmern zu lernen, unter www.Mein-Unternehmen-Zukunft.de

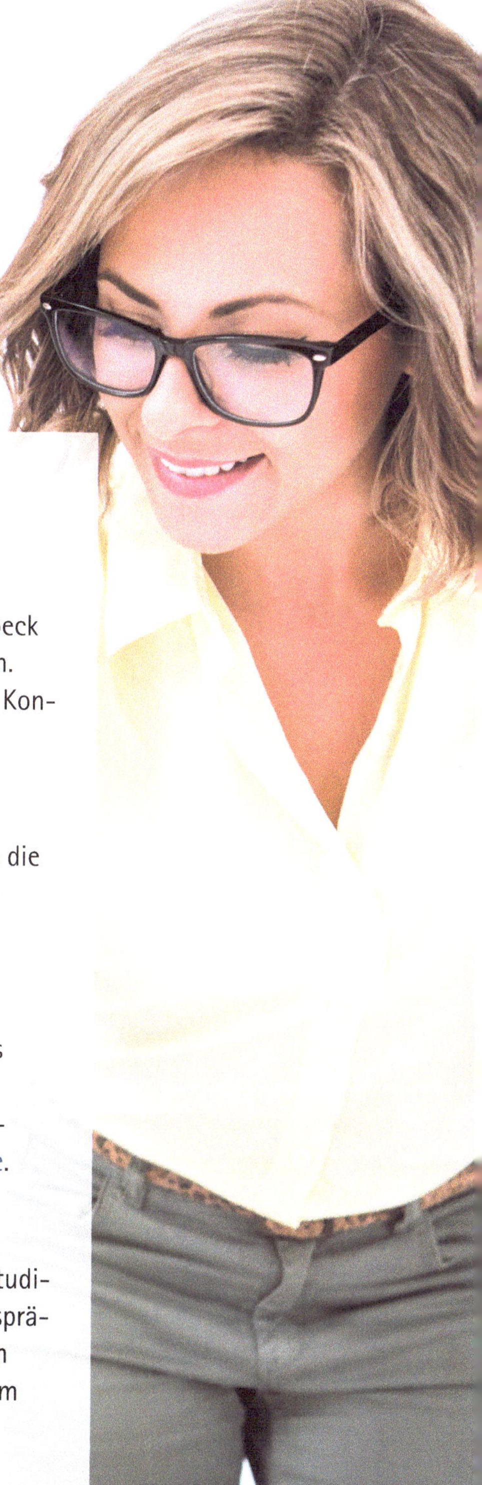

Best of ...

Karrieretag

Der Karrieretag ist eine gemeinsame Veranstaltung der Universität zu Lübeck, der Fachhochschule Lübeck, der BioMedTec Management GmbH sowie der IHK zu Lübeck und richtet sich an Studentinnen und Studenten, Absolventinnen und Absolventen. Beim Karrieretag stellen Sie die Weichen für Ihre berufliche Zukunft und knüpfen Kontakte zu zukünftigen Arbeitgebern. Mehr unter: www.ihk-sh.de/karrieretag

Praktikumsbörse

Die IHK-Praktikumsbörse www.praktikum-sh.de bietet Schülern und Studierenden die kostenlose Möglichkeit, Praktikumsplätze bei Unternehmen in Schleswig-Holstein zu finden.

Beratung StudiLe

Das Studium mit integrierter Lehre (StudiLe) ist ein duales Studienmodell, welches eine betriebliche Ausbildung mit einem Bachelorstudium an der Fachhochschule Lübeck verbindet. In circa vier Jahren können somit zwei berufsqualifizierende Abschlüsse erworben werden. Nähere Informationen finden Sie unter www.studile.de.

Informationen für Studienabbrecher

Die IHK zu Lübeck engagiert sich im Netzwerk „Zweifel am Studium?" und steht Studienabbrecherinnen und Studienabbrechern für individuelle berufliche Beratungsgespräche zur Verfügung. Eine erste Orientierungsberatung zur Studiensituation und dem persönlichen Profil bietet die Handwerkskammer Lübeck im Studentenwerk auf dem Campus der Lübecker Hochschulen an: www.kursaenderung-ins-handwerk.de

Existenzgründung und Unternehmensförderung

Ob Sie eine Firma gründen, die Wettbewerbsfähigkeit Ihres Unternehmens für die Zukunft sichern oder die Nachfolge regeln möchten – wir stehen von Anfang an an Ihrer Seite und begleiten Ihr Unternehmen von der Gründung bis zur Übergabe. Mehr zu finden unter: www.ihk-sh.de/basisinfos

Speziell für Studierende bieten wir Beratungstage zur Existenzgründung auf dem Campus und eine individuell zugeschnittene Finanzierungs- und Förderberatung.

Student Conference Proceedings 2020

Scientific Program Committee

Barth, Prof. Dr. Erhardt	Institute for Neuro- and Bioinformatics, Universität zu Lübeck
Birngruber, Prof. Dr. Reginald	Institute of Biomedical Optics, Universität zu Lübeck
Brandenburger, Dr. Matthias	Fraunhofer Research Institution for Marine Biotechnology and Cell Technology, Lübeck
Buntrock, PD Dr. Gerhard	Institute for Software Engineering and Programming Languages, Universität zu Lübeck
Bunzeck, Prof. Dr. Nico	Institute of Psychology I, Universität zu Lübeck
Busch, Prof. Dr. Hauke	Lübeck Institute of Experimental Dermatology, Universität zu Lübeck
Buzug, Prof. Dr. Thorsten M.	Institute of Medical Engineering, Universität zu Lübeck
Damiani, Dr. Christian	Medical Sensors and Devices Laboratory, Lübeck University of Applied Sciences
Ernst, Prof. Dr. Floris	Institute for Robotics and Cognitive Systems, Universität zu Lübeck
Gehring, Prof. Dr. Hartmut	Department of Anesthesiology, University Medical Center Schleswig-Holstein, Lübeck
Grzegorzek, Prof. Dr. Marcin	Institute of Medical Informatics, Universität zu Lübeck
Gutsmann, Prof. Dr. Thomas	Research Center Borstel, Leibniz Lungcenter
Handels, Prof. Dr. Heinz	Institute of Medical Informatics, Universität zu Lübeck
Heinrich, Prof. Dr. Mattias	Institute of Medical Informatics, Universität zu Lübeck
Hellbrück, Prof. Dr. Horst	Department of Electrical Engineering and Computer Science, Lübeck University of Applied Sciences and Institute of Telematics, Universität zu Lübeck
Huber, Prof. Dr. Robert	Institute of Biomedical Optics, Universität zu Lübeck
Hüttmann, Prof. Dr. Gereon	Institute of Biomedical Optics, Universität zu Lübeck
Ingenerf, Prof. Dr. Josef	Institute of Medical Informatics, IT Center for Clinical Research, Universität zu Lübeck
Jürgens, Prof. Dr. Tim	Department of Applied Natural Sciences, Lübeck University of Applied Sciences
Karpf, Prof. Dr. Sebastian	Institute of Biomedical Optics, Universität zu Lübeck
Klein, Prof. Dr. Stephan	Medical Sensors and Devices Laboratory, Lübeck University of Applied Sciences
Lühmann, Dr. Dagmar	Department of Primary Medical Care, Center for Psychosocial Medicine, University Medical Center Hamburg-Eppendorf, Hamburg
Mastmeyer, Prof. Dr. Andre	Faculty Optics & Mechatronics, Aalen University of Applied Sciences
Meier, Sigrid	Academy of Hearing Aid Acoustics, Lübeck
Mertins, Prof. Dr. Alfred	Institute for Signal Processing, Universität zu Lübeck
Müller, Prof. Dr. Stefan	Medical Sensors and Devices Laboratory, Lübeck University of Applied Sciences
Obleser, Prof. Dr. Jonas	Department of Psychology, Universität zu Lübeck
Özçep, PD Dr. Özgür L.	Institute of Information Systems, Universität zu Lübeck
Paulsen, PD Dr. Hauke	Institute of Physics, Universität zu Lübeck
Rafecas, Prof. Dr. Magdalena	Institute of Medical Engineering, Universität zu Lübeck
Reinholz, Dr. Fred	Institute of Biomedical Optics, Universität zu Lübeck
Rostalski, Prof. Dr. Philipp	Institute for Electrical Engineering in Medicine, Universität zu Lübeck
Rückert, Prof. Dr. Elmar	Institute for Robotics and Cognitive Systems, Universität zu Lübeck
Schildbach, Prof. Dr. Georg	Institute for Electrical Engineering in Medicine, Universität zu Lübeck
Schulz, Prof. Dr. Arndt-Peter	Clinic for Orthopedic and Trauma Surgery, Biomechanics Laboratory, University Medical Center Schleswig-Holstein, Lübeck
Schröter, Jörg	Medical Sensors and Devices Laboratory, Lübeck University of Applied Sciences
Song, Dr. Young-Hwa	Institute of Physics, Universität zu Lübeck
Spitzenberger, Prof. Dr. Folker	Centre for Regulatory Affairs in Biomedical Sciences, Lübeck University of Applied Sciences
Tchorz, Prof. Dr. Jürgen	Lübeck University of Applied Sciences
Wendlandt, Dr. Robert	Clinic for Orthopedic and Trauma Surgery, Biomechanics Laboratory, University Medical Center Schleswig-Holstein, Lübeck

Preface and Acknowledgements

After the great success of the previous meetings from 2012 to 2019, the Student Conference 2020 shows continuing growth both in quality and quantity of scientific contributions. In this year, the 9th Student Conference on Medical Engineering Science is held together with the 5th Student Conference on Medical Informatics, the 3rd Student Conference on Biomedical Engineering and the 2nd Student Conference on Auditory Technology.

The organization team of the Institute of Medical Engineering, the Institute of Medical Informatics, the Institute for Signal Processing and the Program Coordination for Biomedical Engineering at Lübeck University of Applied Sciences in cooperation with the Chamber of Industry and Commerce (IHK) Lübeck, has spared no effort to provide an excellent conference, where master students of the campus present their recent research results to a broad public of academics and industry.

The contributions show how new approaches and methods in medical engineering and medical informatics can advance medicine, health, and health care. Moreover, this conference offers a good opportunity for both students and companies to get in touch at the Recruiticon, i.e. a satellite recruiting fair with industrial exhibition. Students from the Life Sciences programs present their results from projects carried out at the laboratories, clinics, and institutes of Lübeck's Universities, in international research facilities, or research-oriented industrial companies. The conference focus has been placed on topics from medical engineering and medical informatics. The interdisciplinary field of medical engineering has been established at the Lübeck University of Applied Sciences for decades, and Medical Engineering Science (Medizinische Ingenieurwissenschaft – MIW) is an important bachelor and master program at the Universität zu Lübeck as well. Both universities jointly offer the international master degree programs Biomedical Engineering (BME) and Auditory Technology (Hörakustik und Audiologische Technik – HAT).

Furthermore, in the master program Medical Informatics (Medizinische Informatik – MI) the 5th Student Conference on Medical Informatics is integrated as an important element where project results in the emerging field of digital medicine are presented by the students.

As Conference Chairs, we want to thank all people who worked with enthusiasm and dedication to make the conference a successful event. We want to thank the companies who support the meeting. Moreover, our thanks go to Infinite Science for producing these proceedings and organizing the Recruiticon meeting supporting the Student Conference. Personally and on behalf of all colleagues of the Student Conference Committee, we especially want to thank Christina Debbeler and Ksenija Gräfe, they have been the central contact points for all questions of students and the program committee members, as well as Jan-Hinrich Wrage for editing the proceedings and furthermore Silke Venker and Anita Möller. Their in-depth overview of all details of this event is the key to the success of the Student Conference 2020.

Lübeck, March 3-5, 2020

Prof. Dr. Thorsten M. Buzug
Chair of the 9th Student Conference
on Medical Engineering Science

Prof. Dr. Heinz Handels
Chair of the 5th Student Conference
on Medical Informatics

Prof. Dr. Stephan Klein
Chair of the 3rd Student Conference
on Biomedical Engineering

Prof. Dr. Alfred Mertins
Chair of the 2nd Student Conference
on Auditory Technology

Sponsors

Infinite Science

Exhibitors

Contents

Auditory Technology

Spectro-Temporal Response Functions of the Human Brain with Clear and Vocoded Speech in a Bimodal Hearing Situation 3
    F. Kraus, M. Wöstmann, S. Puschmann, J. Obleser

Development of a knowledgebase and Learning Management System for Corona E3 7
    N. Rybacki, S. Meier, K. Langschwager

The effect of experimental design on reverberation preferences in normal hearing and hearing impaired listeners using hearing aids 11
    K. Mohnlein, J. Tchorz, C. Pischel, N. Acs

Skin Conductance as a Measure of Listening Effort: A Pilot Study 15
    M. Kemper, N. Herbert, J. Tchorz

Measurement of Spectro-Temporal Modulation Detection Thresholds with Hearing Impaired Listeners 19
    S. Griepentrog, J. Rehmann, S. Klockgether, V. Kuehnel, J. Tchorz

Assembling and verifying a hearing-aid prototype, consisting of a Raspberry Pi and the opensource-software openMHA 23
    M. Geisen, S. Meier, T. Jürgens

Biomechanics

Using two in-house developed test methods to characterise medical tapes regarding their resistance to shear and friction under load 29
    J. Tiedemann, N. Röhner, A. P. Schulz

Mechanical 4-Point-Bending test and simulation of known material 33
    D. Grollmisch, F. Mutter, R. Wendlandt

Structural changes in the design of a device that allows the formation of artificial lipid Montal & Mueller membranes combined with electrical and optical measurements 37
    R. Meyer, C. Nehls, T. Gutsmann

Case study: Influence of electrical thigh stimulation on physiological and pathological gait 41
    F. Allerding, A. Diercks, N. Bade, L. Röhrich, R. Wendlandt

Development of an in-vitro brain model for diffuse axonal injury occurrence simulation 45
    T. Burkhanova, S. Klein, C. Damiani, K. Krajewski

Biomedical Engineering

Development of a test stand to qualify the endurance life time of the Murata MZB1001T02 microblower 51
    A. C. Hennigs, O. Wallnewitz

Conversion of an existing lung demand valve to a pressure reducer with an output of 50 mbar 55
    J. Beidatsch, A. Freund, H.-U. Hansmann

Conception of a Test Setup to Determine the Service Life of an Infrared Emitter Used for CO_2 Measurement . . 59
    A. Leßmann, M. Kroh, P. Rostalski

Development of a Dosing Unit Using Venturi Injector 63
    M. Badran, W. Janik, S. Müller

Carbon dioxide measurement of different gas concentrations concerning its heat conductivity considering the system humidity 67
    A. Freund, J. Beidatsch, H.-U. Hansmann

Reduction of Condensate in Electrically Heated Breathing Hoses 71
    M. Hünert, L. David, L. Tappehorn

Optimization of a Fish Holder for the Immobilization of Zebrafish 75
    S. Melikov, M. Frerkes, S. Seeger, M. Zvolský, M. Rafecas

Development of a mobile sensory device to trace ambient and plasma treatment conditions 79
    I. Chaerony Siffa, T. Gerling, R. Wendlandt

Towards measurement of the deformation energy of softpellets: Design and Development of a Micro-Force-Spectroscope 83
    C. Pottkämper, C. Nehls, T. Gutsmann

Simulation of Field Free Line Translation for a Single-Sided Magnetic Particle Imaging-Scanner 87
    C. Chinchilla, A. Negash, A. Bakenecker, A. Tonyushkin

A non-invasive core body temperature measurement system for firefighters applied on the sternum 91
    N. Kowalczyk, F. Sattler, J. Koch, S. Müller

Usage of the optical front tracking methods for the measurement of fast-changing micro flows of liquids 95
    M. Ali, J. Schröter

Biomedical Optics

Visible light optical coherence tomography and fluorescence imaging with dual-stage aberration correction . . . 101
    J. Franke, W. Newberry, E. F. Durech, M. V. Sarunic

Non-linear Raman microscopy with an experimental approach on resolution enhancement 105
    V. Lutz, B. B. Jensen, S. Karpf, J. R. Brewer

Infrared raytracing simulation for measuring beam mixing in multi gas sensors 109
    L. Schnelle, B.-M. Dicks, M. Kroh, F. Reinholz

Simulation of Universal Laser Beam Parameter Correction of a High Power UV Laser System 113
    M. Seitz, R. von Elm, K. Seger, R. Huber

Improved semi-automatic quantification of experimental parameters after laser treatment for presbyopia 117
    J. K. Luther, S. Kassumeh, R. Birngruber

E-Health

Analysis of Data Loss when using the Archiving and Exchange Interface for Practice Management Systems . . 123
L. Barkow, O. Meincke, H. Ulrich, J. Ingenerf

Multi-Label Learning with a Cone Based Geometric Model 127
M. Leemhuis, Ö. L. Özçep

Generation of message transformers based on HL7 FHIR StructureMaps within the interface engine Mirth Connect 131
K. M. Vogl, H. Ulrich, J. Ingenerf

Creating a model instance for detection of *de novo* germline mutations using a neural network 135
K. Schultz, A. Künstner, M. Olbrich, H. Busch

Deploying a molecular tumor board on an open platform 139
N. Reimer, H. Busch, J. Ingenerf

MetaCheck – a Web Application for the Automatic Evaluation of Metadata in Healthcare 143
A. Banach, A.-K. Kock-Schoppenhauer, H. Ulrich, J. Ingenerf

Transforming Nutrition Databases Towards Semantically Consistent Content Classification 147
L. Liebenow, C. Twesten, M. Grzegorzek

Image Processing

Generation of 3D Body Models from Single-View Range Images for Robotic Applications in Medicine 153
T.-C. Çallar, S. Böttger, E. Rückert

Automatic anatomical structure detection in aortic CT angiography 157
R. Pallenberg, M. Fleitmann, K. Soika, H. Handels

Automatic layer segmentation of corneal μOCT images for subbasal nerve tracking 161
V. Bühler, A. Wartak, M. S. Schenk, G. J. Tearney, R. Birngruber

Gesture Recognition with Deep Learning 165
J. Zillmann, F. Coleca, E. Barth

Automatic segmentation of prostate zones in MRI data using different variants of the U-Net 169
S. Rüttgers, J. Hocke, M. Heinrich

Estimation of Healthy Aortic Root Shapes from Pathological Images with Conditional Variational Autoencoder 173
M. Mehdi, J. Hagenah, F. Ernst

Frequency based detection of regions of interest in videodata to estimate human breathing frequency 177
J. Sirocko, J. Diesel, M. Heinrich

A Combination of Normalized Metal Artefact Reduction and the Use of Multiple Prior Images in Computed
Tomography 181
P. A. Gunawan

Workflow development to create and analyse QA parameter from MRI images 185
S. Postel, A. Mastmeyer, J. Hirsch

Efficient Self-Supervised Context Learning on MRI data 189
C. Großbröhmer, M. Blendowski, M. P. Heinrich

Development of a Deep Learning based classification of respirator masks for a mobile application 193
L. Bartram, B. Dedersen, M. Heinrich

Automatic multi-object organ detection and segmentation in abdominal CT data 197
O. Mietzner, A. Mastmeyer

Fast Pulmonary Fissure Detection in CT Scans using Deep Learning on Point Clouds 201
M. L. Gillner, L. Hansen, M. P. Heinrich

Medical Electronics

Analysis of current regulation circuits for solenoid valves in a bus communication architecture for a dialysis machine 207
D. A. Briceño Peniche, A. Rieß, C. Schleicher, T. Jürgens
Development of a GUI-based Interface for the Optical Determination of Blood Flow Velocity 211
A. Dixit, R. Behroozian, S. Müller
Design of a Temperature Control System for the Three-Dimensional Magnetic Particle Spectrometer 215
X. Zhang, X. Chen, T. M. Buzug
Experimental investigation of the sEMG-force relationship of the biceps brachii 219
C. Kaußler, E. Petersen, J. Sauer, P. Rostalski
Modelling of a PEEP-Valve for an Anaesthesia Workstation 223
C. Ehlers, G. Männel

Medical Imaging

Segmentation of fluorescently labelled axons in a model of brain hemorrhage-induced axonal degeneration using convolutional neural networks 229
N. K. Menon, A. Palumbo, P. Grüning, S. K. Landt, L. Heckmann, L. Bartram, A. Madany Mamlouk, M. Zille
A low-cost, open-source device for characterizing super-paramagnetic nanoparticles 233
K. Brandt, E. Mattingly, E. E. Mason, M. Śliwiak, L. L. Wald, T. M. Buzug
The semantic congruence effect is not impaired in older subjects but associated with changes in neural processing 237
R. J. Alejandro, P. A. Packard, N. Bunzeck
Development of a Monitoring System for the Team Foundation Server-Build and Test Infrastructure 241
J. Magonov, C.-K. Mahadevaswamy, M. Schett, T. M. Buzug
Monte-Carlo simulation of positron emission tomography combined with Compton-camera imaging 245
A. Bolke, M. Zvolský, M. Rafecas

Regulatory Affairs

Development of a digital tool for Risk Management of Medical Devices 251
S. Shrestha
The benefit and specifications of a digital tool to optimize Clinical Evaluation of Medical Devices 255
U. Satyal

Development of a digital tool for Post-Market Surveillance of Medical Devices 259
    R. Pradhan
Approval of safety valves and globe valves in South Korea 263
    S. Henn, C. Kunath, M. Zaubitzer, T. Cordes
Regulatory requirements for laboratory developed tests (LDTs) – International comparison and conclusions for best-practice requirements for safety and performance of "in-house" IVD medical devices 267
    A. Safi

Safety and Quality

Design and implementation of a quality awareness process to reduce the cost of non-quality 273
    B. Abel, G. Buntrock
Connecting Old Devices Safely To The Cloud 277
    N. Kath, A. Seidel, J. Ingenerf
Medical Packaging Recyclability and Environmental Impact Analysis 281
    G. Chavez Gomez, M. Timmermann
Trajectory Control with Lane Boundary Constraints for Autonomous Vehicles Using Model Predictive Control . 285
    R. Kensbock, P. Huß, G. Schildbach
Modeling of a Failure Mode and Effects Analysis in SysML on a respiratory alcohol detector 289
    J. M. Goldyn, G. Männel, C. Brendle
Evaluation of Measurement Methods for Determining the Humidification Output of a Humidifier System . . . 293
    L. Luers, T. Matthews
Literature analysis of the competitive landscape of minimally invasive treatments for benign prostate hyperplasia 297
    A. Momin
Systematic review on comparative effectiveness of high cut-off membrane haemodialysis to haemodialysis with other types of membrane 301
    W. Łuczak
Selection and implementation of a data pipeline to display the test results of automated tests in a user-defined dashboard 305
    H. Schönmuth, L. Aulmann

Signal Processing

Acoustic scene classification with neural networks 311
    M. Kuttner, J. Tchorz
Measurement of Room Impulse Responses and Comparison of Different Interpolation Methods 315
    G. Röwekamp, F. Katzberg, A. Mertins
Single-Channel EEG - Detection of the Event-Related Potential (ERP) P300 Using an Acoustic Stimuli Synchronization 319
    K. Mrotzeck, H. Gamboa, A. Mertins

Removal of ECG artifacts caused by mechanical CPR using externally recorded compression markers 323
  L. Boudnik, J. Hochreiter, G. Müllegger, M. Heinrich
Interpolation of individual HRTF datasets using principal component analysis 327
  A. Biel, F. Katzberg, A. Mertins
Optimisation of support vector regression for the determination of haemoglobin derivatives in whole blood . . . 331
  S. Karthikeyan, B. Redmer, B. Nestler, S. Müller
Aortic Detection in Image Sequences of Electrical Impedance Tomography 335
  S. Herrmann, D. Schneider, J. Mrongowius, M. Heinrich
Model-based evaluation of magneto-mechanical oscillator signals for wireless sensing and localization 339
  J. Ackers, B. Gleich, J. Rahmer, I. Schmale, T. M. Buzug
Convolutional and recurrent neural networks for surface electromyography-based respiratory diagnostics and therapy 343
  N. S. Brügge, M. Eger, T. Handzsuj, S. Walterspacher, P. Rostalski

1

Auditory Technology

Spectro-Temporal Response Functions of the Human Brain with Clear and Vocoded Speech in a Bimodal Hearing Situation

Frauke Kraus [1], Malte Wöstmann [2], Sebastian Puschmann [2], and Jonas Obleser [2]

[1] Auditory Technology, Universität zu Lübeck, frauke.kraus@student.uni-luebeck.de

[2] Auditory Cognition, Department of Psychology , Universität zu Lübeck, malte.woestmann@uni-luebeck.de, sebastian.puschmann@uni-luebeck.de, jonas.obleser@uni-luebeck.de

Abstract

We know that it is possible to understand spectrally degraded speech in a laboratory setup. But in complex acoustic situations we need as well the spectral information of sounds to separate and ignore irrelevant sound. To improve hearing aid technologies we need a better understanding of this complex accomplishment in the human brain. We presented 5 normal hearing subjects audiobooks with clear and with 8-band and 4-band vocoded speech without a distractor while measuring electroencephalography (EEG). We estimated the spectro-temporal response functions (STRF) for all three conditions with regularized ridge regression. By comparing the STRFs of the three conditions we found that their deflections were smaller for the vocoded conditions. This effect was most dominant in the P2 component and the frequencies up to 2000 Hz.

1 Introduction

Cochlear Implants (CI) are used if the sensorineural hearing loss is too intense to supply it with a non-implanted hearing aid. A CI uses 12 to 22 electrodes to code the acoustic sound to the auditory nerve. In the healthy inner ear there are around 3000 inner hair cells for coding the temporal and spectral information of sounds to the auditory nerve. The coding strategies of modern CIs use the envelopes of the sounds [10]. This means there is no information about the temporal and spectral fine structure available for the CI-users. Though it is possible to reach a very high speech intelligibility wearing a CI [5]. Shannon et. al [9] already found out that a very high speech recognition is possible with sound degraded down to three bands of modulated noise if the temporal envelope is intact. Therefore it is of high interest to understand how the brain restore the loss of spectral information due to a CI or in which hearing situations this loss of information makes speech intelligibility considerably difficult.

Another point of interest is bimodal hearing. This means that a person has a CI implanted on one side and the other ear is normal hearing or supplied with a hearing aid. Only since the end of 90s there are publications of successful use of bimodal hearing [2]. Before the conviction was that there is no benefit for bimodal hearing. Only bilateral deaf patients were implanted with a CI.

To study how the brain deals with the loss of spectral information there is often the use of speech vocoding to simulate a CI using situation with normal hearing participants. The speech is spectrally degraded while using filter banks with different numbers of bandpass filters and multiplying them with white noise [9].

It is not very easy to analyse how the complex brain processes complex stimuli like human speech. One possibility is to estimate a response function of the human brain to get an idea which features of an auditory stimulus are neurally tracked. For this estimation electroencephalography (EEG) or (magnetoencephalography) MEG is measured during the presentation of an acoustic stimuli. Afterwards by mapping the EEG/MEG-signal with the acoustic stimulus the response function of the human brain is obtainable [3], [4], [6]. In this study we resolved the response function in the temporal and in the spectral domain and therefore it is called spectro-temporal response function (STRF). We compared the STRFs of the vocoded and the clear speech condition to analyse in which time and frequency range they differ or not.

We first started with a paradigm without a distractor. With this simple design we wanted to determine first if the STRF-estimation with vocoded stimuli is possible and how the vocoding affects the STRFs before using a more complex hearing situation. We hypothesize that it is possible to estimate STRFs after a stimulation with vocoded stimuli but that STRFs are weaker compared to stimuli with clear speech.

2 Material and Methods

2.1 Participants and Task

Participants were 5 young adults (3 female, age: 24.2 ± 3.5 years). Pure tone audiometry ensured that the PTA (Pure Tone Average) was below 20 dB HL. The PTA on the right side was 5.8 ± 7.3 dB HL and on the left side 1.4 ± 3.9 dB HL. One participant had a moderate hearing loss up to 40 dB HL at the frequencies 500, 750 and 1000 Hz on the right side. Due to the good performance (see Table 1) of this subject in the task and the low number of participants we decided to not exclude the data of this participant. The task was to listen carefully via in-ear headphones to a german audiobook and afterwards answering some questions about the content of the audiobook. We used the audiobook "Das zweite Lebens des Monsieur Moustier" by Aude le Corff, translated by Anne Bran, published in 2016 by *Der Hörverlag*, spoken by a female narrator. The participants were able to set the presentation level to a comfortable volume.

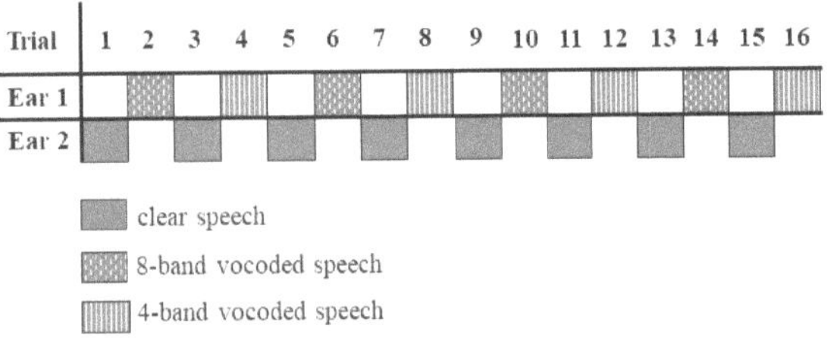

Figure 1: Task Design. In 8 of the 16 trials the participants heard audiobook trials with clear speech on ear one, in 4 trials with 8-band vocoded speech and in 4 trials with 4-band vocoded speech on ear 2. The three conditions were presented in an alternating scheme.

The first 64 min of the audiobook were divided into 16 trials of 4 min each. After each trial the participant had to answer 4 closed questions with 4 answer possibilities about the story (s)he heard in this trial. To simulate the situation of a bimodal CI-patient the participants heard clear speech during the whole experiment on one ear and on the other the audiobook with vocoded speech. While the stimulation was on one side on the other was silence. Three of the participants heard the vocoded speech on the right ear and two of them on the left ear. The experimental design is illustrated in Fig. 1. Half of the trials with vocoded speech were vocoded with 8 bands and the other half with 4 bands.

2.2 Auditory Stimuli

For the vocoded stimuli of the audiobook we used a gamma tone filterbank with 4 filters between 0 and 8 kHz for the 4-band vocoding and one with 8 filters for the 8-band vocoding. After filtering the sounds with these filters each band was multiplied with white noise filtered with the same filterbank and a RMS-correction has been done. Then the 4 or 8 bands were added and again corrected to the RMS-value from the not vocoded audiobook parts [9].

2.3 EEG Recording and Preprocessing

The EEG was recorded from 64 electrodes (ActiChamp, Brain Products, München) at a sampling rate of 1000 Hz,

referenced to the electrode Fz. After recording the EEG data were preprocessed with the FieldTrip toolbox [8] for MATLAB (MathWorks, Inc.). The data were re-referenced to the average of all electrodes and filtered between 1 and 100 Hz. With an Independent Component Analysis (ICA) components clearly corresponding to eye blinks, eye movements, muscle activity and heartbeat were rejected. The data were again low-pass filtered to 10 Hz, cut into trials of 4 min starting at the onset of the auditory stimulus and resampled to 125 Hz. For the further estimation of the STRFs these trials were again cut into trials with a duration of 1 min and z-transformed due to the optimization of the STRF-estimation later [3].

2.4 STRF Estimation

For the estimation of the STRFs of all the conditions (clear, 8-band and 4-band vocoded speech) we used the clear stimuli and not the vocoded ones. To get a spectrally resolved representation of the stimuli we used the NSL toolbox [1] for MATLAB (MathWorks, Inc.). Each of the 16 audiobook trials were split up into 128 frequency bands. To get the onset envelope in each frequency band, the envelope was computed, the first derivation was taken and the negative values were zeroed. After downsampling in time and frequency domain, the envelopes of the auditory stimuli included 16 frequency bands with the following center frequencies in Hz: 199, 291, 367, 462, 582, 733, 924, 1164, 1466, 1847, 2328, 2933, 3695, 4655, 5865, 6772 and the same sample frequency fs=125 Hz as the EEG data.

The STRFs were estimated with the mTRF Toolbox [3] for MATLAB (MathWorks, Inc.). This toolbox uses regularized linear regression as the method for the STRF-estimation. First for every of the 1 min trials a model m was estimated as in (1), where S is a Matrix containing the envelopes. The matrix R contains the corresponding EEG signal of all 64 electrodes and I is the identity matrix. To find the best regularization parameter λ the STRF estimation was carried out with different λ-values leaving out one of the 1 min trials. With the estimated model the left-out trial was predicted. Then the λ-values, which maximize the correlation between the predicted and the left-out trial, was chosen for the STRF-estimation. For our data this was the case for λ=1.

$$m = (S^T S + \lambda I)^{-1} S^T R \qquad (1)$$

For optimization reasons the ouput data of this estimation were normalized before the next steps [3]. To get the STRF for every subject and separated for the three conditions (clear speech, 8-band vocoded, 4-band vocoded) the models for one subject and one condition were averaged over trials. This means the STRF for the clear speech condition was averaged over 32 trials, the one for the both vocoded conditions over 16 trials each. Due to technical problems for subject 2 there were only 28 trials in the clear speech condition available for the estimation. For all conditions the STRFs were than averaged over 9 frontocentral electrodes

(FC1, FCz, FC2, C1, Cz, C2, CP1, CPz, CP2). Fig. 2 shows an example for a STRF.

2.5 Statistical Analysis

To analyse possible differences between the STRFs of the three conditions we used cluster permutation tests in the FieldTrip toolbox [7] for MATLAB (MathWorks, Inc.). For this statistical analysis we used the data of all 64 electrodes, not only the nine described in 2.4. The idea is to compute dependent-samples t-tests between the clear speech and the vocoded conditions for each time,frequency,electrode-triplet of the STRFs. Resulting t-values were used to construct clusters by connecting neighbouring time-frequency-samples with p<0.05 for at least 3 neighbouring electrodes. In each observed cluster the t-values were summed and compared to a permutation distribution. For this permutation distribution clusters were generated by randomly assigning time,frequency,electrode-triplets to one of the conditions through 100 iterations and their t-values summed [7]. Due to the small number of subjects we show all clusters with p<0.1.

3 Results and Discussion

3.1 Behavioural Results

Table 1 shows the percent correct of all content-specific questions separated for the three conditons. With a chance level of 25 % all participants were clearly above this. Additionally there was no single trial in which one of the participants answered less than two answers correctly (chance level). This means they could understand the audiobook in all conditions and were attentively listening to the story. For this reason we did not have to exclude one of the trials for the STRF estimation.

Table 1: Performance in percent correct.

participant	clear speech	8-band voc	4-band voc
1	93.75 %	100 %	81.25 %
2	90.63 %	87.5 %	93.75 %
3	93.75 %	93.75 %	100 %
4	100 %	87.5 %	100 %
5	96.88 %	87.5 %	87.5 %

3.2 STRF

An example for a STRF is shown in Fig.2. There is the first positive component at around 50 ms, called the P1-component. Around 100 ms appears the first negative component N1 and around 180 ms the second positive component P2 [4], [6]. This general pattern is existing in the clear speech condition as well as in the vocoded one. We show here that this P1,N1,P2 pattern is not only present in an attended-ignore-task as in Kong et al. [6] but as well in a task without a distractor as in the present study.

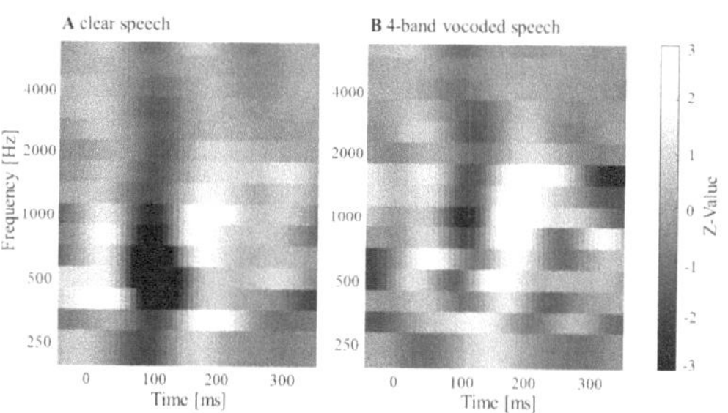

Figure 2: Example for a STRF. **A**, the STRF of a single subject for the condition with clear speech. **B**, the STRF for a single subject for the condition with 4-band vocoding.

3.3 Spectral Vocoding Reduces Neural Speech Tracking

To statistically assess the differences between the three conditions we used cluster permutation tests as described in 2.5. We estimated one test for each of the distinct time periods of the three components (P1: 0.024-0.072 s, N1: 0.072-0.12 s, P2: 0.12-0.2 s) and the three possible combinations of the conditions. Fig.3 shows the significant clusters we observed. For illustrating purpose the cluster plots were interpolated. We could not find any significant differences between the two vocoded conditions in any time period. There are significant clusters between the clear speech and the 4-band vocoding (see Fig. 3A and B). This means the N1 component of the clear STRF is more negative than the one of the 4-band vocoding and the P2 component is more positive in the clear condition. Comparing the clear and the 8-band vocoded condition we could only find significant clusters in the time range of the P2 component (see Fig. 3C), which were similar to the clusters we observed for the 4-band vocoding in Fig. 3B. The topographies below show that all these differences were most dominant in the frontocentral electrodes. These findings are in line with Kong et al. [6] that the amplitude of N1 and P2 are getting smaller the more the speech is degraded.

Most of the significant clusters were observed in the frequencies up to 2000 Hz. We explain this with the fact that speech has the most power in these frequencies. Therefore the STRFs have greatest deflections in these frequencies which leads to the biggest differences between the conditions as well. Ding et al. [4] showed in a MEG-study that the STRFs peaked most between 400 Hz and 2000 Hz with highest peaks below 1000 Hz. The clusters in the very low frequencies (P2a in B and P2c in C) could be explained with the gamma tone filter banks we used for the vocoding. These filter banks filter the frequencies up to 400 Hz in a very intense way. Due to this low power in the vocoded stimuli the STRFs are less powerful in this frequency range for the vocoded conditions compared to the clear condition. These results let us assume that the neural tracking in the human brain for spectrally degraded speech is available but less effective compared to clear speech which suggests higher listening effort for spectral degraded speech.

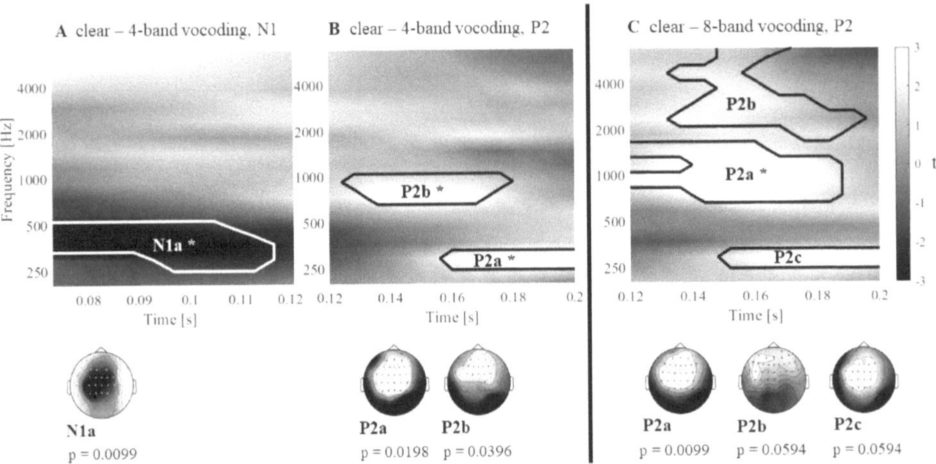

Figure 3: Comparison of the clear STRFs to the vocoded ones in distinct time periods. **A**, one negative cluster for clear − 4-band vocoding in the time period of the N1-component. **B**, two positive clusters for clear − 4-band vocoding in the time period of the P2-component. **C**, three positive clusters for clear − 8-band vocoding in the time period of the P2-component. Below the clusters are the corresponding topographies pictured. The colorbar on the right belongs to the cluster figures and the topographies as well. All clusters with p<0.05 are marked with *. Negative clusters are marked with white lines and positive clusters with black lines.

4 Conclusion

In this study we showed that it is possible to estimate STRFs with vocoded speech (as a model for CIs) in a task without a distractor. The finding that the STRFs decrease for the vocoded condition in the frequencies up to 2000 Hz should be confirmed with more than 5 participants. Additionally it would be of interest to expand this setup to an attend-ignore task with bimodal hearing to see how the differences between the clear and vocoded conditions are present in a complex hearing situation. As a next step, further investigations should study this task with bimodal CI users to see how maybe over time the STRFs between the conditions differ less due to training effects wearing the CI.

Acknowledgement

The work has been carried out at and supervised by the Group Auditory Cognition at the Department of Psychology, Universität zu Lübeck.

5 References

[1] T. Chi, P. Ru, Powen and S.A. Shamma, *Multiresolution spectrotemporal analysis of complex sounds*. The Journal of the Acoustical Society of America, vol.118, pp.887–906, 2005.

[2] T. Ching, P. Incerti, and M. Hill, Mandy *Binaural benefits for adults who use hearing aids and cochlear implants in opposite ears*. Ear and hearing, vol.25, pp.9–21, 2004.

[3] M. Crosse, G. Di Liberto, A. Bednar and E. Lalor *The Multivariate Temporal Response Function (mTRF) Toolbox: A MATLAB Toolbox for relating neural signals to continuous stimuli*. Frontiers in human neuroscience, vol.10, pp.604, 2016.

[4] N. Ding and J. Simon *Neural coding of continuous speech in auditory cortex during monaural and dichotic listening*. Journal of neurophysiology, vol.107, pp.78–89, 2012.

[5] M. Hey, T. Hocke, J. Hedderich, and J. Müller-Deile *Investigation of a matrix sentence test in noise: reproducibility and discrimination function in cochlear implant patients*. International journal of audiology, vol.53, pp.895–902, 2014.

[6] Y.-Y. Kong, A. Somarowthu and N. Ding *Effects of spectral degradation on attentional modulation of cortical auditory responses to continuous speech*. Journal of the Association for Research in Otolaryngology, vol.16, pp.783–796, 2015.

[7] E. Maris and R. Oostenveld *Nonparametric statistical testing of EEG- and MEG-data*. Journal of neuroscience methods, vol.164, pp.177–190, 2007.

[8] R. Oostenveld, P. Fries, E. Maris and J.-M. Schoffelen *FieldTrip: Open source software for advanced analysis of MEG, EEG, and invasive electrophysiological data*. Computational intelligence and neuroscience, 2011.

[9] R. Shannon, F.-G. Zeng, V. Kamath, J. Wygonski and M. Ekelid *Speech recognition with primarily temporal cues*.Science, New York, vol.270, pp.303–304, 1995.

[10] F.-G. Zeng, S. Rebscher, W. Harrison, X. Sun and H. Feng *Cochlear implants: system design, integration, and evaluation*. IEEE reviews in biomedical engineering,vol.1, pp.115–142, 2008.

Development of a knowledgebase and Learning Management System for Corona E3

Nicolai Rybacki [1], Siegrid Meier [2], Kevin Langschwager [3]

[1]Auditory Diagnosis, Universität zu Lübeck, nicolai.rybacki@student.uni-luebeck.de
[2]Auditory Diagnosis, Akademie für Hörakustik, s.meier@afh-luebeck.de
[3]Meidizinprodukte, Pilot Blankenfelde, kevin.langschwager@pilot-blankenfelde.de

Abstract

Due to the age of digitalization, learning management systems (LMS) optimize complex teaching processes through fast and individual learning. This also includes an understanding of unknown technologies. If teaching processes are simplified, time, money and personnel are saved. This paper deals with the establishment of an LMS for the audiological differential diagnostic measuring system Corona E3. For this purpose, an LMS with Moodle is created as a basis for knowledge generation. In order to let the LMS grow and connect knowledge, users are needed who share their experience with the system.

1 Introduction

Due to the complexity and the short innovation cycles in medical technology, knowledge management and the exchange of experience is important for the operators. The distribution of a user manual is not sufficient even for experts. They need e.g. assistance for the interpretation of the measured data. For beginners it is helpful to have education material to avoid typical errors and pitfalls. In order to adopt the users to the frequently changes in hard- and software, a knowledge management system should be developed that meets the following objectives: manufacturer independent, flexible, interactive, scalable, and adoptable to similar systems [1]. This also allows users to actively provide input, e.g. ask questions, research, request interactive learning units or give "best practice tips". Additionally, the manufacturer can quickly distribute updates and background information to the users.

The following project was divided into two sections. In the first part, a knowledge base for the Corona E3 was created and in the second section interactive learning units are produced in the sense of empowerment. The aim of this project is to establish a learning management system (LMS) in which users of the Corona E3 can teach themselves how to use the measuring unit and answer their questions together with other users. As a result, the operators need a mixture of a computer supported collaborative learning (CSCL) and a computer supported alone learning (CSAL) system [2].

2 Material and Methods

2.1 System: Corona E3

Table 1: Features and specification of the CoronaE3

Features	Specification
Medical device classification International Standard	Typ IIa, B and BF, IPX0 EN 60645-1 for audiometer EN 60645-6 for OAE EN 60645-7 for ABR
Transducer	Headphone: HDA 300, Insertphone: ER-3C, Boneconducter: KLH 96, laudspeaker
Masking	Narrow band, broadband, speech masking
Level/frequency range for audiometry	Air conduction: -120dB HL Bone conduction: - 80dB HL Freefield: - 85dB 125Hz – 8kHz
Level and frequency range for OAE	40-85dB SPL 0,5 kHz – 8kHz
Level/frequency range AEP	100dBHL 110dBHL for ASSR
Stimuli for AEP	Click, gauss, toneburst, tonebeep, ASSR
Possible measurements	Electrochleography, FastCERA (cortical evoked potentials), SlowCERA and vestibularistests

The Corona E3 is an audiological diagnostic system [3], which is distributed in the clinical, audiological sector and is developed to get detailed audiological information specifically for the pediatric sector and not for standard examinations. The system enables the following measurements: pure tone audiometry, measurement of the otoacoustic emissions (OAE) and the monitoring of the acoustic evoked potentials (AEP). More detailed data about the Corona E3 is represented in Table 1. This makes it possible to obtain information for the diagnosis of the pathology and the hearing loss of the patients. For this purpose, the system allows a lot of individual test procedures. It takes a lot of experience to find the right settings for the filters, the signals, the levels e.g.

2.1.1 Auditory Brainstem Responses (ABR)

Brain activity can be visualized using various imaging techniques. One of them is the principle of the EEG. The Auditory Brainstem Response is an Evoked Potential that originates at the auditory nerve (Cranial Nerve VIII). This test is used to assess the auditory system's function from the cochlea through the brainstem [4]. The response is identified by "peaks" that occur typically between 1 and 15 milliseconds from the stimulus onset. The ABR peaks are measured and marked traditionally as I, II, III, IV, and V. Each peak has an expected latency to be considered "normal". Delayed or missing peaks are consistent with abnormal auditory function. The amplitudes and latencies (both absolute and inter-peak) are used to diagnose certain auditory pathologies. The test signal in the standard setting is a click, that represent a broadband frequency area. The ABRs are measured with electrodes that are attached to several places at the head (Fig.1). The preparation is very important for this.

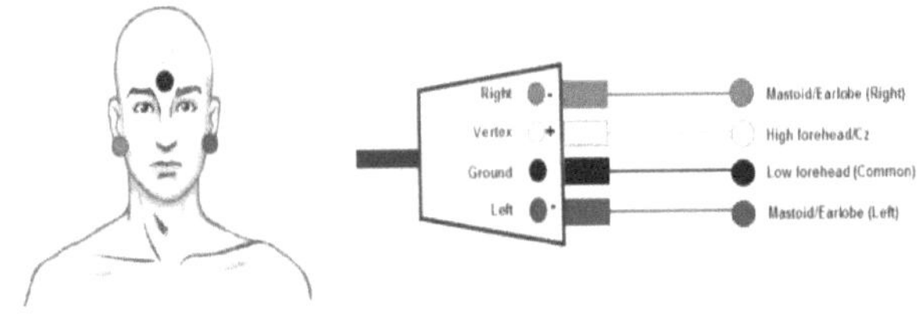

Figure 1: Electrode placement and preamplifier

2.1.2 Measurement with Chirp

If the cochlea is stimulated with a click, the signal responses are also changing in time because of the tonotopy and the resulting transit-time differences. The chirp signal bypasses this mechanism by adapting the time for the frequency of the signal to stimulate the hair cells on the cochlea at the same time. The time manipulation of the chirp (low chirp, high chirp or mid chirp) can also excite certain locations of the cochlea in order to record specifically frequencies In Fig. 2 you can see a setting for a chirp measurement.

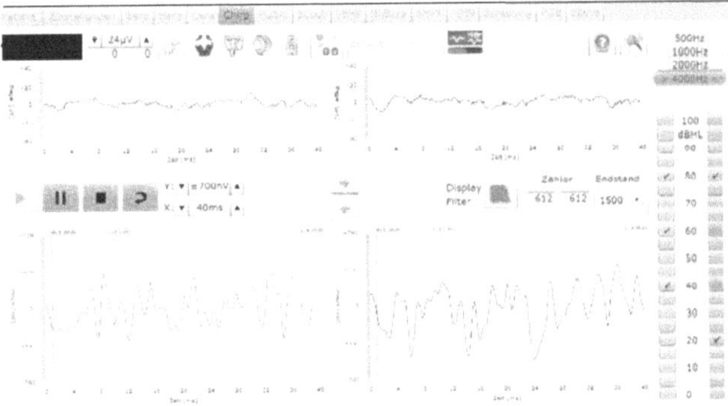

Figure 2: Measurement with Chirps

2.1.3 Measurement with Notched Noise

This test signal is a sound pulse that lies in a notch of a filtered broadband noise. As a result, the frequency of the cochlea is isolated and can be tested alone. This method is important for pediatric audiometry in order to localize a frequency specific hearing loss.

2.1.4 Measurement with Auditory Steady-State Responses

The auditory steady-state response is an amplitude modulated sinusoidal signal which, unlike the ABR, can represent the derived voltage changes of the test signal over the spectrum.

2.2 Knowledgebase

Due to the technical development, the significance of the measurement became better, but the systems get more sophisticated. To this end, some manufacturers are making more and more training manuals available on their websites. Although the Corona E3 is a finished product, pitfalls have been noticed in the operation and use of the software, which in unfavorable cases could also arise during an operation. The knowledge database is primarily a collection of the experiences and reports of Corona E3 users who explain how to use the measuring unit. Not just how it works, but also how to avoid errors, system weaknesses, pitfalls and the like. For this purpose, all users of the knowledge base have the opportunity further feed the knowledge base. This includes interesting facts and useful information, user tips and the pathological interpretation of results. All the knowledge is then documented in a consensus area and for the second section using LMS.

2.2.1 Software

The structure of the knowledge database was created using a Moodle platform. The knowledge is documented and processed in Moodle.

For better coordination and structure, AdobeConnect was used with this virtual classroom. This applies to web conferences and online lessons when the interactive learning units raise questions.

2.2.2 Participants

To generate participants, the german working group AG-ERA and known user were contacted. In addition, the distrubuter of the system was added to the LMS. They should share their experiences and exchange knowledge. They also got instructions for using the software through a web conference, which was recorded for later repetition. To this point twelve participants were signed for Knowledge base.

2.2.3 Structure and functions

To make the interface user-friendly and clear, Moodle provides several functions for coordinating, evaluating and generating knowledge, as you can see in Table 2. To start with a base, the product description is accessible to everyone [5].

Table 1: Features and specification of the LMS

Features	Specification
Responsiveness Interface	easy to navigate on both desktop and mobile devices.
Personalised Dashboard	customisable page for providing users with details of their progress and upcoming deadlines
Collaborative tools and activities	Work and learn in forums, wikis, glossaries, database activities, assignments, chats, choices, feedbacks, lessons, external tools, quiz, survey, workshops
File management	Drag and drop files from cloud storage services including MS OneDrive, Dropbox and Google Drive and file upload, folder, label, page, url
HTML text editor	Format text and conveniently add media and images with an editor that works across all web browsers and devices
Notifications	When enabled, users can receive automatic alerts on new assignments and deadlines, forum posts and also send private messages to one another.
Supports open standards	Readily import and export IMS-LTI, SCORM courses
Track progress and evaluation tools like logfiles	Educators and learners can track progress and completion with an array of options for tracking individual activities or resources and at course level
Webconference	Adobe connect integration as a plugin

2.3 Procedure

To realize and build the database, a simple, clear structure had to form the core. The functionality had to be described for all users and a fast communication had to be possible. The forum function was used for all users to describe their experiences, not only to document knowledge, but also to respond to the assigned participants. Due to the abovementioned occurrences of pitfalls of the Corona E3, a forum for the description of the errors and solutions with exact settings and procedures was set up. At the request of the participants and with regard to urgent cases, the wiki was added so that clinical users could communicate current cases and have a common basis for visualizing results. On this basis, a book function was created so that knowledge that is generally discussed and found to be correct is not changed and documented again. Finally, a feedback system was set up so that the participants can make suggestions for optimization. In Fig. 3 you can see the Knowledge base.

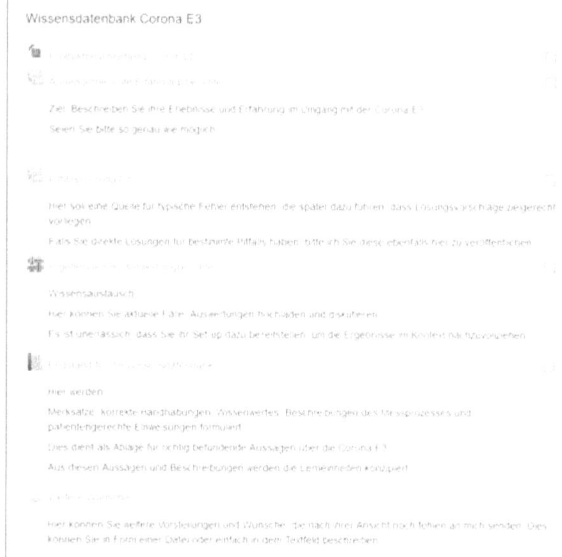

Figure 3: Examples for the CSCL units

3 Results and Discussion

3.1 Learning Management Systems (LMS)

The LMS is an interactive learning platform in which you answer structured topics on the use of corona E3 questions in different chapters, are taught the application step by step and check your self-acquired knowledge in a feedback system.

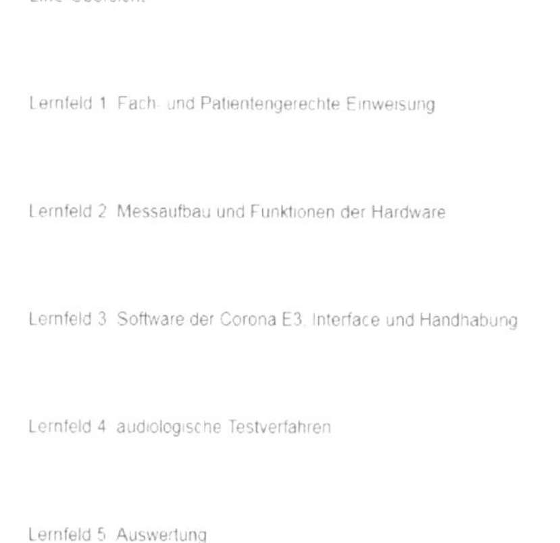

Figure 4: Chapters for the CSAL units

The LMS should be intuitive and motivating for beginners. Therefore, the structure and the choice of the learning fields are based on the chronological beginner's situation questions. The framework around learner questions first covers all external factors and then all internal factors for knowledge acquisition. This means that before the learner gets to know the Corona software better, he is familiarized with the patient's needs, such as the briefing for the upcoming measurement (Fig. 4). This is followed by the hardware description and the explanation of the handling. This also includes any typical mistakes that the beginner will no longer make in practice. One of the last topics for the time being is the evaluation and interpretation of the measurement results, although these are not important for all users of the Corona E3. Each learning field contains different descriptions, exercises, interactive knowledge queries and a review of what has been learned. Each candidate receives information about his or her level of knowledge via an integrated feedback system and can view it at any time.

3.2 Discussion

The LMS is a good method to collect and generate knowledge. It has the advantage over conventional sources of knowledge that you can repeat the exercises and have direct and quick feedback on what you have learned. In addition, the LMS is not a constant system that always remains the same, with more knowledge and new insights it changes. Technical support is also a prerequisite for this generation of knowledge. That means not only the internet-enabled devices, but also maintenance and administrator services. The resulting costs would have to be compared to the savings in technical support from the manufacturer without the LMS.

4 Conclusion

At this stage of the knowledge base it must be clear that the effective benefit depends on the activity of the participants. Such a system has to be fed with information so that the link between the knowledge grows for all participants. Successful implementation can only be promoted by user activation. The implementation and continuation of this project creates clear competitive advantages over other manufacturers. It is therefore conceivable that other manufacturers can design similar systems for their products. In addition, any system requiring explanation with a user-oriented operation or learning unit could be adopted in the future.

Acknowledgement

The work has been carried out and supervised by the Akademie für Hörakustik.

5 References

[1] P. Nikodemus, *Lernprozessorientiertes Wissensmanagement und kooperatives Lernen. Konfiguration und Koordination der Prozesse*. AKAD University Edition, 2017.

[2] U. Hinze, *Computergestütztes kooperatives Lernen (CSCL)*. Lehrbrief. Modul: Didaktiktisches Design. Hg. v. Wissenschaftliche Weiterbildung der Universität Rostock, 2015.

[3] Pilot Blankenfelde, *Gebrauchsanweisung für das Diagnostiksystem Corona e3*. Pilot Blankenfelde, Blankenfelde, 2018.

[4] S. Hoth, K. Neumann, R. Mühler, and M. Walger, *Objektive Audiometrie im Kindesalter*. s.l.: Springer-Verlag, 2014.

[5] Moodle Community, *Moodle Developer Documentation*. Available: https://docs.moodle.org/dev/Main_Page [last accessed on 2020-01-20].

The effect of experimental design on reverberation preferences in normal hearing and hearing impaired listeners using hearing aids

Katharina Mohnlein [1], Jürgen Tchorz [2], Claudia Pischel [3] and Nicola Acs [3]

[1] Auditory Technology, Universität zu Lübeck, katharina.mohnlein@student.uni-luebeck.de
[2] Institut für Akustik, Technische Hochschule Lübeck, juergen.tchorz@th-luebeck.de
[3] WSAudiology, Erlangen, {claudia.pischel,nicola.acs}@wsa.com

Abstract

Audiological studies normally take place in lab conditions, so experimental situations can be reproduced accurately. This study investigates how lab and real life situations can be compared to each other in reverberant conditions. Further, it considers how assessments of normal hearing experienced listeners can be compared to those of hearing impaired listeners. Two programs were prepared, one with and one without a dereverberation algorithm. The study was comprised of three setups: two lab and one real life setup. The two lab presentations were done via headphones or via loudspeakers. Three sound examples were presented in the lab setups. The results indicate that the hearing impaired and normal hearing listeners differ in their subjective evaluation of all three setups. More precisely, the two lab setup preferences differ from the real situation preferences for both participant groups. Finally, for the same sound examples, participants' responses differ between the two lab setups.

1 Introduction

Audiological research often includes listening experiments with normal hearing and hearing impaired people. Generally, it is of great interest to evaluate the perception of hearing aid amplification strategies and features. Hearing aid manufactures for example aim to prove benefits of their products and features in addressed listening situations. This also applies to reverberant environments, which represent a tiring situation especially for hearing impaired listeners [1], [2], [3]. Reverberant situations can be improved with a hearing aid. The *Signia Primax*, for example, can improve speech intelligibility and reduced listening effort in reverberant situations [4]. For research purposes, most objective and subjective measurements are performed in a lab with loudspeakers or headphones and not in realistic conditions. Although hearing aid features are developed for hearing impaired people, research and development also make use of tests with normal hearing listeners. Therefore, in this study we compare the effects of different experimental designs on listeners perception. More precisely, we compare the differences between lab setups and real situations.

2 Material and Methods

The study took place in a soundproof room with three setups (see 2.3). The first setup was tested using loudspeakers in a lab (LSL) wearing hearing aids (HAs). The second setup was tested with headphones (HPH). For the presentations via LSL and HPH, three sound examples were presented. In these six situations, the participants evaluated two program settings (A and B) by answering eight questions (see 2.4). These two programs were prepared in advance (see 2.2). The ordering of program presentation was randomised and the participants were aware of program changes. In the LSL part, the instructor changed the program via the app *touch-Control*. The signal tones in the HAs were deactivated, so there was only a visual sign from the instructor each time the instructor changed the program. A and B were assigned to the two programs each time at the beginning of each example. While doing the HPH part, the participants selected the program via a GUI themselves. Here, the program order was also random and unknown. The real life part (RL) took place in a church during a sermon, attended by the hearing impaired participant (HI) on an individual date. All of the normal hearing experienced listeners (EL) went to the same church, but not to the same sermon. In the RL the change from program 1 to program 2 was done by the participants using the rocker switch or *touchControl* app.

2.1 Participants

The study was conducted with seven HI and seven EL. The seven HI had a bilateral, symmetric, sensorineural hearing loss. The median pure tone average (PTA) of the HI was 50 dB (44 dB - 62.5 dB). The EL were fitted to a PTA of 48.5 dB. All were used to wearing HAs. Each participant took part in all setups, except for the RL part. Here, only six EL participated. There were two female and five male HI. Their age ranged from 70 to 80 years (mean: 75 years). There were four female and three male EL. Their age ranged from 27 to 58 years (mean: 35 years).

2.2   Hearing aids and settings

During the first appointment the audiogram was checked and renewed if older than six months. As a function of their current audiogram, each participant was provided with one of two possible fittings. Fitting 1 and fitting 2 differ in high-frequency gain (see Fig. 1). All participants were equipped with released *Signia Pure 312 7X M* devices with closed *click sleeves*. The EL were always fitted with fitting 2. All the HAs were fitted with two hearing programs. One program was *Universal* (U) with an appropriate fitting and with default feature settings (without Own Voice Processing or individual fine tuning). The other program was fitted with the *Reverberant Room* (RR) algorithm. RR is a dereverberation program to improve subjective perception in reverberant environments. Because of the *Connexx*-setting, U was always the first program in the RL setup. As already stated, program change could be done via rocker switch or via App.

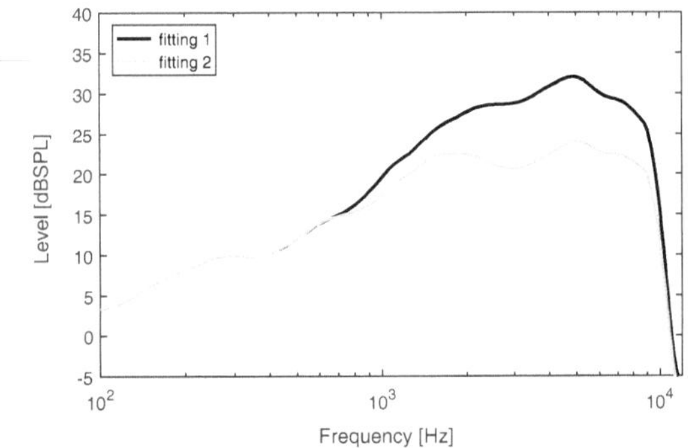

Figure 1: Insertion Gain for the two fittings. The curves were determined with 65 dB input level.

2.3   Experimental Setup

The study contained three setups: LSL, HPH, RL. Both LSL and HPH contained three sound examples, as shown in Table 1. The sermon sound example was always the first, followed by the lecture and then the poem example. The reverberation time was estimated based on the *Altiverb* plug-in, and indicates a rough value only.

Table 1: Details for the three sound examples

S1: Sermon	Speech in quiet	RT60: ~3.5 sec
S2: Lecture	Speech with murmuring	RT60: ~2.5 sec
S3: Poem	Speech in quiet	RT60: ~1.5 sec

In the first part of the study, the three sound examples were presented in LSL. The participants wore the fitted HAs. For every sound example, the two programs were compared. In the next part, the same three sound examples were presented via HPH without wearing HAs. These sound examples were previously recorded in LSL in both programs. RR and U were recorded with fitted HAs (fitting 1 and 2) with a Kemar in the center of the LSL setup. The third part was done in

a church in RL. For more details, see Table 2. The three sound examples were made by a special digital reverberant plug-in.

Table 2: Details for the three setups

LSL	HPH	RL
5 loudspeakers (Genelec 8030)	Fireface UCX, Sennheiser HD650	different loudspeaker(s) in church
S1, S2 and S3 as in Table 1	S1, S2 and S3 as in Table 1 (recorded)	real sermon
loops of whole sound examples	some loops and repetition optional	no repetition
switching via App by the instructor	switching via GUI by the participant	switching via App or switch rocker by the participant

2.4   Questionnaire

For every sound example, the participants were asked to compare U and RR blindly with respect to subjective criteria: the reverberation intensity (Q1), the stability of the sound (Q2), the presence of unnecessary sound (Q3), the speech intelligibility (Q4), the listening effort (Q5), the speech clarity (Q6), the sound quality (Q7) and the overall preference (Q8). Ratings could be given on a seven point scale with a division into seven levels ranging, from *totally agree with U* to *totally agree with RR*. The construction of the response options were equivalent to the abscissa in Fig. 2. Each participant completed seven questionnaires (three for LSL, three for HPH and one for RL).

3   Results and Discussion

3.1   Comparison between LSL and HPH

Fig. 2 shows a comparison between the median responses, given in LSL and HPH. In both lab situations the participants had on average a preference for RR. In LSL, the more reverberant the sound example was, the higher the preference was. In the HPH part, the ratings were closer together than in LSL. Here, ratings for the sermon and the lecture examples were almost identical.

3.2   Comparison between LSL, HPH and RL

To compare all three setups, only the sermon example and question one were considered; *In which program do you perceive less reverberation?* For this comparison, the participant groups were compared based on the setup type and response frequency for the subjective listening criteria. The results are shown in Fig. 3.

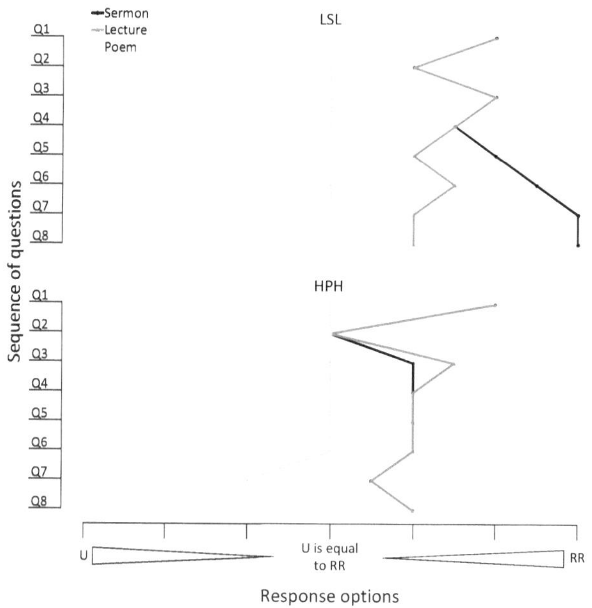

Figure 2: Medians of the response options to the questions, for combined participant groups, all sound examples in the two lab setups.

With the setup LSL, all EL and HI had a lower perception of reverberation with RR. There was only a slight difference between the participant groups: the EL tended a little more to RR. With HPH, EL and HI had not such a clear perception as for LSL. On average the HI had a slight preference for RR, but could not decide easily. Some HI even mentioned this while performing the test. With HPH the EL preferred RR again, but a little less than with LSL. In RL, ratings of HI and EL were the most different; the EL perceived the least reverberation in RR again with a slighter preference than in LSL and even HPH. The RL part was only performed by six EL. Most of the HI perceived less reverberation for U.

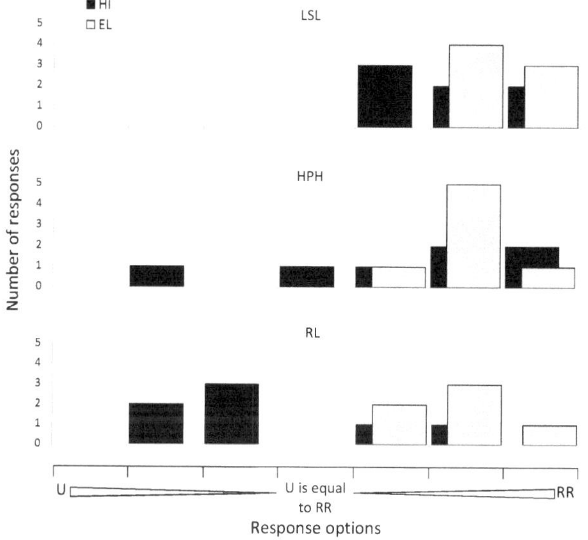

Figure 3: Response options for question one (*less reverberation intensity*), depending on a setup and participant group.

Overall, Fig. 3 indicates a difference between the setups - although a Wilcoxon signed-rank test showed no significant difference (see 3.3). EL preferences differed slightly between the setups, but showed a trend towards RR. Comparing the LSL part to HPH part, one HI perceived less reverberation with U compared to RR. So the preference towards RR was weaker among the HI. In RL the HI participants generally preferred the U program.

3.3   Comparison between HI and EL

For a better overview Fig. 4 was made using the median responses from all questions. The boxplots are for HI and EL in every setup. Here, every sound example was included.

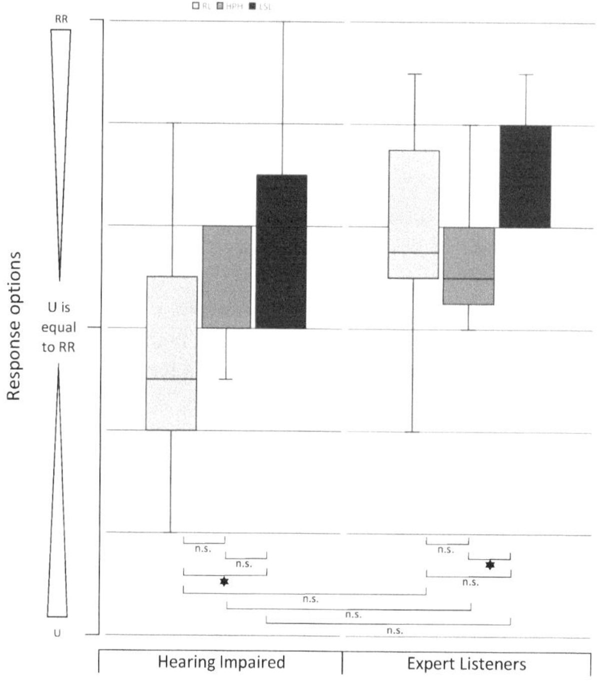

Figure 4: Medians of response options for HI and EL, for all three setups.

The HI preferred RR in both lab setups. With LSL no median showed a trend to U. In RL the HI had no preferences. Only the RL and LSL setups showed a significant difference. Regarding responses of EL in different setups, nearly all of the responses indicated a preference towards RR. Like the HI, in the LSL, the EL chose RR most often. The biggest difference between the HI and the EL became visible in RL. Here the EL preferred for RR while the HI were undecided. Comparing the responses of EL depending on setup, only HPH and LSL showed a significant difference. Altogether, the ratings differed significantly between setups, but not between participant groups.

3.4 Discussion

For the HI, hearing loss was compensated based on a cluster fit, without any individual fine tuning of HAs. Thus, for the HI, the sound differed from their own HAs, which may have caused them to unintentionally compare the experimental settings with their own hearing aids. To alleviate this issue in the future a fine tuning of study HAs could be performed. Because of the different reverberation times, and their noticeable influence on perception (see Fig. 2), it would be better in future to consider the effect of only one reverberation time at the time. Additionally, listening with HPH should not involve a parallel task like handling a GUI. It is known that parallel tasks and more reverberation puts higher demands on the working memory for elderly and hearing impaired people. In this study the HI were also elderly and so the impact on their working memory was increased. LSL and HPH contained the same sound examples, but the ratings were not the same (see Fig. 2). A possible explanation is the loss of individual HRTFs using HPH. Listening to an averaged and not an individual HRTF may harm naturalness in such a way that sensitivity for reverberation is reduced [3], [5].

As shown in Fig. 3, subjective ratings differ between setups. An obvious reason is the fact that all sound examples were optimized for RR in the LSL situation and thus, ratings for this condition were expected to be good. The reason that there was no preference for RR in RL given by the HI (see Fig. 3) could be due to the architecture of the different churches. The chosen churches do not have the same reflections, which results in different acoustic characteristics. Another reason for the lack of RR preference here is that reverberation in real environments is easier to tolerate. Even the participants who felt uncomfortable with the reverberation in the lab did not find it annoying in the RL. The EL preference for RR was maintained in RL. We need to keep in mind that all EL went to the same pre-selected church. Ideally, the HI should have also gone to the same pre-selected church. The perception of reverberation in RL was not the same as the perception of reverberation in lab.

When comparing the ratings of EL and HI, it is clear that the setup had a higher impact on the ratings of HI than on the EL (see Fig. 4). Furthermore, the variance of ratings was higher for the HI than the EL. The suspected reason for this is due to the variance of the hearing losses of the HI. In the results, the different hearing thresholds of the HI could not have been analysed separately. Additionally, the EL were used to such studies, knew the intentions of and interpreted the questions differently from the HI. With more participants, the results could be clearer. For further studies a higher number of participants would be preferable.

In addition, it should be noted that this study focused on subjective outcome measures. In further studies the memory effect could be measured, in order to obtain some objective results, too. The focus in this study was on the reverberation time only. However, the factors of target level, noise level and signal-to-noise-ratio are equally important in judging reverberation preferences and should be considered in the future [3], [5], [6].

4 Conclusion

The aim of this study was to assess how different participant groups, sound examples and setups affect subjective responses. Altogether, we can conclude the following three conclusions. First and foremost, the ratings of HI and EL participants should not be considered equivalent. The EL show a greater preference for dereverberation than the HI. The largest such difference between EL and HI was measured in RL. At the same time the responses within each group had biggest variances. Secondly, the more reverberant a situation, the more participants (whether or not hearing impaired) preferred a dereverberation (an exception was HI in RL). Thirdly, the benefit from RR in both lab situations and RL was not the same. All in all, this study emphasises the importance of RL tests with HI participants when evaluating hearing aid algorithms.

Acknowledgement

The study was carried out at Sivantos GmbH, Erlangen and was supported by the Institut für Akustik, TH Lübeck.

5 References

[1] B. W. Hornsby, G. Naylor and F. H. Bess, *A taxonomy of fatigue concepts and their relation to hearing loss.* Ear and hearing, 37(Suppl 1), 136S, 2016.

[2] R. Ljung, P. Sörqvist, A. Kjellberg and A. M. Green, *Poor listening conditions impair memory for intelligible lectures: implications for acoustic classroom standards.* Building Acoustics, 16(3), 257-265, 2009.

[3] A. Kjellberg, *Effects of reverberation time on the cognitive load in speech communication: Theoretical considerations.* Noise and Health, 7(25), 11, 2004.

[4] P. Folkeard, V. Littmann and S. Scollie, *Using a Dereverberation Program to Improve Speech Intelligibility and Reduce Perceived Listening Effort.* Hearing Review, 24(4), 32-33, 2017.

[5] R. Ljung and A. Kjellberg, *Recall of spoken words presented with a prolonged reverberation time.* Building Acoustics, 16, 301-312, 2009.

[6] I. Holube, K. Haeder, C. Imbery and R. Weber, *Subjective listening effort and electrodermal activity in listening situations with reverberation and noise.* Trends in hearing, 20, 1-15 doi:2331216516667734, 2016.

Skin Conductance as a Measure of Listening Effort: A Pilot Study

Markus Kemper [1], Nicholas Herbert [2] and Jürgen Tchorz [3]

[1] Auditory Technology, Universität zu Lübeck, markus.kemper@student.uni-luebeck.de
[2] Sonova AG, Stäfa, nicholas.herbert@sonova.com
[3] Institut für Akustik, Technische Hochschule Lübeck, juergen.tchorz@th-luebeck.de

Abstract

Listening effort is an important factor for hearing aid fitting and might be useful for evaluating applications such as noise reduction algorithms. One physiological measure of listening effort is electrodermal activity (EDA) which increases during stressful situations. The aim of this pilot study was to investigate potentially influencing factors (speech intelligibility, absolute sound pressure level, artefacts due to speech production/movement and dual-tasks) on EDA. 11 subjects performed a speech in noise task at different sound pressure levels (in appointment 1 closed vs. open response and in appointment 2 with vs. without dual-task) while their EDA was recorded. To answer the research questions, EDA amplitudes in all 8 conditions were compared. Results showed a dependence of EDA amplitudes on absolute sound pressure level and condition-order. Furthermore, subjects exhibited reactions to both speech material and the secondary task. Future studies using EDA to measure listening effort should be mindful of these dependencies.

1 Introduction

In everyday life, speech intelligibility (SI) is often influenced by disturbing factors like ambient noise or reverberation. These can have a negative effect on speech intelligibility, which is regarded as the most important component of communication [1]. Ref. [1] describe the effort for understanding as another element susceptible to these factors. This „listening effort" can vary even if SI remain constant (Fig. 1 (dashed lines)). It is assumed that listeners

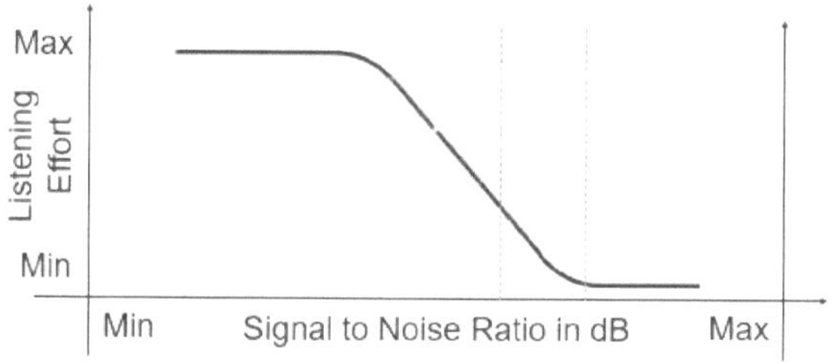

Figure 1: Estimated relationship between speech intelligibility and listening effort. The grey line shows the SI and the black line the listening effort.

feel increased stress in situations with high listening effort [2]. Stress affects the autonomic nervous system (ANS). In stressful situations (or during periods of effort) the sympathetic nervous system, the stimulating part of the ANS, influences many physiological processes. Variations in these can be quantified, e.g. by changes in pupil dilation or electrodermal activity (EDA, also known as skin conductance) [3]. The measurement of listening effort, e.g. via physiological measurements, has been investigated more closely

in recent years, as it is assumed that it is another factor that could be important for the fitting of hearing aids [2]. For example, such measurements could be used in the evaluation of noise reduction algorithms, because they do not effect SI but are perceived positively by hearing aid users [4]. Standardized measures of listening effort are not currently agreed upon and due to varied study methodologies it is difficult to compare outcomes. When measuring listening effort, dual tasks are used often, as they measure the remaining cognitive resources which are bigger when the subject is less stressed [6]. In this study EDA was used as an indicator for determining listening effort. EDA consists of two components: A slowly changing curve (tonic) with quickly changing phasic activity superposed. The phasic response reflects sympathetic nervous activity and could result from increased listening effort. However, the phasic part is very sensitive and can be contaminated by movements or speech production artefacts. Whether the reaction is caused by a stimulus of interest can only be determined on the basis of a defined time point [3] (Fig. 2). The focus in this pilot

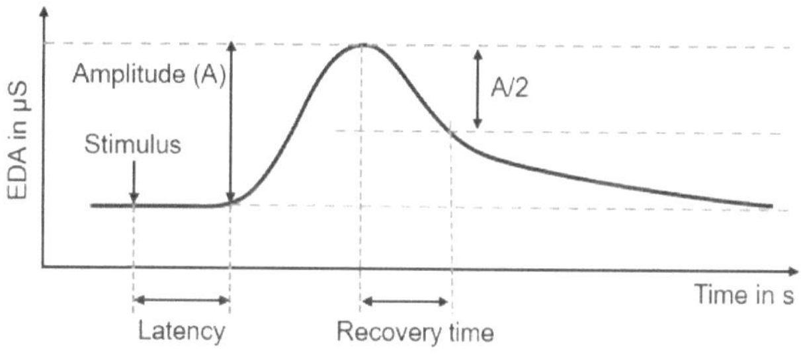

Figure 2: Phasic reaction after a given stimulus at a defined time point.

study was to generate more knowledge about influences of SI, sound pressure levels (SPL), impact of a dual task and artefacts due to movement and speech production on EDA measures.

2 Material and Methods

2.1 Subjects

All subjects were employees of Sonova AG, who voluntarily attended this study. They were native (Swiss-)German speakers, above 40 years old and without age-related atypical hearing loss (the limit was a pure tone average (PTA) of 25 dB HL for the poorer ear). In total eleven subjects (ten men and one woman) participated in this study at the age of 48 to 59 years ($\bar{x} = 53.73$ years $\sigma = 3.36$ years) with a PTA between 3.75 dB HL and 20 dB HL ($\bar{x} = 12.78$ dB HL $\sigma = 4.99$ dB HL).

2.2 Study Material

The study was executed in a sound-proof booth at Sonova AG (Stäfa). All physiological data were recorded with a Nexus 10-MKII (Mind Media B.V.). Signals were played via twelve MM-4XP loudspeakers (Meyer Sound) controlled by a Lynx Aurora 16 (Lynx Studio Technology, Inc.) sound card. The experiment was run using custom software written for MatLAB R2017b (MathWorks, Natick, MA) running on a PC (64-bit Windows 10 Professional). Statistical analysis was undertaken using R Studio 3.6.1 (RStudio Inc.).

2.3 Stimuli

A speech-in-noise task was used as the auditory task. Speech material consisted of sentences from the Oldenburg Sentence Test (OLSA)[5]. Each sentence is constructed according to the scheme „Name-Verb-Number-Adjective-Substantive", e.g. „Ulrich gives seven heavy seats". There are not always meaningful sentences, which has the advantage that it is more difficult to memorise them. Noise material was a modified cafeteria noise (noise recorded in a cafeteria and superposed with the long-term spectrum of the OLSA speech material). Furthermore a 250 Hz sinus tone was created by using Matlab, which was used as a warning tone. For each OLSA sentences a twelve channel audio file was created, consisting of one OLSA sentence and eleven decorrelated modified cafeteria noise channels with a constant Signal-to-Noise Ratio of - 4 dB. In addition, the warning tones were implemented in two points to determine the startpoint of one trial and to give a signal for repeating the OLSA sentence. Subjects sat in the middle of a 12 loudspeaker ring, arranged 30 degrees apart from each other and 1.2 m meter away from subjects. A screen was placed in front of participants and they were instructed to look at it. Speech material was played from loudspeaker at 0 degrees, all other loudspeakers played the noise material.

2.4 First Appointment

The first appointment (APP 1) was divided into screening and actual measurements. Screening included audiometry, an anamnesis and otoscopy to be sure, that there were no unexpected factors which might impact EDA or the auditory task. If no exclusion criteria were found, the actual measurement started. Therefore the palmar surfaces of subjects' non-dominant hands were connected to Nexus 10-MKII via electrodes. To investigate whether different SPLs (76 dB or 66 dB) resulted in EDA variations and which artefacts (movement or speech production) had bigger impacts, four conditions were tested in this APP (Table 1). To con-

Table 1: Test conditions APP 1

Condition	Sum level	Response
1	76 dB SPL	closed
2	66 dB SPL	closed
3	76 dB SPL	open
4	66 dB SPL	open

trol position effects over all subjects, condition-order was pseudorandomized using a latin square. Independent from test condition, the measurement always took the same form (Fig. 3). At the beginning of each trial a warning tone was played for 0.25 s from loudspeaker at 0 degrees (1). This was followed by a 0.75 s break (2). Then each loudspeaker, except from 0 degrees, started playing the modified OLSA noise while the OLSA sentence was played from the loudspeaker at 0 degrees (3). The next step was a break of nearly 9 seconds, which depended on the individual length of the OLSA sentence (4). Then a second warning tone appeared for 0.25 s to indicate, that it was time to answer (5). In the last time slot the subjects gave their answer (6). When the warning tones were played at (1) and (5) time stamps were placed in the EDA data. For the closed con-

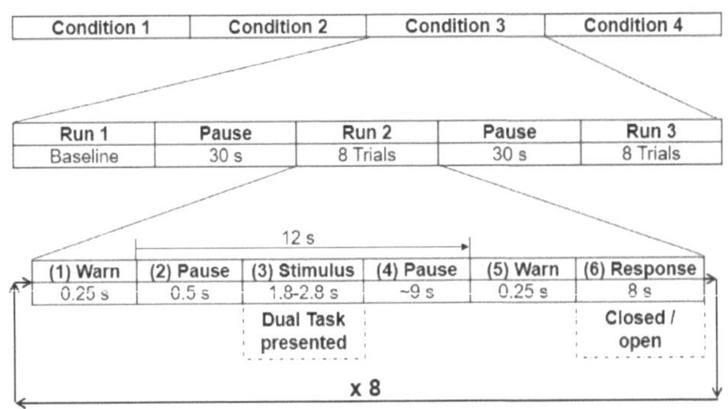

Figure 3: Schematic test procedure. The numbers (1-6) show the different sections within one trial.

dition a matrix (with all OLSA-words) appeared for 25 s on the screen and for the open condition there was a time window of 8 s for participants to answer. After „answering time" the paradigm was repeated. To prevent confusions, the screen in front of the subjects indicated all current tasks using picture icons. One condition consisted of two runs of eight trials with a between-run break of 30 s and a baseline measurement at the beginning which was made up of 8 trials without speech material and followed by a break of 30 s. Subjects were instructed to move as little as possible and to speak only in the between-run break, between conditions or at indicated times. Before the first measurement, subjects were introduced to speech material and the test procedure with 20 sentences at an SNR of 0 dB and a sum level of 65 dB. In addition to physiological measures, speech intelligibility for OLSA was also determined.

2.5　Second Appointment

APP 2 served for quantifying the influence of a dual task on EDA. Test procedure was equivalent to APP 1, but the screening part was omitted and for familiarisation the new test procedure with dual task was also practised before the first condition. As in APP 1, four conditions were tested in randomized order (Table 2). The trial- and baseline-measurements were equivalent to APP 1 (Fig. 3), except conditions 3 and 4, where a dual task was simultaneously presented with speech material within trials (Fig. 3 (4)). During the baseline measurement subjects also undertook the dual task. All responses were given as „open".

Table 2: Test conditions APP 2

Condition	Sum level	Dual Task
1	76 dB SPL	no
2	66 dB SPL	no
3	76 dB SPL	yes
4	66 dB SPL	yes

A visual-coordinative reaction test was used as a measure of spare cognitive capacity (i.e. to gauge listening effort). The task was presented using a custom script written in MatLAB R2017b using Psychophysics Toolbox extensions. It consisted of nine black fields, which were arranged in 3 rows and 3 columns. Two randomly selected fields changed their colour from black to white. In one a digit was displayed, and in the other an „X" was shown. Digits between one and eight were used to ensure an equal chance of even and odd numbers occurring. If an even digit appeared subjects were asked to click on the digit-field. For odd digits they were asked to click on the X-field. Subjects were instructed to carry out this task as quickly as possible.

2.6　Data Processing

EDA data were imported into Matlab 2019b, down-sampled to 128 Hz and then processed. To separate tonic and phasic components, signals were filtered in Matlab 2019b using a direct-form II, second-order, Chebyshev high-pass filter (cutoff frequency 0.07 Hz, pass-band ripple 0.5 dB) [7]. For each condition data were divided in baseline-data and trial-data. The mean response patterns for baseline- and trial- measurements (within-subjects) were calculated because similar responses were expected to each stimulus. Furthermore the calculation of these mean responses had the advantage of mitigating measurement noise (e.g.

movement artefacts). Amplitudes for each condition were computed from these mean response patterns using semi-automatic peak and trough detection in Matlab. All identified peaks and troughs were visually verified by the experimenter.

2.7　Statistical Method

To control whether the subjects reacted to the speech material a paired two sample t-test was performed on the amplitude of baseline-data and trial-data for each APP. To inform the main study questions, a linear mixed-effects model was fitted to the data in the form: $\text{Amplitude} \sim \text{Condition} + \text{SPL} + \text{SI} + \text{Order} + (1|\text{ID})$ where ID describes the individual stubject. To eliminate non-relevant variables a backward regression was performed. Post-hoc-analysis included an ANOVA to identify main-effects and subsequent pairwise comparison using Least-Squares Means.

3　Results and Discussion

Analysis of speech production and movement artefacts are incomplete and is not covered in this article. Due to substantial artefacts over the entire experimental session, the data of one subject was excluded leaving a total number of 10 participants.

3.1　Differences between Baseline and Trials

Fig. 4 shows the processed averaged baseline- and trial-data of all subjects in a single condition for APP 1 and APP 2. The trial-data exhibits an increase in EDA following both stimuli presentation and speech production. APP 1 paired t-test showed significant differences between amplitudes of baseline and trials ($p = <0.01$) which demonstrated subjects' reaction to the speech material (Fig. 4 left and right). In APP 2, amplitudes during baseline and trials did not significantly differ from each other ($p = 0.25$) (Fig. 4 right). This indicates, that the dual task impacts EDA measurements and in such a paradigm listening to speech material had no additionally measurable effect on EDA amplitudes. For further analysis only trial-data were used.

3.2　Linear Mixed-Effects Model

A linear mixed-effects model was fitted for each APP. Backward stepwise regression suggested the same reduced model for both the first and the second APP: $\text{Amplitude} \sim \text{SPL} + \text{Order} + (1|\text{ID})$. This indicates that speech intelligibility and condition did not have a significant impact on EDA amplitudes and are not considered further. No significant differences were found between the full and reduced models. Thus the reduced (simplified) models were retained for further analysis to keep the model as simple as possible. For APP 1 the model showed an increase of $0.08\,\mu S$, for the higher sum level of 76 dB. Also a main effect of order was indicated (higher order results

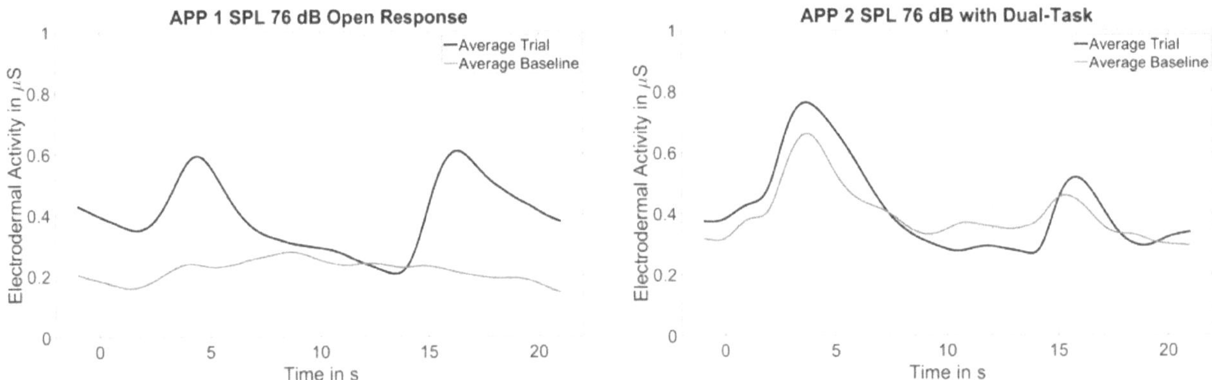

Figure 4: Averaged data from all subjects of one condition for APP 1 (left) and APP 2 (right) with time stamps at 0 s and 12 s.

in lower amplitudes). Post-hoc-analysis identified a significant order-effect $(Pr(> F) = 0.02)$ and a significant effect for SPL $(Pr(> F) = 0.03)$. Pairwise comparisons for the different orders showed a significant difference between orders 1 and 3 $(p = 0.03)$ and orders 1 and 4 $(p = 0.03)$. In further studies the SPL should therefore be kept constant to compare different conditions. The order-effect indicates a potential influence of fatigue and this should be monitored in further investigations for example by keeping the testing time to a minimum. For APP 2 the model showed an increase of $0.11\,\mu S$ for the higher sum level of 76 dB. Again a main effect of order was indicated. Post-hoc-analysis identified a significant order-effect $(Pr(> F) = 0.03)$ and a significant effect of SPL $(Pr(> F) < 0.01)$. Pairwise comparison for different orders showed a significant difference between orders 1 and 2 $(p = 0.04)$ and orders 1 and 3 $(p = 0.03)$. As seen in Fig. 4 (right) the dual task influences EDA measurement and could also leads to fatigue. This might be an explanation for the large order-effects.

4 Conclusion

The aim of this pilot study was to gain more knowledge about factors that can influence EDA measurements. Comparisons between trial- and baseline-data showed that subjects did react to the speech material, suggesting it may be a promising paradigm for testing listening effort. Furthermore subjects should not respond directly after speech material, as EDA-responses also appear to be triggered by speech production (Fig. 4). However, the dual-task also had an influence on measurements, so it is questionable whether amplitude rose due to auditory stimuli or as a result of the dual-task in certain measurements. SPL and condition-order had a significant effect on amplitude. The order effect might indicate an influence of fatigue and so in future appointments testing times should be kept to a minimum and the SPL should be kept constant to compare different conditions as EDA amplitudes depend on absolute SPL. In the following main study, the present paradigm will used as a basis concept to evaluate the benefit from noise reduction

algorithms with respect to the shown dependencies from this pilot study.

Acknowledgement

The work was conducted at Sonova AG, Stäfa and supervised by the Institut für Akustik, TH Lübeck.

5 References

[1] B. Gabriel and M. Meis, *Optimierung eines Messverfahrens für die Höranstrengung.* Zeitschrift für Audiologie, vol. 4, 2001.

[2] C. L. Mackersie and H. Cones, *Subjective and Psychophysiological Indexes of Listening Effort in a Competing-Talker Task.* Journal of the American Academy of Audiologie, vol. 22(2), pp 113–122, 2011.

[3] K. Gramann and R. Schandry, *Psychophysiologie: Körperliche Indikatoren psychischen Geschehens.* 4. Auflage, Beltz PVU Weinheim.

[4] I. Brons, R. Houben and W. A. Dreschler, *Effects of Noise Reduction on Speech Intelligibility, Perceived Listening Effort, and Personal Preference in Hearing-Impaired Listeners.* Trends in Hearing, vol. 18, pp. 1–10. 2014.

[5] K. Wagener, V. Kühnel and B. Kollmeier, *Entwicklung und Evaluation eines Satztests für die deutsche Sprache I.* Zeitschrift für Audiologie, vol. 38, pp. 5–15, 1999.

[6] S. Fraser, J. -P. Gagné and M. Alepins *Evaluating the effort expended to understand speech in noise using a dual-task paradigm.* Journal of Speech, Language, and Hearing Research, vol. 53, pp 18–33, 2010.

[7] H. S. Arnold, M. K. MacPherson and A. Smith, *Autonomic Correlates of Speech Versus Nonspeech Tasks in Children and Adults.* Journal of Speech, Language, and Hearing Research, vol. 57, pp 1296–1307, 2014.

Measurement of Spectro-Temporal Modulation Detection Thresholds with Hearing Impaired Listeners

Sebastian Griepentrog [1], Julia Rehmann [2], Stefan Klockgether [2], Volker Kuehnel [2], and Jürgen Tchorz [3]

[1] Hörakustik und Audiologische Technik, Universität zu Lübeck, sebastian.griepentrog@student.uni-luebeck.de
[2] Sonova AG, Laubisrütistr. 28, CH-8712 Stäfa, (julia.rehmann, stefan.klockgether, volker.kuehnel)@sonova.com
[3] Institut für Akustik, Technische Hochschule Lübeck, jürgen.tchorz@th-luebeck.de

Abstract

Hearing impairment can partially be compensated by using hearing aids. The fitting of hearing aids is based on the individual hearing threshold, measured via audiometry. With comparable hearing losses, hearing impaired people still experience differences in speech understanding, although the audiograms are similar and the hearing aid fitting seems to be sufficient. For further information on the individual hearing loss, spectro-temporal modulation (STM) detection thresholds were examined as an additional psycho-acoustic assessment paradigm. A study with 10 hearing impaired subjects was conducted. As measurement method, an alternative forced choice test was used and the subjects were invited to a training block and to two test appointments. The results show, that hearing impaired subjects are able to detect STMs and that frequency dependent detection thresholds are measurable. Further investigation might show the benefits of a hearing aid feature fitting, based on the extra information obtained from the STM measurements.

1   Introduction

Hearing impaired people require amplification to make sounds audible and to overcome their hearing loss. Therefore they can be fitted with hearing aids. For fitting the hearing aid, it is necessary to know the degree of hearing loss of the hearing impaired person. This measurement is covered by standard audiometry, which results in an individual's hearing threshold. This information will be used by the fitting software to calculate the prescriptive gain, which is the current standard procedure [1].

In reality, the end-user feedback and several studies show that the benefit gained from a hearing aid fitted with regard to the individual hearing threshold can vary strongly [2], [3]. Even within groups of people with almost identical audiograms, the individual benefit from wearing a "well fitted" hearing aid varies strongly. Especially when it comes to the fitting of parameters for particular hearing aid features like a noise canceller, a standard audiogram seems to provide insufficient information on individual hearing loss or the respective residual hearing. One explanation for the varied success of fittings may be a reduction of the frequency resolution capability in the impaired ear.

The frequency resolution capability is a measure of the spectral sensitivity of a listener. The ability is mostly linked to the inner ear's hair-cells and it is often reduced in an impaired auditory system [4]. The determination of the spectral sensitivity of an individual is usually done with a notched noise experiment, which measures the widths of particular auditory filters. The filter width gives an indirect measurement of the frequency resolution capability at the tested frequency. This method is elaborate, slow and only uses the indirect connection between filter width and frequency resolution capability [5]. The approach we use is the perception of spectro-temporal modulation (STM), which is considered to be connected to the frequency resolution capacity [6]. In this way, the availability of more complete, individual data for the fitting would allow a more precise parameterization of various hearing aid features, especially concerning the individual affection by artifacts or distortions which might be generated by a strong hearing aid feature setting. This could result in better adaptation of the hearing aid to individual abilities and needs of the hearing impaired individuals and thus lead to more benefit and satisfaction with the hearing aids. Due to recent findings, an immediate improvement in fitting for the hearing aid features could be expected [7], [8]. With the individual frequency resolution capability as an indicator for e.g. a hearing impaired listener being an artifact hater, a weaker noise canceller setting would be recommended. To obtain knowledge on the extent, to which STM measurements may produce reliable results and if octave wide STM detection thresholds can be measured with hearing impaired listeners, as well as to gain information on the individual frequency resolution capability, a study with hearing impaired people was conducted.

2   Material and Methods

2.1   Subjects

10 hearing impaired listeners (6 male, 4 female) were tested. The age range was 28–86 years (average: 74.3 years, standard deviation: 18.3 years). Since the first question to be answered was whether such thresholds are generally measurable with hearing impaired people, no specific hearing loss was requested. Rather, subjects with a wide range of hearing losses were tested to see if there are limitations of the paradigm, due to the individual hearing loss. As shown in Fig. 1, the individual hearing impairments were widely spread. It was also not necessary for the subjects to be experienced hearing aid users, as the measurements are all conducted without the use of hearing aids. Limited cognitive abilities and normal hearing were defined as exclusion criteria. The subjects were invited for three appointments, to which all of them also attended to.

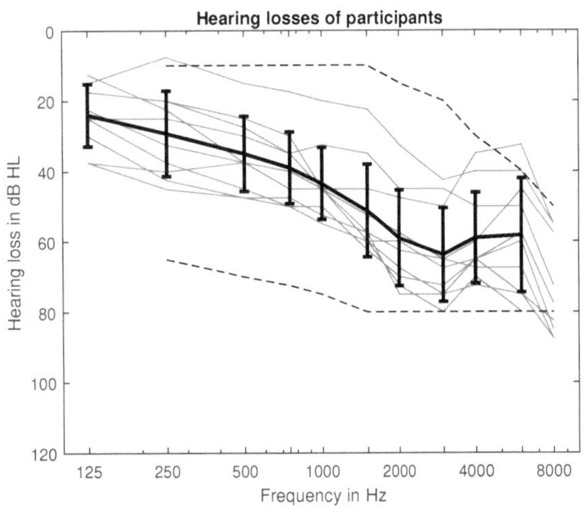

Figure 1: Means of right and left audiograms of all participating subjects (grey lines) and mean of all audiograms with standard deviation (black line with errorbars). The dashed lines show the range between normal (upper dashed line) and severe (lower dashed line) hearing loss, as defined in [9].

2.2   Stimuli

The target stimuli for the STM measurements, spectro-temporal ripples, are linked to the individual frequency resolution capability. Such ripples consist of a white noise carrier which is modulated simultaneously in the time domain (amplitude modulation) and the frequency domain. Besides the temporal modulation frequency and the frequency modulation repetition rate, one could use either the bandwidth of the modulated noise carrier or the modulation depths as parameters in a detection task. Fig. 2 shows some examples of STM stimuli with different temporal and spectral modulation frequencies. As there are many different combinations of modulation frequencies possible, there is a big

difference in detection capability. For this study a temporal modulation frequency of 4 Hz and a spectral modulation rate of 2 cycles per octave was chosen for all tested conditions, since this combination seems to be the easiest to detect [6]. The modulated bandwidth was set to either one or four octaves, logarithmically centered around the particular test-frequencies. An octave wide modulation with two cycles per octave modulation rate would mean the frequency sweep could be perceived twice per stimulus.

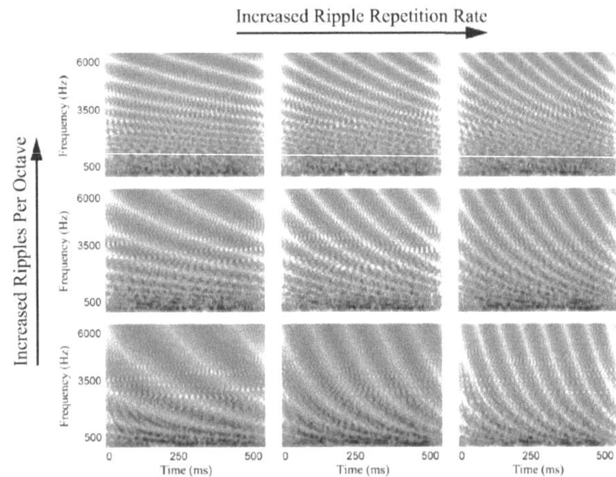

Figure 2: Exemplary depiction of the effects of an increased ripple repetition rate and/or increased ripples per octave on STM ripples [10].

2.3   Alternative Forced Choice Measurement

The measurements were carried out in the form of an alternative forced choice (AFC) paradigm using an 1-up-2-down method. In this type of measurement, the subject is forced to choose a response option, even though the target might be below the detection threshold and the subject has to guess. This makes the AFC-paradigm criteria free. In this study, three intervals were provided, which all could randomly contain the target stimulus.

Three intervals of noise signals lasting one second each were presented, of which one included the target stimulus. The other two intervalls contained another unmodulated random white noise carrier. The task was to identify and name the number of the target interval. With a false response, the modulation index of the target was increased by one step (1-up); however, only after two correct responses, the modulation index was decreased by one step (2-down). For each run, the starting condition was fully modulated with the modulation index set to "0". The lower the modulation index is, the harder gets the task. A lower detection threshold would correspond to a better frequency resolution capability. As there is a guess probability of 1/3, the psychometric function lies between 33% and 100% with the steepest point close to 67%. The detection threshold probability is p = 0.5, meaning 50% on the psychometric function, which lies in a rather flat area. The 1-up-2-down procedure increases the detection threshold probability to

$p^2 = 0.5$, as there are two target hits required to lower the modulation index. The detection threshold probability achieved in this way is $p = 0.707$, thus about 71% on the psychometric function and very close to its steepest part [11]. The modulation index stepsize is adaptive and started with 4 dB. After two reversals (false responses), the stepsize decreased to 2 dB and after two additional reversals, the stepsize decreased to 1 dB. As soon, as the smallest stepsize was reached, the next 6 reversals were used to calculate the detection threshold as a mean with standard deviation.

2.4 Measurement Procedure

All measurements were carried out in a psychoacoustic research lab and the following equipment was used: An *Aurical* audiometer by *Otometrics* with standard audiometry headphones and a PC with *Matlab 2019a* installed and connected to the headphones via a *RME Fireface* soundcard. The subjects were measured on three appointments in total. The first session included a screening and a training block on the target stimuli. For this purpose, three conditions with modulated broadband stimuli in different octave widths (4, 2, 1 octaves) were measured with the AFC method; each condition was measured in two runs. At the second and third appointment the AFC procedure was performed as test and retest measurement. One such experiment consisted of seven conditions. The modulated stimuli were adjusted to a four-octave-wide broadband condition around the center frequency of 2750 Hz and 5 one-octave-wide narrowband conditions, logarithmically centered around the following centerfrequencies: 900 Hz, 1400 Hz, 2000 Hz, 3000 Hz and 5500 Hz.

To ensure a better detection and according to literature, the unmodulated frequency areas of the carrier noise were set off 15 dB lower than the modulation frequencies [6]. The broadband stimulus was fixed for the first and last run, in the remaining 5 runs the narrowband conditions were presented in a randomized order. All stimuli were presented binaurally (dichotic, with a different random white noise carrier on both ears) over headphones with a level of at least 70 dB SPL and additional 15 dB sensation level depending on the individual hearing threshold (within a range up to 100 dB SPL). In detail, if a subject's hearing threshold would be at 65 dB SPL, the stimulus is presented with additional 15 dB - in this case 80 dB SPL. That also means, at the higher frequencies, the hearing level limit might be rather low.

3 Results and Discussion

For all participants but one, individual and octave wide STM detection thresholds could be assessed. The results show, that there are not only learning effects within the broadband conditions of the same appointment but also when comparing test and retest. The median retest broadband STM detection thresholds are up to 3 dB better, than the test broadband STM detection thresholds. Fig. 3 shows the boxplots of test (upper plot) and retest (lower plot) measurements, the broadband (BB) conditions each on the left

side of the plots and separated from the octaveband (OB) conditions by the solid black line. The BB conditions were fixed for the first and last run and are displayed as such. There are also learning effects in the octaveband conditions, as the range of the boxes/whiskers is decreased in the retest measurements and also the median STM detection thresholds for the two high frequency conditions are up to 2 dB better than the thresholds measured on the first appointment. The results also show a tendency for easier detection of modulations at higher center frequencies, which was expected, as even for normal hearing subjects, modulations at lower center frequencies are harder to detect [12].

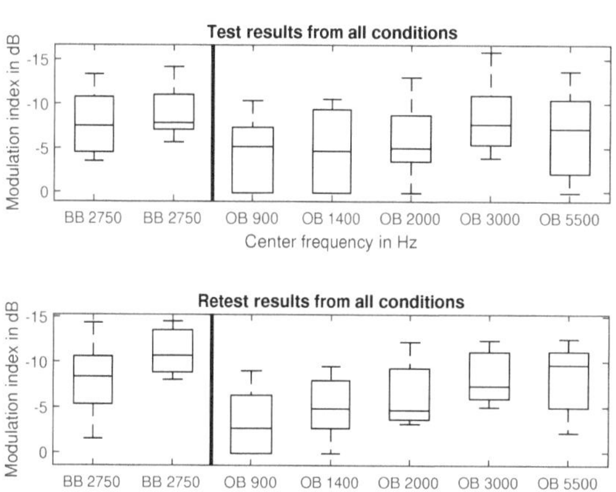

Figure 3: Boxplots of STM detection thresholds from test and retest. The broadband conditions on the left were measured as first and last run for each subject. The boxes show the median and the interquartile range.

Fig. 4 shows the test versus retest accuracy of the modulation index from all measured frequencies and all participants. The closer the scatterdots lie to the indentity line (dashed line), the more accurate is the test-retest result. With a correlation coefficient of 0.83 (83%) and a p-value < 0.001, a highly significant correlation between the test and retest samples is implied (solid line).

With all participants but one, octave wide STM detection thresholds could be achieved. The one subject, not being able to detect the modulations simply may have lost most of the individual frequency resolution capability. This is also a possible and expected outcome of the experiment, as one of the goals was to assess the individual frequency resolution capability by measuring STM detection thresholds. For this one subject, the results could imply, that the individual frequency selectivity is no longer available. This subject's hearing loss also was not the strongest one amongst the participants, that means this occurance cannot be explained by a very poor hearing threshold. Such cases might generally not be affected very much by artifacts or distortions generated by the hearing aid's feature settings. The STM thresholds of the remaining subjects vary individually and were not only measurable, but also differed in their magnitude. This means that there is not only a result in terms of thresh-

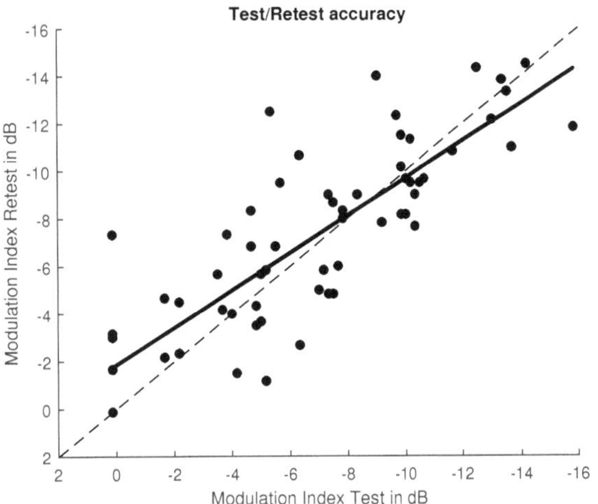

Figure 4: Scatterplot of Test versus Retest modulation index for each participant in all measured frequencies. The dashed line is the identity line and the solid line is the regression line of all samples.

olds being measurable or not, but it also implies that the paradigm allows a feasible dynamic range.

As this study was a pilot, the sample size is rather small and no further statistical evaluation was done. Further studies with a larger number of subjects might produce more statistically interpretable results, such as correlation between age, hearing loss and STM detection thresholds.

4 Conclusion

Measurement of STM detection thresholds is possible with hearing impaired listeners. Only one out of the ten participants was not able to pass the threshold measurement. As with all other participants thresholds could be measured, future studies might concentrate on the benefit from certain hearing aid feature settings, based on the individual STM detection threshold. Different noise canceller settings (at least one based on knowledge from STM threshold measurements) could be fitted and subjective insight via home trial, as well as the relation to other measures such as speech reception thresholds in noise, might be a criterion for spontaneous acceptance and/or listening comfort. With a positive outcome of such measurements, an additional psychoacoustic measurement tool for assessment of the individual hearing loss could be developed and provided.

Acknowledgement

The study was conducted at Sonova AG, Stäfa (CH) and supervised by the Institut für Akustik, Technische Hochschule Lübeck (D).

5 References

[1] Bagatto et al. *Clinical protocols for hearing instrument fitting in the desired sensation level method.* Trends in Amplification 9(4), pp. 199–226, 2005.

[2] R. Bentler, E. Walker, R. McCreery, R.M. Arenas and P. Roush *Nonlinear frequency compressionin hearing aids: Impact on speech and language development.* Ear Hear, pp. 143–152, 2014.

[3] Wolfe et al. *Preliminary evaluation of a novel nonlinear frequency compression scheme for use in children.* International Journal of Audiology, pp. 976–988, 2017.

[4] K. Hopkins and B.C.J. Moore *The effects of age and cochlear hearing loss on temporal fine structure sensitivity, frequency selectivity, and speech reception in noise.* The Journal of the Acoustical Society of America, vol. 130 pp. 334–349, 2011.

[5] B.R. Glasberg and B.C.J. Moore, *Derivation of auditory filter shapes from notched-noise data.* Hearing Research, vol. 47, no. 1–2, pp. 103–138, 1990.

[6] G. Mehraei, F.J. Gallun, M.R. Leek and J.G.W. Bernstein, *Spectrotemporal modulation sensitivity for hearing-impaired listeners: Dependence on carrier center frequency and the relationship to speech intelligibility.* The Journal of the Acoustical Society of America, vol. 136, pp. 301–316, 2014.

[7] J. Zaar, L.B. Simonsen, T. Behrens, T. Dau and S. Laugesen, *Towards a clinically viable spectrotemporal modulation test.* Tahoe City, USA, International Hearing Aid Research Conference (IHCON), 2018.

[8] R. Sanchez-Lopez, T. Dau and M.L. Jepsen, *Hearing-aid settings in connection to supra-threshold auditory processing deficits.* Nyborg, Denmark, International Symposium on Auditory and Audiological Research (ISAAR), 2019.

[9] IEC 60118-15:2012, *Electroacoustics - Hearing aids - Part 15: Methods for characterising signal processing in hearing aids with a speech-like signal.*

[10] J.M. Aronoff and D.M. Landsberger, *The development of a modified spectral ripple test.* The Journal of the Acoustical Society of America, vol. 134, pp. 217–222, 2013.

[11] H. Levitt, *Transformed up-down methods in psychoacoustics.* The Journal of the Acoustical Society of America, vol. 49, pp. 467–477, 1971.

[12] S. Klockgether, J. Rehmann, P. Derleth *Rapid measurement of frequency depending spectro-temporal modulation detection thresholds.* The Journal of the Acoustical Society of America, vol. 145, p.1720, 2019.

Assembling and verifying a hearing-aid prototype, consisting of a Raspberry Pi and the opensource-software openMHA

Marten Geisen [1], Siegrid Meier [2], and Tim Jürgens [3]

[1] Auditory Technology, Universität zu Lübeck, marten.geisen@student.uni-luebeck.de
[2] Akademie für Hörakustik, S.Meier@afh-luebeck.de
[3] Institute of Acoustics, Technische Hochschule Lübeck, tim.juergens@th-luebeck.de

Abstract

In this research project a corporate independent hearing aid prototype working with a 'Raspberry PI' Computer and the opensource software *OpenMHA* was build and technically verified. The assembled mobile hearing device and a state-of-the-art hearing aid were compared via an acoustic coupler according to the standard 'DIN EN 60118-0'. The results show that the prototype provides stable gains up to 34 dB. However, maximal output level and overall gain were lower and total harmonic distortion higher than in the state-of-the-art hearing aid. The construction of a cost-effective and company-independent hearing aid is interesting as a device for research or self-construction to test developed signal processing algorithms in real-world situations. Moreover, it can lower the entry barrier for hearing aid algorithm development and evaluation, because interested people can actively contribute to testing and improving hearing devices.

1 Introduction

Hearing aids help the wearer to follow conversations and acoustic events in different situations. Specialized for speech understanding, hearings aids are the most common form to treat mild to moderate hearing losses. To treat hearing-impaired people accordingly, effective signal processing algorithms are needed. For this reason, the development and improvement of signal processing algorithms in hearing aids is an important research topic. Unfortunately, signal processing in hearing devices is a company secret. For some time now research has been underway to develop a cheap hearing aid technology for do-it-yourself construction. The open-access software *open Master Hearing Aid(openMHA)* [1] is an algorithm development and evaluation platform for real-time hearing aid software development. This platform essentially consists of three elements: a software development environment with a signal processing library, a real-time environment for standard PC-hardware and ARM-based mobile platforms, and a standard set of reference algorithms to simulate a complete hearing aid such as directional filters and algorithms for noise reduction as well as compression. The real-time system can be configured to behave as a hearing device. In combination with an ARM-based portable single-board Computer like the *Raspberry Pi*, it is possible to build a cheap hearing aid. Though this setup does not provide the processing power of a normal PC it offers a lot more mobility. This portable device can be utilized to conduct field tests of hearing aid processing methods, which are necessary to assess the usefulness of new and individual hearing aid algorithms in real-

istic situations. Even if the construction of the prototype is primarily intended for testing signal processing algorithms, it makes sense to test it first objectively using a verification. Hearing aids from companies must be electro-acoustically verified before they are put on the market. The purpose of this verification is to determine if the hearing instruments meet a particular standard and are performing as expected. Moreover, verification is a necessary step to ensure that the patient has maximized the full value of features of the hearing instrument because it provides professional audiologist with more details and accuracy when it comes to the fitting process. Several researchers have suggested that verification can lead to a reduced number of return visits and increased patient satisfaction [2]. However, this hearing aid prototype has not yet been verified. In this work this set-up will be assembled, electro-acoustically verified and compared with a state-of-the-art hearing aid, the Signia Motion 13 7 Nx. This can be considered as a first step to test the suitability of the prototype for everyday use and the signal processing algorithms.

2 Material and Methods

2.1 Structure of the prototype

The assembled mobile hearing device can be seen in Fig. 1. It consists of a *Raspberry Pi*, an external soundcard, a microphone preamplifier, and a binaural stereo microphone headset. The set-up is taken from [3] [3]. A sound card especially developed for the *Raspberry Pi* is used that offers six input channels and eight output channels. Although the

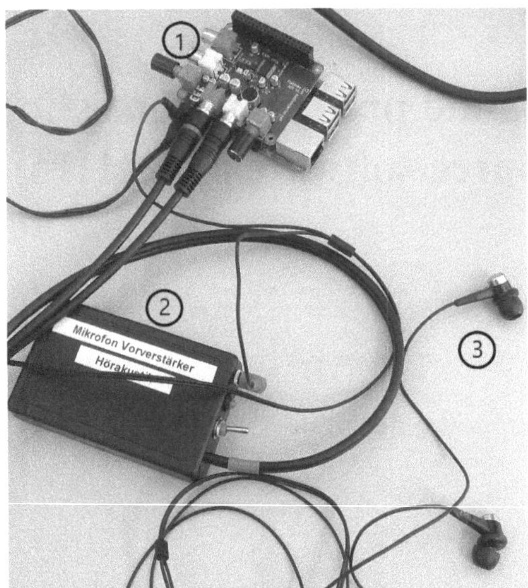

Figure 1: Structure of the prototype. (1) Raspberry Pi with attached sound card (2) Microphone preamplifier (3) Binaural stereo microphone headset

Raspberry Pi offers no hardware support for more than two sound channels, this sound card manages to offer enough input channels to connect two hearing aids with three microphones each [4]. To raise the microphone signals to line level, external microphone preamplifiers are needed. This set-up aims to maintain as much mobility and suitability as possible while being available for any interested person at a low cost of about 250 €. To make details of the individual components, the used software and an instruction manual, more easily accessible a Github wiki was created [3]. The prototype can be controlled via a graphical user interface(GUI) implemented in *MATLAB*. This involves especially settings in the context of hearing aid fitting.

2.2 Measurement Setup

The assembled mobile hearing device and a state-of-the-art hearing aid fabricated by Signia were tested and verified using an acoustic coupler included in the hearing instrument testing module *Aurical Hit*, which is integrated with the *OTOsuite* hearing care platform by *Otometrics*. A seal was applied using *Blu-Tack* adhesive compound to ensure that the headphones were correctly seated in the in-ear coupler connector. Furthermore, the seal is used to reduce feedback and background noise. Both hearing devices were tested in agreement with the standard *DIN EN 60118-0* [5]. This standard gives practice-orientated recommendations for measuring the performance characteristics of air-conduction hearing aids. Using the standard for both devices ensures comparability and reproducibility. Before the measurement, the *Aurical Hit* was first calibrated, so that an offset can be accounted for afterward. As a control, a 60 dB sine wave was played before each measurement to see how strong the offset is. Since the standard for the various

measurements requires the hearing aids to be set to either in *reference test setting of the gain control (RTS)* or the *full-on gain (FOG)* setting, these had to be determined first. The RTS is the maximum position of the gain control at an input SPL of 60 dB. The FOG-setting is the setting for maximum gain. Even before the actual measurements, it was noticed that internal feedback quickly occurred within the prototype. Feedback occurred from a linear gain of 14 dB and could be limited to a frequency of 4 kHz. Therefore, both the FOG and the RTS values in this frequency were chosen below 14 dB. This also prevented the use of compressive signal processing methods, since these immediately led to feedback, so that all results were based on measurements with linear gain algorithms at least for the prototype. For the measurements, a hearing loss of 40 dB over the whole frequency range was assumed for both devices. All frequency response curves were measured in the frequency range of 200 Hz to 8000 Hz for the prototype and 125 Hz to 8000 Hz for the hearing aid. The OSPL90 frequency response curve was recorded with an input SPL of 90 dB in the FOG-setting, to determine the maximum output SPL. The normal frequency response curve was recorded with an input SPL of 60 dB in the RTS, this curve best reflects the behavior of the hearing aid in everyday life. The frequency response curve of acoustic amplification was recorded with an input SPL of 50 dB in the FOG-setting, to determine the maximum gain. The maximum gain in the high-frequency is calculated, as HFA-FOG, by taking the HFA-value from the frequency response curve minus the input SPL of 50 dB. The *high-frequency average (HFA)* is determined by reading and averaging the output levels values at 1000 Hz, 1600 Hz and 2500 Hz from the different frequency response curves. The *Reference test gain (RTG)* is the HFA with an input SPL of 60 dB. The total harmonic distortion (THD) and input noise were measured with the settings specified in the standard. The power consumption of the prototype was measured with a USB measuring device, the power consumption of the hearing aid was measured with the *Aurical HIT*.

3 Results and Discussion

On the left earphone, the FOG-setting was determined with 24 dB gain in the frequencies 250 Hz, 500 Hz, 1000 Hz, and 2000 Hz, as well as 13 dB gain at 4 kHz and 14 dB at 8 kHz. The RTS was 1 dB lower over the whole frequency range. On the right earphone, the FOG-setting was determined with 27 dB gain at the frequencies 250 Hz, 500 Hz, and 1000 Hz, as well as 24 dB at 2 kHz and 13 dB at 4 and 8 kHz. The RTS was 2 dB lower over the whole frequency range. The FOG-setting of the hearing aid amplifies with 43 dB and the RTS with 39 dB master gain. The latency of the prototype measured to be about 3 ms. Fig. 2 and Fig. 3 show the results of the measurements of the hearing aid prototype as acoustic frequency response curves. Fig. 4 shows the results of the measurements of the hearing aid as an acoustic frequency response output level as a function of frequency. While the curves are similar for the earphones

Table 1: Data sheet comparison of the two hearing aids. The results were measured with a 2ccm coupler. For the prototype the results are given for the right and left earphone.

Highest achievable output SPL	Hearingaid Prototype (right, left earphone)	Signia Motion 13 7 Nx
Peak value	114 dB SPL, 112 dB SPL	
Peak value (mean value)	113 dB SPL	131 dB SPL
HFA-OSPL90	105 dB SPL, 102 dB SPL	
HFA-OSPL90 (mean value)	104 dB SPL	126 dB SPL
Acoustic amplification (Input level 50 dB)		
Peak value	34 dB	58 dB
HFA-FOG	28 dB	54 dB
RTG	26 dB	50 dB
Noise behaviour, frequency range, battery		
Equivalent input noise	29 dB SPL	13 dB SPL
Right, Left	30 dB, 28 dB	
f1, f2 (mean)	<200 Hz , 7000 Hz	<200 Hz, 7500 Hz
f2	6700 Hz, 7300 Hz	
Power consumption in RTS	520 mA	1.4 mA
Total harmonic distortion factor in percent		
THD 500Hz	1.4 , 0.9	1.6
THD 800Hz	2.6 , 2.2	0.7
THD 1600Hz	1.5 , 4.7	0.3

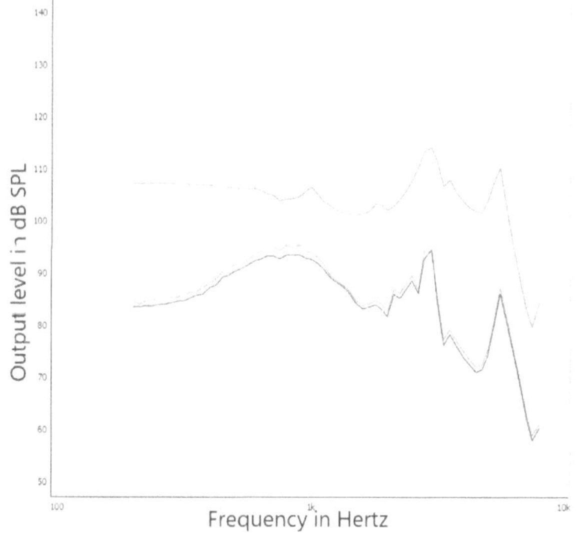
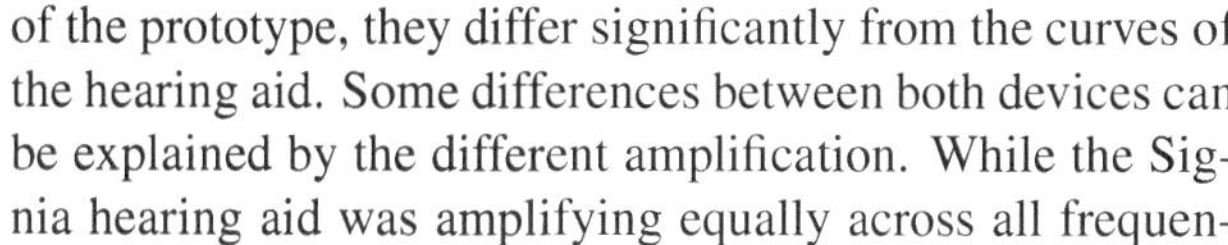

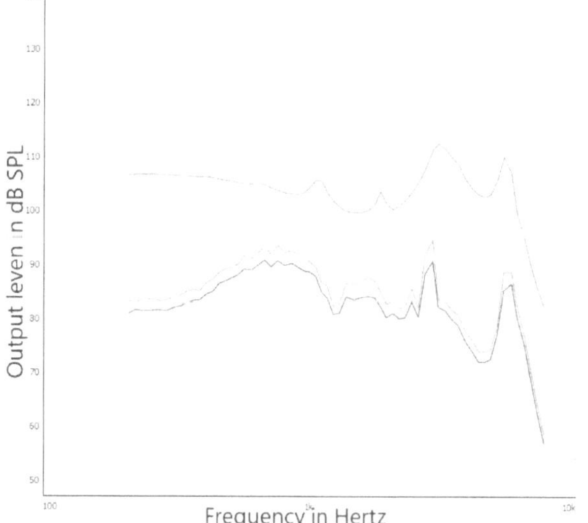

Figure 2: Acoustic frequency response curves of the right earphone of the hearing aid prototype. The top line shows the OSPL90 Frequency Response Curve. The middle line shows the Frequency Curve of Acoustic Amplification at maximum gain. The bottom line shows the Normal Acoustic Frequency Response Curve.

Figure 3: Acoustic frequency response curves of the left earphone of the hearing aid prototype. The display is the same as in Fig. 2

of the prototype, they differ significantly from the curves of the hearing aid. Some differences between both devices can be explained by the different amplification. While the Signia hearing aid was amplifying equally across all frequen- cies, the frequencies between 4000 Hz and 8000 Hz of the prototype were amplified only about half as much as the fre- quency range from 250 Hz to 2000 Hz due to the occurring feedback. This, of course, explains the strong level drop in the frequency range from 3000 Hz to 4000 Hz. The level in- crease between 2000 Hz and 3000 Hz and the maximum at 5500 Hz can only be explained by the behavior of the pro-

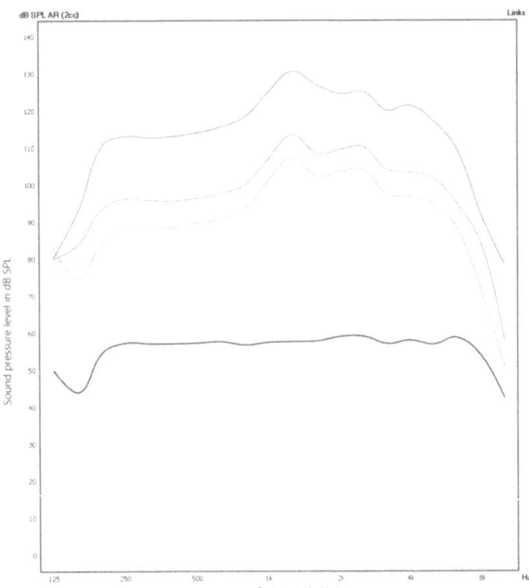

Figure 4: Acoustic frequency response curves of the state-of-the-art hearing aid. The top line shows the OSPL90 Frequency Response Curve. The second line shows the Normal Acoustic Frequency Response Curve. The third line shows the Frequency Curve of Acoustic Amplification at maximum gain. The bottom line shows a calibration line with an input SPL level of 60 dB.

totype. The horizontal course in the low-frequency range of the OSPL90 frequency response curve of the prototype can be explained by reaching the maximum output level. Despite the different amplification levels and the lower gains in the higher frequency range, the level maxima of the prototype are in the high-frequency range between 3000 and 5500 Hz. In contrast, the hearing aid mainly amplifies the frequency range between 750 Hz and 4000 Hz. The hearing aid amplifies, therefore, better in the frequency range of speech. Table 1 shows the most important measured values in the form of a data sheet commonly used for hearing aids. The maximum gain of the prototype is about 34 dB, which is almost 20 dB less than the gain of the hearing aid. Still, the prototype offers stable gains over the whole frequency range. The frequency range of both devices was determined as the cut-off frequencies f1 and f2. The frequencies f1 and f2 can be read by drawing an auxiliary line at the HFA-OSPL90 value minus 37 dB. At 12.9 dB SPL, the input noise of the Signia hearing aid is about 15 dB lower than the input noise of the prototype. The high input noise of the prototype considerably impairs the listening experience and should be overcome as soon as possible. The battery consumption of the prototype was measured to 520 mA. The result was independent of whether the hearing aid was in the RTS setting, as required by the standard, or in the FOG setting. In contrast, the hearing aid runs at significantly lower consumption of 1.4 mA.

4   Conclusion

Due to the resulting internal feedback, the gain of the prototype, especially at the higher frequencies, could not be set high enough to allow the selection of compressive signal processing strategies. In addition to the feedback, the high amount of equivalent input noise is also a problem that has to be overcome to achieve a device for everyday use. Otherwise, the prototype provides stable gains in the suitable frequency range and is therefore appropriate to be used as a test device. To lower the feedback the first step is to change the headphones. An approach would be to take the housing of a company hearing aid including only loudspeaker and receiver. Only after the feedback has been corrected and the input noise has been adjusted it is possible to compare the hearing aid prototype again. In these measurements, it is best to take a hearing aid for comparison, which gives about the same amplification, i.e. about 20 dB. If better results are then available, further measurements, such as percentile analysis or in-situ measurements using a dummy head, could be made. Afterward, the prototype can be tested in real-world situations by initially normal hearing and later also hearing-impaired people.

Acknowledgement

The work has been carried out at the Technische Hochschule Lübeck supervised by the Institute of Acoustic, Technische Hochschule Lübeck.

5   References

[1] Tobias Herzke, Hendrik Kayser, Frasher Loshaj, Giso Grimm, and Volker Hohmann. 2017. Open signal processing software platform for hearing aid research (openMHA). In Proceedings of the Linux Audio Conference, pages 35–42, Saint-Etienne.

[2] Lindsey E. Jorgensen.Verification and validation of hearing aids: Opportunity not an obstacle. J Otol. 2016 Jun; 11(2): 57–62

[3] Marc Rene Schaedler, Hendrik Kayser, and Tobias Herzke. 2018. Pi hearing aid. The MagPi (Raspberry Pi Magazine), 67:34–35. Wiki on Github: https://github.com/m-r-s/hearingaid-prototype/wiki.

[4] Hertzke et al. 2018 Open Hardware Multichannel Sound Interface for Hearing Aid Research on Beagle-Bone Black with openMHA: Cape4all

[5] Electroacoustics - Hearing aids - Part 0: Measurement of the performance characteristics of hearing aids (IEC 60118-0:2015); German version EN 60118-0:2015

2

Biomechanics

Using two in-house developed test methods to characterise medical tapes regarding their resistance to shear and friction under load

Juliana Tiedemann [1], Niko Röhner [2] and Arndt Peter Schulz[3]

[1] Medical Engineering Science, Universität zu Lübeck, juliana.tiedemann@student.uni-luebeck.de

[2] Essity, Hamburg, niko.roehner@essity.com

[3] Department of Orthopaedics and Trauma Surgery of the Biomechanics Laboratories, Universität zu Lübeck, schulz@biomechatronics.de

Abstract

Mechanical influences such as abrasion, scraping or shear stress during sports or daily work may lead to wounds. Medical tapes could extend their medical purpose by protecting the skin. In order to verify their resistance to friction and shear forces two methods have been developed in this research project for testing medical tapes. These methods have been applied to three tapes consisting of a rubber, acrylate and silicone-based adhesive respectively. For a shear strength measurement, the tapes were fixed to steel, PE (polyethylene) and a skin surrogate called Lorica Soft and pulled off afterwards. To measure the resistance against friction, the tapes were fixated onto skin surrogate and pulled over three different textiles. Tape A (rubber) shows the highest while tape C (silicone) the lowest shear strength. Both tapes are also able to withstand the loads whereas tape B (acrylate) shows a low frictional resistance.

1 Introduction

Abrasions belong to the most common injuries, which can occur during sports activities. To prevent such injuries, many athletes use to wear bandages, protective clothing or even tapes [2]. However, mechanical wounds such as abrasions or blisters can also occur in everyday life or during heavy physical work. These mechanical wounds are the result of strong abrasive forces and shearing [1]. Withstanding these forces and protecting the skin is an essential requirement for the preventive function of a medical tape. Friction μ can be described by the following formula [5]:

$$\mu = \frac{F_R}{F_N} \qquad (1)$$

F_N is the normal force and describes the weight force of an object. There are two types of F_R during a friction test: static friction force and dynamic friction force. The static friction force defines the force, which is required to set the object in motion and the dynamic friction force defines the force, which is required to move it after static friction. μ represents a unitless coefficient in (1) [5]. The other force, which is responsible for skin damage, irritations or abrasions, is the shear force. This force F_Z acts parallel to the surface A and causes a shear stress τ_S, which can be described by the following formula [4]:

$$\tau_S = \frac{F_Z}{A} \qquad (2)$$

In this project, test methods were developed to test the properties of three different medical tapes for friction and shear strength. The aim of this research is to measure the characteristics of medical tapes against friction and shear forces in order to investigate and confirm their use as a preventive measure against skin injuries. In contrast to peak loads with high forces, this work considers friction and shear forces with smaller force peaks under longer action.

2 Material and Methods

Three medical tapes were selected for the two test methods: tape A, tape B and tape C. Table 1 lists the respective coatings and adhesives. To imitate the skin tissue, a skin surrogate has been used, which consists of polyurethane and polyamide [3]. Cottenden and Cottenden (2013) describe this surrogate in a comparison of skin replacement materials in a research article. Lorica Soft is most similar to skin in terms of mechanical properties such as friction coefficients [3].

Two methods were developed to simulate a load due to shear and tension as well as abrasion. The first method is a shear test based on a tensile shear test. The second method simulates a load test by friction.

Table 1: Backing and adhesives of the tapes

Tapes	Backing	Adhesive
Tape A	cotton with acrylate	rubber
Tape B	polyacetate	polyacrylate
Tape C	polyacetate	silicone

2.1 General setup of the shear strength method

The shear test was developed as a test method on the basis of a tensile shear test in DIN EN 1465 [6]. According to DIN EN 1465, two bonded surfaces are pulled apart in vertical direction. The tape (A = 2.5 x 2.5 cm) was fixed to steel, PE (polyethylene) or a Lorica Soft covered plate (The artificial synthetic leather was fixed with a double-sided tape). To obtain a reproducible measurement and an uniform adhesion, the tape was attached onto the plate and rolled over with a 5 kg steel cylinder. Afterwards, the part of the plate, which is not covered by the adhesive, was fastened between the clamps in the lower part of the Zwick universal testing machine to the 1 kN load cell and the upper side of the tape was clamped in the upper part of the testing machine (see Fig. 1). The maximum tensile force, releasing the tape off the plate, was determined, using a constant speed of 200 $\frac{mm}{min}$. The pulling direction of the machine corresponded to the adhesive direction of the tape, so that the shear stress can be calculated from the measured maximum tensile force using the adhesive surface according to (2).

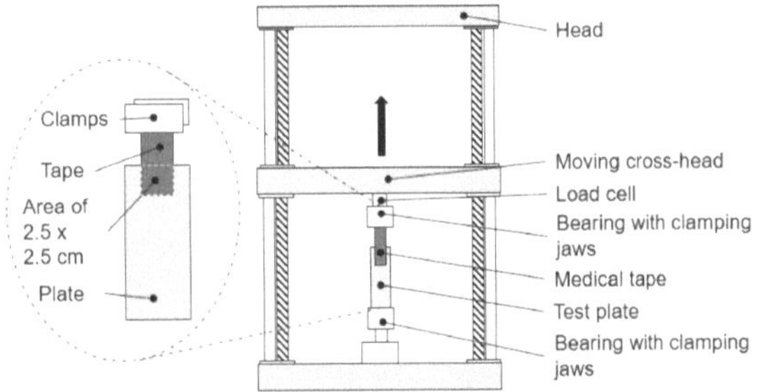

Figure 1: Setup of the shear force method. The medical tape was attached to the clamping jaws at the top and was fixed at the test plate.

In this method, the tapes A, B and C were each measured ten times on the materials steel, PE and Lorica Soft. The maximum tensile force, breaking force and elongation were determined.

2.2 General setup of the load test

The second method was developed to simulate friction and shear load in horizontal direction. Lorica Soft was attached onto a slide and a sample of the tape (with an area of 2.5 x 2.5 cm) was fixed on it. In addition, the slide was loaded with a weight of 2 kg to increase the normal force. The sample was rolled on with a weight of 3 kg by hand. A rope was attached to the slide, guided along a pulley and was fixated to a 200 N load cell of the Zwick testing machine. The

slide was pulled 150 mm over a test surface (Lorica soft, denim or cotton fabric) attached to the table at a constant speed of 2000 $\frac{mm}{min}$ with the aid of the Zwick (see Fig.2). The tape (with adhesive direction to the slide) was loaded by the test surface in terms of shear and friction. In this method, the initial force required to set the slide in motion and the frictional force during the steady motion was measured. The coefficient of static friction μ_s and dynamic friction μ_d can be determined using the normal force according to (1). Ten measuring cycles were run through, unless the tape detached itself during one of the cycles. If this was the case, the area still adhering was determined with a grid. Therefore, lines were drawn horizontally and vertically with a distance of 0.5 cm on a glass plate using a calibrated ruler. Once the plate was placed on the tape, the detached pixels could be counted to allow an objective observation (see Fig. 7).

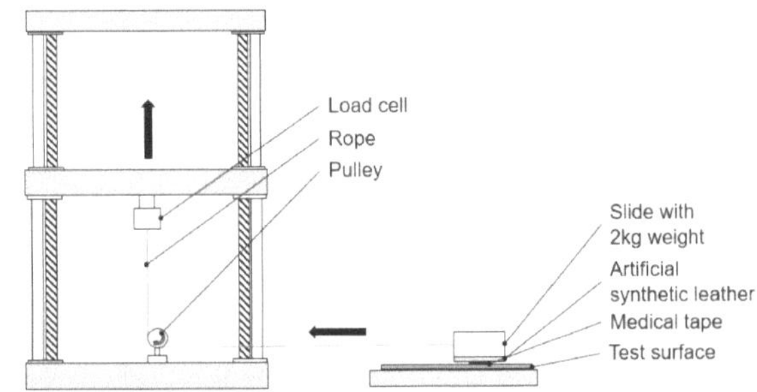

Figure 2: Setup of the shear force method. The slide was pulled over the test material at a constant speed over a predefined path.

3 Results and Discussion

Table 2: Results of the shear force method

Tapes	shear stress $[N/cm]$	standard deviation
Tape A on steel	47.50	2.14
Tape A on PE	34.11	0.81
Tape A on Lorica	12.50	0.90
Tape B on steel	25.86	0.36
Tape B on PE	18.21	0.78
Tape B on Lorica	8.07	0.52
Tape C on steel	13.58	1.18
Tape C on PE	14.64	0.69
Tape C on Lorica	8.84	0.65

Table 2 lists the results of the shear test of the three tapes. Tape A thus shows the highest shear strength, tape B a medium shear strength and tape C the lowest (see Fig. 3). During the measurement, the tensile force was recorded against the elongation (see Fig. 4, 5 and 6). Tape A initially shows a linear course and then begins to decrease after the maximum. This curve is obtained for all three material pairings. Tape B initially shows a linear increase on steel and PE, but this quickly decreases and begins to fluctuate. For

Lorica Soft, the curve looks linear. For eight measurements, the backing material of tape B tore apart before the adhesive could detach from the plate. Tape C shows a linear increase for steel and PE, followed by a strong decrease. For Lorica Soft, the trend looks more linear.

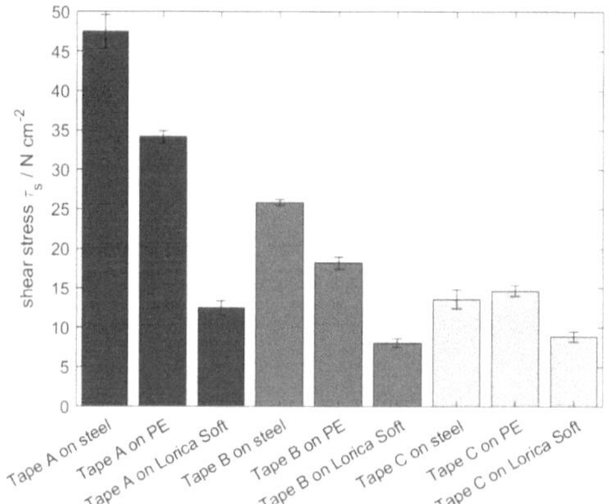

Figure 3: Bar chart depiction with standard deviation for the measured values of the shear strength. Ten measurements per tape and test plate were performed.

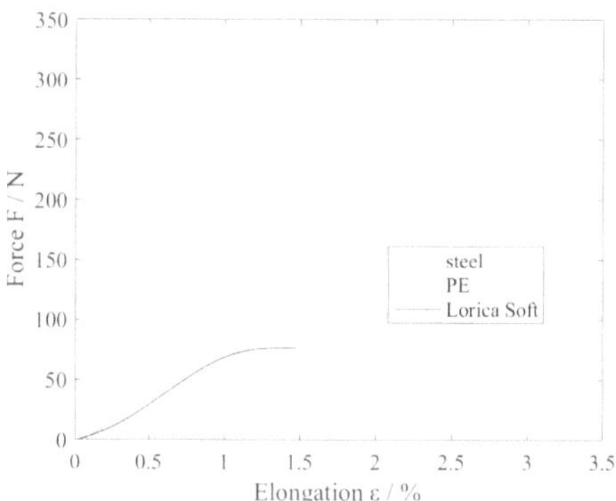

Figure 4: Force - elongation diagram of tape A. A trend line was used to approximate the course of all measured data.

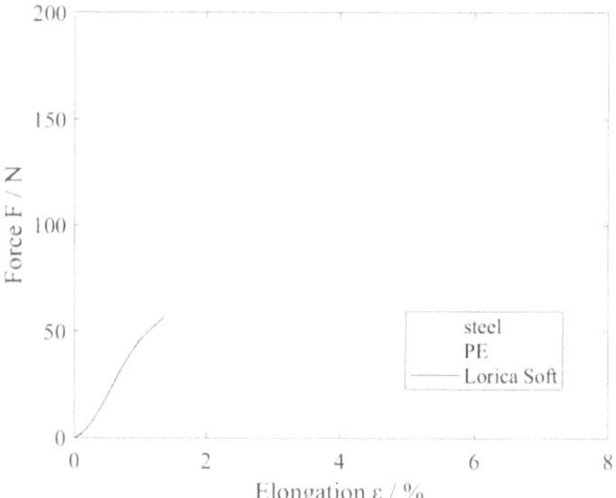

Figure 5: Force - elongation diagram of tape B. A trend line was used to approximate the course of all measured data.

In the second method, the frictional forces of the material pairings Lorica Soft/Lorica Soft, Lorica Soft/denim and Lorica Soft/Cotton were determined. The coefficients of the

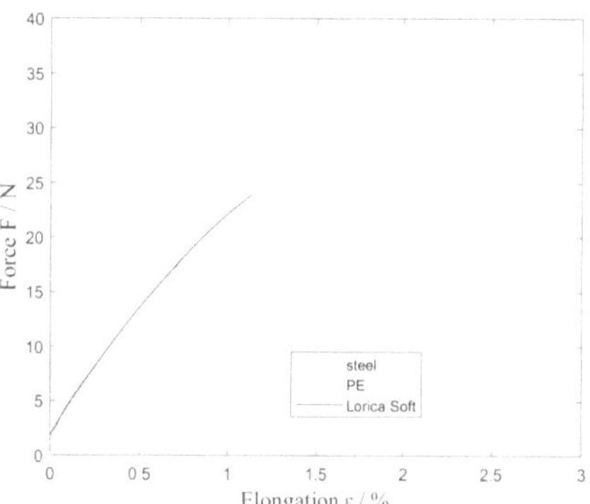

Figure 6: Force - elongation diagram of tape C. A trend line was used to approximate the course of all measured data.

Table 3: Friction coefficients without tape

	Lorica Soft	denim	cotton
μ_s	0,44	0,37	0,34
μ_d	0,22	0,23	0,22

static friction were determined from the friction forces and yield 0.44, 0.37 and 0.34 respectively (see table 3). Lorica Soft on Lorica Soft shows a high static friction. Cotton and denim pairing lead to similar coefficients. The coefficients of the sliding friction are 0.22, 0.23 and 0.22 and, therefore, quite similar for all test methods.

For tape A and tape C, five measurements each with ten measuring cycles were carried out. The static friction force and sliding friction force show no significant differences compared to the measurement without tape. Both tapes did not detach from the plates over ten measuring cycles for all materials. For tape B, five measurements out of ten cycles each were measured with Lorica Soft and no detachment was observed (see table 4). With denim and cotton pairings, on the other hand, the tape always came off during the first measurement, although the friction coefficients show similar values. The tape rolled up completely on denim and on cotton it rolled up partially. 100 % of the pixels were detached for denim and 78 % of the pixels were detached in average for cotton (see Fig. 8). Therefore, ten measurements were made per pairing.

In the first method, the force-elongation curve of tape A initially rises linearly to the yield point and flattens out

Table 4: Results of the load test

pairing	Amount intact	Amount loose
Tape A / Lorica Soft	5	0
Tape A / denim	5	0
Tape A / cotton	5	0
Tape B / Lorica Soft	5	0
Tape B / denim	0	5
Tape B / cotton	0	5
Tape C / Lorica Soft	5	0
Tape C / denim	5	0
Tape C / cotton	5	0

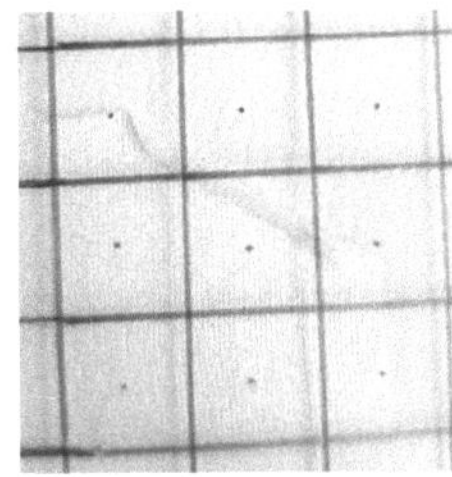

Figure 7: Evaluation of the measurement using a grid. A grid with 0.5 cm x 0.5 cm pixels was used to count the detached pixels quantitatively.

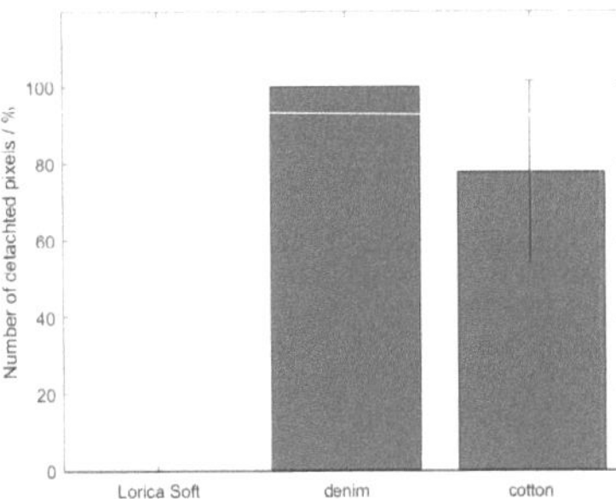

Figure 8: Bar chart depiction with standard deviation for the number of pixels in percent that detached itself during the load test with tape B.

slightly until it breaks. This course indicates a relatively rigid backing. The backing shows only a slight elongation and all measurement series show only a slightly different course. The curve progression of tape B on steel and PE shows greater deviation within the data series. This can be explained by the fact that the backing of some samples is torn apart before the adhesive got loose. Most likely, this event occurs because the shear strength of the adhesive mass is close to the tensile strength of the backing. On Lorica Soft, the shear strength of the backing is significantly higher than the adhesive strength, so that this curve does not indicate a slight slope. Tape C shows a completely different course. After a steady increase, the force to achieve a further elongation decreases, which indicates a creep behaviour of the adhesive. Tape A and tape C showed no detachment from the artificial synthetic leather on all materials. Tape B, on the other hand, peeled off when tested with denim and cotton. It is assumed that the adhesive mass stuck to small fibres of the fabric. Due to the even movement of the slide, the tape rolled up. With Lorica Soft, there were no free fibres to be found and the tapes did not roll up at all. Despite these results, it should be mentioned that the number of samples is small and that despite similar mechanical properties to the skin according to ref. [3], Lorica Soft is very different to skin. Lorica Soft has no sweat or sebaceous glands neither hair. For a realistic simulation, measurements should be made on real skin. Besides, some flaws could be identified using these two test methods from above. The load measurement only takes place in one spatial direction. For realistic measurements, dynamically changing measurements in all spatial directions need to be considered. Furthermore, a speed of 2000 $\frac{mm}{min}$ is quite slow. Higher speeds could be achieved with more specialized testing machines.

4 Conclusion

By measuring the tensile strength and shear stress of medical adhesive tapes applied to plates of different materials, first insights can be gained into the adhesive behaviour and their resistance to wound-causing abrasions: tape A (rubber) has the highest shear strength, but due to the rubber adhesive mass it sticks strongly to the patient's hair and is therefore less skin friendly. Tape C (silicone) shows low shear strengths on the three materials. Tape B (polyacrylate) shows the smallest load measurement, because it rolled up already during the first test. Tape C and tape A, on the other hand, withstood ten repetitions without the tape came off. Under more realistic conditions on the skin, these two tapes could behave similarly. Therefore, further tests on human skin should be carried out. Especially the skin friendly property of tape C makes the tape a promising candidate for the preventive use on the skin against wounds caused by abrasion and has therefore an advantage over tape A.

Acknowledgement

The work has been carried out at and funded by BSN medical GmbH, an Essity company, Hamburg, and supervised by the Department of Orthopaedics and Trauma Surgery of the Biomechanics Laboratories, Universität zu Lübeck.

5 References

[1] M. Grönheim and C. Kemperdick, *Sanitätsdienst: Erstversorgung - Notfallmedizin - Schnittstelle Rettungsdienst*, Urban & Fischer Verlag/ Elsevier GmbH, München, 2019

[2] B. B. Adams, *Friction Injuries to the Skin*, In: Sports Dermatology, Springer, New York, pp. 203-240, 2006

[3] D. J. Cottenden and A. M. Cottenden, *A study of friction mechanisms between a surrogate skin (Lorica soft) and nonwoven fabrics*, in Journal of the mechanical behaviour of biomedical materials, Vol. 28, pp. 410-426, 2013

[4] V. L. Popov, *Das Coulombsche Reibungsgesetz*, In: Kontaktmechanik und Reibung, Springer, Berlin, Heidelberg, pp. 131-152, 2009

[5] G. Habenicht, *Kleben - erfolgreich und fehlerfrei*, Springer, Wörthsee, 2016

[6] DIN EN 1465:2009, *Bestimmung der Zugscherfestigkeit von Überlappungsklebungen; Deutsche Fassung EN 1465:2009*, 2009

Mechanical 4-Point-Bending test and simulation of known material

Daniel Grollmisch[1], Felix Mutter[2] and Robert Wendlandt[3]

[1] Biomedical Engeneering, Luebeck University of Applied Sciences, daniel.grollmisch@stud.th-luebeck.de

[2] Biomedizintechnik, Luebeck University of Applied Sciences, felix.mutter@stud.th-luebeck.de

[3] Clinic for Orthopedics and Trauma Surgery, Laboratory for Biomechanics and Biomechatronics, University of Lübeck, wendlandt@biomechatronics.de

Abstract

Fracture plate implants represent a big field in today's biomedical industry. Pre-clinical mechanical testing is required to obtain a CE certificate for this type of medical product. A 4-point-bending test allows to assess the loading capacity in bending. Finite Element Analysis (FEA) simulation allows to predict the bending behaviour during the design process or for laboratory purposes. No standard procedure for simulation is established in the biomechanical laboratory of University of Lübeck, Germany. Hence, to create such a standard procedure samples in different geometries but with the same material are tested with a universal testing machine in a 4-point-bending setup. Data from FEA simulation is checked against results from mechanical testing. After all, the simulation can only show the behaviour of a material fitting the minimal required properties.

1   Introduction

Different materials and designs of bone plates for fracture treatment are available [1] since every fracture is individual and has to be treated individually. Those implants are loaded in bending often while in use.

The 4-point-bending test as described in ASTM F382 is the standard used to assess the sample parameter of single cycle and fatigue tests. Parameters as bending strength, bending stiffness and fracture load can be derived from the test [2]. Nevertheless, this standard is mainly intended for straight, diaphyseal bone plates [1]. Implants with more complex geometries, which are increasingly used in today's fracture care are not covered by the ASTM F382 but there is a recommendation given for testing plate-screw constructs of osteosyntesis [3]. Different materials and coatings are used for fracture bone plates. Because of the biocompatibility, mainly titanium alloys like Ti-6Al-4V and Ti-6Al-7Nb [4], commercially pure titanium [5] and stainless steel [6] are used for fracture implants nowadays. For design and laboratory purposes of bone plates, Finite Element Analysis (FEA) simulations are required. Not only the geometry of the implant but also the complete material parameters in the elastic and plastic domain are required. Those parameters are usually derived from a tension test of the material since the stress distribution in tension is homogeneous over a cross section of material. Yet, tensile testing requires a specialised test setup, additional sensors and an adapted sample geometry to yield accurate results which also might largely differ from material to material.

A full tension test dataset is only available if one has access to a good material data base or access to a sufficient tension testing machine. Hence, it is practical to come up with a sufficient model of the material behaviour.

The Ramberg-Osgood model [7] is used most often to describe the material behaviour. By knowing some specific parameters of the material, the engineering stress-strain diagram can be modelled and consequently the ›true stress‹-›plastic true strain‹ diagram can be calculated, which is necessary for the material simulation for FEA.

The stress distribution in bending behaves differently. Once plastic deformation starts to occur the stress distribution is S-shaped and is a combination of compression and tension [8]. Thus, bending test is usually not applied to obtain material parameters. Still, using simple bending tests for determination of material parameters for FEA simulations would simplify the workflow for development of new plate designs as no additional test setup or third party data are required.

2   Material and Methods

2.1   Material and test-setup

For obtaining data to compare the FEA simulation, some plates made from aluminium alloy EN AW-6060-T66 (3.3206) were ordered from two different sources: www.aluminium-online-shop.de and www.alu-messing-shop.de. The samples were ordered in simple ge-

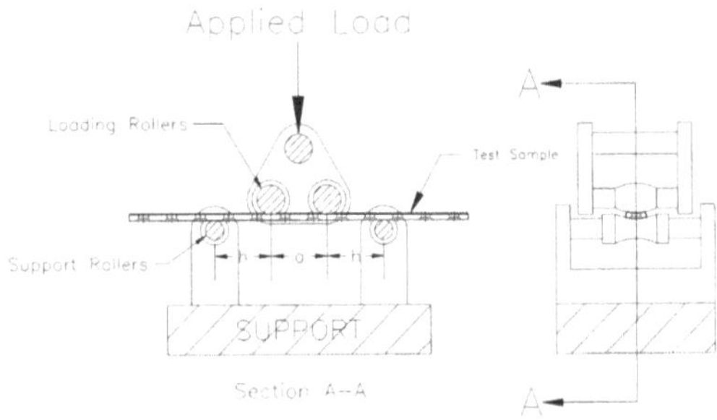

Figure 1: 4-point-bending test setup [2]

ometries: 3x15x150mm and 5x15x150mm. Additionally, some samples were modified by extruding some material for observing the behaviour of slightly more complex geometry in comparison with the simple samples. The samples where tested in a 4-point bending setup as shown in figure 1 made by Felix Mutter based on the ASTM [2]. Samples were tested using a material testing machine (Zwick 1456, Zwick, Ulm with a $\pm 2kN$ force sensor). The tests have delivered data of loading and deformation for each sample (n = x per dimension plus n = y per dimension for the modified sample geometry) tested. The program that controls the testing device needs the information about geometry parameters of the sample like width B and high H in mm. It also needs some parameters of the setup which are the distance of the support roles and the distance of the loading rolls. These parameters were measured for each tested sample individually.

Knowing the setup parameters loading span distance h and centre span distance a as shown in figure 1 as well as the bending stiffness K and the area moment of inertia I_y the young's modulus E can be calculated [2]:

$$E = \frac{I_y(2h + 3a)Kh^2}{12} \qquad (1)$$

Knowing the centre span distance a the 0.2%-offset displacement q can be calculated with

$$q = 0.002 \cdot a. \qquad (2)$$

The point of intersection of the measured data with the 0.2%-line is known as the bending yield point or proof load [2].

Two methods were used to verify the usability for FEA simulation as bending yield point prediction.

2.2 Linear material behaviour

The first methods follows the idea that the notable plastic deformation just starts at the bending yield point and therefore the behaviour of the material is linear until that specific point regardless of the geometry of the sample. So, just by knowing the bending stiffness K and the sample geometry as well as the setup parameters we could possibly just simulate the sample until the bending yield point and get the load that is needed to get the yield stress from literature. Having tested a sample the bending stiffness K can be derived from the slope in the linear part of the load-deformation-diagram:

$$K = \frac{F_2 - F_1}{f_2 - f_1} = \frac{\Delta F}{\Delta f} \qquad (3)$$

For setting up the FE simulation in Ansys Workbench 19.2, simple isotropic elasticity can be used by knowing the poisson ratio from literature and calculating the young's modulus of the sample with (1). After simulating the sample testing under the static structural environment taking each parameter of setup and sample geometry into account, the load-Von Miese stress-diagram can be derived from the program. The stress distribution in tension is homogeneous but in bending it is inhomogeneous. While bending, the highest stress shows up at the highest distance from the neutral fibre when considering the homogenous geometries. This most commonly is the border of the sample in the loading direction. The moment of bending while 4-point-bending test can be calculated as:

$$M_b = F \cdot \frac{h}{2}. \qquad (4)$$

The stress showing up in the border is calculated as:

$$\sigma_r = \pm M_b \cdot \frac{H}{2 \cdot I_y}. \qquad (5)$$

Assuming that the bending stress in infinitesimal elements is equal to tension stress in each element, it can be stated that the needed force to reach the yield stress from literature can be calculated using (4) and (5):

$$F_{0.2\%} = \frac{4 \cdot I_y \sigma_r}{Hh} \qquad (6)$$

The simulated load (6) at Von Miese stress equal to the yield point can be compared with the load at intersection with the 0.2%-line in the real testing data.

2.3 Nonlinear material behaviour

The second method follows the idea that the behaviour is not just linear until the bending yield point but nonlinear. So, the plasticity of the material has to be taken into account. This can be done in the FEA simulation in Ansys by adding multilinear isotropic hardening. For that, the ›true stress‹-›plastic true strain‹ data has to be known and implemented in Ansys as tabular data.

To come up with such data, we have to have access to the material data base which can be found on www.totalmaterial.com, for example. Having access to such data base has the advantage of direct and fast implementation of material behaviour in Ansys. The disadvantage is the high costs of getting access.

Another way is to model the required data which was the method used here. To come up with the Ramberg-Osgood Model (RO-Model), we have to know some material parameters available in material data books or on the internet, for example www.makeitfrom.de. Those parameters would be young's modulus E, ultimate strength F_{tu}, bending yield point F_{ty} and the strain at rupture ϵ_r.

The standard MMPDS-01 [9] shows that the uniform stress ϵ_{us} and the Ramberg-Osgood-Coefficient can be calculated as:

$$\epsilon_{us} = 100 \cdot (\epsilon_r - \frac{F_{tu}}{E}) \tag{7}$$

and

$$n = \frac{ln(\frac{\epsilon_{us}}{0.2})}{ln(\frac{F_{tu}}{F_{ty}})}. \tag{8}$$

With all these parameters, the Ramberg-Osgood equation can be solved to obtain the engeneering stress-strain-curve as follows:

$$\epsilon = \frac{\omega}{E} + 0.002 \cdot (\frac{\omega}{F_{ty}})^n. \tag{9}$$

To come up with the true stress-strain data, RO stress-strain has to be modyfied:

$$\epsilon_{true} = ln(1 + \epsilon) \tag{10}$$

and

$$\sigma_{true} = \sigma(1 + \epsilon). \tag{11}$$

Some programs like Ansys just need the plastic part of the true strain and true stress starting with the proof load. So, the plastic strain has to be calculated from the true strain as:

$$\epsilon_{true,plast} = \epsilon_{true} - \frac{\sigma_{true}}{E} \tag{12}$$

From the simulation, the load-deformation data can be derived easily. Just like the data derived from the 4-point testing of the original sample, the 0.2%-method is used. The intersection points of simulation and real testing can be compared this way.

3   Results and Discussion

3.1   Linear material behaviour

After testing several samples in different geometries regarding high and complexity using the linear material behaviour, comparison of the results has been done.

Due to the huge difference in results, this method has been

Table 1: Extract from results with linear material behaviour compared to real testing in 4-point-bending-test

sample high in mm	0.2%-load in N from real testing	simulated load in N at given yield point
2	93,04	51,99
3	209,61	122,85

abandoned in the early state of testing. Since the material properties are given as a minimum limit and not as a total value, the simulated results have to be considered the same. So the simulation shows the load that is required for 0.2% of plastic deformation that is guaranteed by the standards

that states the regulatory for manufacturing the special material and objects made from it. Since the real testing results are higher than the simulated value, the material is built in a sufficient way. Still, this is not the required outcome of the test since the goal is to predict the real yield point.

The variation of material behaviour is shown in figure 2. Sample 1 and 13 are from one company, sample 21 and 22 from another company. All four samples are made with the same geometry and ordered under the very same material description EN AW-6060-T66. Still, it is to observe that the samples from each shop are fitting good with each other but differ in stiffness, proof load and ultimate load comparing the shops themself.

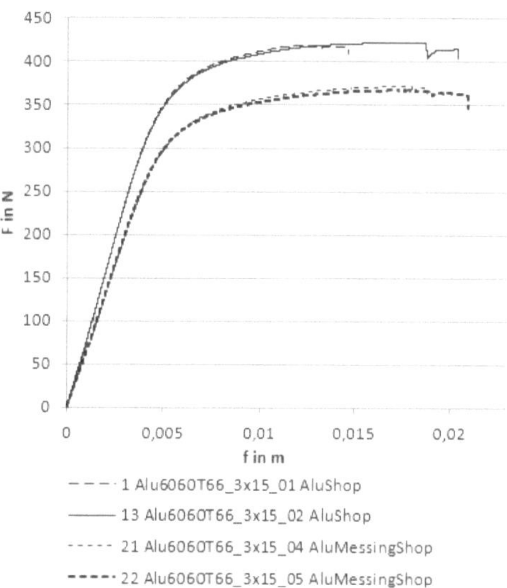

Figure 2: similar samples delivered from different companys

3.2   Nonlinear material behaviour

Since the observation of the first method shows the unreliability on material behaviour for different material sources, we remain relying on just one material source to have comparable material behaviour. To overcome the difference in behaviour of the material and the behavior of the simulation, the simulation was modified in a way that the simulation and a real sample testing fits to each other. Therefore the tabular data of the plastic deformation was modified with an offset for the reason that both curves - the real data and the simulated data - shall intersect in the same point with the 0.2%-line of the real testing as shown in figure 3. By increasing the offset stepwise, it's possible to match the 0.2%-load of the simulation and the mechanical testing. Since the goal is to predict the proof load of more complex geometries, some samples from the very same material were modified by drilling a 3 mm hole in the center of the plates. These were tested the same way the normal samples were tested. The simulation was done with an averaged young's modulus and an averaged offset for the plastic material behaviour in Ansys multilinear isotropic hardening. As shown in figure 3, the curves of testing and simulation

Table 2: Extract from comparison of proof load matching with offset with (indicated by _L) and without hole

sample *highxwidth_-* nr in mm	0.2%-load from real testing in N	simulated 0.2%-load in N	offset 0.2%- in N	ΔF_-0.2% in N
3x15_04	232,066	232,068	37,75	0,002
3x15_05	230,505	230,452	35,45	-0,053
3x15_06_L	229,823	167,417	36,6	-62,406
3x15_07_L	222,220	177,895	36,6	-44,325

perfectly match at the 0.2%-point for simple samples after the modification is done but they still does not fit in the plastic domain. For the drilled sample, the bending stiffness of tested and simulated sample does slightly differ from each other and again the plastic part of the curve does not fit at all. Additionally the results for the drilled samples shown in table 2 ending up with a significant difference compared to the homogeneous samples. This is an indication for the method is still erroneous.

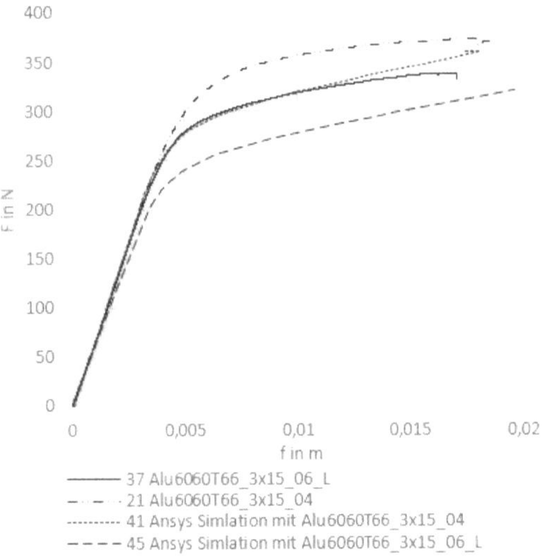

Figure 3: sample with and without hole in real testing and simulation

4 Conclusion and Outlook

Neither the method with linear material behaviour nor the method with modified nonlinear behaviour is sufficient to predict the proof load for bending in a 4-point bending setup. One problem is the unreliability of the literature data of the material properties with this special purpose. Another problem is the failure that occurs when prediction is made on a simulation which is based on a model of material behaviour.
One further possible step could be to actually test a sample in a tension test knowing that all samples are from the same batch. This way all needed and reliable data for the simulation can be gathered and a prediction of more complex geometries would be more likely correct.

Acknowledgement

The work has been carried out at the Laboratory for Biomechanics and Biomechatronics, Universität zu Lübeck and was supervised by Dr.-Ing. R. Wendlandt. The work and the measurements were done by Daniel Grollmisch. The work of building a new setup for 4-point-bending done by Felix Mutter shall be acknowledged here too since this work wouldn't be possible otherwise. Also J. Schroeter shall be mentioned here since his constructive questions about the proceeding were helping a lot to not go in wrong directions.

5 References

[1] H. Schorler, F. Capanni, M. Gaashan, R. Wendlandt, C. Jürgens, A.-P. Schulz, *Bone plates for osteosynthesis - a systematic review of test methods and parameters for biomechanical testing*. DE GRUYTER, 2016.

[2] *ASTM F382-17, Standard Specification and Test Method for Metallic Bone Plates*. ASTM International, West Conshohocken, PA, 2017, www.astm.org, DOI:10.1520/F0382-17.

[3] H. Schorler, R. Wendlandt, C. Jürgens, A.-P. Schulz, C. Kaddick, F. Capanni, *Bone plate-screw constructs for ostosynthesis - recommendation sfor standardized mechanical torsion and bending tests*. DE GRUYTER, 2017.

[4] VK. Ganesh, K. Ramakrishna, Dhanjoo N Ghista, *Biomechanics of bone-fracture fixation by stiffness-graded plates in comparison with stainless-steel plates*. 2005, Available: https://www.ncbi.nlm.nih.gov/pmc/articles/PMC1192810/ [last accessed on 03.02.2020].

[5] Ortrun E.M. Pohler, *Unalloyed titanium for implants in bone surgery*. elsevier, 2000, DOI: doi.org/10.1016/S0020-1383(00)80016-9.

[6] J.A. Disegi, L. Eschbach, *Stainless steel in bone surgery*. 2000, DOI: doi.org/10.1016/S0020-1383(00)80015-7.

[7] G. Gadamchetty, A. Pandey, M. Gawture, *On Practical Implementation of the Ramberg-Osgood Model for FE Simulation*. SAE-international, 2016, DOI: 10.4271/2015-01-9086.

[8] R.C.Hibbeler, *Menchanics of Materials*. Pearson, 2018, tenth Edition in SI Units.

[9] R.C. Rice, J.L. Jackson, J. Bakuckas, S. Thompson, *Metallic Materials Properties Development and Standardization - MMPDS-01*. 2003.

Structural changes in the design of a device that allows the formation of artificial lipid Montal & Mueller membranes combined with electrical and optical measurements

Robin Meyer [1], Christian Nehls [2], Thomas Gutsmann [2]
[1]Medical Engineering Science, Universität zu Lübeck, robin.meyer@student.uni-luebeck.de
[2]Biophysics Group, Forschungszentrum Borstel, {cnehls, tgutsmann}@fz-borstel.de

Abstract

Natural membranes are very complex and as such difficult to understand. The Fluorescence Montal & Mueller Microscope (FMMM) can be used to build a simplified model of a natural membrane and measure its electrical properties while simultaneously observe fluorescence labeled particles passing through and interacting with the membrane. There are three important parts to the FMMM: with part I the membrane is build, part II carries out electrical measurements and part III does fluorescence observations. The starting point of this project is a prototype of a FMMM in which none of the components described above work. The goal is to optimize step-by-step or create new components. At first the sample chamber was exchanged and free-standing membranes were produced. The first electrical measurements could already be performed as proof-of-principle. Additionally a microscope was added to help with positioning the membrane for the optical setup. In the future a different light source will be added and the beam path will be optimized.

1 Introduction

In today's times the evolution of drug resistant bacteria is a problem [1]. One approach to better understand the effects and the attributes of bacteria, is to study processes at the cell membrane. However, since natural membranes are very complex, it is difficult to build them in an experimental setup. To do this, simplified model membranes can be used. It is in this context the Fluorescence Montal & Mueller Microscope finds its use. The FMMM can be used to set up membranes, measure their electrical properties and observe fluorescence labeled particles simultaneously. It consists of three parts. Part I is used to build a simplified model of natural membranes with the Montal & Mueller method. This part consists of the sample chamber which itself consist of two troughs between which a Teflon film is pinched. The Teflon film has a aperture with a diameter of 100-250 µm. The membrane is build inside this aperture. Part II is used to measure the electrical properties of a membrane. This can provide information about the structure and whether or not other particles are embedded in the membrane. This part consist of the Faraday cage in which the sample chamber is placed and two electrodes. Each electrode placed inside one of the troughs. Both electrodes are connected to a signal amplifier which in turn is connected to the software. Part III is the optical setup with which fluorescence labeled peptides and membrane lipids can be observed. This provides information about interactions and accumulations of particles. This part consists of two lasers, the beam path and a camera to observe the fluorescence. The beam path goes inside the Faraday cage, through the sample chamber and ends at the beam stop. A model of I, II and the end of III can be seen in Fig. 1.

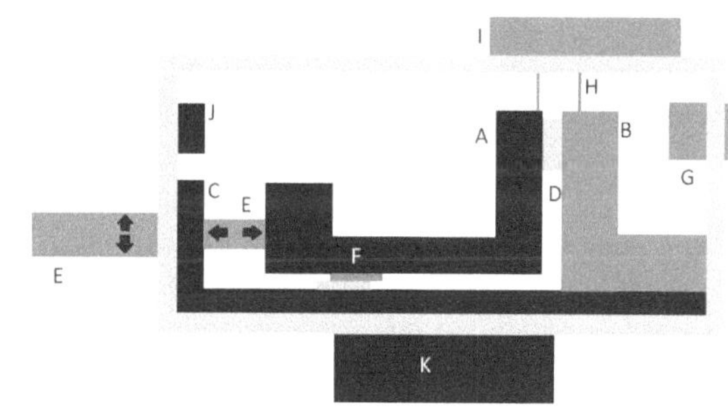

Figure 1: Outline of I and II. A: left chamber holder; B: right chamber holder; C: connection to Faraday cage; D: sample chamber; E: micrometer slider; F: rail; G: end of optical setup; H: electrodes; I: signal amplifier; J: beam stop; K: tool for X-Y-Z positioning

1.1 Cell membrane

Because of the complexity of natural membranes, here is a short introduction to the topic. The main structural elements of membranes are phospholipids and proteins [2] as shown in Fig. 2. The membrane is responsible for the shape of the cell, the diffusion into and out of the cell as well as signal transmission. Phospholipids define the structure of the cellular membrane. They consist of two hydrophobic hydrocarbon chains (called *tail*) and a phosphoric acid diester (hydrophilic *head* group). In case of this project the lipid DOPC was used to build membranes. In a natural membrane the lipids of two monolayers connect to each other at the tails.

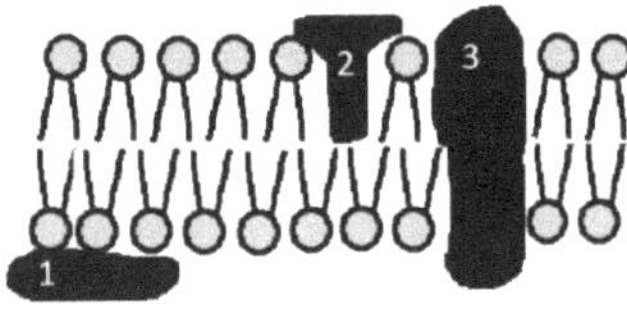

Figure 2: The model of a lipid bilayer membrane. It is formed from phospholipids (gray) with a head group, a tail group. From left to right a peripheral (1), an integral (2) and a trans membrane protein (3) are shown.

1.2 Montal & Mueller method

The Montal & Mueller method is used to build a simplified model of natural membranes for part I. The basic structure of a Montal & Mueller [3] apparatus is that of one box with two chambers inside. Each Chamber is filled with buffer and a single layer of lipids on the surface. The head group of the lipid faces the solution with the tail group facing the air. This way, later when both monolayers come into contact, the hydrocarbon chains of both layers form a connection. An electrode reaches into each chamber. The two chambers are separated by a septum. The septum is made of a thin Teflon film which has a small aperture in its center (100-250 µm diameter).

To build the bilayer membrane more polar solution is slowly added to one chamber through a syringe, which makes the level rise above the hole in the Teflon film. Because of that, a lipid monolayer is build inside the aperture of the Teflon film. When the same is done with the second buffer level, both lipid monolayers connect and form a lipid bilayer. This can be observed through an increase in the electrical capacitance measured between the electrodes in both chambers. The basic principle can be seen in Fig. 3.

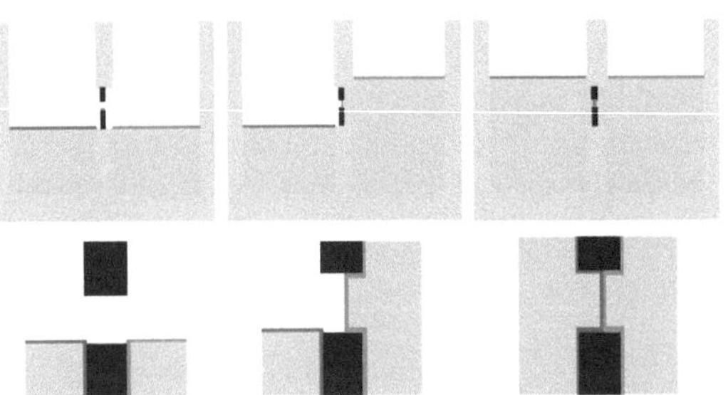

Figure 3: Formation of a planar lipid bilayer. Two chambers are used and filled with lipids on top of buffer.Left: Buffer and lipids are prepared. Middle: More buffer is added into one of the chambers. The lipids form a monolayer in the aperture between the chambers. Right: Buffer is added into the other chamber and a bilayer membrane is formed from the two monolayer in the gap.

1.3 Previous Structure

The structure of the Montal & Mueller device (section I; Fig. 4) used in the FMMM is different from the one described in Chapter 1.2. The box with the two chambers is made from silicone rubber, which is biologically inert, temperature resistant and mechanically flexible [4]. To build the lipid bilayer membrane instead of adding more solution to the chambers, they are slowly compressed. The solution level rises and the membrane is build. The upside to this composition is that there is no need to have syringes sticking inside the sample chambers at all times, but there are flaws, too. The first is that the brackets start to bend backwards when compressing the silicone chambers and therefore making them not usable for further experiments. The second is that the silicone chambers can only be used once and must be replaced after every measurement. The last and most important one is that the membrane does not stay in focus of the beam path (part III) when compressing the chamber [4]. Because of the points described above the decision was made to forsake this design and make a completely new one. One that is easier to use, takes less time to prepare and does not take damage during experiments.

This is an ongoing project divided into three sub projects. The first sub projects deals with designing a new sample chamber with which membranes can be build for I. The second and third sub projects deal with optimizing II and III. This paper concludes the first part with the presentation of the new structure.

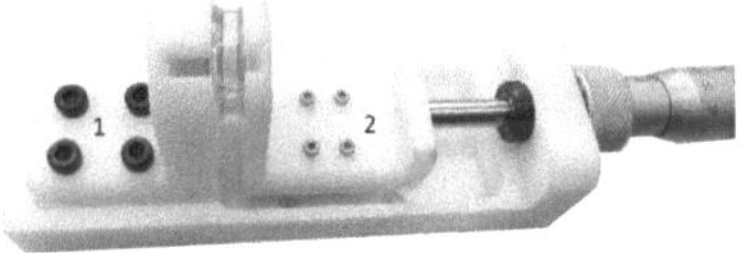

Figure 4: Montal & Mueller setup. The silicone chambers are positioned between the two brackets. One bracket (1) is fastened to the ground while the other (2) is fastened to a rail. Bracket 2 can be moved to compress the silicone chambers and thus building the lipid bilayer membrane. The opening in the top of both brackets are used for the insertion of the electrodes.

2 Material and Methods

2.1 Materials

The requirements for the materials for the sample chamber are that they need to be heat resistant up to 323 K (50 degree Celsius), are not electrically conductive, are chemically inert, rigid and they need to fit inside the Faraday cage. The new sample chamber was taken from another setup in the lab. It consist of 2 troughs made from Teflon which are put in an aluminum frame to hold them together. The chamber of these troughs is delimited by a glass plate (Fig. 7). These sample chambers are not tall enough to reach into the beam line, so a holder needed to be made. The structure of the holder is further described in Chapter 3.1 and Fig. 5. It is made by 3D printer with the printing material being PLA. PLA is heat resistant up to at least 333 K (60 degree Celsius) and rigid.

2.2 Capacitance Measurement

The capacitance of the membrane is described by

$$C = \varepsilon_0 \varepsilon_r \frac{A}{d}, \tag{1}$$

with ε_0 being the electrical field constant, ε_r the relative permittivity, A the area of the membrane and d the thickness. The electrical capacitance of the membrane is measured by applying a voltage through the electrodes in each chamber (part II) according to the stress jump method [5]. To build the membrane 20 µl of the lipid DOPC was used and a 1500 µl solution of 300 mM NaCl to 50 mM HEPES as buffer for each chamber. The Teflon film (200 µm diameter) was first cleaned with methanol and then hydrophobicised with a 10+1 Hexan-Hexadecan solution. No heater was used. The data acquisition was done through a software implemented under LabView.

3 Results and Discussion

3.1 New Structure

Because of the complexity of the drawbacks described in Chapter 1.3, it was decided to not improve the existing structure of part I and to build a new one. The new structure resembles the structure described in Chapter 1.2 more closely. Now there is a aluminum frame for the compartments. Each compartment consist of a Teflon trough and a glass plate and has an opening for a syringe. The Teflon film is impacted between the two Teflon troughs (Fig. 7).

To hold the sample chamber inside the beam path three parts were designed and printed with a 3D printer. The first part (Fig. 6a) is used to fasten a glass plate to the Teflon trough. To form one chamber. The second part (Fig. 6b) connects the sample chambers to a rail, which allows the sample chambers to be moved back and forth. The last part (Fig. 6c) connects the rail, on which the sample chambers moves, to the outer aluminum box. An Outline of this description can be seen in Fig. 5.

Additionally a DM-200 digital microscope from Somikon, Buggingen, Germany was added to the setup. This microscope has a resolution ranging from 1280x720 Pixel to 640x480 Pixel depending on the magnification. It is possible to see the aperture in the Teflon film with this microscope (Fig. 8). It will later be used, when optimizing the optical setup (III), to help position the aperture inside the beam path.

3.2 Results of Capacitance Measurement

Using the experimental setup described in Chapter 2.2 and 3.1 a membrane has been build as seen in Fig. 9. The signal was measured every quarter of a second. The capacitance

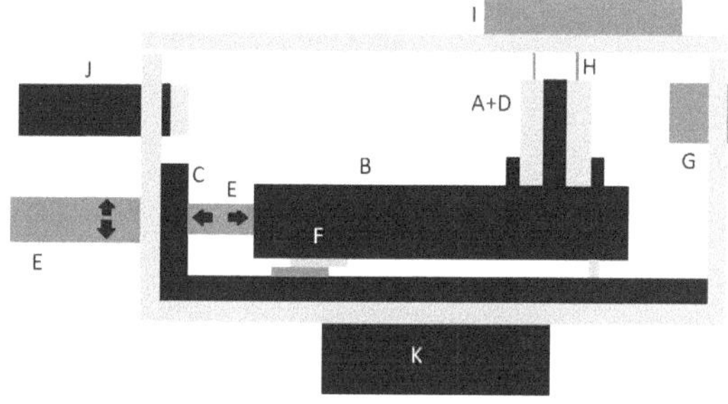

Figure 5: Outline of the Montal & Mueller setup. A: Fig. 6a is inside Fig. 6d; B: Fig. 6b; C: Fig. 6c; D: Fig. 6a+d; E: micrometer slider; F: rail; G: end of optical setup; H: electrodes; I: signal amplifier; J: camera to help with positioning; K: tool for X-Y-Z positioning

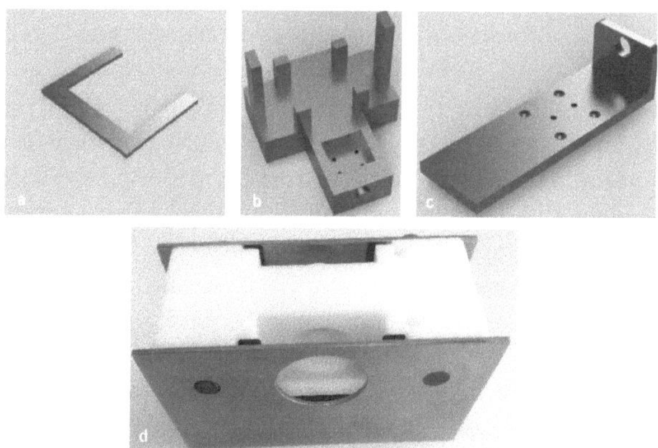

Figure 6: The new sample chamber. From top left to top right: 3D printed frame to press the glass plates against the compartments; 3D printed bracket for Montal & Mueller chamber; 3D printed framework to connect the bracket and the aluminum box. Bottom: Montal & Mueller chamber.

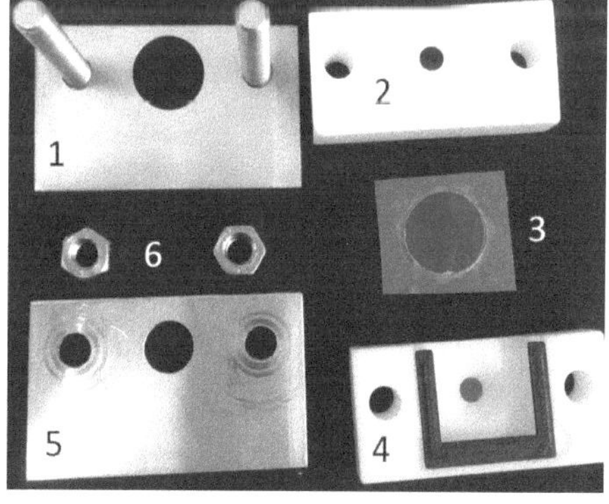

Figure 7: Assembly of the new sample chamber (Fig. 6d). The numeration describes the sequence in which the parts are assembled.

for a membrane depends mainly on the size of the aperture in the Teflon film and the type of lipid [3]. In this experiment the aperture size was $A = 150$ µm and DOPC was used as the lipid. This means when the capacitance C in pF stays between the values of 20-100 for a prolonged period of time, a membrane has successfully formed.

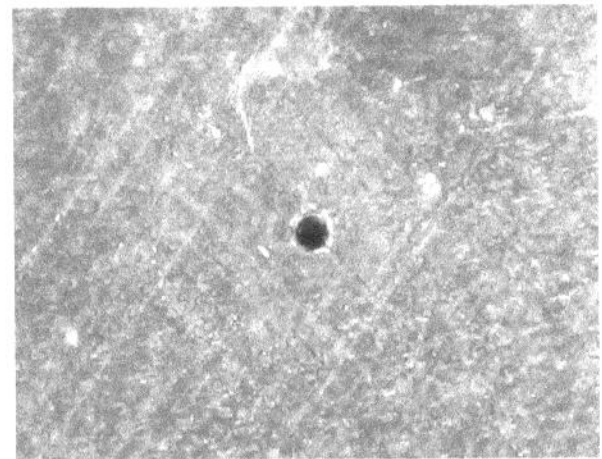

Figure 8: A picture taken the DM-200 microscope showing the aperture in a Teflon film which has a diameter of 150 μm.

In Fig. 9 it can be seen that at first, between 0 and 16 seconds, there is no influence of the membrane, because it hasn't formed yet. After 16 seconds the buffer level is raised one chamber after the other and a membrane is build. The membrane stays intact for 47 seconds. After that it becomes increasingly unstable and breaks. The high values for the capacity after 65 seconds mean that there is a short circuit and electrons flow freely from one electrode to the other without being interrupted by the membrane or the Teflon film. The membrane stood for 47 seconds. Although this is only a small window of time, it proves the new setup works. The time the membrane stood needs to be further improved to use the setup in current research. Ideally the person doing the experiment should be able to freely control how long a membrane stands. To do this it is important that there is enough room to move instruments without touching the two sample chambers in the aluminum box. Greatest care must be taken when preparing and during the experiment. Movements, dust particles or air bubbles in the sample chamber can break the membrane. Therefore it is important for the experimenter to practice the procedures and gain experience.

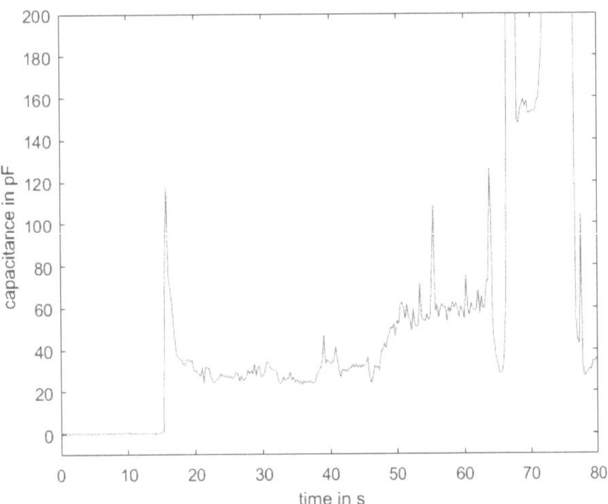

Figure 9: Measurement of a DOPC bilayer membrane with NaCl-HEPES as a buffer. The chambers are filled with buffer and a membrane is build after 17 seconds. After 65 seconds the membrane becomes unstable and breaks.

4 Conclusion

The redesign of the Montal & Mueller setup for part I has had the following effects: 1. Instead of needing to change the sample chamber brackets, the silicone chamber and the Teflon film after every experiment, only the Teflon film needs to be replaced. Every other component can be cleaned and reused. 2. After 5 experiments there was no observable damage done to any component. 3. Before the Teflon film and with it the membrane would move when building the membrane, now the Teflon film is firmly pinched between the two troughs and unmovable, which should make the fluorescence measurement more reliable. 4. Now there is a syringe inside each sample chamber. Big movements of the syringes might rupture the membrane.

With this as the basis, optimizations will continue to be made and improvements to the optical setup (III) of the Fluorescence Montal & Mueller Microscope will be worked on.

Acknowledgement

The work has been carried out at Forschungszentrum Borstel, department of Biophysics and supervised by Prof. Thomas Gutsmann and Dr. Christian Nehls.

I want to thank Prof. Thomas Gutsmann and Dr. Christian Nehls for their support and encouragement throughout this project.

The previous setup was made by Niklas Stein [4].

5 References

[1] H. S. Gold and R. C. Moellering, *Anitmicrobial-Drug Resistance*. In: N Engl J Med; 335:1445-1453, 1996

[2] Y. Shan and H. Wang, *The structure and function of cell membranes examined by atomic force microscopy and single-molecule force spectroscopy*. In: Chem.Soc.Rev., 44, 3617, 2015

[3] M. Montal and P. Mueller, *Formation of Bimolecular Membranes from Lipid Monolayers and a Study of Their Electrical Properties*. In: PNAS 69 (1972), P. 3561–3566.

[4] N. A. Stein, *Vergleich biophysikalischer Methoden zur Charakterisierung der Membran-Peptid-Interaktion* Christian Albrechts-Universität zu Kiel, Diplomarbeit, 2015

[5] O. Alvarez and R. Latorre, *Voltage-dependent capacitance in lipid bilayers made from monolayers*. In: Biophysical journal, 21(1):1–17, 1978.

Case study: Influence of electrical thigh stimulation on physiological and pathological gait

Fiona Allerding [1], Aljoscha Diercks [2], Nils Bade [3], Lotta Röhrich [4] and Robert Wendlandt [5]

[1] Medical Engineering Science, Universität zu Lübeck, fiona.allerding@student.uni-luebeck.de
[2] Evomotion GmbH, Lüneburg, diercks@evomotion.de
[3] Evomotion GmbH, Lüneburg, bade@evomotion.de
[4] Biomedical Engineering, HAW Hamburg, lotta.roehrich@haw-hamburg.de
[5] Clinic for Orthopaedics and Trauma Surgery, Universität zu Lübeck, robert.wendlandt@uksh.de

Abstract

Functional electrical stimulation (FES) on the shank while walking has proven to be a big support for patients with central nervous diseases [5]. The aim of this study is to investigate the effects of FES of the thigh muscles on human gait. One person with pathological and one person with a normal gait participated in this study. FES is applied to the M. quadriceps femoris for extension and to the hamstring group for flexion of the knee. The test persons walked five times along a walking distance of 10 m with a self-selected walking speed. The gait was recorded with cameras at 120 Hz and LED markers were placed on the leg, to track the knee angle. The FES on the hamstring group had a clear effect on the knee. The subject was again able to adopt a more normal gait and inflect the knee.

1 Introduction

If one suffers from neurological diseases, which primarily affects the functionality of the lower extremity, mobility and thus locomotion is often difficult. A consequence is a development of a pathological gait pattern whereby an evasive movement in order to compensate for a lack of knee flexion is adopted. Without functioning knee flexion, the swing-leg forward movement is significantly more difficult.

The patient's limited mobility concerning the lower leg can partially be restored by using functional electrical stimulation (FES) on dorsiflexors and plantarflexors [5]. Muscles are triggered through stimuli applied via surface electrodes on the skin. The electrical impulse is transmitted to the muscle via the nerve, causing a muscle contraction. The strength of the muscle contraction can be influenced by adjusting the stimulation parameters (intensity, frequency and pulse width). The prerequisite for FES is a functioning motor neuron of the muscle which has to be stimulated. Therefore, diseases of the central nervous system can be well treated while periphery nervous diseases can not [1].

The main focus of this work is on the stimulation of the thigh. Therefore, M. quadriceps femoris and the hamstring group were stimulated since they are extensor and flexor of the knee. Furthermore, the M. gastrocnemius being located in the lower leg is also responsible for knee flexion [4]. Therefore, it was tested whether the stimulation of the lower leg and the combined stimulation of the lower leg and the thigh has an effect on the maximum knee angle in the correct gait phase, too. One double step is described as

gait cycle (GC). During normal physiological walking, the stance phase accounts for 60%, the swing phase for 40% of the total GC. For a physiological gait the main knee flexion of about $60°$ takes place during the swing phase of the GC [2]. Therefore, it is expected to see a difference to the gait without stimulation mainly during swing phase.

2 Material and Methods

2.1 Material

For the FES of the thigh the product evomove of the company Evomotion GmbH was used.
The evomove has two channels for FES which can be controlled individually. Thus, two muscle groups can be stimulated at the same time. In addition, the evomove works with an adaptive algorithm in order to guarantee that the stimulation is triggered in the correct gait phase. The stimulation parameters and the timing can be configured by experts via the Medical App "Evomotion". Thus, it is possible to configure the stimulation period in percent of the GC. The stimulation parameters: the intensity in mA, the frequency in Hz and the pulse width in µs are individually adjusted.

2.2 Methods

Four electrodes per evomove were used for the stimulation. The electrodes for the control of the thigh were attached to the beginning of the muscle belly of the M. quadriceps femoris while for the stimulation of the hamstring group the electrodes were attached to the muscle belly of M. biceps femoris and M. semitendinosus.

For the stimulation of the lower leg the electrodes were attached to the muscle belly of the M. gastrocnemius in order to stimulate the knee flexor. The stimulation of the muscle groups was first validated on a physiological (phys.) gait pattern before being tested on the pathological (path.) subject.

To stimulate the three muscle groups, two evomoves were used, one was placed on the thigh, one on the shank with a retaining case and belt. The evomoves were connected via cables to the positioned electrodes. After the evomoves were placed, they need to be calibrated in order to make the gait detection algorithm work properly. The plane in which the patient is walking (sagittal plane) is calculated by a horizontal and vertical position of the leg.

In order to stimulate and support the knee flexion the stimulation parameters are first adjusted while the test person is sitting. The leg is thus not loaded, and a contraction can be easily observed and adjusted. If the parameters are set satisfactorily, the timing of the stimulation is set while the subject is walking. Thereby, the stimulation parameters can still be improved, for example if the selected intensity while sitting is too strong or too weak while walking.

The stimulating unit was configured for thigh and lower leg stimulation. The aim was to achieve a pleasant not painful feeling and a good contraction of the desired muscle group. As it turned out, for the path. test person only the hamstring group needs to be stimulated in order to achieve the greatest possible effect. The additional stimulation of the M. quadriceps femoris was felt to be disturbing.

Light-emitting diodes (LED) were used as markers for the video-based analysis. The LED markers were attached to the joint axes of the leg. The hip joint is marked above the greater trochanter, the knee joint above the lateral condyle, the ankle joint at the lateral malleolus and the forefoot at the level of the metatarsophalangeal joint. The video is recorded with an iPhone 7 at a frequency of 120 Hz. In addition, a fisheye lens (olloclip, Wide-angle + Macro, 2-in-1 Essential Lenses for most Devices) is placed in front of the camera to record a larger spacial angle. Thus, one is able to evaluate more steps of the gait.

Fisheye lenses allow a picture to be taken over 180°, but this results in a barrel-shaped distortion of the image. In order to reproduce the physical reality and thus calculate the correct angles while walking, the video must be equalized. This is realized with the camera calibrator app of MATLAB.

First, the videos of walking without stimulation (noStim) is recorded to avoid an influence of possible aftereffects of the stimulation. Afterwards, the data of stimulating the thigh (StimTh), stimulating thigh and lower leg (StimThLL) and stimulating only the lower leg (StimLL) were recorded.

At the beginning of the measurement the video recording is started in slow-motion. The test person has to cover a walking distance of 10 m recorded by the camera. After the 10 m walking distance, the recording of the video is stopped and saved. The length of the running distance depends amongst others on the distance the gait algorithm needs to calculate the individual walking speed and the initial contact (IC) of each step in order to ensure a stimulation of the right gait phase. The remaining distance is used for the video analysis. The measurement is repeated five times per test person to enable averaging of the data. By comparing the measurements with and without stimulation it is possible to determine the effect of stimulation on the path. gait pattern.

The main focus of the measurements' analysis is on the knee angle since the studies' assumption expects to observe a great effect on it.

The software MATLAB was used for the video-based analysis. At the beginning of the video evaluation, the markers are selected manually once to ensure a suitable starting position in the GC. As the markers are illuminated in cyan, they can be tracked by searching for pixels in each video frame with only green and blue image components. Then each frame of the video is checked for markers by selecting the central coordinates of these tracked pixels. This is used as new starting point of markers in the next frame. With the help of a threshold value, incorrectly assigned green and blue components that do not represent a marker can also be suppressed. For lost markers a new start position of the marker is calculated by extrapolation of the last coordinates. If no marker has been found for more than 5 frames and thus no marker coordinate is available, the mean value for the x and y value of the last coordinates is calculated. This is used as starting point of the marker for the next frame.

Due to lost markers there are missing data points which have to be interpolated in order to get a complete set of measured angles. The empty data points are removed from the array and interpolated over all missing values. For the interpolation the position of the empty data points within the array must be smaller than the array position of the last tracked data point. This last tracked data point of the marker is the reference for the program how far the interpolation has to go. All tracked angles together with the interpolated data points are stored in a new variable being used to form the vectors for the angle calculation.

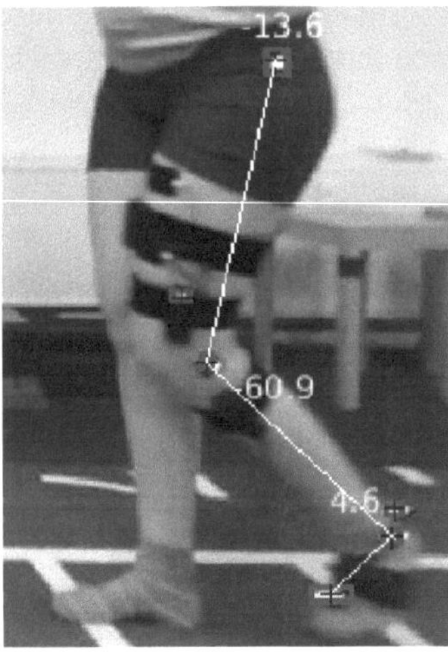

Figure 1: Knee angle (60.9°) calculation of the joints using the vector between hip and knee markers and the vector between knee and ankle markers.

Looking at Figure 1 the angle calculation is performed using the vector between hip and knee markers and the vector between knee and ankle markers. With the use of the standard scalar product and the euclidean norm the angle is calculated and signed within the video.

3 Results and Discussion

3.1 Results

For this study, a phys. and path. subject was measured using video-based analysis. The data is separated in double steps and the mean of all recorded steps (n=3) was build. Figure 2 shows the knee curves with and without stimulation of the thigh of the phys. subject.

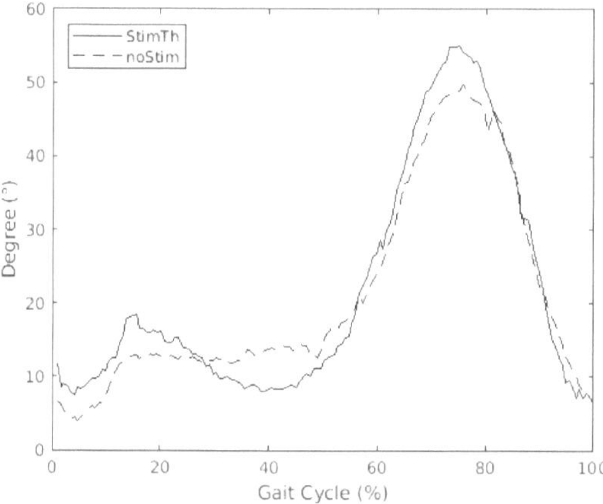

Figure 2: Averaged course of the phys. knee angle (n=3) with and without stimulation of the thigh.

The curves of noStim and StimTh have a similar course. The mean standard deviation $(\overline{SD})$ of StimTh is about $4°$ while the $\overline{SD}$ of no stimulation is about $5°$. According to literature, for a phys. gait the maximum knee angle is about $50°$ to $60°$ [3]. Without stimulation the phys. gait pattern shows a maximum knee angle of about $50°$ during the initial swing (ISw) which is consistent with the literature. With stimulation, a higher angle is seen in the swing phase. During ISw, there is an increase of the maximum knee angle of about $5°$ to $55°$. Without as well as with stimulation of the thigh this angle is at about 75% of the GC. To test whether there is a systematic difference between StimTh and noStim, a t-test was performed (α=0.05, Table 1). The null hypothesis assumes no significant difference between the mean values of the measurements. For the entire GC the t-test results in a p-value higher than the significance level α. Therefore, the null hypothesis can be accepted and the mean value differences are not significant. Thus, the probability of accepting a right null hypothesis is at least 95%.

Furthermore, the step length and speed (Table 2) were calculated. Looking at $\Delta\overline{s}$, there is an increase of $0.27\,\mathrm{m}$ while $\Delta\overline{v}$ rises about $0.13\,\frac{\mathrm{m}}{\mathrm{s}}$ compared to the measurement without stimulation.

Table 1: Evaluation in which phase of the GC, the knee angle with stimulation differs significantly from the one without FES. The t-test (α=0.05) is significant if the null hypothesis is (1). Therefore the probability that the visible change is due to randomness is less than 5%.

GC	phys.	path. Th	path. ThLL	path. LL
10%	0	0	0	0
20%	0	0	0	0
30%	0	0	0	0
40%	0	0	0	0
50%	0	0	0	0
60%	0	1	0	0
70%	0	1	1	0
80%	0	1	1	1
90%	0	0	1	0
100%	0	0	0	0

Table 2: Mean step length and speed of the gait patterns.

param.	noStim	StimTh	StimThLL	StimLL
$\Delta\overline{s}$ phys.	$1.52\,\mathrm{m}$	$1.79\,\mathrm{m}$		
$\Delta\overline{v}$ phys.	$1.38\,\frac{\mathrm{m}}{\mathrm{s}}$	$1.51\,\frac{\mathrm{m}}{\mathrm{s}}$		
$\Delta\overline{s}$ path.	$1.55\,\mathrm{m}$	$1.55\,\mathrm{m}$	$1.38\,\mathrm{m}$	$1.44\,\mathrm{m}$
$\Delta\overline{v}$ path.	$1.29\,\frac{\mathrm{m}}{\mathrm{s}}$	$1.28\,\frac{\mathrm{m}}{\mathrm{s}}$	$1.12\,\frac{\mathrm{m}}{\mathrm{s}}$	$1.19\,\frac{\mathrm{m}}{\mathrm{s}}$

Figure 3 shows the knee angle course of noStim, StimTh, StimThLL and StimLL averaged from 5 double steps (n=5). All four curves have a similar course except of the swing phase showing an increase of the knee angle with stimulation. The $\overline{SD}$ was also calculated for these measurements showing that noStim has about $3.4°$, StimTh about $2.5°$, StimThLL $1.6°$ and StimLL $2.8°$.

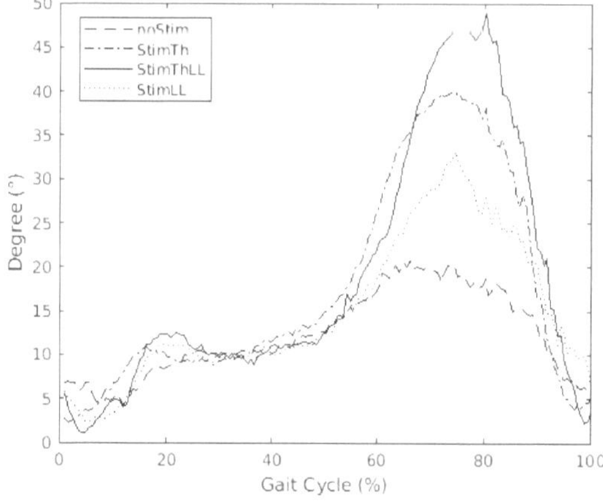

Figure 3: Averaged course of the path. knee angle (n=5) without stimulation, with stimulation of the thigh, the lower leg and the combination of the thigh and the lower leg.

Without stimulation of the thigh, the maximum knee angle is only about $21°$. During ISw, there is an increase of the maximum knee angle (75% GC) of about $19°$ to $40°$ when stimulating the thigh. The t-test (α=0.05, Table 1) shows a significant mean value difference in the swing phase corresponding to ISw (60-80% GC). The calculated p-value describes the probability to achieve the observed results just by chance. Since the p-value is clearly less than α during 60-80% of the GC, there is a systematic difference between the measurements and the null hypothesis can be rejected during this section of the gait.

For StimThLL, there is a maximum knee angle in ISw of about 47°. This is 7° higher than StimTh and 26° higher than without stimulation. Therefore, with StimThLL there is the best result for the knee angle of the path. test person. Furthermore, the t-test (α=0.05, Tab. 1) shows a significant mean value difference during 70-90% of the GC. Therefore, the null hypothesis can be rejected for this gait section.

With StimLL the maximum knee angle is about 33° being 12° more than without stimulation, 7° less than StimTh, and 14° less than with StimThLL. Again the maximum knee angle is at about 75% of the GC during the ISw. The t-test (α=0.05, Table 1) results in a rejection of the null hypothesis for 80% of the GC since there is a significant mean value difference.

Looking at the step length and speed (Table 2), there are small differences between the measurements. The greatest value of $\Delta\overline{v}$ and $\Delta\overline{s}$ is visible during noStim. When stimulating the thigh $\Delta\overline{s}$ remains the same while $\Delta\overline{v}$ begins to decrease.

3.2 Discussion

As shown in Figure 2, the stimulation has just a little effect on the phys. gait pattern. With stimulation of the thigh the maximum knee angle is about 55°. Therefore, compared to noStim the knee angle in ISw rises about 5°. By looking at $\Delta\overline{s}$ and $\Delta\overline{v}$ an acceleration of the gait compared to noStim can be detected. Nevertheless, the t-test does not approves a significance of the curves' mean value differences since p is higher than α during the entire GC.

Looking at the path. gait pattern (Figure 3), the curve of no stimulation clearly shows the pathology of the subject. The knee angle course in the swing phase is significantly smaller compared to the phys. gait pattern.

Already by stimulating the knee flexor in the shank a correction of the gait is achieved. By stimulating the knee flexors in the thigh a further improvement of the angle can be observed. Comparing noStim and StimThLL the maximum knee angle during ISw is more than doubled. Therefore, with StimThLL the path. knee angle could be significantly approached to the phys. knee angle of 50° to 60°. These angle increases are supported by the t-test with a significance of the mean differences during the swing phase. Looking at the step length and speed of the path. test person, $\Delta\overline{s}$ and $\Delta\overline{v}$ decreases when using FES. Especially when using FES at the shank (StimThLL, StimLL) there is a reduction concerning both parameters. This contradicts the expectations as well as the subject's perception. The test person stated an accelerating effect of FES, especially when thigh and shank were stimulated simultaneously. For a valid statement more data should be recorded and evaluated in the future.

All in all, the path. test person experienced the thigh stimulation as supportive. The gait pattern was more fluid and the subject perceived the gait to be safer. In addition, the test person judged the walking distance to be easier to cover when walking with stimulation. Overall, the treatment was perceived as a great support.

Based on the results, the effect of thigh stimulation on the maximum knee angle and thus on the gait pattern of the test person with path. gait pattern could be clearly concluded.

4 Conclusion

On the basis of the results, it can be clearly concluded that thigh stimulation had an effect on the maximum knee angle and thus on the gait pattern of the test person with path. gait pattern. Especially the combined stimulation of the thigh and the lower leg had a great effect. For a valid statement about the influence of thigh stimulation on the knee angle, further test persons should be measured and analysed.

Acknowledgement

The work has been carried out at the company Evomotion GmbH, and supervised by the Clinic for Orthopaedics and Trauma Surgery of the Biomechanics laboratory, Universität zu Lübeck.

Author's Statement

Ethical approval: The research related to human use complies with all the relevant national regulations, institutional policies and was performed in accordance with the tenets of the Helsinki Declaration, and has been approved by the authors' institutional review board or equivalent committee.

5 References

[1] Evomotion GmbH (2019b) *Technische Dokumentation.* Lüneburg, Evomotion GmbH. Interne Dokumente.

[2] J. Perry, *Ganganalyse- Norm und Pathologie des Gehens.* Urban & Fischer Verlag/Elseviert GmbH, 1. Auflage, 2003.

[3] K. Götz-Neumann, *Gehen verstehen. Ganganalyse in der Physiotherapie.* Thieme Verlag, 4. Auflage, Stuttgart, New York, 2016.

[4] N. Palastanga, *Anatomie und menschliche Bewegung: Strukturen und Funktionen.* Urban & Fischer Verlag/Elsevier GmbH, 1. Auflage, München, 2015.

[5] D. G. Embrey, *Functional electrical stimulation to dorsiflexors and plantar flexors during gait to improve walking in adults with chronic hemiplegia.* Archives of physical medicine and rehabilitation, 2010, 91; S. 687–696.

Development of an in-vitro brain model for diffuse axonal injury occurrence simulation

Tamara Burkhanova [1], Stephan Klein [2], Christian Damiani [2,a], and Kara Krajewski [3,a]

[a] These authors share senior authorship
[1] Biomedical Engineering, Luebeck University of Applied Sciences, tamara.burkhanova@stud.th-luebeck.de
[2] Medical Sensors and Device Lab, University of Applied Sciences Lübeck, {stephan.klein, christian.damiani}@th-luebeck.de
[3] UKSH Universitätsklinikum Schleswig-Holstein, KaraLeigh.Krajewski@uksh.de

Abstract

Diffuse axonal injury (DAI) is a common type of brain tissue damage after trauma that often leads to disability and death. DAI is not easily detected with standard medical imaging techniques. In order to investigate and simulate DAI occurrence, a complex in-vitro brain model is necessary. In this work, we studied different compositions of phytagel (PHY) hydrogels as well as mix hydrogels made of polyvinyl alcohol (PVA) 3,4% and PHY 1.7% as model materials. Also unconfined compression – relaxation tests and indentation tests of the hydrogel specimens were done. To enable future measurements of post-traumatic deformations of the model, iodine contrast droplets were embedded within the hydrogel structure and investigated, using MicroCT scans. We achieved hydrogels with E = 0.9 kPa and E = 1.7 kPa for the white matter and corpus callosum models, respectively. These values closely resemble the stiffness of brain tissue according to other studies.

1   Introduction

The mechanical properties of the brain vary due to the complex system of neuronal networks in the brain [1], [2]. Mechanical impact during trauma can have severe consequences for brain function. One of the most dangerous type of brain injury is known as diffuse axonal injury (DAI). This type of injury is caused by sudden rotational movements within the head [2]. During rotation, large shearing forces are applied to the internal brain tissue, leading to elongation and ultimately rupture of axons [2]. Based on clinical studies [2], DAI is primarily observed in regions with a large number of axon fibers, such as cerebral white matter, the brainstem and the corpus callosum. DAI occurrence happens within the first 6-10 ms of the impact [3]. In this paper, we will focus on the corpus callosum region in order to measure the residual deformation and strain from mechanical impact. The corpus callosum is a tract with uniaxially oriented axons from left to right that connects both hemispheres of the brain.

Since brain tissue is hyper-viscoelastic, inhomogeneous and anisotropic [1], the exact value of stiffness is undefined. For example, the range of white matter stiffness goes from $0.59{\pm}0.19$ kPa to $2.30{\pm}0.64$ kPa [4]. The reasons behind the large variability of the reported values could be due to different sizes of tissue samples, the method of measurement, etc. The most cited values are presented in Table 1. Further, the stiffness of corpus callosum vary depends on the measurable plane of the sample: in transverse plane it is 1.1 kPa, in coronal plane 1.3 kPa and in sagital plane 3.32 kPa [1].

Regarding the article [6], the authors achieved hydrogels with mechanical properties close to that of white matter of human brains and created a homogeneous in-vitro model. The composition of the hydrogel was polyvinyl alcohol (PVA) 6% and phytagel (PHY) 0.85%. In this work, we created hydrogel models for corpus callosum, white and gray matter in order to later measure the strain after a mechanical impact. For this, we created and characterized small droplets of iodine contrast agent inside the hydrogels, which should be used as markers.

2   Material and Methods

2.1   Hydrogel manufacturing

The hydrogel consists of polyvinyl alcohol (PVA, MW 146000 − 186000, 87-89% hydrolyzed, Sigma − Aldrich, Germany), phytagel (PHY, Sigma − Aldrich, Germany) and deionized water. The interaction between the two polymers is due to the intramolecular hydrogen bonds in OH groups of PHY and PVA. *PVA* [-CH$_2$CHOH-]n is a synthetic polymer with density of 1290 kg/m^3, Poisson's ratio of 0.42-0.48 [6]. It is non-toxic, biocompatible [6] and has strain-rate dependent characteristics. *PHY* is similar to agar, made from a bacterial substrate based on glucuronic acid,

Table 1: Mechanical properties of brain regions

Brain region	Density, kg/m^3	Poisson's ratio	Young's modulus, kPa	Stiffness, kPa	Bulk modulus, GPa	Shear modulus, kPa
Grey matter	1040 [2]	0.49 [5]	0.8 [1]	0.68 [4]	2.19 [8]	G_0 = 10, G_∞ = 2 [7]
White matter	1040 [2]	0.49 [5]	1 [1]	1.33 [4]	2.19 [2]	G_0 = 12.5, G_∞ = 2.5[7]
Corpus callosum	1080 [5]	0.49 [5]	-	2.67 [9]	2.19 [8]	G= 3.32[9]
Cerebellum	1060 [7]	0.48 [5]	-	0.75 [4]	2.19 [2]	G_0 = 10, G_∞ = 2 [7]
Brain stem	1060 [2]	0.4996 [5]	-	-	2.19 [8]	G_0 = 22.5, G_∞ = 4.5[7]

rhamnose and glucose. It is a tetrasaccharide biocompatible polymer with high strength and a viscoelastic behavior. *Lipiodol* (Ethiodized oil, 480 mgI/mL, Guerbet LLC, provided by the department of neuroradiology, Universitätsklinikum Schleswig-Holstein, Campus Lübeck) was chosen as a contrast agent marker due to its density 1,28 g/cm^3 at 20°C and viscosity 34 – 70mPa·s. It is an injectable radio-opaque agent used in neuroradiological studies. Each milliliter of oil contains 480 mg of iodine organically combined with ethyl esters of fatty acids of poppyseed oil. The precise structure is not known.

The procedure of manufacturing the gel for the model was adapted from [6]. Two different types of hydrogel were prepared: pure PHY and PVA+PHY. For *PVA+PHY* hydrogel, deionized water was warmed up in a microwave mid-low power to achieve 70°C. Then, the solution was put on the magnetic stirrer and covered with a cardboard isolation construction in order to reduce heat loss. The PHY powder or PVA crystals was then poured into the flasks in small quantities to prevent clumping. For PVA, stirring the mixture at 90°C for 1 hour was required. While the PHY stirring was carried out at 90°C for 30 minutes. In order to prevent evaporation, the flasks were covered with aluminum foil. Then, the PHY and PVA solutions were mixed together in a 4:1 ratio and stirred for 30 min at 70°C. The solution was then poured into plastic molds and allowed to cool down. After 15-30 min, the molds were placed in a freezer at -20°C for 5 hours. Subsequently, the molds were taken out to thaw at room temperature. The *pure PHY* hydrogel was fabricated as described above, but the final solution was stirred at 90°C for only 30 min.

Pure PHY hydrogels with different concentrations were manufactured, as well as a mixed hydrogel of PVA 3.4% and PHY 1.7% (*MH*). The composite hydrogel of PVA 6% and PHY 0.85% with ratio 1:1 (*CH*) were prepared according to [6].

In addition, hydrogel samples with embedded Lipiodol as a contrast agent were prepared. The hydrogel was manufactured in a similar way as the PHY and allowed to cool for 10-15 min. Three to four drops with less than a milliliter of Lipiodol each was poured with a syringe into the liquid solution. The mixture was slightly shaken in order to distribute droplets around the gel, reduce their size and keep them away from the bottom of the dish. The hydrogel and the droplets solidified after 10 minutes.

2.2 Mechanical testing

Unconfined compression – relaxation tests were carried out on the hydrogel specimens using the experimental setup schematically shown in Fig.1. The hydrogel samples were prepared according to ISO 3383 and had a cylindrical shape, 13±0.5 mm in diameter and 6.3±0.3 mm high. Prior to the measurements they were kept at room temperature for at least 16 h. A constant deformation of 1 mm was applied to

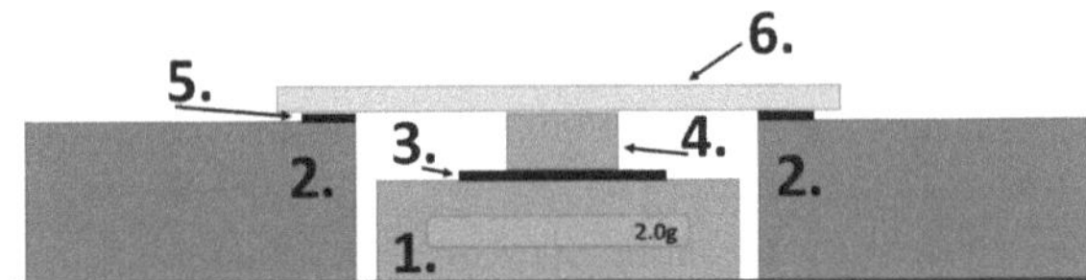

Figure 1: Schematic representation of unconfined compression – relaxation test setup.

the samples and the load-relaxation curve was recorded over time for 300s using a precision balance (PCE-BSH 10000, PCE Instruments, Germany) connected to a Standard PC. The sampling rate was 1 Hz.

The *unconfined compression – relaxation tests* allowed us to calculate relaxation modulus of the fabricated hydrogels. The calculation of the relaxation modulus was performed by equations (1) - (5).

$$E(t) = \frac{\sigma(t)}{\varepsilon_0}, [Pa] \tag{1}$$

where

$$\sigma(t) = \frac{F(t)}{A}, \tag{2}$$

$$\varepsilon_0 = \frac{L_\Delta}{L_0}, [Pa] \tag{3}$$

where

$$F(t) = m_t \cdot 9, 8, [N] \tag{4}$$

$$A = \pi \cdot R^2, [m^2]. \tag{5}$$

Where $E(t)$ - relaxation modulus, $\sigma(t)$ - time dependent stress, ε_0 - constant strain, $F(t)$ - inner force of the specimen, A - cross-sectional area of specimen, L_Δ - constant deformation, L_0 - specimen height, m_t - load, recorded on the scales.

Also, *Indentation tests* were conducted with a Mach-1^{TM}

mechanical testing system from Biomomentum, Canada. These tests were carried out on pure PHY hydrogels of different concentations, MH with different mix ratios and in three equal samples of CH [6]. Each of the tests consisted of two different measurements: force vs. deformation and force vs. time. Then the elastic and shear moduli were calculated by the Mach-1^{TM} Analysis software using internal algorithms of Elastic Model in Indentation. The specimen had a height of 2 mm, and their diameter was at least 5 times larger than that of the indenter. The indenter moved down with the speed of 0.1 mm/s until a force of 0.3 gf (gram - force) was recorded. Then, the indenter was stopped and held for 30 sec while measuring the relaxation.

2.3 MicroCT measurements

In order to characterize the embedded contrast droplets, the hydrogels were imaged using a *MicroCT* (nanotom m, GE, Germany) at 90kV and 100 μA. The voxel size was 20 μm and the distance between the object and the window was 13 cm in order to simulate the conditions needed for a real-sized brain model. The samples were also analyzed using an *optical digital microscope* (VHX-600, Keyence, Germany), in order to detect the difference between the gel, Lipiodol droplets and air bubbles, to measure the average size and distance between the droplets.

3 Results and Discussion

3.1 Mechanical testing

The relaxation curves from the *Unconfined compression – relaxation tests* are shown in Fig.2. The calculated moduli of elasticity are listed in Table.2. The highest modulus of elasticity was found for PHY 1.7% and the lowest for PHY 1.25% and MH 1:1. According to these results, only PHY 1.25% and MH 1:1 can be used for an in-vitro model of stiff regions of the brain.

Table 2: Mechanical properties of hydrogels from unconfined compression – relaxation tests

Hydrogel	Radius*, mm	(E_0),kPa	(E_∞),kPa
MH 1:1	13	1.93	0.5
PHY 1.25 %	11.9	1.92	0.94
PHY 1.42 %	12	9.57	5.42
PHY 1.7 %	13	21.28	7.36

** - Radius of the specimen*

The results from *Indentation tests* are shown in Fig.3 and Fig.4. Fig.3 shows the gel relaxation curves while keeping the indenter inside the gel. A fast relaxation behavior can be seen for all hydrogels except for MH 1:4. The larger relaxation time of MH could be due to the PVA component of the hydrogel.

In order to compare the mechanical properties of the surface of the samples with those of the bulk, the central part of the specimens was cut out and measured using the same

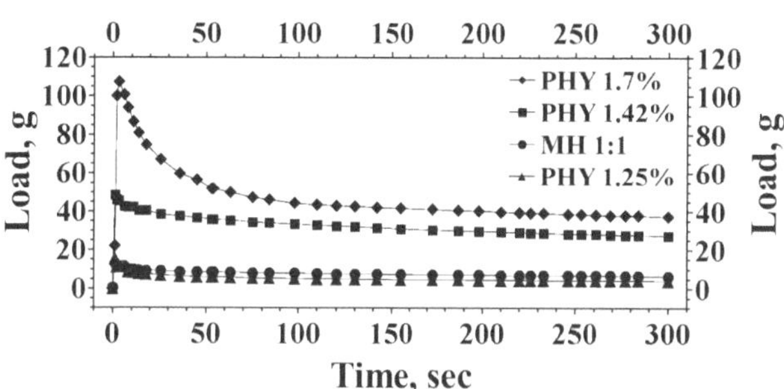

Figure 2: Unconfined compression – relaxation test curves of hydrogels.

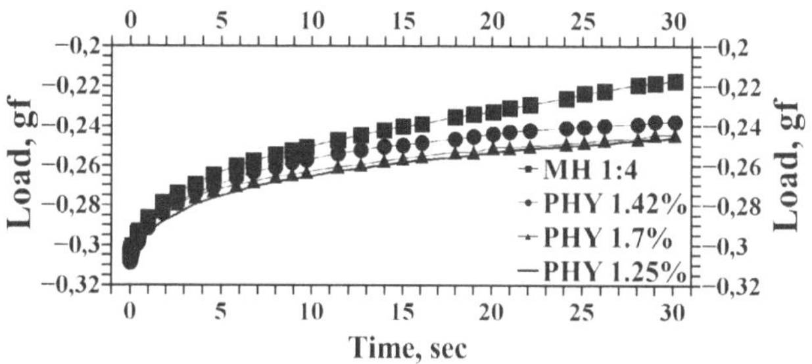

Figure 3: Indentation - relaxation tests of hydrogels.

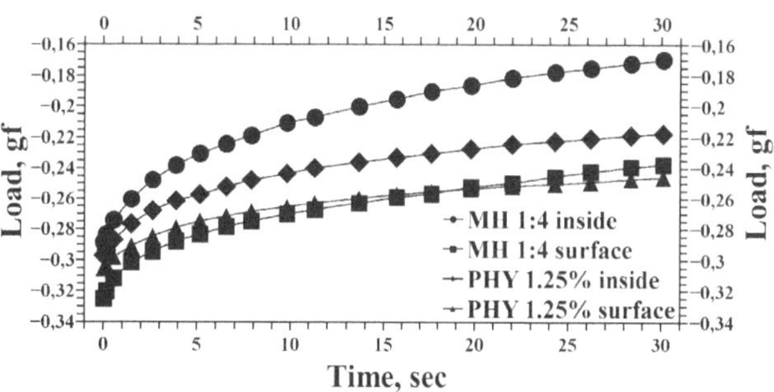

Figure 4: Indentation - relaxation test of hydrogels surface and inner sides.

identation tests as before. Fig.4 shows the relaxation curves for the surface and bulk of the PHY 1.25% and MH 1:4. The viscoelastic behaviors at the surface of both hydrogels are similar, with a relaxation load from -0.3 gf to -0.24 gf. However, the bulk for MH is softer than for pure PHY, ending up with -0.16 gf and -0.22 gf, respectively. These differences can be explained by the fact that oxygen influences the polymerization process on the surface of the hydrogel, but also water evaporation at the surface can lead to a higher stiffness.

The elastic modulus was measured in the first stage of the indentation test. Table 3. shows the elastic moduli of the hydrogels from the indentation test. The elastic modulus of the CH was 0.5 ± 0.05 kPa, which is half as stiff as cortex tissue. The elastic modulus was 10.8kPa for MH 4:1 and 0.72 kPa for MH 1:1, where the latest is softer than healthy cortex brain tissue [1].We achieved measurements on bulk PHY 1.25 % hydrogel of E = 0.9 kPa, which can

Table 3: Mechanical properties of hydrogels from the indentation test Mach-1

Hydrogel	Elastic modulus (E), kPa
CH	0.5
PVA 1% & PHY 2% 1:4	15.7
MH 1:1	0.72
MH 1:4	10.8
MH 1:4 inside	1.7
PHY 1.7 %	14.5
PHY 1.42 %	5.5
PHY 1.25 %	2.9
PHY 1.25 % inside	0.9

be an in-vitro white matter model. According to [7], corpus callosum is stiffer than white matter. Thus, the percentage of PHY for these regions should be increased slightly and the modulus of elasticity should be slightly higher. Based on [9], one can assume that bulk MH 1:4 hydrogel with E= 1.7 kPa can be used for a corpus callosum model.

3.2 MicroCT analysis

The *MicroCT* scan was performed through the horizontal side of the cylinder model. The Lipiodol drops can be seen in Fig.5, right image. The scan allowed us also to obtain a 3D image of the whole hydrogel phantom. Also, the inner part of the PHY 2.2 % hydrogel model with the embedded Lipiodol droplets was analyzed with an *optical microscope.* The droplets in optical microscope image and the circles

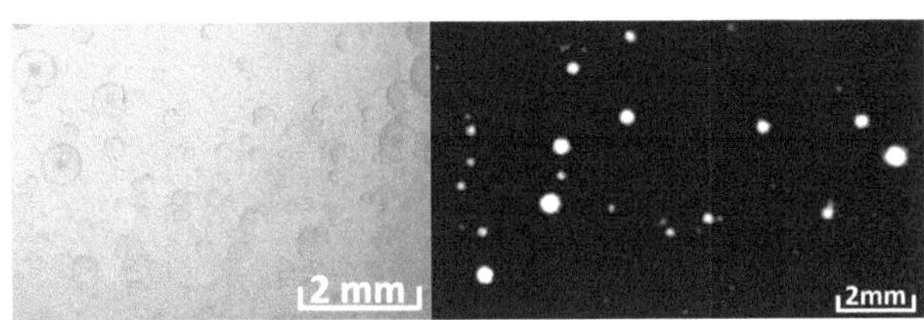

Figure 5: Optical microscope image (left) and NanoCT scan (right) of PHY 2.2% with embedded Lipiodol. The two pictures do not correspond to the same region of the sample.

in MicroCT scan correspond to Lipiodol contrast droplets. This "mesh of markers" with a unique shape and location should be used in future measurements to detect local deformation after simulated trauma impacts. The droplets could be tracked in 3D images to calculate the difference in their coordinates before and after mechanical impact. If the elastic modulus of the hydrogel is known and the deformation is measured, the corresponding stresses can be calculated. Even though this values are of limited use without a validation (e.g. ex vivo), they could help identify the regions of the brain which are at higher risk of DAI.

4   Conclusion

We were able to manufacture hydrogel models of the white matter and corpus callosum with similar mechanical proper-

ties to those of real tissue [1], [9]. Also, tiny droplets of an X-ray contrast agent were successfully embedded into the samples, which could be used to track local deformations inside the model using microCT imaging together with image registration methods. In the near future, tactile tests will be conducted by doctors on hydrogel models to confirm the mechanical similarity of these materials to human brain tissue. A further goal of this project will be the development of custom sensors for registering local deformations inside the brain model in real time.

Acknowledgement

The work has been carried out and supervised by Luebeck University of Applied Sciences. We thank Dr. Moritz Matthiae for the fruitful discussions and experimental support and Dr. Alexander Neumann for providing the contrast agent materials and support in the experiments.

5   References

[1] S. Budday, G. Sommer, et al, *Mechanical characterization of human brain tissue.* Acta Biomaterialia, 48, 319–340, 2017.

[2] A. Madhukar, M. Ostoja-Starzewski, *Finite Element Methods in Human Head Impact Simulations: A Review.* Annals of Biomedical Engineering, 47(9), 1832–1854, 2019.

[3] A.M. Nahum, R.W. Smith, *An experimental model for closed head impact injury.* SAE Technical Papers, 1976

[4] J. Weickenmeier, R. de Rooij, S. Budday, et al, *Brain stiffness increases with myelin content.* Acta Biomaterialia, 42, 265–272, 2016

[5] B. Yang, K.M. Tse, et al, *Development of a finite element head model for the study of impact head injury.* BioMed Research International, 2014.

[6] A.E. Forte, S. Galvan, F. Manieri, F. Rodriguez y Baena, D. Dini, *A composite hydrogel for brain tissue phantoms.* Materials and Design, 112, 227–238, 2016.

[7] T.J. Horgan, M.D. Gilchrist, *Influence of Fe model variability in predicting brain motion and intracranial pressure changes in head impact simulations.* International Journal of Crashworthiness, 9(4), 401–418, 2004.

[8] T. El Sayed, A. Mota, F. Fraternali, M. Ortiz, *Biomechanics of traumatic brain injury.* Computer Methods in Applied Mechanics and Engineering, 197(51–52), 4692–4701, 2008.

[9] L. Zhang, K.H. Yang, A.I. King, *Comparison of brain responses between frontal and lateral impacts by finite element modeling.* Journal of Neurotrauma, 18(1), 21–30, 2001

3

Biomedical Engineering

Development of a test stand to qualify the endurance life time of the Murata MZB1001T02 microblower

Anna Carlotta Hennigs [1], Oliver Wallnewitz[2]

[1] Medical Engineering Science, Universität zu Lübeck, anna.hennigs@student.uni-luebeck.de
[2] Drägerwerk AG Co. KGaA, oliver.wallnewitz@draeger.com

Abstract

As part of a new valve development for a ventilator, qualification of the MZB1001T02 microblower manufactured by "Murata - Innovator in Electronics" is required. In the ventilator, it is the task of the microblower to switch inspiration and expiration valves. These microblowers will be used outside the manufacturer's specifications. As a consequence, a complete qualification of the endurance life time is required.

The aim of the project is to develop a test tool that can qualify the endurance life time of the microblower. To speed up the testing process, the microblowers are tested for 22 hours instead of the two hours per day predicted operating life. In addition, the microblowers operate at a frequency of 2 Hz that exceeds a person's respiratory rate of 18 breaths per minute. So after 59 days of testing six years of use can be simulated. The result of the lifetime test is that no microblower has failed. This means that the microblowers are suitable for use in the inspiration and expiration valve of a medical device.

1 Introduction

The pumping of the microblower MZB1001T02 is created by a thin ceramic, which is oscillating by the converse piezoelectric effect and can be seen in Fig. 1 in the right picture. This effect describes that a certain material, such as specific ceramics or crystals, is deformed by applying an electrical voltage [1]. This effect is used in the microblower by an alternating voltage of about 26 kHz [2]. The amplitude of the deflection is proportional to the supply voltage. An increasing amplitude is accompanied by an increasing air flow and output pressure. As the ceramic swings up and down, the air is drawn into the microblower from below and discharged through the nozzle at the top. The operating voltage assigned by Murata should be between 10 and 20 V. The microblower can operate in an environment of $0\,^\circ$C to $70\,^\circ$C [3]. The microblower MZB1001T02 developed by

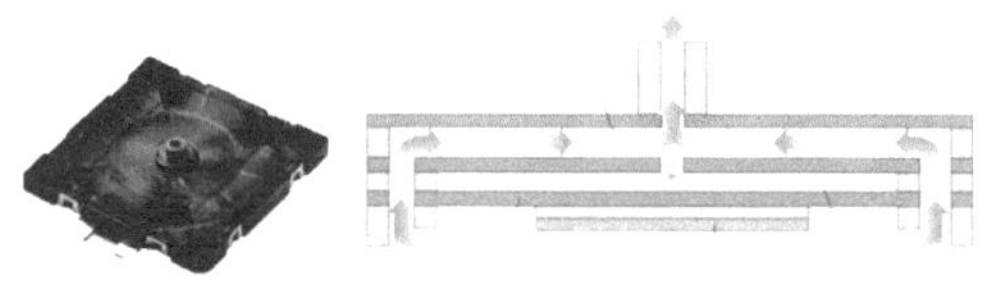

Figure 1: Picture of the microblower (left) and a schematic representation of the microblower (right) [3].

"Murata – Innovator in Electronics" is to be operated in this particular application outside the specifications specified by the manufacturer (24 V and $-20\,^\circ$C to $50\,^\circ$C). The piezo ceramic is very sensitive and the main cause for a defect of the microblower. The operating voltage determines the amplitude at the deflection of the ceramic. The higher the amplitude, the greater the mechanical load on the component. The service life of the microblower is expected to decrease proportionally to the operating voltage [3].

2 Material and Methods

2.1 Calculation of the acceleration factor

The maximum operating time of the microblower in the new application is two hours per day, which corresponds to 730 operating hours per year. Children have a higher respiratory rate than adults, so the average frequency assumed in the lifetime test is 18 breaths per minute [4]. The acceleration factor of the endurance test results from the extended operating time per day. The microblowers are always operated continuously for six hours, followed by a 30-minute break. This results in an acceleration factor of 11.25. Furthermore, the microblower are operated at a frequency of 2 Hz, which corresponds to 60 breaths per minute. This results in an acceleration factor of 3.3 and 37.2 overall, i.e. to simulate six years, the test must run for approx. 59 days.

2.2 Calculation of the minimum number of microblower

For the endurance test the failure rate of the microblower should not exceed 1 % per year. With an expected service life of six years, this results in a permissible failure rate of 6 % ($p = 6$). This hypotheses should be confirmed with a probability of 90 %. The search is for n (minimum number of objects to be tested) [5]:

$$G_{(x,n,p)} = \sum_{i=0}^{x} p^i (1-p)^{n-i} \qquad (1)$$

If $x = 0$ is now set (assuming no errors occur), $p = F_{(t)}$ is the maximum error rate, $G_{(x,n,p)} = 1 - P_A$ (P_A: confidence level). It results in [5]

$$n = \frac{ln(1 - P_A)}{ln(1 - F_{(t)})}, \qquad (2)$$

$$n = 37. \qquad (3)$$

In order to perform this test, a test stand needs to be developed.

2.3 Requirements of the test stand

In the following the requirements of the test stand are summarized in Table 1. The requirements result from the set test objectives, the best possible data acquisition and evaluation and the specified geometry of the new valve, which cannot be explained further here.

2.4 System Design

The following sections is separated into hardware and software design and includes all information to reproduce, maintain and repair the test tool. The basic hardware design can be seen in Fig. 2 and consists of:
48 microblower, 48 mountings each equipped with a membrane that is fixed with an O-ring, 48 oscillators, two relays, one Arduino Uno, two voltage sources, one LCD-Display.

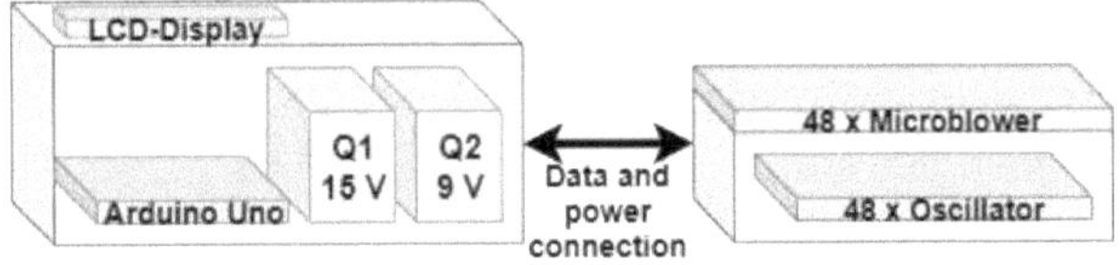

Figure 2: Overall hardware structure of the test tool.

2.4.1 Microblower mountings

The microblower mounting is executed in two parts. The CAD drawing can be seen in Fig. 3. The microblower is glued airtight into the upper part so that it can build up a pressure to be measured. In the system, the valves switch by the microblower, closing a crater with a diaphragm that is inflated. In order for this to work reliably, a minimum pressure of 25 ± 1 mbar needs to be achieved. The pressure can be tapped via a 6 mm Festo hose below the membrane. The upper part needs to be firmly attached to the lower part. This is realized by two lateral recesses so that the brackets can be attached flexibly via a rail system. The mountings are 3D-printed by Dräger from polyamide-12-fine powder.

Table 1: Requirements

	Requirement text
1	**Non-functional Requirements**
1.1	In the test tool all cable of connections are precisely labelled
1.2	In the test tool all settings (e.g. voltage source) are precisely labelled
1.3	In the test tool the electronics are fixated, so that no individual part can be removed
1.4	The electronics of the test tool are inaccessible to the user, so that faulty operation is impossible. The housing of the electronics is accessible by standard tools for repairs or for setting new parameters
2	**Process Requirements**
2.1	Detailed documentation of the test stand structure
2.2	Instructions for installing the microblower and putting the test into operation
2.3	A template is available for data analysis
3	**Sensors**
3.1	All sensors shall be calibrateable
4	**Technical Requirements**
4.1	48 Murata Microblowers MZB100102 fit into the test tool
4.2	The test tool operates the microblower with an oscillating voltage of 25 kHz
4.3	The test tool alternates the voltage between 15 and 24 V
4.4	The test tool exerts no mechanical pressure on the microblower
4.5	The test tool records continuously the pressure of each individual microblower
4.6	The test tool shows on a display the number of cycles, number of pauses and simulated test time
4.7	The electronics of the test tool and the microblower and driver housing must be separated
5	**Maximization of Portability**
5.1	Integration of all system components inside a case
5.2	Reduction of data and supply cables
6	**Leakage**
6.1	Each gas leading part of the test tool shall not leak more than 25 ml/min at 60 mbar
7	**Relay Control**
7.1	The Software shall control the relays for pump cycle control and operating hours
7.2	The Software shall count the pump cycle and the breaks between operation

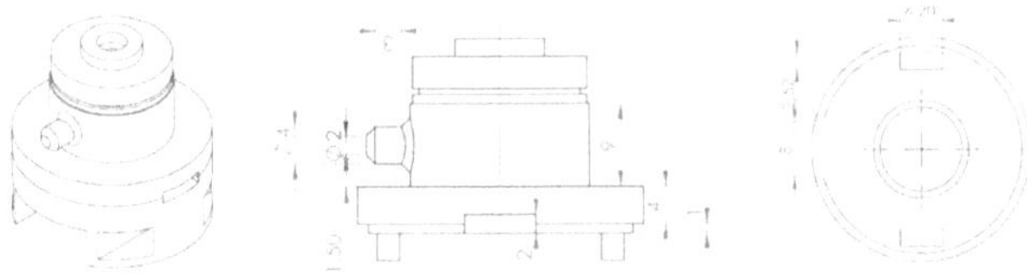

Figure 3: CAD drawing of the microblower mounting: left: total, middle: upper part, right: lower part.

2.4.2 Power supply control

In the ventilator, it is the task of the microblower to switch inspiration and expiration valves. To do this, the microblower should build up a pressure of 25 mbar in the diaphragm so that it inflates and closes the valve. Therefore, a realistic test requires switching between two voltages (valve open and closed) at pump rate. This is achieved by means of two relays which alternately connect a second voltage source in series. The circuit is shown in the Fig. 4. The

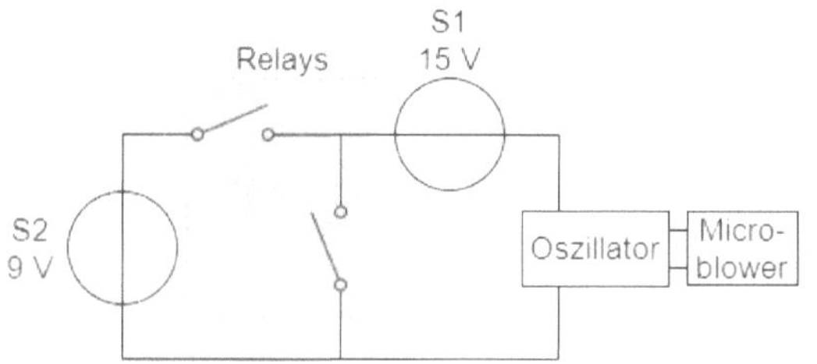

Figure 4: Circuit diagram: The two relays periodically connect source 2 in series with source 1 to add the voltages .

relays are controlled by an Arduino Uno and switches between 15 V and 24 V with a frequency of 2 Hz. The two voltage sources are connected in series, so that the voltages add when the switch is closed, but the current remains constant. The oscillators are all connected in parallel so that they receive the same voltage. About every 6 hours all microblower are switched off for about 30 minutes to simulate the break times and a "cold start".

2.4.3 Software design

The Arduino counts the number of cycles and pauses and displays them on an LCD display. The flowchart of the Arduino software for controlling the relays is shown in Fig. 5.

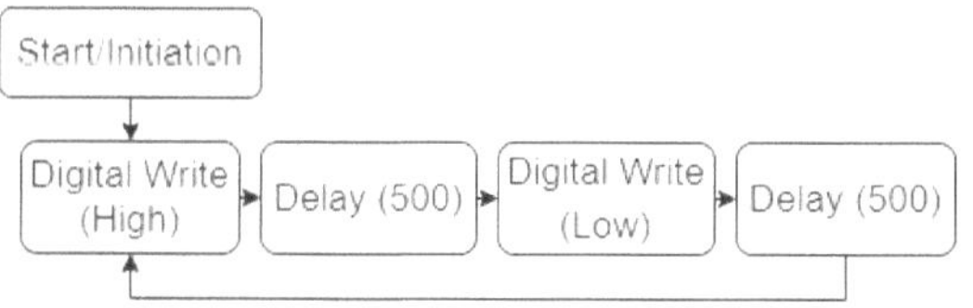

Figure 5: Flowchart of the Arduino software: Digital-Write(High) opens and DigitalWrite(Low) closes the relay.

2.5 Test execution

2.5.1 Gluing the microblower into their mountings

- Soldering two thin cables to the contacts of the Murata Microblower - MZB1001T02

- Fasten the upper part of the housing to the table upside down with dough

- Mix Quick Set Epoxy Adhesive on and quickly apply a thin adhesive ring with a toothpick

- Carefully insert the microblower and weight it

- The adhesive must dry out for five minutes

- Fix the diaphragm with the O-ring to the core of the bracket

- Repeat for all 48 microblower

2.5.2 Commissioning process

When all the microblowers have been wired and glued, they can be connected to the oscillators. The voltage source 1 is preset to 15 V and the voltage source 2 to 9 V. The test start automatically when the power connection is established. Once a week, the pressure of each microblower is manually recorded using a pressure gauge and an oscilloscope. Both the minimum and maximum pressure are documented. A screenshot of the pressure curve is recorded for each microblower. It is important to note that the scaling of the oscilloscope is always the same to allow comparison. The completed test stand can be seen in Fig. 6.

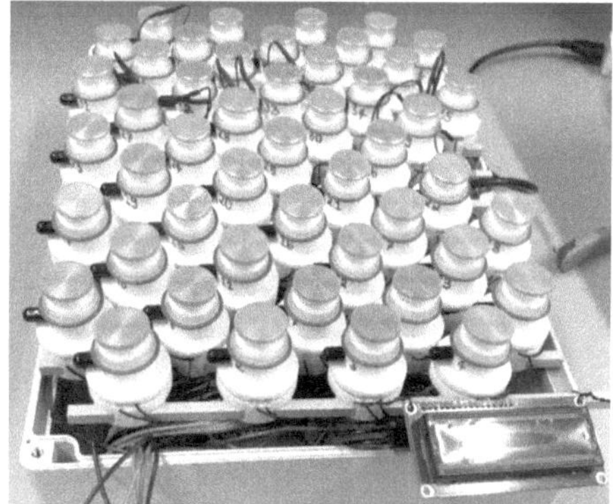

Figure 6: Picture of the completed test stand.

3 Results and Discussion

3.1 Results

After commissioning of the test stand, the output pressure of each microblower (# 1 - 48) was measured once a week. I.e. after 59 days of measurement, the mean value of the pressure (m. pressure in mbar) and the standard deviation (σ in mbar) were calculated for each microblower. The mean results of each microblower can be seen in Table 2. The test is passed if the average pressure does not fall below 25 ± 1 mbar. All microblowers have passed the test. A graphical overview of the results with standard deviation is shown in Fig. 7.

3.2 Discussion

All microblowers are still operating after 59 days of simulation and the goal of the test was achieved. Nevertheless, there are very strong differences within the microblower performance (see Table 2). Some microblowers have just

Table 2: Test results of the endurance test after six years of simulation. #: Number of the microblower, m.pressure: mean value of the output pressure in mbar

#	m.pressure	#	m.pressure	#	m.pressure
1	24,48	17	28,72	33	30,24
2	29,20	18	28,28	34	30,04
3	26,92	19	27,56	35	30,44
4	28,80	20	29,84	36	32,44
5	28,48	21	30,44	37	31,40
6	26,64	22	32,32	38	31,48
7	28,44	23	30,56	39	29,36
8	26,88	24	30,20	40	31,44
9	30,44	25	29,92	41	29,12
10	27,80	26	30,48	42	30,84
11	26,08	27	29,80	43	30,68
12	27,12	28	31,56	44	30,88
13	28,04	29	30,88	45	24,92
14	29,84	30	28,52	46	30,52
15	27,80	31	30,96	47	29,52
16	30,24	32	29,68	48	29,52

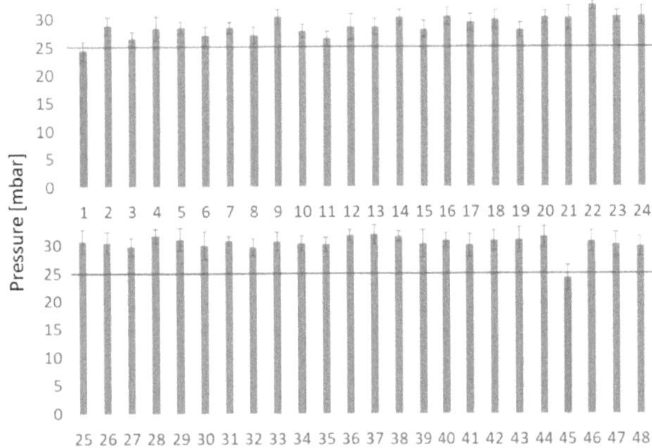

Figure 7: Graphical overview of the results (average pressure with standard deviation) of the microblower # 1-48.

reached the target of 25 ± 1 mbar and others have an average pressure well above 30 mbar. It can be observed that especially the microblower with a lower number (< # 20) show a weaker performance. Since these microblower are from an earlier delivery and therefore stored longer in the office, it can be assumed that either the duration of the storage has an influence on the microblower performance or a production variance can be observed. Otherwise, all microblower were treated equally.

The microblower # 1 and # 45 are consistently the weakest and in many measurements just below 25 mbar. An initial measurement was carried out on each microblower before it was installed in the test bench. In this test, microblower # 1 had a pressure of 28.40 mbar and microblower # 45 a pressure of 33.60 mbar.

After reaching the six-year service life, the test was not switched off, but continued. The maximum service life is intended to be determined. In the event of failure of several microblower, precise fault analyses of the cause could

be carried out in order to better predict the behaviour of the microblower.

4 Conclusion

A test stand was successfully developed and deployed to determine the service life of the Murata Microblower MZB1001Z02 with increased supply voltage. It has been shown that the microblowers achieve and far exceed the desired service life of six years (> 10 years).

The further development of the test stand is recommended as the commissioning and execution of the test is very extensive and time-consuming. The most demanding tasks are the soldering and gluing of the microblower into the holder and the weekly manual measurement of the microblower. In addition, the additional requirement arose that if a microblower fails, the exact time was important. The gluing and soldering can be done with an extended bracket. The microblower must be hermetically sealed without mechanical pressure being exerted on its casing. In addition, the cables must also be led airtight through the bracket to the microblower. Spring contacts that are glued in are recommended. This guarantees a safe, flexible power supply and minimal leakage.

Both the weekly manual measurement and the additional requirement to know the exact time of the failure is realized by an automated and permanent pressure measurement of all microblower simultaneously. A microcontroller must be programmed so that the information can be read from 48 pressure sensors.

Acknowledgement

The work has been carried out at Drägerwerk AG Co. KGaA, Lübeck and supervised by Prof. Dr. Philipp Rostalski, Institute of Electrical Engineering in Medicine, Universität zu Lübeck.

5 References

[1] W. Roddeck, *Einführung in die Mechatronik*. Springer Vieweg, 2012.

[2] Murata - Innovator in Elentronics, *Microblower Driver Information*. Available: https://www.murata.com/en-us/products/mechatronics/fluid/microblower_mzb1001t02. [last accessed on 2019-11-30].

[3] Murata - Innovator in Elentronics, *Murata MZB1001T02*. Available: https://www.murata.com/en-us/products/mechatronics/fluid/microblower_mzb1001t02. [last accessed on 2019-11-30].

[4] W. Oczenski, A. Werba und H. Andel, *Atem-Atemhilfen*. Blackwell Wissenschaft, 1996.

[5] J. Shao, C. Shein-Chung und W. Hansheng, *Sample size calculation*. Taylor & Francis, 2018.

Conversion of an existing lung demand valve to a pressure reducer with an output of 50 mbar

Jan Beidatsch [1], Annika Freund [2] and Hans-Ullrich Hansmann [3]

[1] Medical Engineering Science, Universität zu Lübeck, jan.beidatsch@student.uni-luebeck.de
[2] Medical Engineering Science, Universität zu Lübeck, an.freund@student.uni-luebeck.de
[3] Drägerwerk AG, Advanced Engineering Solutions, hans-ullrich.hansmann@draeger.com

Abstract

The aim of this article is to investigate whether an existing lung demand valve (LDV) with an outlet of 3 mbar can be converted into a pressure reducer with an outlet of 50 mbar for a new lung ventilator. This ventilator operated with a blower shall be expanded to include the function of being able to operated it via an oxygen cylinder. For this purpose, a dimensioning of a spring located in the LDV was carried out. By adjusting the compression of the spring, the inlet pressure could be reduced from 5 bar to 50 mbar. After an appropriate setting was found, it was tested whether the pressure could still be maintained with increasing flow. It turned out that the pressure was sufficiently high in the flow range relevant for ventilation. Since the spring is longer due to the higher force required, the housing of the LDV must be adapted.

1   Introduction

In the event that a patient is unable to breathe or can only breathe insufficiently, ventilation must be initiated immediately to support or replace the breathing. The aim is to ensure a sufficient pulmonary gas exchange, i.e. an oxygenation of the blood and alveolar ventilation (elimination of CO_2). Furthermore, the ventilation promotes the lung and respiratory healing and the weaning of respiratory aids [1], [2].

1.1   Respiration

Artificial respiration, which imitates a normal respiratory function, is based on positive pressure ventilation. This is a contradiction to the actual physiology of the lung, which fills with air through contraction of the inspiratory muscles. The diaphragm is most importantly responsible for the ventilation of the lung. Due to its contraction, a pressure reduction in the thorax is achieved, which leads the alveolar pressure to be lower than the ambient pressure. In order to ventilate the patient's lung protectively with positive pressure ventilation, pressure-controlled and volume-controlled ventilation is used. In case of serious lung diseases, 30 mbar should be applied as pressure-limited or 6 - 8 ml per kg body weight as volume-limited procedure in order to avoid damaging the lungs [1], [3], [4].

A new and small lung ventilator (Fig.1) which is operated with a blower is currently in its development phase and is expected to be enhanced with a further function. The blower, which aspirates ambient air and thus ventilates the patient, shall be extended by an additional connection. The

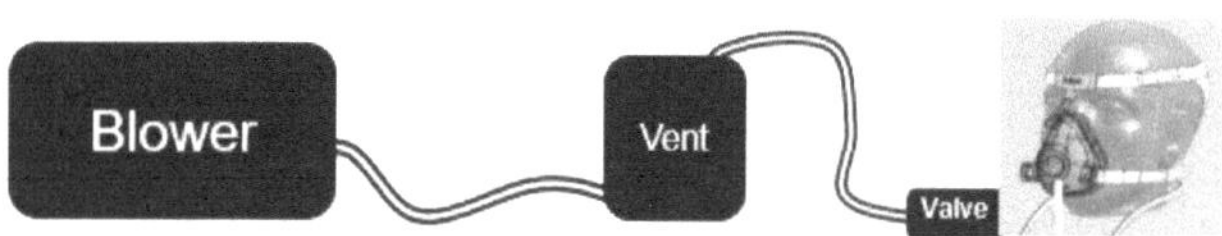

Figure 1: Schematic illustration of the lung ventilator. The blower supplies the room air required for ventilation by drawing it from the ambient air. The Vent system is used for regulation and thus enables pressure-controlled ventilation. The valve in front of the mask is an expiration valve. Blower, Vent and valve are connected via tubes.

central oxygen supply from a hospital or an oxygen cylinder shall be connectable to this input of the blower or directly to the ventilator [1], [3]. This would increase the application possibilities of the device enormously, especially in the field of emergency operations, where pure oxygen is administered to the patient [5], [6].

1.2   Requirements

In order to ensure the operation of the device, it is necessary to add a pressure reducer between the oxygen source and the blower. This is required as the blower only generates an outlet pressure of 50 mbar. However, an oxygen cylinder, which is under a pressure of 200 - 300 bar, still delivers a pressure of 5 to 9 bar despite the cylinder pressure reducer. This pressure would be so high that it would lead to the failure of the device. The pressure reducer is therefore required to convert an inlet pressure of approximately 9 bar into an outlet pressure of 50 mbar. To ensure that it can be used in emergency medical services in addition to everyday clinical applications, a robust design with regard to shock strength, humidity and temperature is required. In addition, an oxygen resistance and an accuracy of ± 5 mbar at constant flow is required. To meet all the requirements listed above, it should be tested whether an existing LDV from the Dräger product family can be converted into a pressure reducer for a new lung ventilator.

2 Material and Methods

In order to be able to convert the existing LDV into a pressure reducer for the blower of the new ventilator, an adjustment of the counteracting force in terms of dimensioning and response time must be found.

2.1 Lung demand valve

The lung demand valve used, shown in Fig. 2, simplified, consists of a spring, a diaphragm, a primary and secondary lever and a balanced piston valve. Without air pressure, the force of the spring pushes the diaphragm and the lever below it downwards. This opens the piston valve. If an inlet pressure of 5 to 9 bar is applied, the slightly opened piston valve causes a pressure drop. The reduced outlet pressure is also present on the underside of the diaphragm due to openings which are not visible in Fig. 2. This pressure thus provides the counterforce to the spring. If the pressure is too high, the diaphragm lifts and the spring is compressed. As a result, the lever closes the valve a and the outlet pressure decreases. If, on the other hand, the pressure is too low, the force of the spring prevails and the valve is opened further, leading to an increase in pressure. This way, the system oscillates until an equilibrium of forces is reached. This ensures that the valve is opened sufficiently to achieve a pressure of 3 mbar at the outlet.

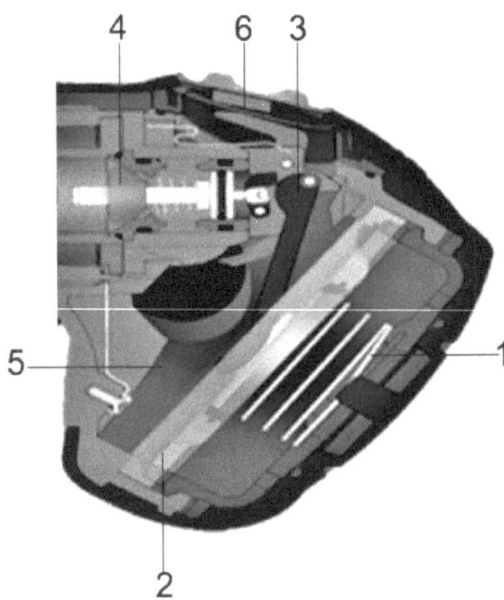

Figure 2: Simplified structure of the lung demand valve: (1) positive pressure spring, (2) diaphragm, (3) lever, (4) balanced piston valve, (5) outlet pressure, (6) reset button by lever. The input of the air pressure cannot be seen in this figure. The inlet opens directly into the area which is closed by the balanced piston valve.

The LDV, which is normally used by firefighters and divers, has an additional reset button. This function is required to prevent the gas from escaping from the compressed air cylinder in an uncontrolled manner. If the gas were to flow out after connection to the compressed air cylinder, the consumption would be too high in some situations. In this case, a lever mechanism prevents the force of the spring from pushing down the lever of the output valve. The gas cannot begin to flow until a connection is established with the mask and breathing begins. With the beginning of the first breath a negative pressure is created at the diaphragm, so that the force of the spring is now sufficient to open the valve [7]. The function of switching on and off is also a requirement of the pressure reducer and must be taken into account when converting it.

2.2 Spring conditions

To achieve a pressure of 50 mbar at the outlet of the LDV, the counteracting force of the spring had to be adjusted. For this purpose, the characteristic curve of the spring belonging to the LDV was first recorded with the aid of a force gauge. By drilling a hole in the cover of the LDV, the spring excursion could be determined with the help of a caliper gauge at an applied inlet pressure of approximately 5 bar. The associated force could be read off the spring characteristic curve. Since the output pressure and the force transmitted from the spring and diaphragm to the lever are proportional, the force, which is required to reach an output of 50 mbar, can be calculated. A spring, which could provide the necessary force, was built into the system. However, since this spring had to apply much more force than the one originally belonging to the LDV, its overall length was much greater. Installing this spring to the system would lead to the consequence that the spring would already be compressed to such an extent that it would generate so much force that the desired result would be exceeded many times over. Therefore, a hole was drilled in the lid so that the spring could protrude from the LDV. The spring then pressed outside the housing against a simple metal plate which was attached to the rest of the cover with two screws. As the plate was fixed along the screws with the help of nuts, the spring could be compressed so far that the desired force was achieved.

2.3 Stabilization of the spring

As the new spring was smaller in diameter than the old one, it is important to note that after applying an inlet pressure, the spring had the possibility to move to the side. Although the length would remain the same, in such cases the force acting on the diaphragm would be different. As the diameter of the new spring was smaller compared to the old one, it no longer fits into the holder fixed on the diaphragm. The fact that the spring may not have been positioned centrally and the additional degree of freedom created by the evasive movement led to an uneven distribution of force acting on the diaphragm. The whole device ultimately behaved like

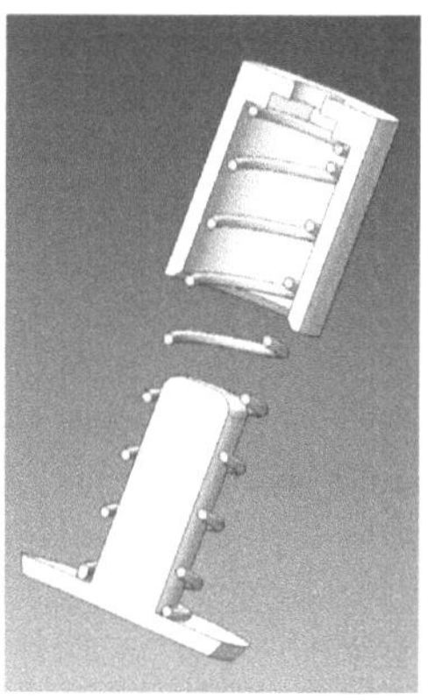 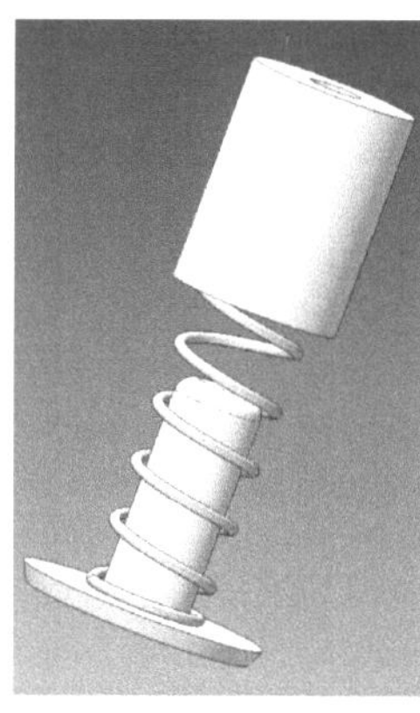

Figure 3: Prop and sleeve for the spring. On the left side as sectional view and on the right side as full view.

a disturbing impulse and caused the system to oscillate. In this case audible resonances occurred.

To solve these problems, two components were designed by using computer-aided design (CAD). On the one hand a prop which ensured the central positioning of the spring and on the other hand a sleeve which together with the prop ensured that the spring could only move in an axial direction were developed. Both components are shown in Fig. 3. The prop was adapted to the holder on the diaphragm for the original spring. Due to the equal distribution of forces over the entire membrane, the system no longer vibrated, causing the audible resonances to disappear. The sleeve, which was placed loosely over the spring, prevented the evasive movement of the spring. For that the sleeve was attached to the metal plate with a nut. When constructing the sleeve, a recess for the nut was therefore inserted on the top side.

2.4 Experimental setup

To set the correct outlet pressure of the LDV, a tube was connected between a manual valve used to regulate the flow and the outlet of the LDV. The manual valve had an additional connection to which a pressure gauge was connected. After applying an inlet air pressure of 5 bar at the input of the LDV with the hand valve closed, the correct compression of the spring could be adjusted. Afterwards the manual valve was opened in increments to increase the flow while recording the pressure. The flow was measured by using a spirolog (an internally existing Dräger product) located behind the hand valve. The technology behind it is based on hot-wire anemometry [8].

Furthermore, the dynamic behavior of the pressure reducer should be assessed. The experimental setup was almost not changed for this purpose. Only the pressure gauge had to be replaced, because the one originally used for our purposes was too slow. Its measuring cycle was over 1 s and was therefore unsuitable for dynamic measurement. For this reason, an HDI sensor from FirstSensor was used to create a measuring device with which a sampling rate of 3125 per second could be achieved. The data of pressure and flow were recorded with an oscilloscope. Due to the

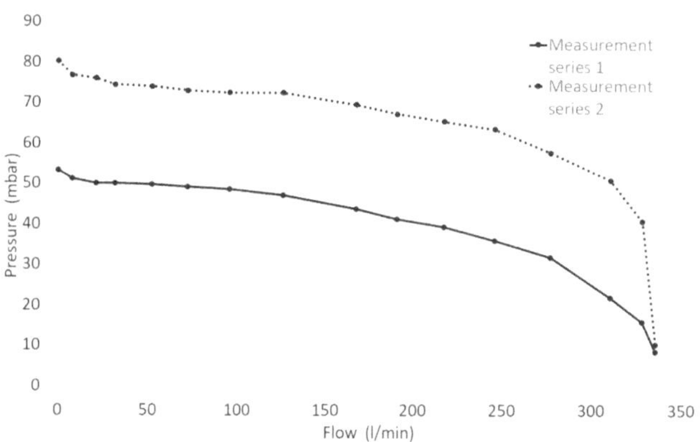

Figure 4: Measurement of outlet pressure from the LDV at an inlet pressure of 5 bar against flow. For measurement series 1, a spring setting was selected so that an outlet pressure of 50 mbar is achieved at a flow of 0 l/min. For test series 2 a higher spring force was used, resulting in a higher initial pressure of 80 mbar.

amount of data the function was smoothed before plotting.

3   Results and Discussion

The results of the measurement are shown in Fig. 4. Shown is the outlet pressure for an inlet pressure of 5 bar depending on the flow. During the first measurement, the spring was compressed so that a pressure of 50 mbar was reached when the valve closed, which means that we had a flow of zero. The solid line in Fig. 4 shows that the pressure was kept constant, especially in the low flow range. As soon as the valve is opened to the point where we have a flow greater than 125 l/min, the pressure begins to decrease more rapidly. Because especially the flow range between 100 and 150 l/min is of interest for the ventilation and we have already lost about 5 mbar of pressure at a flow of 95 l/min, the experimental setup was revised again so that a pressure of 80 mbar was achieved with a closed manual valve. For this purpose the spring force had to be increased. These results are represented by the dotted line in Fig. 4. The course of this measurement is very similar to the measurement with lower spring force, but the pressure values with the increased spring force are in a higher range. Again, above 125 l/min the pressure starts to fall faster and at a flow of 310 l/min it drops rapidly towards zero. As the pressure only drops to a value below 50 mbar at a flow of more than 300 l/min, this spring setting is much more suitable for the ventilation of a person, as the range of more than 200 l/min is no longer relevant for our field of application.

It could be shown that an input pressure of 5 bar could be reduced to a pressure range relevant for ventilation and that it could be kept sufficiently constant in the relevant flow ranges.

To see how the outlet pressure behaves with a changing flow, a dynamic measurement was started. The result of the dynamic behavior from the pressure reducer can be seen in Fig. 5. In the diagram the time is plotted on the abscissa and

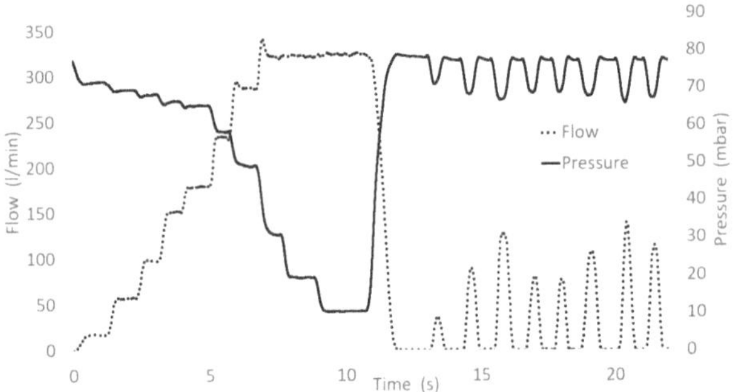

Figure 5: Dynamic pressure measurement: Shown is outlet pressure from the LDV at an inlet pressure of 5 bar and flow against time. The flow was controlled by manually opening and closing the valve and the corresponding pressure was recorded.

the flow and pressure on the ordinate. The flow is applied on the left and the pressure on the right ordinate.

A flow curve (dotted line) can be seen, which was created by opening and closing the manual valve over a period of about 22 s. The solid line shows the pressures that match in time. In the first half of the diagram, the flow was increased in steps over a period of 7 s. It can be seen that the pressure is above 50 mbar for almost the entire period. Only when the flow rises above 300 l/min within this time interval can the pressure no longer be maintained and it drops. However, this is not problematic for our application scenario, as a flow of more than 200 l/min is not common for ventilation. As soon as the valve is closed and the flow drops to 0 l/min, the pressure increases again abruptly. The pressure increases just as fast as the flow decreases. In the second half of the diagram, the valve was opened several times in succession for only a short time of about 1 s. During this time a flow range was reached which is important for ventilation. You can see that the pressure always collapsed when the valve was opened, but it was always greater than 65 mbar. In addition, it always returned to its original pressure as soon as the valve was closed. The pressure was therefore always sufficiently high for the ventilation of a patient.

4   Conclusion

In principle, it could be shown that an existing lung demand valve can be converted into a pressure reducer for a new ventilator. Above all, the correct dimensioning of the spring was necessary to achieve this result. After this has been found, an inlet pressure of 5 bar can be reduced to an outlet pressure of at least 50 mbar.

The next step is the production of a spring by the company "GUTEKUNST FEDERN" according to our specifications. This will also be installed and tested in the existing experimental setup and, if necessary, reworked. After consultation with the company, the length of the new spring does not fit into the pressure reducer, so it is necessary to redesign the cover. Since the final length of the spring is not yet fixed, a variable adjustment option in the cover, which compresses the spring more or less, is necessary. It must also be ensured that the switch-on and switch-off function of the device is maintained. Switching off is still done with the reset button on the side of the pressure reducer. However, switching on must be implemented differently for the ventilator. Previously, the switching on was realized by the negative pressure of the first breath. However, since the ventilator does not generate negative pressure, an over pressure at the top of the diaphragm (i.e. on the side of the spring) should enable switching on. For this purpose, the cover must have an additional button that can create this over pressure.

After implementation, the next step would be to test the pressure reducer directly on the ventilator.

Acknowledgement

The work has been carried out at Drägerwerk AG & Co. KGaA and supervised by PD Dr. H. Paulsen, Institute of Physics, Universität zu Lübeck.

Special thanks are due to Dipl.-Ing. Hans-Ullrich Hansmann, who has supported me throughout with his expertise.

5   References

[1] R. Larsen, T. Ziegenfuß, and A. Mathes, *Beatmung: Indikationen - Techniken - Krankheitsbilder*, 6th ed. Berlin, Heidelberg: Springer Berlin Heidelberg, 2018.

[2] F. Bremer, *1x1 der Beatmung*, 4th ed., ser. Medizin. Berlin: Lehmanns Media, 2014.

[3] H. Lang, Ed., *Außerklinische Beatmung: Basisqualifikation für die Pflege heimbeatmeter Menschen*. Berlin, Heidelberg and s.l.: Springer Berlin Heidelberg, 2017.

[4] F.-J. Kretz and F. Teufel, *Anästhesie und Intensivmedizin*. Berlin, Heidelberg: Springer Medizin Verlag Heidelberg, 2006.

[5] F. Flake, S. Dönitz, S. Goller, H. Hasel, and N. Menche, Eds., *Mensch Körper Krankheit für den Rettungsdienst*, 2nd ed. München: Elsevier, 2018.

[6] M. Linnemann and M. Kühl, *Biochemie für Mediziner: Ein Lern- und Arbeitsbuch mit klinischem Bezug*, 6th ed., ser. Springer-Lehrbuch. Berlin, Heidelberg and s.l.: Springer Berlin Heidelberg, 2003.

[7] Dräger Safety UK Limited, "Instructions for use: Lung demand valve," vol. 2019, pp. 1–2, 2012.

[8] "Flowsensor," 18.12.2019. [Online]. Available: https://www.draeger.com/de_de/Hospital/Products/Accessories-and-Consumables/Sensors/Flow-Sensor-Accurate-and-intelligent-flow-measurement#benefits

Conception of a Test Setup to Determine the Service Life of an Infrared Emitter Used for CO_2 Measurement

Annette Leßmann [1], Martin Kroh [2], and Philipp Rostalski [3]

[1] Medical Engineering Science, Universität zu Lübeck, annette.lessmann@student.uni-luebeck.de
[2] Connect & Develop, Gas Sensors, Drägerwerk AG & Co. KGaA, martin.kroh@draeger.com
[3] Institute for Electrical Engineering in Medicine, Universität zu Lübeck, philipp.rostalski@uni-luebeck.de

Abstract

In ventilation a close meshed control of vital parameter is essential. For getting information about the patient's health, one important parameter is the CO_2 amount in the expiratory gas mixture. Based on a transmission measurement, the special absorbance properties of the carbon dioxide can be utilised. To verify a new CO_2 sensor, a service life measurement is necessary. Due to the fact, that the infrared emitter is the most failure-sensitive component, the durability of the emitter has to be investigated first. In this work, the conceptual design and setup of the endurance test is described. This contains the control and the acquisition of the sensor data. Concluding, a forecast of the long term data is given.

1 Introduction

Patient ventilation plays a decisive role in intensive and emergency care. Often the venting of the lungs is completely or partially controlled by the device. In addition to the ventilation unit, monitoring is also essential to control important vital and respiratory parameters. Furthermore, the medical ventilator measures the compound of the expiratory gas mixture [1].

In metabolic processes, a part of the inhaled oxygen is transformed to carbon dioxide, typically the expiratory air contains 17 vol.-% oxygen (from previously 21 vol.-% in the ambient air) and 4 vol.-% CO_2. If the carbon dioxide concentration deviates, a disorder of the organism or a failure in the connection of the patient could be the reason. Therefore, CO_2 measurement, called capnography, is an essential part of the ventilation unit. In contrast to oxygen measurement, capnography is characterized by the high measurement speed. This provides a time-resolved detection during each respiratory phase [2]. The concentration curve as a function of time, a so-called capnogram, is exemplary shown in Fig. 1 [1].

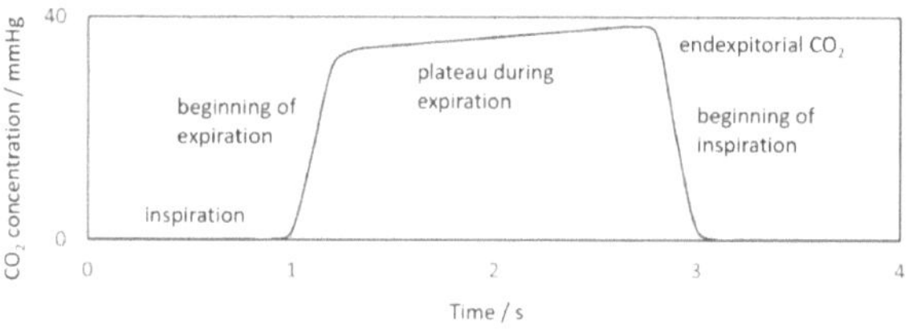

Figure 1: Capnogram of a healthy person [1].

The graph provides information about ventilation, the respiratory condition, dead space ventilation and the function of the medical ventilator. In emergency care, the correct tube position or the effectivity of cardiopulmonary resuscitation can also be observed.

The examination of gas concentration can be done by two different methods. The mainstream method integrates the measuring cuvette directly into the patient's respiratory flow to provide a measurement without delay. But due to its proximity to the patient, the cuvette is often contaminated with blood, secretions and other substances. Additionally, the sensor's weight on the cuvette can exert undesired pulling forces on the tracheal tube. The second possibility is provided by the sidestream method. A small amount of the respiratory gas is continuously aspirated and examined remotely from the patient. The gas is led to the sensor via a water trap integrated in the tube system. With this method, the measurement is not possible without a certain delay [1].

A common CO_2 measuring principle is based on determination of the concentration using an optical method. For this purpose, the specific absorption behaviour of carbon dioxide towards electromagnetic radiation of a broadband infrared (IR) emitter is utilised. The gas concentration can be calculated according to Lambert-Beer's law. The radiation intensity transmitted through the cuvette depends linearly on the initial intensity I_0 and exponentially on the travelled path length l and on a substance-specific and wavelength-dependent extinction coefficient ε^*:

$$I_{measured} = I_0 \cdot e^{-l \cdot \varepsilon^* [C(\lambda)]}. \tag{1}$$

The following relationship applies to the concentration, using the spectral absorption coefficient ε_λ with

$$[C(\lambda)] = \frac{1}{l \cdot \varepsilon_\lambda} \cdot lg \left(\frac{I_0}{I_{measured}} \right). \qquad (2)$$

However, this method is error-prone, since the exponential relationship only applies to an ideal setting. A bandpass filter in front of the detector does not allow the passing of just one wavelength, contrary to the ideal solution. In realistic measurements, the logarithmic relationship $-lg(I_{measured}/I_0)$ has to be transformed into a linear context by using a previous calibration to convert transmission to concentration [3]. To be used in everyday clinical practice, the sensor has to meet a specific reliability target. The most sensitive component of this design is the infrared emitter. Due to insufficient manufacturer qualification, a service life test for using the radiation source should be designed. The hardware of a CO_2 sensor prototype is used to run the emitter in a suitable manner. To operate the emitter appropriately, the hardware of a sensor prototype is used, which is why the functionality is discussed in detail below.

2 Material and Methods

2.1 Sensor

Typical mainstream CO_2 sensors consist of an infrared emitter, a cuvette through which the breathing air is passed, a beam splitter and two detectors, as shown in Fig. 2.

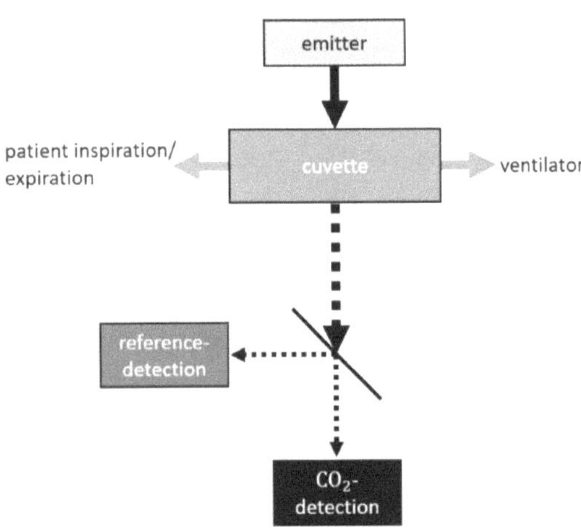

Figure 2: Schematic of the sensor setup, consisting of the IR emitter, a cuvette with patient gas, a beam splitter and two detectors.

The radiation produced by the infrared emitter passes through the cuvette to the beam splitter and to both detectors. Depending on the concentration of the carbon dioxide in the cuvette, a certain amount of the radiation is attenuated.

The measuring channel observes the intensity at a wavelength of 4.26 μm, given that broadband radiation is most strongly absorbed by CO_2 here. Interference filters with bandpass characteristics are used in front of the detectors

for spectral confinement. The reference channel, which detects radiation of a slightly lower wavelength, is used to minimize the influence of external noise factors such as contamination. To keep the structure abacterial, the cuvette can be changed. In addition, the windows of the cuvette are heated to a certain temperature to prevent condensation [2].

2.2 IR emitter

Usually, thermal radiation sources are used to generate infrared light. These are emitters whose spectral radiation density dP_λ of area A can be calculated according to Planck's radiation law with

$$dP_\lambda = \frac{2\pi hc^2}{\lambda^5} \cdot \frac{A}{e\left(\frac{hc}{k\lambda T}\right) - 1} d\lambda. \qquad (3)$$

The calculation assumed an ideal black body with an emissivity of one. In reality, however, there is no surface coating providing this property. The strong dependence on temperature and wavelength is shown in Fig. 3. As the temperature rises, the maximum intensity shifts into the short-wave spectral range. Despite a high surface temperature like 700 °C, the radiation remains invisible to the human eye.

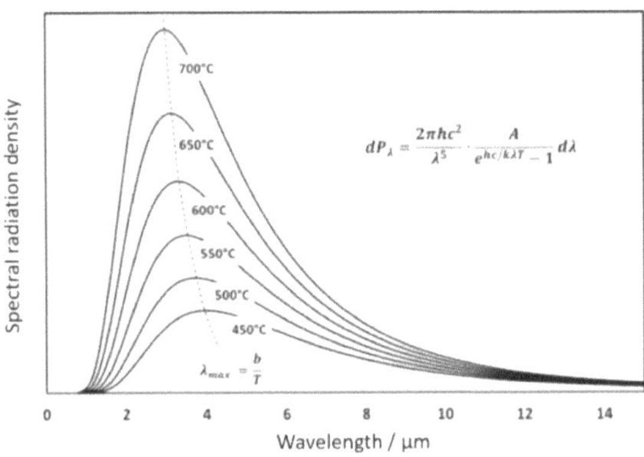

Figure 3: Temperature-dependent spectrum of black body radiation [2].

In order to be able to determine the possible types of failure of the infrared emitter more precisely, a deeper understanding of the operation mode is essential. Specifically, it is a micromechanical surface emitter, described in Fig. 4.

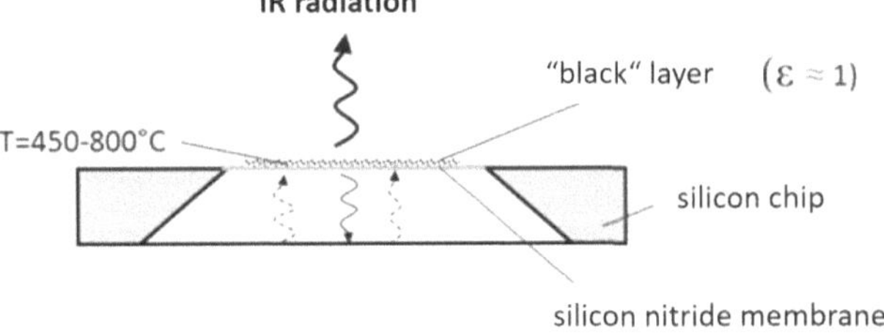

Figure 4: Membrane layer of an infrared emitter [2].

A silicon microchip forms the fundament for the radiation source. An opening on the upper side contains a very thin silicon nitride membrane, on which a platinum structure is applied. Electrical current flows through the metal structure, heating the membrane and emitting IR radiation. The emitter surface has a further, very porous, three-dimensional layer through which the ideal emission coefficient is approximated. This enables efficient energy conversion from electrical to optical power. The radiation source in Fig. 5 can be used with a reflector for radiation bundling. Due to the fact, that the emitter has an adjoining window to reduce external disturbances, it has a ventilation opening on the bottom. This prevents outgassing from destroying the layer structure of the membrane. Since a more efficient specific reflector solution was found for the sensor design used in this project, the emitter operates without a window.

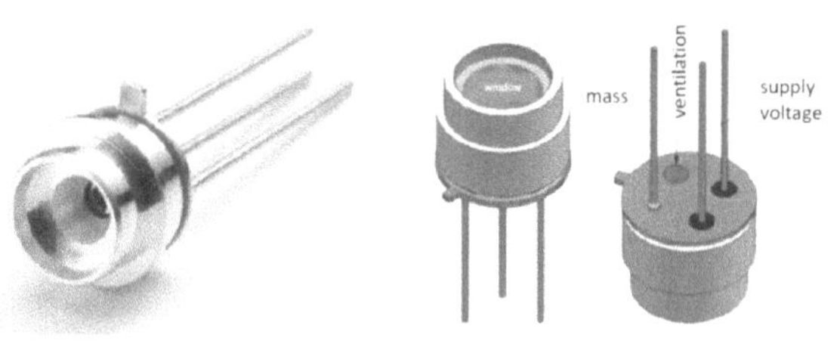

Figure 5: A typical micromechanical surface emitter with reflector and window [2].

A decisive advantage of the membrane emitter is the low thermal membrane mass. Switching the power supply on and off is accompanied by heating and cooling within a few milliseconds. At low frequencies (approx. 3 Hz) the maximum radiation emission is reached during one switching cycle as well as the minimum value. If the operating frequency is increased, the temperature maximum and minimum is no longer achieved. This means that the amplitude is reduced and the modulation level decreases with increasing frequency. At 100 Hz, for example, it constitutes only 40 %. However, the temporal sensitivity of the detector in the sensor design usually limits the number of measurement cycles per second. Typically the emitter operates at membrane temperatures of approximately 450 °C decreasing to the sides. The desired temperature is achieved by adjusting the electrical power, being the decisive parameter for lifetime prediction of the emitter [2].

2.3　Weibull distribution

The following Weibull distribution is used to calculate the component's reliability most precisely as a function of time with

$$R(t) = e^{-\left(\frac{t}{\tau}\right)^{\beta}}. \tag{4}$$

After the characteristic durability τ, 63 % of all test items are failed. β describes the failure steepness and has to be determined experimentally by durability tests [2]. Because of its shape, a typical failure rate is called the bathtub curve. The shape parameter β can be used to draw conclusions about the type of failure. An early failure is for example

most likely triggered by production errors. In further process, random failures may occur and after the product has already been in operation for a long time, the so-called late failures usually predominate [4].

3　Results and Discussion

3.1　Conception

Since the infrared emitter needs to last several years and the test should be completed earlier, the measurement takes place under accelerated conditions. By means of a power control, the test items are constantly driven at defined power values, independent of the membrane resistance. As a control group, some emitters are operated with the fixed sensor parameters. In this first setup, 60 test items are used and no failures are expected. However, in order to obtain a suitable statement on the durability, errors should be provoked. An acceleration is achieved in other setups by raising the power and the ambient temperature. Both factors are contributing to increase the membrane temperature. 20 test items with slightly raised peak power, in which the power were increased by 10 %, and further 20 emitters with strongly increased peak power should be operated in the climatic chamber. The frequency could also be reduced. The increased modulation level would further increase the temperature gradient. To draw conclusions about the realistic running process with sensor parameters, an acceleration factor can be calculated. The goal is a failure rate of 1 % per 20000 hours of operation. This is based on the specification of error-free running for at least 8 years.

3.2　Test setting

The first idea to operate the emitters on self-designed circuit boards and to control them individually through a Raspberry Pi was rejected again due to the complex handling. Instead, the hardware of a sensor prototype was used. The advantage here is the supply of power control and the possibility of communication via a specific sensor protocol which has already been provided on the board. A housing was designed in order to build the test setup repeatable and to keep the sensor as independent as possible of external factors. A cover plate protects against breathing artifacts from the operator and against dust particles that could falsify the measurement result. Using the computer-aided design program SolidWorks by *Dassault Systemes Deutschland GmbH*, the primary sensor mounting was adapted, as shown in Fig. 6. Due to the requirement to be able to examine individual IR emitter even during test operation, the setup should not require adhesives to allow an easy removal. The reflectors on top of the IR emitter and also above the detectors are inserted with a guide rail and a round opening in the middle ensures the suitable radiation of the gas mixture. This ensures a stable optical path. By using rapid prototyping technology, the polyamide housing gets produced quickly by a 3D printer.

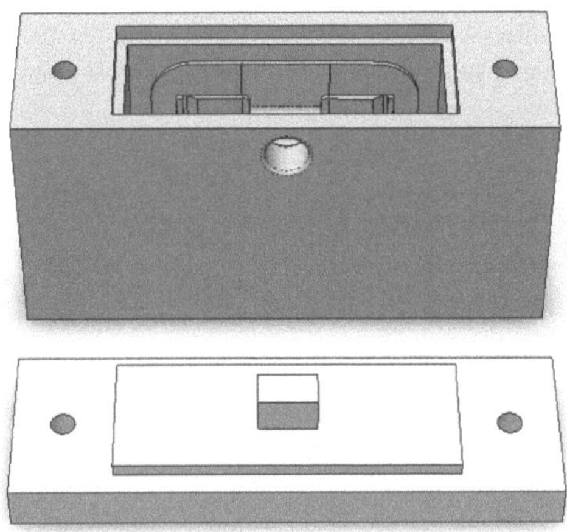

Figure 6: Construction of a non-adhesive sensor housing for the test setup.

3.3 Sensor control

The internal sensor protocol STORM provides a means for sending necessary commands and reading out relevant sensor data. The transmitting and receiving of the appropriate commands is carried out by an automated test environment, which controls all sensors serially. This function is implemented in the test management software TestStand by *National Instruments Corporation*. Python source code was integrated here, which runs the communication protocol. The main program starts with setting the sensor mode from "Standby" to "Measurement". In the next step, the operating parameters can be adapted to the test conditions. In addition, the values required for evaluation of the emitter operation can be read out via this interface, including emitter voltage, emitter current measured via a shunt resistance and the transmission values of both detectors. If the ratio of the two remains the same, this is interpreted as nominal operation. However, if one of the values differs from the other, a spectral shift of the emitter is to be expected. Since this is a planned endurance test that should only be stopped after a certain failure rate, no test results are available at this project. After completion, the evaluation of Weibull plots will allow a detailed conclusion.

First measurement results of emitter operation over 13 days with realistic sensor parameters already showed a trend to the desired stability. The three visible spikes can be explained by noise. Since the test had not yet taken place in the designed sensor housing, external influences, such as the approximate number of people in the laboratory, can be distinguished in the measurement data in Fig. 7. The aberration of the detector ratio approximates 0.0125 a.u., which corresponds to a CO_2 concentration deviation of 0.03 vol.-%.

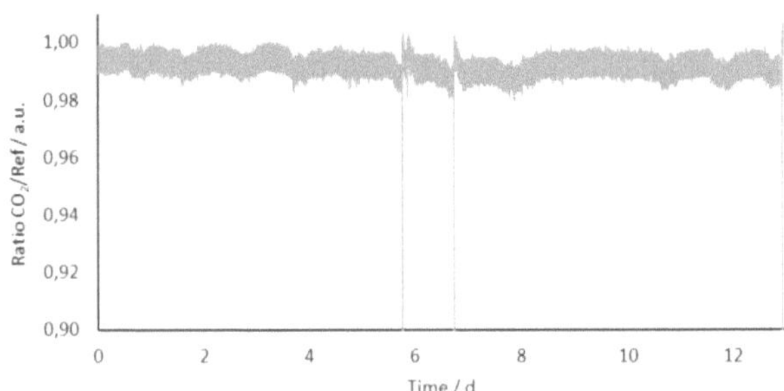

Figure 7: Detector ratio of an emitter stability measurement over 13 days.

4 Conclusion

We have shown that a newly developed setup for measurements of properties of infrared emitters delivers information not available so far such as quick prediction of the long-term failure rates. For a more exact consideration of the physical processes, also in connection with possible damages, a further test setup has to be developed. An infrared camera will be used to measure the time resolved shift of the maximum membrane temperature. If the product behaves as expected, the emitter can be installed in the upcoming CO_2 sensor, prototypes of which were already used for testing. With its reduced power consumption and smaller design, the choice will have a positive effect on the costs associated with the emitter. These advantages also speak in favour of replacing the emitter in another already developed sensor.

Acknowledgement

The work has been carried out at Drägerwerk AG & Co. KGaA, Gas Sensors. We'd like to express special gratitude to Bernd-Michael Dicks for his support. The project is supervised at Universität zu Lübeck, Institute for Electrical Engineering in Medicine.

5 References

[1] R. Larsen, T. Ziegenfuß and A. Mathes, *Beatmung*. Springer, Berlin, 2018.

[2] G. Wiegleb, *Gasmesstechnik in Theorie und Praxis*. Springer Vieweg, Wiesbaden, 2016.

[3] S. Leonhardt and M. Walter, *Medizintechnische Systeme*. Springer Vieweg, Berlin/Heidelberg, 2016.

[4] Universität Bremen, *Die Weibullverteilung*. Available: http://www.math.uni-bremen.de/~dickhaus/downloads/LehrerInnen-SoSe2013/Weibullverteilung.pdf [last accessed on 2019-12-11]

Development of a Dosing Unit Using Venturi Injector

Mahmoud Badran [1], Waldemar Janik [2], and Stefan Mueller [3],

[1] Biomedical Engineering, Luebeck University of Applied Sciences, Mahmoud.badran@stud.th-luebeck.de
[2] B. Braun Avitum AG, Department of New Technologies, Waldemar.janik@bbraun.com
[3] Department of Applied Natural Sciences, Luebeck University of Applied Sciences, Stefan.mueller@th-luebeck.de

Abstract

As part of many industrial applications, mixing of fluids is needed. In most of these cases, pumps are used. However, this increases the initial and maintenance costs. With the use of Venturi Injectors (VI), an expensive dosing unit can be replaced by a cost-effective and low maintenance system. This work studies the effect of different parameters on the performance of the system and compares the theoretical calculations with the actual performance. The findings in this work are important for different fields including the biomedical field where pumps are being used.

1 Introduction

The main task in this paper is to suck in fluid B from a certain elevation and add it to fluid A in the main flow line. For such applications, it is typical to use pumps because they are easy to handle and it is a common practice. However, the use of pumps has many disadvantages including energy consumption, wear, and required maintenance due to moving parts. In addition, Printed Circuit Board (PCB) is needed to control the pump operation, and pumps may be stuck. For these reasons, venturi injectors are becoming more useful in such applications in different fields. Venturi injectors (VI) are build up of three main parts, a nozzle, a throat and a diffuser. Because of the reduction in the cross-sectional area, a negative pressure is created which can be used to suck in another fluid and mix it with the fluid in the main flow line. This concept is used for example in the car carburetors where air is the fluid in the main flow line and due to the area reduction in the throat of VI, fuel is sucked in and mixed with air. VI does not contain any active parts, and its cost is cheaper compared to the price of the pump. In the medical field, mixing of fluids is necessary, for this reason it is important to study the effectiveness of using VI in this field. High delivery precision is very important and low flow rates and small tubes diameter are typical in the medical applications. This work starts with defining the concept of VI, followed by the choosing of the VI's dimensions for the defined application and ends with a comparison between the theoretical and experimental performance.

2 Material and Methods

2.1 System Requirements

As mentioned previously, the application is to suck in fluid B and add it to fluid A flowing in the main flow line. There

are several requirements for this application for which the VI is being designed and tested. The elevation difference between the VI and the bottom of fluid B canister is -50 cm, and the height of the canister is 15 cm. Fig. 1 shows the system that was used in the experiment. In our experiment, fluid B in the canister has a density of 1161 $\mathrm{kg/m^3}$. In the main flow line, fluid A is degassed permeate water at room temperature of 23 °C. The operating flow rates of the pump is between 100 and 800 $\mathrm{ml/min}$.

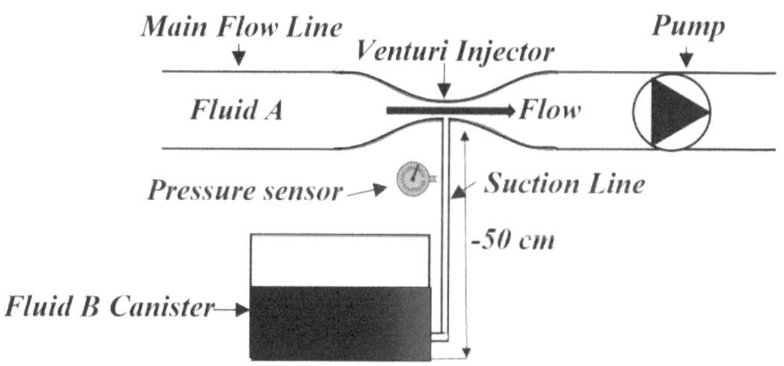

Figure 1: Test setup, where the pump is placed after the VI and an elevation of -50 cm between VI and fluid B canister. The elevation of the pressure sensor is the same as the VI .

As proven by [4], positioning a pump after the VI will create a higher suction pressure (negative pressure) compared to positioning the pump before the VI. In addition, in the first case, the relative pressure at the inlet and exit of *VI* will be less than 0 mbar compared to a greater than 0 mbar and 0 mbar respectively in the second case. For this reason, the position of the pump is chosen as shown in Fig. 1.

2.2 Venturi Injectors Concept and Design

VI were developed based on the same idea of the venturi meter. In Fig. 2 the schematic of VI is shown. The main idea of VI is that with the reduction of the cross-sectional area the velocity of fluid increase and thus the static pressure decreases generating a suction pressure for the fluid in the canister. The total pressure is the sum of three types of pressure: static pressure, dynamic pressure and hydrostatic pressure. In the case of VI, the hydrostatic pressure is constant between the throat and flow line because the elevation is the same. The dynamic pressure is generated from the flow velocity thus the dynamic pressure will increase in the throat, and the static pressure is pressure when the fluid is static and thus it will decrease at the throat.

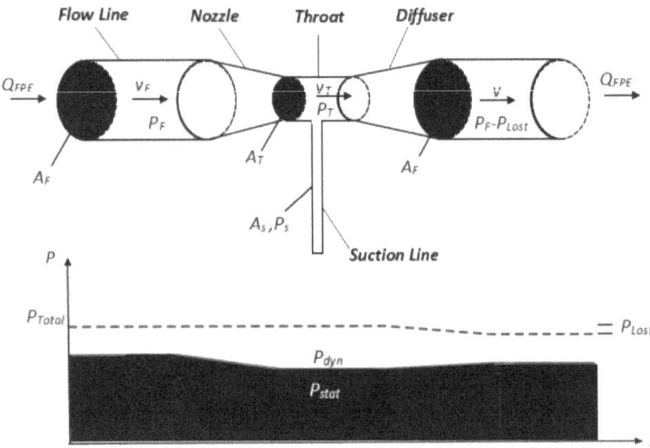

Figure 2: Schematic for venturi injectors with parameters and pressure distribution along the VI. P_{dyn} is the dynamic pressure, P_{stat} is the static pressure and P_{Total} is the total pressure.(modified to [4]).

In Fig. 2, A is the cross-sectional area, V is the flow velocity, Q is the Volumetric flow rate and P is the pressure. The subscripts f is for the flow line, T for the throat line, and s for the suction line. Q_{FPE} represent the flow rate created by the pump.

To start with, from the conservation of mass for incompressible fluid, steady and non-viscous flow, the continuity equation can be determined between the main flow line and the throat as shown in (1)

$$Q = A_F \cdot V_F = A_T \cdot V_T. \tag{1}$$

From (1) it can be seen that at the throat the velocity will increase because of the reduction in the cross-sectional area. As mentioned previously the pressure is divided into three components of pressure which can be seen clearly in the Bernoulli equation as in [3], and shown in (2)

$$P_F + \frac{\rho_w \cdot V_F^2}{2} + \rho_w \cdot g \cdot h_F = P_T + \frac{\rho_w \cdot V_T^2}{2} + \rho_w \cdot g \cdot h_T \tag{2}$$

Where ρ_w the density of the fluid in the main flow line (Fluid A, which in this case is permeate water), g the gravity and h the elevation. The first term corresponds to static pressure, the second to dynamic pressure and third term corresponds to hydrostatic pressure. The VI is placed horizontally and thus h_F and h_T are the same. Then the Bernoulli equation is simplified to (3)

$$P_F + \frac{\rho V_F^2}{2} = P_T + \frac{\rho V_T^2}{2} \tag{3}$$

As tested by [4] and shown in Fig.2, the total pressure is constant at the main flow line and the throat but is less after the diffuser, where theoretically it should be constant. This reduction in pressure is affected by different parameters including the ratio of the main flow line and throat areas, the Reynolds numbers and installation conditions especially the position of the pump. For the flow line and throat line the total pressure is constant, thus from (1) the flow velocity increases and as a result from (3), the static pressure decreases generating a pulling force for the fluid in the suction line. It is possible to define $\Delta P = P_F - P_T$, due to the fact that the suction line is connected directly to the throat with no pressure reduction, thus the static pressure is shown in (4)

$$P_s = P_T = P_F - \Delta P. \tag{4}$$

Moreover, the cross-section is circular, thus $A = \pi \cdot \frac{D^2}{4}$. By substituting the area in (1), the flow velocity can be determined as in (5).

$$V_{T,F} = \frac{4 \cdot Q_{FPE}}{\pi \cdot d_{T,F}^2}. \tag{5}$$

By Substituting (4) and (5) in (3), the pressure change can be obtained as shown in (6):

$$\Delta P = \frac{8 \cdot Q_{FPE}^2 \cdot \rho}{\pi^2} \cdot \left(\frac{1}{d_T^4} - \frac{1}{d_F^4} \right). \tag{6}$$

Equation (6) gives a relation between the pressure difference (between the main flow line and the throat), the dimensions of the venturi injector, the flow rate and the density of the fluid in the main flow line. This equation can be used to determine the required dimensions of the venturi injector to fulfil the requirements of pulling the fluid from the canister. The negative pressure needed to pull fluid from - 50 cm is shown in (7), where ρ_s is the density of the fluid that is flowing in the suction line (Fluid B).

$$P_s = \rho_s \cdot g \cdot h. \tag{7}$$

P_s = 1161 kg/m^3 $\cdot$ 9.81 m/s^2 $\cdot$ - 50 cm = - 5694.71 Pa. Thus the required pressure is 5694.71 Pa under the assumption that $P_F = 0$ $mbar$. Then $\Delta P = -P_T$ and a fixed diameter $d_F = 5$ mm and a flow rate of 300 ml/min. The density of fluid A is taken to be 992 kg/m^3 , the required diameter can be determined by arrangement of (6) to get (8).

$$d_T = \left(\frac{8 \cdot \rho \cdot Q^2 \cdot d_F^4}{\Delta P \cdot \pi^2 \cdot d_F^4 + 8 \cdot \rho \cdot Q^2} \right)^{\frac{1}{4}} \tag{8}$$

This will lead to a required diameter of $d_T = 1.368 \times 10^{-3}$ $m \simeq 1.3$ mm because in this case rounding down will create higher suction pressure. For a fixed flow rate the only factor that affects the generated suction pressure is the area ratio between the main flow line and the throat line. Another important parameter is the design of the nozzle and diffuser where it is recommended by [5] to have an opening and closing angle of less than 15 ° in order to avoid flow separation. Flow separation mainly occur due to an adverse pressure gradient in the direction of flow such that the boundary layer cannot follow the wall, because fluid sudden deceleration occur or when the wall velocity is zero and an inflection occur in the velocity profile due to sudden geometry change [1]. This lead to reduction in flow and might lead under certain circumstances to back flow as shown in Fig. 3. For this reason, a maximum angle of 14 ° was used.

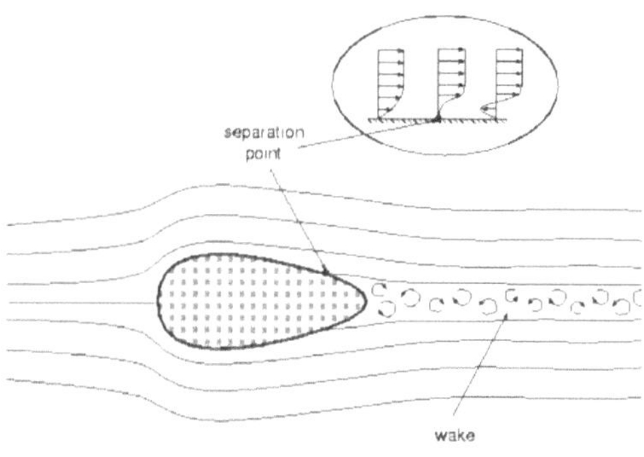

Figure 3: Flow separation behavior in the area of sudden change in geometry [1] .

Theoretically, the suction diameter should not have much influence on the generated pressure and this is what was tested by [2] and then by [4], where it had only influenced the volumetric flow rates of the fluid being sucked in from the canister. For higher volumetric flow rates of fluid B, a higher suction diameter is required and vice versa. Where this behavior can be predicted by the rearrangement of Bernoulli equation for the tube between the bottom of the canister which indicated in the equation as c and the suction point in the VI indicated as s, then the volumetric flow rate of B is obtained as in (9).

$$Q_s = \sqrt{\frac{(\Delta P - \rho_s \cdot g \cdot (h_s - h_c)) \cdot \frac{\Pi^2}{8 \cdot \rho_s}}{(\frac{1}{d_s})^4 - (\frac{1}{d_c})^4}} \qquad (9)$$

In (9), d_c was chosen to be 5 mm, thus for the same pressure difference between the bottom of the canister and the throat area in the venturi injector, a higher diameter of the suction line of VI leads to a smaller denominator in (9), thus higher suction flow rate. So depending on the application and the required flow rate of fluid B, d_s of the VI can be determined using (9).

Different dimensions of polyamide VI were tested. Those VI produced using Selective Laser Sintering (SLS) with a layer thickness of 0.1 mm. Some of these VI also contain sealing to make sure that there will be no effect of the surrounding on the system behavior and no particles can pass through the membrane of the VI to avoid the introducing of gas from VI wall.

2.3　Experiments

Three sets of VI were tested with the same nozzle and diffuser angles and main flow and exit diameter, but different suction and throat diameter as shown in Table 1.

Table 1: Tested VI Dimensions

VI	Throat Diameter (mm)	Suction Diameter (mm)
1313	1.3	1.3
1410	1.4	1.0
1414	1.4	1.4

The VI listed were tested using the setup in Fig.1. The flow rate varied between 100 and 800 ml/min with a step of 100 ml/min, each flow rate had a duration of 1 min. Both the increasing and reduction of flow rate was done in order to make sure that the measured data is not affected by the order of flow rate variation. The design of VI 1410 and VI 1414 is to make sure that the suction pressure generated will be independent of the suction diameter. The suction line was closed and a pressure sensor was installed to measure the pressure in the suction line as in Fig. 1.

3　Results and Discussion

3.1　Results

The resulted pressure for the testing of the three VI under different pump flow rates can be seen in Fig. 4. This data was recorded using Labview and then analyzed using Matlab.

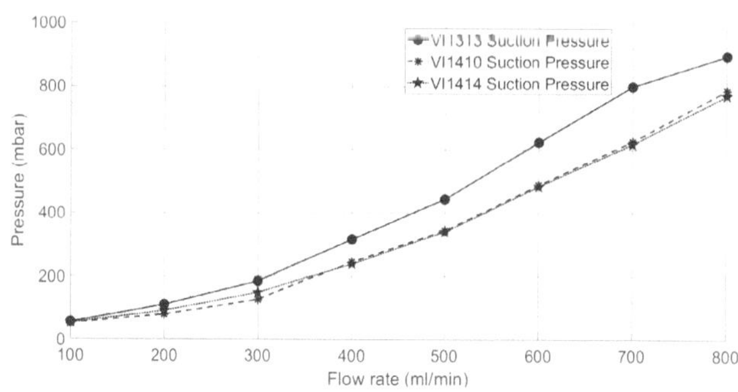

Figure 4: Absolute values of the measured suction pressure for each VI for different flow rates

In Fig. 4 it can be seen that for a higher flow rates, the generated suction pressure is higher. Besides, the generated pressure for VI1313 is higher than that for VI1410 and VI1414. Moreover, the generated suction pressure for VI1414 and VI1410 is the same. In order to compare the results with the theoretically calculated suction pressure, (6) was used. Fig. 5 shows the calculated pressure for different flow rates for both throat diameters of 1.3 mm and 1.4 mm for the same parameters used before.

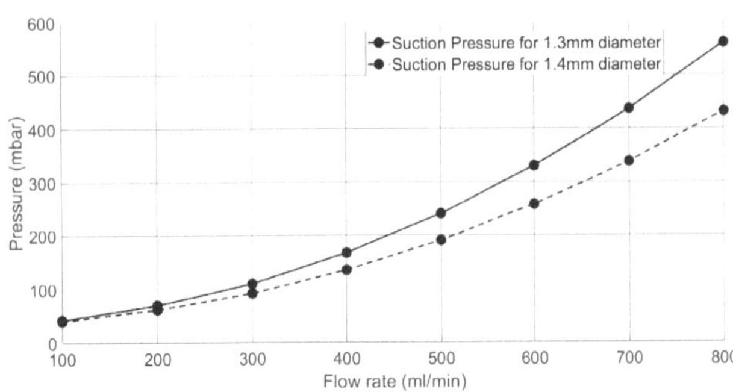

Figure 5: Calculated suction pressure for VI with 1.3 mm and 1.4 mm throat diameters

3.2 Discussion

From Fig. 4, it can be seen that the generated suction pressure for both cases of VI1414 and VI1410 are the same and thus it is a validation of the conclusion made by both [2] and [4] that the suction diameter has no effect on the generated suction pressure. In addition, as expected the smaller the throat diameter, the larger is the suction pressure. Fig. 5 shows the calculated suction pressure. In Fig. 6 the difference between the measured and calculated suction pressure for different suction diameters can be seen. The calculated suction pressure takes into account that P_F is not zero. Where TH (throat diameter) represent the calculated suction pressure for the respective diameter using (9).

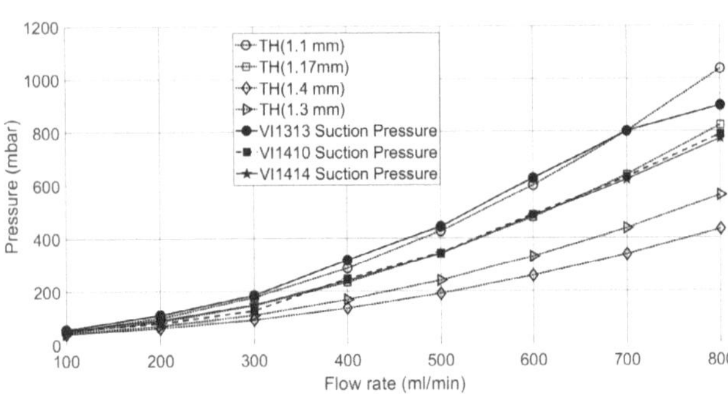

Figure 6: Deviation between calculated and measured suction pressure

From Fig. 6, it can be seen that the performance of VI1313 is close to the expected performance of TH (1.1 mm), and the performance of VI1414 is close to TH (1.17 mm). Where a suction pressure of 185 mbar for VI1313 was measured and a suction pressure of 110 mbar was expected for 1.3 mm throat diameter and 177 mbar was expected for 1.1 mm throat diameter. Thus, the throat diameter is smaller than expected. Theoretically, the throat diameter of VI1313 is 1.1 mm and not 1.3 mm but this was not measured. This is due to the SLS method with a layer thickness of 0.1 mm that was used in the production of the VI. Moreover, the area is not perfectly circular which affects the performance. In addition, there is the effect of reduction in the pressure after the constriction area as explained previously in Fig.2, where theoretically the total pressure should be constant over all the VI parts, so this system will not behave the same as the ideal case taken into account in the calculated pressure. The generated pressure is higher than expected, thus the flow rate for the sucked in fluid will be higher and thus the system is operating very well to pull the fluid and to mix two solutions together in a continuous way.

4 Conclusion

This paper discussed the concept of using VI in order to pull fluid from a certain elevation and mix it with another fluid. It covered the concept of VI, how to design them, and what are the important parameters that affect the performance of VI. Even though the performance deviates from the theoretical analysis, the system is operating well concerning introducing an additional fluid to the fluid in the main line. This can be of interest in many applications especially in the biomedical field where a lot of pumps are being used and thus initial and maintenance cost increase. However, in case of VI the initial cost is low especially in the case of mass production. Depending on the application, an additional sensor can be added to detect other properties of the mixture from fluid A and fluid B, such as pH sensor or temperature sensor. Then control the addition of fluid B by controlling a valve installed in the suction line.

Acknowledgement

The work has been carried out at B. Braun Avitum AG, New Technologies and supervised by the Department of Applied Natural Sciences, Lübeck University of Applied Sciences.

5 References

[1] A. Bakker, *Applied Computational Fluid Dynamics*. Available: http://www.bakker.org/dartmouth06/engs150/11-bl.pdf [last accessed on 2018-12-10]

[2] B. Ahmet, O. Fahri and O. Mualla, *Experimental Investigations of air and liquid injection by Venturi tubes,* Water and Environment Journal 20, Nr. 3, 2006.

[3] D. Warnack, *Computer Aided Techniques in Design* [lecture Notes]. Available: https://lernraum.th-luebeck.de/course/view.php?id=16 [last accessed on 2019-11-10]

[4] M. Moritzen, *" Konzeption einer durch leitfähigkeit geregelten dosiereinheit einer dialysemaschine ",* Master's Thesis, University of Kassel, Department of Electrical Engineering, 2018.

[5] S. Herbert, *Technische Fluidmechanik.* Springer Vieweg, Berlin/Heidelberg, 2017. ISBN 978-3-662-54467-9

Carbon dioxide measurement of different gas concentrations concerning its heat conductivity considering the system humidity

Annika Freund [1], Jan Beidatsch [2] and Hans-Ullrich Hansmann [3]

[1] Medical Engineering Science, Universität zu Lübeck, an.freund@student.uni-luebeck.de
[2] Medical Engineering Science, Universität zu Lübeck, jan.beidatsch@student.uni-luebeck.de
[3] Drägerwerk AG, Process Engineering, hans-ullrich.hansmann@draeger.com

Abstract

The company Drägerwerk AG & Co. KGaA is aimed at the development of new technologies in the section of human ventilation. One of the current projects is the processing of a new small ventilation system. For checking the correct intubated ventilation, the system device requires a sensor to measure the CO_2 level. The goal is to use an internally existing Dräger module. Usually this module measures oxygen in the patient's respiration. A possibility of measuring the CO_2 concentration via heat conductivity is sought. For implementation it is required to use different gas concentrations in an experimental setup to check the dependence of humidity. The conclusion shows the linearity between the signal of the module and the CO_2 concentration of the gas. The result is an effective working of the module for detecting CO_2 indirectly through its heat conductivity.

1   Introduction

Respiration is the most important factor for the human body to survive and is divided into inspiration and respiration. A healthy body independently respires the required amount of O_2 and emits the resultant amount of CO_2. But there are many use cases where people need support with their respiration. One possibility for assisting respiration is the use of a ventilator. Each ventilation is aimed at the elimination of dyspnoea, improvement of the gas exchange, amendment of the pressure-volume relation, promotion of lung and respiratory healing and the facilitation of respiratory equipment remove. From a medical point of view, ventilation should eliminate hypoxaemia, correct respiratory acidosis and treat respiratory distress.

The respiration occurs in two different parts. There is a pathway for the transportation of gas and a part where the gas exchange happens. Ventilation can only be administered during the transport of gas. In general, air can only flow from a higher to a lower pressure. During inspiration, the pressure in the alveoli must be lower than the ambient pressure. Inspiration ends when the alveoli pressure is the same as the ambience pressure and no airflow exists. Respiration takes place according to the same working principle only the other way round. With human respiration the pressure inside the lung is higher than in the ambient [1], [2]. If using ambient air, a gas mixture is applied for ventilation. Ambient air consists of 20.93 % oxygen, 78.10 % nitrogen, 0.9325 % argon, 0.03 – 0.04 % carbon dioxide, 0.01 % hydrogen and some inert gases. The mixture of inspiration and expiration is shown in table 1.

Table 1: components of respiration air [3]

	inspiration	expiration
nitrogen	78 %	78 %
oxygen	21 %	16 %
carbon dioxide	0.03 %	4 %
other inert gases	1 %	1 %

The table only serves as an overview, since the sum does not add up to 100 %. A healthy body has in its expiration gas a 4 % CO_2 concentration, also based on standard conditions. During intubated breathing special attention must be paid to the tracheal position of the intubator. This is checked by a constant CO_2 release of approximately 4 %. If this is not the case it means that the intubator is located at the esophagus and no ventilation takes place [3]. In general, humidity is water vapour in a closed volume. When the gas is air, it is called air humidity or air moisture. Absolute air humidity means the water vapour in g/m^3 depends on temperature. Air can only absorb a specific amount of water vapour increasing by higher temperature up to maximal humidity. The relative humidity in % is the relation between absolute and maximal humidity depending on temperature. In the following equation 1, the relative humidity describes the water vapour saturation in %.

$$relative\ humidity\ [\%] = \frac{absolute\ humidity}{maximal\ humidity} * 100 \quad (1)$$

Variations of relative humidity can occur due to modification of the actual amount of water vapour or temperature changes [4], [5]. Among other things, changes in relative humidity are caused by changes in the actual amount of water vapour or temperature. The maximum value of water vapour also changes. The same actual amounts of water vapour then correspond to different relative humidities at different temperatures [6], [7].

The aim is to identify whether an internal existing module with its included sensor can measure CO_2 via heat conductivity. Normally this module measures O_2 and there is no direct possibility to detect the CO_2 value.

2 Material and Methods

An experimental setup was used to carry out the test. The target is to check the possibility to measure the CO_2 concentration and the measurement dependence of humidity. The internally existing software "gseos" for data acquisition was used to support the test.

2.1 Experimental setup

To simulate the human respiration a breadboard as shown in Fig. 1 was used. Two different inflows are required for simulating the concentration of human respiration. They are shown in part one of Fig. 1. One inflow is used for the compartment air, like inspiration air, and the other inflow needs different CO_2 concentrations for the simulation of human expiration. The level of CO_2 concentration was chosen in the area of more and less than the human respiration CO_2 release based on standard conditions (4 %). To achieve the humidity found in human expiration gas, an impinger was used. In order to achieve the smallest possible dead space volume when humidifying, a small impinger is required.

An electro-magnetic valve switches between the two different inflows with a frequency of 0.2 Hz. That means 2.5 s for inspiration and 2.5 s for expiration. One breath per 5 s equates to a normal respiratory frequency of 12 breaths of air per minute for an adult without physical effort [8].

The heat and moisture exchanger (HME) filter is required for moisture compensation and is independent of external heat- and moisture feeding. The HME saves the moisture and heat of expiration air during expiration and during inspiration the air takes the moisture back. There is no loss of heat and moisture but instead a constant moisture over a breathing cycle [4], [9].

The complete first part in Fig. 1 describes the fraction outside of the ventilation system, the mask and patient. A one-meter tube connects the interior and exterior. The connecting tube is later the connection between the patient front and the ventilation system. At first, a connection length of 1 m was assumed. Variations are still possible within a certain framework. The second part includes different measurement modules for checking flow and moisture measures and a small pump. The Paramagnetic Independent Analyser (PIA) is internally the standard measure module for oxygen in air. The working principle is based on various values of heat conductivity at different temperatures of several gases, e.g. oxygen. There is no direct possibility to obtain the CO_2 value. The voltage value U_{th} of PIA's measurement cell is then compared to the CO_2 value of a second module, the Infrared Low Cost Analyzer (ILCA), to confirm whether the values match. For getting the air and gas mixture into the system, a pump is necessary. Sensors were attached to the end of the breadboard to check flow and humidity. It is required to reach a flow of 70 ml/min, which corresponds to the flow of the exit of the breadboard. Flow measurement is especially used for checking the pressure.

All test parts are connected to each other via tubes (PVC and silicon). The length of the tubes corresponds to the optimum connection, so that a possible change or exchange of components is easily possible.

2.2 Test performance

The gas mixture was joined through a gas bag on the system. It is possible to fill the gas bags exactly with the right concentration so that they serve as a reference. To simulate human breathing, a CO_2 concentration of 4 % and several values around that (2 %, 3 % and 5 %) were chosen. Furthermore, four different humidity levels were considered. In the first attempt, compressed air instead of ambient air was used to achieve a system with average room humidity. All different CO_2 concentrations were checked with these different moisture values. In Fig. 2 the results of the PIA and the ILCA at room humidity are shown which correspond to about 50 %, in this measurement 52.6 %. The measurement occurs with four different CO_2 concentrations, 2.005 %, 2.997 %, 4.01 % and 4.951 % bagged in gas bags shown at the abscissa. On the left ordinate the measured CO_2 concentration of the ILCA is shown. The concentration is indicated in Vol.-%, the unit for concentration with regard to Volume [10]. On the right the voltage hub difference divided by 100 of the PIA is shown. The measured CO_2 level and the measured U_{th} have a similar linearity. A slight divergence in the curves occurs only at higher concentrations. Overall, the curve of the PIA shows more linearity than the one of the ILCA. All measurements with different CO_2 concentrations were repeated by various relative humidity values, about 75 %, 85 % and 95 %. For reaching approximately

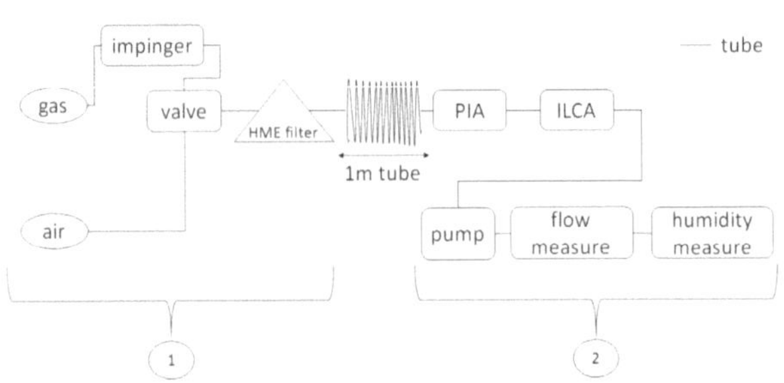

Figure 1: Schematic breadboard for the measurement of CO_2 via measurement modules. The first part describes the part outside the ventilation system. The second part shows the interior compartments. Both are connected via one-meter tube. The single parts are also linked via tubes.

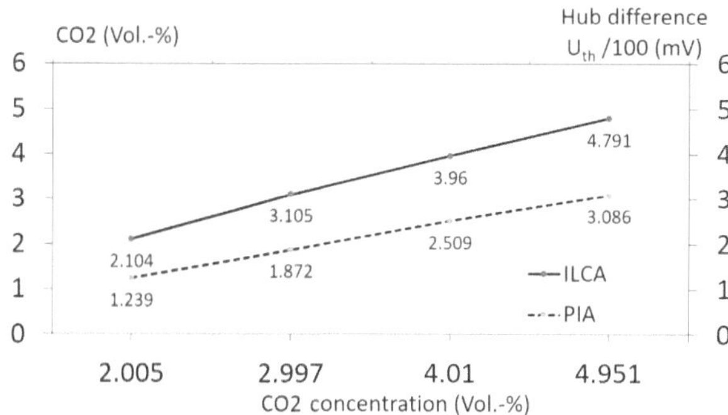

Figure 2: The PIA and the ILCA results at room humidity. On the left ordinate the measured CO_2 value of the ILCA is shown. On the right side the hub differences (voltages changes) divided by 100 of the PIA is presented and the abscissa displayed the reference (different CO_2 concentrations in bags).

85 % humidity, another frequency was used to raise the humidity in the system. A switching frequency of 2.5 s to 5 s was chosen. At up to 95 % humidity it is necessary that both inflows will be humidified. It is not possible to humidify both by one impinger because there is no possibility to receive a breathing simulation signal. It will not reach the expiration and inspiration values because of too much dead space volume. In general, it is important to give the system enough time until the humidity is constant. Any moisture above room humidity needs time to become a constant in the system. At higher humidities it can take up to half an hour.

The software "gseos" was used to evaluate the measuring modules. This software enables communication with the PIA and the ILCA. The measuring cells each have their own user interface in the software. With the ILCA, the CO_2 value can be read directly, whereas with the PIA, the value is based on the U_{th}. The ILCA was always running as a check system during the measurements. To see both user interfaces simultaneously, working with two computers was necessary. The ILCA surface also provides the O_2 value, which is a good comparison to see if the ILCA is running correctly, especially in indoor air. Before the measurement is started, the ILCA needs a reset, which is automatically requested after about 20 minutes. The pump of the ILCA must be activated so that it absorbs the ambient air to reach a zero value.

In order to use the PIA for this test setup, it is important to deactivate the integrated heating. Otherwise the heating will cause the gases to behave differently than at ambient temperature.

3 Results and Discussion

It has been shown that it is possible to use the measurement module PIA to indirectly detect the CO_2 concentration to verify the correct ventilation. By determining a factor, the measured voltage U_{th} of the PIA measuring cell can be used to calculate the CO_2 value.

PVC-tubes have proven to be suitable. Tubes made of silicone on the other hand proved to be unsuitable. With silicone, there was a loss of water vapour and precise humidity measurement was no longer possible.

3.1 Simulated breathing

First the simulated breathing was evaluated. Only after the first satisfactory results were obtained, spontaneous breathing was also evaluated. In Fig. 3 all PIA values for all different CO_2 concentrations are on the abscissa and the hub difference in % on the ordinate. Test 1 is the measurement of CO_2 at 52.6 % (10.7 g/m^3) humidity, test 2 is measured at 69.678 % (14.25 g/m^3) humidity, test 3 is measured at 74.6 % (15.33 g/m^3) humidity and test 4 included 90.075 % humidity (18.7525 g/m^3). In each case it was measured with 2.005 %, 2.997 %, 4.01 % and 4.951 % CO_2. During the first test run the humidity results do not show desired humidity values. Even after a long waiting period only deviating humidity could be measured. When checking, two tubes were found to be insufficient. In general, more humidity is lost through silicone tubes, so PVC-tubes should be preferred. In this case, the source of error was directly in front of the humidity sensor, so that it can be assumed that the correct humidity was present in the system itself. The result in Fig. 3 pictures all different humidity values of 16 measurements in one diagram. There is nearly the same hub difference between the diverse humidity values. So, the PIA was found to be constant in measuring the same values for varying humidity.

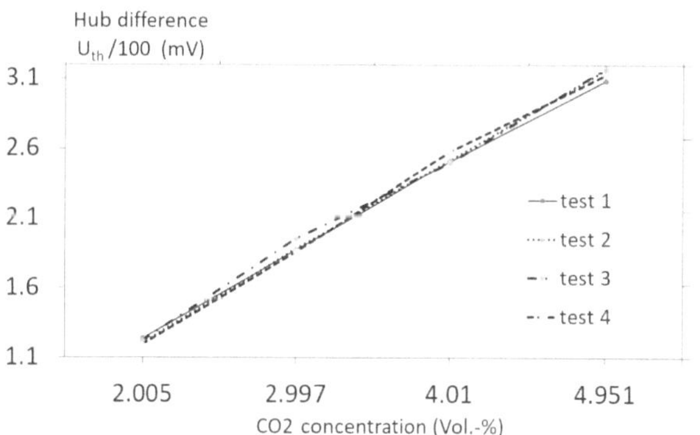

Figure 3: Paramagnetic Independent Analyzer values of four tests with four different humidity values and various CO_2 concentrations. On the ordinate the hub differences divided by 100 are shown and the abscissa displayed the reference.

3.2 Spontaneous breathing

Due to the positive results, first spontaneous breathing experiments were carried out. First, average respiratory frequencies of an adult at rest were examined. For this purpose, the test person was given a breathing mask and breathed directly into the test set-up by mouth. Beforehand, 10 breaths per minute were measured by the test person in rest as a reference. The exhaled air of the test per-

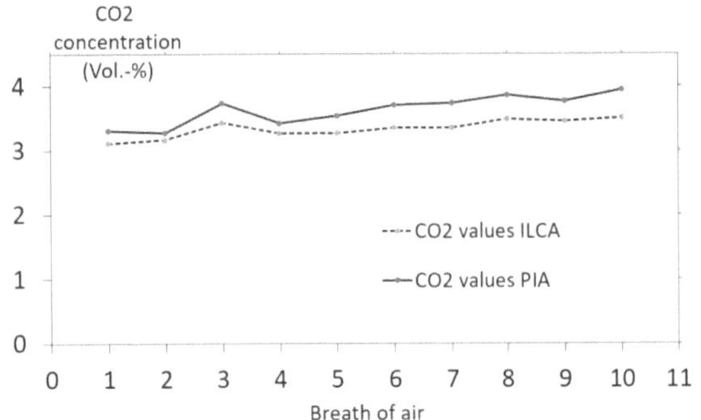

Figure 4: CO_2 concentration of expiration per breath of air by normal breathing frequency (10-12 breaths per minute). On the ordinate the CO_2 concentration of the PIA and the ILCA is shown and the abscissa displayed the breaths of air.

son had a humidity of 71 % as an average over inspiration and expiration. Fig. 4 shows the first attempt at spontaneous breathing through the mouth. The abscissa represents the breaths of air and the ordinate shows the CO_2 concentrations in Vol.-%. There is an offset between the PIA and ILCA values. A similar offset can also be seen when breathing through the nose. Further breathing experiments with different breathing frequencies were performed. During rapid breathing, irregularities in signal frequencies occurred. It must be checked whether the measuring modules have enough time to detect the full concentration during rapid breathing.

4   Conclusion

It is possible to use the PIA for CO_2 measurements. The next step is the implementation in a new ventilation system. Therefore, the electronic measuring device in the PIA must be removed from the measuring module. Additionally, the best position of the sensors for measuring the expiration air must be found. Different options are possible, in the front of the mask near the HME or on the side of the mask, for example. Furthermore, it is necessary to check the complete system at different temperature levels and variable heating and cooling. Also the succession from warm to cold and the other way round will be checked which is important for further system use cases. Temperature levels around the dew point (the temperature at which the first condensate of the gas occurs [11]) and how the system works with condensation inside are of special interests. It is still necessary to carry out breathing tests to determine the reason for the offset in rapid breathing and to check the general validity of the system.

Acknowledgement

The work has been carried out at Drägerwerk AG & Co. KGaA and supervised by PD Dr. H. Paulsen, Institute of Physics, Universität zu Lübeck. I am thankful for the cooperation with Drägerwerk AG & Co. KGaA and for being included on the project team. My thanks go to all workmates on the team and let me express a special thanks to my supervisor Dipl.-Ing. Hans-Ullrich Hansmann.

5   References

[1] W. Melanie, "Machbarkeitsuntersuchung zur Integration und Validierung eines C02-Sensors unter den Rahmunbedingungen eines bestehenden Beatmungssystems," Master's thesis, University of Applied Sciences Westfälische Hochschule Gelsenkirchen, 2019.

[2] R. Larsen, T. Ziegenfuß, and A. Mathes, *Beatmung: Indikationen - Techniken - Krankheitsbilder*, 6th ed. Berlin Heidelberg: Springer-Verlag, 2018. [Online]. Available: https://www.springer.com/de/book/9783662548523

[3] G. Baumbach, *Luftreinhaltung: Entstehung, Ausbreitung und Wirkung von Luftverunreinigungen — Meßtechnik, Emissionsminderung und Vorschriften*. Springer-Verlag, 2013, google-Books-ID: P3fPBgAAQBAJ.

[4] W. Oczenski, *Atmen - Atemhilfen: Atemphysiologie und Beatmungstechnik ; 53 Tabellen*. Georg Thieme Verlag, 2008, google-Books-ID: Mif8hp0KW1MC.

[5] D. Weber, *Technische Feuchtemessung: in Gasen und Festkörpern*. Vulkan-Verlag GmbH, 2002.

[6] E. Heyer, *Witterung und Klima: eine allgemeine Klimatologie*. Teubner, 1963, google-Books-ID: j64MAQAAIAAJ.

[7] E. Nürnberg and P. Surmann, *Hagers Handbuch der pharmazeutischen Praxis: Band 2: Methoden*. Springer-Verlag, 2013, google-Books-ID: IRSDBwAAQBAJ.

[8] M. Seel and E. Hurling, *Die Pflege des Menschen im Alter: Ressourcenorientierte Unterstützung nach den AEDL*. Schlütersche, 2010, google-Books-ID: aAQuBQAAQBAJ.

[9] H. W. Striebel, *Operative Intensivmedizin: Sicherheit in der klinischen Praxis*. Schattauer Verlag, 2014, google-Books-ID: _Jy0BAAAQBAJ.

[10] J. Schatz and R. Tammer, *Erste Hilfe - Chemie und Physik für Mediziner*. Springer-Verlag, 2015, google-Books-ID: DJq4CQAAQBAJ.

[11] F. Tiemann, *Einfluß des Gasschwefels auf den Taupunkt von Verbrennungsgasen im Hinblick auf die Abgas-Korrosionswirkung*. Springer-Verlag, 2013, google-Books-ID: sO2GBwAAQBAJ.

Reduction of Condensate in Electrically Heated Breathing Hoses

Maximilian Hünert [1], Lasita David [2] and Ludger Tappehorn [2]

[1] Medical Engineering Science, Universität zu Lübeck, m.huenert@student.uni-luebeck.de
[2] Drägerwerk AG & Co. KGaA, Lübeck, {lasita.david; , ludger.tappehorn}@draeger.com

Abstract

The project work deals with the task of gaining an understanding of the parameters that influence the formation of condensate in heated breathing hoses. The focus is on the heated neonatal and adult hose systems with an active humidification of the breathing air. The placement of the temperature sensor outside of the incubator leads to a decrease of the total amount of condensate by 275 %. The use of an additional expiratory incubator extension resulted in a reduced expiratory condensate quantity from 3.15 ml to 0.63 ml. By using the incubator extension, the risk of newborn babies suffering skin damage can be prevented. The surface temperature 25 cm away from the Y-Piece can be reduced by 9.6 °C. The adult breathing tube system cannot handle high flow changes and the humidification cannot be maintained. A high flow ventilation change from 10 l/min to 60 l/min causes the humidifier to switch off during the warm-up phase of 30 min.

1   Introduction

For long-term ventilated patients, the dry and cold respiratory gases lead to a damage of the lungs. The use of a heated air tubing system and an active humidifier supports the preservation of the self-cleaning process of the lung and the patient is spared [1]. Active humidification is used in the neonatal, paediatric and adult clinical areas. The field of active humidification, with an additional heated breathing tube system is gaining more significance over the last years. The warm and humid breathing air is guided through the heated tubes to the patient and back to the ventilator. Condensate can occur on this path of the breathing air, which can lead to a restricted ventilation performance. The formation of condensate must therefore be avoided. In order to compare the standardized conditions and the environments that may occur in reality, condensate formation on already available heated breathing tubes will be investigated.

active humidification, the heated hoses increase the respiratory gas temperature between 31 °C to 40 °C. The lower ambient temperature cools the tubing during ventilation and the breathing gas. Low respiratory gas temperatures can store less humidity. The remaining humidity is released in form of condensate.

Condensation describes the transition of a vapor (breathing gas) into the liquid state (water). Condensation occurs if the gas temperature falls below the saturation temperature or if the absolute humidity is increased and therefore an over-watering takes place. Energy is released into the environment. The opposite of condensation is boiling or evaporation [2]. This effect is used with the active humidification systems.

Evaporation describes the phase transition of a liquid into the gaseous state. If this phase transition takes place at boiling temperature, it is called vaporation. Vaporation is much faster than evaporation [3].

2   Material and Methods

The water content of the air is referred as humidity and is expressed as absolute humidity AH [mg/l] [2]. By converting the ideal gas equation, the absolute humidity can be calculated with the mass of the water in the air m_w [mg], and the volume of the gas V [l]. More specifically it can be described with the temperatur T [°C], the specific gas constant R_s [J/kg · K], and the water vapour pressure p_{rel} [hPa]:

$$AH_{max} = \frac{m_w}{V} = \frac{p_{rel}}{R_s \cdot T} \qquad (1)$$

The water absorption capacity of the gas mixture is limited by a maximum humidity AH_{max} [mg/l] [2]. During

2.1   Heated Breathing Hose Systems

In order to maintain the required amount of water vapour within the breathing hose, heated hoses keep the air at a constant temperature. This provides also the advantage of avoiding condensation by cooling the breathing air. A setup of an active humidification is illustrated in Fig. 1.

Dry gas from the central gas supply is conducted with a temperature of approx. 10 - 15 °C to the ventilator. The breahting gas is heated up and has an inspiratory temperature of 20 - 25 °C. The inspiratory nozzle is connected via an unheated hose (A) with one opening of the water chamber (H) of the active humidifier MR850 of the company Fisher & Paykel. (G). By heating up the water, active humidification takes place in the water chamber. The warm and humid

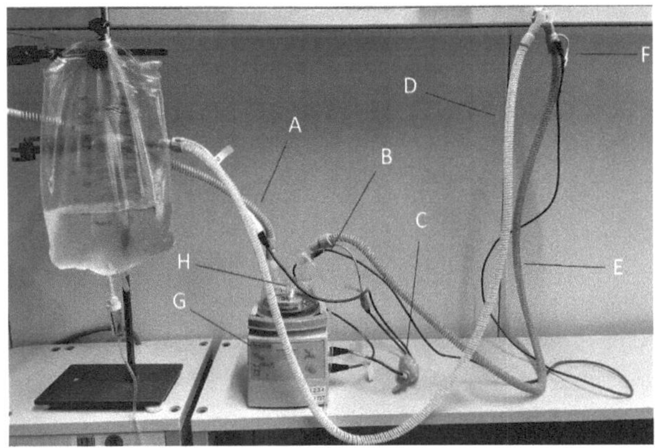

Figure 1: Setup of the active humidifier. It includes a ventilation system, the humidifier with the water chamber, inspiratory and expiratory heated hoses and sensors to measure the breathing gas temperature of the water chamber and right before the Y-Piece (connector between Hoses and artificial lung).

air flows through the heated inspiratory hose (E) into the patient and back to the ventilator via the heated expiratory hose (D). In addition, a temperature sensor (B) is located at the water chamber and at the end of the inspiratory hose (F). The sensors are connected via the heater wire adapter (C). The humidifier regulates the temperature at the end of the inspiratory hose and the chamber outlet temperature. Only the temperature is measured and regulated, but not the humidification. Depending on the application, a temperature of 34 - 40 °C is reached at the Y-Piece [4]. The heated and humidified air reaches the lung of the patient and is also returned via an expiratory heated tube to the ventilator. An unheated expiratory hose causes more condensate due to the colder ambient temperature. The condensate that occurs could flow back into the patient's lungs and into the ventilator. This must be avoided with an additional water trap in the unheated expiratory hose. If condensate should occur nevertheless, another water trap is placed behind the expiratory nozzle of the ventilator.

2.2 Changes in the Measurment Setup

The ventilation of premature babies takes place in an incubator. In most cases the temperature inside the incubator is between 34 °C and 39 °C [5]. The breathing hoses must be guided through the openings of the incubator. This can lead to a different positioning of the temperature sensor as shown in Fig 2.

Measurements have been taken to examine the influence of the ambient temperature on the temperature sensor and the resulting formation of condensate in the ventilation hoses and in the water trap of the device. The temperature sensor on the Y-Piece is located inside or outside the incubator. For both partial measurements, the same set of hoses is used with an additional unheated inspiratory incubator extension. The used devices are the Dräger Incubator Caleo, with a set incubator temperature of 33 °C and the VN500 ventila-

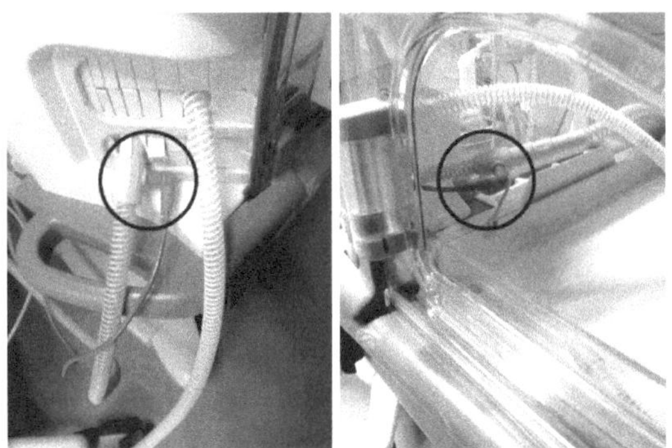

Figure 2: Placement of the temperature sensor outside of the incubator (left photo) and inside of the incubator (right photo). If the sensor is inside, a part of the heated inspiratoy hose leads into the incubator.

tor in Pressure Controlled - Continuous Mandatory Ventilation (PC-CMV) mode. The parameters of ventilation are: Respiratory Rate RR = 60/min, Inspiratory Pressure P_{ins} = 15 mbar, Inspiratory Time T_{insp} = 0.33 s and Positive End Expiratory Pressure PEEP = 5 mbar. Before each test run, the breathing hoses and the water trap are weighed. After a running time of 4 hours, the weight is again documented. The neonatal ventilation system is sold with an unheated inspiratory incubator extension. The heated expiratory tube is thus located in the inside of the incubator. Another consideration of the formation of condensate in the breathing hose system is the influence of an additional unheated expiratory incubator extension. The measuring setup consists of the humidifier MR850 (Auto HC mode), the Dräger incubator Caleo (33 °C incubator temperature) and the same ventilation settings from above. The temperature sensor is located outside the incubator. After a measuring time of 4 hours, the condensate formation is monitored with and without an additional expiratory incubator extension.

Depending on the age of the newborn, the maximum incubator temperature is 36.5 ± 1,8 °C. This warm environment, in combination with a heated expiratory hose, leads to an increased surface temperature of the hose. This could be dangerous due to the thin skin of newborns. In addition to the formation of condensate, the surface temperature of the expiratory hose with and without an incubator extension is measured. The limit value of the surface temperature of a tube, for a longer skin contact than 10 min, is at 43 °C for adults [6]. The value for newborns is below that threshold. For the measurement of the surface temperature the incubator temperature is 38 °C. After a running time of 3 hours, the surface temperature 10 cm, 20 cm and 25 cm away from the Y-Piece is recorded.

2.3 Regulation of Respiratory Gas Temperature

Further investigations consider the formation of condensate and temperatue differences of the respiratory gas temperature based on changes of the environmental temperature and changes in the ventilation parameters. The regulation of the breathing gas temperature of the hose systems can be examined by measuring the environmental temperature and the temperature at the end of the inspiratory hose. For this, a neonatal and adult breathing hose system is tested.

During conventional ventilation, high-frequency ventilation and constant flow, the breathing gas temperature measured at the end of the inspiratory hose shall be between 33 °C and 41 °C and alternate only about ± 2 °C after the warm-up time of 30 min.

Additional temperature values will be added during constant flow changes from 4 l/min to 30 l/min for the neonatal system and flow changes from 10 l/min to 60 l/min for the adult version. The temperature should also remain constant if the environmental temperature changes from 26 °C to 18 °C.

3　Results and Discussion

Table 1 shows the results of the condensate measurement depending on the position of the temperature sensor with a temperature of 33 °C inside the incubator.

Table 1: Results of the condensate measurement depending on the position of the temperature sensor. The temperature sensor is placed inside or rather outside of the incubator. After a runtime of 4 hours, the quantity of condensate in the expiratory hose, the inspiratory hose and in the water trap is noted. Also the breahting gas temperature of the water chamber and the Y-Piece is listed at the end of the measurement.

Parameters	Sensor Outside	Sensor Inside
Condensate Expiratory [ml]	0.51	1.93
Condensate Inspiratory[ml]	0.98	1.96
Condensate Trap [ml]	-	0.22
Chamber Temperature [°C]	36.9	37.0
Y-Piece Temperature [°C]	40.0	40.0
Environmental Temperature [°C]	20.5	20.4

If the temperature sensor is located outside the incubator, the resulting water quantity after 4 hours in the expiratory tube is 0.51 ml and in the inspiratory hose 0.98 ml. The water trap of the ventilator is dry during the entire measurement. The placement of the temperature sensor in the inside of the incubator leads to a significant increase of condensate. The total amount of water produced during the measurement increased by 275 %. In summary the temperature sensor inside the incubator is negatively influenced by the high ambient temperature of the incubator. The placement of the temperature sensor into the interior of the incubator should be avoided. Another measurement was to investigate the dependency from the usage of an additional expiratory

incubator extension to the formation of condensate. The results are listed in the table 2.

Table 2: Results of the condensate measurement depending on an additional expiratory incubator extension. After a runtime of 4 hours, the quantity of condensate in the expiratory and inspiratory hose with and without an additional expiratory incubator extension is noted. Also the gas temperature of the water chamber and the Y-Piece is listed at the end of the measurement.

Parameters	Inspiratory Extension	Plus Expiratory Extension
Condensate Expiratory [ml]	3.15	0.63
Condensate Inspiratory [ml]	2.27	1.04
Chamber Temperature [°C]	36.9	37.0
Y-Piece Temperature [°C]	40.0	40.0
Environmental Temperature [°C]	20.4	22.5

The use of an additional expiratory incubator extension resulted in a reduced expiratory condensate quantity from 3.15 ml to 0.63 ml. Inspiratory, the formation of condensate could be reduced from 2.27 ml to 1.04 ml. However, it should be noted that the ambient temperature was 2.1 °C lower during the second measurement. The cooler ambient temperature promotes the formation of condensate. Further tests should be carried out at the same ambient temperature and different incubator temperatures. The results of consideration of the surface temperature of the expiratory hose inside the incubator with and without an additional unheated expiratory extension are listed in Table 3.

Table 3: Results of the measurement of the surface temperature of the expiratory hose with and without an additional expiratory extension and a temperature of 38 °C inside the incubator. The ventilation mode is PC-CMV, with the values: Ti = 0.33 s; RR = 60/min; Slope = 0.1 s; Pinsp = 15 mbar and PEEP = 5 mbar. After a running time of 180 min the surface temperature 10 cm, 20 cm and 25 cm away from the Y-Piece is displayed.

Setup	10 cm	20 cm	25 cm
Expiratory Surface Temperature with Extension [°C]	34.0	34.2	34.4
Expiratory Surface Temperature without Extension [°C]	41.5	42.9	44.0

If no expiratory incubator extension is used, the surface temperature determined 25 cm away from the Y-Piece is above the limit of 43 °C. The additional unheated expiratory incubator extension reduces the surface temperature to a maximum of 34.4 °C. To investigate the regulation of respiratory gas temperature, the specifications listed in chapter 2.3 have been carried out. The neonatal breathing system has passed all requirements. During the different ventilation methods of constant flow changes from 4 l/min to 30 l/min and the environmental temperature change from 26 °C to 18 °C, the temperature at the end of the inspiratory hose has reached a constant value between 33 °C and 41 °C. An example is shown in Fig. 3.

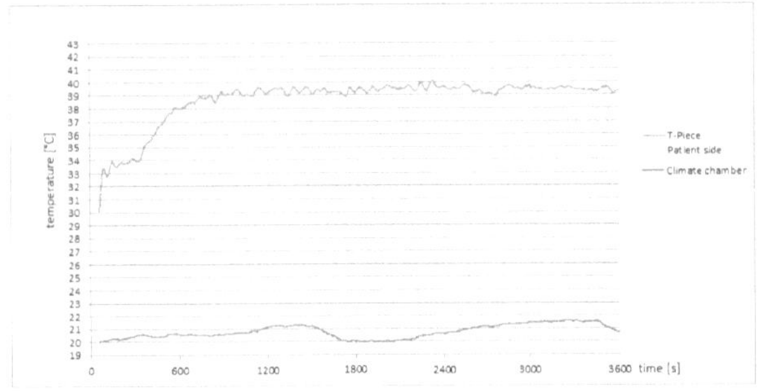

Figure 3: Temperature over 30 min of the temperature at the Y-Piece (top curve) and the environmental temperature (bottom curve). The measurement was started at power up the humidifier and the conventional ventilation.

After 8.88 min, the defined tolerance range is reached and the temperature is almost kept constant. Since no significant temperature fluctuations occur, no condensate is formed. The adult version of the breathing tube system has not passed the change of the constant flow from 10 l/min to 60 l/min.

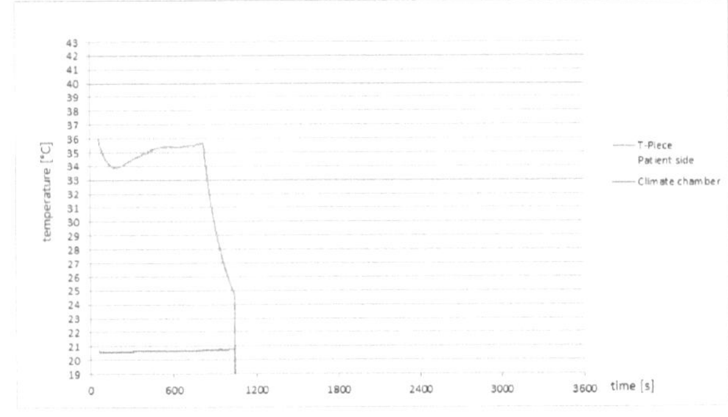

Figure 4: Temperature over 16.65 min of the gas at the end of the inspiratory hose (top curve) and the environmental temperature (bottom curve). The measurement was started at the constant flow change from 10 l/min to 60 l/min.

As shown in Figure 4, the humidifier cannot compensate the high flow change and thus the cooling of the respiratory gas. After 13 minutes, the water chamber temperature is too low and the humidifier switches off. The cause can be found in the larger diameter (19 mm) of the adult hose. The flow change to 60 l/min represents an extreme situation in high-flow ventilation. For future applications, the sudden increase of the flow should be discouraged. An intermediate step of 30 l/min should be installed.

4   Conclusion

In this project, several changes were made to the setup of heated breathing systems with active humidification. The use of an additional expiratory incubator extension leads to a lower formation of condensate and to lower surface temperature of the hose inside the incubator. In order to avoid the risk of premature infants being burned by too high incubator temperatures (> 37 °C), it should be considered to use an additional unheated expiratory incubator extension. In summary, it should be noted that low ambient temperatures, sudden and strong flow changes have the most significant influence on the formation of condensate.

A further consideration is to develop a method that allows the humidification of the air as close to the patient as possible. This would reduce the cooling distances and the formation of condensate.

Acknowledgement

This work has been carried out at the Drägerwerk AG & Co. KGaA, Lübeck and supervised by the Institute of Physics, University of Lübeck.

5   References

[1] Draegerwerk AG & Co. KGaA, "Gebrauchsanweisung - VentStar Helix dual heated (N)," 2018.

[2] Wiegleb, G, "Gasmesstechnik in Theorie und Praxis: Messgeraete, Sensoren, Anwendungen.," *Springer-Verlag*, 2016.

[3] Kuchling,H, "Taschenbuch der Physik," *Carl Hanser Fachbuchverlag*, 2014.

[4] Fisher & Paykel, "Technical Manual - MR850 Respiratory Humidifier," 2005.

[5] Agren J., "Transepidermal water loss in infant born at 24 and 25 weeks of gestation," *Acta Paediatr*, 1998.

[6] IEC 60601-1, "Allgemeine Festlegungen fuer die Sicherheit einschliesslich der wesentlichen Leistungsmerkmale, Seite 172, Tabelle 24," 2012.

Optimization of a Fish Holder for the Immobilization of Zebrafish

S. Melikov [1], M. Frerkes [2], S. Seeger [3], M. Zvolsky [3] and M. Rafecas[3]

[1] Medizinische Ingeniuerwissenschaft, University of Luebeck, salim.melikov@student.uni-luebeck.de
[2] Biophysik, University of Luebeck, maja.frerkes@student.uni-luebeck.de
[3] Institute of Medical Engineering, University of Luebeck, {seeger,zvolsky, rafecas}@imt.uni-luebeck.de

Abstract

The importance of fish, in particular zebrafish, as model organism for medical and biological research is constantly increasing. Positron Emission Tomography (PET) is usually used on laboratory animals to better understand the modelled diseases on the basis of metabolism. Nevertheless, only very few PET studies of fish have been published. Fish immobilization is an indispensable and at the same time demanding step on the way to a PET scan of the fish. This work aims to optimize an immobilization system with the focus on praticability for a damage-free fixation of the test animal. An elastic fish holder was designed and compared to an existing holder in terms of fixation, usability, pressure on the fish and flow simulations. The new prototype allows a smoother and more time-efficient immobilization procedure of the zebrafish. After testing the elastic material for biocompatibility, the new prototype can be used for the immobilization of anesthesized zebrafish.

1 Introduction

The zebrafish (lat. *Danio rerio*) plays an important role in biomedical research. In recent years, the zebrafish has gained increasing interest as a model organism and is now the third most frequently used animal model after the mouse [1]. The suitability of the zebrafish results from its completely sequenced genome, easy genetic modifyability and the financial feasibility due to its high fertility rate as well as fast development. Zebrafish embryos are nearly transparent, so that internal morphology and metabolism can be easily studied e.g., using microscopy or other optical imaging modalities. In contrast an adult zebrafish is opaque and requires alternative imaging techniques.

PET, which is a modality of nuclear imaging, offers the possibility to visualize various metabolic processes using radiolabelled substances. PET imaging of zebrafish could therefore provide a better understanding of certain metabolic processes associated with disease models. While PET is already regularly used on mice, it has hardly been used on fish. The zebrafish are only 4-5 cm long; therefore, conventional small-animal-PET systems with a resolution of about two millimetres are not able to visualize the metabolism of zebrafish sufficiently well. A further difficulty for the realization of PET on zebrafish is the immobilization of the fish in an aquatic environment. Multi-Emission Radioisotopes Marine Animal Imaging Device (MERMAID) is the first project worldwide that aims to design the first dedicated PET prototype for fish, especially zebrafish. The current MERMAID prototype and the existing fish holder are characterized by a flexible configuration which can accommodate fish of different sizes. A first attempt to immobilize the fish was made by embedding the fish in agarose within a water-filled vessel [2]. This work aims to optimize the immobilization approach.

1.1 Agarose-filled Holder

The entire immobilization prototype, shown in Fig. 1 consists of a flow chamber and the fish holder inside, which is filled with agarose. A 1 % agarose solution is prepared and then gelled on the inner surface of the fish holder. In the gelled agarose, a cavity for the fish is then scraped out and the fish is clamped in the agarose between the two halves of the fish holder. Finally, the fish holder is retracted into the

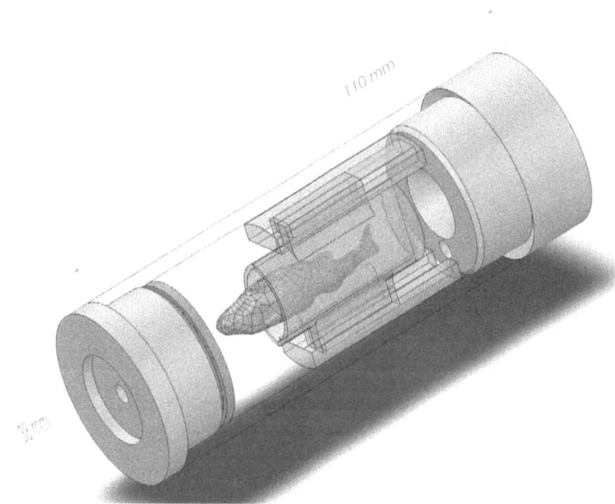

Figure 1: Flow chamber with an immobilized fish phantom in the agarose-filled holder.

flow chamber by means of two rails and fixed in place.

1.2 Requirements and Challenges of the Immobilization

The mentioned approach presents some shortcomings: Firstly, the preparation of the hollow for the fish becomes a demanding and not precisely reproducible task due to the crumbling property of the agarose. Secondly, during the scan, the fish can excrete toxins such as ammonia and possibly radioactive decomposition products of the radiotracer through its skin and gills. A previous work has shown that excreted radioactivity could diffuse into the agarose [2], which in turn could affect the image quality. Essentially the small body size and very sensitive skin of zebrafish pose further challenges. During the image acquisition, the fish must be supplied with oxygen and sedated with an anaesthetic. The latter is necessary because otherwise the fish will be exposed to severe stress during the immobilization process, which could result in physiological changes and risking the animal welfare. Additionally, anaesthesia minimizes the risk of movement-related skin damage and image artifacts. As a consequence of all these facts, the entire animal housing have to be integrated in a pump system supplying fresh water. This is ensured by the already existing flow chamber. Thus, this work focuses on improving of the fish holder, which is located inside the flow chamber and in direct contact with the fish.

Another important requirement is to keep the radiation attenuation by the flow chamber and the fish holder as low as possible. Hence the used materials need to have a low mass density.

2 Material and Methods

The structure of the animal fish holder was designed using SOLIDWORKS CAD software. Due to the very sensitive skin and the mechanical vulnerability of the zebrafish, only soft materials were considered. Since the anaesthetic relaxes the muscles of the fish, the fish holder dimensions were chosen that the entire fish trunk is supported except for the area between the head and the gills.

2.1 Construction

Advanced 3D printing methods allow printing from soft and stretchable materials. The new immobilization structure was made with a Form 2 Desktop SLA 3D-printer [3] from a silicone-like material called "Elastic" [4]. After printing, the mass density of the fish holder taking into account its total mass and volume was determined to be $0.66\,\text{g/cm}^3$.

2.2 Fixation

The suitability of the respective fish holder regarding the fixation was investigated primarily by a visual and tactile test. Special attention was paid to the fact, whether the fish remains in its position after the water is pumped. For this

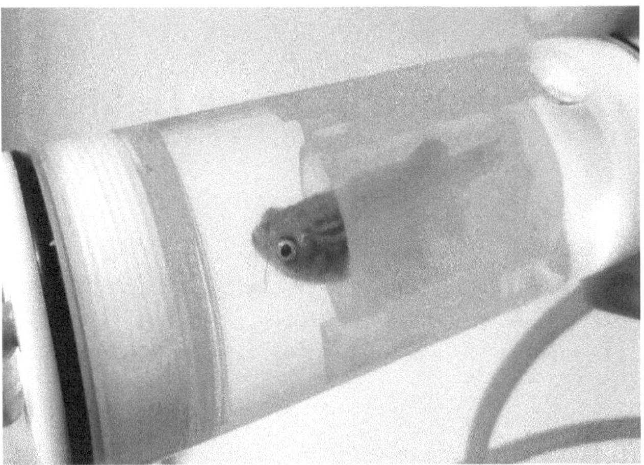

Figure 2: The flow chamber with the elastic holder, including a fixated dead zebrafish in first water pump tests.

purpose, a zebrafish phantom was printed from the "Elastic" material using Form 2.

2.3 Pressure Measurements

In order to be able to relate the forces acting on the fish from the respective fish holder, an experimental setup for force measurement, illustrated in Fig. 3, was implemented. To this aim, a cylinder with an oval base corresponding to the largest sagittal cross-section of the fish was used. The cylinder was printed as two halves and fixed in the respective fish holder. Each of the halves was attached at each end to an approximately weightless thread. Each thread was deflected towards the ground, so that if the weight at the end of the threads was sufficiently high, the two halves of the cylinder are pulled apart. At the end of the threads the weights were increased simultaneously. The resulting gap between the two halves was measured at both ends using a blade gauge with a step of 0.1 mm. In order to consider a possible hysteresis effect, expressed as residual deformation of the fish holders, the gap size was also measured backwards by reducing the weights.

The measured values were then processed to estimate the correlation between the fish diameter and the pressure acting on a fish. In this regard, the gap width for each weight was measured at both ends of the cylinder and then averaged. The diameter of the cylinder was then added to the averaged gap width. Finally, the pressure was estimated by taking into account the area of the exposed cylinder halve. The respective pressure were then visualized as a function of the fish diameter in a plot for each fish holder.

2.4 Flow Simulation

To maintain reliable supply of oxygen and anesthesia, as well as to drain the toxic excretion products, simulations of the fluid mechanics, carried out using the flow simulation feature in SOLIDWORKS, were evaluated. A volume velocity of 10 ml/s was simulated. The simulated data were then visualized by means of the graphical representation

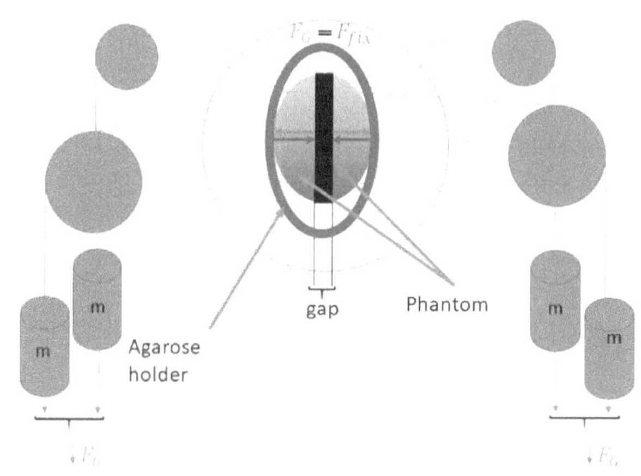

Figure 3: Setup for determination of the pressure acting on the fish by the fish holder.

of the flow velocity. The graphical representation was checked for turbulence, and the flow velocities within the flow chamber for both fish holders were compared quantitatively.

2.5 Practicability

The process of immobilization should be as fast and easy as possible. In order to compare the time required for immobilization using both fish holders, two trials of immobilization were made in each case, and the time required was firstly measured and then averaged. The time for filling the agarose-holder with agarose and its following hardening was measured separately, as no contact with the fish is necessary for this, allowing its preparation to be carried out independently of the time of the image acquisition. A dead zebrafish was provided by the Fraunhofer Research Institution for Marine Biotechnology and Cell Technology (EMB) for this experiment.

3 Results and Discussion

The visual and tactile tests have shown that both fish holders provide sufficient fixation of the zebrafish in the pumped flow chamber. The silicon fish maintained for a period of 10 minutes and showed no movement or vibration due to the water flow or oscillation of the pumping system, even when the pump was operated at its full power of 10 ml/s. The simulations (see Fig. 4) showed that none of the fish holders has any significant adverse effect on fluid flow dynamics. Small local turbulences on the surfaces of the fish holders and above the pressure openings cannot be excluded, but are very unlikely to cause a problem for the outflow of the excreted products. However, in the simulation with the elastic holder, there were some small turbulences just in front of the fish phantom, but they have not been observed in real flow test. In a real study with a living fish, the turbulences might distort the correct supply of oxygen and anesthesia, and higher concentration of both could be required.

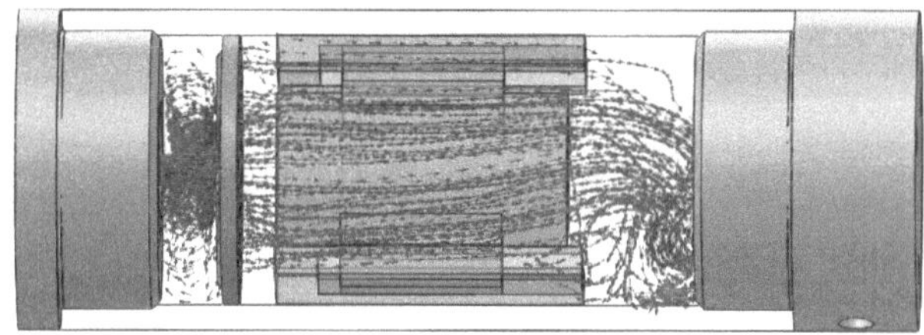

Figure 4: Flow chamber with the agarose holder and flow simulation at 10 ml/s. The direction of the arrows represents the direction of the local velocity vector. The flow within the cylinder is laminar.

Fig. 5 and 6 depict the approximate pressure exerted on the fish by the respective fish holder as function of its maximum diameter. As the amount of weights needed to create the smallest measurable gap in the fish holders varied, the plots do not contain the same number of data points. The maximum deviation of the weights of $\pm 5\%$ was considered as a relative error for the pressure. The error of the diameter corresponds to the thickness of the blade gauge (± 0.1 mm). In order to interpret these results, we have considered the following facts: zebrafish are bred in containers at a depth of up to 25 cm [5]. Bodies submerged in fluids experience hydrostatic pressure that increases approximately by 1 bar $= 1 \times 10^5$ Pa per meter. This would imply that zebrafish can tolerate a pressure of up to 2500 Pa. This reference value is shown as a horizontal line in Fig. 5 and Fig. 6. The figures show that the pressure exerted by the agarose holder is almost a factor two higher than the pressure exerted by the elastic holder. Fig. 5 shows that the measurement in the reverse direction has provided a constant value for the agarose holder. This can be explained by the fact that the agarose has escaped through the openings of the fish holder due to the pressure applied to it. The elastic holder, on the other hand, shows almost no hysteresis effect, except for low diameters. This effect, however, could also be attributed to measurement uncertainties.

Due to the overall higher pressure exposure by the agarose holder, the referenced pressure is already exceeded for phantoms with 7.8 mm diameter. However, in the case of the elastic holder, it occurs above a diameter of 8.9 mm. In summary, it can be concluded that that for fish of the same size, the agarose holder exposes the fish to a significantly higher pressure than the elastic holder.

With regard to the time required for the immobilization, time measurements were carried out. Filling the agarose holder with agarose solution lasted a total for 26 minutes. The most time-consuming and challenging process was cutting out a suitable cavity for the fish in the agarose. Immobilization using the agarose holder took on average 3 minutes, while immobilization using the elastic holder took 1.5 minutes. Obviously, the latter is a more time-efficient method.

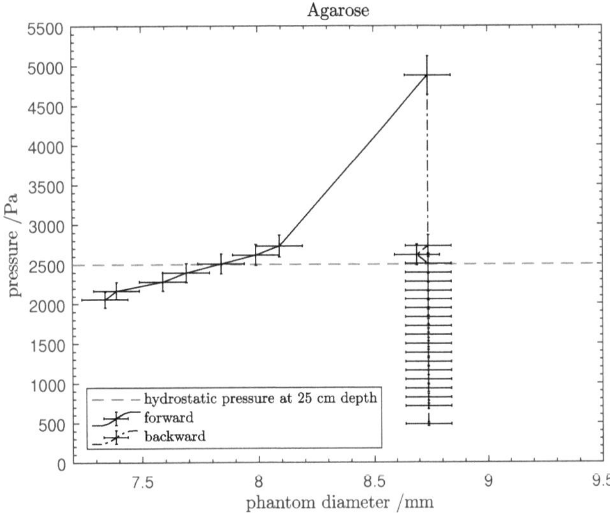

Figure 5: Pressure acting on the phantom as a function of the phantom diameter for the agarose holder.

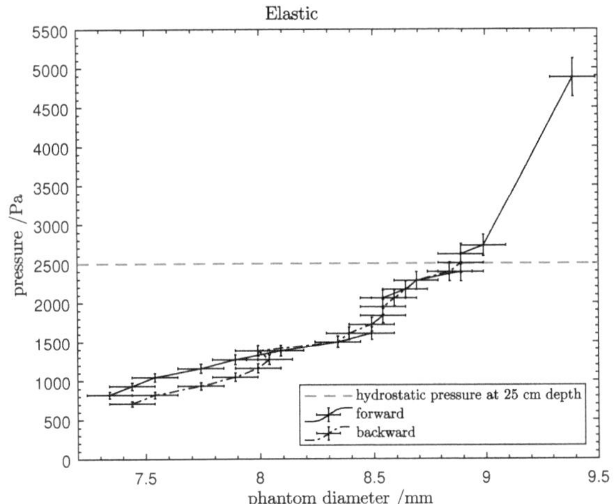

Figure 6: Pressure acting on the phantom as a function of the phantom diameter for the elastic holder.

4   Conclusion

In-vivo nuclear imaging of zebrafish requires the immobilization of the animal. Within the frame of the MER-MAID project, different methods for this immobilisation have been developed. In this study, an agarose-based holder was compared to a 3d-printed elastic holder in terms of usability, pressure on the fish and water flow behaviour. It is shown that the elastic holder offers a more time-efficient and smooth immobilization of the zebrafish than the agarose holder.

The next step should be to check the elastic material for biocompatibility and replace the material if needed. The agarose holder could be improved significantly by a more practical and precise method of forming the fish cavity. The findings of this study will help to further optimize the immobilization procedure before performing animal trials.

Acknowledgement

The work has been carried out at the Institute of Medical Engineering, University of Lübeck.

Special thanks are dedicated to Dirk Steinhagen and Cindy Läken, who supported us in the technical realization of this work. We would also like to thank the Fraunhofer EMB for providing the dead zebrafish.

References

[1]   PS Narayana, D Varalakshmi, T Pullaiah. *Research methodology in Zoology*. Scientific Publishers, 2018.

[2]   J Burbat. "Development of methods and processes for nuclear in-vivo-imaging of zebrafishes". Bachelor's Thesis. University of Luebeck, 2019.

[3]   Formlabs Inc. *Form 2: Affordable Desktop SLA 3D Printer*. [last accesses on 2020-01-14]. URL: https:// formlabs.com/3d-printers/form-2/.

[4]   Formlabs Inc. *MATERIAL DATA SHEET,FLDUCL02*. [last accesses on 2020-01-14]. URL: https://formlabs-media . formlabs . com / datasheets / Durable _ Resin _ Technical.pdf.

[5]   R Barney and J Maggy. "Guidance on the housing and care of Zebrafish". In: *Southwater: Royal Society for the Prevention of Cruelty to Animals* (2011).

Development of a mobile sensory device to trace ambient and plasma treatment conditions

Ihda Chaerony Siffa [1], Torsten Gerling [2,3,4], and Robert Wendlandt [5]

[1] Biomedical Engineering, Luebeck University of Applied Sciences, ihda.chaerony.siffa@stud.th-luebeck.de
[2] Leibniz Institute for Plasma Science and Technology (INP), Greifswald, gerling@inp-greifswald.de
[3] Centre for Innovation Competency ZIK plasmatis, Greifswald
[4] Competency Centre for Diabetes KDK Karlsburg, Karlsburg
[5] Clinic for Orthopedic and Trauma Surgery, University Medical Center Schleswig-Holstein, robert.wendlandt@uksh.de

Abstract

Possibilities of using plasma on an outpatient basis are being explored, for this purpose a mobile device that can house plasma devices is being developed. An additional mobile sensory device for recording of the ambient and plasma treatment conditions is needed. A prototype of the mobile sensory device was developed with Raspberry Pi 3 model B as the main processing unit, a sensor for the measurement of ambient temperature and humidity, and a microphone to detect and trace on/off time of the plasma device as well as a detection algorithm for this. A touchscreen display is used for display and input mechanism—for this, a GUI system was developed. Collected data are compiled into a report which can be viewed on the display or exported to a USB-stick. A sound detection reliability testing at 5 cm distance showed that 1 sensitivity, 0.971 specificity, and 0.986 accuracy were achieved.

1 Introduction

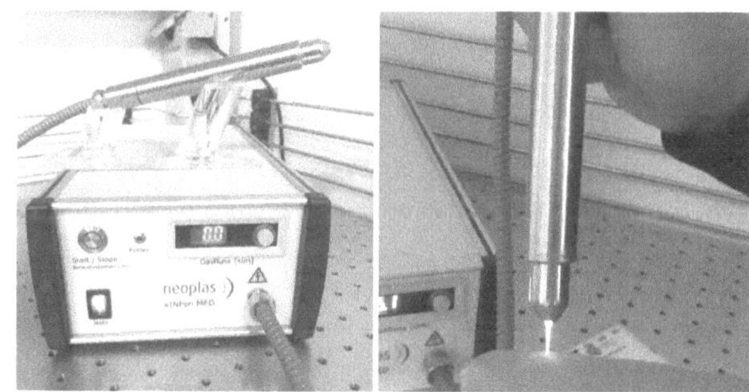

Figure 1: kINPen® MED device (left) and its plasma jet (right) in contact with skin surface.

Up to this day, plasma is a subject of continuous research and applied in many different fields, one of them is medicine. Plasma medicine, unlike other applications of plasma generally doesn't use high temperature plasma, rather cold atmospheric plasma (CAP). Low temperature plasma like CAP can be achieved by applying electrical energy to a not directly biologically effective gas (argon, helium, oxygen, nitrogen, air, or mixtures thereof). The main part of the applied energy goes into the production of energetic ('hot') electrons whereas the majority of gas atoms, ions and molecules remain in a low-energetic state resulting in a low plasma temperature. Additionally, excitation processes result in emission of electromagnetic radiation (UV/VUV, visible light, IR/heat, electric fields) [1]. CAP can be applied in therapeutic purposes, such as in wound healing and in treatment of infectious skin diseases due to its interaction with biological tissues [2]. The application of CAP in therapeutic purposes is generally achieved with dielectric barrier discharge (DBD), and atmospheric pressure plasma jet (APPJ). Plasma devices like kINPen® MED (Leibniz Institute for Plasma Science and Technology (INP-Greifswald)/neoplas tools GmbH, Greifswald, Germany)—an APPJ, and PlasmaDerm®VU-2010 (CINOGY Technologies GmbH, Duderstadt, Germany)—a DBD source, have been CE-certified in Germany and used clinically for the past few years [2]. Application of CAP for wound treatment is currently the most common use of these plasma devices, which has shown its effectiveness [2]. Researchers and engineers at INP-Greifswald are developing the possibilities of using plasma on an outpatient (mobile) basis. For this purpose, a mobile device is being developed that can house both devices (kINPen® MED and PlasmaDerm®VU-2010). *In addition to that*, a mobile sensory device to detect and record ambient conditions such as humidity, temperature and plasma source activity is needed to help in assessing the effectiveness of the treatment. This paper is focused on the development of a prototype of a mobile sensory device for kINPen® MED device. Fig. 1 shows the kINPen® MED device and its plasma jet application on skin surface.

2   Material and Methods

2.1   Design and construction of the prototype

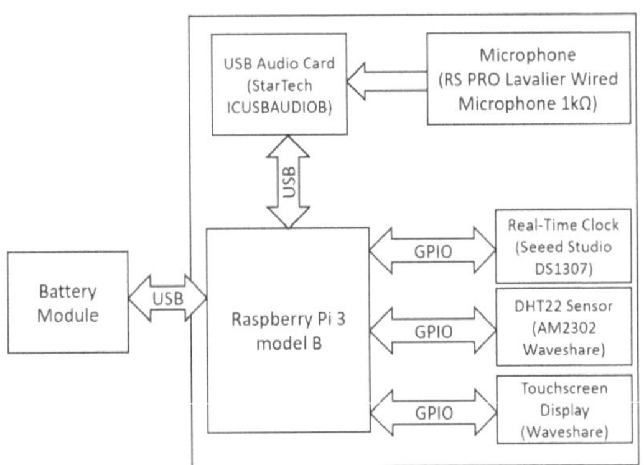

Figure 2: Block diagram of the hardware system.

The need of a small and lightweight device is crucial for high mobility. Raspberry Pi 3 model B was chosen as the main processing unit of this system due to its size, price and ease of use. A condenser microphone is used to detect the sound signal produced by the plasma source and interfaced with Raspberry Pi through a USB Audio Card. A DHT22 sensor was chosen for measurement of ambient temperature and humidity due to its simple interface protocol with Raspberry Pi and its compactness—two measurements in one sensor package. A display and user interface module is also needed for operation and control of the device, a 5" HDMI touchscreen display module was chosen to keep the design as small as reasonably practicable. A battery module was also considered to further increase the mobility. As logging of treatment and ambient conditions is part of the objectives, the correctness of logging date and time is imperative, to ensure this, a real-time clock module was also added to the system. The block diagram of this system is shown in Fig. 2. Virtual components were structured carefully in Autodesk Inventor® to help with estimating the size of the casing. The casing was designed accordingly, encapsulating the virtual components, then 3d-printed and assembled together with the components (see Fig. 3). The

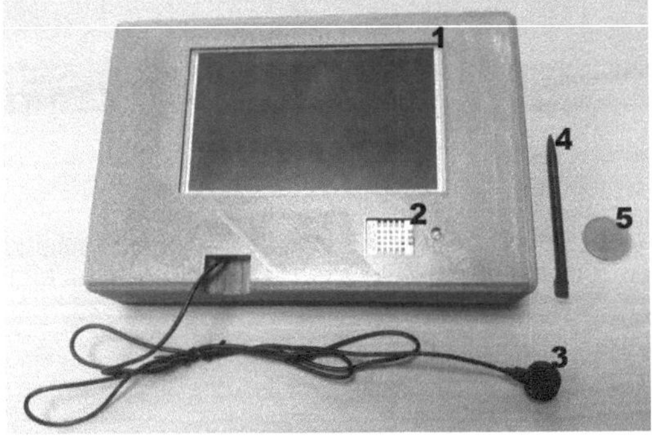

Figure 3: Constructed prototype: **1.** Touchscreen display; **2.** DHT22 sensor; **3.** Microphone; **4.** Stylus; **5.** One euro coin (for scale).

microphone is planned to be attached to the kINPen® MED plasma source or placed closely to it for optimal reading of the sound signal. Fig. 4 shows a heuristic explanation of the working principle of the prototype.

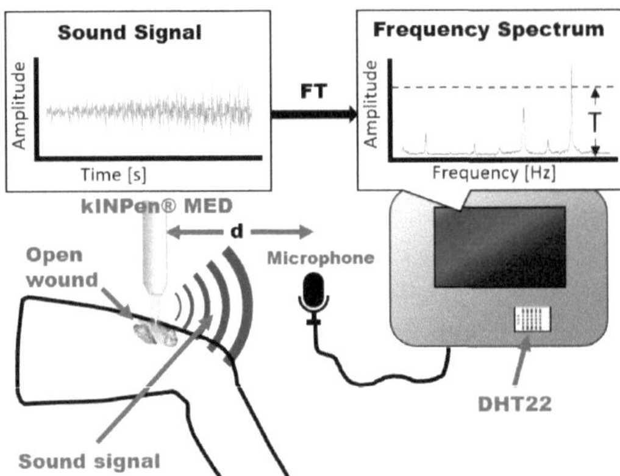

Figure 4: Diagram of the working principle of the prototype.

2.2   Software Overview

The software consists of several programs written in Python (v3.7.3), utilizing open-source libraries like NumPy (v1.18.1), SciPy (v1.4.1), and Kivy (v1.11.1). The programs were developed in a Raspberry Pi linux-based operating system (OS)—Raspbian Buster 10. The programs are managed by one main program which takes the advantage of multiprocessing capability of the OS, where multiple programs can be executed concurrently and scheduled properly. A graphical user interface (GUI) system was also developed using Kivy library and embedded to the main program. Since the display also provides touchscreen capability, the main program also takes care of the input mechanism. Inputs from the user will trigger the execution of the sub-programs. In addition to that, a plasma device sound detection program was also conceived which is explained in the next subsection.

2.3   kINPen® MED sound detection

The kINPen® MED is a CAP-argon based plasma jet device. To generate its plasma, a sinusoidal voltage (2–6 kVpp) with a frequency of 1.0–1.1 MHz is applied to the feed gas (argon gas) and pulsed with a frequency of 2.5 kHz (plasma on to off ratio = 1:1) [3]. We assumed the 2.5 kHz pulse also corresponds to a sound event of the same frequency. Given the distinct sound signature, a detection algorithm can be developed. The idea is to detect the frequency specific to the one generated by kINPen® MED plasma source. Frequency decomposition of sound signals can be done generally with Fourier transform. Since the process is done in a digital system, discretization of the sound signals is needed. In this system, the sound signal is sampled at 44.1 kHz which is a common sampling frequency f_s for sound signals [4]. Discrete Fourier transform

(DFT) can be used for discrete signals as shown in (1).

$$X_k = \sum_{n=0}^{N-1} x_n \cdot e^{-\frac{2\pi jkn}{N}} \tag{1}$$

where $X_k \in \mathbb{C}$ is the Fourier amplitude at frequency bin k, $N \in \mathbb{N}$ is number of samples, and x_n is n-th sample value (amplitude of the sample).

The algorithm for kINPen® MED sound detection is structured as follows:

Input: an array of sampled sound signal $\mathbf{x} = \langle x_0, ..., x_{N-1} \rangle \in \mathbb{R}$.

Output: a return value P is an integer value of either 1 or 0. Where 1 indicates the plasma device is on and 0 is off.

Set parameter of detection: length of recording time $t_r \in \mathbb{R}$, target detection frequency $f_d \in \mathbb{N}$, sampling frequency $f_s \in \mathbb{N}$ (following Nyquist-Shannon sampling theorem, $f_s \geq 2f_d$ [5]), and threshold value $T = [0, 1] \in \mathbb{R}$.

Main loop of the algorithm:

1. Record sound signal for t_r long. Sampled sound signal $\mathbf{x}$ is obtained, with $N = t_r \cdot f_s$.

2. Compute the DFT for $\mathbf{x}$ at all available frequencies $\mathbf{k} = \langle k_0, ..., k_{f_s-1} \rangle \in \mathbb{N}_0$, complex number results are obtained $X_k = a_k + b_k j$, with $a, b \in \mathbb{R}$. Compute the absolute value of X_k for each result with $A_k = |X_k| = |a_k + b_k j| = \sqrt{a_k^2 + b_k^2}$. An array of frequency amplitude in frequency domain is obtained $\mathbf{A} = \langle A_0, ..., A_{f_s-1} \rangle \in \mathbb{R}$.

3. Min-max normalization of $\mathbf{A}$ values with

$$Z_k = \frac{A_k - \min(\mathbf{A})}{\max(\mathbf{A}) - \min(\mathbf{A})} \tag{2}$$

where Z_k represents k-th normalized frequency amplitude value. An array of normalized frequency amplitude values is obtained $\mathbf{Z} = \langle Z_0, ..., Z_{f_s-1} \rangle \in \mathbb{R}$, $Z_k = [0, 1] \in \mathbb{R}$.

4. Detection of the target frequency

$$P(Z_{f_d}) = \begin{cases} 1 & if \ \ Z_{f_d} \geq T \\ 0 & if \ \ Z_{f_d} < T \end{cases} . \tag{3}$$

5. **Return** the value of $P(Z_{f_d})$.

The implementation of this algorithm was done in Python with *scipy.fftpack* to compute the DFT. In addition to sound detection, the system should also be able to track how long the plasma device is on. This can be done with profiling the detection program for each loop. The computation time of each loop is stored, and the total detection time is the summation of computation times of all loops where the sound is detected.

3 Results and Discussion

3.1 GUI and sensor integration

A GUI system which is integrated with the whole system was developed. The GUI consists of several screens, which some of them are interfaced with sub-programs such as ambient condition detection program, data compilation program, and report generation program. The main screen

shows the latest humidity and temperature readings confirming the communication between the sensor module and the system is established, and additionally, current time and date is also shown (Fig. 5. **a**). From the main screen, the user can navigate to logging of plasma treatment (sound detection), collection of last reports, or to a wound assessment screen using the provided buttons. A turn off button is provided to terminate the system safely. A German language version of the GUI is also provided. Virtual keyboard provided by Kivy is also added enabling the user to input information of the patient (Fig. 5. **b**). After the treatment

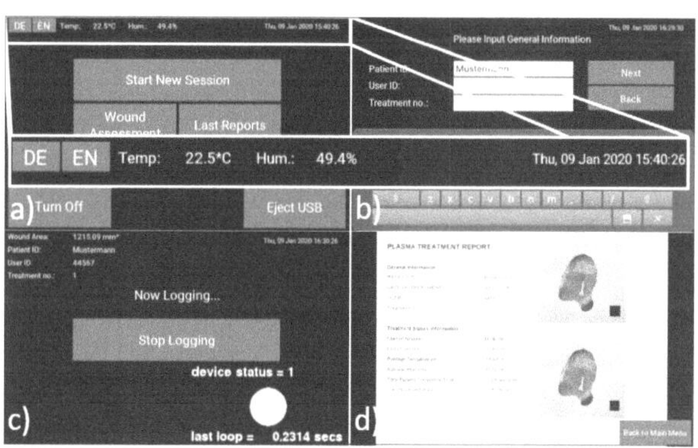

Figure 5: Some examples of the GUI screens. **a)** Main screen of the GUI; **b)** Patient information input screen with a virtual keyboard; **c)** Logging/detection screen; **d)** Report display screen.

is done, the user can choose whether to generate a report, which summarizes general information, average temperature and humidity, and the length of plasma treatment (Fig. 5. **d**). The report is saved in the system storage and can be viewed on the display or exported to a USB-stick. The current software is still in alpha stage (version 1.0.0-alpha), other features like wound assessment (2d wound area measurement) and integration with 3rd party wound assessment device, are also being developed, which are reserved for the next iteration of the software.

3.2 Testing of sound detection

kINPen® MED sound signal was recorded continuously for 1 second long, at 44.1 kHz sampling frequency, and 10 cm away from the sound source. The DFT of this signal was computed using an fft function from SciPy library and then plotted from 0 to 5100 Hz frequency bins (Fig. 6). The plot shows that a peak frequency happens at around 2550 Hz which is assumed to be the sound frequency produced by kINPen® MED. A simple indication GUI was added to show when a detection happens (see Fig. 5. **c**). As for an initial reliability testing, the detection system was tested in a noisy condition. A noise source, an online newscast streamed using Samsung Galaxy J7 Prime phone at maximum volume with an average loudness of 49 dB, was added into the testing environment. A kINPen® Med plasma source and the noise source were placed at the same distance from the microphone. Several measurements were

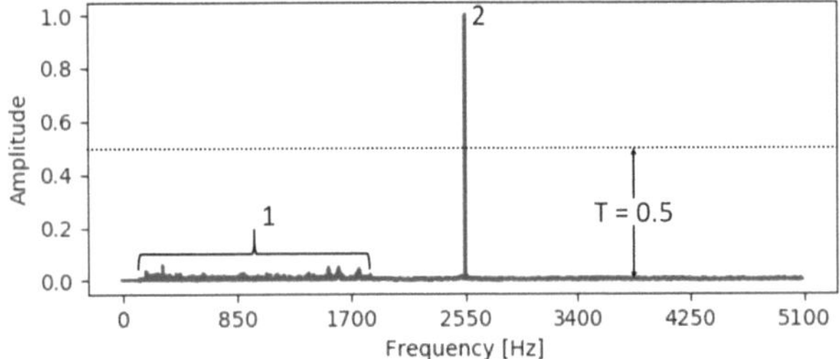

Figure 6: Fourier analysis of kINPen® MED sound signal ($t_r = 1s$, $f_s = 44.1\,kHz$) shows the frequency amplitudes of the recorded signal, with **1** being the ambient noise and **2** the assumed kINPen® MED sound signal peaked above T = 0.5.

conducted at different distances d from 5 cm to 30 cm with 5 cm increment, and 1 m and 2 m for extreme case. Automatic measurement time was used (programmed measurement time, measurement stops automatically after set-time) and each measurement was conducted for 30 seconds. A threshold value of 0.5 was used. The sensitivity/True Positive Rate (TPR), specificity/True Negative Rate (TNR), and accuracy (ACC) of the detection were calculated, and plotted (log scale x-axis) as shown in Fig. 7. We achieved the

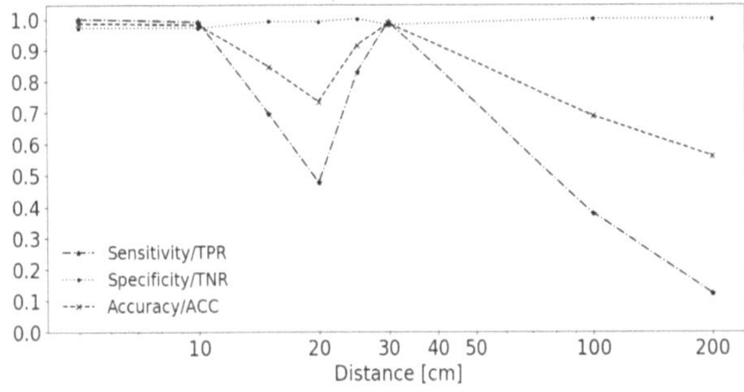

Figure 7: TPR, TNR, and ACC of the detection at different distances with a presence of 49 dB noise from a newscast.

best detection performance at 5 cm measurement distance with TPR of 1, TNR of 0.971, and ACC of 0.986. We assumed the performance of the detection degrades as the measurement distance increases which was confirmed until 20 cm measurement distance. However, the performance of the detection increased again at 25 cm and 30 cm. We assume this is due to optimal interference with the noise signals, though a further investigation is needed. At 100 cm and 200 cm measurement distances, the detection performance followed back the decreasing trend.

3.3   Device classification consideration

In its completion, the mobile sensory device will be part of a mobile housing device that can house clinically certified plasma devices to be used together for an outpatient basis plasma treatment. *When integrated together with the mobile housing device* as one mobile device, this device as well as the software within is projected to be a medical device

and hence subjected to the medical device regulation of a medical device class I [6].

4   Conclusion and next steps

The proposed prototype presented in this paper has shown its potential for ambient and plasma treatment parameter detection. The logging of ambient and plasma treatment conditions by this device may also help in assessing the effectiveness of the treatment. The same sound detection algorithm can be used with different plasma source devices, with minimal modification due to their specific pulsing frequencies. Though the detection of kINPen® MED sound signal has shown to be successful at 5 - 10 cm measurement distance, a further evaluation is needed to address the performance of the detection with the presence of different kinds of noises, and to find the best threshold value for better reliability. An integration with a 3rd party wound assessment system is also needed for a better treatment work-flow, which is part of the ongoing development. Field testing is the next natural step of the development to evaluate the user experience and to find a better treatment work-flow.

Acknowledgement

The work has been carried out at and funded by INP-Greifswald and supervised by Luebeck University of Applied Sciences. We would like to thank Hannes Bendt, Michael Timm, and others at INP-Greifswald for their assistance.

5   References

[1] K-D. Weltmann and Th. von Woedtke, *Plasma medicine—current state of research and medical application.* Plasma Phys. Control. Fusion, 59 014031, 2017.

[2] T. Bernhardt, M. L. Semmler, M. Schäfer, S. Bekeschus, S. Emmert, and L. Boeckmann, *Plasma Medicine: Applications of Cold Atmospheric Pressure Plasma in Dermatology.* Oxidative Medicine and Cellular Longevity, vol. 2019, Article ID 3873928, 10 pages, 2019.

[3] S. Bekeschus, A. Schmidt, K-D. Weltmann, and Th. von Woedtke, *The plasma jet kINPen–A powerful tool for wound healing.* Clinical Plasma Medicine. Volume 4, Issue 1, Pages 19-28, 2016.

[4] J. Watkinson, *The Art of Digital Audio, 2nd edition.* Focal Press, Oxford, 1994.

[5] J. Blackledge, *Digital Signal Processing (Second Edition).* Horwood Publishing, vol: ISBN: 1-904275-26-5, 2006.

[6] Official Journal of the European Union, *REGULATION (EU) 2017/745 OF THE EUROPEAN PARLIAMENT AND OF THE COUNCIL of 5 April 2017.* Available: https://eur-lex.europa.eu/homepage.html [last accessed on 2020-02-05].

Towards measurement of the deformation energy of softpellets: Design and Development of a Micro-Force-Spectroscope

Christian Pottkämper [1], Christian Nehls [2], Thomas Gutsmann [2]
[1] Medical Engineering Science, Universität zu Lübeck, christian.pottkaemper@student.uni-luebeck.de
[2] Biophysics Group, Research Center Borstel, {cnehls, tgutsmann}@fz-borstel.de

Abstract

In the case of lung diseases, it is advisable to take medication by inhalation. This allows the active ingredients to reach their destination directly and to act locally. In the treatment of acute tuberculosis, the usual therapy is oral administration. In order to enable inhalation application, the active substance is being processed into so-called softpellets. These can be crushed by using a suitable device, such as the Turbohaler®, and will be absorbed in the lungs by inhalation. To measure the energy required to use the Turbohaler efficiently, the deformation energy of the pellets must be measured. Since it was not possible to measure with an atomic force microscope (AFM), a micro-force-spectroscope has been designed and constructed for this purpose. A subsequent validation of the device showed that the design could allow force measurements. However, further adjustments are pending for this purpose and a suitable filter needs to be selected.

1 Introduction

In the treatment of many lung diseases, Dry Powder Inhalers (DPI) are used increasingly in practical application. This is caused by a high-dose and successful administration via the respiratory tract. A cost-effective and simple method to create this kind of agglomerated micronized powder is the production of softpellets. Through mechanical effects, for example a vibrating chute, these particles become a larger structure in the micrometer range consisting of pure active substance. At this size, a conventional capsule inhaler that punctures capsules is unsuitable because the softpellets would not be released. To ensure precise dosing of the powder, the softpellets can be crushed by using a Turbohaler. The deformation energy of the softpellets must be known, so that the DPI system can enable efficient disaggregation. The correct dosing is carried out by rotating the dosing unit in the storage unit towards a pressure plate below the dosing unit. The material discharge of the dosing system is exclusively volume-related and thus quantity-related and is based on the volumetric principle [1].

Active tuberculosis is a lung disease for which inhalation application of medication is recommended. The standard therapy lasts at least six months and the administration of the active ingredient is taken orally [2]. Christian Etschmann, from Department of Pharmaceutics and Biopharmaceutics, Kiel University, presents a method in which the antibiotic rifampicin can be inhaled in the case of acute tuberculosis. For this purpose the active substance is agglomerated to produce softpellets consisting of the pure active substance.

To determine the deformation energy of these manufactured softpellets, an Atomic-Force-Microscope can be used to ob-

tain a force-distance curve [3]. However, since the tip of the largest cantilever available for the AFM is too small, only a local area of the surface was dented. The image is shown in Fig. 1.

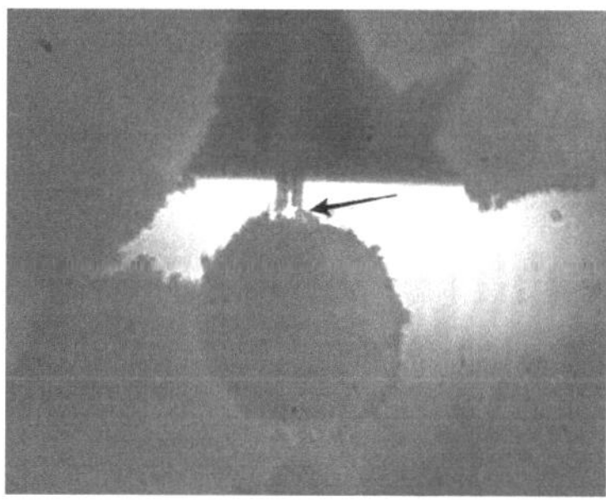

Figure 1: AFM-measurements on the softpellet. The arrow marks the point at which the tip of the cantilever presses through the softpellet locally.

The aim of the present project is to construct a device that can measure the deformation energy of the whole softpellet. For this purpose the measuring should be in the micro-meter range. Afterwards the device should be validated to ensure a subsequent force-measurement on the softpellets.

2 Material and Methods

In order to fulfill the requirements for an experimental setup metioned in the previous, it is necessary to select suitable materials and components. A further criterion is to keep the construction as simple as possible while allowing a mobile application. In addition, the construction should be relatively inexpensive and at best be realized with the available tools.

2.1 Material

When selecting the materials, the basic framework of the device was produced with a 3D-printer. The used material is polylactide (PLA), which was available as filament with a diameter of 1.7 mm.

To meet the requirements for generating a force-distance curve in the desired value range, the device must consist of the following components. On one hand, a tactile sensor, constructed in the manner of a cantilever of an AFM, which is pressed onto the sample (Fig. 4A). With a suitable displacement sensor (Fig. 4B), the force can be deduced from the deflection of the cantilever. On other hand, a stepper motor is required so that an axial movement including the mounting of the touch sensor can take place (Fig. 4C). The smallest step should be in the micrometer range to ensure precise measurement. The position of the cantilever on the sample can be checked by observing with a suitable microscope (Fig. 4D).

2.2 Construction

Due to the limited print size, the entire structure had to be cut into pieces so that they could be printed. The challenge here was to create a construction that was stable enough to withstand the ambient vibration and the movement of the components. At the same time the design had to be as compact as possible. Figure 2 shows the final prototype of the basic framework. In addition, it was developed completely by the *Autodesk Fusion 360 Software* and printed with the *Prusa i3 MK3 3D* Printer.

Figure 2: Final prototype of the framework, divided into the individual parts.

2.2.1 Touch sensor and displacement measurement

To exert pressure on the softpellets to measure the deformation energy, a cantilever with a tip design was developed. The cantilever is made of copper and measures 0.1 mm (thickness t), 8 mm (width w) and 42 mm (length l). At the end, a tip has been attached vertically downwards, which has a planar side with a diameter of 0.5 mm and consists of polylactides. To the upper side, a ferromagnetic core is attached. To measure the deflection of the end of the

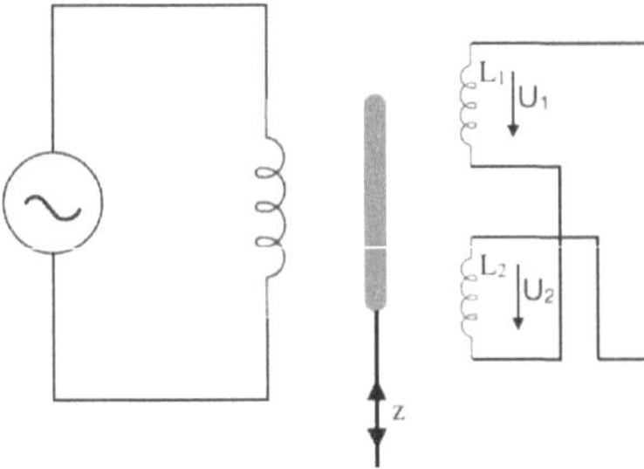

Figure 3: The functional principle of an LVDT. Left is the main coil, right the two secondary coils. When the iron core moves parallel z, a voltage is induced depending on its position to the secondary coils..

cantilever Δz, a Linear Variable Differential Transformer (LVDT) was installed above it. This is an analogue sensor for displacement measurements. As shown in Fig. 3, it consists of a primary coil on one side and two secondary coils on the other side. If the ferromagnetic core is inserted into the interior of the tubular coil, a voltage is induced depending on its position relative to the secondary coils [4]. The bending of the cantilever is:

$$\Delta z = \frac{4}{E_w} \left(\frac{l}{t}\right)^3 F_n$$

$$\Rightarrow k_n = \frac{F_n}{\Delta z} = \frac{Ew}{4} \left(\frac{t}{l}\right),$$

(1)

where k_n is the spring constant on a rectangular spring beam. The angle $\Delta\alpha$, around which the bar end moves, is calculated from the theory of elasticity [5]:

$$\Delta\alpha = \frac{3\Delta z}{2l} = \frac{6}{Ewt} \left(\frac{l}{t}\right)^3 F_n$$

(2)

$$\Rightarrow F_n = \frac{3\Delta z}{2l} \frac{Ewt}{6} \left(\frac{t}{l}\right)^3 .$$

Taking into account the modulus of elasticity E, the effective force F_n can now be concluded.

2.2.2 Implementation of the axial movement.

The movement of the cantilever and the LVDT onto the sample is done by a holder connected to a stepper motor (NEMA17) with an step angle of 1.8 degrees. To achieve even smaller steps, a motor driver (DRV8825) was additionally installed. This enables a control of 1/32 steps per motion unit. Furthermore, a rail guide with a ball screw was connected so that the movement can take place in a vertical direction. The existing spindle pitch is 5 mm. In total, this results in a full step of 25 micrometers and a 1/32 step of 0.78 micrometers. With a softpellet size of 500 micrometers there is a maximum value range of 641 steps.

Figure 4: Final device with all installed components. A is the holder for the cantilever and B the LVDT. C is the vertical movement apparatus by means of a stepper motor, D the USB microscope camera and E the cross stage for the slide.

2.3 Other components

The finished device also consists of other components, as can be seen in Fig. 4. A USB microscope camera is installed to ensure that the sample can be aligned correctly and a cross table allows the movement of the slide. In addition, the device has a cross stage for the specimen slide, an auxiliary rail for the motor and push buttons that define the maximum range of movement. The electrical parts are controlled by an Arduino Atmega 2560. Because the internal A/D converter has only a value range of 10 bits, an i2c chip (ADS1115) to extend the range to 16 bits.

2.4 Performing the validation measurements

To validate the accuracy, the components can measure data values in relation to each other, two experiments were carried out, as shown in Fig. 5. In the first one the soft iron core was fixed on the slide. Afterwards, the LVDT was driven onto the core via the stepper motor until a certain value has been reached. Subsequently, the device should remain in this position. In the second experiment the core was attached to a cantilever and aligned in the LVDT. Then

a solid material, a block of polylactide, was placed on the slide about 0.45 mm below the spring bar. The LVDT was moved towards the block and should also hold its position after reaching a certain distance.

3   Results and Discussion

The aim is to give an estimation whether the device in its current state allows measurements on the softpellets or whether it needs to be revised. Fig. 5a shows a measurement in which the LVDT detects a certain length of the fixed ferromagnetic core and has to hold the position at a depth of 0.75 mm. On the Y-axis the displacement measurement at the LVDT is plotted and on the X-axis the relative axial movement of the stepper motor is depicted.

Please note that for a better representation the axial movement (X-axis) reaches to 0.75 mm. Afterwards the stepper motor remains in its position.

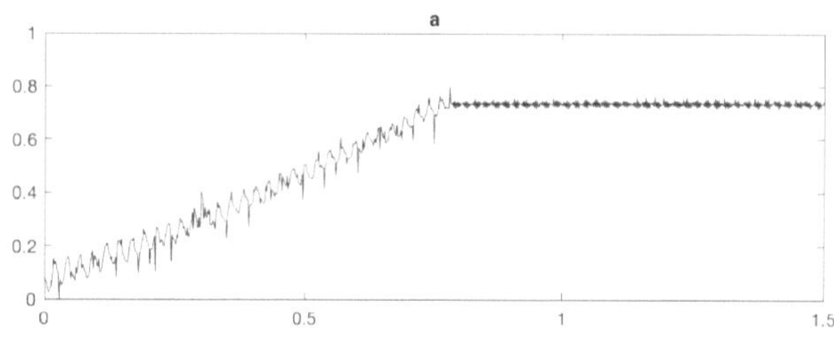

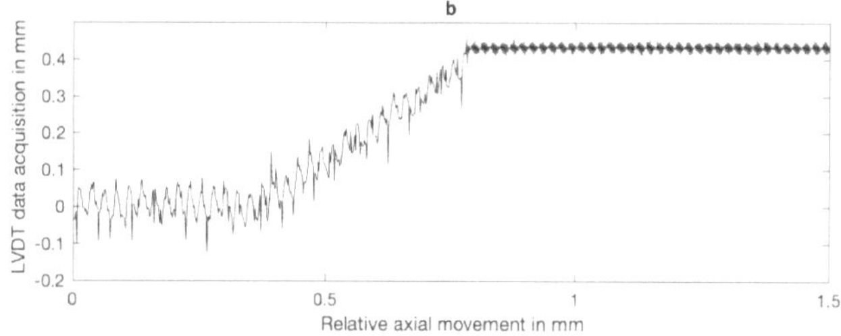

Figure 5: Displacement measurements on the LVDT as a function of vertical movement by the stepper motor. The measurement at the fixed core is shown in (a). In comparison, the measurement with attached cantilevers is shown in (b).

Thereby a reference measurement is to be created to evaluate the precision of the displacement measurement at the cantilever (Fig. 5b). Here it can be seen that a higher noise is present before the spring beam reaches the polylactide block. The deflection of the cantilever and the remaining in one position can indeed be measured. As shown in Fig. 6a and b, to compare the results statistically, the two measurements were each performed 10 times and the data was then averaged. In order to evaluate the noise level and the accuracy of the cantilever (b), the standard deviation during movement and while holding the position, was calculated. The resulting total standard deviation without cantilever is +/- 0.234 mm while the stepper motor is moving. If the position is held, it has a value of +/- 0.0138 mm. Comparing the data with cantilever measurements shows that the values are relatively similar. These results have an overall higher

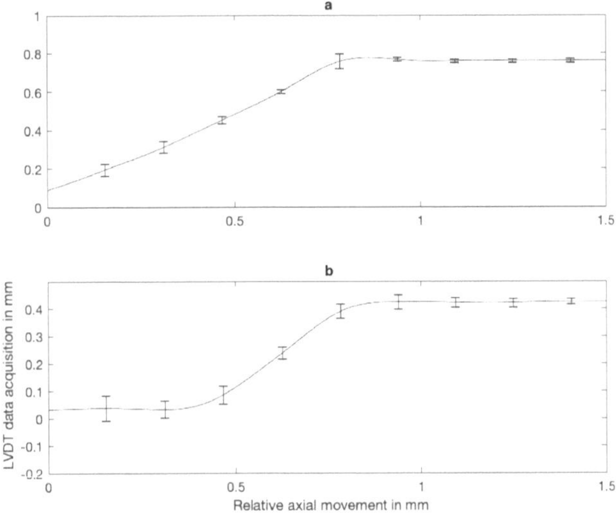

Figure 6: The averaged measurement data (10 runs) without cantilever (a), with cantilever (b) and their corresponding standard deviations displayed as vertical bars.

averaged interval range of +/- 0.0278 mm during movement and +/- 0.0218 mm in the rest position. In addition, isolated distinct peaks can be recognized in the measurements.

4 Conclusion

In summary, the construction could be suitable for measuring the force on the softpellet. However, the resulting measuring error, caused by the standard deviation of about +/- 22-28 micrometers, is still relatively high and has to be reduced. For this purpose further adjustments will be useful. Firstly, a stepper motor with a smaller step angle should be used. This would allow a larger measuring range. Furthermore, several averaged measurements between steps would lead to a more accurate result. Finally, a filtering should be considered. Here, for example, a suitable low-pass filter would be a good choice, since higher frequencies should be eliminated accordingly.

Acknowledgement

The work was developed in cooperation with the Biophysics Group at Reseach Center Borstel, Leibniz Lung Center and the Department of Pharmaceutics and Biopharmaceutics of the University of Kiel. Special thanks to Regina Scherließ and Christian Etschmann, who is researching at the rifampicin softpellets and produced and provided it.

5 References

[1] Wetterlin, K., *Turbuhaler: A New Powder Inhaler for Administration of Drugs to the Airways. Pharmaceutical Research*, 05(8), pp. 506–508, 1988.

[2] Bley, C. H., Centgraf, M., Cieslik, A., Hack, J., & Hell, T., *I care Krankheitslehre* Berlin: Thieme, pp. 357-360, 2015.

[3] Fauster, T., Hammer, L., Heinz, K., & Schneider, A. M., *Oberflächenphysik: Grundlagen und Methoden (German Edition)*, de Gruyter Studium, Berlin, pp. 165-169, 2019.

[4] Hering, E., & Schönfelder, G., *Sensoren in Wissenschaft und Technik: Funktionsweise und Einsatzgebiete (German Edition)*, Vieweg+Teubner Verlag, Wiesbaden, pp. 140-144, 2012.

[5] Soergel, E., *Untersuchung von Oberflächenladungen auf photorefraktiven Kristallen mit dem Rasterkraftmikroskop*. München: Utz, Wiss., 1988.

Simulation of Field Free Line Translation for a Single-Sided Magnetic Particle Imaging-Scanner

Carlos Chinchilla[1,2], Amanuel Negash[2], Anna Bakenecker[3], Alexey Tonyushkin[2]

[1] Medical Engineering Science, Universität zu Lübeck, carlos.chinchilla@student.uni-luebeck.de
[2] Physics Department , University of Massachusetts Boston, MA, USA, alexey.tonyushkin@umb.edu
[3] Institute of Medical Engineering, Universität zu Lübeck

Abstract

Magnetic particle imaging (MPI) is a tomographic imaging modality that avails of the nonlinear response of magnetic nanoparticles to an oscillating magnetic field. In the past years, various scanner topologies have been proposed. Among these, a single-sided scanner was presented in 2008. This particular scanner features the advantage that all its hardware is located on one side, offering accessibility without limitations due to the size of the object of interest. In this paper, the selection field of such a single-sided scanner geometry is simulated. Hereby, the field-free line (FFL) generated by the selection coils for the specific scanner topology is simulated. Time-varying currents are applied to the selection coils to translate the FFL. With the changing ratio of the currents for the selection coils, the FFL can be moved to discrete positions enabling two dimensional spatial encoding of the nanoparticle distribution. Additionally, a numerical simulation delivers information about the magnetic field profile at various heights.

1 Introduction

Magnetic Particle Imaging (MPI) is an emerging medical imaging technology, free of ionizing radiation, which detects and visualizes the three-dimensional distribution of superparamagnetic iron oxide nanoparticles (SPIONs). Invented by Gleich and Weizenecker in 2005 at the Philips Research Laboratories, MPI uses the nonlinear magnetic response of the SPIONs to generate images [1]. The advantages of MPI include such features as high sensitivity, sub millimeter resolution and a recording time estimated to be lower than 0.1 s [2]. Although it features various promising aspects, it should be stressed that MPI does not deliver anatomical reference as in other modalities, e.g. magnetic resonance imaging. MPI can only image the concentration distribution of the SPIONs.

Once the SPIONs have been injected into the object of interest, a time varying magnetic field excites the SPIONs. To achieve spatial resolution, a gradient magnetic field, which is also called a selection field, is superimposed on the excitation field. The selection field generates an area of zero magnetic field strength, producing a field free point (FFP) or a field free line (FFL) [2]. Only signal from SPIONs within or close to the FFP or FFL can be detected by receive coils. The signal's frequency spectrum is then used for image reconstruction. The basic MPI theory is depicted in Fig. 1. To date, several scanner topologies have been proposed [3],[4],[5],[6]. Among these is the single-sided MPI scanner, which was first introduced for FFP field topology by Sattel, et al. [4]. The single-sided MPI scanner topology confines all the hardware to one side of the device, which

is beneficial because it provides the sample, or the patient, unrestricted access to the scanning area [2],[7],[8]. Given the nature of the measurements by the FFL scanner in discrete angle positions a sinogram can be generated, which can be further used to reconstruct images of the measured data with filtered backprojection method [9],[10].

In this paper, a single-sided scanner topology is used as shown in Fig. 2, which uses a FFL to spatially encode the measured signal. In this approach, the sample is placed upon a rotating platform, which is located above the coils generating the selection and excitation fields. Rotation of the FFL is done by rotating the platform mechanically to discrete angle positions. The excitation coil at the bottom in Fig. 3 is used to generate the excitation field, which is superimposed to the selection field. Calculations on magnetic fields generated through the currents in the selection coils have been simulated using Biot-Savart Law. As currents are applied in opposite directions to the selection coils, the generated FFL varies its position, achieving spatial encoding.

2 Theory

The simulations presented later on for the selection field are based on Biot-Savart Law and the calculations for the applied currents. Hence, the theory is briefly introduced in the following sections.

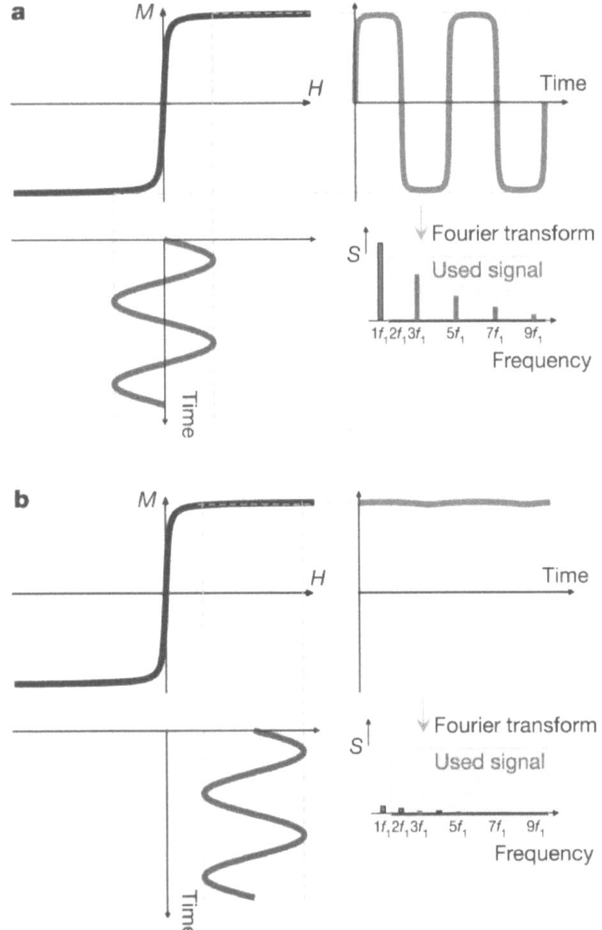

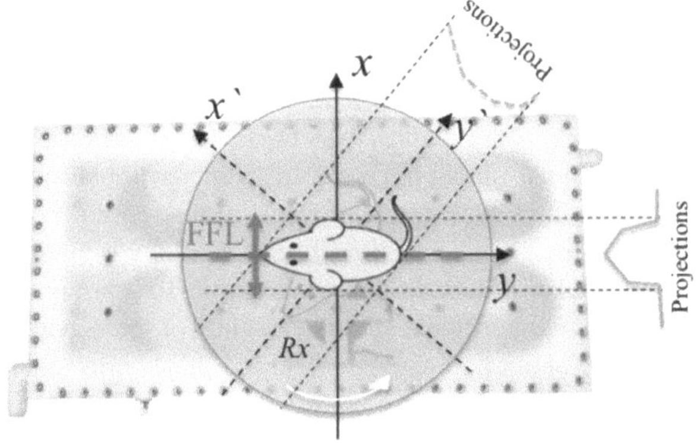

Figure 1: Theory of Magnetic Particle Imaging.
(a): SPIONs within the field free region. Left: The excitation field magnetizes the SPIONs according to their magnetization curve, which can be observed on the upper left. SPIONs that are within and near the FFP/FFL generate a signal (time-varying magnetization illustrated on the upper right). This signal can be detected and represented as a frequency spectrum (lower right). **(b)**: SPIONs outside the field free region. SPIONs that are magnetically saturated, i.e. oustide the FFP/FFL, present no response. Hence, do not contribute to the signal of the SPIONs. Image taken from [1].

2.1 Biot-Savart Law

Biot-savart law calculates the magnetic field **B** generated by a uniform current I flowing through a conducting wire. The differential magnetic field dB is calculated for every infinitesimal element dl of the coil, whose direction is the same as the direction of the current

$$d\mathbf{B} = \frac{\mu_0}{4\pi} \frac{I d\ell \times \hat{\mathbf{r}}}{r^2}. \qquad (1)$$

r is defined as the distance from the infinitesimal element dl to the point in space where the magnetic field is calculated. μ_0 is a constant known as the permeability of free space.

2.2 Currents

The currents applied to the selection coils to generate the FFL along the y-direction and its translation along the x-direction are obeyed by the following equations:

$$I_1 = I_0(1 + \alpha \cos(\omega t)), \qquad (2)$$

$$I_2 = I_0(1 - \alpha \cos(\omega t)), \qquad (3)$$

where I_0 is the current amplitude that defines the gradient field generated. The α factor determines the spatial range covered by the FFL, as it determines the gap between the maximum values of the currents.

Figure 2: Single-sided MPI scanner. The FFL is generated through the selection coils along the y-axis and translated along the x-direction for 2-D spatial encoding at a fixed height. Image taken from [11].

3 Materials and Methods

The two racetrack coil arrangements simulated to generate the FFL have a length of 24.5 cm, width of 54 mm, height of 18.3 mm, 26 turns, core gap of 10 mm and a gap between the selection coils of 11 mm. The conductor wire is made of copper with a cross section of 1 mm. The upper surface of the excitation coil, is located 5 mm from the lower surface of the selection coils centered between them. Fig. 4 shows the coil arrangements used for this single-sided scanner.

To compute the Biot-Savart law (see Eq. 1), numerical integration was implemented in Matlab. For simplicity the shape of the coils were approximated to a rectangular shape. For this simulation $I_0 = 100$ A and $\alpha = 0.65$ was used (see Eq. 2 and Eq. 3).

The currents used for the selection coils are displayed in Fig. 5. Here we simulated the translation of the FFL with Matlab applying currents for the selection coils as depicted in Fig. 5, showing that 2-D encoding is feasible. Consequently, that allows data to be acquired and reconstructed to generate images.

Furthermore, the selection coils were modeled and numerically simulated in COMSOL Multiphysics® (see Fig. 3). For the simulations a "Magnetic Fields" Interface was used

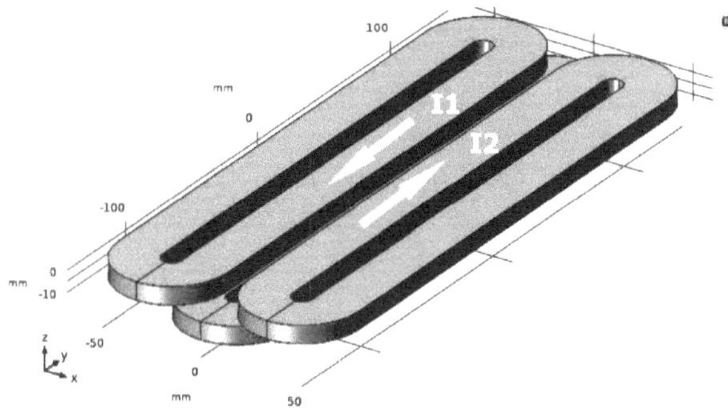

Figure 3: Coils simulated with COMSOL Multiphysics®. Upper coils are used to generate a selection field. The lower coil is used for excitation of particles. Currents I_1 and I_2 are applied in opposite directions to generate an area of zero magnetic field (FFL).

Figure 4: Actual coils from the single-sided MPI scanner modeled for the simulations of the system.

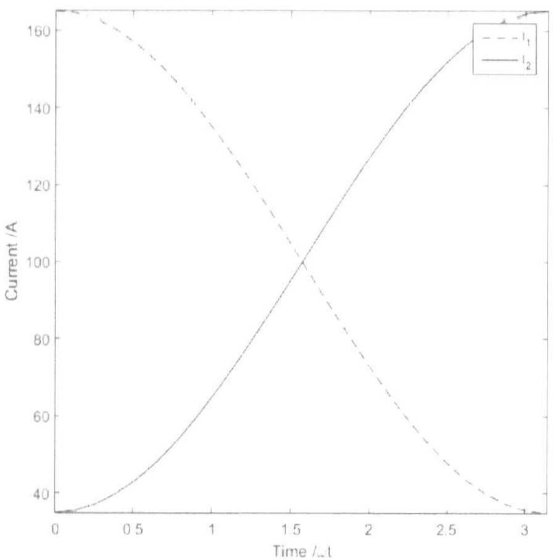

Figure 5: Currents applied to the two coil arrangements.

with $I = 100$ A flowing in opposite directions. The coils were simulated as a homogenized multi-turn coils. Simulations provided information about the magnetic field strength for different heights above the surface of the selection coils.

4   Results

In the images the magnetic field density is shown in the xy-plane at a height of 20 mm above the surface of the coils displaying the FFL along the y-direction. Fig. 6 displays the simulated translation of the FFL for this single-sided MPI scanner geometry from -20 mm to 20 mm.

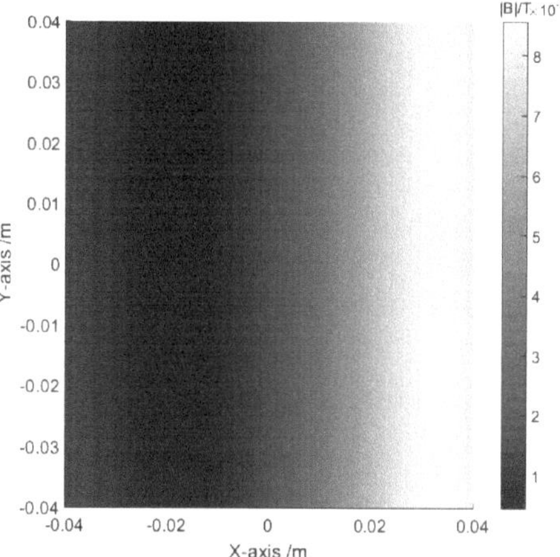

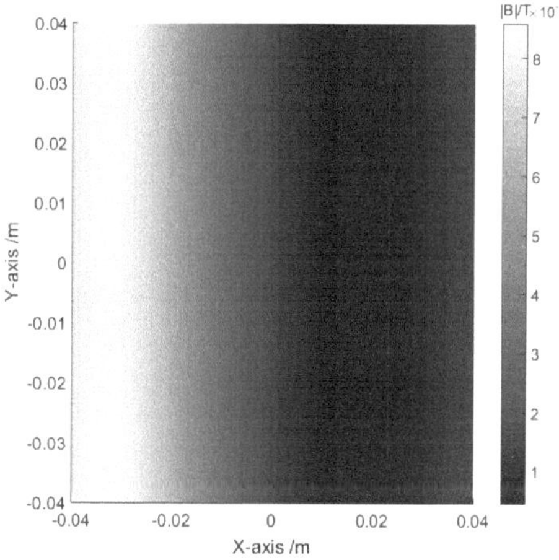

Figure 6: Simulations of the FFL translation along the x-axis. The upper figure displays the FFL at -20 mm, followed by the displacement to 20 mm.

Fig. 7 visualizes the results for the simulations of the magnetic field generated in the xz-plane. The profile of the magnetic field shows the formation of the FFL at 26 mm above the surface of the coils. The simulated coils generate a higher magnetic flux density on locations were the coils are placed. Fig. 8 shows a plot of the magnetic field strength along the z-axis. This plot shows that for increasing height, the magnetic field strength decreases. The curve reaches 0 at 26 mm hence forming the FFL.

4.1   Discussion

The simulated coil geometry for the selection field provides insight into the region for potential 2-D encoding for this single-sided scanner through the translation of the FFL. The

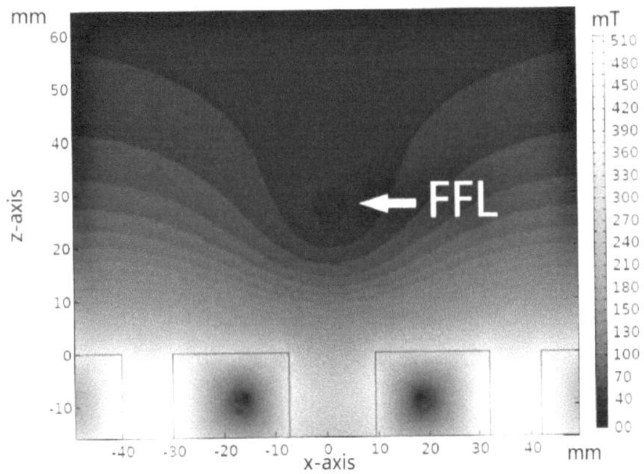

Figure 7: Simulations of the FFL generated by the selection coils with COMSOL Multiphysics® display the magnetic field strength profile on the zx-plane.

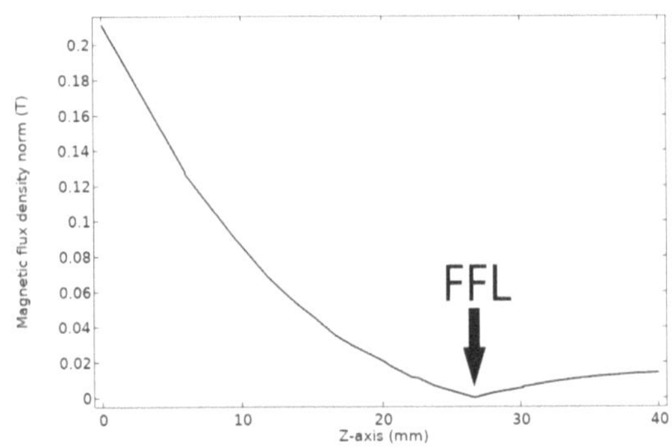

Figure 8: Magnetic field density along the z-axis visualizing the FFL at a height of approximately 26 mm.

results for the COMSOL simulations at different heights show the FFL forming at approximately 26 mm. Discrepancy among the values obtained for magnetic flux density from both simulations may be due to approximations made for the calculation of the Biot-Savart law. Furthermore, the simplification of the geometrical shape for faster computational time may also cause a difference in the magnetic fields.

5 Conclusion

Numerical simulations for a single-sided MPI scanner in Matlab and COMSOL Multiphysics® were achieved, which demonstrate the magnetic field configuration of a FFL. The simulations depicted in Fig. 6 show that with the selected currents 2-D spatial encoding of a field of view of 40 mm is feasible. Furthermore, the simulation results shown in Fig. 7 and Fig. 8 demonstrate the formation of the FFL at 26 mm above the coil surface.

Acknowledgement

The work has been carried out at University of Massachusetts Boston and supervised by Alexey Tonyushkin from the Physics Department, University of Massachusetts Boston and the Institute of Medical Engineering, Universität zu Lübeck.

6 References

[1] Gleich, Bernhard and Jürgen Weizenecker. "Tomographic imaging using the nonlinear response of magnetic particles." Nature 435.7046 (2005): 1214.

[2] Knopp, Tobias and Thorsten M. Buzug. Magnetic particle imaging: an introduction to imaging principles and scanner instrumentation. Springer Science Business Media, 2012.

[3] Weizenecker, Juergen; Bernhard, Gleich and Joern Borgert. "Magnetic particle imaging using a field free line." Journal of Physics D: Applied Physics 41.10 (2008): 105009.

[4] Sattel, Timo F., et al. "Single-sided device for magnetic particle imaging." Journal of Physics D: Applied Physics 42.2 (2008): 022001.

[5] Sattel, T. F., et al. "Magnetic field generation for multidimensional single-sided magnetic particle imaging." Proc. Int. Soc. Magn. Reson. Med. Vol. 18. 2010.

[6] Panagiotopoulos, Nikolaos, et al. "Magnetic particle imaging: current developments and future directions." International journal of nanomedicine 10 (2015): 3097.

[7] Tonyushkin, Alexey. "Single-Sided Field-Free Line Generator Magnet for Multi-Dimensional Magnetic Particle Imaging." IEEE Transactions on Magnetics 53.9 (2017): 1-6.

[8] Tonyushkin, Alexey. "Single-sided hybrid selection coils for field-free line magnetic particle imaging." International Journal on Magnetic Particle Imaging 3, no. 1 (2017).

[9] Mason, Erica E.; Clarissa Z. Coole; Stephen F. Cauley; Mark A. Griswold; Steven M. Conolly and Lawrence L. Wald. "Design analysis of an MPI human functional brain scanner." International journal on magnetic particle imaging 3, no. 1 (2017).

[10] Murase, Kenya. "A simulation study on image reconstruction in magnetic particle imaging with field-free-line encoding." arXiv preprint arXiv:1606.03188 (2016).

[11] Mason, Erica and Tonyushkin, Alexey. "Towards a Single-Sided FFL MPI Scanner for in vivo Breast Cancer Imaging", IWMPI, New York, NY (2019).

A non-invasive core body temperature measurement system for firefighters applied on the sternum

Natalia Kowalczyk [1], Frank Sattler [2], Jochim Koch [2] Stefan Müller [3]

[1] Biomedical Engineering, Lübeck University of Applied Sciences, natalia.kowalczyk@stud.th-luebeck.de
[2] Technology & Intellectual Property, Drägerwerk AG & Co. KGaA Lübeck, {frank.sattler,jochim.koch}@draeger.com
[3] Department of Applied Natural Sciences, Lübeck University of Applied Sciences, stefan.mueller@th-luebeck.de

Abstract

Thermal strain and hypothermia can have a large impact on human body. A system exists which may help to prevent it by measuring core body temperature in a non-invasive way during operation of patients and in an ICU (Dräger Tcore®). A new attempt is to adopt this sensor on the sternum where it can be used for firefighters to monitor their temperature and heat strain during work. Two different prototypes were compared in their accuracy for the measured temperature. A main point was to develop a prototype which resists movement artefacts and is robust and waterproof. Results show that by using one of the evaluated prototypes, the system gives a sufficient accuracy for core body temperature, movement artefacts are removed in an adequate way and the system has the potential to work even more accurately, if the algorithm of the used sensor gets further optimized.

1 Introduction

People who professionally work under high thermal strain often suffer from heavy consequences of these working conditions. According to NFPA 64 firefighters died in 2018 in the United States, 44 % due to overextension or stress [1]. Meanwhile, there exists a system as a prototypical demonstrator at Drägerwerk AG & Co. KGaA (in the following called Dräger), which measures the core body temperature (CBT) in a non-invasive way and with which strain of the body can be observed objectively. To evaluate the stress of a working person, Moran's Physiological Strain Index (PSI) is used. It was published and developed in 1998 in the United States with young healthy soldiers and allows for a rating of heat stress based on a dimensionless scale between 0 and 10. Though it is calculated by temperature and also heart rate [2], PSI and heart rate analysis will not be considered in detail. The main focus of this document remains on the temperature sensor embedding and performance in the used demonstrator. For the purpose of temperature measurement, part of an already existent product, the Dräger Tcore® is used. It is build up of a so called Double Sensor. This product was however developed as a disposable. It is designed for the application at a person's forehead, under work and climatic conditions of an operating room as well as ICU [3]. The new field of application shall now be fire service and rescue teams in mining. This is the reason why the positioning of the Double Sensor is changed from forehead to sternum, where it can be shielded from rough environmental conditions. This position was declared as reliable in former studies [4]. Moreover, the embedding of the

Double Sensor is changed, so that it does not work exactly like the Dräger Tcore®.

2 Materials and Methods

2.1 Materials

The technical functionality of the double sensor is based on the measurement of heat flow between two temperature elements as known from thermodynamics, where one of them is placed directly on the skin (T_{h1}) and the other one is situated within its spitting distance of some millimeters towards the environment (T_{h2}). Between both elements, there is a defined insulation. Based on the thermal balance of the thermal flux between the environment and skin temperature, the CBT (T_{core}) is estimated as follows:

$$T_{core} = T_{h1} + \frac{k_s}{k_g} \cdot (T_{h1} - T_{h2}) \qquad (1)$$

with k_s and k_g being the insulation and tissue heat transition coefficient, respectively. Moran's physiological strain index

$$PSI = 5 \cdot \frac{(T_{ret} - T_{re0})}{(39.5\,°C - T_{re0})} + 5 \cdot \frac{(HR_t - HR_0)}{(180\,bpm - HR_0)} \qquad (2)$$

with T_{re0} being the starting rectal temperature, T_{ret} the temperature at time t and HR_0 the starting heart rate, evaluates the equally weighted influences of both CBT and heart rate of the measurement. To do so, heart rate signals are collected by BioHarness™ pulse chest belt. For fixation of the double sensor on sternum and measurement of heart rate, the double sensor and its embedding are integrated is the chest belt. The chest belt is extended by additional straps for better fixation on the body and worn directly on the skin. Two different embedding designs are evaluated, design A and design B. They work as a connection between double sensor and skin and are shaped in such a way that the double sensor is embedded in the middle of their front side and the shape is compressible and soft so that it fits a regular chest shape. The difference of both designs lies in their mechanical and material properties. Whereas both designs are made out of silicone, design A is rather flat and relatively hard and design B is softer and thicker.

2.2 Methods

2.2.1 Data Acquisition and Interpretation

The chest belt has to be worn one hour prior to each measurement, so that thermal balance of the sensor is reached and no dynamic error of data is recorded. Real-time data transmission is made possible by a bluetooth connection between double sensor and a smartphone with a corresponding application. The collected data is further analyzed with a written routine in R. Bland-Altman plots serve as a substantial tool to analyze data later on [7].

2.2.2 Design A

Preliminary tests with design A were first carried out at volunteers at Dräger. After wearing the chest belt with its sensor for a sufficient time so that thermobalance was reached, the subjects had to stretch and bend their back and rest in this position, while data was recorded. Various chest types were evaluated, such as the so called funnel and chicken breast, as well as obese and lean people. These tests were performed without a reference temperature, but the subjects did neither heat up or cool down actively and the ambient conditions were hold constant, so that in this way changes in the temperature signal were definetively an indicator for movement artefacts.

Also a heat tolerance test (HTT) [5] was carried out with design A at one of Dräger's business partners. In this setup, 10 male subjects (mean age: $35{\pm}8$ years) had to ingest a telemetric temperature pill delivered by BodyCap which records data from inside of the body and serves as a reference value. The temperature pill was ingested four hours before measurement to ensure a reproducible position in the human body. Eating and drinking was not allowed up to two hours before the measurements. The heat tolerance test is a test, during which mine workers and rescuers are put under heat stress for around one hour. In a closed room with a temperature of approximately 33 °C and a relative humidity of app. 98 %, the tested people dressed only in shorts must perform steps in a rhythm of "up-up-down-down" on a step in front of them. The rhythm is given by a clock. To adjust the power of stepping up and down in an adequate way for everybody, different platforms are put in front of the step. The height of the platform depends on the person's weight. The work protocol contains half an hour stepping,

one minute break while walking around in the room and drinking 250 ml water and again half an hour stepping. Before and after the test, rectal temperature measurements are carried out as a reference to the temperature measurement of the chest belt.

The outcome of both tests with design A were evaluated and use of the gained knowledge was made to improve the design and develop a new one, design B.

2.2.3 Design B

The study executed at Dräger with design B included 23 subjects, 7 female and 16 male (mean age: $31{\pm}22$ years). They had to ingest the same temperature pill four hours before measurement to measure reference CBT. Eating and drinking was not allowed up to two hours before and during the whole study. Before and after the test, sublingual temperature measurements were carried out as a reference to the temperature measurement of the chest belt. A forehead sensor was also used as a reference. First, subjects were cooled down for 15 minutes by spraying water on front of their naked torso and standing in front of a ventilator (distance was app. 0.4 m). After precooling, the subjects had to wait for another 30 minutes so that the double sensor could calibrate itself on the cold skin. As a next step the subjects heated up actively by riding on an ergometer in short clothing (t-shirt and shorts) for 40 mintes. The power of the ergometer was adjusted in regular steps each 5 minutes beginning from 25 W to the subject's G26-maximum [6]. The G26 is a medical suitability test for firefighters and their personal protection equipment. At the end of working phase, data was recorded further on for 15 minutes as the subject was resting. After the resting phase each subject was asked to bend over to the front for five minutes so that data of a short movement test could be gained.

3 Results and Discussion

3.1 Results

3.1.1 Design A

First tests performed with design A show strong movement artefacts as subjects were bending and stretching their chest while holding the position for some minutes. This can be seen in Fig. 1.

A poor adaption between double sensor and skin can be seen in changing skin temperature (T_{h1}, dotted line). As a consequence, CBT is estimated wrong (T_{core}, solid line). Changes such as mechanical and material modifications were implemented in design B and resulted in measurements in Fig. 2.

The following Fig. 3 gives an example of the results of HTT. Here, the solid line represents data collected by telemetric pill and the dotted line corresponds to data gained by double sensor. Rectal temperature measurements are implemented as crosses. In this example, they illustrate a high correlation between them and the telemetric pill, whileas double sensor

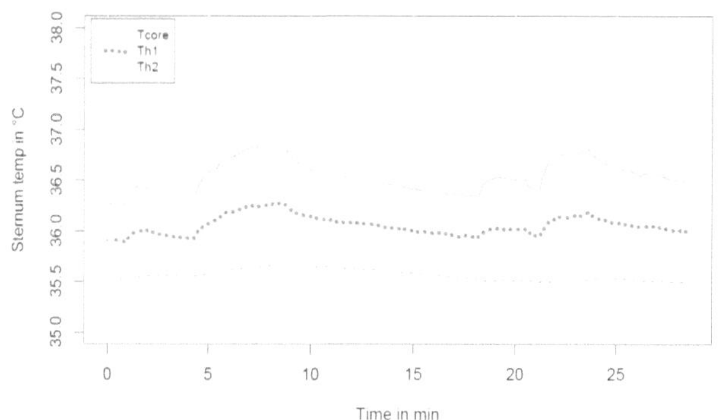

Figure 1: example of a measurement with movement arte-facts

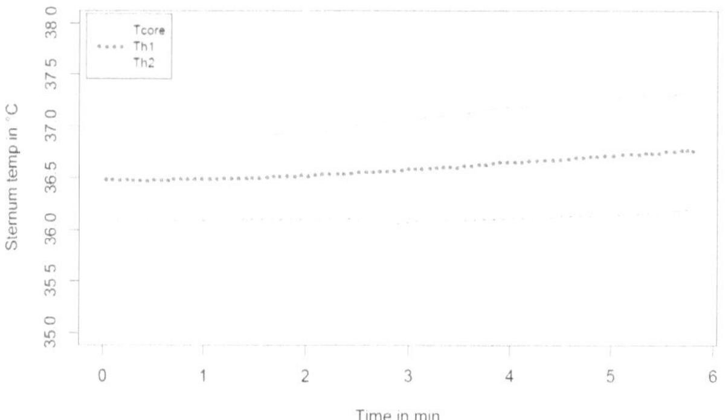

Figure 2: example of a measurement without movement artefacts

shows a correlation with telemetric pill above a threshold temperature above app. 38 °C.

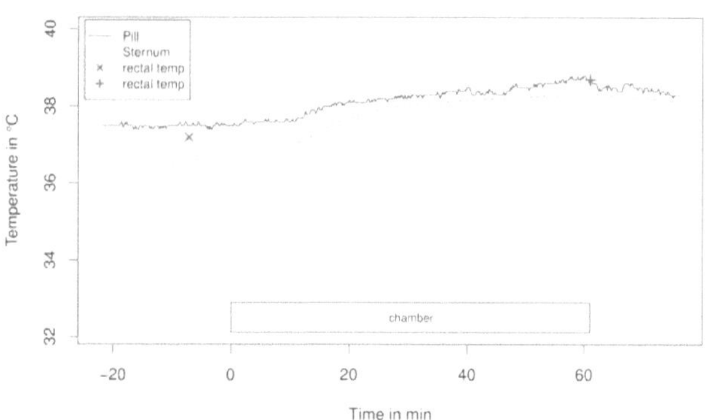

Figure 3: example of a HTT measurement

3.1.2 Design B

An exemplary measurement resulting from the study is shown in Fig. 4. Sternum, forehead and pill temperature are shown as curves and spot measurements as crosses. Due to invalid data, only 7 datasets could be considered. Data gained from the forehead was declared as invalid due to poor adaption of the double sensor to skin while the sub-

jects were sweating. Also the spot measurements result in less useful data. Due to these restrictions the statistical results have limited accuracy.

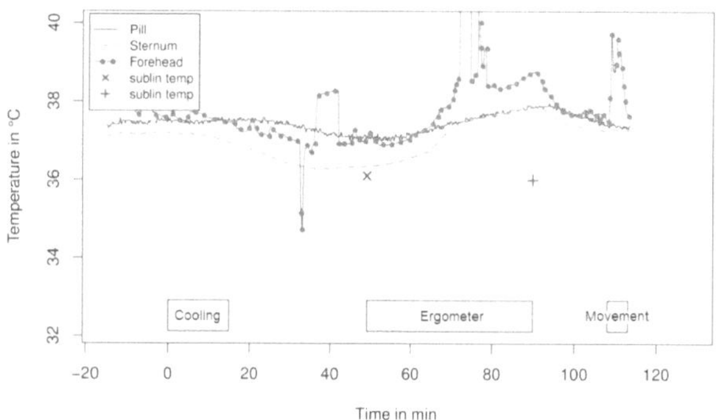

Figure 4: example of a measurement during the internal study

Data from the internal study was analyzed by the help of a Bland-Altman plot, see Fig. 5, to evaluate the correlation beetween the telemetric pill and the double sensor.

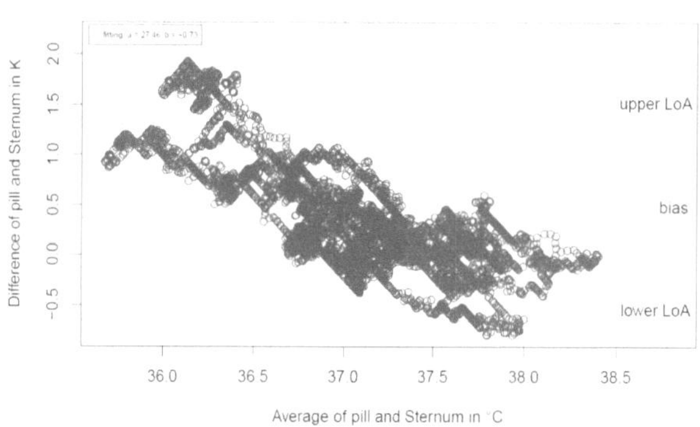

Figure 5: comparison of telemetric pill and double sensor

To improve the accuracy of the double sensor and approach the accuracy of the telemetric pill, especially in temperature range lower than 38 °C, one attempt was to slightly change the algorithm mentioned in eq. 1. This resulted in the following Fig. 6.

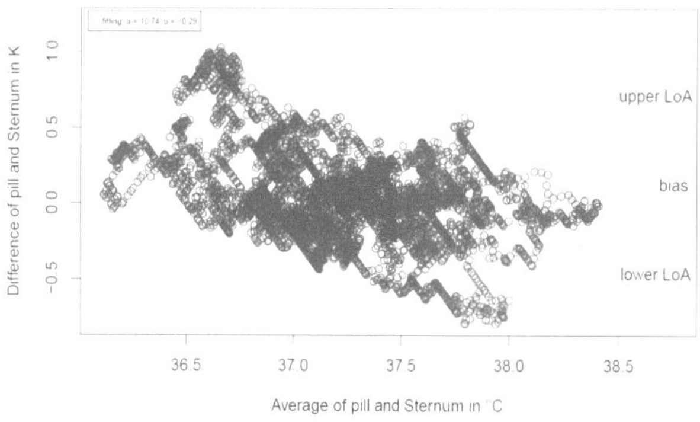

Figure 6: comparison of telemetric pill and double sensor with changed algorithm

Table 1 summarizes the outcome and comparison of HTT and the performed study considering the following targets set:

1. standard deviation between double sensor and pill ≤ 0.35 K

2. 85 % of data within a range of ± 0.5 K

3. bias between double sensor and pill ≤ 0.1 K

Table 1: Comparison of Measurements

Target	HTT	Internal Study, normal algorithm	Internal Study, changed algorithm
1	0.37 K	0.53 K	0.30 K
2	74 %	64 %	89 %
3	0.31 K	0.37 K	0.05 K

3.2 Discussion

Comparing both designs of the embedding adaption for double sensor between chest belt and skin, first measurements with design A deliver big movement artefacts. This is due to its mechanical properties: the embedding is too stiff, so that the module does not follow the shape of the wearer's chest but allows for an air gap between double sensor and skin as the wearer bends over to the front. This could be observed especially with skinny subjects, where the skin and tissue does not move to the front while bending. Considering this fact, design B was created softer and also higher, so that in case of a funnel chest the occuring air gap can be filled by silicone, while on a flat chest the module gets compressed and a good adaption between double sensor and skin is still present. Moreover, design B was developed more round than design A, so that it fits better between breast muscles and does not get pressed away by them.

The comparison of the results of the HTT and the internal study shows that at first sight, there are still inaccuracies between temperature pill and double sensor below 38 °C. However, the HTT performed with design A reaches better values than a study with improvement of design B. As measurements shall be recorded without movement artefacts, it is not possible to renounce design B. A change of coefficients in the algorithm leads to excellent values in table 1 comparing the set target. This is plausible with regard to literature.

Unfortumately, not all datasets could be considered in the internal study. This is why the statistical accuracy of these measurements is limited.

4 Conclusion

Concluding all results and after discussing the outcome of this research project, one can say that the Double Sensor has the potential to perform a sufficiently accurate CBT estimation. Design B adapted with a chest belt guarantees a more stable positioning of the double sensor on the desired spot. The double sensor is also shielded from rough outer influences by the embedding. Further research in the algorithm explained in equaltion 1 would probably improve the performance of double sensor adopted at sternum, so that the accuracy is given for CBT below 38 or 37 °C.

Acknowledgement

The work has been carried out at Drägerwerk AG & Co. KGaA Lübeck, Technology and Intellectual Property and supervised by Prof. Dr.-Ing. Dipl.-Ing. Stefan Müller, Department of Applied Natural Sciences, Lübeck University of Applied Sciences. Reasearch on this topic also would not be able without the engagement of Mrs. Cornelia Schrader, who was responsible for the modification of the used chest belts as well as all volunteers who volunteeringly contributed to the tests and studies.

5 References

[1] R. F. Fahy and J L. Molis *Firefighter Fatalities in the US - 2018*. National Fire Protection Association, Quincy, Massachusetts, USA, 2019.

[2] D. Moran, A. Shitzer and K. Pandolf *A physiological strain index to evaluate heat stress*. In: American Journal of Physiology 275, 1998.

[3] M. Soehle, H. Dehne, A. Hoeft and S. Zenker *Accuracy of the non-invasive $Tcore^{TM}$ temperature monitoring system to measure body core temperature in abdominal surgery*. In: Journal of Clinical Monitoring and Computing 28, 2019.

[4] H.-C. Gunga et.al *The Double Sensor—A non-invasive device to continuously monitor core temperature in humans on earth and in space*. In: Respiratory physiology & neurobiology, vol. 169 Suppl 1, pp. 63-68, 2009.

[5] T. Nabi et al. *Assessment of cardiovascular fitness [VO2 max] among medical students by Queens College step test*. In: International Journal of Biomedical and Advance Research, 2015.

[6] Dr. med. A. Argo *Die Untersuchung nach G 26.3 aus arbeitsmedizinischer Sicht*. Available: https://www.atemschutzunfaelle.de [last accessed on 2020-01-10].

[7] D. G. Altman and J. M. Bland *Measuring agreement in method comparison studies*. In: Statistical Methods in Medical Research 8, 1999.

Usage of the optical front tracking methods for the measurement of fast-changing micro flows of liquids.

Mariya Ali[1], Jörg Schroeter[2]

[1] Biomedical Engineering, University of Applied Sciences Lübeck, mariya.ali@stud.th-luebeck.de

[2] Medizinische Sensor und Gerätetechnik, University of Applied Sciences Lübeck, joerg.schroeter@th-luebeck.de

Abstract

Due to the underestimation of human errors in drug infusion therapy, research is carried out as a joint project with the European Metrology Institute to to minimalize the ratio of errors. The goal is to lessen the errors caused by external or internal factors. A previous system was reassembled and put into operation to examine and conduct an experiment. Two types of liquid flow speeds: 0.7 and 0.1 microliter/min were used. Data obtained from the two different flow sets were compared with the actual and measured values from the sensor and the camera's software. A continuous flow rate was observed with an exception of a deviation in one of the flow rates due to the sensor's working method. Hence the measured and actual values were observed to be the same.

1　Introduction

A drug delivery is a process used to administer a medicinal drug into a patient for a pharmaceutical purpose [1]. Modern medical technology regards the different liquid flows taking place in the human body as a significant matter. The credibility of the liquid doses should be taken into account in order to ensure the safe usage of medical devices. A great amount of deviations is experienced with flow adjustments during infusion therapy in the complex flow systems. The type of drug and its amount are the important things while delivering the drug into a patient. Moreover, there is a great number of drugs whose actual flow should be taken into great attention during their infusion. Usually accidents are prone to take place when inadequate amount of the drugs are delivered, involving the low flow rates. Some of these accidents occur due to a lack of understanding of the characteristics of a drug delivery system, lack of awareness or an improper metrological infra-structure for low flow rates [2].

In order to analyze and control complex flow systems, there must be a measuring technique introduced for the dynamic limitations of conventional flow meters. This will prove to benefit the flow sensors in medical applications like thermal flow sensors and differential pressure sensors and the dynamic flow rates in medical application like peristaltic, disposable, implantable, membrane and pulsating infusion pumps. An earlier project's model was therefore taken as a reference and further on reassembled to be put into operation. The goal of this project was to improve the sufficiency and improve the chances for less errors. This project is in collaboration with the European Metrology Institute.

2　Material and Methods

The system comprised of the following components:

2.1　Materials

- High-Speed Camera: The personal workstation computer had to have its IP settings inputted for the functionality of the high-speed camera. The high-speed camera used, was the DMK 22AUC03, which was manufactured by Imaging Source. Some of its properties are displayed in Table 1.

- Magnifying telecentric lens: The magnifying telecentric lense T45/4.0L was manufactured by Vision and Control. A few of its mechanical properties are depicted in Table 2.

- neMESYS Syringe Pump 290 N: The infusion pump used was the Low Pressure Syringe Pump neMESYS 290 N. It was also manufactured by Cetoni. It is widely used for industry and research experiments/projects. The pump has great precision for miniscule uids which is good enough for fluid speeds of millimeters and nanoliters per second. The piston moves forward smoothly due to its syringe pump's PID controlled drive system, which doesn't allow stick slip effects and allowing high degree of precision. It comes with a comprehensive software package with LabVIEW support. [5]

- LG16 liquid flow meter series sensor and USB cable: The USB RS485 and the USB cable was manufactured by the Sensirion sensor company. The sensor cable permits the easy installation for the Sensirion's flow meters.

- Hamilton Syringe: A syringe with a 0.25-micrometer diameter.

- Capillary: The capillary was with an inner diameter 310.77 micrometer with a +- 2.16 micrometer possible maximum error.

2.2 Softwares

- neMESYS User Interface: The IC sensor interface allows complex logic like data buffer for asynchronous readout, dispense volume controller and automatic dispense detection [4]. The sensor viewer is by the same company for the simple use of testing the digital sensors. It allows plug and play testing and testing of all Sensiron's sensors with IC interface. The software has different set of parameters and can be easily exported to Excel format.

- IC Measure and IC Capture: Used for live capturing of the images from the High-speed camera

- Liquid Flow Viewer: Used for the data collected from the sensor. It was provided by the campany along with the sensor.

- LabView: Used for the processing of raw data.

- DIAdem: The DIAdem 2018 was downloaded from the National Instrument's website and was used for the analyzation and further smoothing of the raw information obtained from the sensor viewer data.

As this project required several different softwares to work with, all of them had to be setup prior to the beginning of any experiments/calculations conducted.

Table 1: Properties of the High Speed Camera [3]

Resolution	744 x 480
Frame rate	76
Sensor format	1/3
Pixel Size	6 micro meter
Lens mount	C/CS mount

Table 2: Properties of the Telecentric Lense T45/4.0L[4]

Maximum object field diagonal	5.4 mm
Field view for 1/3 inch sensor (mm)	1.2 x 0.9
Working distance	$67\pm1mm$
Total Length	188 mm
Weight	330g

2.3 Preparation

2.3.1 Hamilton Syringe

The Hamilton syringe and the capillary needed to be disinfected and dried with an air pressure gun in order to eliminate any foreign body particles ie.: air bubbles and dust. The experiments were usually hindered due to them.

2.3.2 Infusion pump

It was connected and tested as demo from its own software. The movement of the different dosing modules and syringes were checked. The measured and the actual values were calculated and compared.

2.3.3 High-Speed camera

The High-speed camera and its software was installed and tested by taking sample pictures of the capillary with liquid passing through it. The high-speed camera was to remain capped after its use had ended.

2.4 Working Method

The system setup can be viewed in Fig.1. The high-speed camera was placed in a vertical position above a light source, an LED collimated. The capillary was positioned between both. Meanwhile, the Hamilton syringe was fixed on the syringe pump.

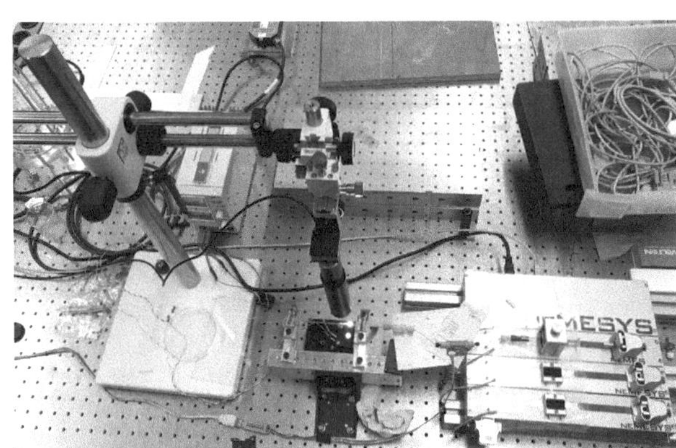

Figure 1: Image of the whole setup on the workstation.

The infusion pump's program interface was started at the beginning of the experiment, along with the sensor viewer software. This was to analyze and compare the flow readings between the syringe and the sensor along with the high-speed camera. The goal was to go below 100 nanoliters/min for the flow rate. Two types of dosage sizes were used:

- 0.1 micro liter/min

- 0.7 micro liter/min

Table 3: Some properties of the Hamilton Capillary

Capillary Inner Diameter	310.77 micrometer
Frames per second	30
Pixel pro micrometer	0.5925

The properties of the Hamilton syringe are shown in Table 3. These properties were entered into the neMESYS interface software [6]. The user interface is user friendly and can be easily configured to apply a dosing system. The syringe pump was started, and the liquid was then pumped into the syringe to the sensors and into the capillary as shown in Fig.2. The camera kept recording live pictures and or video, capturing the moment when the water flow was detected.

Figure 2: An image of the high speed camera capturing the flow movement of 0.7 microlitre/min inside the capillary

The pump was stopped when the desired position had been reached. The pictures were stored in the location chosen before the experiment was started. The location of the images can be exported to the Virtual Instrument LabVIEW in order to process them. The Virtual Instrumentation had been created in LabVIEW for the evaluation of the camera images. A series of data had to be entered in order for the LabVIEW to process the data i.e: kernel size, width, edge strength and other parameters wished to be changed.

Table 4: Values for the flows 0.7 and 0.1 microliter/min in the Virtual Instrument Labview.

Flow	Kernel	Width	Edge Strength	Process
0.7	3	3	10	Best Edge
0.1	3	3	10	Best Edge

After the complete export, the image of the capillary captured by the camera was displayed in LabVIEW. A straight line was drawn horizontally across the capillary on the image. After continuing, LabVIEW displayed the volume and flow of the experiment, an example is shown in Fig.3 for 0.1 microliter/min. This file could be exported to diadem for converting the raw data into processed data. The time signal was inverted by linearly scaling it and synchronized. Similarly the sensirion data for the flow rate was also imported into the DIAdem by installing the sensirion data plug in in the software. The sensirion data was also linearly scaled and synchronized.

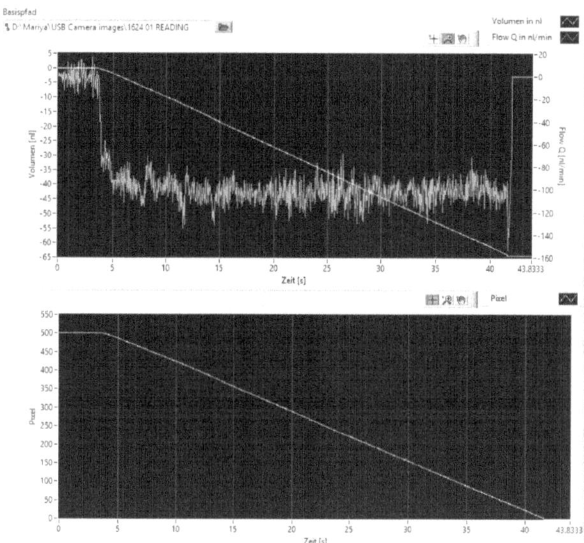

Figure 3: An image of the labVIEW processing of the raw data for the flow of 0.1 microlitre/min

3 Results and Discussion

The post processing of the raw data obtained from the two flows are shown in the following Figures 4 and 5. Fig.4 shows the High-Speed camera's and the sensor's flow rate signals. The 'linearly scaled 1' represents the camera's flow while the other represents the sensor's flow data.

It can be seen clearly seen the flow is continuous by both of the systems except a shoot at 4th second by the camera. This could be due to the internal measurement method of the sensor, the optical measurement method for the camera, the capillary and because it was measured in different positions. The length of this deviation was 400 millisecond. Other than that, the flow rate was always continuous of 700 nl/min with no other deviations for both of the systems.

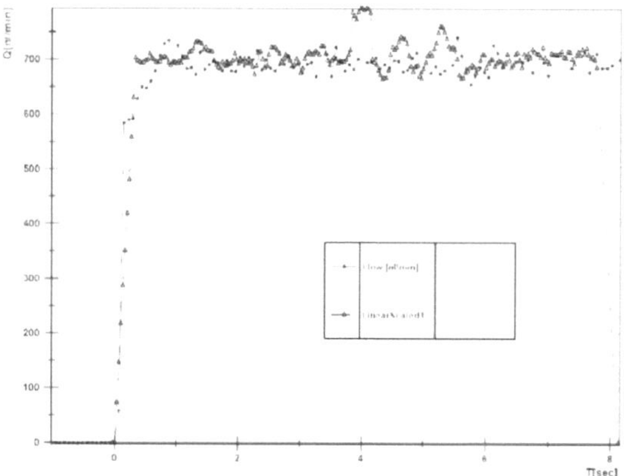

Figure 4: The comparision for the flow of 0.7 microliter/min between the High-Speed Camera and the Sensor.

Figure[5] displays the data for the flow rate of 0.1 microliter/min for the High-Speed Camera and the sensor. The graph shows a very clear flow rate going up then being continuous before going towards a gradual decrease. The sensirion sensor had overshooted in the beginning. This is because it is an internal sensor with a heater and two thermal sensors. The heater is heating the area with no flow, hence it is heating up and if the flow starts from zero, the heated part is washed off and the sensor recognizes it as a higher tempertaure area. With a continuous flow rate, the heater will be heating the water a little bit, and the sensor registers it as a higher temperature. This is calibrated to the design flow rate.. In the beginning, we have a high temp diff between the two sensors with different flow rates as it keeps heating with no flow rate. The hot spot is washed with the second sensor but not the first one therefore deviation occurs. This is characteristical for the flow sensor.

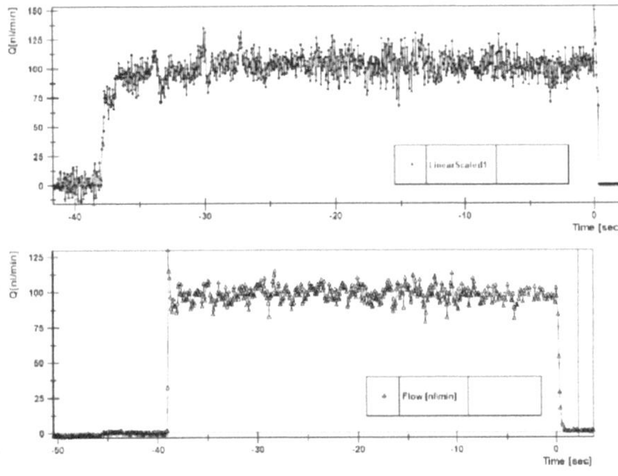

Figure 5: The comparision for the flow of 0.1 microliter/min between the High-Speed Camera and the Sensor.

4   Conclusion

Infusion devices need to be reliable for it to be utilized in drug infusion delivery. Nevertheless, it can not always be calibrated in a perfect manner. This is due to the different conditions and environments these devices are being used along with different assort of accessories. Low flow rates are prone to a great amount of inaccuracies leading to serious consequences.

It is believed that the camera is collecting 30 frames per second and sending every frame to the computer by the sensor, and the computer storing and collecting next picture. This could be a delay due to that. We can not assume that the camera is collecting with exactly 30 fps. With this experiment it gave us the result that the camera could be capturing the images for 29 frames per second instead of 30. This project hence gave an idea to how the measured value and the actual values are calculated and the possibility of its accuracy.

Acknowledgement

The work has been carried out and supervised by the Institute of Medical Sensor and Device Technology of University of Applied Sciences Luebeck.

5   References

[1] Collins Dictionary, *Available*: http://bit.ly/2R7KVy0 [last accessed on 18-01-2020]

[2] *Biomedical Engineering Biomedizinische Technik*, vol 60, no 4 , pp . 269–275, 2015.

[3] Source for the equipment manual Available: https://www.theimagingsource.com/products/industrial-cameras/usb-2.0-monochrome/dmk22auc03/   [last accessed on 18-01-2020]

[4] *Sensirion datasheets* Available: http://bit.ly/2RvhHrT [last accessed 18-01-2020]

[5] Available: https://www.cetoni.com/products/low-pressure-syringe-pump-nemesys-290n/ [last accessed 18-01-2020]

[6] Available: https://www.cetoni.com/products/nemesys-userinterface/ [last accessed 18-01-2020]

4

Biomedical Optics

Visible light optical coherence tomography and fluorescence imaging with dual-stage aberration correction

Jonas Franke [1], William Newberry [2], Eduard F. Durech [2], and Marinko V. Sarunic [2]

[1] Medical Engineering Science, Universität zu Lübeck, jonas.franke@student.uni-luebeck.de
[2] Engineering Science, Simon Fraser University, Burnaby, BC, Canada, msarunic@sfu.ca

Abstract

The combination of Optical Coherence Tomography (OCT) and fluorescence Scanning Laser Ophthalmoscopy (fSLO) allows the addition of functional imaging provided by fluorescence imaging to the detailed structural OCT images. Both these imaging modalities suffer from decreased image quality due to aberrations. In order to reduce aberration artifacts, the implementation of Sensorless Adaptive Optics (SAO) has been successfully tested for both imaging techniques [1]. In this paper, a modified version of the multi-modal system discussed in [2] is presented, that is capable of a more sophisticated aberration correction by using two deformable mirrors.

1   Introduction

Optical Coherence Tomography (OCT) has proven to be a powerful imaging modality in medical applications. Modern Fourier Domain OCT systems are capable of high imaging speeds combined with a sub-micrometer lateral resolution and an axial resolution of close to one micrometer. Thus, these systems are well suited for applications especially in ophthalmology, where they are widely used for detailed imaging of retinal structures. By combining OCT systems with fluorescence Scanning Laser Ophthalmoscopy (fSLO), the structural imaging can be extended by a fuctional imaging modality. Fluorescence imaging enables the system to show labeled reporter cells within the imaged tissue, therefore allowing the visualization of biological processes.

Most commercial OCT systems use a near-infrared light source. However, the OCT described in this publication uses visible light with a tunable centre-wavelength. Results discussed here were acquired with either a centre-wavelength of $470nm$ or $560nm$ The use of light in the visible spectrum allows a high axial resolution due to shorter wavelength while using a narrower bandwidth [3]. However, the lateral resolution remains limited by diffraction. The use of visible light enables the acquisition of spectroscopic information due to the high absorption of visible light within blood vessels. Furthermore, the wavelength used for visible light OCT (VIS-OCT) are capable of exciting the fluorescence markers used for fSLO. Therefore, in theory the same light source and wavelength can be used for both imaging modalities, the OCT as well as the fSLO, allowing parallel acquisition.

In order to integrate the fSLO system into the OCT assembly, it is necessary to implement an appropriate detector for the low-power fluorescence signal emitted by the fluorophores of the sample. For this purpose a Photo Multiplier Tube was introduced. Since for the presented setup the intensity in the excitation band of the used fluorophores (i.e. Fluorescein) was realtively low using the supercontinuum laser ($60\mu W$), an additional light source with a centre wavelength of 488nm is coupled into the sample arm for parallel use to the supercontinuum light source. In principle, both wavelengths, $560nm$ and $488nm$, can be used for both OCT and fSLO

Both imaging modalities, the fSLO and the VIS-OCT, suffer from decreased image quality due to aberrations. For in-vivo imaging, these aberrations are mostly introduced by various optical imperfections of the eye as well as, to a certain extend, of the imaging system itself. In the setup presented here, the aberrations are compensated by the use of sensorless adaptive optics (SAO). The benefits of SAO for aberration correction in OCT have been successfully demonstrated before [1]. Also, the usability in VIS-OCT was shown in [2]. The concept is based on the use of deformable mirrors (DM) in order to modify the wavefront in a way that compensates for the deformations caused by aberrations within the sample. For evaluating the wavefront usually a wavefront sensor is used. For sensorless adaptive optics, the optimization using the DMs is not done based on the wavefront itself, but by the image quality. Different aberration modes are aplied to the DM, until the best image quality is achieved.

The systems from [1] [2] were relying on the use of a single DM to correct aberrations and a variable focus lens to allow compensation for defocus. In the system discussed here, a second DM has been introduced. This allows aberration compensation in a two-step manner: While one DM is used to compensate for lower order aberrations, such as defocus and astigmatism, the second DM can independently be used to correct higher order aberrations, caused by varying refractive indices of different components of the eye. These higher order aberrations include spherical aberrations and coma, and can be described mathematically by higher order Zernike polynomials which are then applied to the shape of the DMs. This concept therefore follows a 'woofer-tweeter' approach: while one DM may have advantages when it comes to maximum stroke, the other DM is better suited for adapting to higher spacial frequencies due to its higher actuator count.

The setup including two DMs is especially useful for evaluating novel SAO methods. By being able to generate known aberrations using one of the DMs, these aberrations can then be corrected by using the second DM. Thus making the system well suited for testing and developing new SAO related aberration correction approaches, in particular those based on Deep Reinforcement Learning. The development of such approaches will be another future application of the system presented here.

In addition to the changes to the OCT and fSLO sample arm, the spectrometer was refined to get a more robust system with a higher sensitivity for better overall OCT imaging quality. More detail on these changes will be discussed below.

2   Material and Methods

The schematic representation of the system in use is shown in Fig.1. As a light source for the VIS-OCT a supercontinuum white laser light source from NKT Photonics was chosen. This light source allows the adjustment of the wavelength in a spectrum from 390 to $840nm$ and a maximum bandwidth of $100nm$. OCT imaging was performed using centre wavelengths between $470nm$ and $560nm$. In all cases a bandwidth with a FWHM of 60 nm was chosen, allowing a power of ca. $70mW$ at the sample. In order to excite fluorescein for concurrent fSLO image aquisition, a wavelength of $470nm$ was chosen. Therefore, with

$$\Delta z = \frac{2ln2}{\pi} * \frac{\lambda_0^2}{\Delta\lambda}. \tag{1}$$

where λ_0 is the centre wavelength and $\Delta\lambda$ is the bandwidth, this allows an axial resolution of $1.6\mu m$. Analogously for $560nm$ follows an axial resolution of $2.3\mu m$.

A Thorlabs $560nm$ 50/50 wide band fibre coupler was used to direct the light from the supercontinuum laser to sample and reference arm and to couple the reflected light from both arms back into the spectrometer. For the excitation of certain fluorophores (i.e. fluorescein) while using $560nm$ wavelength for the OCT, a second light source was added

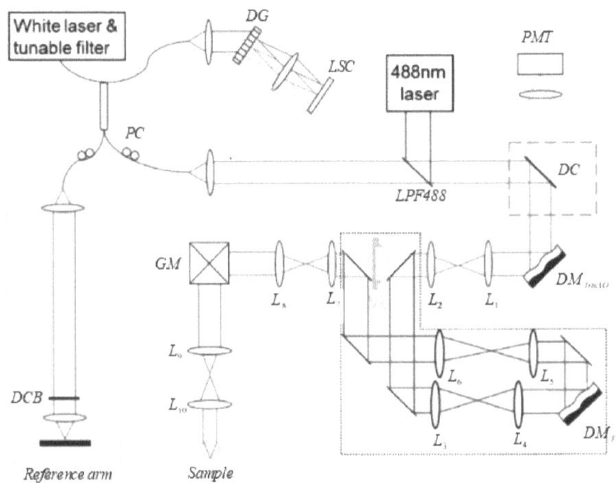

Figure 1: Schematic representation of the system used. GM, 2-D galvanometer scanner; VL, variable focus lens; DG, diffraction grating; PC, polarization controller; LSC, line scan camera; DCB, dispersion compensation block; MEB, multi-edge beam splitter; MEF, multi-edge filter; M, mirror; PMT, photo-multiplier tube), $L_1, L_2, L_7, L_8 = 200mm$; $L_3 = 6mm$; $L_4 = 15mm$; $L_5 = 300mm$; $L_6 = 380mm$; $L_9 = 75mm$; $L_{10} = 25mm$.

to the system. A $488nm$ laser (Coherent OBIS 488) was chosen. To couple the light from both light sources into the sample arm of the system, a set of two dichroic mirrors was set up as shown in Fig.1, one beeing a LPF488 for reflecting the light from the $488nm$ source but transmitting longer wavelengths, the other being a multi-edge filter for transmitting the fluorescence light to the PMT, but reflecting the light used for the VIS-OCT into the optical relays of the sample arm. After beeing collimated by a Thorlabs reflective collimator to a 4mm beam diameter, the light hits the first DM, an IrisAO segmented deformable mirror(PTT-111, IrisAO, Inc) capable of adapting to higher order zernike polynoms. For restricting the beam diameter to the size of the DM, an adjustable aperture was put in place in front of the DM.

In order to complement the existing system from [2] by a second DM, further modification to the sample arm was necessary. The variable focus lens was bypassed with two additional sets of optical relays. The first relay consists out of a $6mm$ and a $15mm$ focal length lens, the second includes two lenses with $300mm$ and $380mm$ focal length. As an additional DM a Thorlabs continuous surface deformable mirror was chosen, which is capable of modifying incoming wavefronts in a way that compensates for aberrations described by lower-order zernike polynomials.

After the second DM, the light is directed to the galvanometric scanning mirrors (GM) via another set of relay lenses ($200mm$ focal length each), then into the last set of relay lenses with a focal length of $75mm$ and $25mm$, and further to the sample. For ex-vivo imaging of phantoms, an additional objective lens with a focal length of $8mm$ and a numerical aperture of 0.60 is added. With

$$\omega_0 = \frac{\lambda_0}{\pi * NA}.$$
(2)

this allows a lateral resolution of $\omega_0 = 249nm$ for $\lambda_0 = 470nm$ and $\omega_0 = 297nm$ for $\lambda_0 = 560nm$

Since the modified sample arm now consists of a set of additional lenses to bypass the variable focus lens, the dispersion had to be re-matched. For this purpose a set of dispersion compensation blocks was brought into the system's reference arm and additional dispersion compensation was performed numerically [4].

The quality and resolution of the VIS-OCT imaging highly depends on the characteristics of the spectrometer. For this setup, a custom build spectrometer was used. The beam coming from the fiber coupler was collimated by a $75mm$ focal length lens. Together with the fibers numerical aperture of $NA = 0.13$ with

$$d = 2 * (NA * f).$$
(3)

the diameter of the collimated beam equals 19.8mm. For diffracting the incoming light according to the wavelength, a transmission holographic grating with a grating constant of 1800 lines per millimeter was placed in front of the detector. The detector is a 2048 pixel line scan tall-pixel OCT camera (Telendyne e2v octoplus) with a pixel size of $10 * 200\mu m$. Thus the axial range l for this setup is as follows [5]:

$$l = \frac{\Delta z}{2} * \frac{N}{2}.$$
(4)

with N being the number of detector elements (in this case 2048) and the axial resolution Δz. This leads to an axial range of $835\mu m$ for $470nm$ centre wavelength and $1.17mm$ for $\lambda_0 = 470nm$.

The spectrometer was designed and simulated using Zemax Optic Studio. Spectral alignment was performed by maximizing the peaks in the fourier transform of the captured spectrum.

Imaging was performed with an adjustable depth scan rate which allowed the selection of values between $10kHz$ and $100kHz$. The aquistion for both modalities, the fSLO as well as the VIS-OCT, was done using a GPU-accelerated custom-build program.

The abberration correction using the AO could then be performed on either the en-face VIS-OCT images or the confocal fSLO images. Therefore an optimization algorithm based on the hightest image sharpeness as a merit function, was used, thus defined by

$$S = \sum [I(x, y)]^2.$$
(5)

where $I(x, y)$ is the intensity of each pixel [1].

3   Results and Discussion

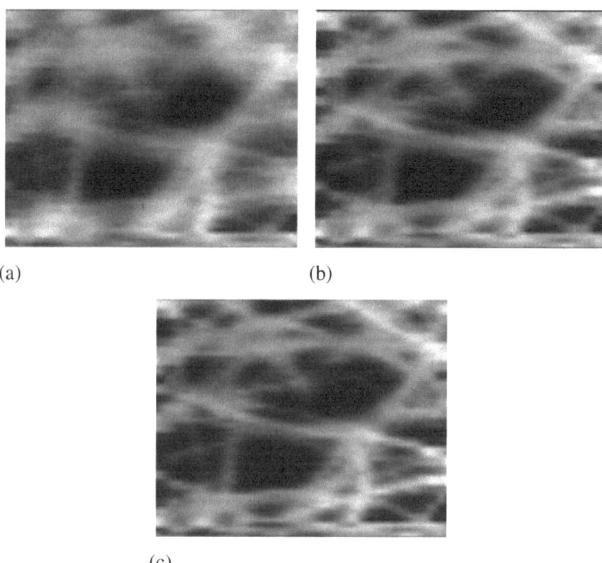

Figure 2: Images acquired using the system's fSLO, in (a) without aberration correction, (b) with optimization via the Thorlabs DM and in (c) with additional optimization using the IrisAO DM.

Fig. 2 shows three images acquired with the fSLO System. As a sample, a piece of lens tissue, to which some fluorescein was applied, was put in front of the objective lens. The shown images were averaged over 10 acquired single images. As light source the $488nm$ Coherent OBIS laser was chosen, since it delivers more power into the samplearm and therefore leads to a higher signal coming from the sample back to the PMT. In order to create aberrations, the sample was placed slightly off-focus. Furthermore, the offset of the galvanometric scanning mirrors was chosen in a way, that the beam hits the last lenses at an angle to the optical axis, thus creating astigmatism. Now, the two DMs where used to compensate for these aberrations.

Fig.2 (a) shows the image without any additional aberration correction, resulting in a blurry image with low brightness. Now, in order to compensate for defocus and astigmatism, the higher stroke of the Thorlabs DM was used. The parameters of the mirror control where therefore manually changed, until both, the brightness and the sharpness of the imaged features, increased. The results can be seen in Fig.2(b). In the last step, the IrisAO DM was used by applying the optimization algorithm based on (5), resulting in the image shown in Fig.2 (c).

It is apparent that the major differences are created by the aberration correction with the high-stroke Thorlabs DM. The image highly increases in sharpness and visibility of the samples features. The optimization with the IrisAO however, in this case, does only have a slight effect on the overall image quality. This result is expected, since the introduced aberrations, astigmatism and defocus, are represented by lower order zernike polynomials, which can be corrected more effectivly with the high stroke DM rather then with the segmented IrisAO, which is more effective for higher spatial frequencies.

The OCT system however suffers from high losses within the sample arm. This results in less light being coupled back into the spectrometer. The reason for this lies in the relatively large amount of silver coated mirrors which are necessary to direct the light into the optical relays that are necessary for setup with two DMs. These mirrors have a reflectivity of approximately 95 to 96% for the used wavelength. In the current setup the sample arm includes seven silver coated mirrors from Thorlabs. Given a measured power at the first DM of $120\mu W$ this, for a reflectivity of 95% for the used wavelength, leads to a power at the sample of $120\mu W * 0.95^7 = 83.8\mu W$. Additional losses appear at the two DMs, the galvanometric scanning mirrors and, although to a much lesser extent, on the lenses. The measured power at the sample therefore is $63\mu W$. The low amount of light being coupled back from the sample leads to a low amplitude of the fringes detected at the spectrometer and makes it hard to distinguish them from noise, making especially the numeric dispersion compensation and the creation of the look-up table required for this, much more difficult. All in all, OCT imaging of multiple layer phantoms or in-vivo imaging was not yet possible. Fig.3 shows the fourier transform of the detected spectrum for multiple depths. It is apparent that with increasing depth the intensity falls off and the width of the peaks increases. This fall-off in sensitivity therefore requires a numerical dispersion compensation.

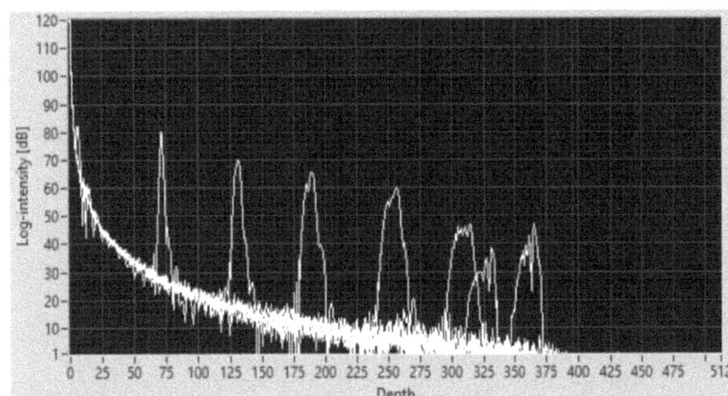

Figure 3: Fourier-transform of the detected spectrum for multiple depths relative to the DC.

4   Conclusion

It was shown that the approach of aberration correction using two difformable mirrors delivers promising results when applied on fSLO images. Especially the high stroke of the Thorlabs DM proved to be capable of effective lower order aberration correction, even when optimization was carried out manually.

Due to the two imaging modalities included, the alignment of the system turned out to be relatively difficult in order to optimize functionality for both, the fSLO and the OCT. The aim here was to minimize system aberrations, such as defocus and astigmatism, as much as possible without compromising on the functional capability of both systems. Since the system had to be used for carrying out parallel research on the DRL network, one modality (the fSLO) had to be functioning at any time, thus restricting the degrees of freedom for OCT alignment. That made the iterative process of alignment more time consuming then expected. The higher complexity of the system including two DMs therefore turned out to be a disadvantage of this approach.

For further work on the system, it is now necessary to increase the power in the sample arm by testing other optical components and improving the alignment of the system. That will decrease the influence of noise and make the numerical dispersion compensation more effective. Also the supercontinuum light source will be replaced by one generating higher output power. Furthermore the use of additional dispersion compensation blocks within the reference arm will be tested.

Acknowledgement

The work has been carried out at Simon Fraser University, Burnaby, Britsh Columbia under guidance of Professor Marinko V. Sarunic and supervised by Professor Gereon Huettmann, Institute of Biomedical Optics, Universität zu Lübeck. Special thanks to Ji Hoon Kwon for creating the look-up-tables for dispersion compensation.

5   References

[1] Daniel J. Wahl, Ringo Ng, Myeong Jin Ju, Yifan Jian and Marinko V. Sarunic, *Sensorless adaptive optics multimodal enface small animal retinal imaging*. Biomedical Optics Express Vol. 10, No. 1, 2018.

[2] Myeong Jin Ju, Christine Huang, Daniel J. Wahl, Yifan Jian and Marinko V. Sarunic , *Visible light sensorless adaptive optics for retinal structure and fluorescence imaging* . Optics Letters Vol. 43 No. 20, 2018.

[3] Shau Poh Chong, Tingwei Zhang, Aaron Kho, Marcel T. Bernucci, Alfredo Dubra, and Vivek J. Srinivasan , *Ultrahigh resolution retinal imaging by visible light OCT with longitudinal achromatization* . Biomedical Optics Express Vol.9, No. 4, 2018.

[4] Wojtkowski M, Srinivasan V, Ko T, Fujimoto J, Kowalczyk A, Duker J, *Ultrahigh-resolution, high-speed, Fourier domain optical coherence tomography and methods for dispersion compensation*. Optics Express, Vol. 12 ,No. 11, 2004.

[5] Kenny K.H. Chan, *Spectral Domain Optical Coherence Tomography System Design: Sensitivity Fall-off and Processing Speed Enhancement, Chapter 4*. The University Of British Columbia, August 2010.

Non-linear Raman microscopy
with an experimental approach on resolution enhancement

Valerie Lutz [1], Brian Bjarke Jensen [2], Sebastian 'Nino' Karpf [3], and Jonathan R. Brewer [2]

[1]Medical Engineering Science, Universität zu Lübeck, valerie.lutz@student.uni-luebeck.de
[2]Department of Biochemistry and Molecular Biology, University of Southern Denmark, brewer@memphys.sdu.dk
[3]Institute of Biomedical Optics, Universität zu Lübeck, sebastian.karpf@uni-luebeck.de

Abstract

We implemented a system, combining coherent anti-Stokes Raman scattering (CARS) and stimulated Raman scattering (SRS) in one home-built setup. Both non-linear Raman techniques are label-free and provide chemical specific images which are shown and compared in the following. Furthermore, a first approach on enhancing the resolution of those techniques was made by adding a third laser-beam, which demonstrated that at least two-photon depletion and resolution enhancement on a small scale is already possible. This builds the foundation to aim for resolution enhanced CARS and SRS in future projects.

1   Introduction

Nowadays there are many methods to enhance the visibility of specific features in probes beyond pure magnification. As a chemically specific and label free method the detection of Raman scattering is widely used, especially for characterising the chemical properties of biological specimens. Contrary to common fluorescence techniques there is no need for external labeling with dyes, which can interfere with the sample properties. In comparison with infrared spectroscopy, the spectra do not suffer from a dominating water signal and are therefore more convenient for the investigation of biological tissue. As the standard Raman technique generates quite weak signals, it is possible to increase the signal intensity by going into the non-linear regime and detecting coherent anti-stokes Raman scattering (CARS) or stimulated Raman scattering (SRS). Both techniques also provide a high intrinsic temporal and spatial resolution.[1]-[3].

Based on the idea of Stimulated Emission Depletion (STED) Microscopy [4], where a donut-shaped light-source is used to gain super-resolution fluorescence images, the system presented here is a first step to improve the resolution of CARS/SRS-Microscopy beyond the diffraction limit. While with STED the samples need to be labeled with specific fluorophores, they are not needed in Raman imaging, as it is based on the same principle as general CARS/SRS-detection. Only the point spread function (PSF) gets modified by the overlay with a vortex-shaped beam, which changes signal generation in the outer rim of the spot. As this technique is still at the beginning of investigation, there are mostly theoretical simulations with different approaches. This system is based on the same theoreti-

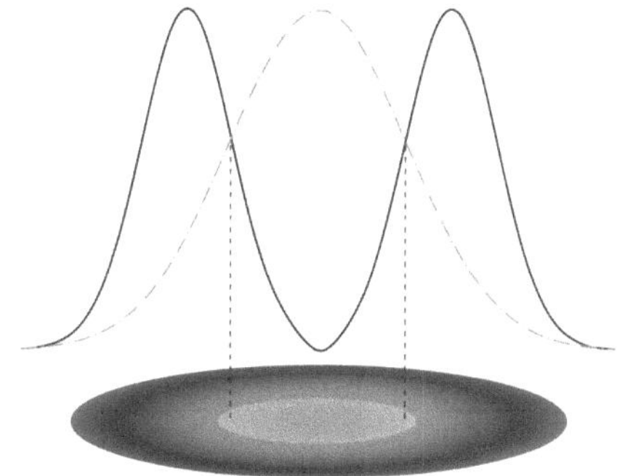

Figure 1: Overlayed point spread functions (PSF) of the pump, Stokes (grey, dotted line) and depletion beam (black, continuous line). Below: resulting spot on the sample. Spatially reduced part, contributing to the image in light grey.

cal approach as Beeker et al. [5] described with density matrix simulations. The two overlaid Gaussian-shaped beams drive the molecules inside the middle of the donut in coherence and build the precondition for anti-Stokes scattering and energy transfer between the two laser frequencies. To prohibit this, shortly in advance, the additional beam depopulates the ground state in higher vibration modes with long lifetimes and destroys the build-up of coherence by introducing an incoherent population transfer to the Raman active vibration state. This leads to the presumption that mainly the laterally minimized central part, in which only the two Gaussian pump- and Stokes-beams interfere, is providing the imaged signal. A representation of the PSFs of the used beams and the resulting spot on the sample is displayed in Fig. 1.

2 Material and Methods

2.1 Technical setup

The imaging setup is based on the *picoEmerald* laser source from *APE*, which consists of an internal laser emitting a 1064 nm beam with a repetition rate of 80 MHz and 2 ps pulse duration. This beam gets frequency doubled and used to pump the optical parametric oscillator (OPO). The resulting signal beam is tuneable from 720 nm to 990 nm. Since it is not further used, the idler beam gets blocked. Both, the IR- and OPO-beam, leave the *picoEmerald* aligned on top of each other and get guided over several mirrors and lenses into the *Nikon TI Eclipse* microscope. For a schematic setup refer to Fig. 2. The scan head consist of two galvanometer mirrors, each for one axis, and a concave mirror which projects the laser beam on the oscillating scan mirrors. They realize the movement of the laser spot on the sample and describe a rectangular scanning-frame which defines the spatial dimensions of the generated image. For the CARS/SRS-excitation the tuneable beam is used as the pump laser and the fixed infrared beam is used as the Stokes laser. Inside the inverted microscope a dichroic mirror separates the anti-Stokes scattering from the exciting laser. In epi-direction the signal gets detected by a *Hamamatsu Photosensor Module H7422P-40* (PMT) in photon counting mode. To differ from fluorescence or second harmonic generation a filter wheel with different wavelength bandpass filters is mounted in front of the PMT. Contrary, for the SRS signal the detection is in forward direction. This is realized with an IR-blocking filter and a photodiode to measure the stimulated Raman loss of the pump beam and analyse it with a lock-in amplifier. The amplifier is the *SRS detection Module* from *APE*. To use this technique the Stokes laser beams gets modulated by a build in electro-optic modulator at 10 MHz. The *picoEmerald* also provides the TTL-reference signal for the lock-in amplifier.

For the resolution enhancement feature a near infrared laser was used. The donut shape of the beam is generated by a 2-π-helical phase plate to get a circularly polarized beam, which is combined with a quarter wave plate to remove longitudinal components in the center. To adjust the focus height in the z-axis a beam expander is inserted. The beam gets coupled into the light path of the other two beams by a dichroic mirror.

2.2 Imaging

The here presented system does not acquire whole Raman spectra automatically, but images at specific wavenumbers to display the presence of molecular bonds with matching vibrational frequencies. With manual tuning of the pump laser a range of different wavenumbers can be scanned and displayed. Due to the possible tuning frequencies of the pump laser and the used PMT it is possible to cover the wavenumbers from 2435 cm^{-1} to 4490 cm^{-1} for CARS and down to 702.51 cm^{-1} for SRS-detection including the fingerprint region below 1500 cm^{-1}.

The samples used for adjusting and aligning the beams

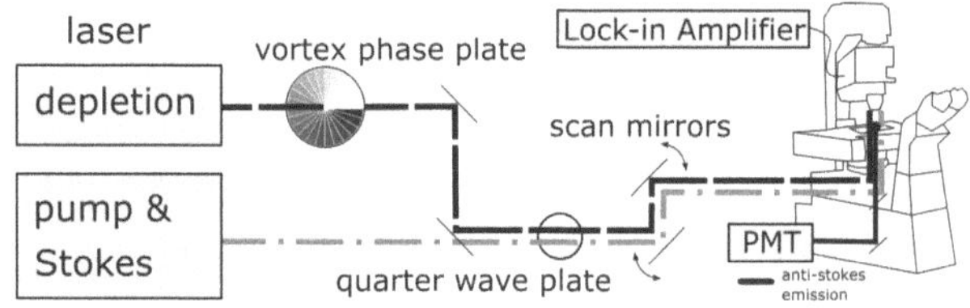

Figure 2: Simplified setup for the generation of CARS, SRS and resolution enhanced images. The depletion laser gets donut-shaped by a vortex phase plate and coupled into the light-path of the pump and Stokes-beams by a dichroic mirror. They pass a quarter-wave plate to correct polarization and over a scanhead into an inverted microscope. In epi-direction the anti-Stokes emission is detected by a photomultiplier tube and in forward-directon the signal loss of the pump-beam gets detected by a photodiode and passed to a lock-in amplifier for stimulated Raman detection.

on top of each other were *Polymere Microspheres - green fluorescing - 1.0 μm* and *Qdot^{TM} 605 ITK^{TM} Carboxyl Quantum Dots*. To measure the depletion *FluoSpheres^{TM} Carboxylate-Modified Microspheres, crimson fluorescent, 0.2 μm* were used. All from *Thermo Fischer Scientific*. Additionally slices of human skin with a thickness of 15 μm were imaged. They were deep-frozen before hydrating them with phosphate buffered saline on a microscope slide. No labeling was carried out. All images were made with a 60x/1.27 apochromat IR-corrected water immersion objective from *Nikon*.

3 Results and Discussion

Figure 3 shows a comparison of the different methods used in this approach by imaging the same area in a piece of butter. The images on the top were created using the SRS-detection and the ones on the bottom are made with the CARS-setup. For the images on the left, the laser was tuned to 2850 cm^{-1} to match the CH_2-stretching present in lipids. Butter consists mostly of fast with small water droplets inside. They don't get excited with the chosen wavelength combination and therefore remain black. To image them, the laser is tuned to excite the OH-bonds (3422 cm^{-1}) in water. Since water is more fluid, the images on the right look more blurred than their counterpart with lipids. The images were taken with equal laser power, except for the CARS-water-image where 1.5x of the power was needed. It can be seen, that the SRS-images have a more equally distributed intensity. They also contain contributions from a broader depth in the z-direction giving it a more three-dimensional look. This probably origin from the lower numerical aperture of the lens collecting the signal for the SRS-detection in forward detection. The CARS-signal gets collected from the same high-NA objective the lasers are focused with.

In order to demonstrate the technique on a biological sample, the laser was tuned again to take pictures of human skin,

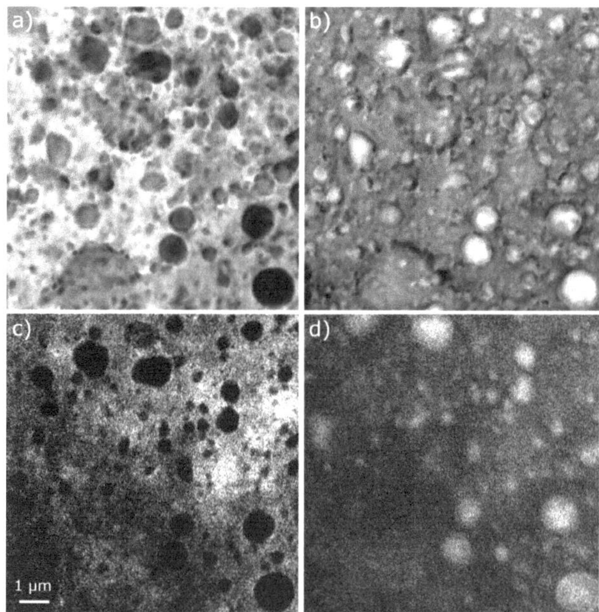

Figure 3: Comparison of SRS- a),b) and CARS c),d)- images from the same area in butter. Detected wavenumber for a) and c): 2850 cm^{-1} (lipids), b) and d): 3260 cm^{-1} (water)

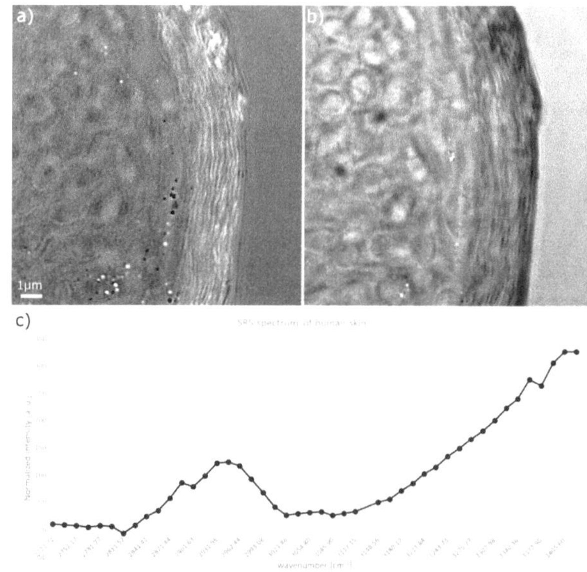

Figure 4: SRS-image of human skin. a) 2841.41 cm^{-1} (lipids), b) 3422.02 cm^{-1} (water) c) SRS-spectrum of the area

once on lipids and once on water. The images are shown in Fig. 4 a) and b). It depicts the filaments on the top layer of the epidermis, the stratum corneum, with a high lipid content and the outlines of the cells forming the layers below with more water content. The surrounding media consists of water. The images were made with a pixel dwell time of 50 μs and 20 μs integration time of the lock-in amplifier. A Raman spectrum was also recorded for this area, shown in Fig. 4 c). To cover the displayed wavenumber range the pump laser needed to be tuned from 825 nm to 780 nm. The two peaks between 2850 cm^{-1} to 3010 cm^{-1} origin from the symmetric and asymmetric CH$_2$-stretching and indicate the presence of lipids. The rising intensity above 3200 cm^{-1} is coming from the OH-bonds in the water.

For the resolution enhancement feature the donut shaped beam needed to be generated and aligned concentrically on the two other beams. Fig. 5 shows the diffraction pattern from carboxyl quantum dots excited with the 775 nm depletion beam at different heights on the z-axis. One can clearly see the donut shape at the top and bottom of the PSF but the intensity within is not distributed evenly and concentrates in two opposed spots. The effect gets stronger in the middle, resulting in two entirely separated spots. Also the focus points of the two sides of the beam are not on the same height. This behaviour probably arises from chromatic aberrations within the optical system.

Despite the fact, that the depletion laser does not have the desired optimal shape, measurements with 200 nm fluorescent beads were made to show two-photon depletion. The beam of the depletion laser was aligned a bit lower in the z-axis to be more donut-shaped around the focus of the excitation beam. For the excitation only the pump beam at

a wavelength of 816 nm was used, resulting in two-photon excitation. When the additional depletion laser is turned the PSF of the fluorescent beads gets smaller. This is displayed in Fig. 6 a). The corresponding intensity distributions across the beads are shown in Fig. 6 b). For better visibility only the normalized, centered Gaussian fits are plotted. However, the signal-to-noise ratio is rather low, making it difficult to come to a clear conclusion about the real extent of spatial reduction. The $\frac{1}{e^2}$-width from the Gaussian fits without depletion are 376 nm and 536 nm and are reduced to 344 nm respectively 424 nm with 300 mW of depletion power. That means a relative reduction to 91 % from the bottom right to the bottom left and to 79% from top left to the bottom right, which is in line with the intensity distribution of the depletion-donut as seen in Fig. 1. Increasing the power even more only leads to a drop in the overall intensity. Summing up one can conclude, that stimulated depletion occurs reducing the lateral PSF. But the intrinsic resolution of two-photon excitation is already high and the initial image was already near the real size of the bead. Therefore it could not get much smaller. Additionally, the PSF of the depletion beam is not optimal which makes the depletion uneven. Due to fast photo-bleaching it was not possible to take conclusive images with the smaller beads which were available.

A precondition for the depletion is the right timing of the two laser pulses. In case of fluorescence, the depletion beam needs to come shortly *after* the excitation to stimulate de-excitation. To realize this the length of the triggering cable between the two lasers was changed accordingly. In the case of resolution enhancement for the CARS/SRS-Signal, ground-state depletion need to happen *before* the two excitation beams come. Since the timing need to be more precise for that, it was not possible to get the right timing by changing the cable length. This was only possible

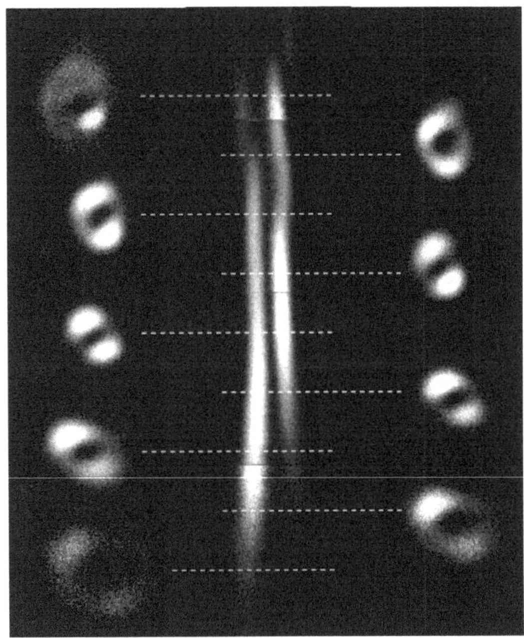

Figure 5: Point spread function of the depletion beam on a carboxyl quantum dot. In the middle is a scan of the z-axis. To cover the whole depth four images needed to be combined. Xy-scans are displayed to the left and right at the indicated positions with a distance of 0.5 μm are displayed.

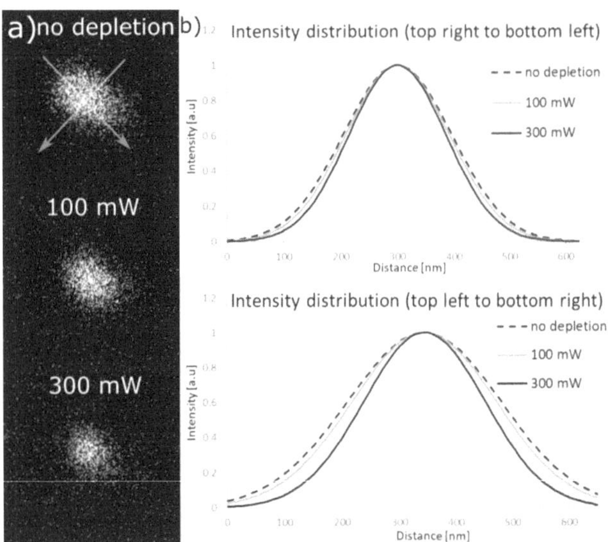

Figure 6: a) Images of fluorescent beads with 200 nm size, two-photon excitation through 816 nm laser with 300 mW initial power. In each image the energy of the depletion laser is indicated. The arrows show the direction of the analyzed intensity distributions. b) corresponding intensity distribution with various intensity of the depletion laser. Normalized and centered to compensate photobleaching and center depletion.

in discrete steps of 10 cm of cable, which lead to 500 ps travel time by signal speed of $2*10^8 \frac{m}{s}$ through the cables. Compared to the pulse duration of a few ps it is not precise enough.

4 Conclusion

It was possible to set up a system to image label-free different characteristic molecular vibrations from a broad range of wavenumbers with the techniques of coherent anti-Stokes Raman scattering and stimulated Raman scattering. Also, a third laser beam was included into the system, which was formed into a donut-like shape. With the correct temporal relation it could be shown that two-photon depletion is possible and resolution enhancement in a small extend was demonstrated. To optimize the depletion beam shape one can consider putting the phase-plate in a conjugate plane of the back-focal plane of the objective. With the current setup this was not possible. Additionally a transmissive liquid crystal devices could compensate the aberrations. It was not possible to observe any reduction in anti-Stokes intensity with the turned on depletion laser. Probably the time delay between the three beams was not optimal, as it couldn't be set any more specific in the present setup. For future experiments a solution for this, may be the use of an optical delay line for the excitation or the depletion beam to realize a continuous delay on a smaller timescale. This might also optimize the two-photon depletion.

Acknowledgement

The work has been carried out at the Danish Molecular Biomedical Imaging Centre, Southern University of Denmark under the supervision of the Institute of Biomedical Optics, Universität zu Lübeck.

5 References

[1] J. Cheng and X. S. Xie, *Coherent Raman Scattering Microscopy*. Boca Raton, 2016.

[2] T. Dieing, O.Hollricher and J. Toporski *Confocal Raman Microscopy*. Springer-Verlag Berlin Heidelberg, 2011.

[3] R.C. Prince, R.R. Frontiera and E.O. Potma *Stimulated Raman Scattering: From Bulk to Nano*. In Chemical reviews 117.7 (2017), pp. 5070-5094.

[4] S.W. Hell and J. Wichmann, *Breaking the diffraction resolution limit by stimulated emission: Stimulated-emission-depletion fluorescence microscopy*. In: Optics letters 19.11 (1994), pp. 780–782.

[5] W.P. Beeker et al., *A route to sub-diffraction-limited CARS Microscopy*. In: Optics express 17.25 (2009), pp. 22632–22638.

Infrared raytracing simulation for measuring beam mixing in multi gas sensors

Lucas Schnelle [1], Bernd-Michael Dicks [2], Martin Kroh [3], and Fred Reinholz [4]

[1] Medical Engineering Science, Universität zu Lübeck, lucas.schnelle@student.uni-luebeck.de
[2,3] R&D Gas Sensors, Drägerwerk AG & Co. KGaA Lübeck, [bernd-michael.dicks; martin.kroh]@draeger.com
[4] Institut für Biomedizinische Optik, Universität zu Lübeck, reinholz@bmo.uni-luebeck.de

Abstract

The reliable measurement of anesthetic gases is one of the most important control mechanisms during surgery. Contamination that occurs inside a sensor can block parts of the detector unit. In order to avoid this problem, most optical gas sensors use expensive beam splitters to uniformly mix the the infrared light beams. To investigate the pathway of the beam distribution, various sensor setups with and without internal beam splitter were simulated in Zemax and individual raytracing experiments were performed. An existing sensor design with beam splitter was used as a reference in these experiments and the simulation of the new design was optimized accordingly. During this work various models of a planar detector structure were investigated and results comparable to the existing gas sensor could be achieved. The newly simulated sensor designs allow beam mixing with uniform intensity distributions on the detector chips without internal beam splitters.

1 Introduction

The *Infrared Low Cost Analyzer* ILCA3 is the most utilized multi gas sensor of the *Drägerwerk AG & Co. KGaA* for the five anesthetic gases halothane, enflurane, isoflurane, sevoflurane and desflurane as well as carbon dioxide CO_2 and nitrous oxide N_2O. Of the used anesthetic gases, only nitrous oxide is available as gas. Under normal pressure and temperature conditions the remaining five anesthetics are in liquid form. These can be delivered into the lungs with an evaporator [1].

The volatile narcotic gases have an inhibitory effect on the transmission of excitation to the nervous connections so that a hypnotic, muscle relaxant and analgesic reaction is induced. On molecular level the exact mechanism of action is unexplained despite decades of everyday application [2].

In order to determine the exact concentration of the respiratory gases, a small probe stream is continuously branched off via the sidestream method and fed into the optical gas sensor. Inside the sensor the exact concentrations can be ascertained by light transmission measurements. In this way a safe dosage of anesthetics can be guaranteed during surgery. During the measurement, the gas molecules are stimulated via vibrational excitation by the infrared light. Consequently, the photons are absorbed in specific ranges of wavelengths between 2.5 - 15μm [3]. By reflecting the light on several detectors with different filter channels, the special absorption characteristic of the similar, but nevertheless individual spectra in the middle infrared range can be recorded. While carbon dioxide and nitrous oxide have

transmission minima in the range between 3 - 5μm, the range between 8 - 10μm is used for the measurement of anesthetic gases. Fig.1 shows how the transmission properties of two anesthetics differ between 8 and 9μm. A transmission value of 100% signifies that all incoming radiation passes through the medium. A transmission of 0% means that the total radiation is absorbed by the gas molecules. In addition, the respective signal strenghts are displayed in three spectral windows [4].

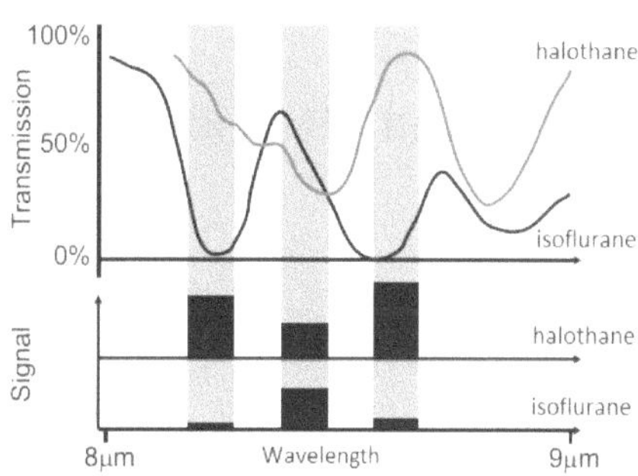

Figure 1: Top: Transmission spectrum of halothane and isoflurane; Bottom: The respective signal strenghts in three spectral windows [4].

The individual transmission properties of an infrared-active gas can then be solved for a specific concentration of the molecule in question via the *Lambert-Beer law*. Equation (1) can be adjusted subsequently to (2) so that the concentration of the substance quantity $C(\lambda)$ is calculated from the light attenuation related to the initial luminous intensity I_0, the travelled path length L and the substance-specific and wavelength-dependent extinction coefficient ϵ.

$$I_M = I_0 e^{-L \cdot \epsilon \cdot C(\lambda)} \tag{1}$$

$$\Rightarrow C(\lambda) = \frac{1}{L\epsilon} \cdot lg(\frac{I_0}{I_M}) \tag{2}$$

However, the exponential relationship between the detected intensity I_M and $C(\lambda)$ is only valid under ideal conditions. The real measuring arrangement does not have infinitely narrow bandpass filters, but finitely wide, Gaussian-shaped profiles. In this case, the logarithmic relationship $-lg(\frac{I_M}{I_0})$ no longer provides a linear correlation. As a result, the relation between transmission and concentration in practice often has to be approximated experimentally by calibration curves [4].

Typically, the light is directed onto the detector surfaces with the aid of a beam splitter. This beam splitter is used to ensure reliable beam mixing so that the light beams fall symmetrically onto the detector chips. In the course of this work it had to be tested whether a detector structure with a planar sensor unit without beam splitter could serve as a more cost efficient, but just as reliable alternative. For this purpose, both the original detector structure and the new planar structure were simulated in *Zemax* and the influence of different beam blockages in the beam path was assessed. Subsequently, various concepts of a possible new detector layout were set up and compared with the original design.

The aim of this paper is to:

- Present the different sensor assemblies

- Explain the *Zemax* simulations of the detectors

- Demonstrate the necessity of beam mixing in the sensor

2   Material and Methods

2.1   Light source

A thermal radiator is used as the light source for the sensor. Infrared emitters of this type are heated to certain temperatures, so that the spectral distribution corresponds approximately to that of a black body. The intensity $\tilde{I}$ of the heat radiation emitted in the wavelength range $d\lambda$ is described by the *Planck radiation law* [5] [6]:

$$\tilde{I}(\lambda)d\lambda = \frac{2\pi hc^2}{\lambda^5} \frac{d\lambda}{e^{\frac{hc}{\lambda k_B T}} - 1} \tag{3}$$

Where h is the *Planck constant*, c the speed of light, T is the temperature and k_B is the *Boltzmann constant*. Thereby the maximum intensity shifts with increasing temperature to shorter wavelengths according to the *Wien displacement law* [5] [6]:

$$\lambda_{max} = \frac{2897,8\mu m \cdot K}{T} \tag{4}$$

The radiation characteristics of the light sources were on the one hand simulated in *Zemax* and on the other hand described by rayfiles of the manufacturer. Rayfiles are a set of data that describe the beam distribution and beam intensity of the infrared sources at their exit window and can directly be implemented into the simulation.

2.2   Cuvette

Typically, the cuvette containing the measuring medium forms the optical path between the emitter and detector. In the case of anesthetic gas measurements, a cylindrical cuvette made of aluminum is used. In the design of the ILCA3, the cuvette is used to separate the sample gas from the ambient air. Through two embedded windows, a constant measuring volume is to be maintained in the sensor, which is protected against contamination from the outside. To ensure optimum transmission in the measured wavelength range the material of the windows is chosen appropriately.

2.3   Detector

The radiation receiver utilised in the measurement setup is a thermal detector. Here, the optical radiation is converted into an electrical sensor signal via the *Pyroelectric Effect*. If the infrared radiation $\phi_S(t)$ emitted by the light source hits the sensor surface A_S, the radiation is almost completely absorbed. The absorption of the radiant energy causes a temperature change ΔT which changes the distance of the grid structure d_S of the pyroelectric material. This leads to an electrical polarization of the surface, whereby realeased charges can be picked up with an outer electrode [5] . The functional principle of the detector is further depicted in Fig. 2.

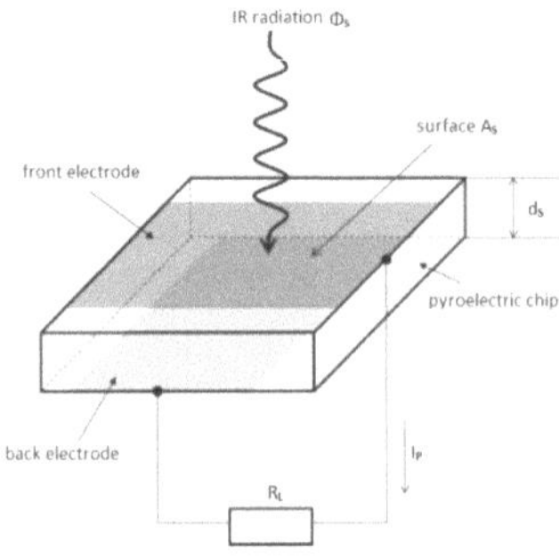

Figure 2: Pyroelectric infrared detector: The time-varying radiation flux leads to an electrical polarization of the the detector surface so that charge carriers are realeased [5].

2.4   Four-channel sensor with internal beam splitter

The detectors are four channel detectors, whose basic structure is illustrated in Fig. 3. During the measurement the infrared radiation hits a beam splitter which ideally reflects the light uniformly in all directions. The deficits of this detector design are on the one hand the costly arrangement of pyramid arrays which provide the beam mixing. On the other hand the angular design of the filter and detector elements is problematic as it can lead to power losses.

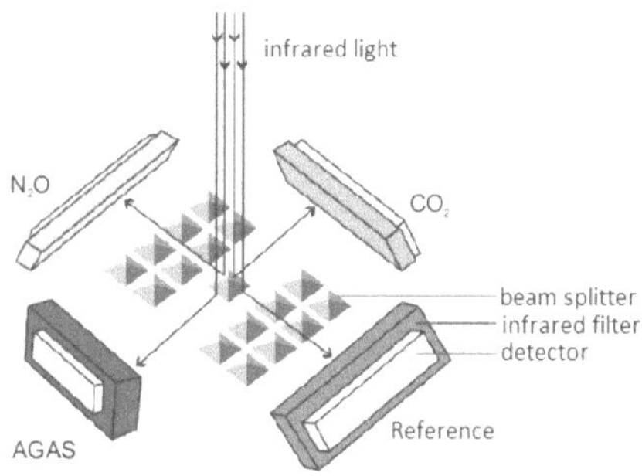

Figure 3: Illustration of the beam mixing produced by the pyramid array. The light beam is then directed onto the various detector channels ([4] modified by [7]).

2.5   Four-channel sensor without beam splitter

Another possibility is the use of a much less complex detector without beam splitter. The four sensor chips are mounted on a planar, two-dimensional plane directly behind the infrared transparent window. In this configuration, specific filters are likewise attached in front of the respective detectors.

2.6   Zemax simulation

In Zemax, the original sensor in Fig.4 (a) with light source, cuvette, beamsplitter and detector as well as the new sensor design Fig.4 (b) with simpler detector setup were simulated. The assumed wavelength of the light source was set to 4.3μm with a nominal output power of one Watt. The detector housing and the detector base were ideally parameterized as completely absorbing and the cuvette and the reflector were assumed to be perfectly reflecting. Finally, silicon substrates were placed in front of the detectors as filter dummies.

By various raytracing experiments it was now determined whether a constant beam mixing without shading and with 50% shading at the detector window can be achieved in both sensor simulations. Since some beam distributions proved to be insufficient, the sensor design was adapted and simulated iteratively.

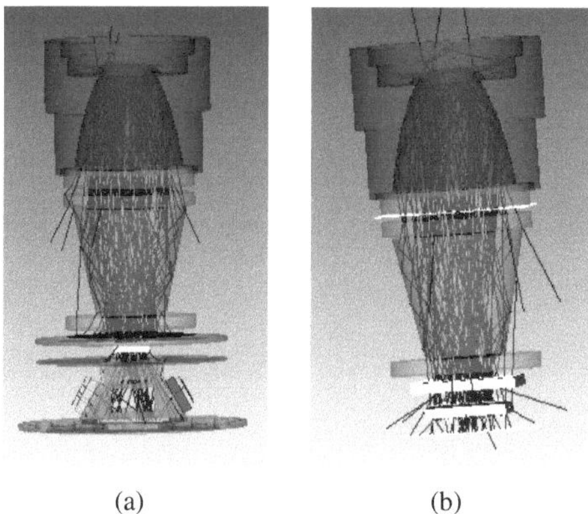

(a)   (b)

Figure 4: *Zemax* simulation of the original sensor (a) and the simulation of the modified detector design with planar detector chips (b).

3   Results and Discussion

The raytrace results of the simulation of the existing detector should serve as a reference. Raytracing without shading showed that beam powers of approx. 3 mW per detector chip could be measured with the ILCA3 detector.

For good beam mixing within the detector, it is necessary for the individual signals of the four chip surfaces to have similar power. If the center of gravity of the beam distribution shifts, the ratio of the signal strengths of the chips to each other also changes. A high asymmetry therefore stands for an uneven illumination of the detector surface, whereas low asymmetries indicate a good beam mixing and uniform illumination.

In the ILCA3 setup, the differences in symmetry in the beam distribution were less than 1%. If 50% of the rear cuvette window is shaded, the incident beam power is also halved to 1.5 mW. The incoming asymmetry of the beam distribution was now at 8%. The next task was to achieve similar results in terms of beam power and symmetry in the new detector design.

The beam distribution arriving at the planar detectors without shading (a) and with shading (b) is shown in Fig. 5.

Raytrace experiments in a planar detector without shading show that significantly higher beam powers of 22 mW per detector chip could be measured. The asymmetry also continually remains below 1%. By inserting the absorber between the cuvette window and the detector, 50% of the light was blocked. It was afterwards observed that the beam power of the chips on the darkend side decreased to approx. 3 mW. The beam power on the unshaded side, on the other hand, only droped to 19 mW per chip. The proportion of radiation was thus 13% on the side of Chip1 and Chip2 and 87% on the detector half of Chip3 and Chip4. In this way, the asymmetry of the beam distribution increased to 74%.

Since the beam distribution of the sensor was insufficient, two new possible sensor designs with plane detectors were additionally implemented in *Zemax*. The new models should improve beam mixing and, in the best case, deliver comparable results to the existing ILCA3 sensor.

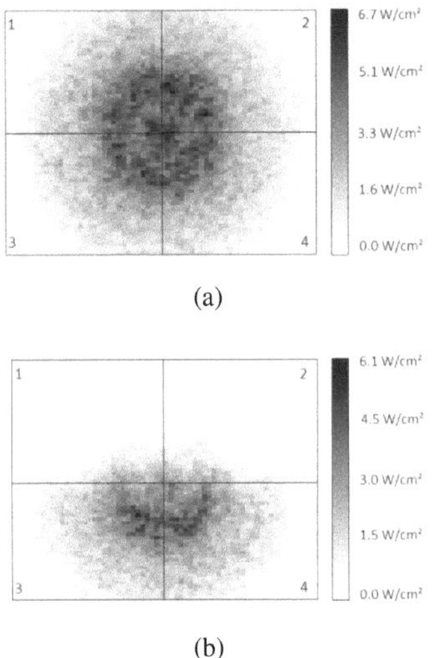

(a)

(b)

Figure 5: Beam distribution onto the planar detector without blocking (a) and with blocking at the rear cuvette window (b).

The new intensity distribution of Model A during a raytrace experiment with 50% shading is displayed in Fig 6 (a).

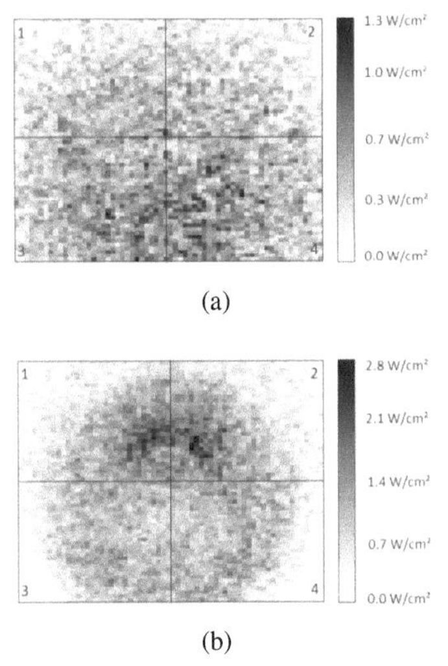

(a)

(b)

Figure 6: Beam distribution onto the planar detector with 50% blocking at the rear cuvette window: Modell A (a) and Modell B (b).

In terms of symmetry, significantly better results than the original planar detector have been achieved with asymmetries of 24%. The detected beam power was 3.6 mW per chip on the darkened side and 5.8 mW per chip on the non-darkened half and was therefore in comparable ranges to the ILCA sensor.

Fig. 6 (b) shows the light distribution onto the detectors of Model B with 50% shading. Despite the absorber material between the rear window and the detector, a reliable beam mixing could be simulated. The asymmetry of light was less than 1% and the detected beam power was between 9.8 and 10 mW per Chip. Compared to the existing sensor with internal beam splitter, the simulation of the planar detector

structure with half shading thus achieves a beam distribution improved by 7% and signal strenghts six times higher.

4   Conclusion

The intention of the *Zemax* simulation of the detectors and the raytrace experiments was to evaluate the necessity of beam mixing inside the sensor. The outcome of the comparison between the existing detector with internal beam splitter and a planar detector structure without beam splitter provided promising results. On the one hand the simulations confirmed that an internal beam mixing is absolutely necessary so that reliable gas measurements can be carried out despite contamination within the sensor unit. On the other hand it was shown that by modifying a planar detector design, uniform beam distributions could be achieved despite 50% shading. Further laboratory tests are planned to confirm whether the sensor simulation and raytrace analysis have delivered valid results. Subsequently, a new sensor concept with a less expensive planar detector without beam splitter can be set up.

Acknowledgement

The work has been carried out at Drägerwerk AG & Co. KGaA, R&D Sensor Systems, Lübeck.

5   References

[1] F.-J. Kretz, J. Schäffer and P. Terboven, *Anästhesie, Intensivmedizin, Notfallmedizin,Schmerztherapie*. Springer-Verlag Berlin Heidelberg, 2016.

[2] M. Heck, M. Fresenius and C. Busch, *Klinikmanual Anästhesie*. Springer-Verlag Berlin Heidelberg, 2008.

[3] R. Kramme *Medizintechnik*. Springer-Verlag Berlin Heidelberg, 2011.

[4] S. Leonhardt and M. Walter, *Medizintechnische Systeme*. Springer Vieweg, 2016.

[5] G. Wiegleb, *Gasmesstechnik in Theorie und Praxis*. Springer Vieweg, 2016.

[6] I.V. Hertel and C.-P. Schulz, *Atome, Moleküle und optische Physik 1*. Springer Spektrum, 2017.

[7] J. Schüttler and H. Schwilden, *Modern Anesthetics*, ser. Handbook of Experimental Pharmacology. Berlin: Springer Verlag, 2008, Vol.182.

Simulation of Universal Laser Beam Parameter Correction of a High Power UV Laser System

Marc Seitz [1,2], Rüdiger von Elm [2], Kai Seger [2], and Robert Huber [3]

[1] Medical Engineering Science, Universität zu Lübeck, marc.seitz@student.uni-luebeck.de
[2] Coherent LaserSystems GmbH & Co. KG, Lübeck, {Ruediger.vonElm, Kai.Seger}@coherent.com
[3] Institute of Biomedical Optics, Universität zu Lübeck, robert.huber@uni-luebeck.de

Abstract

In order to generate ultraviolet (UV) light, a source beam of a certain wavelength is directed into a nonlinear crystal. There, the incoming beam frequency is changed through frequency conversion. During this non-linear process, undesirable changes occur within the nonlinear crystal, causing the output characteristic of the laser beam to change. The most common manifestations of beam parameter changes include beam waist position and beam waist radius. In order to compensate for the deviations in the nonlinear crystal that occur over time, the aim of this work is to develop a control software for a beam parameter correction system with adaptive optics. For this purpose the programming language Python is used. In order to solve the problem, simulations of the beam caustics as a function of different input parameters as well as of the control behavior were performed as preliminary tasks.

1   Introduction

Diode pumped solid-state lasers, such as the used Paladin Compact 355-4000 UV laser, represent the technology of choice for a wide range of different applications due to their high reliability, low operating costs and superb output characteristics. In order to generate UV laser radiation, laser systems usually direct a source beam to a nonlinear crystal. By traversing this nonlinear crystal under proper phase-matching conditions, the source beam undergoes a frequency conversion, which creates a beam with a shorter wavelength. Directing a 532 nm beam to a nonlinear crystal leads to generation of a 266 nm beam, which would be the second harmonic. Furthermore, a 1064 nm beam and a 532 nm beam can create a 355 nm beam, the third harmonic of the 1064 nm beam. This is called sum frequency generation. Typically several different crystals are required to generate a desired UV wavelength [1].

One example for a nonlinear crystal is cesium lithium borate (CLBO), which has a large thermal acceptance bandwidth and is also used in the employed Paladin Compact laser system. The Paladin Series from Coherent are mode-locked oscillators with 20 W to 30 W average power, which emit pulses of 10 ps to 20 ps length with a repetition rate of 80 MHz or 120 MHz. Since high peak powers on the range $> 10\,\text{kW}$ occur in the pulses, an efficient frequency conversion can be performed outside the resonator. During the generation of light with short wavelengths, the high photon energies cause changes in the nonlinear crystal, resulting in changes in the beam parameters of the UV laser beam. The potential of degradation is increasing with higher power density and shorter wavelengths. As a result of the early commencement of degradation in the harmonic conversion efficiency, nonlinear crystals used in high power UV laser systems are particularly prone to such cumulative changes. Among others, the most common manifestations of beam parameter instability are beam quality M^2, axial beam waist location z_0 and beam waist radius ω_0 [2], where waist location and waist radius are the most sensitive parameters.

UV lasers are used in critical applications with especially high reliability requirements such as microelectronics, material processing, original equipment manufacturer (OEM) components and instrumentation as well as scientific research and government programs [3]. Thats why the stability of the beam parameters is of central importance. Therefore the goal of the present work is to implement a control software for an adaptive beam parameter correction system that can compensate for the occurring changes in the nonlinear crystal.

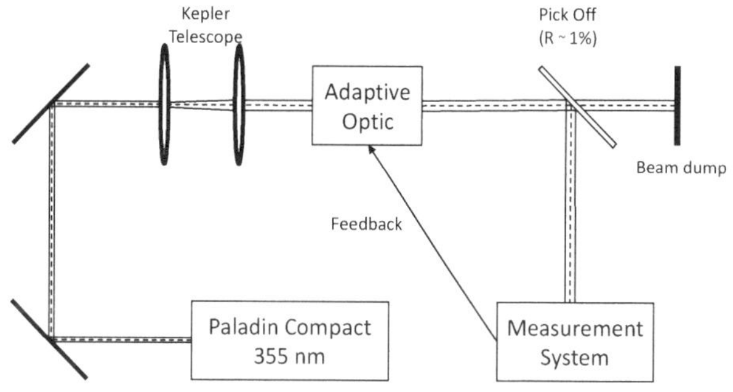

Figure 1: Schematic experimental setup consisting of laser, telescope, adaptive optic component, pick off for beam sampling and measurement system.

2   Material and Methods

2.1   Experimental Setup

The employed experimental setup, that is shown in Fig. 1, contains a Coherent Paladin Compact 355-4000 UV laser, which emits light with the wavelength 355 nm at a power of up to 4 W and an output beam radius of 0.5 mm. Via two mirrors, the laser beam is directed onto a Kepler telescope and enlarged to double its size. After passing the telescope, the laser beam is directed onto the adaptive optic components, which carry out the beam parameter correction. The adaptive optic components are therefore actuated via an Adafruit 16-Channel 12-bit Pulse Width Modulator (PWM)/Servo Driver over I^{2}C and controlled by an algorithm implemented in Python. The central component of the control circuit is a Raspberry Pi, which connects the measuring system and the Adafruit PWM board. The input data of the control loop is generated by scanning the beam waist radius ω_0 at different positions along the direction of propagation of the laser beam. Preliminary work at Coherent on this topic has shown, that an observation of the phase front detected by a camera-based detection system, is particularly suitable for the measurement and evaluation of the beam caustic.

The beam waist radii are hence continuously measured and transferred to the control algorithm running on the Raspberry Pi. During the algorithm run, the waist radius and waist location parameters are calculated and compared to the desired set value. Depending on the difference between the reference variable and the measured value, a control variable is determined and transmitted to the PWM drivers, triggering an according change of the adaptive optic components. This process is continued until the reference value is reached or the remaining error becomes negligible.

2.2   Theoretical Considerations on the Caustic of a Gaussian Beam

The Gaussian beam at fundamental mode embodies the theoretical optimum of focusing that can be achieved with a laser beam. In reality however, many lasers emit radiation with complicated intensity distributions. The propagation of a laser beam is determined by the radius of the beam waist ω_0 and by the wavelength. In large distances from the waist, the far field for $z >> z_R$, the expansion of the beam radius is approximately linear and is given by the far field divergence angle [4] [5]

$$\theta_0 = \frac{\lambda}{\pi \cdot \omega_0} = \frac{\omega_0}{z_R}. \tag{1}$$

At the point $z = 0$ on the optical axis, the beam has the beam waist radius ω_0, at which it has the smallest beam cross section and a plane wave front. In increasing distance from the beam waist in z-direction, the beam radius $\omega(z)$ increases according to

$$\omega(z) = \omega_0 \cdot \sqrt{1 + (\frac{z}{z_R})^2} \tag{2}$$

with the Rayleigh length z_R, in which $\omega(z)$ for $z = z_R$ is $\sqrt{2}$-times bigger than $\omega(0) = \omega_0$ [7]. The Rayleigh length z_R is the exact distance, at which the wavefront has its strongest curvature and can be described by

$$z_R = \frac{\omega_0^2 \cdot \pi}{\lambda}. \tag{3}$$

Effects, such as the oscillation of higher transverse modes or amplitude and phase disturbances due to inhomogeneous amplification in the laser medium, can cause the real beam profile to deviate from the ideal Gaussian profile. In order to quantify these deviations and to evaluate the generated beam quality, the beam parameter product (BPP)

$$BPP = \theta_0 \cdot \omega_0 = \frac{\lambda}{\pi} \cdot M^2 \tag{4}$$

with the beam waist radius ω_0, the far field divergence angle θ_0, the wavelength λ and the beam quality factor M^2 is calculated [5]. The smaller the BPP, the higher is the quality of the focused laser beam. Due to the fact that BPP = $\frac{\lambda}{\pi}$ applies to the Gaussian beam, M^2 is usually given as the ratio of calculated BPP and the optimum value, resulting in values $M^2 \geq 1$ [4]. The Paladin Compact used in this project has a beam quality factor of $M^2 < 1.2$ [6]. In order to concretize the points raised, Eq. 1 and Eq. 4 are used to determine the Rayleigh length for the parameters present in the experiment. The beam is observed after passing through the Kepler telescope, with the enlarged beam waist radius $\omega_0 = 1$ mm:

$$z_R = \frac{\omega_0^2 \cdot \pi}{\lambda \cdot M^2} = \frac{(1 \times 10^{-3}\,\text{m})^2 \cdot \pi}{355 \times 10^{-9}\,\text{m} \cdot 1^2} = 8.85\,\text{m}. \tag{5}$$

In the far field, the beam radius expands approximately linearly along the divergence angle θ_0. This and a simulated Gaussian beam width evolution for the parameters in Eq. 5 is shown in Fig. 2.

As mentioned in Sec. 1, nonlinear crystals are particularly prone to cumulative changes, that are caused by the frequency conversion process in high power UV laser systems. The most common manifestations of beam parameter instability are, among others, the axial beam waist location z_0

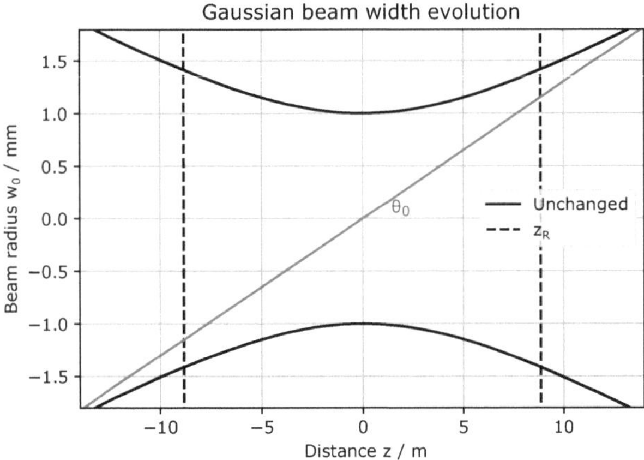

Figure 2: Laser beam caustic for $\lambda = 355$ nm, $\omega_0 = 1$ mm, $z_0 = 0$ and $M^2 = 1$ simulated in Python.

and the beam waist radius ω_0 [2]. Fig. 3 displays the most important possible effects on the beam caustic after the diversions in the nonlinear crystal fall into place. Either the beam waist location z_0 is shifted in z-direction, or the beam waist radius ω_0 changes. In Fig. 3 (a) the comparison between an unchanged beam caustic and a beam caustic, that is shifted to $z_0 = \frac{z_R}{10} = 0.885\,\mathrm{m}$, is displayed. The effects on the beam caustic for smaller or bigger beam waist radii are shown in Fig. 3 (b) and (c). In both cases it is obvious, that the respective Rayleigh lengths z_R are altered. Combinations of both effects can occur as well.

2.3 Control Simulation in Python

Adaptive optics cannot function without a control system that continuously monitors the parameters of the laser beam via the detection system, evaluates them and transmits control commands to the actuators of the adaptive optic components. Therefore, a control system was designed, which is adjusted to the requirements of the adaptive optics. The goal of the control algorithm is to adjust the laser beam with the adaptive optics in such a way, that a shifted beam waist position or a changed beam waist radius can be compensated.

The first approach to the problem is a control simulation programmed in Python. The input parameters, which in reality are supplied by the detection system, are generated as random variables in each program run. In this algorithm, a PI controller is implemented. The error $e(t)$ of the reference element is fed to the controller. The error is formed by the difference between the reference variable $w(t)$ and the control variable $x(t)$. In the following, the reference variable is referred to as the set value and the control variable as the actual value. The controller should ensure, that the actual value is adjusted to the set value, if it deviates from it. The procedure can be quick, slow or slowing down, depending on the control parameters of the proportional and integral component(k_p, k_i). The intervention is called time response. In this thesis, the time response is artificially generated by multiplication and integration [8]. In order to choose the suitable control parameters, the Ziegler-Nichols method is used. After the calculation by the controller, the manipulated variable $y(t)$, is thereupon transmitted to the control system to be influenced. A known control system element, which is utilized in the created simulation, is called PT1 element. Characteristic for the PT1 element is a proportional transmission behaviour with a first order delay. If input $y(t)$ is applied, this element generates the control variable $x(t)$ with a delay of time constant τ [9].

3 Results and Discussion

In reality, the calculated output variable $y(t)$ is transmitted to the PWM modulators as a voltage signal. Depending on the deviation between the set value and the actual value, the adaptive optics are then activated and ensure, that the altered beam caustic is gradually moved back into the desired position. Such typical transient oscillation processes

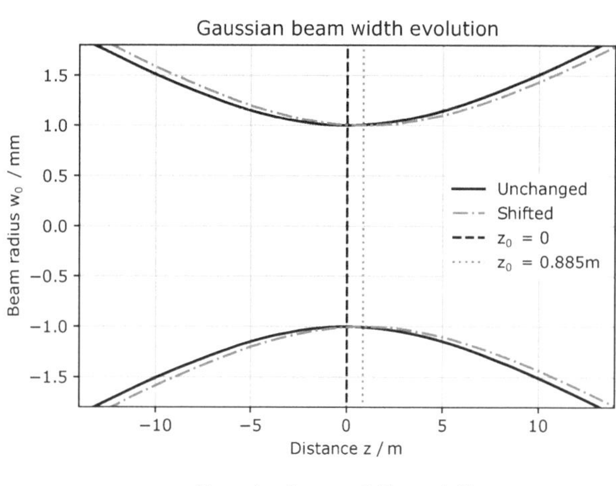

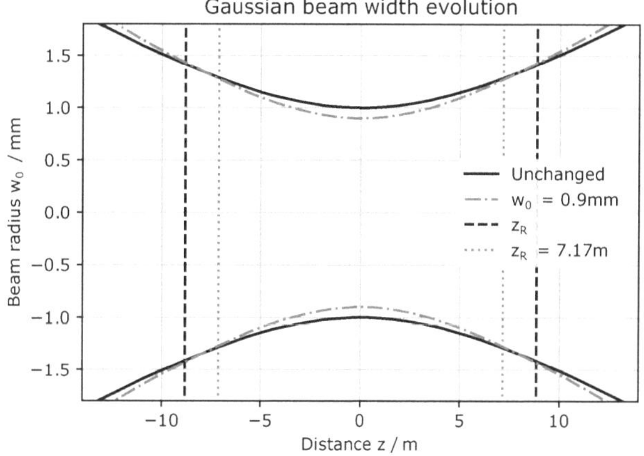

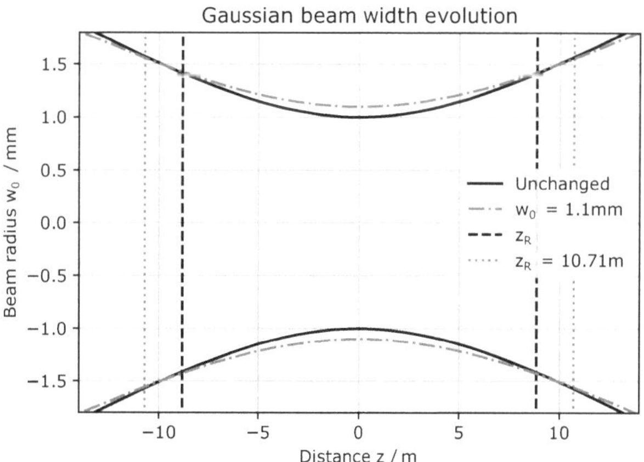

Figure 3: Comparison of beam caustic (Python): (a) Shift to $z_0 = \frac{z_R}{10} = 0.885\,\mathrm{m}$. (b) Smaller beam waist radius $\omega_0 = 0.9\,\mathrm{mm}$. (c) Bigger beam waist radius $\omega_0 = 1.1\,\mathrm{mm}$.

are shown in Fig. 4. The displayed curves describe the desired transient response of the actual value over a period of $t = 100\,\text{s}$. The shown time response already corresponds relatively closely to the set expectations on the PWM modulators. In case of Fig 4 (a), the gain constants were not adapted by the Ziegler-Nichols method, whereas in Fig 4 (b), the proportional gain constant is determined to $k_p = 0.855$ and the integral gain parameter can be defined as $k_i = 0.167$. While the starting input value is determined randomly with each program run, as described above, $w(t) = 1.2$ was selected as the desired set value. In the further course of the project, the detection system and beam parameter evaluation algorithm, that have been developed in preliminary works, are to be added on the basis of the implemented control simulation.

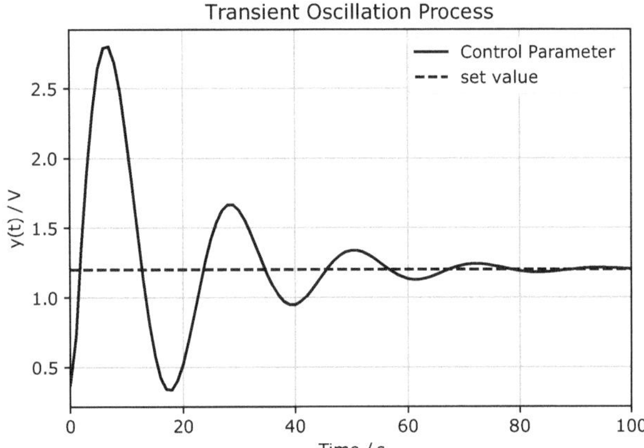

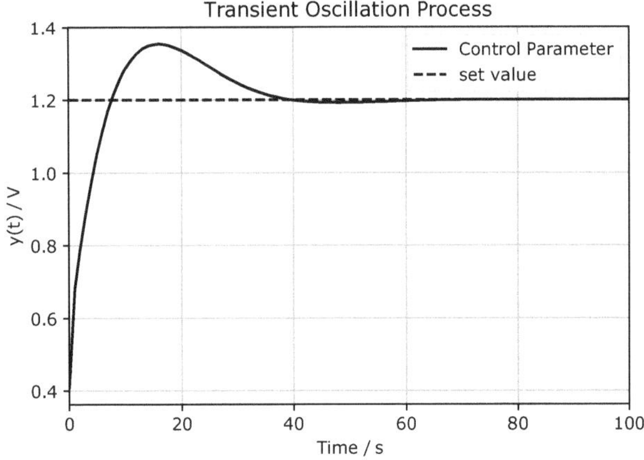

Figure 4: Simulated transient oscillation processes of the implemented PI control unit for one random actual value, respectively. (a) No further adaption of k_p or k_i. (b) Adaption with the Ziegler-Nichols method resulting in $k_p = 0.855$ as well as $k_i = 0.167$.

4 Conclusion

In the present work, simulations of different laser beam caustics with different input parameters were programmed and carried out in Python. In the further course of the project, the goal is to realize a universal beam parameter correction system, based on adaptive optics. To counter-act the degradations occuring in the nonlinear crystals, a control loop should therefore be implemented. The control algorithm independently records the necessary beam parameter values and modifies and corrects the undesired output characteristics according to the desired set values. In this paper, first simulations of a control program were presented, which serve as a basis for the extension of the already existing detection system and evaluation algorithm. Furthermore, measurements are to be carried out, in which the functionality of the implemented control system is to be tested. Additionally, the already existing experimental setup is to be realized in a reduced form, for example in a housing or on a breadboard.

Acknowledgement

The work has been carried out at Coherent LaserSystems GmbH & Co KG , Lübeck and supervised by Prof Dr. rer. nat. Robert Huber, Institute of Biomedical Optics, Universität zu Lübeck.

5 References

[1] L. Gruber, N. Hodgson, H. Hoffman, J. Angstenberger, *Algorithm for enhancing the lifetime of critical components in a laser system.* Patent US6890474B2, Newport Corp, 2002.

[2] G. Hollemann, A. Kneip, *Controlling laser beam parameters by crystal shifting.* Patent US10312659B1, Coherent Lasersystems GmbH & Co KG, 2019.

[3] Coherent LaserSystems GmbH & Co KG, *Coherent Laser Anwendungen.* Available: https://de.coherent.com/applications/ [last accessed on 2020-01-16].

[4] H.J. Eichler, J. Eichler, *Laser - Bauformen, Strahlführung, Anwendungen.* 8. Auflage, Springer, 2015.

[5] D. Meschede, *Optik, Licht und Laser.* 2. Auflage, Springer, 2005

[6] Coherent LaserSystems GmbH & Co KG, *Coherent Paladin Compact 355- 2000/4000 Datasheet.* Available: https://de.coherent.com/lasers/laser/paladin-compact-355 [last accessed on 2020-01-18].

[7] F. Pedrotti, L. Pedrotti, W. Bausch, H. Schmidt, *Optik für Ingenieure - Grundlagen.* 3. Auflage, Springer, 2005.

[8] W. Schneider, B. Heinrich, *Praktische Regelungstechnik*, 4.Auflage, Springer, 2017.

[9] J. Lunze, *Regelungstechnik 1 - Systemtheoretische Grundlagen, Analyse und Entwurf einschleifiger Regelungen.* 10. Auflage, Springer, 2014.

Improved semi-automatic quantification of experimental parameters after laser treatment for presbyopia

Jannik K. Luther [1,2,4], Stefan Kassumeh [3,4], and Reginald Birngruber [2,4]

[1] Medical Engineering Science, University of Lübeck, Jannik.Luther@student.uni-luebeck.de
[2] Institute for Biomedical Optics, University of Lübeck, bgb@bmo.uni-luebeck.de
[3] Department of Ophthalmology, Ludwig-Maximilians-University München, Stefan.Kassumeh@med.uni-muenchen.de
[4] Wellman Center for Photomedicine, Massachusetts General Hospital, Harvard Medical School, Boston MA, USA

Abstract

Presbyopia affects about a quarter of the global population and its high prevalence is estimated to rise due to an aging society. The trend is towards permanent surgical correction, that leads to a life without the need for glasses, while achieving optimal visual correction at the same time. To accomplish this, the quantification of relevant therapeutic parameters is crucial for a predictable procedural outcome. Our approach is to improve and extend the analytic processing of corneal OCT data on the basis of an existing MATLAB system of functionalities that allows the calculation of the refractive power of the cornea [1]. The adapted software system was applied to 85 experimental OCT datasets.
During this work, the software could be extended by a volume calculation method and the automatic determination of the optical zone. Nevertheless, further adaptations and improvements have to be realized to mitigate associated problems for a potential clinical use.

1   Introduction

Presbyopia is the irreversible loss of accommodative power leading to problems with reading and focusing objects nearby. Common, non-invasive correction options include the use of reading glasses or bifocal contact lenses. Due to the rising desire of people to live a life without glasses, different surgical correction methods, such as corneal implants and laser refractive procedures, were developed [2]. However, most of them can cause visual disturbances, e.g. halos, glare or might lead to a reduced visual acuity and contrast sensitivity [3]. Our new surgical approach to correct presbyopia is based on the compensation of the loss of accommodation capabilities by modifying the inner curvature of the cornea centrally. This is realized by injecting a transparent, liquid filler material into a laser-induced "pocket" inside the corneal center. The aim of this is to create bifocality by flattening the central posterior corneal curvature, reducing the negative refraction of the cornea leading to an increase of refractive power, without affecting the anterior or peripheral posterior curvature. The proof of concept of this kind of treatment was shown in other projects, where a similar attempt to treat hyperopia and astigmatism was evaluated [4]. Since this approach is innovative and experimental, current commercial software solutions are not applicable to the problem of quantifying the effect and distinct parameters of this new form of therapy. Therefore, there is an unmet need of an analytic software that calculates the induced change in refractive power of the cornea, the geometrical properties of the injected filler material and the resulting optical zone, to evaluate the experimental data.

2   Material and Methods

For the experiments, *in vitro* rabbit eyes (Pel-Freez Arkansas LLC, Rogers, AR, USA) were treated with a VisuMax® femtosecond laser (Carl Zeiss Meditec AG, Jena, Germany) operating at 1043 nm with a pulse duration between 220 - 580 fs and a pulse rate of 500 kHz. Circular corneal pockets of different diameters in a predetermined depth were cut into the corneal stroma. Following pocket creation, a solution of hyaluronic acid 1 % w/v (Millipore Sigma, Burlington, MA, USA) in phosphate-buffered saline (Thermo Fisher Scientific Inc., Waltham, MA, USA) was injected, with a 28-Gauge mouse femoral vein catheter (SAI Infusion Technologies, Lake Villa, IL, USA) attached to a 28.5-Gauge insulin syringe (Becton Dickinson, Franklin Lakes, NJ, USA). Three-dimensional OCT scans were acquired before and after the filler injection, using a TEL320C1 Spectral Domain OCT System (Thorlabs, Newton, NJ, USA) with a central wavelength of 1300 nm, an axial resolution of 3.5 μm and a lateral resolution of 23.6 μm in air. The A-Scan averaging was set to 2 and scanning speed was 76 kHz. The Field of View (FOV) was 8 mm $\times$ 8 mm $\times$ 3.15 mm (X $\times$ Y $\times$ Z) covering a sufficiently large area of the cornea for the analysis. Fig.1 shows an exemplary B-Scan of a cornea before and after the filler injection. All image processing steps were solely

realized using the numerical computing environment MAT-LAB 2018a (The Mathworks, Inc., Natick, MA, USA).
In previous works, a validated system of MATLAB functionalities merged into a workflow was implemented for the semi-automatic analysis that is used as a basis. This existing system enables the picking of surface points, the application of a spheric and a toroidal fit using a least-squares algorithm, the three dimensional raytracing of all optical surfaces and the correction of OCT- induced scanning errors [1].

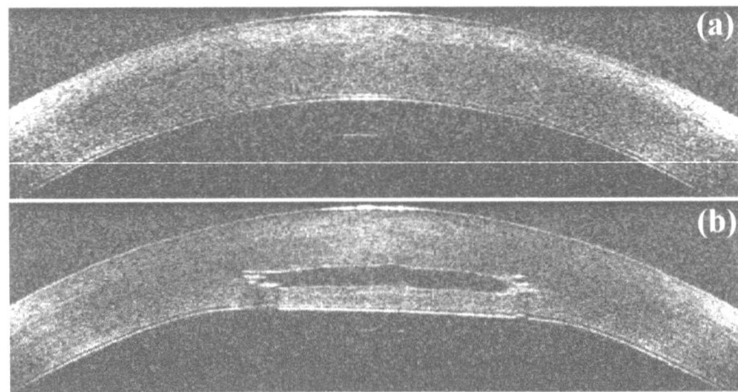

Figure 1: OCT B-Scan of a Cornea before (a) and after (b) the filler injection.

2.1 Quantification of the filler volume

For the quantification of the filler volume a new approach was implemented and evaluated, to compensate the uncertainty experimental data brings with themselves inherently. Ideally, this should be done without extending the runtime. The valid determination of the filler volume is essential for a real-time dosimetry during surgery because the injected filler volume correlates with the induced change in refractive power. To identify the anterior and posterior surface of the filler material, a concatenation of thresholding, morphological operators and the biggest connected component algorithm is used. The basic principle of the volume calculation approach is presented as a flowchart in Fig. 2. During this process, a whole stack of the filler volume is calculated and each voxel inside the volumetric dataset is classified either belonging to the filler or not. Therefore, a simple interception of this binary stack, counting of all voxel belonging to the detected filler structure and multiplying the sum with the given voxel volume, as proposed in [5], is our basic step for the volume calculation. The previous approach to calculate the filler volume was to determine the intersection of the geometric fits of the anterior and posterior curvature of the filler pocket, counting all the voxels enclosed by both geometric fits and then multiplying the sum with the given voxel volume. Current results of volume calculation using the previous method suggest that the approach used delivers incorrect results and leads to overstimation. A direct comparison between the actual injected filler volume with the proposed methods is not feasible because the surgeon removes part of the injected volume by applying pressure onto the cornea for a precise tuning. This removed amount of filler material cannot be measured.

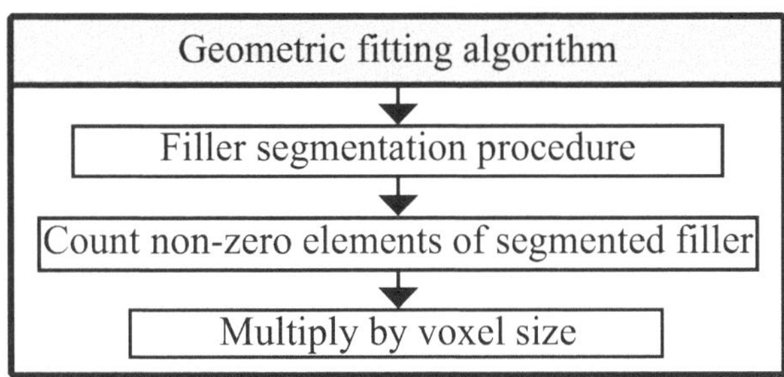

Figure 2: Flowchart of the voxel-based filler volume calculation method.

2.2 Determination and visualization of the optical zone of the central near-correction

For assessing the optical properties after treatment of presbyopia, the optical zone of the correction is an important measure. There are different definitions of the term "optical zone" [3], but the fundamental concept can be divided into the optical zone (OZ), which is the modified area in the cornea directly after treatment and the functional optical zone (FOZ), which depicts the diameter after alterations due to biological healing processes. Since refractive surgeries can affect the visual acuity differently, depending on the actual intervention, an adequate criterion to determine the optical zone is crucial. In our case, we wanted to specify the flattened area of the central posterior curvature. In line with flattening the central posterior curvature, inevitably a kink between the peripheral and the central corneal curvature is measurable, that is used as a measure of the optical zone. During the fitting procedure, both, the central and peripheral corneal curvature were approximated by spherical fits. Therefore, it is clearly noticeable, that those fits intersect in the center of the kink, defining the optical zone. The basic principle of the implemented procedure is presented in Fig. 3 below.

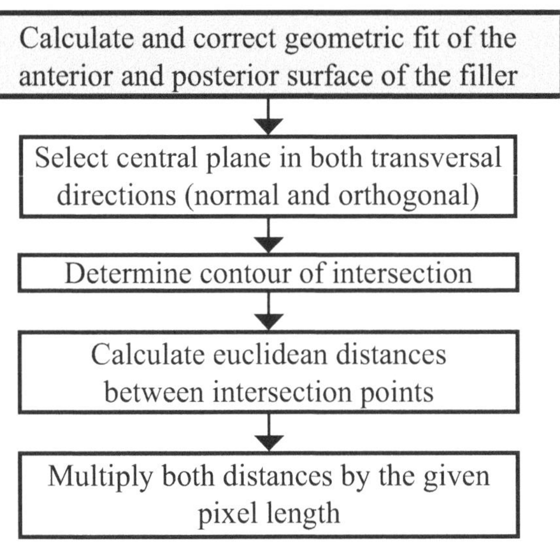

Figure 3: Flowchart of the determination of the optical zone.

3 Results and Discussion

In the following section, the calculations of the filler volume and the determination of the optical zone are exemplary evaluated for the 46 filler pockets with a diameter of 3 mm.

3.1 Calculation of the filler volume

After the implementation of the new filler volume calculation method, the outcome of both methods were compared visually and inspected on the basis of the OCT image data. In the previous examination of the filler volume, unexpected high values were obtained which could not be explained phenomenologically. Also, logical correlations and presumptive dependencies of the volume could not be shown with the current volume calculation method. In a direct visual comparison of a typical filler volume calculated using both methods as shown in Fig. 4, the difference between the geometric model and the experimental reality is apparent. The geometric fit of the filler surfaces encloses volume obtained using the voxel-based method. This difference leads in most cases to an overestimation of the filler volume, if the geometric fit approach is used for the volume calculation. In

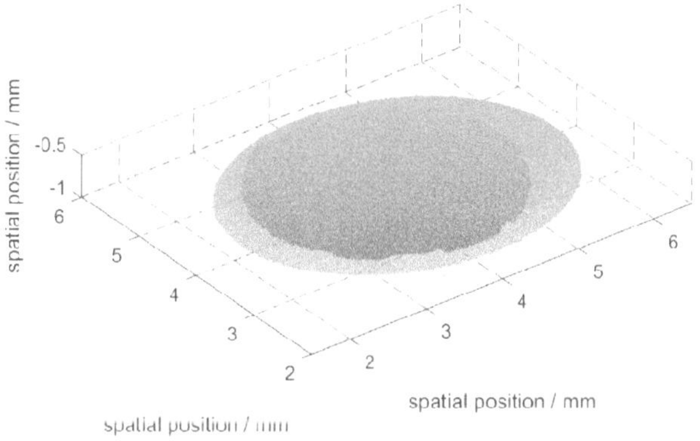

Figure 4: Modeled volume of the filler based on the geometric fit (the rounded shape). This shape reflects the geometric fit of the contour of the anterior and the posterior surface of a filler from an exemplary dataset. Inside this geometric shape, the modeled volume of the same filler based on a voxel-based thresholding and classification approach is displayed.

theory, the central thickness of the filler should be dependent on the amount of injected filler material since there is only one opening of the pocket and the cornea showed similar expansive behavior in [4]. It is therefore surprising and incomprehensible that we cannot detect a clear correlation at this stage with our geometric volume calculation method in Fig. 5 (upper plot). The voxel-based approach seems to enable a confirmation of the obvious assumption, that with increasing amounts of injected filler material, the central thickness of the filler increases too. The correlation coefficient of both variables of the voxel-based approach in Fig. 5, in the plot below, indicates a distinct correlation between both variables, with a correlation coefficient of 0.819. The

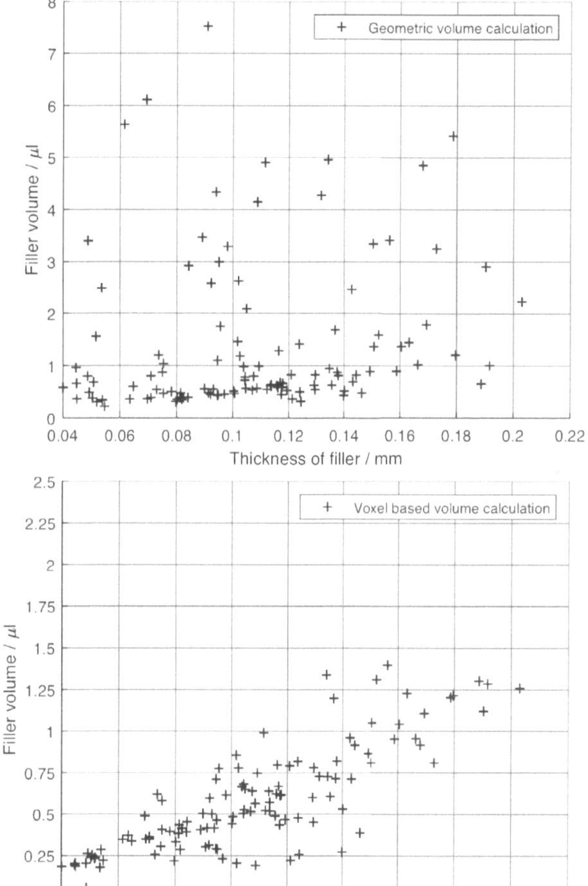

Figure 5: Calculated volume distributions, based on the geometric approximation (above) and the voxel-based approach (below), in dependency of the filler thickness. Varying filler volumes were injected, whereas the same datasets were evaluated with both volume calculation methods.

disadvantage of the fit-based volume calculation method is particularly visually apparent in Fig. 6.

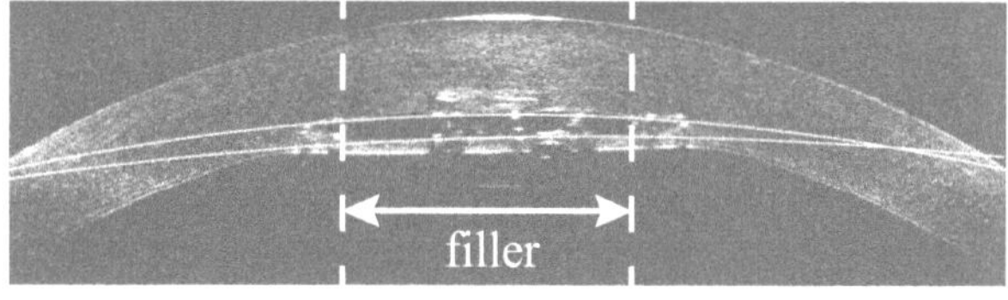

Figure 6: In the case of a near parallel geometric fit of the pocket surfaces in thin pockets, both lines do not intersect in the edges of the filler. Therefore the volume would be overestimated, although the fit reflects the surfaces adequately.

3.2 Determination of the optical zone

The determination of the optical zone could be successfully implemented by applying the explained principle in Fig. 3 to calculate the intersected area between the geometric fits of the peripheral and the central posterior curvature. The resulting optical zones for 46 different experimental OCT datasets were evaluated. In general, the optical zone is larger than the resulting pocket diameter.

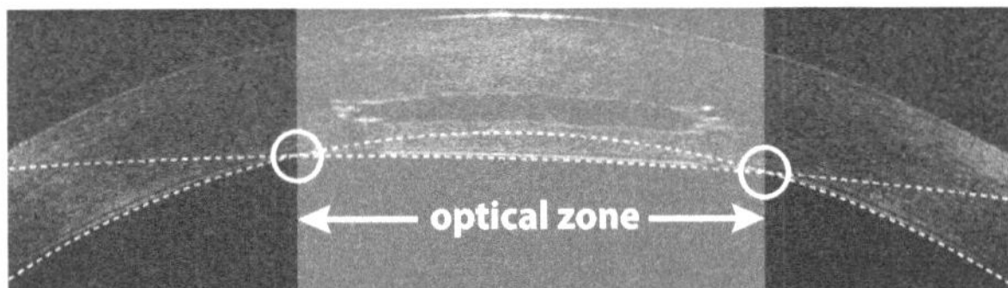

Figure 7: Exemplary visualization of the determined optical zone. The intersection between the geometric fits of the peripheral and the central posterior curvature is marked with circles.

3.3 Discussion

3.3.1 Real-time dosimetry

For a reliable and precise future treatment of presbyopia with the explained filler injection procedure, the existence of an on-line algorithm enabling a real-time dosimetry for the surgeon is necessary. The complete runtime for an entire image processing pipeline as it is utilized at the moment, requires about 340 seconds in total for a single eye. This time span is far too long for a useful on-line application in clinical practice, where the surgeon is dependent on quick feedback. Moreover, a semi-automatic approach for the point picking procedure of the surfaces needs to be avoided. In [6], an approach using the shortest-path graph search method to segment layers of the macular region automatically is proposed, that only requires 26.15 seconds to segment a whole OCT C-scan. Since the fundamental problem, the segmentation of certain layers, or in our case optical surfaces, is similar, it might be used to replace our semi-automatic and time-consuming current method of the picking procedure. In addition, an approach of training-based algorithms which learns the detection of boundaries by extraction of abstract descriptors while reducing the runtime could be conceivable, various examples of such methods can be found in current literature.

3.3.2 *In vivo* experiments

The presented implementations and functionalities were solely applied to the datasets acquired from *in vitro* eyes, whereas in clinical practice, the ideal condition of immobility is not met. It is necessary to investigate systematic mitigation of motion artifacts and to validate new methods during *in vivo* experiments, especially with regard to the current analysis software.

4 Conclusion

During the course of the work, the two new functionalities such as the calculation of the amount of the injected filler material and the determination of the optical zone, could be included into the image processing pipeline, adding value to the experimental data analysis. Those changes allowed a more detailed quantification of relevant experimental parameters. Besides the OCT imaging limitations, air bubbles due to the fs-laser-cutting mechanism are complicating the analysis. The implemented calculation of the filler volume is more precise, because the estimate is only based on truly classified image data, instead of the idealized assumption of fitting geometric models into the experimental data. A suitable criterion for the determination of the optical zone was found and realized.

In the course of this work, another step towards a more reliable and precise analysis of experimental corneal OCT-data was made.

Acknowledgement

The work has been carried out at the Wellman Center for Photomedicine at the Massachusetts General Hospital, Harvard Medical School and supervised by the Institute of Biomedical Optics, Universität zu Lübeck. The utilized fs-laser was kindly provided by the Carl Zeiss Meditec AG. The research team of the Wellman Center for Photomedicine provided exceptional support to this project. Furthermore, the previous work of Steffen Kaminsky and Katharina Brandt developing the basis of the used software system, deserves appreciation.

5 References

[1] K. Brandt, *Quantifizierung der Hornhautbrechkraft anhand von 3D-OCT-Bildern*. Bachelor's thesis (confidential), University of Lübeck, 2019

[2] G.L. Mancil et. al., *Optometric clinical practice guideline - care of the patient with presbyopia*. American Optometric Association, St. Louis, 2011

[3] J. Tabernero, S.D. Klyce, E.J. Sarver, P. Artal, *Functional optical zone of the cornea*. Investigative Ophthalmology & Visual Science, vol. 48, no. 3, Arvo Journals, Rockville, 2007

[4] S. Freidank, A. Vogel, R.R. Anderson, R. Birngruber, N. Linz, *Correction of hyperopia by intrastromal cutting and liquid filler injection*. Journal of Biomedical Optics, vol. 24, no. 5 SPIE, 2019

[5] H. Handels, *Medizinische Bildverarbeitung*. 2nd edition, Vieweg + Teubner, Wiesbaden, 2009

[6] J. Tian, B. Varga, G.M. Somfai, W.-H. Lee, W.E. Smiddy, D.C. DeBuc, *Real-Time automatic segmentation of optical coherence tomography volume data of the macular region*. PLoS ONE vol. 10, no. 8, Public Library of Science, San Francisco, 2015

5

E-Health

Analysis of Data Loss when using the Archiving and Exchange Interface for Practice Management Systems

Linus Barkow [1], Oliver Meincke [2], Hannes Ulrich [3], and Josef Ingenerf [3,4]

[1] Medical Informatics, Universität zu Lübeck, linus.barkow@student.uni-luebeck.de
[2] Epikur Software & IT-Service GmbH & Co. KG, Berlin, oliver.meincke@epikur.de
[3] IT Center for Clinical Research, Universität zu Lübeck, {h.ulrich, josef.ingenerf}@uni-luebeck.de
[4] Institute of Medical Informatics, Universität zu Lübeck, ingenerf@imi.uni-luebeck.de

Abstract

The archiving and exchange interface for practice management systems of the *Kassenärztliche Bundesvereinigung*, defined by FHIR (Fast Healthcare Interoperability Ressources) profiles with extensions, describes a new opportunity for medical practice owners to change the system provider. The expectation is to transfer an entire database of the legacy system to the new system without data loss. This paper analyzes the potential data loss performed by a parameter comparison. The results show that during an import on average 75 % of the parameters per profile are supported and on average only 49% of the reviewed parameters, existing in the exporting system, could be included in the interface representation.

1   Introduction

As a step of digitalization, the German "E-Health Act" regulates that all medical practices must be connected to the telematics infrastructure (TI) which contains the management of insured persons' master data with a medical practice management system (*Praxisverwaltungssystem* - PVS). Many medical institutions are already using such a system to manage daily tasks and quarterly accounting [1].

There are more than 200 different providers of practice management systems in Germany that offer different systems for the diverse specializations of the ambulatory health care like psychotherapy [1].

If the system can not be fully integrated into the workflow of the users or errors occur, changing the provider of the PVS normally results in major problems. The data transfer process is error-prone and time-consuming because the entire database needs to be transferred to the new system. However, the architectures and data models of different PVS are very heterogeneous so that a complete transfer and transformation is only possible with great effort and costs [2].

Therefore, the new SGB V §291d motivates and regulates the use of open interfaces in medical IT-systems in 2019:

> "In information technology systems that are used to collect, process and use personal patient data in 1. the contract physician care, 2. the contract dentist care and 3. hospitals, open and standardized interfaces for the system-neutral archiving of patient data as well as for the transfer of patient data during a system change are to be integrated."

The National Association of Statutory Health Insurance Physicians (*Kassenärztliche Bundesvereinigung* - KBV) is assigned for defining the interface specification. This new interface is called PVS archiving and exchange interface (*PVS-Archivierungs- und Wechselschnittstelle* - AWS) and is supposed to allow the human-readable archiving of patient related data and exchange between different systems [2].

The interface structure was defined to represent the most common data stored in a PVS. In the following, it will be analyzed how the interface can represent special features and parameters of a certain popular practice management system and examine the differences between both representations.

2   Material and Methods

2.1   Fast Healthcare Interoperability Ressources (FHIR)

The interface definition provided by the KBV describes various profiles which are based on the Fast Healthcare Interoperability Ressources (FHIR) standard. FHIR is an exchange format for medical data which was introduced by Health Level 7 (HL7) in 2013. The focus is to simplify the HL7 adoption significantly and to use existing open internet standards.

By using the Representational State Transfer (REST) architecture style described by Roy Fielding, the architecture becomes platform independent, scalable and usable on small devices due to the lightweight interfaces [3]. The objects can be loaded and manipulated via the HTTP methods: GET, POST, PUT/PATCH and DELETE. The unique resource identifiers (URIs) can be generated by the system or set manually. A versioning can be realized by adding a history number to the URI. Following the REST principle of different resource representations, it is possible to build the structures in XML, JSON or RDF format.

The profiles defined using FHIR represent the most important units of health information during data exchange. The specification considers these profiles as distinguished building blocks, which can be integrated into many processes as a logical entity. One profile instance refers to other profiles by using the unambiguous identifier so that a data representation can be composed by making use of various building blocks. Due to general applicability and complexity, the modules of the standard are limited to the most necessary but relevant parameters. To be able to integrate all information, it is possible to define and integrate further attributes via extensions. Depending on their content, the profiles are divided into different packages as well as sub-packages.

In addition to the profiles, Value Sets and Code Systems are used to aggregate codes from different ontologies, terminologies or enumerations. To guarantee interoperability, only specified codes are permitted in a defined context. By using the bundle resource, any number of profiles can be packaged and transferred in one object [4].

Multiple reference implementations for the FHIR standard already exist. The current version is R4 (4.0.1) which was released on 30 October 2019. The previous version STU 3, released in 2016, is currently available in version (3.0.2) and marked as "historical" [5].

2.2 Specification of the interface

The KBV specified a catalog of requirements to define and constrain the interface. The specified profiles constrain general STU 3 FHIR resources. To be able to represent all peculiarities of the German health system, many extensions are integrated and Code Systems are defined. The resulting profiles are grouped into five different bundles: address book, treatment modules, patient files, office supplies and appointments. For example, the patient file bundle is created for each patient and contains all patient-relevant data. The basic requirement of the KBV is that the complete data set of the PVS is meant to be exported by using all possible profiles. To enable parameter importance, some elements in the Structure Definition are marked with the property "mustSupport". These elements are supposed to be supported independently of cardinality if they are available in the information system.

When importing data records, all exported data must be transferred according to the scope of the importing system. The context of the information is meant to be preserved correctly [2], [6].

2.3 Interface Implementation

The AWS definition is available in XML files which contain the structure definitions of the various profiles. According to these definitions, objects were created using the open source Java reference implementation HAPI-FHIR [7] and filled with all available data from the database of a commercially available PVS.

For the mapping of the various coding systems, the given systems were extended by the tag <pvsEnumClass> and <pvsDate> to establish a reference to the enum in the PVS implementation. For each code at least one internal code was added in the tag <pvsCode> (see Fig. 1). Thus, it is possible to load the mapping of the codes automatically in the program and changes to the systems can be performed easily via the XML files. This feature is important if the interface version changes and new codes must be supported.

```xml
<CodeSystem xmlns="http://hl7.org/fhir">
    <url value="https://fhir.kbv.de/CodeSystem/
74_CS_AW_Medikamenten_Abgabeart"/>
    <version value="1.10.002.pvs.1"/>
    <name value="74_CS_AW_Medikamenten_Abgabeart"/>
    <status value="active"/>
    <date value="2018-10-05"/>
    <publisher value="Kassenärztliche
Bundesvereinigung"/>
    [...]
    <concept>
        <code value="Gruenes_Rezept"/>
        <display value="Grünes_Rezept"/>
        <pvsCode value="REZEPT_GREEN"/>
    </concept>
    <concept>
        <code value="Privatrezept"/>
        <display value="Privatrezept"/>
        <pvsCode value="REZEPT_PRIVAT"/>
        <pvsCode value="REZEPT_PRIVAT_LANG"/>
    </concept>
    [...]
    <pvsEnumClass value="FormEnum"/>
    <pvsDate value="2019-11-26"/>
</CodeSystem>
```

Figure 1: Example of a modified CodingSystem for mapping interface defined codes to the PVS codes.

As the structure of the objects in the PVS often does not match the profiles in the AWS interface, it is challenging to ensure that the data transfer is correct and as lossless as possible. Since a complete transfer of the data into the structured elements is not possible, further data is stored in free text fields or as a PDF attachment. This is often the case for documents such as orders, letters, recipes, labaratory data and medical documentation.

2.4 Calculation of parameter difference

For the quantitative and statistical evaluation of each AWS profile it is examined whether a parameter is supported by the reviewed PVS and vice versa. Each parameter was manually assigned to one of three labels. In addition to "Yes" and "No" for "is supported"/ "not supported", a parameter can be marked with "support unknown" if the support of this parameter is unclear. The occurrence of the labels for a profile is counted and the percentage in relation to the total parameter number is calculated. Note that only parameters which contain variable data were analyzed. If a parameter is predefined or does not contain any information, this parameter is not included in the statistic. To display the results, profiles were summarized to packages and sub-packages.

Additionally, all used profiles were analyzed to see how many of the "mustSupport" elements are supported. To find out which and how many profiles are supported by the PVS, the support was documented and evaluated.

Furthermore, the parameters stored in the tables of the PVS database were examined. The evaluation is supposed to determine how many of the stored parameters can be represented by the structured elements of the modified FHIR profiles. This analysis was only carried out for selected tables.

3 Results and Discussion

3.1 Import: Interface to PVS

The first statistical evaluation shows how many parameters of the interface definition the PVS supports. The Fig. 2 visualizes overall packages from that at least 62% of the parameters per object are supported and maximally 26% are not supported. On average 78% of the parameters per profile are provided. The evaluation included on the whole 1073 different parameters.

Examples of parameters that are not supported are the prefix qualifier and suffix part of names, the dagger-asterisk coding and version of ICD 10 codes and the coded dose information of drugs. Furthermore, profiles such as allergy, cure, vaccination, preventive power of attorney are not provided by the PVS because these information are not stored in the database.

Additional evaluation shows that the PVS supports 93% of the "mustSupport" parameters. All other information could be written in an unstructured parameter or the information is not available in the dataset. Therefore, the data loss during an import out of the FHIR structure is relatively small. In this context the quality of data is relevant for the result of the import. Only when given bundle data are validated against the structure definitions of the interface, an import can be performed.

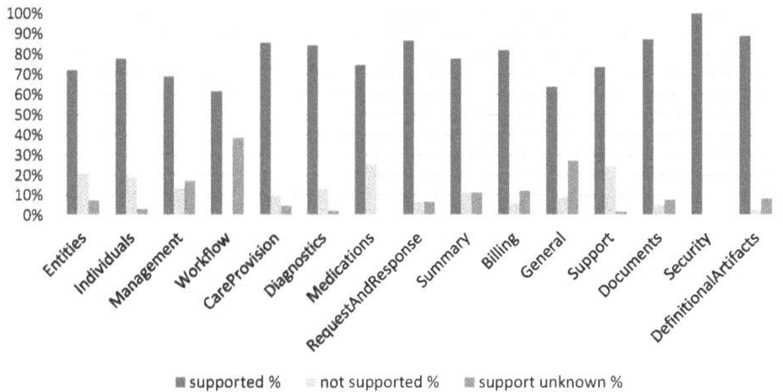

Figure 2: Percentage per package of supported and not supported FHIR interface parameters by the PVS.

3.2 Export: PVS to interface

The more relevant part is the export of all data stored in the database of the PVS. Fig. 3 presents that many parameters are not supported by the interface representation.

The diagram differentiates between five object groups and is based on the analysis of 637 parameter. For patient data only 49% could be exported into the structured parameters of the interface. These parameters often contain special and additional information like the blood type, living environment or company of the patient, data about relatives, information about births of children, assignments to groups or psychotherapy sessions that have already taken place. The patient package also includes many individual patient settings like the preferred contact method, therapeutic schemata, private service factors or session counter.

For the "user" class more parameters (54%) are supported but 35% are not supported. On the one hand information like the name, the address or the doctor number can be exported. On the other hand data like the relation between a doctor and their assistances and the assignment to user-profiles and the settings in the PVS are not directly provided in the interface. Additionally, it is neither possible to export data about the doctor's clearance like the type of confession or if the doctor is allowed to process children groups and so on. These data are not directly relevant to archive but for reimport, because these information are relevant for accounting.

The result of the "service" package is quite good because of 57% supported and only 24% not supported parameters. This is due to the fact that the interface is particularly specialized in billing objects. The billing-relevant data can be transferred with few losses. Own fee schedules, as well as service items, can only be integrated into the interface structure in a very complicated way. Information like the tax profiles, visibilities or internal marks can not be integrated into the FHIR structure. In this case, an interoperable exchange of additional data is difficult. The listed aspects also explain the good results for the analyzed "TimeLineElements" like diagnosis, therapy or observation elements.

The "QuickButton" class – which corresponds to the "DefinitionalArtifacts" in Fig. 2 – supports 43% and does not support 53%. The rough structure of the treatment modules can be represented well, but as soon as the parameters become more specific and individually adaptable, they can no longer be adopted. For example, key combinations, color settings or conditions for the individual components get lost. The occurring problems during the export of set-

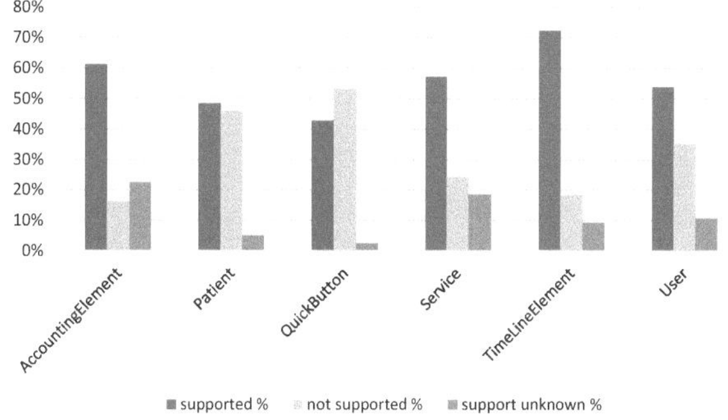

Figure 3: Percentage per package of supported and not supported PVS parameter by the interface structure.

tings highlight the focus of the interface. On the one hand, the doctors and therapist expect to transfer all data and settings to the new system. One the other hand it is impossible to transfer all settings and system specifics to permit a change without a long downtime of the systems. Therefore, the practice owner must spend some time to prepare and individualize the new system and then import all relevant and supported patient, accounting and documentation data. Finally, the most important data can be transferred without additional work.

3.3 Compatibility challenges

In addition to the described differences between the PVS database and the interface structure, there are further challenges regarding the compatibility.

The Code Systems and Value Sets show clear differences to the internal mappings. For example, in the interface specified Code System for the family status a distinction is made between more species than in the PVS system. The necessity of the codes for annulled, interlocutary, polygamous, never married and domestic partner may be arguable, but this example demonstrates that no lossless data exchange is possible. Another example is the attachment types of a document. Here the KBV specifies fewer codes than used in the PVS system. This could be due to the fact that many document types and form templates are no longer used. However, these are still available in the exporting PVS and are meant to be exported without information loss. Thus, it is difficult to integrate old forms into the given structures.

Another major problem is that the PVS saves the parameters in its own context and not in the context of the FHIR structures. That results in missing references in the FHIR instances. For example, the parameter diagnosis of a formula is only stored as an unstructured text in the PVS. The corresponding FHIR object includes the same information but contains references to other profiles, in this example the diagnosis object. This object cannot be generated from the String parameter. Therefore, it works for data archiving but a lossless data exchange is not possible.

One important disadvantage is that the interface is not downward compatible. An import can only be performed on data with the same interface version as the software component. A change of the interface version, as well as the update to the current FHIR version R4 and following updates, will entail major changes in the PVS.

4 Conclusion

We analyzed the data loss during data export and import with the archiving and exchange interface for practice management systems of the KBV. The results show that there is a considerable difference between the PVS and the interface representation. An export of all exportable data and a reimport in the same system will result in a significant data loss. Even if all parameters that are not supported could be integrated in unstructured text fields, a reimport is much more complicated and a loss of information not avoidable.

For archiving patient-related data, the interface is a good option because of the structured data representation and the generated human-readable output of each profile. Therefore, the medicine and therapist can access their data even if the system which generated the data is outdated.

The current version of the interface does not fulfill the requirements of the user because of the listed problems. Through a collaboration between software manufacturers and KBV the interface will be updated and currently not supported parameters will be added. It remains to be seen whether this interface will simplify the change between PVSs and how far it will be used for the archiving of practice data.

Acknowledgment

The work has been carried out at Epikur Software & IT-Service GmbH & Co. KG, Berlin and supervised by the Institute of Medical Informatics, Universität zu Lübeck.

5 References

[1] Kassenärztliche Bundesvereinigung (KBV), *Praxisverwaltungssysteme*. Available: https://www.kbv.de/html/pvs.php [last accessed on 2019-12-18].

[2] Dezernat Digitalisierung und IT in der Arztpraxis, *Festlegung der Archiv- und Wechselschnittstelle nach § 291d Abs. 1 SGB V*, 1st ed. Kassenärztliche Bundesvereinigung, May 2019. Available: ftp://ftp.kbv.de/ita-update/291d-Schnittstellen/PVS-Archivierungs-_Wechsel-Schnittstelle/ [last accessed on 2019-11-06].

[3] R. T. Fielding, *Architectural Styles and the Design of Network-based Software Architectures*. Ph.D. dissertation, University of California, Irvine, 2000.

[4] D. Bender and K. Sartipi, *HL7 FHIR: An Agile and RESTful approach to healthcare information exchange*. In: Proceedings of the 26th IEEE International Symposium on Computer-Based Medical Systems, IEEE, 2013.

[5] HL7.org, *FHIR: Fast healthcare interoperability resources*. 2019. Available: http://www.hl7.org/fhir [last accessed on 2019-12-16].

[6] Dezernat Digitalisierung und IT in der Arztpraxis, *Anforderungskatalog der Archiv- und Wechselschnittstelle (AW-SST)*, 1st ed. Kassenärztliche Bundesvereinigung, May 2019. Available: ftp://ftp.kbv.de/ita-update/291d-Schnittstellen/PVS-Archivierungs-_Wechsel-Schnittstelle/ [last accessed on 2019-11-06].

[7] University Health Network, *HAPI-FHIR fhir made simple*. 2019. Available: https://hapifhir.io/ [last accessed on 2019-12-16].

Multi-Label Learning with a Cone Based Geometric Model

Mena Leemhuis [1] and Özgür L. Özçep [2]

[1] Medical Informatics, Universität zu Lübeck, mena.leemhuis@student.uni-luebeck.de
[2] Institute of Information Systems, Universität zu Lübeck, oezcep@ifis.uni-luebeck.de

Abstract

Ontology information is in many fields available, particularly in medicine. The use of the thereby provided information can facilitate classification tasks, improve the results and guarantees ontologically correct results. But the incorporation of ontology information is challenging, especially for not hierarchical structured ontologies. Therefore an approach is presented which can be used for an arbitrary propositional $\mathcal{ALC}$-ontology. The method creates a representation of an ontology in the vector space. This geometric model is used for classification. The classification is based on a decomposition of the problem in a minor problem for each dimension of the geometric model. In this way ontological induced connections between elements are considered and the ontological correctness of the results is guaranteed. A test with the Gene Ontology results in a performance similar to a comparative approach. This shows its general applicability and underscores the necessity of further investigation.

1 Introduction

Multi-label classification problems, i.e. problems where each element can have multiple labels, are hard to compute and are a currently widely researched topic [1].

There are many strategies available for solving this kind of problem, such as decomposition into several minor problems, e.g. a separation of each class versus each other class, or a separation of each class from all others [1]. There are also combination methods such as K-nearest-neighbour or neural networks possible [1].

Another approach is to use ontology information to improve the classification result. An ontology is a structured representation of concepts and their relations. The use of ontologies is highly relevant for medical topics because many ontologies exist in this field, for example SNOMED CT [2] and the Gene Ontology [3]. Using an ontology for classification guarantees a result that conforms to the ontology, which is necessary for example for disease prediction.

The aim is to get a classifier which is trained with annotated data and results in annotations for test elements. In the annotated training data there could be information in different granularities, in the medical context e.g. that one person is known to have a broken leg and the other one is known to have a broken lower leg. With this classifier such relations and the thereof resulting knowledge are considered.

There are different approaches for learning with ontology information. In this paper the interpretation of the ontology as a cone based geometric model which was described by Özçep et al. [4] is used for the training of the classifier.

A geometric model is based on a transfer of an ontology into a high-dimensional Euclidian vector space, where every concept is placed in a convex region. The geometric model is fully expressive and can cover the description logic $\mathcal{ALC}$, thus it can also map negation and disjunction, which is not possible in many other ontology based approaches [4]. This interpretation is used because of its simple architecture and high expressiveness.

In this paper a multi-label classification approach is introduced, which is based on the decomposition of the main problem in many minor problems. The training set is separated into different sets of elements with the same properties. Classifiers are trained for different combinations of these elements. In this way not only obvious relations are learned but also relations which are based on ontology information. It follows that the number of elements usable for training and the separability increases.

The paper is structured as follows: The geometric model and its generation is presented in section 2. Next the approach for the geometric model based multi-label classification is shown. In section 3 the test results of the method on the Gene Ontology are given and compared to a similar but not ontology-based approach.

2 Material and Methods

2.1 Propositional $\mathcal{ALC}$

Description logic provides a formal language for knowledge representation including tools for drawing conclusions. Propositional $\mathcal{ALC}$ is a subtype of the description logic and is described by the grammar

$$C \rightarrow A \mid \bot \mid \top \mid C \sqcup C \mid C \sqcap C \mid \neg C, \qquad (1)$$

with an atomic concept A and an arbitrary concept descriptions C, but no relation symbols [5]. The $\mathcal{ALC}$ interpreta-

Name	Syntax	Semantics
top	$\top$	$\Delta^{\mathcal{I}}$
bottom	$\bot$	$\emptyset$
conjunction	$C \sqcap D$	$C^{\mathcal{I}} \cap D^{\mathcal{I}}$
disjunction	$C \sqcup D$	$C^{\mathcal{I}} \cup D^{\mathcal{I}}$
negation	$\neg C$	$\Delta^{\mathcal{I}} \setminus C^{\mathcal{I}}$

Table 1: Syntax and semantics for the DL $\mathcal{ALC}$ for an arbitrary interpretation $\mathcal{I}$

tion is $(\Delta, ()^{\mathcal{I}})$ consisting of the domain Δ (the space of possible elements) and an interpretation function $()^{\mathcal{I}}$ which maps constants to elements in Δ and concept names to subsets of Δ. The semantics of arbitrary concept descriptions is given in Table 1.

An ontology $\mathcal{O}$ consists of a tuple $(\mathcal{T}; \mathcal{A})$. The terminological-box ($\mathcal{T}$-box) contains general concept inclusions of the form $C \sqsubseteq D$, where C and D are arbitrary concepts, the assertional-box ($\mathcal{A}$-box) consists of facts of the form $C(a)$, which shows the affiliation of an constant a to a concept C.

2.2 Geometric models

A geometric model is a representation of an ontology in a Euclidian vector space. It is applicable to $\mathcal{ALC}$, but here only propositional $\mathcal{ALC}$ is used. The region-based idea represents the $\mathcal{T}$-box axioms in a geometric way, that means when $A \sqsubseteq B$, then A is a subspace of B in the model.

The main idea is to split the vector space into convex regions, because handling convex objects is computationally efficient. To preserve the convexity under disjunction and negation, a special convex structure - namely an axis aligned cone (al-cone) - is used [4].

Definition 1. *An al-cone is a special case of a closed convex cone. An al-cone in the n-dimensional space is of the form*

$$(X_1, ..., X_n) \text{ where each } X_i \in \{\mathbb{R}, \mathbb{R}_+, \mathbb{R}_-, \{0\}\}. \quad (2)$$

For example the al-cone $(X_1 = \mathbb{R}_+, X_2 = \mathbb{R}_+)$ would be the upper right quadrant in $\mathbb{R}^2$. The negation of an al-cone is defined by its polar cone, which is

$$X^\circ = \{v \in \mathbb{R}^n | \forall w \in X : \langle v, w \rangle \leq 0\}. \quad (3)$$

For the example al-cone it is $(X_1 = \mathbb{R}_-, X_2 = \mathbb{R}_-)$. Al-cones are consistent under intersection, conjunction, negation and disjunction and the distributive law is valid [4]. For better readability subsequently $\mathbb{R}, \mathbb{R}_+, \mathbb{R}_-, \{0\}$ are replaced by $u, +, -, 0$.

Every concept of an ontology is assigned to an al-cone. An operation on the al-cone of a concept, e.g. a conjunction, is executed per dimension. The concepts need to be assigned with respect to the $\mathcal{T}$-box axioms. The constants are placed in a region were the corresponding $\mathcal{A}$-box axioms are valid. The exact position depends on the approach that is being used and is explained in the next chapter.

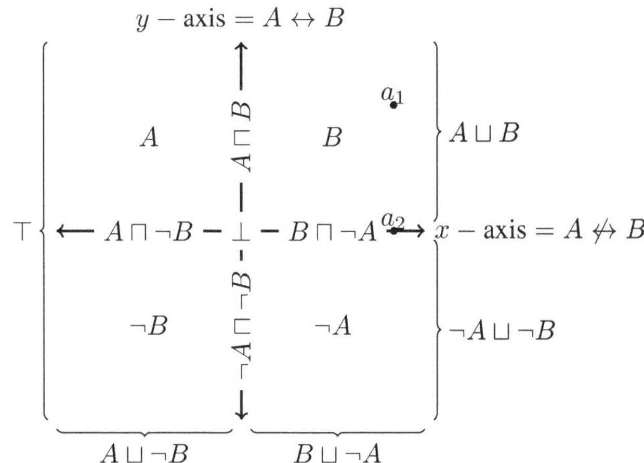

Figure 1: Example of a geometric model for an empty $\mathcal{T}$-box and two concepts A and B. The $\mathcal{A}$-box consists of $B(a_1), B(a_2)$ and $\neg A(a_2)$ [4]

Special cases are the top-concept $\top$ which is $\{u\}^n$ and thus cover the whole vector space and the bottom concept $\bot$, which is at the point of origin $\{0\}^n$.

It has been shown that the geometric model of an ontology is satisfiable iff the ontology is satisfiable [4].

A special feature of the geometric model is its ability to model partial knowledge. It is not obligatory that an element is an instance of a concept or of its negation, its assignment can also be unknown. Then, in the geometric model the element can be placed in a region which belongs neither to the concept nor to its negation. The element a_1 in Fig. 1 is in a region where it is neither in A nor in $\neg A$ and is therefore an example for the representation of partial knowledge.

Construction of a geometric model The geometric model for a given $\mathcal{T}$-box is constructed based on the set K of all possible fully specified concepts k of the ontology [4]. A concept is fully specified when it contains every atomic concept or its negation. The geometric model has the dimension $d = \left\lceil \frac{|K|}{2} \right\rceil$. A conjunction between fully specified concepts is not possible, so every k is placed on one half-axis, e.g. $B \sqcap \neg A \rightarrow (+, 0)$. The al-cone for each concept can be determined by disjunction of all k in which the concept appears positive. The corresponding negative concept can be found by negating the positive concept. With an empty $\mathcal{T}$-box and n concepts this results in 2^n fully specified concepts and thereby in $d = 2^{n-1}$ dimensions.

With a non-empty $\mathcal{T}$-box the number of possible k decreases, but it is still exponential in the most cases. The construction of the model is similar to the empty case.

The arbitrary assignment of the half-axes results in different geometric models, which are equivalent in the sense that they all model the same $\mathcal{T}$-box, but have different properties. Therefore some are more suitable for classification than others. This is analysed in the result section.

2.3 Multi-label classification with a geometric model

The geometric model can be used in combination with the $\mathcal{A}$-box axioms from the training data to train a classifier.

The main idea is to allocate every element x of the training data to a subspace of the vector space by creating a code vector $cv(x) = \{+, -, 0, u\}^d$. In this way a vector is not represented by an individual point in space but by an al-cone. In every al-cone there could be several elements. The creation of the vectors depends on whether the closed- or the open-world assumption is used. This is discussed later. For every dimension of the code-vectors a classifier is trained separately to divide the code-vectors in classes determined by their symbol in this dimension.

Closed vs. open-world-assumption Two interpretation variants for the elements as code vectors are possible: the closed- and the open-world assumption (CWA and OWA). In the CWA, full knowledge about the world is assumed, meaning that a concept which is not known to be true must be false. The OWA allows incomplete knowledge, so that a unknown concept can be either true or false [5].

The code vector for an element with respect to the OWA is a conjunction of the geometric model representation of every known concept and every known negated concept for the element. Unknown concepts are ignored.

In the CWA a code vector is a conjunction of the representations of all known and of the negations of all unknown concepts. Hence the representation of an element is a fully specified concept and has only one non-zero element.

The geometric interpretation of the elements is based on latent features and shows information about similarities and differences between them.

In each dimension each element could be $+, -, 0$ or u. An element with an u in this dimension doesn't incorporate any knowledge because its exact position is unknown. The space of 0 is part of $+$ and $-$, thus it is possible in the OWA, that an element assigned with $+$ or $-$ could also be 0. So elements are not strongly separable, but non-separable elements must have incomplete knowledge and belong to class 0.

This reasoning is viable for every dimension. This means that the geometric model interpretation of the elements allow to train a classifier for each dimension separately.

The classifier for each dimension For each dimension $1 \leq i \leq d$ all elements are separated into classes as follows:

$$X_{pos,i} = \{x | cv(x)_i = +\} \tag{4}$$

$$X_{neg,i} = \{x | cv(x)_i = -\} \tag{5}$$

$$X_{zero,i} = \{x | cv(x)_i = 0\}. \tag{6}$$

A code vector with an u at dimension i is ignored.

The training of the classifier follows Wan and Xu [6]. For each dimension i the separation of $X_{pos,i}, X_{neg,i}$ and $X_{zero,i}$ is learned: When only one of the three classes in dimension i is used, then a training of the classifier is not possible and all elements are assigned to the existing class. For two used classes a binary classifier is trained and the third class is ignored. When all three classes are used, a ternary classifier has to be trained. Therefore two binary classifiers are trained. The first one trains $X_{pos,i}$ against $X_{zero,i}$, the second one $X_{neg,i}$ against $X_{zero,i}$.

Classification of data For classification for each dimension the classification result for the test element is determined. For the ternary classifier the classification result is $+$, when one classifier results in 0 and the other one in $+$, analogue for $-$. In all other cases the result is 0.

The results of every dimension are concatenated and produce a code vector for the test element. This code vector is then placed in the geometric model. The element which needs to be annotated belongs to the concepts which are subsumed by its code vector in the model.

2.4 Experiments

The presented method can be used for every propositional $\mathcal{ALC}$ ontology. Here the Gene Ontology (GO) [3] is used. It does not contain negation or union and is hence a directed acyclic graph (DAG). The relations of GO have not been considered.

Data The data set of Saccharomyces cerevisiae (baker's or brewer's yeast) [7] is used. The geometric model needs exponential space. This was the reason for restricting the data to the first eight concepts of the GO to test the general functionality of the approach.

First the concepts of the training elements are extended in the way that all ancestor concepts of the given concepts are contained. Then every concept without enough elements representing it was deleted to facilitate the training process.

Implementation For classification a support vector machine with a polynomial kernel is used.

For comparison the approach of Wan and Xu [6] is implemented. It is a similar approach but without the use of ontology information and is based on a variant of the one-vs-one-classifier. Any two concepts are compared to each other in a ternary way. A separation of elements of concept A, of concept B and of concept $A \sqcap B$ is learned for all concepts A, B. Via a voting-scheme and a threshold the concepts of an element are obtained. For better comparability instead of the used Tri-class SVM the presented ternary SVM is used.

3 Results and Discussion

The classification of the test set results in similar performance measures for the OWA and the comparison approach and for worse results for the CWA (see Table 2).

The bad results of the CWA are caused by the full specification of the elements. Every element has only one

	Accuracy	Precision	Recall
OWA	0.522	0.827	0.620
CWA	0.029	0.039	0.031
Wan, Xu [6]	0.529	0.825	0.631

Table 2: Results for the approach with OWA, CWA and the approach of Wan and Xu [6]

non-zero dimension. This means that a special variant of an one-vs-all classifier is trained which compares every possible combination of classes with the rest. This results in a bad generalisation quality because of the small training sizes for the non-zero elements.

It follows that this approach is only suitable for problems where the OWA can be used.

An advantage of the presented approach for OWA is that it can only have ontological correct results, while the approach of Wan and Xu [6] can result in contradictions. Another advantage is the error tolerance for concepts at the top of the DAG. They cover a great space in the geometric model, so they are chosen even when there are some errors in the classification.

The OWA achieves no better results than the comparison method. This is caused by the simple structure of GO, which incorporates no negation or disjunction. Only subsumptions need to be considered. By the construction of the method of Wan and Xu [6] and with the extension of the data set to incorporate ancestors, subsumption is also learned in their approach. For example, when A is subsumed by B than A never appears alone and thus the subsumption relation is also learned. Another reason could be the fact, that classifying the GO is no classical weak supervised learning problem and thus the interpretation as an open-world problem is perhaps not appropriate.

The classification result is highly dependent on the choice of the geometric model. When a geometric model is chosen that has a high amount of $+$ and $-$ in the concept representations, than it results in the results given in Table 2. When a representation is chosen which has more 0 and u, then the recall decreases up to 15% in line with a simultaneous increase of precision. This is because with many u there are less possibilities to classify in a dimension and more elements with different properties in the same class, thus the recall is higher but the correctness of the answer is lower.

4   Conclusion

A method was presented which uses a given ontology for classification. An example application for the use of the geometric model for multi-label learning has been shown. Though the approach has no benefit for the classification of the GO, it has a similar performance, but ontological correct results. It needs to be tested for more complicated $\mathcal{ALC}$-ontologies were a simple approach such as the approach of Wan and Xu [6] can not incorporate the ontological know-ledge automatically.

It could also be tested for a classical weak supervised learning problem for which the possibility of the geometric model to model partial information could be useful.

For efficient usage of the approach a method for dimension reduction has to be found to reduce the exponential size of the geometric model.

The approach could be improved by using a different classifier for the dimensions instead of the ternary SVM to include that non-separable elements have to be 0 in the specific dimension and to incorporate information about the certainty of the classification to avoid misclassification. It is also possible to use the geometric model in a totally different way, e.g. as an output of a neural network. Another possibility is to expand the geometric model so as to learn not only concepts but also relations.

The experiment shows the general usability of the presented ontology-based multi-label classification approach and underscores the necessity of further investigation.

Acknowledgement

The work has been carried out and supervised by the Institute of Information Systems, Universität zu Lübeck.

5   References

[1] E. Gibaja and S. Ventura, "Multilabel learning: A review of the state of the art and ongoing research," *Wiley Interdisciplinary Reviews: Data Mining and Knowledge Discovery*, 11 2014.

[2] K. Donnelly, "Snomed-ct: The advanced terminology and coding system for ehealth," *Studies in health technology and informatics*, vol. 121, p. 279, 2006.

[3] M. Ashburner, C. A. Ball, J. A. Blake, D. Botstein, H. Butler, J. M. Cherry, A. P. Davis, K. Dolinski, S. S. Dwight, J. T. Eppig *et al.*, "Gene ontology: tool for the unification of biology," *Nature genetics*, vol. 25, no. 1, p. 25, 2000.

[4] Ö. Özçep, M. Leemhuis, and D. Wolter, "Cone semantics for logics with negation," *Unpublished.*

[5] F. Baader, D. Calvanese, D. McGuinness, D. Nardi, and P. F. Patel-Schneider, Eds., *The Description Logic Handbook: Theory, Implementation, and Applications.* Cambridge University Press, 2003.

[6] S.-P. Wan and J.-H. Xu, "A multi-label classification algorithm based on triple class support vector machine," in *2007 International Conference on Wavelet Analysis and Pattern Recognition*, vol. 4, Nov 2007, pp. 1447–1452.

[7] C. Vens, J. Struyf, L. Schietgat, S. Džeroski, and H. Blockeel, "Decision trees for hierarchical multi-label classification," *Machine Learning*, vol. 73, no. 2, p. 185, Aug 2008.

Generation of message transformers based on HL7 FHIR StructureMaps within the interface engine Mirth Connect

Kai Marten Vogl [1], Hannes Ulrich [2], Josef Ingenerf [3]
[1] Medical Informatics, Universität zu Lübeck, kai.vogl@student.uni-luebeck.de
[2] IT Center for Clinical Research (ITCR-L), Universität zu Lübeck, h.ulrich@uni-luebeck.de
[3] Institute of Medical Informatics, Universität zu Lübeck, josef.ingenerf@uni-luebeck.de

Abstract

Mirth Connect is an interface engine developed by Mirth Corporation. In its most basic form it can be utilized to transform incoming messages via user-defined functions contained as Javascript in channel structures. In order to test the feasibility of generating and deploying these transformations automatically, the container service Docker and the *StructureMap* resource of the FHIR health care data exchange standard were successfully leveraged and adapted to serve as the basis of this endeavor. The actual conversion mechanism was implemented in Java. Moreover, the efforts of a previous bachelor project concerning the *Unified Code for Units of Measure* were integrated as an advanced use case. It was possible to show that Mirth Connect can, in conjunction with FHIR and Docker, be leveraged for deploying common up to more complex transformation scenarios based on the mapping descriptions contained in *StructureMaps*.

1   Introduction

According to a study conducted by Germany's leading consulting company Roland Berger, it is only the biggest health institutes that are working with overarching IT solutions while the majority of smaller facilities lack even the basic framework of a hospital information system [1]. A solution for this set of problems lies within utilizing a middleware (commonly referred to as an interface engine) in order to ease the effects of heterogeneity of interfaces and to simplify the maintenance, definition and logging of communications and file transfers by transforming, routing, cloning and translating messages within a central application. The process of *Extract, Transfer, Load* (ETL) is a common approach for fusing and transforming data from multiple heterogeneous sources but it is often based on partly manual mappings or scripts with a complex overhead in terms of maintenance and responsibilities of each component considering the size of the overarching systems they are part of [2].

This research paper aims at showing that it is possible to automatically generate and deploy simple up to advanced message transformation scenarios based on mapping rules contained in FHIR's *StructureMaps* via the transformer component of the interface engine Mirth Connect. By leveraging this simple but powerful technology stack, many exemplary scenarios were implemented with one of them being based on previous research efforts concerning the *Unified Code for Units of Measure*(UCUM) that facilitates unambiguous electronic communication of quantities together with their physical units.

2   Material and Methods

2.1   Mirth Connect

Mirth Connect is an interface engine used in the healthcare industry that enables managing information via two-way sending of many types of messages. Communication in Mirth Connect is implemented through so called channels. Channels are one-to-many pipe structures that send and receive data via connectors that are distinguished as reader and writer respectively based on their direction [3]. Users specify the type of reader by choosing from a set of protocols such as TCP but it is also possible to read files locally or to connect several channels. As depicted in Fig. 1, they are then routed into a filter mechanism that allows to reject or accept incoming messages based on rules implemented by the user. They are then transformed and sent to the specified destinations via the writer.

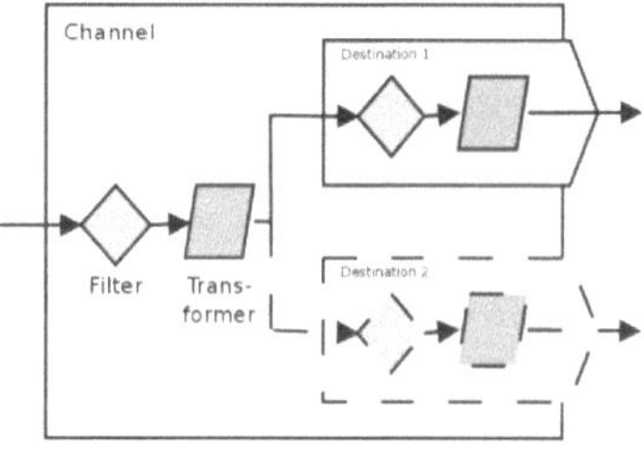

Figure 1: The basic structure of a channel in Mirth Connect with Destination 2 being optional [3].

The heart of a channel lies within its message transformer [4]. Despite the usage of common standards, it is unrealistic to deploy a uniform solution for the needs that arise within the context of heterogeneous systems and the specialized requirements of the healthcare world. Through the use of Javascript it is possible to transform incoming messages as needed. One of the key components in realizing the goal of this project are so called *code templates*. These predefined Javascript functions increase the reusability of code segments by enabling the user to call them by their function name, optionally with transfer parameters. Moreover they can be arranged into libraries for easy importing and exporting just like channels themselves. As a matter of fact, it is very simple to access and manipulate any of these components as they are stored as XML files which made manipulating them effortless. To further increase the capabilities of Mirth Connect there is an extension mechanism that enables a user to import external Java libraries [5]. It was utilized to provide simple HTTP capabilities to make use of the functionalities of a microservice from a previous research effort that successfully implemented the specifications of the Unified Code for Units of Measure.

2.2   StructureMaps

Fast Healthcare Interoperability Resources (FHIR) is a standard for health care data exchange developed by HL7. The FHIR specification includes a mapping language which consists of a concrete and an abstract syntax. It describes how one set of directed acyclic graphs can be transformed into another. For implementing the project, only the abstract part of the syntax is relevant which is outlined by the so called *StructureMaps*. *StructureMaps* are intended to enable experts in formats and intercompatibility to formalize their knowledge in order specify the entirety of the relations between two FHIR structures. This is accomplished through a detailed set of rules that are sufficient to carry out automated transformations of instance data in a *unidirectional* way from source to target [6].

Although *StuctureMaps* were devised for the use within the FHIR ecosystem, there is nothing stopping them from being used outside of it by separating them from the greater context of the associated mapping language. It is therefore unproblematic to disregard most of the contents of the *StructureMap* for this project as only the actual set of rules will be utilized in order to accomplish the automatic generation of fully valid Javascript code for use within Mirth Connect's message transformer. Each rule is identified by a name and consists of two basic components that outline which elements of the source are mapped to which element in the destination and how this transformation is accomplished. *StructureMaps* contain a set of basic operations like *copy*, *delete* and *truncate* but it is also possible to forward additional transformation parameters [6]. This aspect was additionally utilized to store, and later forward Javascript code to Mirth Connect.

2.3   Unified Code for Units of Measure

The Unified Code for Units of Measure (UCUM) is a coding system which encompasses all units of measure that are used internationally in contemporary science. By using this coding system, quantities and their accompanying physical unit can be transmitted unambiguously. The UCUM coding system consists of a simple set of terminal symbols that can be combined with one another according to a syntax to form an infinite amount of highly complex and expressive so called atomic units. Single symbols get linked via the algebraic operations of division and multiplication. Integral exponents signify the dimensional properties and metric units can additionally be assigned metric prefixes. In this way, each expression that is valid within UCUM has a precisely defined meaning. Modifications or additions to the tables of terminal symbols are therefore explicitly prohibited in order to promote *intercompatibility* [7].

UCUM is only provided as a specification and not as standardized implementation in software. Due to the fact that the openly accessible solutions were of variable quality and conformity compared to the official requirements, an effort was made to develop a microservice in a previous research project. This endeavour was highly successful and now resulted in including this service as part of the extension mechanism that accepts external Java libraries for use within Mirth Connect.

3   Results and Discussion

3.1   The basic architecture

As a side note, there is also the container service Docker. Docker is an open source software that realizes a virtualization mechanism on the operating system level. All Docker applications are based on images that bundle runtime and system tools aswell as dependencies. They only have a slim performance overhead and a single image can spawn numerous independent and parallel instances which are highly portable.

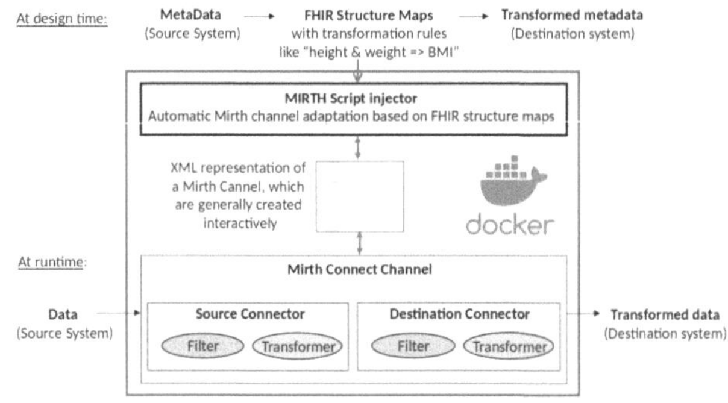

Figure 2: The basic boundaries of the Docker container.

As shown in Fig. 2, *StructureMaps* make up the metadata while Docker bundles Mirth Connect and its dependencies in a container that deals with the instance data as specified by them.

3.2 Generating the channel file

The software that generates the message transformer was implemented in Java. It expects three inputs, the actual *StructureMap* and two JSON files that conform to the structure of the expected input and output data. Mirth Connect refers to the latter two as the inbound and outbound template respectively. By storing these within the channels XML representation, it is easy to directly reference the elements in the transformation code. As stated, Mirth Connect stores channels as XML files which can be imported, exported and easily manipulated. Channels are not generated uniformly which becomes evident when comparing them to one another. There exists a set of basic modules which are brought together upon creation. Each method of input and output generally has its own structure with some shared overlap between them, most noteably the message transformer section and metadata. Additionally there is a special segment added should libraries for custom code be associated with the channel in question.

This project used a basic file reader and file writer structure. A dummy channel was created and associated with one of the default code libraries in order to generate the respective segment. The resulting channel then became the starting point for creating a *channel template* by stripping away the content in XML segments that would later be replaced based on the *StructureMap*. The idea behind this process was simple. Important elements were initially assessed by manually scouring the dummy file before replacing them with a unique identifier in place of its former content. This identifier could then get automatically replaced via a regular expression.

The most obvious segment to be replaced is of course the message transformer which contains the transformation commands as Javascript code but the input and output segments are open to customization efforts too, for example by setting the file location. Despite the fact that Mirth Connect generates IDs for each channel, they are apparently not used to distinguish between them. It was therefore important to find out that it is actually the channel's name that is used not only for display purposes but also for detecting collisions. In order to accomodate the *inbound* and *outbound* templates, they first needed to be encoded by Base64 to be processible in Mirth Connect.

As mentioned in the previous chapter, Mirth Connect allows users to define and store their own reusable functions in a *code template library*. By storing the command *evaluate* within the transform field of a *StructureMap*, it is possible to signify that an entire block of Javascript code needs to be processed and transferred. A regular expression fetches the function name and adds a unique identifier in order to avoid collisions as single code templates are distinguished by their name and not by their assigned ID just like channels themselves. The functions are then stored until each rule has been processed after which they get added into the library. It is important to note that escaping and unescaping is vital in order to go from JSON to plaintext to XML. After the *StructureMap* has been processed, the Java channel generator will output an XML file that is a valid Mirth Connect channel. It should be noted that there is no additional user input needed. The generated channels are not specific to this project either, meaning that the generated channels could be imported into any Mirth Connect application.

3.3 Working with StructureMaps

FHIR supports both JSON and XML formats for representing data and *StructureMaps* are no exception. This implementation focused on utilizing the JSON format specifically due to the fact that the project is part of a bigger software pipeline that deals with *StructureMaps* specified in the JSON format. The open source Java library *Gson* was used in order to extract the necessary data from a *StructureMap* via deserialization [8]. As stated in the previous chapter, *StructureMaps* contain a lot of data that is optional or not necessary for carrying out transformations outside of the FHIR context. Therefore, utilizing the rules section in particular was the central aspect.

```
{
    "name":"example_rule",          "target":[
    "source":[                          {
        {                                   "context":"target",
            "context":"source",             "contextType":"variable",
            "element":"first_name"          "element":"full_name",
            "variable":"a"                  "transform":"concat"
        },                              }
        {
            "context":"source",
            "element":"last_name",
            "variable":"b"
        },
    ],
```

Figure 3: An example of a rule which combines the first and last name of the source into a single name in the destination.

Fig. 3 depicts a common use case for a transformation. The *StructureMap* describes that the information contained within the elements *first_name* and *last_name* should be transformed via the basic command *concat*. They should be concatenated and the newly created string is then to be placed into the *full_name* element in the target. It should be noted that the *StructureMap* contains no information about the actual contents of instance data. It can merely point towards the location of the source and target elements. Fig. 2 already illustrated the divide between metadata and instance data more formally. Transfer parameters in the actual Javascript code can therefore not be arguments from instance data. They instead point towards the location in the instance data from which the data needs to be fetched during runtime.

```
tmp.full_name =
        concat('["first_name", "last_name"]', "full_name")
```

Figure 4: The *StructureMap* leads to generating this line of Javascript code.

Fig. 4 shows the actual rule that was generated. The signifier *tmp* refers to calling the instance data of the output. As there can be multiple sources but only one destination, it is necessary to store all source elements in an array. There is no way of predicting the amount of elements in the instance data but there needs to be a way of uniformly generating these assignments based on a single input argument.

3.4 Advanced use cases using extensions

The option to utilize Javascript code within the message transformer already offers a user many possibilities to solve the most common transformation problems but the already mentioned *extension* mechanism enables Mirth Connect to go beyond that and to make use of entire Java libraries. As an exemplary advanced use case, the UCUM microservice from a previous research effort was imported in order to transform the weight and height from the imperial units pound and weight into the metric system. In addition, the UCUM service was used to calculate the body mass index in the destination. Of course both this advanced scenario and the simple showcase of concatenating two strings are only two exemplary uses for this powerful technology stack. Mapping between differing value sets that mark genders by numbers instead of letters or sanitizing input data in regard to illegal characters are just a few of many more conceivable applications.

3.5 Discussion

The integration of a microservice taken from previous research efforts on UCUM went beyond the initial goals and carries with it two important implications. For one, extending the capabilites of Mirth Connect beyond what is provided by default and possible via Javascript functions, does not necessarily mean that the Java libraries to be included would have to be tailored for with that purpose in mind, meaning that already existing Java code can be included in a modular and clean fashion. Secondly, these existing solutions can be integrated into the Docker image in order to form a portable container that is quick to deploy (even in numerous parallel instances) and simpler to maintain with all dependencies bundled. It was not necessary to modify the basic architecture of the *StructureMap* reference in order to accommodate the injection of custom Javascript functions for utilization within the Mirth Connect message transformer and no further manual tweaking was needed in the process of generating channels.

Unfortunately the freely available documentation on the more niche aspects of Mirth Connect is rather lacking. It is apparent, that special use cases like extensions are intended to be reserved for a market that creates revenue through receiving support and buying software components such as the Core Extension Bundle. The unofficial developer guides by Shamil Nizamov were of great help in this regard.

4 Conclusion

It could successfully be shown that the *StructureMap* resource provided by FHIR could indeed be detached from the greater FHIR ecosystem in order to be leveraged for transformation purposes regarding the automatic generation of Mirth Connect channel structures. Compared to the ETL approach, this project is based on a standardized reference and the processing pipeline is highly automized with the necessary tools being freely available. Docker enables a quick and clean deployment based on specific transformation scenarios dictated by the initial *StructureMap*. It also bundles all necessary components for easy maintenance. Moreover, it should be highlighted that this project was a first attempt at utilizing the FHIR *StructureMap* resource and Mirth Connect in this way. The results of this research should therefore be considered as preliminary groundwork. Future research could focus on incorporating references to rules from other *StructureMaps* to possibly form transformations based on basic rule modules. Additionally, the default template for a Mirth Connect channel was assumed to be a file reader and writer but the transformer logic is not bound to the connector type. An obvious idea for future modifications lies in working with network connectors.

Acknowledgement

The work was supervised at the Institute for Medical Informatics Lübeck under the direction of Prof. Dr. Josef Ingenerf. I would like to thank Hannes Ulrich for providing his expertise on the FHIR standard.

5 References

[1] A. Sunyaev and J.M. Leimeister and A. Schweiger and H. Krcmar, *Integrationsarchitekturen für das Krankenhaus*. IMC - Information Management & Consulting, vol. 21, no. 1, pp. 28–35, 2006.

[2] F. Naumann and U. Leser, *Informationsintegration : Architekturen und Methoden zur Integration verteilter und heterogener Datenquellen (German)*. dpunkt.verlag, Heidelberg, pp.382–386, 2007

[3] P. Köppen and S. Langenberg, *Erfahrungsbericht über den Einsatz des open source Kommunikationsservers Mirth Connect am Universitätsklinikum Bonn*. Informatik 2014, Bonn, pp. 1465–1474, 2014.

[4] S. Nizamov, *Unofficial Developer's Guide to FHIR on Mirth Connect*. Shamil Publishing, pp. 14–18, 2016

[5] S. Nizamov, *Unofficial Mirth Connect v3.6 Developer's Guide*. Shamil Publishing, pp. 201, 2013

[6] *Implementation Support StructureMap*. Available: https://www.hl7.org/fhir/structuremap.html [last accessed on 2020-01-19].

[7] G. Schadow and C. McDonald, *Unified Code for Units of Measure Specification*. Available: http://unitsofmeasure.org/ucum.html [last accessed on 2020-01-19].

[8] I. Singh and J. Leitch and J. Wilson, *Gson User Guide*. Available: https://sites.google.com/site/gson/gson-user-guide [last accessed on 2020-02-04].

Creating a model instance for detection of *de novo* germline mutations using a neural network

Kristin Schultz [1], Axel Künstner [2], Michael Olbrich [2] and Hauke Busch [2]

[1] Medical Informatics, Universität zu Lübeck, kristin.schultz@student.uni-luebeck.de

[2] Medical Systems Biology, Institute of Experimental Dermatology, Universität zu Lübeck, axel.kuenstner@uni-luebeck.de, michael.olbrich@uksh.de, hauke.busch@uni-luebeck.de

Abstract

The key element in genetic analysis is the detection of variations. This *variant calling* can be performed by a number of different tools. *Variant callers* are constantly reevaluated and new approaches are being published. In this project the neural network *NeuSomatic* was put to test. The goal was to see how employable neural networks are, how they compare to most commonly used variant callers and also whether they can adapt to different data sets. First, pre-trained model instances were used to call somatic variants. Second, a model instance was trained to detect germline variants in synthesized samples. While the model instance did adapt to new input characteristics, it performed not as good as other variant callers. The high computational requirements did limit the overall performance.

1   Introduction

In recent years genomic sequencing and analysis has started to become integrated into clinical practice. The DNA sequence of any two people is 99.5 % identical. When looking at an individual's DNA analysis one is most commonly interested in these 0.5 % variations. Variations that affect all cells are called *germline* mutations and can be passed to offsprings. Germline mutations that are not inherited are called *de novo* variations. In contrast, variations that develop during lifetime are called *somatic* mutations. They occur in single cells and are often handled by DNA repair mechanisms. However, if they affect cell regulation genes they can cause diseases such as cancer. There are specialized callers for somatic and germline calling. Moreover, there are callers that only call *single nucleotide polymorphisms* (SNP). Calling structure variants, such as *deletions* and *insertions* (indels), is done by other tools and is generally considered a more difficult task. However, some SNP caller also call small indels from up to 50 nucleobases.

This project aims to evaluate whether an artificial neural network published for somatic variant calling is flexible enough to be trained on and applied to detect de novo germline mutations.

2   Material and Methods

2.1   Tools and Data in Genomic Analysis

Modern DNA sequencing produces *FASTA* files that contain a collection of fragment reads. Each read comes with information about the *base quality*, a per-base estimate of error emitted by the sequencing machine. For analysis the fragments are mapped to a human reference genome. There are many algorithms that perform read mapping, each resulting in a different alignment. Therefore the aligned output *BAM* file contains a *mapping quality* score for each base match. Aligned read data is handled with command-line tools, like SAMtools [1], that complete simple tasks such as translating from binary to human readable, but also more complex tasks like indexing and merging files. Reference genome, index files and aligned reads are the basis for every variant caller. Detected genetic variants are stored in a *Variant Call Format* (VCF) file, a schema in text form.

2.2   Choice and Execution of Variant Callers

In December of 2017 Google published *DeepVariant* [2], a deep convolutional neural network for variant calling. It gained a lot of attention and in the following years two other variant caller *Clairvoyante* [3] and *NeuSomatic* [4] were published, that also utilized neural networks. DeepVariant was excluded for this project because it required connection to the Google Cloud Platform. Clairvoyante was deliberately designed for somatic calling. NeuSomatic, while also designed for somatic calling, was inspired by a network for image recognition and was therefore considered a more general architecture and more likely to adapt to germline variant detection. NeuSomatic is available as a docker [5] image, a software container that creates a reproducible environment. Since it is not advised to run docker images on computer clusters the image was translated into a singularity [6] image. For competing variant callers, MuTect2 [7] was acquired through the *Genome Analysis Toolkit* (GATK) java executive. The somatic caller is part of the GATK best practice workflow [8]. For germline calling GATK provides *HaplotypeCaller*. *VarScan 2* [9] is a well established variant caller available as java executive, that only needs pre-processing with SAMtools' pileup function. *Strelka2* [10] works by configuring and running a python workflow.

2.3   Acquisition of Data Sets

DREAM Challenges are an open science effort to collectively find solutions to biomedical questions. The *ICGC-TCGA DREAM Genomic Mutation Calling Challenge* was proposed to identify the most accurate somatic mutation detection algorithms [11]. Six synthetic whole genome data sets, consisting of tumor-normal pairs, were published during the challenge. The tumor sample has added mutations, some randomly selected and some known in cancer-associated genes. The synthetic data sets 1, 2, 3 and 6 are accessible without ICGC approval. Set 3 was also available as *exome only* (WES) raw files. Since *whole genome sequencing* (WGS) requires way more computational power than WES, the exome version of set 3 was selected as test set for somatic calling.

The *RealTimeGenomics* (RTG) [12] software offers tools for genomic analysis and simulation commands. Training a model instance of NeuSomatic requires a set of ground truths for each training set. There are real world data sets that have been sequenced intensely and are often used for research purposes. While these sets offer a truth file for variants they come with uncertainties. For reproducibility and uniformity of the data sets, as well as the possibility to train with a larger number of sets, synthetic data sets are used for training. The simulation of a suitable read coverage for the synthetic WGS data lead to a memory overflow even when using 360 GB of RAM, which was at the time the maximum of available resources for students on the OMICS cluster. It was not possible to limit the reference genome to the exome only due to the internal indexing of the SDF file system RTG uses. The human reference genome *hg19* was therefore reduced to chromosome one. Based on this new reference population specific variants are simulated and collected in a VCF, mimicking the *Single Nucleotide Polymorphism Database* (dbSNP) for the synthetic samples. Two genomes, one for the father and one for the mother, were simulated using the reference genome and random picks from the synthetic dbSNP. The genome of the child is generated by combining the parents' genomes in a diploid fashion. For the synthetic data of chromosome one, 555 random de novo variants were added to the 15874 germline mutations already generated. To simulate the reads with sequencing errors an error model for the sequencing technology *Illumina paired end sequencing* was selected. Table 1 shows the values of all read relevant parameters. The values were chosen by recommendations of the official *Illumina* website.

All 13 synthetic trio data sets generated in this project were synthesized with the same parameters and references but random seeds for generating variations. A truth table VCF file was generated for each individual in the process. A VCF file containing only de novo mutations of each child was created by subtracting the child VCF file from every annotation found in the parents' VCF files.

Table 1: Parameters for Illumina paired end read simulation.

Parameter	Value
Coverage	50
Error rate	0.01
Left read length	250
Right read length	250
Min. fragment length	500
Max. fragment length	1000

2.4   Testing of NeuSomatic with Somatic Variant Calling

Included in the NeuSomatic docker image are two pre-trained models for variant calling in tumor-normal sample pairs. Both come in a standalone and an ensemble version. The ensemble version includes a pre-processing step that further filters the candidates with external variant callers, including Strelka and MuTect2. Since NeuSomatic is later compared to these variant callers, only the standalone version is used. One model was trained with the WGS Dream Challenge Stage 3 data set, herein after referred to as *dream3 model*. The data set was manipulated to emulate multiple contamination levels, which were used as input for training. The other model was trained with data sets of two tumor-normal-paired breast cancer cell lines (HCC1395 and HCC1395BL), provided by the *Sequencing Quality Control Phase II* (SEQC-II) consortium. Therefore this model is herein called *SEQC-II model*. It was trained with 20 WGS replicate pairs with different coverage, tumor impurities and library preparations.

In this project the raw FASTA files from the WES Dream Challenge Set 3 were pre-processed and aligned following the GATK guideline for best practice [8]. NeuSomatic required additional MD tags which were added with SAMtools' *calmd* function. The resulting BAM files were forwarded to dream3 and SEQC-II separately.

2.5   Training of NeuSomatic for Germline Variant Calling

Of the 13 synthesized trio data sets, number one to ten were used to train NeuSomatic for germline variant calling, creating the *germline model*. Sets 11 to 13 are considered test sets. The child genome read sample is committed to the network as tumor-sample, while the parents' read samples are piled up using SAMtools' *merge* function and committed as normal-sample. The training of the neural network required extensive resources. The pre-processing for each data set was split into ten jobs that each run with 12 cores and 75 GB RAM and required 11.3 hours on average. The

Table 2: Somatic variant calling of the Dream Challenge Stage 3 data set 3 (exome only) with common variant callers.

	MuTect2	Strelka2	VarScan 2
Precision	0.9324	0.9135	0.6195
Sensitivity	0.8817	0.4616	0.4670
F1 Score	0.9063	0.6133	0.5325

Table 3: Somatic variant calling of the Dream Challenge Stage 3 data set 3 (exome only) with NeuSomatic pre-trained models.

	dream3	SEQC-II
Precision	0.9160	0.7330
Sensitivity	0.7793	0.5618
F1 Score	0.8422	0.6361

model training exceeded memory limits when run on default settings. The number of variation candidates loaded into memory had to be reduced by a tenfold. Ten training epochs took 5 days to complete. The default number of training epochs is 1000, which would have taken around 500 days to complete. Since this exceeded the scope of this project the number of training epochs had to be reduced to 40. By default the learning rate decreases at 400 epochs, indicating that the system is expected to converge towards an solution at this point.

3 Results and Discussion

NeuSomatic as well as the comparing variant caller output VCF files. The comparison of these variant call sets is not trivial. When looking at the same locations in ground truth and called variant set complications arise due to possible differences in representation of variants. Therefore the utilities package *RTG tools* is used for evaluation. The RTG command *vcfeval* minimizes false positives and negatives for a set prediction score field in the VCF file. The selected field was *QUAL*, which states a scaled probability that a polymorphism exists in the sequencing data. The vcfeval algorithm optimizes the call set's interpretation by scaling up a threshold for QUAL fields to consider valid. All common variant callers, however, already filter their results for high QUAL fields and thus the algorithm is not able to further minimize false positives. They did, however, benefit from the a dynamic-programming algorithm which is used by vcfeval to deal with call representation discrepancies.

3.1 Somatic Variant Calling

Table 2 shows that MuTect2 performs best in all metrics. Strelka2 has good precision with > 0.9 but only moderate sensitivity. VarScan 2 performs moderately in both sensitivity and precision. MuTect2's good performance may be due to its fine tuned filters for somatic SNP calling. Strelka2 and VarScan 2 both output indel calling along SNP calling and offer germline calling, MuTect2 does neither. It can be assumed that Strelka2 and VarScan 2 trade performance in SNP calling for broader application possibilities. MuTect2 may also benefit from it being part of the GATK pipeline, whose recommendations were followed during pre-processing. NeuSomatic's performance differs depending on the model as seen in Fig. 1 and Table 3.
Model SEQC-II was trained with more tumor-normal pairs than dream3 and is recommended to use by NeuSomatic's

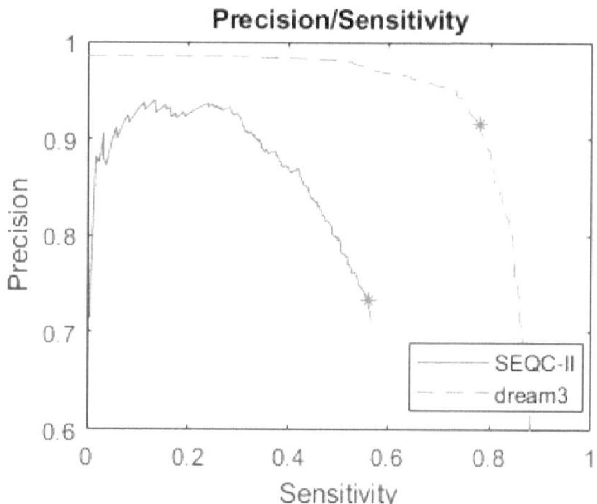

Figure 1: Sensitivity and precision curve of dream3 and SEQC-II NeuSomatic models. The best QUAL threshold for dream3 was considered 22.447. The best QUAL threshold for SEQC-II was considered 5.472. The best F1 scores are marked with *.

authors. However, it performs worse than model dream3, which does well on all metrics. While MuTect2 still performs better, dream3 outperforms VarScan 2 and Strelka2. The bad performance of SEQC-II might come from the difference in tumor type. SEQC-II was trained with two breast cancer samples while Dream Challenge data sets are simulated to compare to pancreatic and prostate cancer samples.

3.2 Germline Variant Calling

This evaluation will consider the sensitivity, precision, and F1 score of the calls for all germline variants. The sensitivity for de novo mutations is considered separately and marked as *sensitivity**. By doing so it can be ruled out that de novo mutations are silently ignored. The germline model was evaluated three times with intermediate steps at ten, 20 and 40 training epochs. The results are shown in Table 5. HaplotypeCaller and Strelka2 achieve very high scores in all metrics, as seen in Table 4. When using NeuSomatic somatic models for germline calling the model is not able to result in any calls. Therefore any of the pre-trained models and germline-trained models are distinct and not interchangeable. The models trained for germline calling achieve only moderate results, but sensitivity and F1 score increase with the number of training epochs, which indicates that the model might be able to achieve good results given enough training. Since it was not possible to train adequately NeuSomatic's capabilities could not be fully

Table 4: Evaluation of different common germline variant callers. Values from data sets 11 to 13 were averaged.

	HaplotypeCaller	Strelka2	VarScan 2
Precision	0.9994	0.9989	0.9847
Sensitivity	0.9941	0.9910	0.9226
Sensitivity*	0.9938	0.9882	0.9216
F1 Score	0.9967	0.9950	0.9526

Table 5: Evaluation of NeuSomatic Germline Model trained with ten, 20 or 40 epochs. Values from data sets 11 to 13 were averaged.

	Epoch 10	Epoch 20	Epoch 40
Precision	0.7628	0.7644	0.7633
Sensitivity	0.7518	0.7542	0.7574
Sensitivity*	0.8729	0.9103	0.9215
F1 Score	0.7572	0.7593	0.7603

ascertained. Still the network has shown the ability to adapt to other data. Instead of dealing with one genome present in a tumor-normal pair it was able to detect germline variants when given three individual genomes of a trio sample.

4   Conclusion

The pre-trained models of the neural network were easy to set up but required high computational resources for pre-processing. In comparison to common variant callers, which were even easier to use and show good results, the network was less practicable. Moreover, pre-trained model instances did perform only moderately well compared to other somatic variant callers.

It was possible to train a model that adapted to the new problem of germline calling, which shows a potential that can be explored further. The model still fell short compared to other germline variant callers but did better than SEQC-II did in the test somatic calling.

Further, one must consider that the quality of a neural network depends on its training sets. Synthesized data sets do not adequately represent real world data. This problem is exacerbated by the lack of transparency in machine learning black box models. However, large real world data sets with known truth variants are not yet available. When dealing with real world genomics data, acquiring ground truth for a single data set is often done in a big project. These projects consider raw data from by multiple different sequencing machines and new technologies are getting available as we are currently in the transition phase from short read to long read sequencing. The library of fully analyzed genomic data sets may increase in the future and enable a reliable training of neural networks for variant calling.

Acknowledgement

The work has been carried out at the research group of medical systems biology, which is part of the Lübeck Institute for Experimental Dermatology (LIED) of the University of Lübeck.

5   References

[1] H., Li, *The Sequence Alignment/Map Format and SAM-tools.* Bioinformatics, vol. 25, pp. 2078–2079, 2009.

[2] R. Poplin et al., *A Universal SNP and Small-Indel Variant Caller Using Deep Neural Networks.* Nature Biotechnology, vol. 36, 2018.

[3] R. Luo, F.J. Sedlazeck, T.-W., Lam and M. Schatz, *Clairvoyante: A Multi-Task Convolutional Deep Neural Network for Variant Calling in Single Molecule Sequencing.* bioRxiv, 2018.

[4] S.M.E. Sahraeian, R. Liu, B. Lau, M. Mohiyuddin and H.Y.K. Lam, *Deep Convolutional Neural Networks for Accurate Somatic Mutation Detection.* bioRxiv, 2018.

[5] D. Merkel, *Docker: Lightweight Linux Containers for Consistent Development and Deployment.* Linux journal, vol. 2014, pp. 2, 2014.

[6] G.M. Kurtzer, V. Sochat and M.W. Bauer, *Singularity: Scientific Containers for Mobility of Compute.* PloS one, vol. 12, pp. e0177459, 2017.

[7] D. Benjamin et al., *Calling Somatic SNVs and Indels with Mutect2.* BioRxiv, 2019.

[8] A. Prabhakaran et al., *Infrastructure for Deploying GATK Best Practices Pipeline.* Intel White Paper, 2016.

[9] D.C., Koboldt, D.E., Larson and R.K. Wilson, *Using VarScan 2 for Germline Variant Calling and Somatic Mutation Detection.* Current protocols in bioinformatics, vol. 44, pp. 15–4, 2013.

[10] C. T. Saunders et al., *Strelka: Accurate Somatic Small-Variant Calling from Sequenced Tumor–Normal Sample Pairs.* Bioinformatics, vol. 28, pp. 1811–1817, 2012.

[11] A. ,Ewing et al., *Combining Tumor Genome Simulation with Crowdsourcing to Benchmark Somatic Single-Nucleotide-Variant Detection.* PloS one, vol. 8, pp. e60204, 2013.

[12] J.G. Cleary, R. Littin,L. Trigg,I. Sean and B. Hilbush, *Quantitative Analysis of Shotgun Metagenomic Data with the Real Time Genomics Platform.* JBT, vol. 24, pp. S33, 2013.

Deploying a molecular tumor board on an open platform

Niklas Reimer [1], Hauke Busch [2], Josef Ingenerf [3,4]

[1] Medical Informatics, Universität zu Lübeck, niklas.reimer@student.uni-luebeck.de
[2] Institute for Experimental Dermatology, Universität zu Lübeck, hauke.busch@uksh.de
[3] Institute of Medical Informatics, Universität zu Lübeck, josef.ingenerf@uni-luebeck.de
[4] IT Center for Clinical Research, Lübeck (ITCR-L), Lübeck, Universität zu Lübeck, josef.ingenerf@uni-luebeck.de

Abstract

Decision support and interoperability have become very popular and important in health care. One way to meet these requirements is to build an open data platform. Here we develop a toolkit with a modular architecture that helps with the data integration based on the openEHR platform. We use it to deploy a molecular tumor board for cancer treatment that uses clinical and genomic data out of this platform.

1 Introduction

During the past few years, decision support tools have become more and more popular in health care, especially when it comes to cancer treatment using genomic data. Using such decision support tools requires interoperability so that structured data can be transferred from the hospital information system (*HIS*) and other locations [1].

In most cases, data is stored in a proprietary format maintained by a single vendor without the ability of direct access. In this case, data can just be accessed using interfaces that serve the data either in standardized or in proprietary formats. To provide more flexible and reliable health platforms as well as tighter coupling between research and care there is a slight movement to open platforms that provide direct access to the data. This is also the intention of the HiGHmed consortium [2] which is part of the Medical Informatics Initiative Germany. HiGHmed decided to use openEHR [3] as their primary platform to store clinical data to enable cross-institutional data analysis on participating locations.

At the same time technologies like genome sequencing have fallen in price. Therefore they have become a more affordable method in cancer research. Using it will require comprehensive documentation about treatment decisions in the patient's health record.

To bring all this together, first the patient's data needs to be extracted from the HIS and together with the analyzed genomic data committed to the openEHR based platform [1]. In the second step, the decision support tools need access to the data in the repository. The last step is to transfer documented decisions from the decision support tool back into the patient's electronic health record.

In the following, we describe how we developed a toolkit that is capable of supporting this workflow. Due to its modular architecture, it will also be possible to extend the toolkit for further applications or use cases.

2 Material and Methods

In the following the software components and the data that we used will be explained.

2.1 Input data

2.1.1 Clinical data

Clinical data is provided in a structured format that is based on a standard like XML or the HL7 family. For cancer patients, they are based on the ADT/GEKID standard [4] providing demographic information like sex and age as well as the ICD-O codes, gradings, histologic results and information about surgeries.

2.1.2 Whole-exome sequencing (*WES*)

While the human genome consists of more than three billion pairs of the four nucleobases adenine, guanine, cytosine, and thymine the exome makes up about 1.1% [5]. This is the part that is potentially protein-coding and contains most of the pathogenetic mutations making the sequencing of the exome a much more reliable method for the diagnostic of genetic disorders. The data is stored in *FASTQ* files that contain an identifier, the sequence of nucleobases and a quality value for each base. The sequencing is done in two reads, having a size of about 4 gigabytes each.

In the special field of oncology, WES is used to analyze differences between a sample from normal tissue and a sample from tumor tissue. The WES data sets used in this work were obtained from the Texas Cancer Research Biobank [6].

2.2 MIRACUM-Pipe

To gain useful information on WES data the research field of bioinformatics has developed a whole set of analysis tools for genetic raw data. In November 2019

MICRACUM-Pipe was released as a part of the Medical Informatics Initiative Germany [7]. It provides a fully automatic pipeline of various analysis tools for alignment, variant calling and annotation of genomic raw data using tools like ANNOVAR [8]. While the results of each step can be used for further processing, the pipeline also generates a human-readable report.

2.3 openEHR

Based on the ISO 13606 standard, openEHR is an open standard for Electronic Health Records (*EHRs*). It defines a Reference Model and the Archetype Definition Language to model basic concepts like blood pressure in archetypes. Those archetypes can be combined to build a template. This architecture allows using the same archetype for multiple purposes as well as sharing them on national or international platforms called *Clinical Knowledge Managers* [9]. A contribution to an EHR is called *Composition* and will be stored via a RESTful interface in the repository of an openEHR server. Data inside the repository can be accessed using Archetype Query Language (*AQL*), a special language for openEHR which is similar to common query languages like SQL.

This work was developed around EHRbase [10], a server that implements the openEHR standard but can also be applied to every other server which is compliant to the openEHR specifications [3].

2.4 cBioPortal

To organize and visualize studies in cancer genomic, the Memorial Sloan Kettering Cancer Center started the cBioPortal project [11] [12]. It provides the functionality to compare different cancer studies and visualizes clinical and mutation data as well as image data. While displaying the data cBioPortal is also capable of using up to date online information from databases like OncoKB to provide additional information about the impact of mutations.

To use it as a decision support tool, recent development has reached out to add features that will allow proper documentation, an administrative structure with users, groups and permission management [13].

3 Developing an openEHR mapping tool

The whole process can be split into a part that is generic for openEHR and a part that is specific to our intended use case. The first part involves the generation of EHR extracts based on templates, managing EHRs and providing access to the repository by using the RESTful interface. The second part is about processing the input data before contribution to the repository and processing of data obtained from the repository. These circumstances lead to a modular architecture as shown in Fig. 1 where the second part was also divided into an input and an output part.

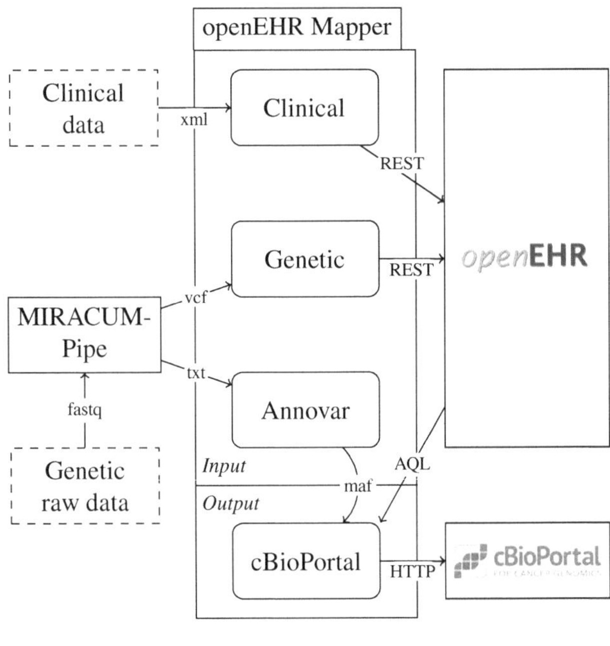

Figure 1: Architecture of openEHR Mapper

It also shows the prior processing of the genetic raw data that is done through the MIRACUM-Pipe whose results are then processed in *openEHR Mapper*.

Initially for a proof of concept, clinical and oncological data is provided in a proprietary XML format, but oriented to the ADT/GEKID standard. To obtain data from it we use XPath queries and map the attributes directly to corresponding openEHR fields. Attributes of the openEHR standard that do not underly any or the same terminology as the attribute from ADT/GEKID can be directly assigned. If the openEHR attribute is bound to specific terminology, we obtain the corresponding value by using a lookup table.

Genetic data comes in different formats and will be processed in two different ways: The output of MICRACUM-Pipe will contain a *vcf* file describing several thousand somatic mutations of the given samples as well as a text file created by ANNOVAR that contains annotations to all of the mutations. As the final target cBioPortal only takes annotated mutations from *extended maf* files, we go through a conversion process of two steps using a library written in *R*.

Committing the genetic data is done using the openEHR template *Molekularpathologischer Befund* that was created by the HiGHmed consortium [14] and is capable of taking most of the attributes in the vcf file from MIRACUM-Pipe. It also maintains additional information about the sample and the way it was processed.

To generate EHR extracts openEHR Mapper takes the mapped data described above and a template file and combines them. The template file contains the structure, datatypes, and terminologies of the contained attributes. We implemented a generator that parses the template on the

Table 1: Attributes that are being mapped from clinical data to openEHR archetypes

Attribute	Mapping status
Age	✓
Gender	✓
TNM status	✓
UICC classification	✓
Histology (ICD-O)	✓

Table 2: Attributes that are being mapped from vcf files to openEHR archetypes

Attribute	Mapping status
Chromosome	✓
Position	✓
ID	empty
Reference base	✓
Alternative alleles	✓
Quality score	empty
Filter	✓
Quality parameters	3/10

fly and builds objects containing the previously mapped attributes. To speed up the process we integrated a simple caching mechanism that prevents excessive memory access. Afterward, the document instance will be serialized as a JSON object and can be committed to the openEHR repository.

To import the data to cBioPortal it requires a *study* to be generated. It contains metadata, the annotated mutations from the Annovar module described above and the clinical data form the openEHR repository in tab-separated files. The clinical data can be obtained by sending AQL queries to the openEHR repository. Those queries can be very specific and select single attributes so that most of the work is done on the server. After the study generation is complete it can be sent to the cBioPortal instance via HTTP using a Python script.

4   Results and Discussion

The following results present the mapped attributes as well as performance reports for two data formats and the impact of threading. Attributes of clinical data that will be mapped are shown in Table 1. They provide the majority of important information for this use case but lack some information about diagnoses and surgeries. Further mapping has been left out because the input source will most likely change due to necessary clarifications regaring the routinely used clinical tumor software ODS Easy.

Genetic variants were also successfully mapped. Table 2 lists the attributes mapped to openEHR. While the chromosome, position, reference base, and alternative alleles are essential, we can also map attributes that give information about the filtering and quality of the data. Especially the quality parameters are currently pretty limited in the corresponding openEHR archetypes and it may be useful to extend them and add more parameters or add slots that provide the option to define them when needed.

As the openEHR specification [3] offers storing contributions either in XML or in JSON we implemented both and compared the performance of our implementations. To generate contributions in reasonable size we generated EHR extracts containing the somatic mutations of four patients from TCRB [6] in both formats and compared the results as shown in Fig. 2. The measurements show that the JSON implementation is at least 21% faster than XML. Unlike XML, the JSON format supports key-value pairs and arrays while XML only uses key-value pairs. This means that

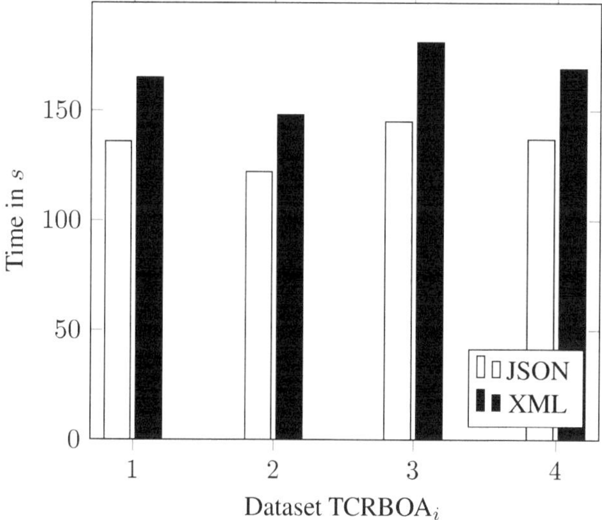

Figure 2: Comparing the amount of time that is needed to genereate a EHR extract when using XML or respectively JSON as target data format

XML generates way more objects and is less efficient for this use case.

Another improvement in performance was achieved by using multiple threads, which required a thread-safe data structure. To achieve this we implemented workers for the openEHR classes *OBSERVATION, ADMIN_ENTRY,* and *CLUSTER.* The mapped data is stored in queues to ensure that each attribute will only be processed once. Again we used mutation data from the TCRB dataset described above. The results in Fig. 3 show that the performance scales almost linear with a rising number of threads.

Storing clinical data inside the repository works and can be a way to ensure data quality and terminology binding. Also having the openEHR repository placed between source systems and cBioPortal means that parts of the whole process could be replaced without changing the other. Though storing genetic data works in principle but has no further use right now than archiving the mutation data as there are no tools that will use this data. This may change when openEHR and the genetic archetypes will be more established.

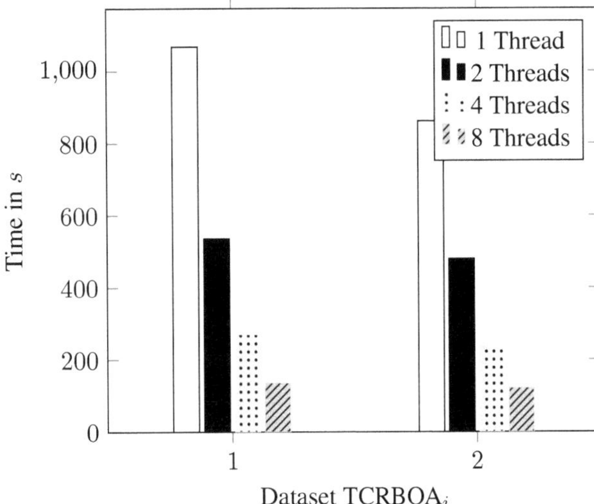

Figure 3: Overview how the generation of clinical EHR extracts (JSON) scales with an increasing number of threads

5 Conclusion

In this paper, we showed the integration of clinical and genomic data into EHRs stored in repositories based on the openEHR standard as well as deploying decision support tools for molecular tumor boards on top of such a repository. The openEHR standard in combination with access via a RESTful API and having AQL as query language turned out to be powerful. Unfortunately due to the lack of applications with openEHR support it does not add very much value yet.

As the input data is already structured the integration process can be configured to run almost fully automated without major abuses in the amount of data or data quality. During the development we also discovered that the generation of EHR extracts can be heavily improved by using JSON instead of XML, caching XPath queries and the usage of threads.

Acknowledgement

The work has been carried out at Institute of Medical Informatics, Lübeck and supervised by the Institute of Medical Informatics, Universität zu Lübeck. We acknowledge both members of the HiGHmed and MIRACUM consortia as part of the Medical Informatics Initiative Germany. The project is funded by the German Federal Ministry of Education and Research (BMBF, grand id: 01ZZ1802Z).

6 References

[1] P. A. Maranhao, G. Bacelar-Silva, D. Goncalves-Ferreira, P. Vieira-Marques, and R. Cruz-Correia, "Challenges in Design and Creation of Genetic openEHR-Archetype," *Stud Health Technol Inform*, vol. 247, pp. 835–839, 2018.

[2] The HiGHmed Consortium, "HiGHmed Medical Informatics," Available: https://www.highmed.org/ [last accessed on 2020-01-09].

[3] openEHR Foundation, "openEHR - Working Baseline," Available: https://specifications.openehr.org/ [last accessed on 2020-01-07].

[4] Arbeitsgemeinschaft Deutscher Tumorzentren e.V. (ADT) and Gesellschaft der epidemiologischen Krebsregister in Deutschland e.V. (GEKID), "Einheitlicher onkologischer Basisdatensatz ADT/GEKID," Available: https://www.gekid.de/adt-gekid-basisdatensatz [last accessed: 2020-01-20].

[5] M. Choi *et al.*, "Genetic diagnosis by whole exome capture and massively parallel DNA sequencing," *Proceedings of the National Academy of Sciences of the United States of America*, vol. 106, no. 45, pp. 19 096–19 101, Nov 2009.

[6] L. B. Becnel *et al.*, "An open access pilot freely sharing cancer genomic data from participants in Texas," *Scientific Data*, vol. 3, p. 160010, Feb 2016.

[7] P. Metzger *et al.*, "MIRACUM-Pipe," Available: https://github.com/AG-Boerries/MIRACUM-Pipe [last accessed on 2020-01-07].

[8] K. Wang, M. Li, and H. Hakonarson, "ANNOVAR: functional annotation of genetic variants from high-throughput sequencing data," *Nucleic Acids Research*, vol. 38, no. 16, 07 2010.

[9] The HiGHmed Consortium, "HiGHmed Clinical Knowledge Manager," Available: https://ckm.highmed.org/ckm/ [last accessed: 2020-01-20].

[10] Vitasystems and Hannover Medical School, "EHRbase," Available: https://ehrbase.org [last accessed on 2020-01-07].

[11] J. Gao *et al.*, "Integrative analysis of complex cancer genomics and clinical profiles using the cBioPortal," *Sci Signal*, vol. 6, no. 269, p. pl1, Apr 2013.

[12] E. Cerami *et al.*, "The cBio Cancer Genomics Portal: An Open Platform for Exploring Multidimensional Cancer Genomics Data," *Cancer Discovery*, vol. 2, no. 5, pp. 401–404, 2012.

[13] P. Unberath, C. Knell, H.-U. Prokosch, and J. Christoph, "Developing New Analysis Functions for a Translational Research Platform: Extending the cBioPortal for Cancer Genomics," *Studies in Health Technology and Informatics*, vol. 258, no. ICT for Health Science Research, pp. 46–50, 2019.

[14] C. Mascia, P. Uva, S. Leo, and G. Zanetti, "OpenEHR modeling for genomics in clinical practice," *Int J Med Inform*, vol. 120, pp. 147–156, 12 2018.

MetaCheck – a Web Application for the Automatic Evaluation of Metadata in Healthcare

Alexandra Banach [1], Ann-Kristin Kock-Schoppenhauer [2], Hannes Ulrich [2] and Josef Ingenerf [2,3]

[1] Medical Informatics, Universität zu Lübeck, alexandra.banach@student.uni-luebeck.de
[2] IT Center for Clinical Research, Lübeck (ITCR-L), Lübeck, Universität zu Lübeck,
{annkristin.kockschoppenhauer, h.ulrich, josef.ingenerf}@uni-luebeck.de
[3] Institute of Medical Informatics, Universität zu Lübeck, josef.ingenerf@uni-luebeck.de

Abstract

High quality of medical data is essential for adequate patient care and meaningful results in clinical trials. Well-specified metadata are an important aspect of data quality. For instance, the choice of correct data types and the specification of value ranges help to avoid errors during data entry. As the amount of metadata is increasing continuously, the automatic evaluation of quality is desirable. Therefore, existing quality criteria were checked for the adaption of ISO 11179-3 conformant metadata and new metrics were derived. Several computable metrics were implemented as a web application *MetaCheck* to evaluate single metadata sets. A comparison of two metadata sets to check which achieved the better quality was also implemented. With the help of computed scores for each metadata set and the computed metrics, it is possible to evaluate the quality and to detect critical factors as the evaluation of 12 research metadata sets shows.

1 Introduction

As medical data can be reused for clinical trials [1], data quality is an important factor not only for patient care but also to achieve meaningful results in research. Reasons for a lack of data quality might be a human failure during manual data entry, time pressure, misinterpretation of specified metadata elements or missing plausibility checks. In this context, metadata are an effective instrument to assure data quality, e.g. by defining appropriate data types and value ranges [2]. Therefore, with the help of well-specified elements, better quality on the instance data level can be achieved.

Generally, metadata can be interpreted as additional data to the actual instance data [3]. Existing metadata elements can be reused for further purposes and do not have to be specified again which reduces effort and costs [4].

As the amount of metadata is increasing continuously and manual data checks would be very complex and time-consuming, the automatic evaluation of metadata quality is desirable. In the past, several criteria to evaluate metadata quality were developed [5], [6], but not all are adaptable to ISO 11179-3 conformant metadata or enable an automatic computation [7].

The purpose of this work was to check whether existing quality criteria were adaptable to metadata sets. Automatically computable metrics for a local metadata repository and for collections were derived and implemented as a web application which allows the evaluation of single metadata sets as well as the quality comparison of two metadata sets.

2 Material and Methods

2.1 ISO 11179-3

ISO 11179-3 [8] is a standard which characterizes the structure and format of a repository and its metadata. It specifies required metadata attributes such as a *designation* of a metadata element or a more detailed *definition*. For each element *data types* have to be defined. It is also possible to define *value ranges*, e.g. for data types such as floats or integer.

An open-source solution for the creation and management of ISO 11179-3 conformant metadata is the metadata repository software *Samply.MDR*. In Samply.MDR a data set of metadata elements is called *namespace*. As metadata sets may have a nested structure based on their semantic meaning, *data element groups* are a feature of Samply.MDR. With the help of these groups, metadata elements can be aggregated. Groups may have subgroups, so a metadata set receives a nested structure.

2.2 Quality criteria

Deppenwiese et al. [7] analyzed existing metadata quality criteria whether they were adaptable for ISO 11179-3 conformant metadata and whether they were automatically computable. Therefore, they checked the criteria developed by Ochoa and Duval [5]. These criteria were based on the metadata standard Dublin Core. For instance, Ochoa and Duval defined a completeness metric Q_{wcomp} shown in (1) to compute the proportion of required fields of a metadata element which do not contain NULL values. They assumed an individual weight factor α_i for a field of a metadata element which must be a positive number. $P(i)$ is 0 if the field contains a NULL value. Otherwise, $P(i)$ is 1. The maximum value of $Qwcomp$ is 1 if all required fields are not empty. The minimum value is 0 in the case that all required fields are empty.

$$Qwcomp = \frac{\sum\limits_{i=1}^{N} \alpha_i \cdot P(i)}{\sum\limits_{i=1}^{N} \alpha_i} \qquad (1)$$

Furthermore, the criteria developed by McMahon and Denaxas [6] were analyzed. Deppenwiese et al. [7] concluded, that some criteria were adaptable and computable such as the completeness metric $Qwcomp$. They also developed additional metrics for ISO 11179-3 conformant metadata elements:

- number of slots: These fields are key-value pairs to capture any additional information. As these fields are free text fields, they are difficult to analyze because of missing standards. So they should not be used too often.

- number of distinctive keys: If slots are used it can be helpful to check the number of distinctive keys in these fields. It might indicate whether there is at least some kind of standardization in using slots in a metadata set.

- proportion of float and integer elements without units of measure: Units of measure are a helpful instrument to ensure data quality. For instance, it is necessary to specify unit of measures for calculation.

- length of definition value: A metadata element with a too short definition might be misinterpreted which causes errors in data collection and leads to low data quality.

- number of string data elements: As these free text fields are difficult to be analyzed automatically, they should not be used too extensively.

- proportion of string data elements without regular expression (REGEX): Regular expressions can help to improve the data quality of string data elements as they are a kind of validation rule. Thus, the use of regular expressions is preferred.

3 Results and Discussion

3.1 Metrics

For the automatic evaluation of a metadata set each element has to be checked. We assume a metadata element "body mass index" which is well-specified as a float with the unit of measure "kg/m^2". We assume another metadata element "BMI" specified as a string without regular expression. For the second element, it is not possible to ensure uniformly

handling in data entry as string data elements without regular expressions can have any content. Therefore, metadata sets containing the second element should achieve a lower rating than metadata sets containing well-specified elements such as the first one.

As the metrics by Deppenwiese et al. have more statistical character which requires human interpretation, additional metrics for automatic evaluation of metadata sets based on their results were derived in this study:

- proportion of boolean elements with too short questions (< 10 words): Closed questions should be carefully expressed to avoid misinterpretation. Therefore, at least 10 words were assumed to get the right context.

- elements with an average ratio of definition and designation values < 2: The definition value should contain more information than the designation value. If the ratio is smaller than two, the definition does not provide an added value of information.

- proportion of groups with number of elements < 2 or > 10: The use of data element groups in Samply.MDR helps to get a well-structured metadata set and to understand the semantic context of its elements.

- proportion of float and integer elements without range: As an analogy to string elements and their regular expressions, value ranges help to improve the data quality of float and integer elements. They can be interpreted as some kind of validation rule.

- proportion of enumeration with number of elements < 3 or > 10: Enumerations with only one key are not appropriate. In addition, enumerations with two keys should be specified as boolean elements. If the number of key values is getting too high, the effort of creating and organizing these metadata elements increases. Therefore, an upper limit of 10 keys was chosen.

All ranges of these metrics are assumptions which are mainly based on personal experiences in metadata management. In addition, the proportion of string elements without REGEX and the proportion of float and integer elements without unit of measure were adapted from Deppenwiese et al. [7] as they suit the purpose.

3.2 Score of a namespace

All metrics presented in the previous chapter can also be formulated in a positive way to characterize the complementary set. Therefore, we assume for each metric i a tuple (x_i, y_i) and x_i denotes the proportion of elements fulfilling the desired criteria (e.g. string with REGEX) and y_i the complementary proportion (e.g. string without REGEX). To compare the quality of two metadata sets the computation of a score for each namespace was necessary. For the computation of the score, the completeness metric $Qwcomp$ developed by Ochoa and Duval was adapted. We assumed α_i to be the computed metric values x_i and y_i.

$P(i)$ is 0 for metric values y_i as they have a negative influence on the metadata quality. Therefore, we only have to add the metric values x_i as $P(i)$ is defined as 1 for them. The score of a namespace is computed as shown in (2):

$$score_{namespace} = \frac{\sum\limits_{i=1}^{N} x_i}{\sum\limits_{i=1}^{N} x_i + y_i} \qquad (2)$$

High metrics values y_i lead to a smaller score.

The possible scores of a namespace are in the range of 0 to 1. A score of 1 means that all elements of the evaluated metadata set meet the desired criteria x_i.

3.3 Web application MetaCheck

Before starting the implementation different public available metadata test sets in healthcare were used and recreated in Samply.MDR to check the implementation continuously during its development. The test metadata set *basismodule* [9] contains metadata elements for patient data and corresponding case data, laboratory results, diagnoses, etc. The metadata set was developed by the Medizininformatik-Initiative to model a core data set for clinical research and patient care [10].

For an automatic evaluation of metadata sets, the web application *MetaCheck* was developed in Java. It was decided to use Samply.MDR as a repository for metadata as it supports ISO 11179-3 and the developed metrics in 3.1 are based on that standard.

When starting the application the user has to choose whether a single namespace of an instance of Samply.MDR should be evaluated or two namespaces should be compared. For data fetching the user has to enter the URL of the Samply.MDR, the ID of each namespace and the language in which the metadata elements are created.

After metadata fetching the developed metrics for a namespace and its score are computed. In addition, some statistical values such as the proportion of different data types are calculated. These information give a complete overview of the metadata set and it also can be used to check which reasons led to a lack of metadata quality. The distribution of different data types in a namespace is also presented as a donut chart.

3.4 Evaluation

Table 1 shows the computed metrics for the metadata set *basismodule*. This metadata set achieved a namespace score of 0.524. The main reasons for the low score are the high proportion of string elements without REGEX and the proportion of elements with an average ratio of definition and designation value < 2 as table 1 shows.

The metadata set also contains string elements for which a REGEX is difficult to define, for instance first or last name. Therefore, it could be considered to add an additional factor

for such metric values to achieve a lower influence on computing the score.

Some metadata elements, e.g. "sex of a patient ", are often defined as an enumeration with the two keys "female" and "male ". With the metric "enumeration with number of elements < 3 or > 10 ", a lower score is achieved. But the example shows that it does not make sense to define all of these elements as boolean. In the meantime, a third key value "diverse" was established for "sex of a patient". This extension could not have been implemented if the metadata element was specified as boolean.

Both examples show that the defined metrics do not cover all particularities. Nevertheless, the comparability of the computed scores of two similar metadata sets is still ensured as both datasets achieve a lower score with such metrics.

MetaCheck was also evaluated with 12 metadata sets originated from a German biomedical research project containing 548 metadata elements. The number of elements reaches from 4 to 154 per metadata set. Table 2 shows the computed scores for each namespace. The best score of 0.614 was achieved by namespace 9, the lowest score of 0.25 by namespace 3. Most of the scores can be found in a range from about 0.4 to 0.5. The main reasons for the low score of namespace 3 are string elements without REGEX and elements with an average ratio of definition and designation value smaller than 2, as the computed metrics showed. Checking the metadata set in Samply.MDR also showed that the namespace contains several groups without any metadata elements. Therefore, the metric "groups with number of elements smaller than 2 or greater than 10" had a negative influence on the score, too. Namespace 9 which achieved the best score contained float and integer elements without ranges. The score was also negatively influenced by the metric "proportion of elements with an average ratio of definition and designation value smaller than 2". It was noticeable that the score of the best namespaces 8, 9 and 11 is decreased by the metric "proportion of elements with an average ratio of definition and designation value smaller than 2". The main reasons for a lower score in the range from 0.4 to 0.5 are string elements without REGEX and also the proportion of elements with an average ratio of definition and designation value smaller than 2. Therefore, it could be considered to reduce the influence of this metric on the score by implementing an additional factor between 0 and 1 as for some elements the designation may be meaningful enough.

Generally, the computed scores and metrics seem plausible as a manual data check in the Samply.MDR has confirmed. The influence of single metrics on the score might be considered and modified by additional factors or by redefining the metric ranges depending on the metadata to be evaluated.

4 Conclusion

The web application MetaCheck allows the automatic evaluation of ISO 11179-3 conformant metadata sets, as well as the direct comparison of two data sets as the evalua-

Table 1: Computed metrics of namespace *basismodule* (rounded to three decimal places)

metric name (desired quality, poor quality)	metric values
proportion of boolean elements with questions ($>= 10$ words, < 10 words)	(0.000, 0.008)
proportion of elements with average ratio of definition and designation value ($>= 2$, < 2)	(0.645, 0.355)
proportion of groups with number of elements ($>= 2$ and $<= 10$, < 2 or > 10)	(0.838, 0.162)
proportion of float and integer elements (with units of measure, without units of measure)	(0.016, 0.073)
proportion of float and integer elements (with range, without range)	(0.000, 0.089)
proportion of enumeration with number of elements (between 3 and 10, < 3 or > 10)	(0.073, 0.073)
proportion of string elements (with REGEX, without REGEX)	(0.000, 0.532)

Table 2: Comparison of 12 namespaces from a German biomedical research project. The lowest and the highest computed scores are highlighted in bold.

namespace	computed score
1	0.487
2	0.468
3	**0.250**
4	0.515
5	0.403
6	0.506
7	0.333
8	0.592
9	**0.614**
10	0.467
11	0.555
12	0.486

tion of the metadata set *basismodule* and the comparison of the medical research metadata sets showed. The score of a namespace enables checking immediately which namespace shows a better data quality. With the help of the implemented metrics and statistics, the factors which could improve metadata quality can be detected. Whenever an instance of the metadata repository Samply.MDR is part of clinical research projects, MetaCheck can be used to check metadata quality.

As the implemented metrics do not focus on semantic factors, the integration of semantic tools such as ontologies for further developments of MetaCheck would be a possible option as well as an automatic evaluation of single metadata elements. So far, all computed metric results are represented by a table. For further implementation, a graphic representation of the computed metrics could be integrated.

Acknowledgement

The work has been carried out at IT Center for Clinical Research, Lübeck (ITCR-L), Lübeck and supervised by the Institute of Medical Informatics, Universität zu Lübeck and is part of the project "Weiterentwicklung und Etablierung des Nationalen Metadata Repositories (NMDR) " funded by the Deutsche Forschungsgesellschaft (IN 50/3-2).

5   References

[1] S. M. Meystre et al., *Clinical Data Reuse or Secondary Use: Current Status and Potential Future Progress.* Yearb Med Inform, 2017.

[2] N. G. Weiskopf and C. Weng, *Methods and dimensions of electronic health record data quality assessment: enabling reuse for clinical research.* Journal of the American Medical Informatics Association, vol. 20, no. 1, pp. 144–151, 2013.

[3] D. Apel, W. Behme, R. Eberlein and C. Merighi, *Datenqualität erfolgreich steuern: Praxislösungen für Business-Intelligence-Projekte.* Dpunkt. verlag, 2015.

[4] S. M. Ngouongo, M. Löbe and J. Stausberg, *The ISO/ IEC 11179 norm for metadata registries: Does it cover healthcare standards in empirical research?.* Journal of Biomedical Informatics, vol. 46, no. 2, pp. 318–327, 2013.

[5] X. Ochoa and E. Duval, *Automatic evaluation of metadata quality in digital repositories.* Springer-Verlag, vol. 10, no. 2-3, pp.67–91, 2009.

[6] C. McMahon and S. Denaxas, *A novel framework for assessing metadata quality in epidemiological and public health research settings.* AMIA Summits on Translational Science Proceedings, 2016.

[7] N. Deppenwiese, A.-K. Kock-Schoppenhauer, H. Ulrich, P. Duhm-Harbeck and J.Ingenerf, *Automatic Evaluation of Metadata Quality in ISO 11179-3 Conformant Healthcare Metadata Repositories.* German Medical Science GMS Publishing House, 2019.

[8] ISO/IEC 11179, *Information Technology – Metadata registries (MDR).* Available: http://metadata-standards.org/11179 [last accessed on 2020-01-03].

[9] art-decor.org, *MII Core Data Set - Datasets.* Available: https://art-decor.org/art-decor/decor-datasets [last accessed on 2020-01-04].

[10] Medizininformatik-Initiative, *MI-I-Kerndatensatz.* Available: https://www.medizininformatik-initiative.de/sites/default/files/inline-files/MII_04_Kerndatensatz_1-0.pdf [last accessed on 2020-01-12].

Transforming Nutrition Databases Towards Semantically Consistent Content Classification

Laura Liebenow [1], Christoph Twesten [2], and Marcin Grzegorzek [3]

[1] Medical Informatics, Universität zu Lübeck, laura.liebenow@student.uni-luebeck.de
[2] Perfood GmbH, MillionFriends, christoph.twesten@perfood.de
[3] Institute of Medical Informatics, Universität zu Lübeck, grzegorzek@imi.uni-luebeck.de

Abstract

The purpose of transforming nutrition databases towards semantically consistent content classification is to support the evaluation of personal nutrition. In this paper, we describe a method to rearrange the entries of the nutrition groups of two databases to a semantically consistent level. Subsequently we present an application that calculates a statistic on the different kinds of nutritions a person has eaten. This information could be used by experts for a simpler, but more specific guidance into personalised nutrition.

1   Introduction

Many diseases are results of an inappropriate nutrition. As reverse conclusion, these diseases can be prevented by an appropriate nutrition [3]. The problem here is, that *right* nutrition can not be generalized, because not every person reacts the same to different kinds of foods. This subjectivity of nutrition leads to the idea of a personalised nutrition [3], [4].

In order to recommend personalised nutrition, information about individual reactions to food is necessary like e. g. the individual glucose reactions or the gut microbiome composition.

It is known that an unsteady blood glucose level, especially a high blood glucose level as response to a consumed meal, increases the risk of adiposity and its resulting complications. A low-glycemic nutrition could prevent this [4].

The gut microbiome is essential for many metabolisms and immune functions in the human body. This means it affects the physiology of a human being, like the blood glucose reaction or the body weight regulation [6]. An individual's microbiome is unique [5]. This fact supports the need of a personalised nutrition.

Based on the information above, the Start Up Perfood GmbH is specialised on delivering personalised nutrition to the costumer.

In order to obtain the information about the glucose reaction, they use a combination of an app and a glucose measurement device, which consists of a sensor that is attached to the body and measures the glucose level and a reading device.

The meals eaten are manually recorded on the app. The timing of recording a meal and what it consists of is important for precise results. Utilising the noted time of the meals, an automatic algorithm finds the corresponding segment of the measured glucose curve. This information is needed to evaluate which meals evoke what kind of glucose reaction.

In order to simplify the recording of a meal, two nutrition databases are connected to the app. These databases are the *Bundeslebensmittelschlüssel* (BLS) [1], and the *Fooddatabase* (FDDB) [2]. The advantage of two databases is that more foods and their micronutrients are available.

One problem is that the nutrition classes of the two databases are on different semantic levels of abstraction. This disparity makes it difficult to automatically combine recorded meal-ingredients of different databases.

The purpose of this paper is to define representative nutrition classes and assign the items of the two databases to them. As a second step, a statistical evaluation should be calculated by an automatic algorithm. More information about this process is provided in section 2.

It is requested that the result eases the more specific guidance into personalised nutrition, as e. g. to check if a person eats enough vegetables and fruits for a balanced nutrition [7]. Section 3 deals with the question of the usefulness of this approach for Perfood GmbH.

Lastly, section 4 gives a short summary and a future outlook as conclusion of this work.

2   Material and Methods

In this section, the databases *Bundeslebensmittelschlüssel* and *Fooddatabase* are summarised for a better understanding of the semantic level differences. Afterwards, the semantic abstraction and the process of the histogram generation are explained.

For gaining information about the database nutrition classes, postgresql is used. The algorithm for the statistic evaluation is written in python 3.7.

2.1 BLS

The BLS is the standardised database for nutrition studies in Germany. It contains around 15 000 foods and their nutritional values. The basis for these values are made by the results of research performed by the Max Rubner Institute and its national co-operation partners. These values are completed by scientific papers and international nutrition databases [1]. For the purpose of qualitative correctness, experts check the entries of the database frequently and only release changes after a good examination. It is not possible to register items to the BLS as a non-qualified person.

The entries of the database can be identified by a semantic ID of seven digits. These seven positions order the database in a hierarchical way. Figure 1 shows an example of the hierarchic classification regarding the BLS class F - *fruits*. Entries with a key beginning with X or Y are meal components of a household or gastronomy.

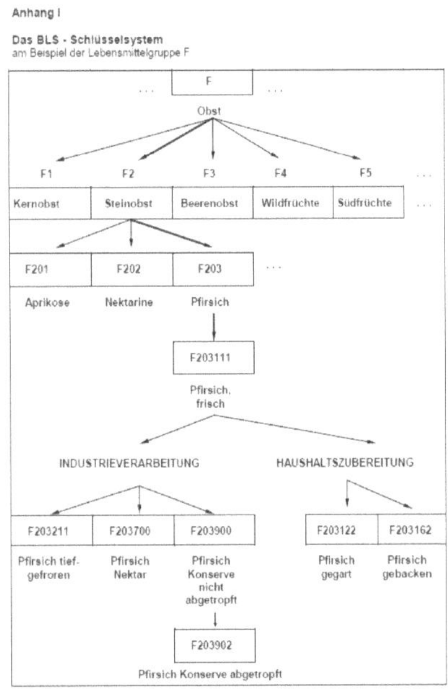

Figure 1: The first position is an alphabetic character, which stands for the nutrition class, e.g. F for *fruit and fruit products*. The following four positions are numbers which identify the nutrition more precisely. The sixth and seventh position refer to the correlation between the way of preparation and the reference weight. [1].

2.2 FDDB

The FDDB is an open database where every user is allowed to enter ingredients or meals and their nutrition values. The database has around 416 800 entries and is not ordered in a hierarchical way. The nonsemantic item ID's simply are increasing numbers, not sorted by nutrition class. But there are ways to categorise the entries. While BLS entries are divided in classes, FDDB entries are categorised by groups, like producer or FDDB nutrition group. An FDDB nutrition group can be *softdrink*, *candy* or some other synonym describing the food item. We used these for our semantic abstraction.

2.3 Semantic Abstraction

The MillionFriends app is connected to the two nutrition databases BLS and FDDB. Since the databases are running in the background, the users are not aware that they are logging items of different sources. Due to the different semantic levels of the entries it is difficult to automatically evaluate them together.

To solve this problem, a semantic abstraction is needed. Therefore both databases are examined manually on their nutrition classes to find out which classes are represented in which database and what items can be found in them. With this information, new classes can be created, representing semantically identical classes of both databases. This is the first part of this approach and visualised by step (1) and (2) of the processing pipeline in Fig. 2.

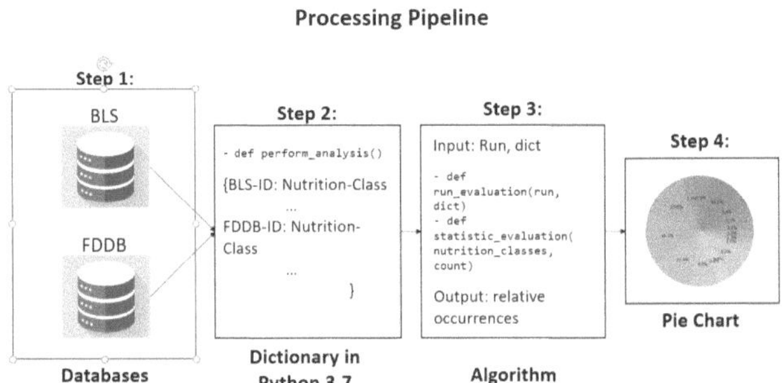

Figure 2: The processing pipeline of this work. First, the nutrition classes of the databases are analysed. Then these classes are reordered to semantically consistent nutrition classes via a dictionary. Third step and second part of this work, is an algorithm, which statistically summarises meal-ingredients by their appearance and shows the result in a pie chart.

Regarding the example class *fruit* of the BLS shown in Fig. 1, fruits are categorised in berries, stone fruits, pome fruits and so on by the second and third number of their BLS ID. In the FDDB fruits are categorised by their nutrition group, which could be *exotic fruits*, *fresh fruits*, *dried fruits*, *canned fruits* or *fruits and fruit products*.

For example you can say, that in BLS class F3, which represents berry fruits, raspberries are included, but also dried raspberries are included in this class, specified by the BLS key as a way raspberries can be eaten. In the FDDB raspberries are *fresh fruits* or *fruits and fruit products* and dried raspberries are *dried fruits*.

There are so many entries in both databases, that an exact mapping with SQL requests is costly. We only want to know if a person has eaten fruits and not what kind of fruits it has been that was eaten. Therefore a representative class *fruit products* is created, which unifies the BLS entries of the class F3 and the FDDB entries of the previously named FDDB nutrition groups. Some adjustments have to be done as well, as the BLS class F3 contains fruit juices. These has to be selected, due to the fact that a more precise categorisation would be the class *drinks* together with the FDDB group *fruit juices* and other matching BLS classes and FDDB nutrition groups.

Moreover the FDDB nutrition group *fruits and fruit products* contains jam, which is not represented by the BLS class F3 rather by the BLS class S13, which represents sweets. Because it is not possible to exclude the jam from the FDDB group due to different spellings and synonyms made by the persons writing the entries of the FDDB, the BLS class S13 also is included in the representative class *fruit products*.

The mapping of each item of the BLS and of the FDDB to their representative class is made in a similar way. Figure 3 visualises this process. After doing this step manually it is made by creating a dictionary in Python, with the BLS ID or the FDDB ID as keys and the corresponding new nutrition class as value. In the end there are thirty-one representative nutrition classes in total.

The items of both databases mapped to their representative classes are now semantically consistent. Only a key-request is needed to find out the consistent nutrition class of an item regardless if it is originally an item of the FDDB or an item of the BLS.

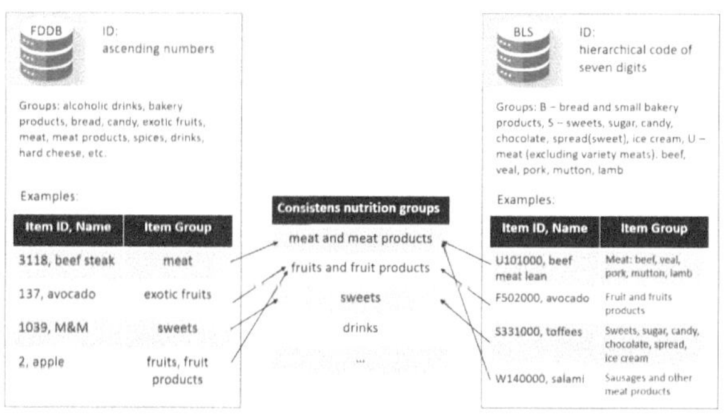

Figure 3: This figure shows that the entries of the two databases FDDB and BLS are classified with different kinds of ID's. For each database, example items with their ID's and their nutrition classes are listed in the tables below. These items are regrouped to the consistent nutrition groups. This process is symbolised by the arrows.

2.4 Histogram Generation

The previously made semantic abstraction is now useful to implement an automatic algorithm for the statistic evaluation. This evaluation should be made over all the meals a person has eaten over a period of time. This information is saved as a *run* in the company's internal database. Every ingredient a person has recorded in their app is connected with the corresponding meal. And every meal a person has eaten is connected with the run ID.

To calculate the statistic, the algorithm iterates over the ingredients which are connected with a run to get their database ID. This is the FDDB ID for items of the FDDB or the BLS ID for items of the BLS. Knowing the ID's, the consistent nutrition classes of the ingredients can be found in the dictionary we constructed in the semantic abstraction step. To summarise the occurring nutrition classes, a new dictionary is created. This dictionary has nutrition classes as keys and their occurrence amount as values. The whole procedure of creating the new dictionary executes as follows: An ingredient ID is checked if it appears in the semantic abstraction dictionary. If that is not the case, the key *undefined* is created in the new dictionary or increased by 1 if it already exists. The key *undefined* sums up every item that can not be assigned to a semantically consistent nutrition class due to different reasons. For example the FDDB nutrition group *new products* can not be generalised, because it includes many different nutrition classes that are not specified correctly. A BLS example for not yet classified items are the items with a BLS ID starting with X or Y, because these items are mostly complete meals.

If an ingredient ID appears in the semantic abstraction dictionary, the corresponding nutrition class is added to the new dictionary or its value is increased by 1 if it already exists in this dictionary. Thus only the amount of nutrition classes that appear in a run are taken into account.

After examining every ingredient of a run this way, we have a dictionary with the appearing nutrition classes and their absolute occurrence. To achieve a relative occurrence, every value is divided by the absolute number of ingredients in a run.

3 Results and Discussion

The dictionary with the relative occurrences of each nutrition class a person has eaten over a period of time can be converted into a pie chart, similar to the one shown in Figure 4. It visualises the occurrence of each appearing nutrition class compared to the others.

This pie chart is easy to read and can help experts to give a more personalised nutrition guidance to an individual person. For example, they can see how often fruits and vegetables have been eaten during a run and maybe give a suggestion to eat more of them due to the *5 per day rule* [7].

Since now, only an evaluation of the consumed amount of carbs, proteins and fats are given in respect to the normal recommendation for a person of a certain age, height and weight by the *Deutsche Gesellschaft für Ernährung e. V.* [8]. This recommendation is modified by the person's nutrition type. A nutrition type in this case can be a fat type, a protein type or a mixed type, which refers to what micronutrients process the individual's glucose metabolism in a beneficial way. One advice based on this evaluation could be that a fat type person should eat the recommended amount of fats by all means.

The alternatives of the personalised nutrition guidances that are available for costumers or nutrition experts can now be improved. Because it is a new development in an ongo-

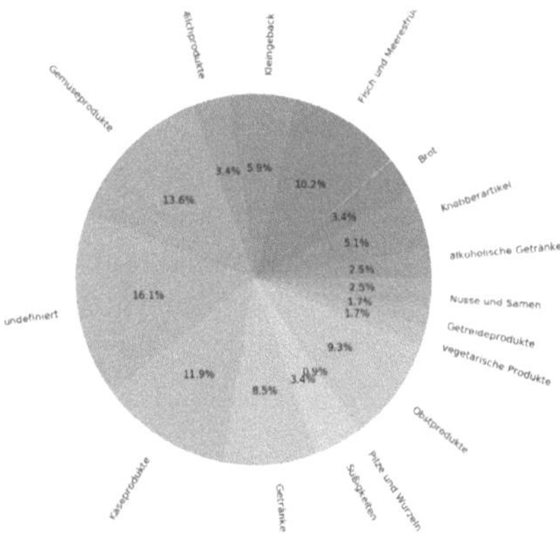

Figure 4: The appearance of each class is represented by its percentage within the fields. The nutrition classes are listed next to their field. The class *undefined* is the most abundant class. Of course this depends on how many unclassified products where logged, but it visualises a problem that complicates the evaluation.

ing process, there is no data for a scenario where the algorithm improves the effect of the provided nutrition guidances. This makes an ad-hoc validation of the approach difficult. A practical examination is recommended. But the previous example with the *5 per day rule* [7], gives the presumption that there will be an improvement if the amounts of how often a nutrition class has been eaten is known.
As future work the not classified ingredients should be classified somehow and the amount of a nutrition class eaten during a specific meal should be taken into account to improve the result.

4   Conclusion

Personalised nutrition is a way to prevent certain diseases. Due to the fact that everybody reacts different to different kinds of nutritions, it is beneficial to give a personalised nutrition guidance. The guidance of Perfood is evaluated via a test phase during which a person registers their meals in an app and measures their glucose level reaction to the meals with a device. A microbiome is needed for a better evaluation. Experts evaluate the measurements corresponding to the logged foods and the microbiome sample to specify if a person is a fat type, a protein type or a mixed type. Then they can make an assumption about what kind of nutrition the person's microbiome can process best.
In our work, we had the goal to transform the nutrition classes of the databases BLS and FDDB to a semantically consistent level. This results in an easier evaluation of the recorded meals. In a second step an automatic algorithm statistically evaluates runs in respect to the amount an ingredient occurs in it. Finally, a pie chart gives a visualised overview over the relative occurrences of a nutrition class.
This approach is a possible innovation in the evaluation process of Perfood GmbH. A more individualised nutrition guidance is something more special than a simple nutrition advice, which may improve the guidances of the concern

and in succession potentially raises consumer demand too. As further steps it would be interesting if a special nutrition pattern appears in the nutrition habits of persons with the same chronical issue, for example migraine. This may be a sort of food which migraine patients eat or do not eat in an extraordinary amount. Or, this may be a nutrition class which patients with less symptoms eat in an extraordinary way compared to other patients with the same issue.

Acknowledgement

The work has been carried out at MillionFriends, Perfood GmbH and supervised by Dr. med. Christoph Twesten of MillionFriends, Perfood GmbH, and Prof. Dr.-Ing. habil. Marcin Grzegorzek of the Institute of Medical Informatics, Universität zu Lübeck.

5   References

[1] Bundesministerium für Ernährung und Landwirtschaft, *Bundeslebensmittelschlüssel*. Available: https://www.blsdb.de/ [© 2005-2017 MRI, Standort Karlsruhe]

[2] Fooddatabase, *FDDB*. Availabe: https://fddb.info/ [since Sept. 2004]

[3] Perfood GmbH, *MillionFriends*. Available: https://www.millionfriends.de [© 2019 Perfood GmbH]

[4] Zeevi D., Korem T., Zmora N., Israeli D., Rothschild D., Weinberger A., et. al., *Personalized Nutrition by Prediction of Glycemic Responses*. Israel, 2015 Nov. 19

[5] V. K. Ridaura, et al., *Cultured gut microbiota from twins discordant for obesity modulate adiposity and metabolic phenotypes in mice*. In: Science(N.Y.)2013 Sept. 06 Available: https://www.ncbi.nlm.nih.gov/pmc/articles/PMC3829625

[6] M. H. Mohajeri, et al., *The role of the microbiom for human health: from basic science to clinical applications*. In: European Journal of Nutrition, 2018, Available: https://www.ncbi.nlm.nih.gov/pmc/articles/PMC5962619/

[7] 5 am Tag e.V., *5 am Tag Obst und Gemüse*. Available: https://www.5amtag.de/wissen/was-ist-5-am-tag/ [Copyright 2020 – 5 am Tag e.V.]

[8] Deutsche Gesellschaft für Ernährung e.V., *D-A-CH-Referenzwerte*. Available: https://www.dge.de/wissenschaft/referenzwerte/

6

Image Processing

Generation of 3D Body Models from Single-View Range Images for Robotic Applications in Medicine

Tolga-Can Çallar [1], Sven Böttger [2], and Elmar Rueckert [2]

[1] Medical Engineering Science, Universität zu Lübeck, tolgacan.callar@student.uni-luebeck.de

[2] Institute for Robotics and Cognitive Systems, Universität zu Lübeck, {boettger, rueckert}@rob.uni-luebeck.de

Abstract

Computer-aided medical systems, e.g. in the fields of medical robotics or image-based assistance, are continuously investigated to overcome human limitations with respect to perception, memory or dexterity. A common requirement of such systems is the availability of a digital model describing the patient's position and morphology during a procedure. Operational complexity and technical limitations of established 3D imaging methods leave the clinical environment in need of a method for the fast acquisition of a three-dimensional patient representation. For this purpose, we propose an automatic and efficient 3D body model generation pipeline based on the elastic registration of single-view range images of patient's with parametric body shape models. Initial results that we obtained for the registration of several bodies show a favourable representative quality of the generated models and indicate the general feasibility of our approach as an instant 3D modelling method for robotic and other applications in medicine.

1   Introduction

The clinical success of surgical, diagnostic and therapeutic procedures exhibits a high degree of dependency upon the interaction and capabilities of medical personnel. Through the emergence of computer-aided medicine and in particular the introduction of robotic systems in clinical environments, many conceptual solutions for negative effects caused by human limitations of perception, cognition, physical strength or dexterity have been developed [1]. Currently studied use cases within medical robotics include robotic surgery as well as interventional and non-interventional diagnostics, e.g. robotic medical ultrasound examinations which indicate improvements of reliability and consistency of outcome compared to conventional medical practice [2]. A commonality of automated medical systems employed for direct patient interaction is the requirement of an accurate and complete virtual representation of the human body. In the context of robotics, this is necessary for motion planning purposes to determine collision-free trajectories to anatomical target sites. Already established 3D imaging modalities with varying image quality, acquisition speed and nocuous potential, e.g. computed tomography and magnetic resonance imaging are seemingly suitable for the generation of individual 3D patient models. However, the associated health risks and procedural complexity of the image acquisition and registration often render these technologies medically and economically unjustifiable for 3D patient modelling. Furthermore, unavoidable positional changes of the patient after CT or MRI scans reduce the obtained body models' capacity to represent the patient at the moment of treatment or examination.

This encouraged us to investigate the applicability of stereoscopic range imaging of the body surface for this purpose, as this technology has several advantages: Image data are obtained instantly, increasing the generated target model's representative quality while also improving the time-efficiency of the whole target generation process. Furthermore, stereo depth cameras are inexpensive and easy to operate, which facilitates their integration into the clinical environment. A technical difficulty inherent to this approach is the occurrence of optical occlusions restricting the retrievable body surface area from a given acquisition perspective which leads to missing surface regions in the range image of the body. The proposed techniques to resolve the issue of incomplete scans with stereo imaging generally employ a type of multi-view acquisition setting to increase the captured surface area of the body. Multiple overlapping partial scans are captured and classically rigidly registered [3] to reconstruct a complete body model. More recently, promising results have been achieved through the non-rigid registration of densely acquired surface data using template models as shape priors to realistically reconstruct missing regions [4]. As such methods involve complex calibration processes and often rely on human input, we present an unsupervised modelling framework based on the elastic registration of single-view range images with parametric templates as a contribution to efficient 3D patient modelling in healthcare environments.

2   Material and Methods

Our proposed approach for the generation of the patient surface model consists of three main elements: First, we create an initial target space representation, i.e. the subspace of the robot's workspace within which the medical procedure is performed, acquiring only one single-view depth image of the patient lying down on the examination couch with the low-cost *Intel RealSense D415* stereo depth camera. This restriction of our image data basis per patient is deliberate to allow easy integration within the clinical workflow. Afterwards, we perform a segmentation of the patient's body and apply several artefact reduction techniques. In the last step, the segmented body surface is elastically registered with a parametric template body model that has been modified in advance to resemble the patient's anthropometric features. The segmentation and registration processes of our framework are implemented utilizing the 3D-processing software *VTK* [5].

2.1   Range Imaging Acquisition

We capture a range image of the complete scenery where the robot will interact with the patient. This serves as the depth image source from which the sections belonging to the patient are extracted and used as an input for the registration of the body surface. The stereo camera is mounted on a tripod and placed lateral to the long side of the examination couch and oriented so that the longitudinal axis of the patient and the x-axis of the camera frame align. The camera's height and vertical tilt are adjusted, until the field of view spans the entire space the patient is expected to occupy, with the image frame being centred around the center of the examination bench. As visualized in Fig. 1 we acquire a frontal and oblique image for every patient in a supine position to generate realistic source data for the subsequent evaluation of our registration process. In preparation for the target segmentation based on background-subtraction, we also capture an image without the patient for every perspective. All of the obtained range images are converted to point clouds and exported as a polygonal mesh through the internal Delaunay-triangulation function of the RealSense Viewer application [6].

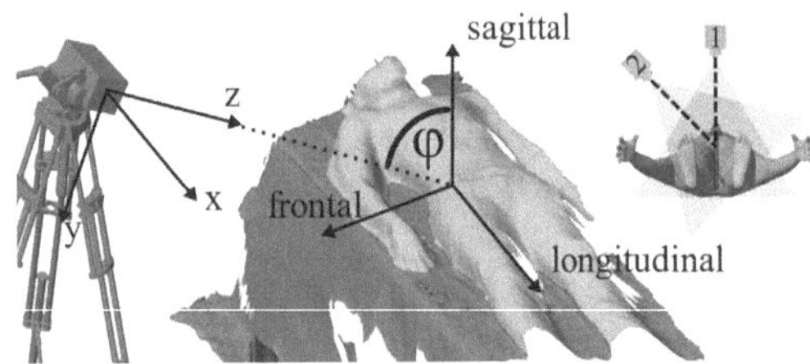

Figure 1: Visualization of the setup together with the segmented range image of the operating field.

2.2   Body Segmentation

To define the area of the range scan we aim to register with the template model, we perform the segmentation of the polygons representing the patient in the triangulated point cloud of the scene. For this purpose we implemented a background subtraction variant utilizing *VTK*: We generate k-d tree based space-partitionings of both the background scenery captured in advance and the scene including the

patient. On these, we perform a nearest-neighbour search to identify those vertices in the patient scene that do not have any vertex from the background mesh in their vicinity. This raw segmentation result contains artefacts caused by the stereo imaging technique. Predominantly, this is visible in the form of a local elongation of the triangular faces of the mesh caused by occlusions in the captured scene which results in steep depth gradients around border regions of the body (Fig. 2). These are sparsely filled in with points that form stretched faces in the exported mesh. We exploit the large positional displacements of the vertices forming these faces to filter them out.

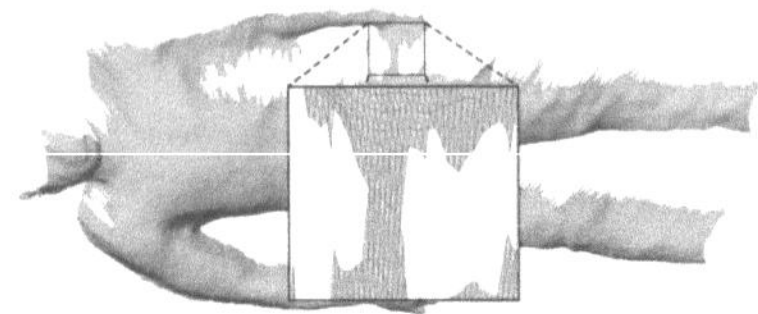

Figure 2: Elongated triangular faces caused by locally steep depth gradients.

2.3   Elastic Registration

Even if we disregard optical occlusions, single-view range imaging can only retrieve depth information of less than half of the body surface as shown in Fig. 2. This renders the mere extraction of the patient from the depth image insufficient for 3D patient modelling, as large portions of the body would remain unmodelled. Therefore, we propose the elastic registration of the segmented range image with a patient-specific body shape template to augment the anatomically incomplete source depth data while accurately reconstructing the captured surface anatomy. To ensure that the final model is descriptive of the patient's real dimensions and surface topology, we employ parametric human body polygon models by courtesy of the *Humanshape* project [7] as template meshes in our registration process. The associated shape generation tool is based on the statistical analysis of whole-body scans and provides realistic human models for a broad spectrum of morphologies. This allows us to create an individual base shape accommodating for the respective anthropometric properties of the patient, consequently improving our registration performance. We implemented a derivation of the registration methods as described in [4] and [8]. Similar to the well-known point set registration algorithm *ICP*, we iteratively search for a transformation that minimizes the distance for all nearest-neighbour correspondences between the template and target meshes. The difference between classical ICP and the method we apply is, that we do not utilize global transformations on our template model. Instead we define n individual affine transformations $\mathbf{X_i}$ for every vertex $\mathbf{v_i} \in V$ of the template mesh $\mathbf{T}$. These are arranged in a matrix $\mathbf{X} := [\mathbf{X_i} \ldots \mathbf{X_n}]$ representing the entirety of the per-vertex transformations which is regularized by the distance to closest points on the target surface $\mathcal{T}$ as expressed by the energy term

$$E_d(\mathbf{X}) := \sum_{\mathbf{v_i} \in V} w_i \, \mathrm{dist}^2(\mathcal{T}, \mathbf{X_i}\mathbf{v_i}) \,, \qquad (1)$$

where w_i denotes a binary reliability weighting for the nearest-neighbour correspondences. Two reliability tests are performed on every candidate vertex pair $(\mathbf{v_j} \in \mathbf{T}, \mathbf{v_j} \in \mathcal{T})$ after finding the initial set of correspondences. A correspondence is rejected, if the template's vertex normal falls below a threshold of collinearity θ in the same direction with the closest target vertex normal, i.e.

$$\mathbf{n_{v_i \in T}} \times \mathbf{n_{v_j \in \mathcal{T}}} < \theta \quad \wedge \quad \mathbf{n_{v_i \in T}} \cdot \mathbf{n_{v_j \in \mathcal{T}}} < 0 \,, \quad (2)$$

and if the nearest template vertex is located on the border of the template mesh. This avoids the mapping of many template vertices onto the same target vertex. Additionally, we penalize the weighted dissimilarity of the transformation matrices of neighbouring template vertices with a weighting matrix $\mathbf{G} := diag(1,1,1,\gamma)$ based on the Frobenius norm $||\cdot||_F$ defined as the square root of the sum of the squares of the matrix elements through an energy term for similiarity:

$$E_s(\mathbf{X}) := \sum_{\{i,j\} \in \mathcal{E}} ||(\mathbf{X_i} - \mathbf{X_j})\mathbf{G}||_F^2 \, \mathrm{dist}^2(\mathcal{T}, \mathbf{X_i v_i}) \,. \quad (3)$$

This ensures the transformations to retain some local rigidity, which results in a globally smooth deformation of the template and causes holes in the target mesh to be filled in realistically, so that the registration result is robust against missing shape topology and resembles a good approximation of the actual body surface. We deliberately omitted a penalty for the distance between landmarks on the mesh surfaces, to not complicate the registration process by setting landmark correspondences. The final cost function assumes the shape

$$E(\mathbf{X}) := E_d(\mathbf{X}) + \alpha E_s(\mathbf{X}) \quad (4)$$

and constrains an iterative sparse solver [9] which we employ to solve for an optimal set of transformations $\mathbf{X}$. As transformations for regions containing rejected correspondences are not subject to the distance constraint $E_d(\mathbf{X})$ and thus would remain unchanged, we enforce smooth transformation of these vertices by emphasizing the similarity constraint by through the parameter α.

ICP-like registration algorithms act solely upon a point to point distance metric and thus strongly depend on the prior coarse alignment of the two point sets to be registered. We handle this issue by first aligning the longitudinal axis of the template mesh with the x-axis of the camera (Fig. 1) to confine the rotational difference to be around this axis. Exploiting that for a patient in a supine position the sagittal axis coincides with the surface normal of the couch underneath, we calculate the couch's surface by averaging the vertex normals of the couch nearest to its intersection point with the camera's z-axis. The template is rotated about its longitudinal axis with the angle that the patient's sagittal axis deviates from the camera's optical axis. The translational misalignment is handled by finding the translation vector between the centres of mass (COMs) of the template and target. Due to missing regions in the patient mesh, the COM of the template and patient do not correspond to the same anatomical location. To compensate this, we recalculate the template's COM for the surface area that would

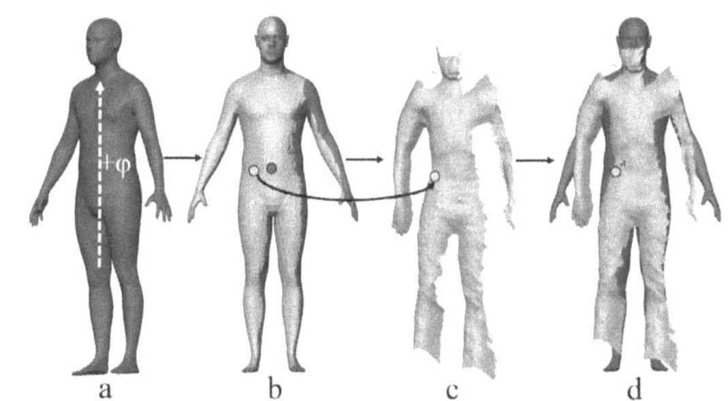

Figure 3: **a** Rotational alignment of the template. **b** Definition of a placeholder COM corresponding with **c** the target's COM. **d** Resulting alignment after translating with the vector between the COMs.

be visible from the camera's viewing angle (Fig. 3). The anatomically corresponding COMs are used to define the translational displacement and obtain an initialization for the elastic registration.

3   Results and Discussion

We evaluate the performance of our method by registering frontal and oblique range images of four male volunteers covering a spectrum of body morphologies (Table 1). The modelling quality is assessed by visual inspection of the obtained meshes for realism and by calculating the residual RMS Euclidean distances between the vertices of the source meshes and their respective closest vertices in the deformed templates (Table 2). As depicted in Fig. 4, our segmentation specifically recovers the data only belonging to the patient while removing major artefacts. Concerning the elastic registration, overlapping regions between template and target, e.g. around the torso, are accurately approximated by the deformed template, due to large amounts of accepted correspondences. With deviations of under 1 mm in these regions, we conclude that the registration of the trunk may be interpreted as a mere reparametrization of the source image. However, the registration quality of articulated regions such as the head and extremities are prone to suffer from unavoidable positioning inconsistencies. Due to the employed template models being fixed in their stance, said regions are only imperfectly recovered. But in most applications, this does constitute a relatively negligible issue, as these regions often are not of direct medical interest and only serve for collision avoidance purposes.

Table 1: Anthropometric averages of the registered participants

	Age	Height (cm)	Weight (kg)
Mean	36.5	180.25	100.75
σ	12.69	5.89	13.63

When comparing the registration quality obtained through either of the two view angles, the similar residual RMS Euclidean distances are misleading. The overall model realism achievable through frontal views is generally higher, as the

Table 2:　Residual RMS Euclidean distance between the target and template for the full meshes and specifically the region of the torso. The results are given as averages for the employed acquisition views.

View	$d_{RMS}(\mathcal{T}, \mathbf{T})$	$d_{RMS}(\mathcal{T}_{torso}, \mathbf{T})$
Frontal	12.1357 mm	0.8015 mm
Oblique	13.3294 mm	0.6562 mm

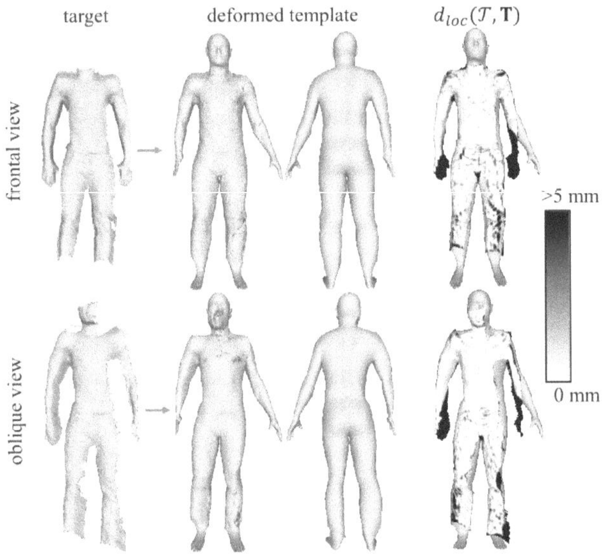

Figure 4: Visual registration results obtained for frontal and oblique range images with residual local minimal Euclidean distances for one subject of our volunteers.

symmetric source data evenly guides the deformation process. Depending on the use case, the view angle capturing most of the respective region of interest is generally preferred. Frontal views only provide source data for registering anterior aspects of the torso. Consequently, for a patient in a supine position, the lateral aspects of the body, cannot be captured. Oblique scans are more suitable for this purpose, as they guide the deformation for both the lateral and anterior aspects, albeit only for one half of the body. As the infilling of holes in the target mesh is solely based on the enforced similarity of neighbouring vertex transformations, the representativeness of a target model produced by such oblique imaging remains questionable for applications demanding higher overall modelling accuracy. A possible intuitive solution of this problem we aim to investigate in our future work would be to exploit the symmetry of the human body and mirror obliquely acquired surface scans about the sagittal plane.

4　Conclusion

We presented an efficient 3D patient modelling framework based on the markerless elastic registration of single-view range images. Our method is designed to be applicable within clinical environments, requiring minimal technical equipment, source data and user interaction for the model generation process. The initial experiments indicate the general feasibility of this technique, validating its robustness and accuracy under variable acquisition settings. This work represents a contribution to alleviate the establishment of automated diagnostic and therapeutic systems in clinical practice, by providing the ability to rapidly obtain complete surface anatomy representations that can be targeted by robotic systems. More generally, we demonstrate an alternative imaging technique that can be utilized for the assessment of the external morphology in patients. Based on these preliminary results we intend to improve our framework's shape completion accuracy and robustness against pose variations.

Acknowledgement

The work has been carried out at the Institute for Robotics and Cognitive Systems, Universität zu Lübeck.

5　References

[1] T. Marc, *Computer-aided medicine and surgery.* In: Digital Medicine. 2.3, p. 81, 2016.

[2] M. Al-Badri, S. Ipsen, S. Böttger and F. Ernst, *Robotic 4D ultrasound solution for real-time visualization and teleoperation.* In: Current Directions in Biomedical Engineering 3.2, pp. 559-561, 2017.

[3] C. Chen, Y. Hung and J. B. Cheng, *A fast automatic method for registration of partially-overlapping range images.* In Sixth International Conference on Computer Vision, IEEE, pp. 242-248, 1998.

[4] B. Allen, B. Curless and Z. Popović, *The space of human body shapes: reconstruction and parameterization from range scans.* In: ACM transactions on graphics (TOG) 22.3, pp. 587-594, 2003.

[5] W. Schroeder, B. Lorensen and K. Martin, *The visualization toolkit: an object-oriented approach to 3D graphics.* Kitware, 2004.

[6] Intel Corporation, *librealsense.* Available: https://github.com/IntelRealSense/librealsense/ [last accessed on 2020-01-20].

[7] M. Reed, U. Raschke, R. Tirumali and M. Parkinson, *Developing and implementing parametric human body shape models in ergonomics software.* In: Proceedings of the 3rd international digital human modeling conference, Tokyo. 2014.

[8] B. Amberg, S. Romdhani and T. Vetter, *Optimal step nonrigid ICP algorithms for surface registration.* In: 2007 IEEE Conference on Computer Vision and Pattern Recognition, IEEE, 2007.

[9] G. Guennebaud, B. Jacob and others, *Eigen v3, 2010.* Available: http://eigen.tuxfamily.org [last accessed on 2020-01-20].

Automatic anatomical structure detection in aortic CT angiography

René Pallenberg [1], Marja Fleitmann [2], Kira Soika [2], and Heinz Handels [2]

[1] Medical Informatics, Universität zu Lübeck, rene.pallenberg@student.uni-luebeck.de

[2] Institute of Medical Informatics, Universität zu Lübeck, (fleitmann, soika, handels)@imi.uni-luebeck.de

Abstract

CT angiography (CTA) is an important tool in modern radiology. The administered contrast agent can cause various adverse effects that can be avoided by dose reduction. For the development of a system that can recommend an adjustment of the commonly used standard dose for individual patients, the dependencies between contrast, patient-individual parameters and contrast agent dose have to be examined. Therefore, an assessment of the image contrast of CTA scans is necessary. The manual assessment is based on determining regions of interest which is time-consuming and not convenient for a large number of scans. We propose a method to automatize the computer-assisted assessment by replacing the manual slice detection with an algorithm. The automatic slice detection achieved an accuracy of 90% on our data compared to the expert's selection. The resulting system allows reducing the time needed for the assessment by proposing the found slices to the expert.

1 Introduction

The visualization of blood vessels and areas well supplied with blood is necessary for the detection of diseases such as aneurysms, blockages, blood clots, and tumors [1]. This visualization can be done with CT angiography (CTA) images. A commonly used contrast agent is Imeron that contains iodine. Severe hives, throat swelling, seizure, and cardiac arrest are occurring adverse effects caused by it [2]. An additional risk relates to generating images with insufficient contrast caused by a too small dose.

To avoid the occurrence of these adverse effects, it is important to reduce the amount of admitted contrast agent. In most cases, a standard dose is applied which does not take differences between individual patients into account. For the search of dependencies between a patient's vital parameters, the contrast dose and the contrast quality, a large amount of annotated images is necessary. The resulting dependencies would allow an adaptation of the dose based on the patient's vital parameters.

Therefore three classes were defined for the contrast quality of an image, insufficient, optimal and excessive. The current contrast quality assessment is based on, the manual segmentation of regions of interest (ROIs) by a radiology expert. This work is time-consuming and not convenient for a large number of images. Therefore an automatic contrast quality assessment is needed. To solve this issue, we have developed a semi-automatic algorithm which automatize the determining of the ROIs and the following assessment. The determination is done by circle detection as in [3] based on the Hough transform [4]. At least a rule-based assessment gets applied based on the threshold used in [5].

In this approach the slice on which the ROIs should be detected has to be selected manually. In this paper, we propose a system for the automatic selection of these slices of interest (SOI), to get a system that can automatic classify the contrast quality of an image. For the localization, a template matching method will be used which has been previously applied on medical images [6]. The algorithm is visualized in Fig. 1.

2 Materials and Methods

2.1 Data

Our data consists of 73 contrast-enhanced CTA volumes. These images were produced by the Department of Radiology and Nuclear Medicine, UKSH Lübeck. The administered contrast-agent was Imeron 300. The images have been generated using an aortic scan protocol. The z-spacing is 5 mm for all slices. A small percentage does not contain all three ROI related anatomic structures. All axial slices have a resolution of 512x512 voxel (sagittal, coronal). The x and y spacing is not equal between different images. All images are stored in the DICOM format. For each image, the SOIs, ROI segmentation, and the contrast quality class have been annotated by radiology experts. These annotations provide the reference standard for the evaluation.

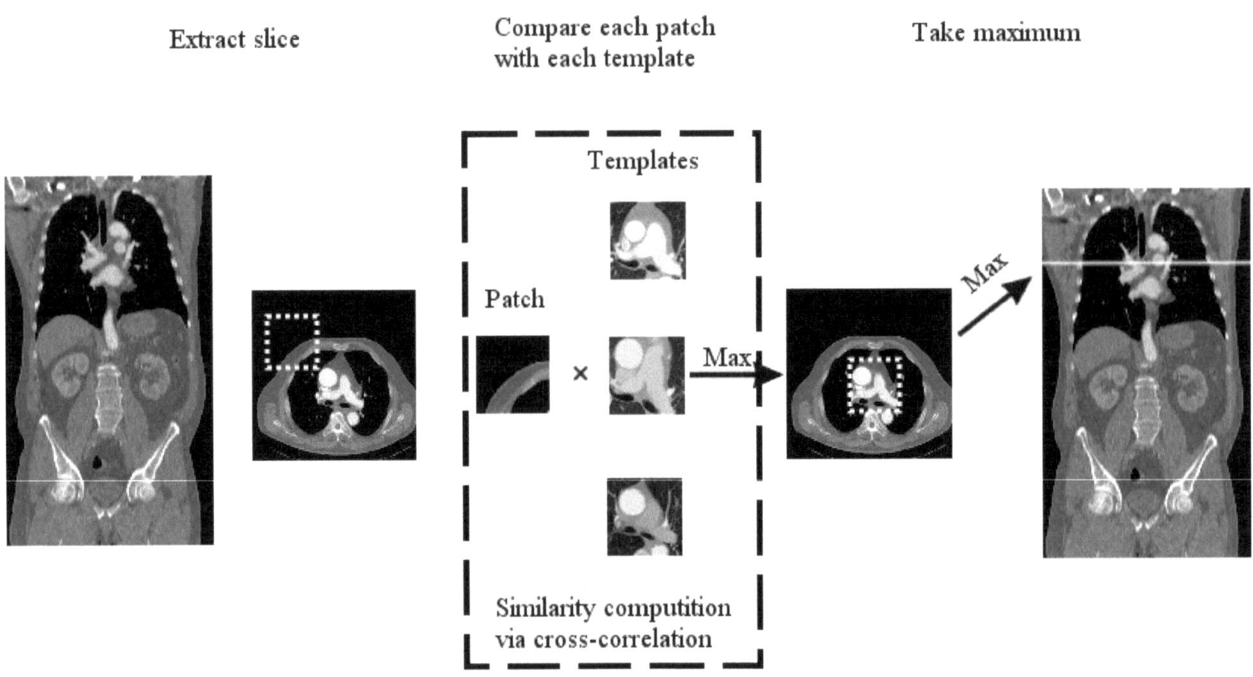

Figure 1: This figure illustrates the flow chart of a SOI 3 detection for an example by using three different templates.

2.2 Template Matching

The template matching algorithm compares a 2D template to a 2D input image. Each possible patch of the same size as the template image gets extracted from the input image. These patches are then compared to the template image by computing a similarity value. There are many possible metrics to compute this value such as the sum of square difference or cross-correlation coefficient. The whole process can be seen as a sliding window operation.

2.3 Slice Detection

For the images of the aorta protocol three ROIs have been selected by radiology experts. These ROIs are:

- ROI 1: Arteria femoralis communis

- ROI 2: The Aorta at kidney level

- ROI 3: The Aorta at pulmonary artery bifurcation level

The first step in determining the ROIs is the detection of the correct axial slice. The aorta and the arteria femoralis communis do not have a significant shape that would allow detecting the slice. Instead of detecting the ROIs directly, the algorithm is searching for significant anatomical structures which are located on the same slices of interest (SOI). These structures are:

- SOI 1: Caput femoris

- SOI 2: The lower end of the Thorax

- SOI 3: Pulmonary artery bifurcation

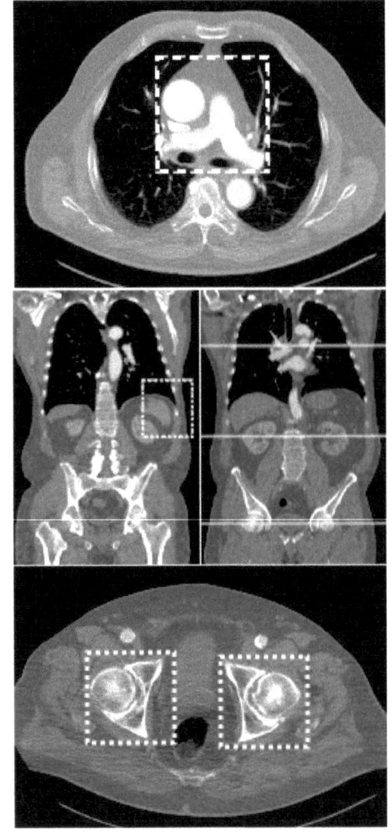

Figure 2: This graphic shows example templates and their location (Right) for SOI 1 (Bottom), 2 (Left) and 3 (Top).

An example for the different templates and their SOI location is presented in Fig. 2. The detection of these structures is done by a template matching [6]. The template matching allows the application of different similarity functions. The Hounsfield units in CTA scans can be highly different for the pulmonary artery. Therefore the normalized cross-correlation function is used, which is the covariance between the template and the image patch divided by the product of the variances. The range of the similarity value is between zero and one where one corresponds to a perfect matching.

To detect the SOIs the template matching is applied on each slice. The slice with the highest similarity value will be selected as SOI. To achieve a higher generalization we applied different templates for the same structure and combined the results by taking the SOI corresponding to the highest value overall templates. The process is illustrated for SOI 3 in Fig. 2.

The templates were generated by extracting patches with ITK-SNAP, which contain the structures for the SOIs. These templates were stored in the Neuroimaging Informatics Technology Initiative (NIFTI) format which contains important information such as voxel spacing. Both the caput femoris dexter and the caput femoris sinister were extracted from the same slice. The size of the caput femoris templates is 115x115 voxel for both sides and 140x140 voxel for the templates of the pulmonary artery bifurcation. The template size was chosen so that no structure of interests in the tested data-set extends beyond the boundaries.

For the automatic selection of SOI 2, we changed the method because the thorax ending does not have a significant structure in axial slices. Instead of selecting the SOI by applying the template matching on all axial slices we apply the template matching on all coronal slices. To get the slice we take the z-coordinate of the best matching directly. The templates were selected so that the last rip is located in the bottom line of the template. The template size is 140x35 voxel. As an alternative, we also compute SOI 2 depending on the two other SOIs. This allows the reduction of the computing time by $\frac{1}{3}$ and is more robust if axial slices of the CTA volume are not available.

After SOI 1 and 3 have been detected, SOI 2 is computed by the function presented in Table 1. The value 46 corresponds to the half of the mean slice difference between SOI 1 and 3 in the tested data-set.

Table 1: The results of SOI 2 for the different conditions.

Condition	SOI 2
ROI 1 and 3 available	mean(SOI 1, SOI 3)
ROI 1 missing	max(0,SOI 3 - 46)
ROI 3 missing	min(maxSlice,SOI 1 + 46)

For the implementation the ITK [7] and the openCV [8] image library have been used. The whole algorithm is implemented in c++. First, the template images and the input image are loaded as ITK image. The Hounsfield units get mapped to values in the range of 0 to 255. Next, the images get converted to the openCV format. The template images

are resized to have the same voxel spacing as the input image. Accordingly, the openCV template matching is applied to each axial slice. To combine the results of both SOI 1 templates the mean of the similarity values is computed.

In some images axial slices are not available, so not each anatomical structure corresponding to an SOI is contained in all CTA volumes. To detect whether all are available two thresholds have been applied. The first one for the number of slices and the second one for the similarity value. If the number of axial slices is below 92 and the highest similarity value is below 0.6 the ROI is labeled as non existing. The value 0.6 was chosen because the lowest similarity value of a correct detected SOI was above 0.65. The value 92 corresponds to the mean slice difference between SOI 1 and 3 in the tested data-set.

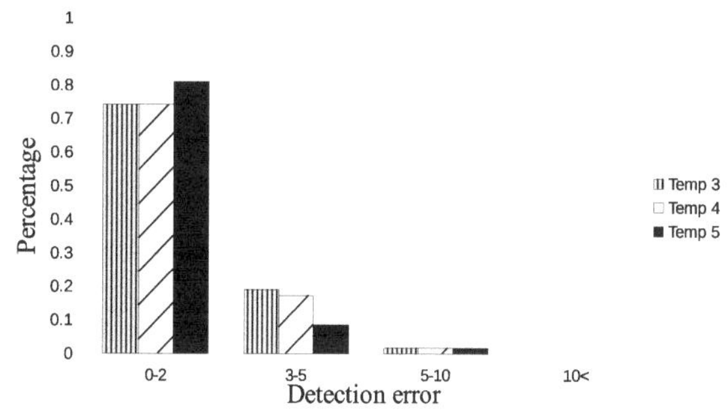

Figure 3: Relative frequency histogram of the detection error of SOI 1.

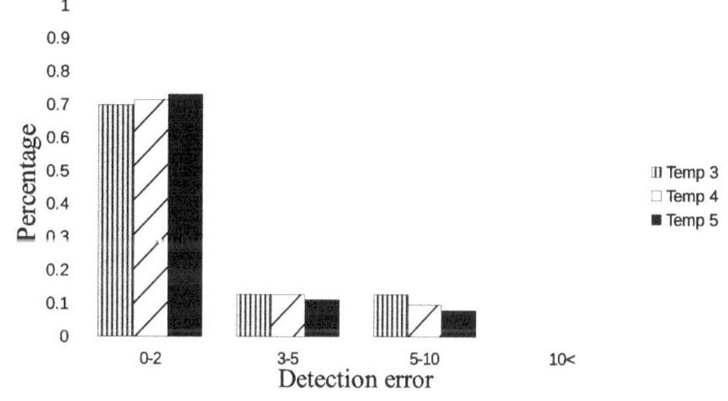

Figure 4: Relative frequency histogram of the detection error of SOI 3.

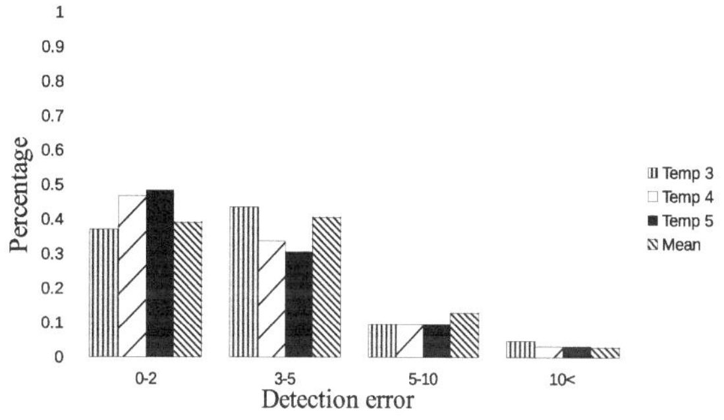

Figure 5: Relative frequency histogram of the detection error of SOI 2.

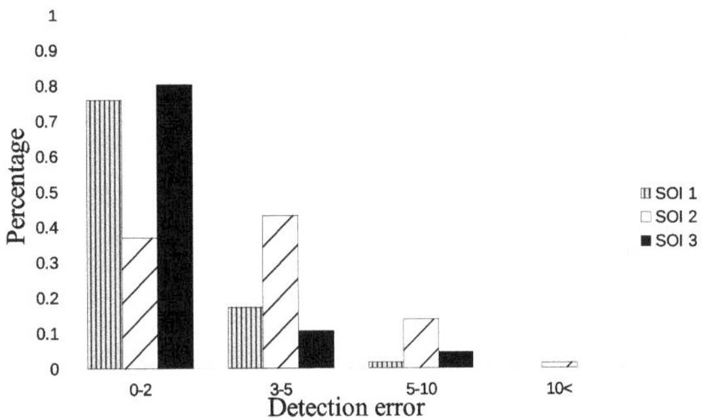

Figure 6: Relative frequency histogram of the detection error for all three SOIs with scaling.

3   Results and Discussion

The detection error is defined as the difference between the slice number of the automatic detected SOI and the expert selection. For each SOI we tested all possible combinations of the templates for three, four and five templates. In the first test, we did not scale the images to get the same spacing. We selected the template combinations with the lowest sum of the squared difference detection error. The images of the used templates are excluded. The relative frequency histograms resulting in computing the difference between the predicted and the expert selected SOIs are in Fig. 3, 5 and 4.

We decided to use three templates for each SOI detection to get a high generalization. The scaling does have a positive impact on the automatic detection. The results for the improved SOI detection with three templates and mean of SOI 1 and 3 for SOI 2 are presented in Fig. 6. The results of the automatic detection of missing SOIs can be seen in Table 2. To evaluate the SOI detection for the task of contrast quality

Table 2: The two fourfold tables show the results of the algorithms existence check for SOI 1 and 3 in comparison to the experts opinion.

| | | Expert Slice | |
		Available	Missing
SOI 1	Available	61	2
	Missing	0	10
SOI 3	Available	71	0
	Missing	0	2

assessment, we compared the contrast quality classification result by segmenting the ROIs on the predicted SOIs with the results by taking the expert-selected SOIs. The classification in the three different contrast quality classes resulting in using these slices achieved an accuracy of 95% over all classes. In considering the slice thickness of 5 mm a difference of at least ten slices is within a certain area.

4   Conclusion

The SOI detection with the template matching achieved high and robust accuracy for the CTA scans of the aorta protocol. The openCV library also includes an GPU supported version of the template matching which can reduce the computation time in the future. The method is the first major step for a fully automatic contrast quality assessment. Using the predicted slices as a preselection for the expert classification reduces the needed time significantly.

Acknowledgment

The work has been supervised by the Institute of Medical Informatics, Universität zu Lübeck. We thank Andreas Martin Stroth, Jan Gerlach, Alexander Fürschke, Jörg Barkhausen and Arpad Bischof from the Department of Radiology and Nuclear Medicine (UKSH) for generating and labeling the CTA scans and their helpful expertise.

5   References

[1] D. Theisen, H. v. Tengg-Kobligk, H. Michaely, K. Nikolaou, M. Reiser, and B. Wintersperger, "Ct angiographie der aorta," *Der Radiologe*, vol. 47, no. 11, pp. 982–992, 2007.

[2] M. Andreucci, R. Solomon, and A. Tasanarong, "Side effects of radiographic contrast media: pathogenesis, risk factors, and prevention," *BioMed research international*, vol. 2014, 2014.

[3] U. Kurkure, O. Avila-Montes, and I. Kakadiaris, "Automated segmentation of thoracic aorta in non-contrast ct images," *IEEE International Symposium on Biomedical Imaging: From Nano to Macro*, pp. 29–32, 05 2008.

[4] C. Kimme, D. Ballard, and J. Sklansky, "Finding circles by an array of accumulators," *Communications of the ACM*, vol. 18, no. 2, pp. 120–122, 2 1975.

[5] H. C. Becker, C. Hong, A. Knez, A. Leber, R. Brüning, U. Schoepf, and M. Reiser, "Optimal contrast application for cardiac 4-detector-row computed tomography," *Investigative radiology*, vol. 38, pp. 690–4, 12 2003.

[6] J. N. Sarvaiya, S. Patnaik, and S. Bombaywala, "Image registration by template matching using normalized cross-correlation," in *2009 International Conference on Advances in Computing, Control, and Telecommunication Technologies*, Dec 2009, pp. 819–822.

[7] L. Ibanez, W. Schroeder, L. Ng, and J. Cates, *The ITK Software Guide*, 1st ed., Kitware, Inc., 2003, iSBN 1-930934-10-6. [Online]. Available: http://www.itk.org/ItkSoftwareGuide.pdf

[8] G. Bradski, "The OpenCV Library," *Dr. Dobb's Journal of Software Tools*, 2000.

Automatic layer segmentation of corneal μOCT images for subbasal nerve tracking

Verena Bühler [1,2], Andreas Wartak [2,3], Merle S. Schenk [2,4], Guillermo J. Tearney [2,5,6] and Reginald Birngruber [2,7]

[1] Medical Engineering Science, Universität zu Lübeck, verena.buehler@student.uni-luebeck.de
[2] Wellman Center for Photomedicine, Boston, {vbuehler, awartak, mschenk1}@mgh.harvard.edu
[3] Harvard Medical School - Department of Dermatology, Harvard University, awartak@mgh.harvard.edu
[4] Department of Ophthalmology, Ludwig-Maximilians-Universität München, mschenk1@mgh.harvard.edu
[5] Harvard-MIT Division of Health Sciences and Technology, gtearney@partners.org
[6] Harvard Medical School - Department of Pathology, Harvard University, gtearney@partners.org
[7] Institute of Biomedical Optics, Universität zu Lübeck, bgb@bmo.uni-luebeck.de

Abstract

The corneal subbasal nerve plexus (SNP) is a network of thin and unmyelinated nerve fibers. It is located between the corneal basal epithelium and the Bowman's membrane (BL). Cornea related diseases such as diabetes can cause changes of nerve fiber density, their respective shape and thickness. The technology of micro optical coherence tomography (μOCT), a non-invasive imaging modality, providing high-speed volumetric and high-resolution cross-sectional imaging, allows for visualization and mapping of the SNP. We present an automated algorithm for segmentation of BL in μOCT image data of excised swine corneas *ex-vivo*. The aim of this project is to facilitate the tracking and quantification of the subbasal nerve plexus to potentially support early diagnosis of peripheral denervation in diseases.

1 Introduction

The cornea is known to be one of the most sensitive tissues in the human body, housing a dense network of nerves [1]. The corneal nerve bundles, that originate from the trigeminal nerve, enter the cornea from the side and divide into a network of many small branches and distinct plexuses, usually parallel to the corneal surface. In the anterior stroma the nerve branches rotate 90° towards the epithelium and branch through Bowman's layer (BL) [1]. At this boundary between basal epithelium and BL the so-called subbasal nerve plexus (SNP), a network of thin (from 0.5 to 1.5 μm in diameter) autonomic nerves, is located [2]. The sensory and unmyelinated nerve fibers of this plexus have a protective role and are important for the unification, such as wound recover and rapid cell reproduction [1].

However, corneal nerves are reported to be affected by diseases and disorders of the peripheral nervous system, of which peripheral neuropathy is the most common one [3]. The causes of peripheral neuropathy are manifold ranging from diseases such as diabetes to autoimmune diseases and from hereditary to infectious diseases [4], [5]. Even though a large number of people suffer from such diseases, early diagnosis of peripheral neuropathy is currently a major clinical need. Since nervous degeneration begins in the most peripheral of nerve branches [3], it is reasonable to conclude that earliest onsets of peripheral neuropathy might be detected precisely at the layer of the micrometer-sized nervous structures – the subbasal plexus of the cornea.

As of today, *in-vivo* confocal microscopy (IVCM) is the gold-standard for corneal nerve imaging in the clinics [4]. This method provides a spatial resolution below 1 μm and speckle-free images, however, entails several shortcomings: *i)* IVCM requires direct contact to the eye which may cause infections and indisposition for the patient. *ii)* IVCM can only show a very small field-of- view (FoV) of roughly 400 x 400 μm because of the required high numerical aperture (NA) of $\sim$0.9. *iii)* IVCM only generates en-face images of a single depth layer [4], [6], [7]. In order to retrieve a depth resolved 3D data set, a multitude of en-face images at different imaging depths need to be acquired.

Optical Coherence Tomography (OCT) is a non-invasive imaging modality, providing high-speed volumetric high-resolution cross-sectional imaging [8]. Its main application is in biomedical imaging, and through its histology-like imaging capabilities of tissues *in-vivo* it revolutionized in particular ophthalmic disease diagnosis. Such, OCT has previously been reported to allow for *in-vivo* imaging of the corneal nerval structures in the subbasal plexus [6], [7].

Although the resolution of conventional OCT is incapable of visualizing the finest nerve fiber structures, recent advances in OCT technology have enabled voxel resolutions of $\sim$1 μm axially and $\sim$2 μm laterally. This form of highest resolution OCT is termed microOCT (μOCT) [9].

In order to estimate the subbasal nerve density, automatic layer and nerve segmentation is needed. Manual segmenta-

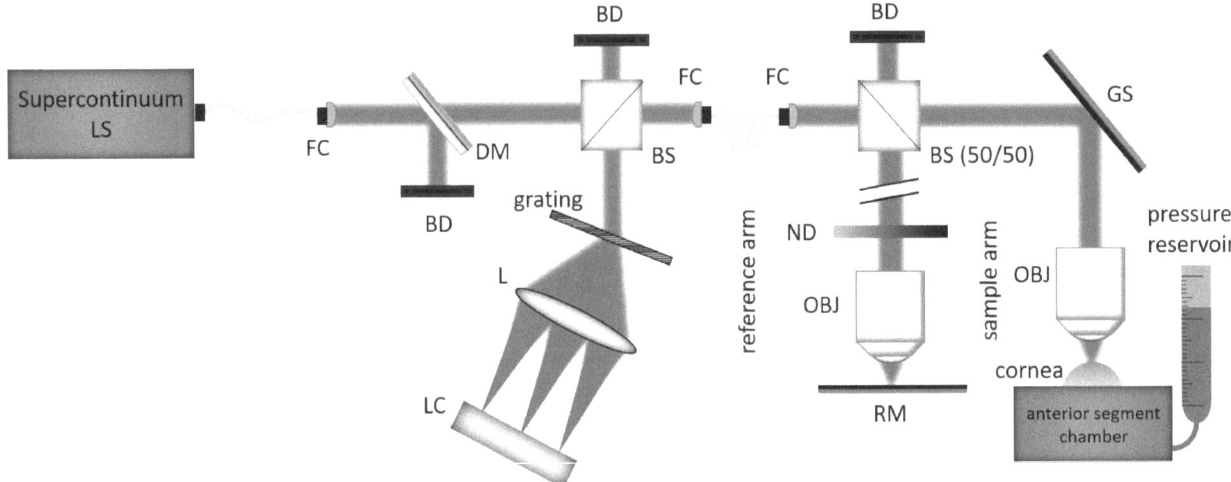

Figure 1: Sketch of the μOCT setup used for corneal imaging. The laser light of the supercontinuum light source (LS) is emitted from a fiber towards the bulk optics setup. The spectral bandwidth is cropped by a dichroic mirror (DM) to the desired wavelength band (650-950 nm), before the light is separated by a beam splitter (BS) to be input to the interferometer and spectrometer. The second BS splits the beam into sample and reference arms. In the latter, the beam is transmitted through a neutral density (ND) filter and is focused onto a reflective mirror through an objective (OBJ, plan apochromat, NA: 0.4). In the sample arm, a 2D galvo scanner (GS) is used to raster scan the cornea before another objective focuses the beam onto the sample. A pressure regulator is applied to maintain a pressure gradient that is similar to that of the physiologic intraocular pressure, and thus keep the cornea's natural shape. The back-scattered light from the sample as well as the back reflection from the reference mirror (RM) are guided back through the system. The first BS redirects part of this light towards the spectrometer unit. Here, the light gets focused on a line-scan camera (LC) by a lens (L) after passing a grating to spatially separate the spectral components.

tion is subjective and, in particular, time-consuming. However, tracking in 3D is a challenge that has not yet been overcome by completely automatic means, and is usually accomplished by semi-automatic algorithms [10].

In this paper we present an automatic segmentation algorithm that segments and flattens Bowman's layer in a 3D μOCT dataset, in order to produce 2D maps of the subbasal nerve plexus. These maps greatly facilitate nerve segmentation by reducing the dimensionality of the data.

2 Material and Methods

2.1 Imaging setup

A bench-top spectral domain (SD-)OCT instrument, featuring a supercontinuum light source (NKT Photonics, Denamrk) was used for imaging. Images were acquired with an 8192-pixel high spectral resolution line-scan camera (Basler, Germany), at a frame rate of 20 kHz. The system operated at a central wavelength of 800 nm with a bandwidth of +/-150 nm. Volumetric images were acquired within a time of $\sim$13 s per volume over a FoV of 1x1 mm at a sampling of 512x512 pixels. The resolutions in tissue were $\sim$1.5 μm axial and $\sim$1.4 μm lateral. On the corneal surface the power of the imaging beam was $\sim$20 mW. A schematic representation of the SD-OCT imaging setup is shown in Fig. 1.

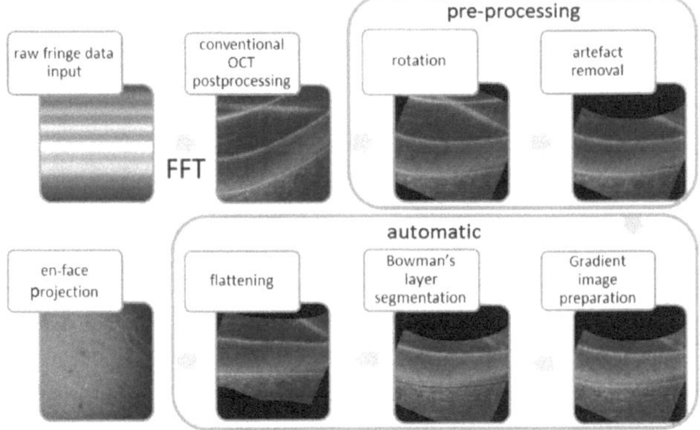

Figure 2: Processing pipeline: layer segmentation and flattening approach including pre-processing and automatic algorithm steps.

2.2 Automatic layer segmentation

After the image acquisition, standard SD-OCT processing steps to generate logarithmic intensity scans from the raw spectral fringe data by means of, mean spectrum subtraction, λ-k-remapping, numerical dispersion compensation and discrete fast Fourier transform (FFT), were performed. ImageJ [11], a Java-based image processing program, was used to rotate the image to create a top view of the cornea and remove an image artefact caused by the highly reflecting surface of liquid Phosphate-buffered saline (PBS) was applied topically to protect the swine cornea from dehydra-

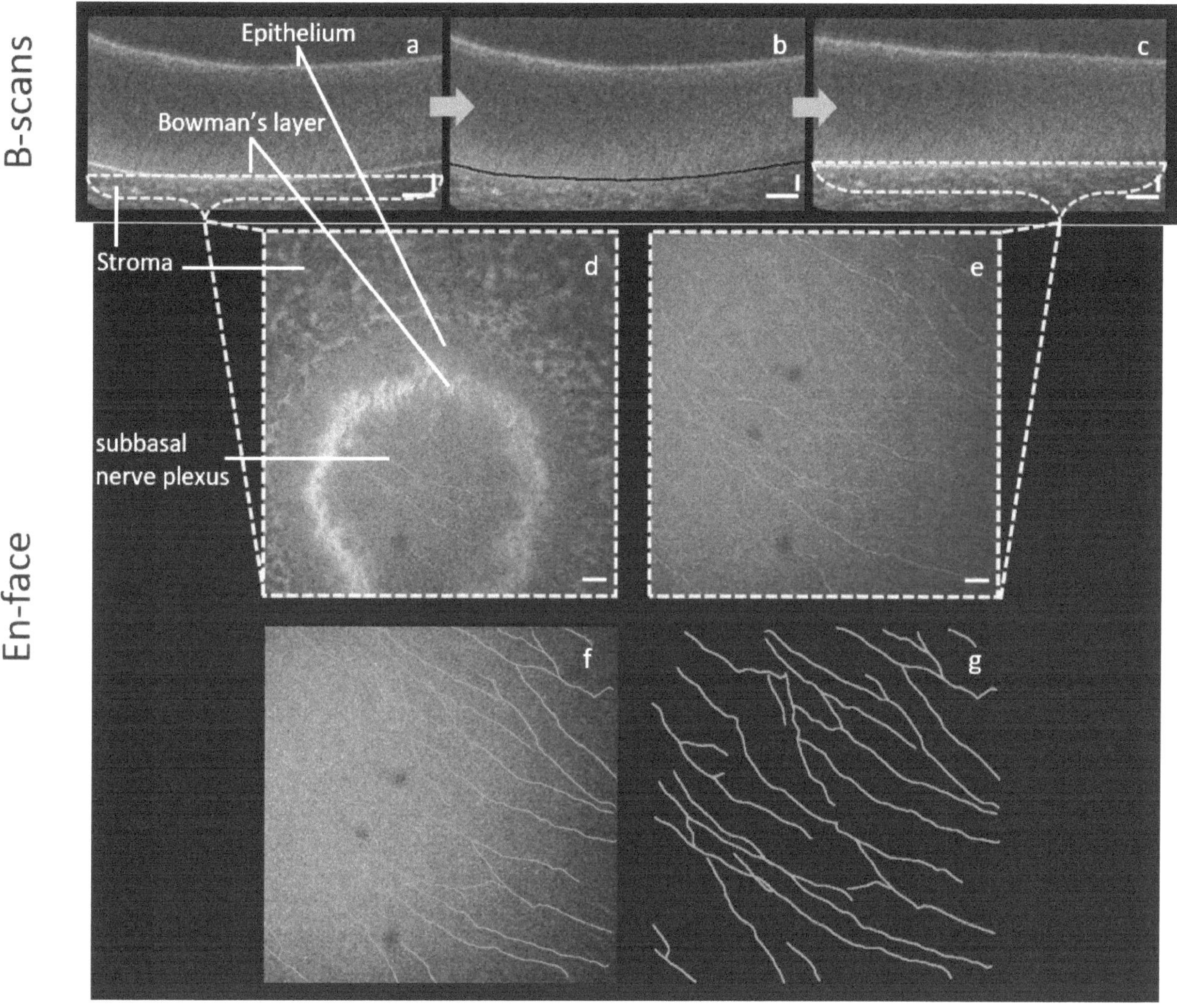

Figure 3: Results of the automatic BL segmentation algorithm and layer flattening applied on an OCT image data of a swine cornea *ex-vivo*. The B-scans (a-c) represent the algorithm steps and the en-face images (e-f) the results. (d) depicts an original en-face projection before segmentation, only enabling a very limited FoV for the visualization of the nerve structures. (e) shows the entire SNP layer. (f) presents the traced nerve skeleton, overlaid over the initial 2D en-face image in magenta. (g) only the nerve skeleton is depicted without overlay. Scale bar= 50 μm.

tion. Figure 2 illustrates the major individual operational steps of the processing pipeline.

The automated segmentation of the subbasal corneal nerve divides into several steps: Fist, the gradient image preparation of general layer segmentation was adapted from previously presented work [12]. Here, each pixel of the B-scan was represented as a node and the edges of the graph connecting adjacent pixels were identified by a costfunction mainly based on the axial image gradient. By keeping the cost function minimized while passing through the B-scan, each layer was segmented iteratively. Besides the original 2D segmentation, an additional 3D adjustment was implemented. The algorithm was modified to only segment BL. Second, after segmenting BL, the algorithm reconstructed every B-scan and sets the layer in each scan to a defined axial position to generate a volumetric flattened dataset. After the image stack was imported, the processing time ranged from between 5 and 15 minutes (depending on the actual size of the dataset) until a new, flattened image

stack was generated and saved. ImageJ was used to further process the scan and generate en-face images over a certain depth (roughly 3-5 μm) of corneal epithelial BL. Using the imaging software plugin "NeuronJ", a fast semiautomatic tracing of the nerves was generated (see Figure 3).

3 Results and Discussion

To test the algorithm, the segmentation was applied to several OCT scans of excised swine eyes *ex-vivo*. Figure 3 presents exemplary results of a swine cornea for the previously described technique. Figures 3a-c depicts the algorithm steps from the original to the flatted image. The upper left figure (Fig. 3a) depicts an exemplary input B-scan, that visualizes the epithelium, BL and part of the stroma. The middle image (Fig. 3b) displays the segmented line of the BL overlaid on top of the μOCT B-scan. The flatted B-scan is shown on the right side (Fig. 3c).

The en-face projection of the original (Fig. 3d) and flattened (Fig. 3e) image are depicted side-by-side to demonstrate the improved subbasal plexus segmentation. With the assistance of ImageJ, a maximum intensity projection (MIP) image in the en-face plain where the subbasal nerves in the entire FoV are visible, was generated.

The produced image enables tracking and analysis of the subbasal nerve plexus in 2D, as illustrated in Fig. 3f, g.

The trace parameters, such as number of traces, individual length, and total length, were determined automatically by the software. In the depicted case, 35 nerves fiber traces were detected with a complete pixel length of 7761.9 μm and an average pixel length of 221.2 μm. The total number of pixels in the image were, as already mentioned, 512x512 corresponding to 706.6x706.56 μm. Thus, the subbasal nerves covered an area ratio of 2.15%.

4 Conclusion

In this paper we present an automatic Bowman's layer (BL) segmentation and flattening algorithm to visualize the corneal sub-basal nerve plexus (SPN) in bench-top μOCT 3D image datasets of swine corneas in a single 2D en-face image. This algorithm enables one to track the nerves and create a comparable area ratio for nerval density measurements. Future improvements include algorithm optimization to shorten the computational time and an automatic nerve tracking procedure.

Acknowledgement

The work has been carried out at Wellman Center for Photomedicine, Boston and supervised by the Institute of Biomedical Optics, Universität zu Lübeck.

5 References

[1] L. Oliveira-Soto and N. Efron, "Morphology of corneal nerves using confocal microscopy", *Cornea*, vol. 20, no. 4, pp. 374–384, 2001.

[2] C. F. Marfurt, J. Cox, S. Deek, and L. Dvorscak, "Anatomy of the human corneal innervation", *Experimental Eye Research*, vol. 90, no. 4, pp. 478–492, 2010.

[3] C. N. Martyn and R. A. Hughes, "Epidemiology of peripheral neuropathy", *J Neurol Neurosurg Psychiatry*, vol. 62, no. 4, pp. 310–8, 1997.

[4] T. Utsunomiya, T. Nagaoka, K. Hanada, T. Omae, H. Yokota, A. Abiko, M. Haneda, and A. Yoshida, "Imaging of the corneal subbasal whorl-like nerve plexus: More accurate depiction of the extent of corneal nerve damage in patients with diabetes", *Investigative Ophthalmology Visual Science*, vol. 56, no. 9, pp. 5417–5423, 2015.

[5] P. F. Chance, M. K. Alderson, K. A. Leppig, M. W. Lensch, N. Matsunami, B. Smith, P. D. Swanson, S. J. Odelberg, C. M. Disteche, and T. D. Bird, "Dna deletion associated with hereditary neuropathy with liability to pressure palsies", *Cell*, vol. 72, no. 1, pp. 143–151, 1993.

[6] B. Tan, Z. Hosseinaee, L. Han, O. Kralj, L. Sorbara, and K. Bizheva, "250 khz, 1.5 μm resolution sd-oct for in-vivo cellular imaging of the human cornea", *Biomedical Optics Express*, vol. 9, no. 12, pp. 6569–6583, 2018.

[7] V. Mazlin, P. Xiao, E. Dalimier, K. Grieve, K. Irsch, J.-A. Sahel, M. Fink, and A. C. Boccara, "In vivo high resolution human corneal imaging using full-field optical coherence tomography", *Biomedical Optics Express*, vol. 9, no. 2, pp. 557–568, 2018.

[8] D. Huang, E. Swanson, C. Lin, J. Schuman, W. Stinson, W. Chang, M. Hee, T. Flotte, K. Gregory, C. Puliafito, and a. et al., "Optical coherence tomography", *Science*, vol. 254, no. 5035, pp. 1178–1181, 1991.

[9] L. Liu, J. A. Gardecki, S. K. Nadkarni, J. D. Toussaint, Y. Yagi, B. E. Bouma, and G. J. Tearney, "Imaging the subcellular structure of human coronary atherosclerosis using micro–optical coherence tomography", *Nature Medicine*, vol. 17, no. 8, pp. 1010–1014, 2011.

[10] C. Canavesi, A. Cogliati, A. Mietus, Y. Qi, J. Schallek, J. P. Rolland, and H. B. Hindman, "In vivo imaging of corneal nerves and cellular structures in mice with gabor-domain optical coherence microscopy", *Biomedical Optics Express*, vol. 11, no. 2, pp. 711–724, 2020.

[11] C. A. Schneider, W. S. Rasband, and K. W. Eliceiri, "Nih image to imagej: 25 years of image analysis", *Nat Methods*, vol. 9, no. 7, pp. 671–5, 2012.

[12] S. J. Chiu, X. T. Li, P. Nicholas, C. A. Toth, J. A. Izatt, and S. Farsiu, "Automatic segmentation of seven retinal layers in sdoct images congruent with expert manual segmentation", *Optics Express*, vol. 18, no. 18, pp. 19 413–19 428, 2010.

Gesture Recognition with Deep Learning

Josephine Zillmann [1], Foti Coleca [2], and Erhardt Barth [3]

[1] Medical Engineering Science, Universität zu Lübeck, josephine.zillmann@student.uni-luebeck.de
[2] gestigon GmbH, foti.coleca@gestigon.de
[3] Institut für Neuro- und Bioinformatik, Universität zu Lübeck, barth@inb.uni-luebeck.de

Abstract

Gesture recognition is an interesting application for human-machine-interaction and an active area of research using deep learning networks. After a research of state-of-the-art approaches for gesture recognition, we have identified the *Temporal Relation Network* as a promising approach and applied it to a novel dataset by means of transfer learning. It resulted in very good outcomes with an accuracy of 96.39 %. The only difficult classes were finger-rotate gestures.

1 Introduction

Gesture recognition is an important part of bringing human-machine-interaction to a higher level. Objects like entertainment systems can be controlled through gestures in a car. Phone calls can be accepted or declined with a swipe or the channel of the radio can be changed with a thumb pointing to the right side. The focus can stay on the road while interacting with the entertainment system. As a result, the gesture control leads to a safer driving experience.

Gesture recognition with deep learning is still a young field of research. Standard image processing approaches like finite state machines already do well on gesture control and are fast [1]. One common problem of gestural interfaces is the detection of false positives when the user gestures without intent for them to be recognized (e.g. during talking or while operating other interfaces). With machine learning these unintentional gestures can be learned as separate classes, which can reduce the false positives. Furthermore, when using machine learning, there is no need for handcrafted feature extraction algorithms, which take a long time to develop.

In this work we want to identify an approach for gesture recognition, that is balanced in accuracy and real-time requirements and apply it to a new, application-specific dataset. Regarding gesture recognition from video clips, mostly 3D-CNNs (Convolutional Neural Network in time and space) are used, some approaches added RNNs (Recurrent Neural Networks) especially LSTMs (Long-Short-Term-Memory Networks) for dealing with different duration of gestures. First, we want to present two approaches and evaluate their difficulties. In the chapter Material and Methods, we then describe the identified network.

One interesting approach is the use of a 3D-CNN followed by a Convolutional-LSTM: The 3D-CNN (in this case four layers from C3D Net) is able to learn local short-term spatio-temporal feature maps and transforms the input frames into a 2D-feature map. Then a convolutional LSTM is used because a simple LSTM network would ruin the spatial structure of the images. The input of an LSTM is one dimensional, an image needs to be flattened. The network learns the global temporal feature map and is appended to a 2D-CNN (mobileNet) to learn deeper features. A Support-Vector-Machine or fully-connected-layer predicts the label. Summarized, the network consists of a 3D-CNN, a ConvLSTM and a 2DCNN. It is trained and tested on the Jester dataset (see Chapter 2), which results in an accuracy of 95.01% [2]. Because this approach has a lot of parameters, inference will be slow on conventional hardware. Factorizing the 3D-convolutional layers into separate spatial and temporal components would lead to lower computational costs [3].

Figure 1: Example of gestures of the Jester dataset. From left to right: Sliding Two Fingers Down, Swiping Left, Thumb Up [4].

Convolutional networks that use optical flow (OF) are a good alternative to 3D-CNNs: One approach divides the video into segments and computes the OF using the frames of the segments. The displacement information is fused to the RGB data as an extra channel, called Motion Fused Frames (MFF). Each MFF is the input of the CNN and the features of the CNN are concatenated and passed to a two-layer MLP, then a softmax layer outputs a classification probability. They show, dividing the video into eight segments is the best choice for their use case and each extra channel of OF boosts the accuracy. On the Jester dataset, they achieve an accuracy of 96.28% [5]. While accurate, this method has the disadvantage of expensive computational costs due to the optical flow.

2 Material and Methods

2.1 Temporal Relation Network

The *Temporal Relation Network* (TRN) was identified as a promising approach in terms of accuracy and real-time requirements because it only uses a two-dimensional CNN. The authors of [6] evaluated the temporal information and found that a sparse set of frames leads to better results even for 3D-CNNs, when there is a good frame selection process. For gesture recognition, some gestures can be recognized from only a few frames (e.g. a swipe), while others need more frames (e.g. finger rotation). For the TRN the video is divided into eight segments. From these segments, one frame is selected and put separately into a CNN. During the training process, the selection is executed randomly. The CNN computes the features, as shown in Fig. 2 (simplified version with four segments instead of eight). First, the features are concatenated and put into a Multi-Layer-Perceptron (MLP). In the next step, only three features are concatenated and put in an MLP. There are several possible combinations of the three features, wherefrom three combinations are chosen. Then the number of features is reduced again (e.g. only two features), and three combinations of these features are put in another three MLPs. The features are reduced until only two features are selected, from which as well three combinations are used. The outputs of the MLPs are summed up and a fully connected layer predicts the classification.

The CNN is the Inception network with batch normalization [7], which is used because of its balance between accuracy and efficiency and pre-trained on ImageNet. It has good classification results on the Jester dataset with 94.78 % [8]. The network does not need an exact starting point of the gesture but has problems with large differences in the execution. The network can classify for example a gesture with five to 40 frames (input 40 frames) or 100 to 150 frames (input 150 frames), but if the network has an input of 150 frames, it could not classify a gesture with only 10 frames.

A technique called partial batch normalization was used when applying transfer learning, as a regularization method. This method has been introduced in [9], which the TRN uses as a starting point. When applying partial batch normalization, only the mean and variance of the first batch normalization layer are updated adaptively. The first batch normalization layer is not frozen because the distribution of the new input is different, so the mean and variance should be re-estimated. The authors argue that batch normalization

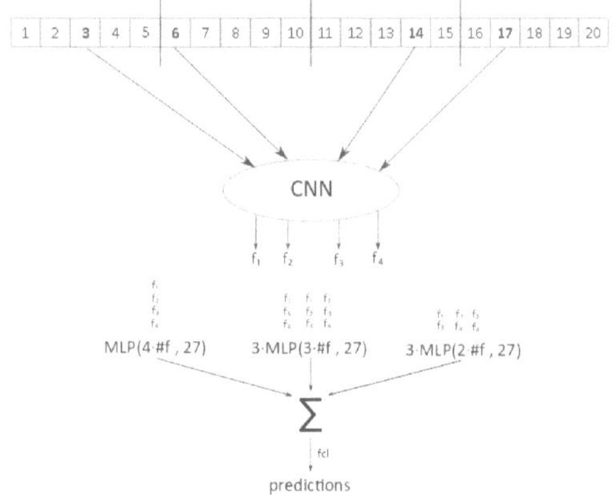

Figure 2: TRN [8] with four segments. The features extracted by the CNN are combined and put into the MLPs. The results are summed up and a fully-connected layer estimates the probabilities of the classes.

usually decreases training time but also increases the risk of over-fitting during transfer learning. This approach boosted the accuracy (from 90% to 92% of a Two-Stream network without TSN), which is why it is used as well in the TRN [9]. To avoid over-fitting, the training uses dataset augmentation techniques. These differ from the ones that are used for one-shot classification, due to the time-series nature of the input data. First, the frames from each segment are randomly chosen. Secondly, random cropping and horizontal flipping with 50% probability are used. This technique appears to have a regularizing effect similar to drop-out, due to the fact that some input frames can be skipped entirely.

2.2 Jester dataset

The company *twentybn* [4] provides various data sets for research purposes. The Jester dataset is a large collection of video clips of people performing predefined gestures in front of their webcam (see Fig. 1). This makes it possible to train robust machine learning systems to recognize hand gestures. The dataset consists of 27 different classes. Each of these classes has around 5000 videos, with each gesture having an approximately similar number of frames, a validation dataset, and a test dataset without labels. Each execution of a gesture needs in between five and 30 frames. There are 68 submissions so far. The list is led by the approach called "Fusion TSN LSTM" with an accuracy of 97.09%. Motion Fused Frames' approach is in 9th place with 96.28%, Guangming Zhu's approach (3DCNN+ConvLSTM+CNN) in 21st place with 95.01% and the TRN in 22nd place with 94.95%. The approaches in between have no meaningful names or papers can't be found.

2.3 Astacus dataset

We have chosen to test the TRN network on our dataset (provided by gestigon GmbH), mostly due to its real-time capabilities. The dataset contains nine classes of gestures

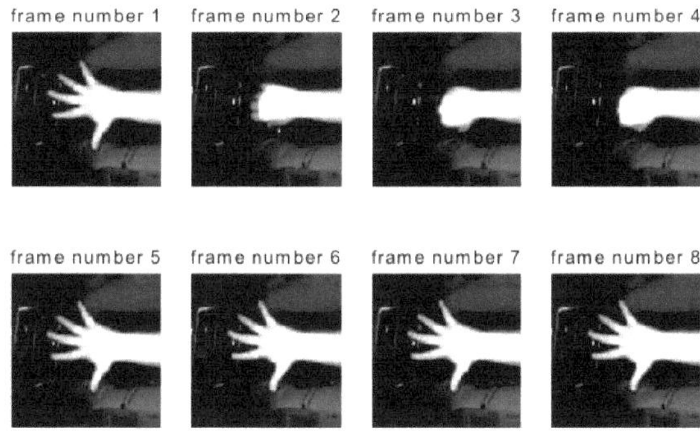

Figure 3: Example of a double-five gesture of the Astacus dataset. The Astacus dataset was provided by gestigon GmbH (www.gestigon.com/home/), and is an in-house dataset recorded for the purpose of gesture recognition in an automotive environment.

(e.g. hand swipes, rotating a finger).
The recordings were done in three different cars, which resulted in an average of 115 videos per gesture. The training data set consisted of 5/6 of the videos and the remaining 1/6 were used for validation. The data is captured by a Time-of-Flight camera, giving amplitude and depth images instead of RGB images. The training was limited to the amplitude data, as these are similar to the images that the TRN network was pretrained on. The images had to be adapted to the Jester dataset. The original amplitude images contain 16-bit values. However, for our use case, there are few regions of the image that go higher than a value of 255. Therefore we clipped the amplitude data to a maximum of 255. Since the network learns the moving gestures rather than the background, the assumption was made that the clipped values (most often seen on shiny background areas) did not affect the classification. Furthermore, the images needed to be preprocessed such that the input fits into the pretrained network. The images were downsampled and interpolated to a dimension of 341 by 256 and afterward cropped to a dimension of 224 by 224.

The range of frames per gestures differed between five and 160 frames, which is a big difference from the gestures of the pretrained dataset. The longest gesture differences have the "Double Five" gesture, see Fig. 3.

The length of gestures was normalized because at the moment the network can't deal with big differences. The video clips in the Jester dataset had a very similar length (35-36 frames each). They were divided into 8 segments because in [9] this partitioning got good results. For the Astacus dataset, we created the clips from the start of the gesture until the end.

3   Results and Discussion

We have applied transfer learning using a pre-trained network (TRN with BNInception as CNN trained on the Jester dataset) on the Astacus dataset. The weights were fine-tuned with a batch size of four because of the small number of gestures per class. The learning rate was 0.001 as before with the Jester dataset.
The training took place without augmentation techniques, no changes have been applied otherwise.

The training was stopped early after 19 epochs, it would have resulted in over-fitting otherwise. This can be seen from the fact that the training error continued to decrease while the test error started to increase again. An accuracy of 96.39 % was achieved. In Fig. 4 it can be seen that only the rotation of a finger in different directions causes problems (see Table 1. This error is explainable because even for a human annotator the differentiation of the rotation gestures by the selection of only eight frames is difficult.
After further analysis of the results, it became clear that the annotations of the gestures for this application were not consistent with our intended usage. The label of rotational gestures did not end after one full revolution but rather ended after a random number of rotations. In order to improve the classification results, further training with shorter inputs of the rotational gestures should be tested. This again shows that a neural net can only be as good as its input data and that exploratory dataset analysis is an important requirement for good results. Besides, it underlines the hypothesis of Huang et al. that the right choice of frames influences the results significantly [6].
One obvious way to improve the dataset is to re-label it and retrain. Another would be to record more data from various classes and with different participants to assure generalization. As well a test dataset, which the network hasn't seen would be useful for further evaluation. By limiting the number of tests done on a third set, we will get a clearer picture of how the algorithm would behave in an unseen real-world scenario.
The inference time is 80ms per clip on an NVIDIA GeForce (GTX 1080 Ti) GPU. This is in line with our real-time requirements, and as such the algorithm would be a good fit for implementation in an automotive embedded ECU.
The background reflections had a minimal effect on the classification accuracy, which confirm our initial assumptions.

4   Conclusion

After state-of-the-art research, we have identified the TRN to be the most promising approach, balancing overall accuracy and real-time requirements. We applied the Network to a new dataset using transfer learning and obtained good results. Rotating gestures were the only difficulties in the classification task. After examination of the data, it was discovered that the annotations need to be adjusted. Another training with better annotation would be interesting.
Furthermore, the network still has to be adapted such that it can cope with larger differences in gesture lengths. A promising idea is the addition of a Recurrent Neural Network after the MLPs.

		1	2	3	4	5	6	7	8	9
swipe left	1	1	0	0	0	0	0	0	0	0
swipe right	2	0	1	0	0	0	0	0	0	0
double five	3	0	0	1	0	0	0	0	0	0
rotate ccw	4	0	0	0	0.73	0.27	0	0	0	0
rotate cw	5	0	0	0	0.05	0.95	0	0	0	0
static pose two	6	0	0	0	0	0	1	0	0	0
thumb left	7	0	0	0	0	0	0	1	0	0
thumb right	8	0	0	0	0	0	0	0	1	0
point click	9	0	0	0	0	0	0	0	0	1
		1	2	3	4	5	6	7	8	9

Table 1: Confusion Matrix with the results of the training on the Astacus dataset (left: True Label, bottom: Predicted Label). Only the finger rotation gestures are problematic.

Acknowledgement

The work has been carried out at gestigon GmbH and supervised by Prof. Dr.-Ing. Erhardt Barth from the Institute for Neuro- and Bioinformatics, Universität zu Lübeck. Hereby I would like to thank gestigon GmbH for the opportunity to work on this topic.

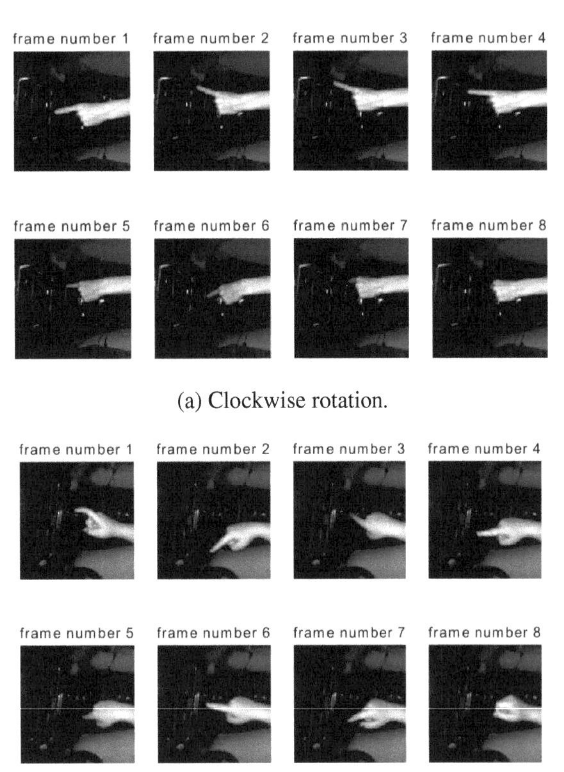

(a) Clockwise rotation.

(b) Counterclockwise rotation.

Figure 4: Even for a human it is difficult to classify the direction of the rotational gestures - the clip contains multiple finger rotations.

References

[1] Pengyu Hong, M. Turk, and T. S. Huang, "Constructing finite state machines for fast gesture recognition", in *ICPR-2000*, Sep. 2000, pp. 691–694.

[2] G. Zhu, L. Zhang, P. Shen, J. Song, S. Shah, and M. Bennamoun, "Continuous gesture segmentation and recognition using 3dcnn and convolutional lstm", English, *IEEE Transactions on Multimedia*, pp. 1011–1021, Apr. 2019.

[3] D. Tran, H. Wang, L. Torresani, J. Ray, Y. LeCun, and M. Paluri, "A closer look at spatiotemporal convolutions for action recognition", *2018 IEEE/CVF*, Jun. 2018.

[4] twentybn. (Oct. 14, 2019). The 20bn-jester dataset v1, [Online]. Available: `https://20bn.com/datasets/jester`.

[5] O. Kopuklu, N. Kose, and G. Rigoll, "Motion fused frames: Data level fusion strategy for hand gesture recognition", *2018 IEEE/CVF*, Jun. 2018.

[6] D.-A. Huang, V. Ramanathan, D. Mahajan, L. Torresani, M. Paluri, F. F. Li, and J. C. Niebles, "What makes a video a video: Analyzing temporal information in video understanding models and datasets", Jun. 2018, pp. 7366–7375.

[7] S. Ioffe and C. Szegedy, "Batch normalization: Accelerating deep network training by reducing internal covariate shift", in *Proceedings of the 32Nd International Conference on International Conference on Machine Learning - Volume 37*, JMLR.org, 2015, pp. 448–456.

[8] B. Zhou, A. Andonian, A. Oliva, and A. Torralba, "Temporal relational reasoning in videos", *Lecture Notes in Computer Science*, pp. 831–846, 2018.

[9] L. Wang, Y. Xiong, Z. Wang, Y. Qiao, D. Lin, X. Tang, and L. Van Gool, "Temporal segment networks for action recognition in videos", *IEEE Transactions on Pattern Analysis and Machine Intelligence*, 2017.

Automatic segmentation of prostate zones in MRI data using different variants of the U-Net

Svenja Rüttgers [1], Jens Hocke [2], and Mattias Heinrich [3]

[1] Medical Informatics, Universität zu Lübeck, svenja.ruettgers@student.uni-luebeck.de
[2] FUSE-AI, Hamburg, jens.hocke@fuse-ai.de
[3] Institute of Medical Informatics, Universität zu Lübeck, heinrich@imi.uni-luebeck.de

Abstract

Fully convolutional neural network architectures have become a popular method for medical image segmentation. Here, we apply U-Net, 3DUNET, and aniso-3DUNET to segment prostate zones in MR images. This automatic segmentation can be used to increase radiologists' efficiency during diagnostics. The difficulties are the high anatomical variability and the low gray value differences. The aniso-3DUNET, an extension of the U-Net for anisotropic data, proved to be the best model. An ensemble of nine aniso-3DUNETs achieves an average dice coefficient of 0.9070 for the Transition Zone and 0.8457 for the Peripheral Zone.

1 Introduction

Prostate cancer is the most common cancer among men in Germany with 60.000 new cases each year [1]. Magnetic resonance imaging (MRI) has been increasingly used to diagnose this disease. A way to increase radiologists' efficiency are automatic suggestions for the diagnosis and prefilled medical reports by investigating the segmentation of the prostate zones in order to localize tumors and to fill out the report. Additionally, the segmentation of the zones could improve tumor classification [2], [3].

The prostate can be divided into different numbers of zones. The following is a segmentation into two zones, the Transition Zone (TZ) and the Peripheral Zone (PZ). The difficulty lies not only in the small differences in gray values but especially in the high anatomical variability of the prostate. This is not only caused by the carcinomas but is also related to the fact that the mean age at which the disease occurs is around 69 years and therefore hyperplasia – a benign enlargement of the tissue through an increase in the cell number – is often present [1].

Deep learning is increasingly used for medical image analysis. The U-Net by Ronneberger et al. is an often used segmentation network [4]. An extension to 3D by Çiçek et al. [5] was for example used for the segmentation of the full prostate by Meyer et al. in 2018 [6]. Another network to obtain a prostate segmentation is the Yu-Net proposed by L. Yu in 2017 [7].

A segmentation of the prostate zones was done with an atlas method by Padgett et al. in 2016 and achieved a dice score of 0.83 for the prostate and 0.57 for the PZ [8]. Moreover, approaches based on the U-Net were presented by Mooij et al. in 2018 to segment the prostate into two zones. With a 3D U-Net a dice score of 0.85 for the TZ and 0.6 for the PZ was reached. By adding an autoencoder to the training the segmentation of the PZ could be improved to a dice score of 0.67 [9]. Furthermore, Mooij et al. presented an approach, called aniso-3DUNET, which achieved the same results as the 3D U-Net for their test data and will be presented in the following [10].

In this work, we use different variants of the U-Net to perform a segmentation of the prostate zones on anisotropic MRI data and analyze the connection between data structure and network architecture. Moreover, we want to find out which improvements could be achieved by ensembles.

2 Material and Methods

2.1 Image Data

We have 86 MRI data sets of the prostate available of which 32 are training samples of the Medical Segmentation Decathlon challenge [11]. These data are multi-parametric, that means several modalities and sometimes different directions of acquisition for each patient are available. We use only the transversal T2-weighted series for the segmentation of the prostate zones because it best illustrates the morphology of the prostate. The pixel spacing is 0.5 mm and 0.6 mm in the x- and y-dimension and between 3 mm to 4 mm in the z-dimension. The ground truth segmentation was created by an experienced radiologist who segmented the two zones of the prostate.

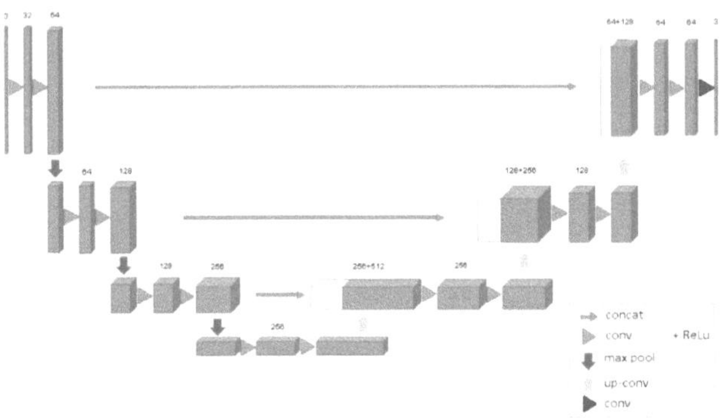

Figure 1: Schematic representation of the network architecture of the 3DUNET by Çiçek et al. (2016) [5].

2.2 Model Architectures

As a baseline we used the U-Net, which was introduced 2015 by Ronneberger et al. [4] and extended 2016 by Çiçek et al. to 3D operations [5]. The network consists of a contracting path that captures the context information and a symmetrical expanding path for precise localization. The network architecture can be seen in Fig. 1.

The U-Net consists of several repeating downsampling steps. First, two consecutive convolutions are performed with a size 3 kernel followed by a rectified linear unit (ReLU). This is followed by max pooling of kernel size 2 with a stride of 2 so that the resolution is halved. In addition, the number of feature channels is doubled at the end of each downsampling step. The expanding path symmetrically includes an up-convolution, which consists of an upsampling convolution with a size 2 kernel, to halve the number of feature channels. This is followed by concatenation with the feature map of the corresponding layer via a residual link and two size 3 convolutions, followed by ReLU. The last layer included a convolution with a kernel size of 1 to map the feature maps to the number of classes.

In the next step an extension of 3DUNET by Mooij et al. (2018), called aniso-3DUNET, shall be considered [10]. This extension of the network architecture was developed in order to acknowledge the issue that the resolution of the MRI series in the z-dimension is significantly lower than in the x-/y-plane and therefore only few layers are available. The difference to the model described above is that the network starts and ends with two blocks of 2D convolutions and 2D max pooling. The image data is therefore not folded as often and reduced in size in the third dimension so the anisotropy of the data set is paid particular attention to.

2.3 Ensemble

Due to the randomness of the training, a slightly different network is trained with each training run, therefore the segmentation result is always slightly different. Especially voxels with a high uncertainty are assigned different labels by different networks. To improve the segmentation result ensembles are used [12]. The idea is the same as in classical learning methods like Random Forest. Not only one network is used for the prediction of segmentation, but several networks. The decision is then made either by the mean of the prediction probabilities or by a majority decision.

3 Training and Testing

Due to the small amount of training data, augmentation is applied in the form of a left-right flip. Moreover, the images are normalized before training the network, since the intensities in the image data vary greatly. The maximum intensity values of the existing image data are between 800 and 2800. This is mainly due to the fact that the MRI, in contrast to computer tomography, has no physical scale on which the values are based. By normalization, where the mean is set to zero and the variance to one, this can be adjusted so that the network converges faster and gives better results. The test data sets are also normalized before they pass through the network for prediction.

The region of interest is determined by the segmentation of the whole prostate by a Yu-Net [7]. A square bounding-box with some pixel margin in x- and y-direction and all slices were extracted from the 3D image so that the network is just trained on this part of the MRI. Furthermore, the segmentation of the prostate is used as an additional input channel for the network using a binary mask with the same size as the input image.

The Keras library was used in Tensorflow to implement the networks. The processing size of the input data is 96×96 for the 2D network and $96 \times 96 \times 32$ for the other networks. The size of the 3D images is limited by the memory size of the graphics processing unit (GPU).

A softmax activation of the final feature map calculates the probability for each class, in this case for each zone. Moreover, the cross entropy is used to calculate the loss. Since significantly more pixels can be assigned to the background than to one of the zones, this imbalance is compensated by weighting the classes c in the loss function L. The loss is then given as

$$L = -\sum_{c=0}^{C} w_c \cdot (y_c \cdot \log(\hat{y}_c)) \tag{1}$$

where $\hat{y}$ is the predicted value. The weight w_c is recalculated before each prediction by:

$$w_c = 1/\log|c| \tag{2}$$

This loss is then optimized via the Adam Optimizer.

The dice coefficient is used to determine the quality of the results and to compare them:

$$Dice = \frac{2|P \cap G|}{|P| + |G|} \tag{3}$$

Where P is the predicted segmentation and G is the ground truth segmentation. A four-fold cross validation with three repetitions each was carried out in order to be able to make representative comparisons and estimate the performance of the network despite the relatively small data set.

4   Results and Discussion

In the following the results of the different U-Net architectures are presented. Image slices and the corresponding ground truth segmentations can be seen in Fig. 2.

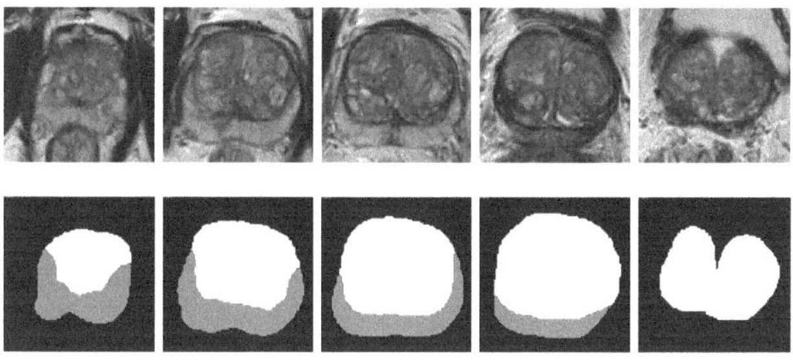

Figure 2: Transversal T2-weighted MRI series and corresponding ground truth segmentation of the two zones (white: TZ, gray: PZ).

First the 2DUNET is tested for the segmentation of the two prostate zones. For training 2D slice images are used and the batches are randomly combined from the images. The batch size is 24 and the learning rate and the decay rate are set to 0.001. Just 25 epochs are trained because further training leads to overfitting. The results are presented in Table 1. The 2DUNET has a dice coefficient of 0.8447 for the TZ and 0.7439 for the PZ. Looking at the result images in Fig. 3, it is noticeable that the zones are less smooth than the ground truth and individual outliers exist. It could be the case that the network is not able to learn all important information because it tends to overfit too quickly. Furthermore, there is no information from the neighboring slices and thus no possibility to determine the position in the prostate.

Table 1: Results for the segmentation of the TZ and the PZ with different U-Net variations as dice scores.

Network	TZ	PZ
2DUNET	0.8447 ± 0.0615	0.7439 ± 0.1241
3DUNET	0.8753 ± 0.0631	0.7903 ± 0.0828
aniso-3DUNET	0.8852 ± 0.0551	0.7998 ± 0.0830

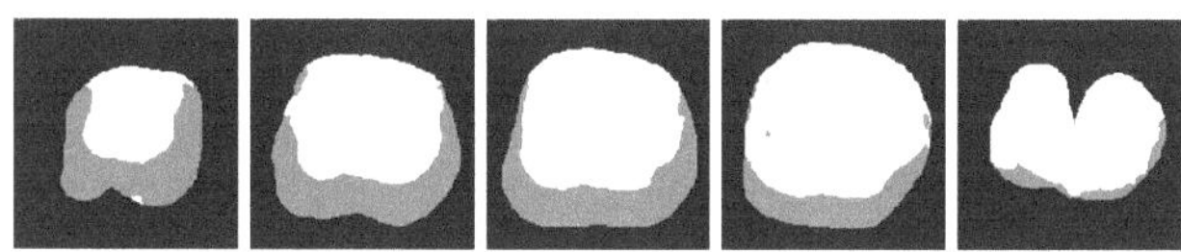

Figure 3: Predicted segmentation by 2DUNET.

In the next step the 3DUNET is used for segmentation. The extension to 3D enables the network to learn correlations of neighboring layers. We trained the network with a learning rate of 0.00005 over 80 epochs and a batch size of 4. The result is a dice coefficient of 0.8753 for the TZ and 0.7903 for the PZ. In comparison to 2DUNET, significant improvements of more than 3 % per zone are attained. The segmentations are shown in Fig. 4 where an improvement is visible. There are fewer outliers and the results are smoother. The

reason for this is that the network not only learns from individual layers, but also from neighboring layers. But this is also the explanation why the segmentation is slightly too symmetrical to the vertical of the image slice.

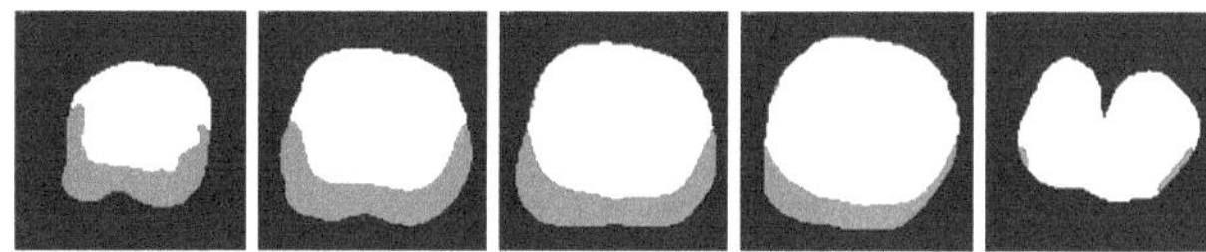

Figure 4: Predicted segmentation by 3DUNET.

The last network to be tested here is the aniso-3DUNET. It is trained with a learning rate of 0.0001 and a batch size of 4 over 200 epochs. For aniso-3DUNET the average dice coefficient is 0.8852 for the TZ and 0.7998 for the PZ. Thus an improvement to the 3DUNET is reached of approx. 1 % per zone. In the result images in Fig. 5 the improvements are visible. The last three layer images barely differ from the ground truth segmentation and there are no more outliers. Hence, the network architecture is best suited to the nature of the MR images. At first detailed information from the 2D layer images is obtained and then the neighboring slices are examined by the convolution in 3D so context information is learned as well.

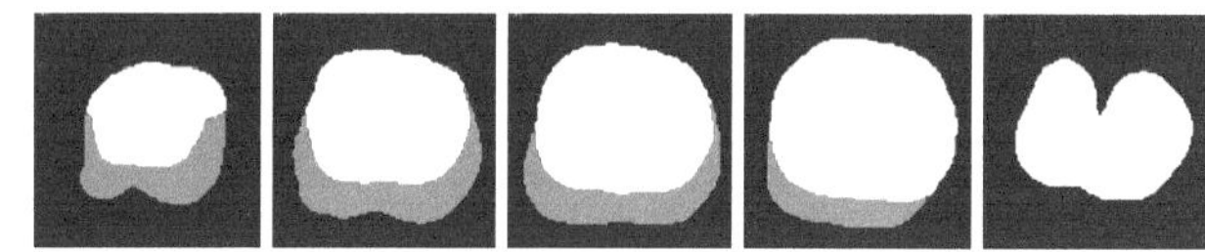

Figure 5: Predicted segmentation of aniso-3DUNET.

In order to further improve the results an ensemble approach is being used. Several aniso 3DUNETs will be trained and evaluated in ensembles of five or nine nets. The decision is made either by the mean value of the prediction probabilities or by a majority voting. The results are presented in Table 2 and Fig. 6. The scores determined by the mean of the prediction probabilities are slightly better than those obtained by majority voting. Moreover, the larger ensemble with nine nets achieved better results than those with only five nets. This is mainly due to the fact that the result segmentations are slightly different due to the small amount of training data and the randomness in training, especially in the difficult areas such as the borders between the zones. With more networks, weaknesses can be compensated for by other networks. The best result is achieved by an ensemble of nine aniso-3DUNETs calculated with the mean of the prediction probabilities. For the TZ the ensemble achieves a dice coefficient of 0.9070 and for the PZ 0.8451. Compared to a single aniso-3DUNET, an improvement of approx. 2 % was achieved for the TZ and approx. 4.5 % for the PZ. The additional expenditure due to the use of an ensemble is thus worthwhile.

172

Table 2: Results for the segmentation of the TZ and the
PZ with an ensemble of five and nine aniso-3DUNETs with
and without majority voting (MV) and for comparison the
results of a single aniso-3DUNET.

Network	TZ	PZ
1 aniso-3DUNET	0.8852	0.7998
5 aniso-3DUNETs	0.9043	0.8445
5 aniso-3DUNETs MV	0.9042	0.8443
9 aniso-3DUNETs	0.9070	0.8457
9 aniso-3DUNETs MV	0.9067	0.8451

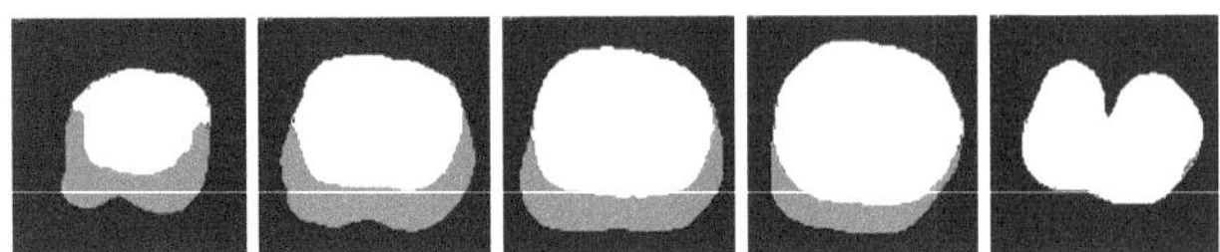

Figure 6: Ensemble of nine aniso-3DUNETs where the final segmentation is predicted by the mean of the prediction
probabilities of each network.

5 Conclusion

In this paper different variants of the U-Net were used to
perform an automatic segmentation of the prostate zones in
MRI data.
The 2DUNET performed the worst for the segmentation
because of the missing information about the neighboring
slides and the tendency to overfitting. The results could be
improved by using 3D convolutions instead of 2D added information also in z-dimension. The aniso-3DUNET proved
to be the best model because the network architecture fits
the anisotropic MRI data best. A single network achieved a
dice coefficient of 0.8852 for the TZ and 0.7998 for the PZ.
An ensemble of aniso-3DUNETs was used for further improvement. The best was an ensemble of nine networks
that made its decision on the basis of the mean of the output
probabilities. The dice coefficient improved to 0.9070 for
the TZ and 0.8457 for the PZ.
An adjustment of the model to the problem and the data
thus turned out to be an advantage. Perhaps further improvements can be reached by using more features of the
data.

Acknowledgment

The work has been carried out at FUSE-AI, Hamburg and
supervised by the Institute of Medical Informatics, Universität zu Lübeck.

6 References

[1] Leitlinienprogramm Onkologie (Deutsche Krebs-
gesellschaft und Deutsche Krebshilfe und
AWMF), *Interdisziplinäre Leitlinie der Qualität
S3 zur Früherkennung, Diagnose und Therapie
der verschiedenen Stadien des Prostatakarzi-
noms*, Langversion 5.1, 2019. Available: `http:
//www.leitlinienprogramm-onkologie.
de/leitlinien/prostatakarzinom/` [last
accessed on 2020-01-16].

[2] S. Yoo, J. K. Kim, and I. G. Jeong, *Multiparametric
magnetic resonance imaging for prostate cancer: a review and update for urologists*, Korean journal of urology, vol. 56(7), pp. 487–497, 2015.

[3] J. C. Weinreba, J. O. Barentszb, P. L. Choykec, F. Cor-
nudd, M. A. Haidere, K. J. Macuraf, D. Margolisg,
M. D. Schnallh, F. Shterni, C. M. Tempanyj, H. C.
Thoenyk, and S. Vermal, *Pi-rads prostate imaging -
reporting and data system: 2015 version 2*, American
College of Radiology, 2015.

[4] O. Ronneberger, P. Fischer, and T. Brox, *U-Net: convolutional networks for biomedical image segmentation*,
arXiv e-prints, 2015.

[5] Ö. Çiçek, A. Abdulkadir, S. S. Lienkamp, T. Brox, and
O. Ronneberger, *3D U-Net: learning dense volumetric
segmentation from sparse annotation*, arXiv e-prints,
2016.

[6] A. Meyer, A. Mehrtash, M. Rak, D. Schindele,
M. Schostak, C. M. Tempany, T. Kapur, P. Abolmae-
sumi, A. Fedorov, and C. Hansen, *Automatic high resolution segmentation of the prostate from multi-planar
mri*, 2018 IEEE 15th International Symposium on
Biomedical Imaging (ISBI 2018), pp. 177–181, 2018.

[7] L. Yu, X. Yang, H. Chen, J. Qin, and P.-A. Heng, *Volumetric convnets with mixed residual connections for
automated prostate segmentation from 3d mr images*,
in Proceedings of the Thirty-First AAAI Conference on
Artificial Intelligence, ser. AAAI'17. AAAI Press, pp.
66–72, 2017.

[8] K. Padgett, A. Swallen, A. Nelson, A. Pollack, and
R. Stoyanova, *SU-F-J-171: robust atlas based segmen-
tation of the prostate and peripheral zone regions on
MRI utilizing multiple MRI System Vendors*, Medical
Physics, vol. 43, no. 6 Part 11, pp. 3447–3447, 2016.

[9] A. de Gelder and H. Huisman, *Autoencoders for multi-
label prostate mr segmentation*, arXiv e-prints, vol.
abs/1806.08216, 2018.

[10] G. Mooij, I. Bagulho, and H. Huisman, *Automatic segmentation of prostate zones*, arXiv e-prints, 2018.

[11] A. L. Simpson, M. Antonelli, S. Bakas, M. Bilello,
K. Farahani et al., *A large annotated medical image
dataset for the development and evaluation of segmentation algorithms*, 2019.

[12] T. G. Dietterich, *Ensemble methods in machine learning*. Springer Berlin Heidelberg, 2000.

Estimation of Healthy Aortic Root Shapes from Pathological Images with Conditional Variational Autoencoder

Mohamad Mehdi [1], Jannis Hagenah [2], and Floris Ernst [2]

[1] Biomedical Engineering, Luebeck University of Applied Sciences, mohamad.mehdi@stud.th-luebeck.de
[2] Institute for Robotics and Cognitive Systems, Universität zu Lübeck, {hagenah, ernst}@rob.uni-luebeck.de

Abstract

Aortic root aneurysm is treated by replacing the dilated root by a grafted prosthesis which mimics the native root morphology of the individual patient. The challenge in predicting the optimal prosthesis size rises from the highly patient-specific geometry as well as the absence of the original information on the healthy root. Therefore, the estimation is only possible based on the available pathological data. In this paper, we show that representation learning with Conditional Variational Autoencoders is capable of turning the distorted geometry of the aortic root into smoother shapes while the information on the individual anatomy is preserved. We evaluated this method using ultrasound images of the porcine aortic root alongside their labels. The observed results show highly realistic resemblance in shape and size to the ground truth images. Furthermore, the similarity index has noticeably improved compared to the pathological images. This provides a promising technique in planning individual aortic root replacement.

1   Introduction

The aortic aneurysm is often accompanied by severe risks such as aortic rupture and cardiac complications. Therefore, patients suffering from aortic insufficiency due to dilated aorta undergo valve-sparing aortic root replacement operation. To prevent aortic root dissection, the aneurysm is treated by the reimplantation of the aortic root with a tubular graft [1]. During the surgery, the native aortic valve is preserved for patients with near-normal cusps in order to maintain smooth outflow and avoid anticoagulation [2]. The size of the implanted prosthesis has a significant influence on the blood flow in the aortic annulus as well as the long-term stability of the aortic valve [3].

However, the estimation of the optimal prosthesis size is still an intricate task. Patient prosthesis mismatch could lead to further complications as the aortic root shape and geometry is highly patient-specific [4]. Additionally, the available data on the patient root is only in the dilated state. Our aim is to assist the surgeons in the decision-making process by developing a preoperative planning tool. For this tool, we need an estimation of the healthy shape and the only input we have is the dilated state.

Previous works based on classic Machine Learning were conducted for personalized prosthesis size prediction. In the early steps, geometrical features were extracted from the ultrasound images as pairs of healthy and dilated features. These features were thereafter used to train a machine-learning algorithm, such as Support Vector Regression (SVR), to learn a mapping between the dilated and healthy states. Then, the SVR-model can be utilized to predict individual healthy features based on dilated ones [5]. A recent study has introduced an alternative solution for reconstructing the healthy state of the aortic root. Instead of handcrafting features from the ultrasound images, a new method, based on representation learning with neural networks, has been developed to learn the features from the input images at the hidden layers of the network. Variational Autoencoders (VAE) provide the possibility to convert the input into a compressed low-dimensional representation, also known as the latent space. After having the latent space for all images extracted, a translation vector is manually defined between the two classes to map between the dilated and the healthy images [6].

Variational Autoencoders have shown considerable performance as generative models, however, with standard VAE, the latent space is randomly sampled. This means that all latent points lie in one distribution. Thus, there is no control over the generated images. As previously mentioned, the mapping between different image classes was performed manually. Within the latent space, a translation vector is defined to map from the cluster mean of the dilated images to the cluster mean of the healthy images. In this work, we propose an alternative approach for generating the healthy aortic root without direct interaction with the latent space. By including a condition on the image class to produce, it is possible to direct the autoencoder to output new reasonable images with specific attributes. In this paper, we demonstrate the study setup as well as the generating process of the aortic root images. The ability of the proposed model to generalize is tested using K-Fold cross-validation. The obtained results are compared with those from different works.

2   Material and Methods

2.1   Preparing the Dataset

In this study, a dataset of 48 2D ultrasound images of porcine valves, obtained from previous work, has been used [7]. The 2D slices were extracted from 3D volumes, in which only the commissure plane was taken. The dataset contains the healthy and dilated states of the aortic root, which mimic the normal as well as the pathological human aortic root in its geometry. The images are available in greyscale with a size of 100×100 pixels.

2.2   Model Architecture

In this experiment, the Conditional Variational Autoencoder (CVAE) has a similar architecture of traditional VAE. The encoder consists of five convolutional layers (with kernel size 3×3) followed by average pooling (with kernel size 2×2). The decoder is obtained by mirroring the architecture of the encoder. Furthermore, a new layer that represents the condition is imposed on the encoder and decoder inputs [8]. The labels are given as one-hot encoding. Figure 1 illustrates the architecture of the proposed model.

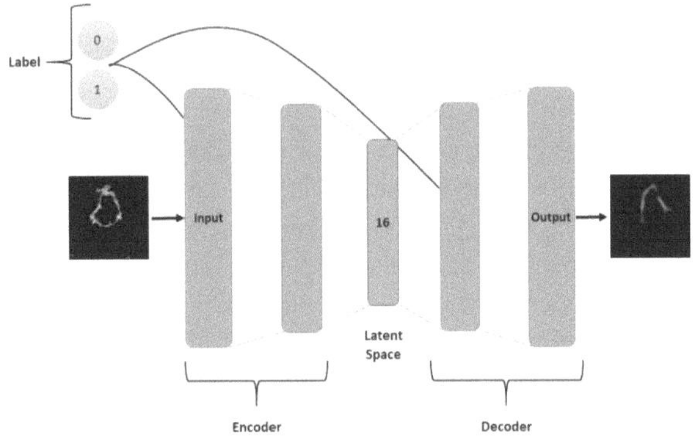

Figure 1: Schematic illustration of the Conditional Variational Autoencoder architecture.

The activation function used after each convolutional layer is "ReLU" activation except for the last layer in the network, which uses "Sigmoid" activation. The size of the latent space is 16. This experiment is carried out using Neural Networks built with the help of Keras library with Tensorflow backend.

2.3   Model Training

The objective of training the VAE is to maximize the variational lower bound [9]. As mentioned above, generating specific data with the VAE is intricate. This is because the encoder does not take into account the input labels when modeling the latent space z, as per the distribution $Q(z|X)$. This is also the case for the decoder when generating outputs under the distribution $P(X|z)$. By introducing the condition c, the latent variable is now distributed under $P(z|c)$. This means that for each condition c, we obtain a distinct

distribution $P(z)$ [8]. So, the loss function can be written as follows:

$$\mathcal{L} = \mathbb{E}[\log P(X|z,c)] - D_{KL}[Q(z|X,c)||P(z|c)] \quad (1)$$

Adam optimizer is chosen as an optimization algorithm. Due to the limited data available, data augmentation makes it possible to increase the size of the dataset by synthetically creating different variations of the images. The techniques used in this study for enhancing the dataset size encompass rotation (max. $\pm 10°$), translation (0.1 height and width shift), zoom (0.1), and horizontal flip.

After training the model for 200 epochs, the model was used in two separate stages. We first used the model to reconstruct the encoded representation of the healthy aortic root images to evaluate the representational capability of the architecture for the given problem. The recovered images were then compared to the original input. In the second stage, we utilized the model for producing the healthy geometry of the aortic root based on the pathological root images. The dilated root images were passed to the encoder to get their latent spaces. To this end, from this reduced representation, the manipulation is done by forcing the decoder to reconstruct the desired geometry by switching the condition on the decoder to healthy. Thereby, the latent space is now reconstructed in the healthy root distribution.

2.4   Evaluation Methodology

To test the reliability of the model when encountering new unseen images, we applied K-Fold cross-validation. Thus, the dataset is divided into six groups. For each group, two sets were created: a training set which contains 40 images (taken from 20 valves), and a test set which contains the remaining eight images (taken from 4 valves).

The similarity between the generated images and the original input is measured with Mean Squared Error (MSE) and Structural Similarity Index (SSIM). However, using these metrics requires performing image registration as the generated and ground truth (healthy) images are not aligned. For image alignment, intensity-based rigid image registration was performed using a built-in function in MATLAB [10].

We further evaluated the model by experimenting with other architectures that have different network depth. We also tested the change in the latent space size on the encoding ability of the model.

3   Results and Discussion

To evaluate the performance of our model, the experiments are conducted as follows. First, we examined the learning process of the CVAE to represent certain input in a lower-dimensional space so that the model is able to reconstruct the output from the reduced representation. Then, we generated aortic root images based on the described method. Afterwards, we checked the effect of changing the depth as well as the latent space size on the accuracy of the results.

Finally, the results of this study were compared with those from other approaches.

3.1   Representation Learning

Table 1 lists the similarity results between the healthy aortic root images propagated through the network and their recovered output. The given results are the average values over each set.

Table 1: Similarity results between the reconstructed and input images based on MSE and SSIM metrics.

	MSE	SSIM
Training set	249 ± 14	0.881 ± 0.002
Test set	475 ± 231	0.83 ± 0.02

The choice of data representation plays a significant role in the success of the performance of machine learning algorithms. In other words, learning proper representations facilitates solving the task by extracting the most relevant and useful feature points to train the desired algorithm. Moreover, representation learning is heavily reliant on the training data. This can be seen from the drop in similarity values when predicting on the test set.

3.2   Aortic Root Reconstruction

Figure 2 shows three reconstructed aortic root images. For qualitative analysis, the images are compared with the ground truth and the pathological (dilated) states.

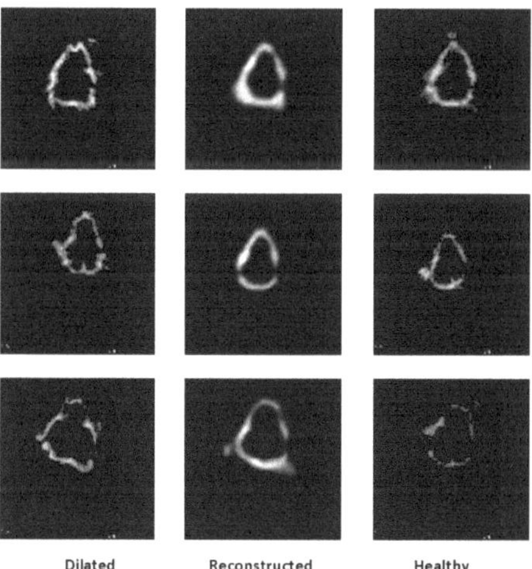

Figure 2: Visualization of three reconstructed images compared with the dilated and healthy states. The first row is obtained from the training set, and the rest is from the test set.

The changes in shape and size attributes can be clearly observed in the reconstructed images. The distorted shape caused by the dilation has disappeared. Instead, the images show smoother geometries without losing the anatomical properties of the aortic root. Despite splitting the dataset, the model still shows potentials in generating meaningful

images when new unseen images are introduced. The similarity results between the predicted aortic root images and the ground truth healthy valves images are recorded in table 2.

Table 2: Quantitative results for the similarity between the generated aortic root and the ground truth based on MSE and SSIM metrics.

	MSE	SSIM
Pathological vs. Healthy	1566	0.766
Reconstruction vs. Healthy (Training set)	1039 ± 129	0.801 ± 0.009
Reconstruction vs. Healthy (Test set)	915 ± 448	0.79 ± 0.03

The achieved results are far from the optimum. However, based on the results between the training and testing sets, the model is still able to maintain a good generalization when provided with new images.

3.3   Evaluation of Various Architectures

Figure 3 demonstrates the relationship between the depth and structural similarity index for different latent space sizes. Four architectures have been used in this comparison. A1 represents a network with 3 convolutional layers in the encoder, A2 has 4 convolutional layers, A3 (the proposed architecture) has 5 convolutional layers, and A4 has 6 convolutional layers. The decoder is the mirrored architecture of each relevant encoder. Increasing the network depth does not seem to improve the accuracy of reconstructing the aortic root. With a small dataset size, taking into account the increase in the number of nodes with depth, adding more layers would only contribute to a higher number of hyperparameters; consequently, the model is more likely to overfit. On the other hand, using shallow architectures, for example, three convolutional layers in the encoder and similarly for the decoder, resulted in the lowest similarity index values compared to other depth choices. In terms of latent space

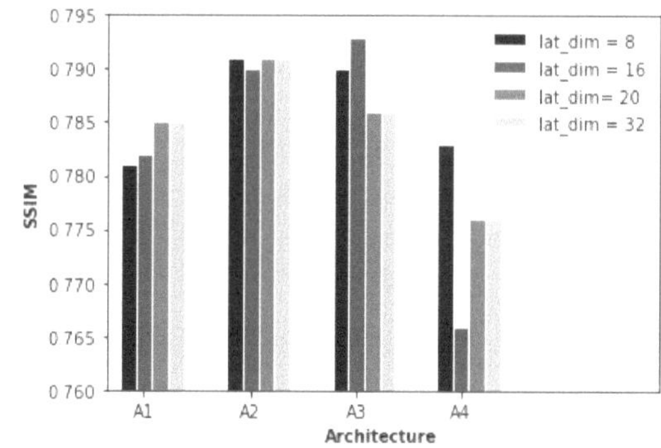

Figure 3: Comparison between various architectures with different latent space sizes

size, using eight nodes was not sufficient to capture the features necessary to provide optimal input retrieval. By fixing the latent space size to 16 neurons as in the proposed architecture, the model provides the highest similarity among other options.

3.4 Comparison to State-of-the-Art Models

The results of this study have also been compared with previous related works. Figure 4 displays a bar graph comparing the SSIM values of four different models when performing the above tasks The first model uses a standard autoencoder AE with convolutional layers. The second one represents a standard variational autoencoder VAE both with a manually designed translation vector in the latent space. The next model is the one proposed by Hagenah et al. [6]. The last model is our proposed model CVAE.

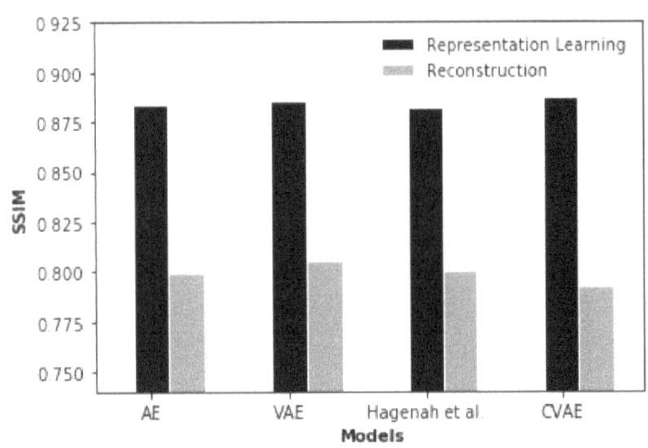

Figure 4: Bar graph depicting the SSIM values of four different models based on their performance in representation learning and reconstructing the healthy aortic root.

The primary downside with standard autoencoders is the noncontinuous distribution of the latent space, which makes it challenging to generate new realistic images when drawing new samples from the latent space. Nevertheless, in variational autoencoder, new images are sampled from a Gaussian distribution over the latent space, therefore, resulting in smooth images. Despite that CVAE does not show ideal performance in reconstructing the healthy aortic root, it still gives the benefit of non-direct manipulation of the latent space as the whole mapping from pathological to healthy shapes is learned implicitly and completely data-driven. The significant difference between the proposed model and the one from Hagenah et al. is the process in which the latent space is modelled. Here, we used a condition on the encoding as well as the generating process. Whereas the other model distinguishes the different classes using a classifier.

4 Conclusion

In this study, we proposed a novel, end-to-end automatized approach for reconstructing the healthy aortic root using a Conditional Variational Autoencoder. The proposed model showed strong performance in generating new images based on changing the label of the latent space of the dilated im-

ages. Further experiments could employ transfer learning to address the problem with the small dataset.

Acknowledgment

This work has been carried out at the Institute for Robotics and Cognitive Systems at the University of Luebeck.

5 References

[1] V. L. Gott, et al., *Replacement of the Aortic Root in Patients with Marfan's Syndrome.* vol. 340, no. 17, pp. 1307–1313, 1999.

[2] K. J. Zehr, M. J. Thubrikar, G. G. Gong, J. R. Headrick, and F. Robicsek, *Clinical introduction of a novel prosthesis for valve-preserving aortic root reconstruction for annuloaortic ectasia.* J. Thorac. Cardiovasc. Surg., vol. 120, no. 4, pp. 692–698, 2000.

[3] T. Kunihara et al., *Preoperative aortic root geometry and postoperative cusp configuration primarily determine long-term outcome after valve-preserving aortic root repair.* J. Thorac. Cardiovasc. Surg., vol. 143, no. 6, pp. 1389-1395.e1, 2012.

[4] V. Paruchuri et al., *Aortic Size Distribution in the General Population: Explaining the Size Paradox in Aortic Dissection.* Cardiol., vol. 131, no. 4, pp. 265–272, 2015.

[5] J. Hagenah, M. Scharfschwerdt, A. Schlaefer, and C. Metzner, *A machine learning approach for planning valve-sparing aortic root reconstruction.* Curr. Dir. Biomed. Eng., vol. 1, no. 1, pp. 361–365, 2015.

[6] J. Hagenah, M. Mehdi, F. Ernst, *Generating Healthy Aortic Root Geometries from Ultrasound Images of the Individual Pathological Morphology using Deep Convolutional Autoencoders.* In computing in cardiology, vol. 46, pp. 1-4, 2019.

[7] J. Hagenah, E. Werrmann, M. Scharfschwerdt, F. Ernst, and C. Metzner, *Prediction of individual prosthesis size for valve-sparing aortic root reconstruction based on geometric features.* Proc. Annu. Int. Conf. IEEE Eng. Med. Biol. Soc. EMBS, vol. 2016-Octob, pp. 3273–3276, 2016.

[8] K. Sohn, X. Yan, and H. Lee, *Learning structured output representation using deep conditional generative models.* In Advances in Neural Information Processing Systems, vol. 2015-Janua, pp. 3483–3491, 2015.

[9] I. Goodfellow, Y. Bengio, and A. Courville, *Deep Learning.* MIT Press 2016.

[10] MATLAB Release 2019a, The MathWorks, Inc., Natick, Massachusetts, United States.

Frequency based detection of regions of interest in videodata to estimate human breathing frequency

Johannes Sirocko [1], Jasper Diesel [2], Mattias Heinrich [3]

[1] Medical Engineering Science, Universität zu Lübeck, j.sirocko@student.uni-luebeck.de
[2] Drägerwerk AG & Co. KGaA, Lübeck, jasper.diesel@draeger.com
[3] Institute of Medical Informatics (IMI), Universität zu Lübeck, heinrich@imi.uni-luebeck.de

Abstract

In order to diagnose sleep related breathing problems, patients sleep in the sleep laboratory. During this night, many vital parameters are monitored for one night by invasive measuring methods. To minimize sleep disturbance, contactless recording of these parameters is favorable. Our non-invasive setup based on recording of infrared and thermal video data. For an analysis of the respiratory rate, it must be determined which pixels in the video carry relevant information. Therefore the aim of this study was to detect regions of interests (ROIs) with respiratory related signals. The results show that in thermal imaging data ROI were located at the boundaries from warm to cold areas that shift due to respiration. Furthermore, breathing induced signals were found at the area of the mouth, which was already stated by [5]. For the infrared data, ROIs were located at the center of the chest, abdomen or close to the mouth.

1 Introduction

So far, various approaches have been tested to realize contactless acquisition of vital parameters. Some experimental setups attempted to exploit the Doppler effect in radar measurements, but the results showed a poor signal to noise ratio at distances greater than one meter from the object. In addition, the antenna had to be aligned with the chest and the results were generally degraded by motion artifacts. Another option is to use pressure sensors that are mounted above and below the bed mattress. But these results were also corrupted by motion artifacts [1]. Another setup already tested with various data-processing algorithms involves the acquisition of depth image information and infrared (IR) image information using a Kinect V2 camera [1]. The main problems with camera based approaches are head movement, motion artifacts, the need to estimate the pose and the fact that the area of the mouth needs to be seen by the cameras. Besides depth and IR data, thermal image data are an option. For thermal image data, research groups focused on facial regions [5]. For example, in [2] the head was tracked by an algorithm and then the information from the head region was analyzed by statistical methods. Nevertheless, the same disadvantages as for the depth and IR data occurred. In order to improve on these restrictions, the aim of this work is to automatically predict at which areas of the patients body breathing related signals can be found in thermal and IR image data. The analyzed data considered various clinical situations with different patients and unlike patient poses. The results show region of interests (ROI) which could be used for subsequent assessments of respiratory associated vital signs.

2 Material and Methods

2.1 Data Acquisition

The experimental setup includes the Kinect V2 camera of Microsoft and the Seek Compact Pro camera, which are installed at a height of two meters at the foot end of the bed on the wall. The Kinect sensor is a time of flight sensor and uses a 850 nm laserdiode as illuminator. The image sensor is a monochrome image sensor. In the following, only the amplitude signal of the time of flight sensor is used. This signal is referred to as IR data. The Seek Thermal Compact Pro is a thermal sensor which is capable of detecting radiation between wavelengths of 7.2 and 13 μm [3].

The sampling rate at which the video data of the Kinect V2 were recorded, was 30 Hz, the sampling rate of the Seek Compact Pro was 17 Hz. The sampling rate of the camera limits the maximal frequency that can be detected according to the Shannon-Nyquist theorem. For the data of the Seek Compact camera, the highest frequency that could be detected was 8.5 Hz. The cut-off frequency of the Kinect V2 was 15 Hz. Since human breathing is substantially below one Hz, our signal is valid. The field of view of the two cameras involved the patient's bed which was filmed obliquely from above. At the boundaries of the frame a bit of the environment of the bed was detected. The distance from the camera to the patient was 3.5 m and the image resolution was 240x320 pixel for the thermal data and 424x512 pixel for the Kinect V2 data.

The dataset included four patients who have slept on suspicion of sleep apnea at the sleep laboratory at the UKSH in Lübeck. The patients are between 45 and 65 years old and weighed between 75 and 110 kg with body heights between 174 and 180 cm. Both male and female patients were included. The measurement time for a patient corresponded to the sleep duration of the patient and was between 5.5 and 6.5 hours long. In addition to the video data, the sleep laboratory measures many vital parameters using various invasive devices. These parameters included a respiratory flow measurement, the recording of an electrocardiogram, as well as the determination of the breathing movements of the upper body by means of two straps that are strapped around the chest and stomach. The straps stretch and shrink according to the breathing of the patient and this signal was used to determine the breathing frequency via Fourier transformation.

2.2 Data processing

The working pipeline of the implementation first considers a downsampling of the data, calculating the average of the pixel intensities in a neighborhood. The size of the neighborhood determined the factor of the downsampling. The purpose of the downsampling is to smooth the images spatially and to reduce the necessary computational effort. Afterwards the pixel values of the video data are extracted for each pixel over time. So the result is a timeseries of each pixel intensity which is analyzed in time intervals of 13 to 45 sec with a real valued Fourier transform. The length of the time interval is short enough, that the patient did not move too much and on the other hand long enough, to yield a reasonable spectral resolution in the frequency analysis. The spectral resolution of the Fourier transform increases with the length of the input data. Because the signal is cut off at two timepoints, one needs to multiply the signal with a hamming window in order to suppress a sinc disturbance in the signal. The real valued Fourier transform is chosen because it takes into account that the input signal is real valued and discards the symmetric imaginary part. The used function is the rfft-function from the numpy package. With the calculated spectra of the timeseries of each pixel, the ROIs could be extracted considering different decision rules.

2.3 Implementation of the ROI detection

Four different ROI selection rules were applied. The main idea regarding the extraction of the ROIs is to sum up the output of the Fourier transform in a short interval around the breathing frequency for the timeseries of each pixel. This is visualized in Fig. 1. The maximal size of the interval is 0.014 Hz.

In decision rule one the spectral sum around the breathing frequency is calculated for each pixel and compared to the maximum of spectral sum which can be found considering all pixels. If the value of the spectral sum for a specific pixel is higher than 50 percent of the maximum of the spectral sum, the pixel is assigned to the ROI. Decision rule two bases on the same rule, but the threshold for assigning pixels to the ROI is set to 85 percent, so that only the most significant pixels are extracted.

Decision rule three uses a different approach by comparing the spectral sum of the small interval around the frequency of breathing to the spectral sum of a frequency interval from zero Hz to close to the breathing frequency. If the ratio of these two values is higher than a threshold, the pixel is assigned.

The criterion in decision rule four is whether the maximum peak of the whole spectrum is located at the breathing frequency, or not. So pixels which have the maximum of the amplitude of the Fourier transform at the breathing frequency are set as ROI.

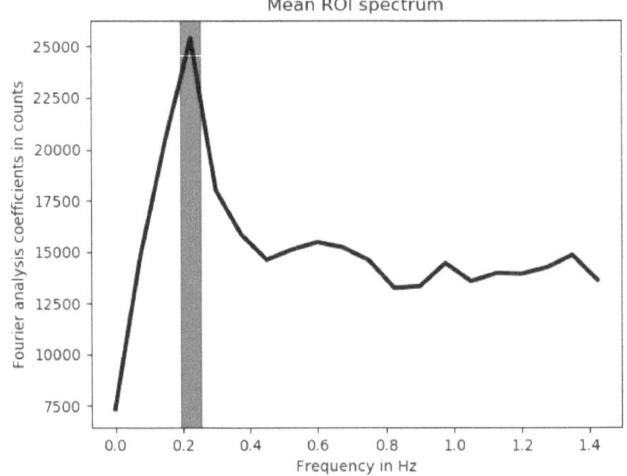

Figure 1: Visualization of how the spectral sum around the breathing frequency is calculated. The shaded area is the interval around the breathing frequency, where the Fourier transform output is accumulated. This spectrum sum is determined for each pixel and used in the decision rules.

3 Results and Discussion

The results showed, that downsampling by a factor of two did not have an impact on the results compared to the data that were not downsampled. The selected ROIs were at the same locations at the patient, only the amount of ROI pixels decreases according to the fact, that the whole smoothed and downsampled image itself had half of the pixel quantity. Regarding the fact, that the necessary computing power is noticeably reduced by downsampling, the computation time can be reduced. Downsampling to a factor of three led to a blurred image that could not be properly analyzed anymore. Furthermore, the time interval length that was chosen for the analysis had an impact. As the patients were all assumed to have sleep apnoe it was most likely that they had short time periods with no breathing. But if the time interval was lasting longer than this time, an ROI could be extracted because the onset of breathing after the apnoe was covered by the time interval. The drawback was that with an increasing time interval the likelihood of patient movement was also rising. This led to confusing ROI pixels when the ROI pixels were scattered on the last frame of the corresponding video sequence.

To analyse the frequency of each pixel timeseries, the function rfft from numpy was used. In order to check, if this function was suitable, the spectra of this function were compared to those of the periodogram-function from the signal package. The results showed little differences in the spectra, but these differences did not follow a structured pattern and the ROI selections did not differ between the functions.

3.1 Results for the different ROI selection rules

In general, all four decision rules yielded good results. There were no outliers at pixels, where no breathing induced signal could be. This was confirmed when analysing a time interval every ten minutes for all patients. Outliers only were found at two specific locations in the IR data, where either a reflection or a defect pixel has been. These defect pixels or reflections are stationary pixel with the maximum value that could be measured.

Decision rule one and two had to be very similar as they both used the same criterion but with different thresholds. The results proved this because all the ROI pixel from rule two were included in rule one. In rule one just a few more pixel in less relevant regions were added.

Comparing decision rule three and four, the most significant difference was, that decision rule three extracted about half of the amount of pixel decision rule four chose. The exact ratio differed from 0.4 to 0.7 depending on the patient and according to the situation. The plausibility of the extracted ROIs for all decision rules were similar as they always extracted pixel at the same regions of the patient. There was no instance, where a decision rule defined ROIs at a different location at the patient, it was always just a matter of how much pixel were chosen.

3.2 Thermal data

Considering that the inhalation entails a cold air flow followed by a warm air flow during the time of exhalation, the chosen ROI in the mouth area in Fig. 3 is intuitiv. That is why the face and especially the mouth area were chosen by [6] and other research groups as ROI. But as shown in Fig. 3, there are more areas showing breathing related signals. Another ROI for the thermal images tended to be at edges from warm to cold areas at the surface of the patient. For example at locations where the cold blanket conjuncts with the warm area of the patients chest. Or at the edges of the warm arm and the cold bed surrounding. This can be explained by the fact, that the sharp edges from warm to cold are moving up and down or sideways due to the breathing induced movement. For sideways movements it is obvious, that breathing related signal can be obtained, but for up and down movements one need to take into account that these movements result in a vertical shift of the edges, as the camera perspective is from above under an oblique angel.

The spectrum of the ROI pixel and of those pixel outside the ROI are shown in Fig. 2. One can see that the Fourier transform amplitude is significantly higher in the spectral range around the breathing frequency for the ROI pixels. For the pixels outside the ROI the Fourier transform amplitude accumulates to a very low value compared to the value of those from the ROI pixels.

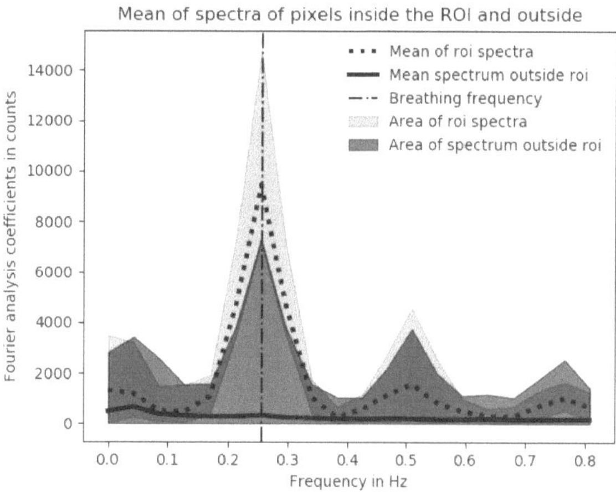

Figure 2: The mean spectrum of all ROI pixels and the mean spectrum of all pixels outside the ROI is plotted. The vertical line shows the breathing frequency where one can clearly distinguish the two spectra. The ROI spectrum is significantly higher at the breathing frequency.

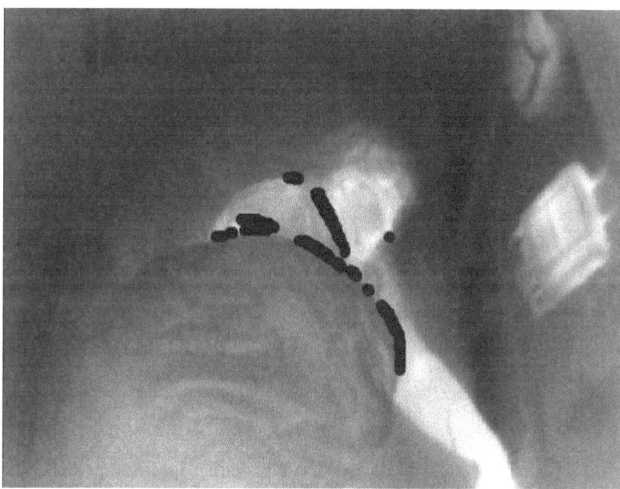

Figure 3: The breathing related ROI selection in a thermal image sequence is shown with the black dots scattered on a frame of the corresponding video sequence. As one would intuitively expect, the area of the mouth is selected as ROI, as we expect nasal air flow to be detected. But additionally, the edge from the warm chest to the cold blanket was chosen, as this edge moves due to breathing induced motion.

3.3 IR data

The results of the IR data were different compared to those of the thermal imaging data. The detected areas mainly included areas of the face, especially at the mouth, the chest and the abdomen which is presented in Fig 4.

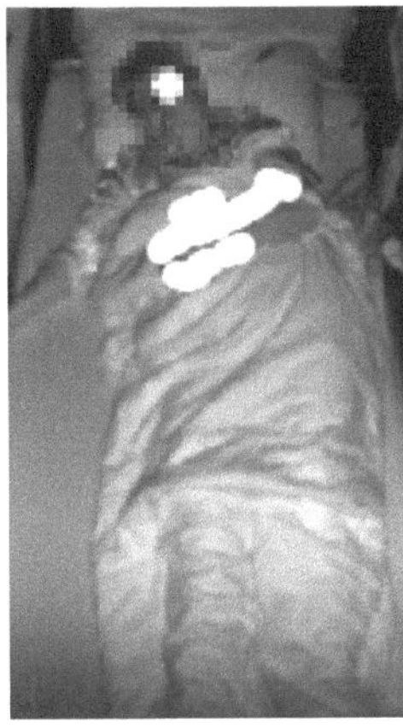

Figure 4: In the near infrared spectrum the ROI selection can be at the area of the head and the chest, as the white dots indicate in the image. In contrast to the thermal data results, the ROI is set to the center of the region and not to the boundaries from patient to surrounding. The area of the head is blurred due to patients privacy. But one can still see that an ROI pixel was chosen there.

Another area which contained breathing related signals was located at the extremities, if they were not covered by a blanket, or if the breathing induced movement was so distinct, that even the blanket moved. Having a closer look the ROI pixel at the extremities, they tended to be at the edges from the patient tissue to the surrounding. Similar to the results which were obtained from the thermal image data.

4 Conclusion and Outlook

The results of this paper show, that a breathing related analysis is possible at different areas of the patient and their close surrounding. These areas differ due to the patients position, the video modality that is used and how much the patient is covered by the blanket. Especially which areas are covered by the blanket.

In [4] three papers were cited which stated, that detecting respiratory rate in thermal data is not possible if the mouth area is not clear to analyze. They stated, that either the temperature flow in the nasal region or the detection of emission of carbon dioxide is the reason for a signal. So this paper proposes new ROIs to get a breathing related signal as there are movements of the edges from warm to cold areas beside the signal in the mouth area. Another problem that was figured out by [1] were head movement or movement in general. This problem can be reduced, considering that for example the rotation of the head will not necessarily result in a movement of the chest area, where an edge from warm to cold at the blanket is likely. So if not the entire body of the patient is moving, but just part of it, its possible that a ROI at another area is not affected and could deliver good results.

Furthermore, the results of the IR data propose that breathing induced signals can be obtained at different areas. The chest and abdomen was manually selected by researching

groups like [7], but to the best of our knowledge no one used the extremities if they are visible.

In the future these results could be used to develop an algorithm to detect breathing apnoe which has a huge clinical importance. This could be done by using deep learning methods to learn the detection of the ROI based on a ground truth represented by the presented data. With these situation adaptive ROIs, algorithms like those presented in [8] could improve their results.

Acknowledgement

The work has been carried out at Drägerwerk AG & Co. KGaA in Lübeck and was supervised by the Institute of Medical Informatics at Universität zu Lübeck.

5 References

[1] Christoph Hoog Antink, Simon Lyra, Michael Paul, *A Broader Look: Camera-Based Vital Sign Estimation across the Spectrum.* Yearbook of medical informatics, 2019.

[2] Marcin Kopaczka, Özcan Özkan, Dorit Merhof, *Face Tracking and Respiratory Signal Analysis for the Detection of Sleep Apnea in Thermal Infrared Videos with Head Movement.* Springer, 2017.

[3] Hardware-Kinect-info, *Kinect-hardware info.* Available:[https://openkinect.org/wiki/Hardware_info [last accessed on 2019-12-17].

[4] Ali Al-Naji, Kim Gibson, Sang-Heon Lee and Javaan Chahl, *Real Time Apnoea Monitoring of Children Using the Microsoft Kinect Sensor: A Pilot Study.* MDPI, 2017.

[5] Murthy R, Pavlidis I, Tsiamyrtzis P., *Touchless monitoring of breathing function.* Annual International Conference of the IEEE Engineering in Medicine and Biology Society 2004

[6] Aleš Procházka, Hana Charvátová, Oldrich Vyšata, Jakub Kopal and Jonathon Chambers, *Breathing Analysis Using Thermal and Depth Imaging Camera Video Records.* MDPI Sensors 2017.

[7] Carmina Coronel et al., *Measurement of respiratory effort in sleep by 3D camera and respiratory inductance plethysmography.* Springer 2017.

[8] Aleš Procházka, Martin Schätz, Oldřich Vyšata and Martin Vališ , *Microsoft Kinect Visual and Depth Sensors for Breathing and Heart Rate Analysis.* MDPI Sensors 2016.

A Combination of Normalized Metal Artefact Reduction and the Use of Multiple Prior Images in Computed Tomography

Patrick Adrian Gunawan [1],

[1] Biomedical Engineering, Technische Hochschule Lübeck, patrick.adrian.gunawan@stud.th-luebeck.de

Abstract

In 2017 Nam et al. proposed a method to reduce metal artifacts in computed tomography (CT) images by using multiple prior images. The method utilizes a combination of priors and a linear interpolation approach. Instead of a linear interpolation, the usage of normalized metal artifact reduction (NMAR) is proposed to improve the quality of the reconstructed image. An Image that was corrected by NMAR is segmented by using Otsu's method with several thresholds to produce multiple prior images. These prior images are forward projected and the corresponding projection data are linearly combined. The newly calculated sinogram is combined with the original measurement in order to replace projection data that passes through a metal object. The process is iteratively repeated to reduce the residual error and improve image quality. The combination of NMAR with multiple prior images shows improved image quality compared to a linear interpolation approach with multiple prior images.

1   Introduction

In the field of computed tomography, image artifacts that are caused by high-density objects can reduce the image quality drastically. In the last decades, the field of metal artifact reduction has proven to be a very active research area with a publication of several correction approaches. These approaches are divided into sinogram inpainting, the adaption of iterative methods and image filtering techniques.

For instance, sinogram inpainting methods treat the projections that pass the metal object as missing data. Generally, these metal projections are then replaced with various interpolation methods. One major problem with most of these interpolation techniques, i.e. the linear interpolation (LI) approach [1], is that the newly calculated projections produce inconsistencies in the raw data. To counteract this problem, a method called normalized metal artifact reduction (NMAR) was presented [2]. Here, the basic idea is to create a prior image that takes the bone and other high contrast structures into account. With the use of this prior image a normalization of the newly calculated projection values if performed to reduce inconsistencies. For the final reconstruction result, the previously generated prior image plays a crucial role in terms of image quality [3].

In order to further increase the image quality when multiple high-density objects are present, an improved method of NMAR was introduced in 2017 [1]. The approach is based on multiple prior images, which result from the segmentations of a pre-corrected reconstruction. In this paper, instead of using recursive active contours as proposed in [1], a segmentation using Otsu's method is performed. Otsu's segmentation approach is a global thresholding technique

that depends on the gray values of the different anatomical structures [4].

In this paper, the combination of NMAR and multiple prior images is investigated. Instead of a LI, NMAR is used to reconstruct a pre-corrected image that is used for the calculation of multiple prior images.

2   Material and Methods

2.1   Normalized Metal Artifact Reduction

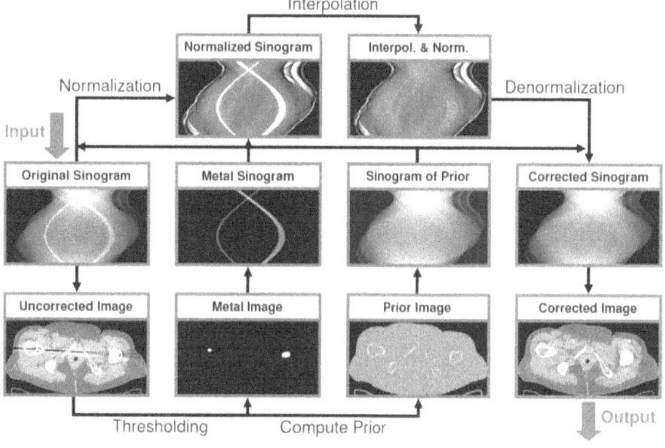

Figure 1: The complete step by step process of NMAR [2].

Initially, a filtered back-projection is used to reconstruct an uncorrected image. The metal image is obtained by segmentation of the uncorrected image. The threshold value of the metal is 1645 HU for Shepp-Logan phantom and 150 HU for Hardware Phantom. To create a prior image for the normalization step, the uncorrected image is smoothed by a gaussian filter and segmented to get the structure of bone and soft tissues. The threshold values for hardware phantom are 118.45 HU for the bone and -343.3 HU for the soft tissues. Whereas for the Shepp-Logan phantom, the threshold values are calculated automatically by Otsu's method. The original sinogram is divided by the forward projection of the prior image. After the normalized sinogram is produced, the sinogram is linearly interpolated in the region of the metal trace. Afterward, the sinogram is denormalized by multiplying it with the forward projection of the prior image. This sinogram is used to reconstruct the final image. The metal object is inserted back into the final image. The approach is shown in Fig. 1.

2.2 Combination of NMAR and Multiple Prior Images

Before the pre-correction step, all the metal implants need to be identified. The uncorrected image is reconstructed by NMAR. NMAR has the advantage to reconstruct the details of the structures, which are missing due to inconsistencies in the raw data produced by LI. With this advantage in the pre-correction, it is possible to create more precise prior images. The contours of bone or soft tissue structures in the close neighborhood of metal objects are pre-corrected with NMAR, which simplifies the creation of prior images. Before the segmentation of the bones and the soft tissues, the NMAR reconstructed image is smoothed with a gaussian filter. The threshold values of the bones and the soft tissues are set by Otsu's method. These subregions represent the multiple prior images. The forward projections of these prior images β are linearly combined. Due to the segmentation process, the values in each pixel of each prior images are either 0 or 1. To get the correct values of each pixel, a weighting factor for each of the prior image is needed. The correct weighting factor is calculated by

$$d = \arg\min_{d_j \geq 0} \| P_o - \sum_{i=1}^{N} d_j \beta_j \|_{T_m^c} . \tag{1}$$

P_o is the original sinogram. $j \in \{1, 2, ..., N\}$ where N is the number of the prior images, d_j are the weighting factors, and β_j is the sinogram of each prior image. T_m^c represents the set of all indices of projections that are not associated with metal objects. The initial values of the weighting factor are the mean of the pixel values of the pre-corrected image in each subregion area.

The combined sinogram $\widehat{P}$ is defined as

$$\widehat{P} = \sum_j d_j \beta_j. \tag{2}$$

The residual errors (i.e. $\delta = P_o - \widehat{P}$), which still exist in the metal trace, need to be compensated. The residual errors δ is corrected with NMAR in the metal trace region and added to the combined sinogram $\widehat{P}$. The corrected image is reconstructed by a filtered back-projection of this sinogram. The process of sub-region segmentation, sinogram combination and residual error compensation is repeated three times.

2.3 Simulation

The numerical simulations were executed with Matlab (version 2019b). The Shepp-Logan phantom and a hardware phantom, which are provided by the Institute of Medical Engineering, University of Lübeck, are used for the evaluation. The images were reconstructed with LI and NMAR in combination with multiple prior images. An implementation of the forward projection and filtered back-projection, are provided by the Institute of Medical Engineering, University of Lübeck.

3 Results and Discussion

The combination of LI and the use of the multiple prior images approach is referred as LI_m and the combination of NMAR and the use of the multiple prior images approach is referred as NMAR_m. There are two phantoms (a hardware phantom and a software phantom) to be simulated and compared. Two metal implants are artificially inserted into the software phantom (see. Fig. 2). The hardware phantom is equipped with two metal rods (see. Fig. 4).

3.1 Software Phantom

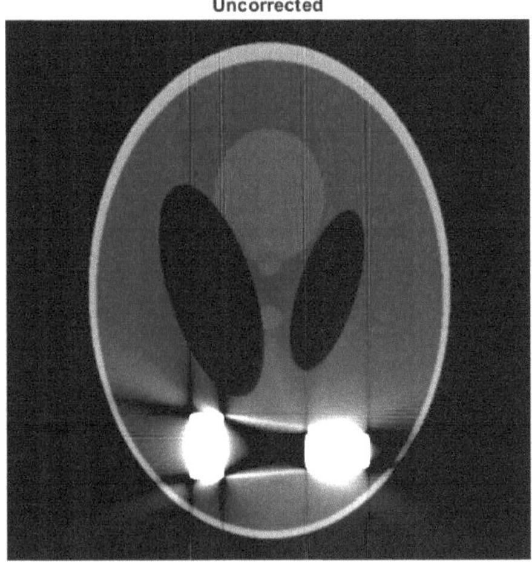

Figure 2: The uncorrected image of the Shepp-Logan phantom with two metal implants.

The uncorrected image is reconstructed by LI, NMAR, LI_m, and NMAR_m. Fig. 3 shows that the NMAR_m is more effective to reduce streak artifacts than NMAR. During the denormalization process of the NMAR algorithm, small interpolation errors in the metal trace region are amplified. The NMAR_m solves this amplification of small interpolation errors by the compensation of the residual error. Therefore, some streak artifacts are seen in the reconstructed image by NMAR. Whereas the difference between LI_m and NMAR_m is shown in the region of interest 1 (ROI1). The reconstructed image by LI_m shows clearer streak artifacts in the ROI1.

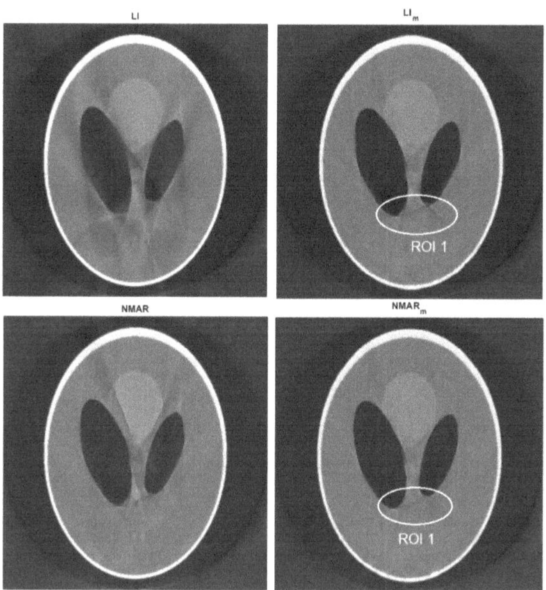

Figure 3: The comparison of reconstructed image by LI (top-left), LI_m (top-right), NMAR (bottom-left) and $NMAR_m$ (bottom-right).

To evaluate the difference between LI, NMAR, LI_m, and $NMAR_m$ more accurately, the mean-squared error (MSE) between a Shepp-Logan phantom as ground truth and the reconstructed image by LI, NMAR, LI_m, and $NMAR_m$ is calculated. The results are shown in Table 1.

Table 1: The results of MSE between the Shepp-Logan phantom and reconstructed image by LI, NMAR, LI_m, and $NMAR_m$. The results are in Hounsfield Unit HU

	LI	NMAR	LI_m	$NMAR_m$
MSE	17,597	17,166	16,608	15,845

Table 1 shows that the reconstructed image by $NMAR_m$ has the lowest MSE. This means that the reconstructed image by $NMAR_m$ has the highest quality of the image.

3.2 Hardware Phantoms

A hardware phantom with two metal rods is reconstructed by LI, NMAR, LI_m, and $NMAR_m$. The hardware phantom without metal rods is used as a reference (see Fig.5).

The reconstruction images by LI, NMAR, LI_m, and $NMAR_m$ are shown in Fig.6. In comparison with the reconstructed image by NMAR, $NMAR_m$ has fewer streak artifacts in the region of the bone. The reconstructed image by LI_m shows an artifact in ROI2, but the reconstructed image by $NMAR_m$ has no artifact in ROI2.

3.3 Discussion

Qualitatively $NMAR_m$ shows a reconstruction image with fewer artifacts than LI, NMAR, and LI_m. The compensa-

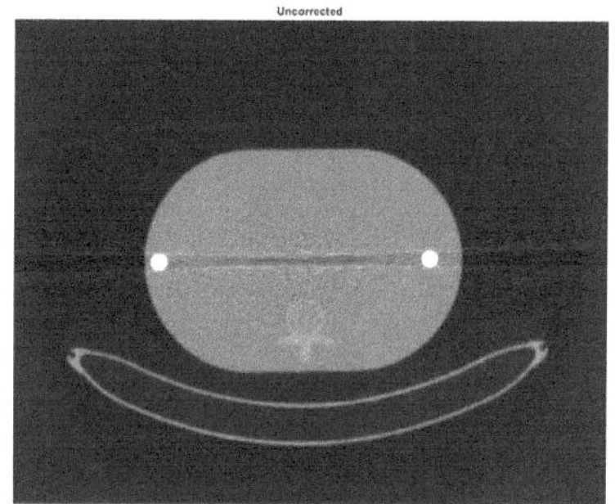

Figure 4: Hardware Phantom with two metal implants. This hardware phantom is going to be reconstructed by LI, NMAR, LI_m, and $NMAR_m$.

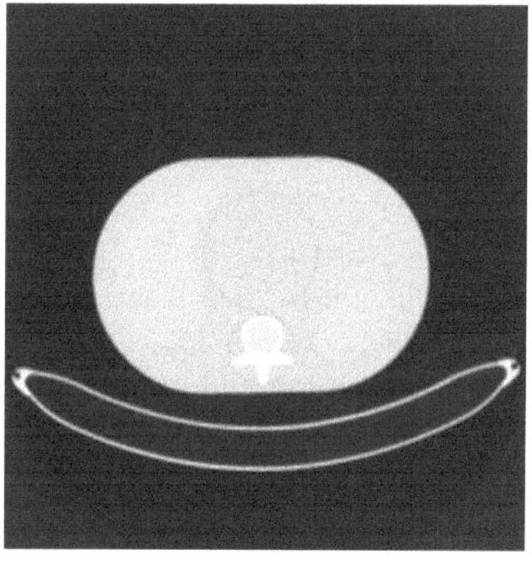

Figure 5: Hardware Phantom without any metal implant. This image is the reference image for the comparison of the reconstructed image by LI, NMAR, LI_m, and $NMAR_m$.

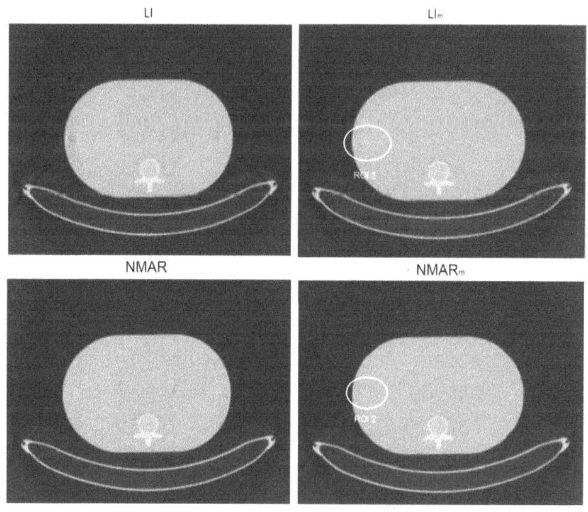

Figure 6: The reconstructed image by LI (top-left), LI_m(top-right), NMAR(bottom-left) and $NMAR_m$(bottom-right).

tion of the residual errors has reduced the streak artifacts. The application of NMAR in the pre-correction step improves the quality of the prior images. An artifact in the reconstructed image of hardware phantom by LI_m is produced because of the prior image (see Fig. 7).

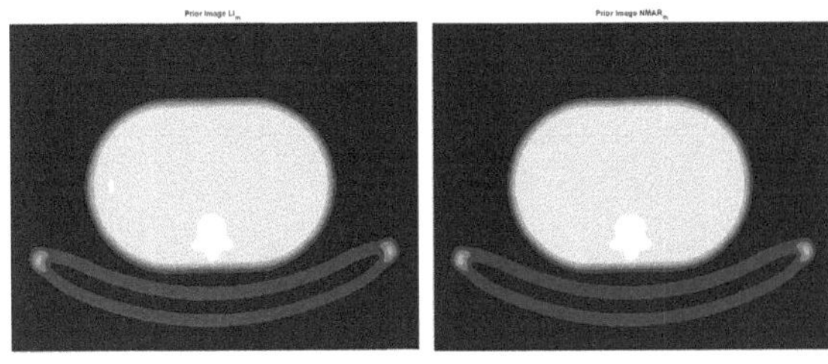

Figure 7: The prior image of LI_m (left) and NMAR_m (right).

4 Conclusion

The reconstructed image by NMAR_m has fever artifacts than the reconstructed image by LI, NMAR, and LI_m. In the simulation of the Shepp-Logan phantom, the streaks artifacts are reduced by NMAR_m. The result of MSE confirms that the quality of the reconstructed image by NMAR_m is better than LI, NMAR, and LI_m. Qualitatively the reconstructed image of hardware phantom by NMAR_m shows fewer artifacts. The reconstructed image by LI and NMAR shows more streak artifacts and LI_m shows an artifact in ROI 2. The NMAR reconstructed images produce a better quality of the prior image. Therefore the quality of the final reconstructed image is improved.

To confirm the conclusion of this article, the simulation shall be performed with clinical CT images with more complexity,i.g. more of bone structures or metal implants inside the bone. In future work, it is suggested to combine Frequency Split Metal Artifact Reduction (FSMAR) with the use of the multiple prior images [5].

Acknowledgement

The work has been carried out at the Institute of Medical Engineering, University of Lübeck.

5 References

[1] H. Nam, J. Baek, *A Metal Artifact Reduction Algorithm in CT Using Multiple Prior Images by Recursive Active Contour Segmentation*, PLOS ONE, vol. 12, e0179022, 2017. Available: https://doi.org/10.1371/journal.pone.0179022 [last accessed on 2020-02-04]

[2] E. Meyer, R. Raupach, M. Lell, B. Schmidt, M. Kachelreiß, *Normalized Metal Artifact Reduction (NMAR) in Computed Tomography*, Medical Physics, vol. 37, pp 5482–5493, 2010

[3] Y. Zhang,H. Yan,X. Jia,J. Yang,S.B. Jiang,X.Mou, *A Hybrid Metal Artifact Reduction Algorithm for X-ray CT*, Medical Physics, vol. 40, 041910, 2013.

[4] H.J. Vala,A. Baxi, *A Review on Otsu Image Segmentation Algorithm*, International Journal of Advanced Research in Computer Engineering Technology, vol. 2,pp 387-389, 2013.

[5] E. Meyer, R. Raupach, M. Lell, B. Schmidt, M. Kachelreiß,*Frequency Split Metal Artifact Reduction (FSMAR) in Computed Tomography*, Medical Physics, vol. 39, pp 1904-1916, 2012.

Workflow development to create and analyse QA parameter from MRI images

Sören Postel [1,3], Andre Mastmeyer[2], and Jochen Hirsch [3]

[1] Medical Informatics, Universität zu Lübeck, soeren.postel@student.uni-luebeck.de
[2] Digital Health Management, Aalen University, andre.mastmeyer@hs-aalen.de
[3] Fraunhofer MEVIS, Fraunhofer Institute for Digital Medicine MEVIS, Bremen, jochen.hirsch@mevis.fraunhofer.de

Abstract

Magnetic Resonance Imaging (MRI) images are difficult to rate with respect to image quality without visually checking the images. This is a very time-consuming task and for image computing, it is important to have image quality information. The development of the software was focused on usability and automated quality assessment (QA). The software process is divided into four steps - data import, QA measures estimation, storing, and visualizing QA measures -, and compatible with external data. The application is implemented in a docker image for easy sharing. The docker image is set up as a container with Elastic Search and Kibana from a single configuration file. The analysis in the application is to compare quality assurance among each other and to similar data from the database. With the help of the system, quality outliers can be detected or data grouping can be done.

1   Introduction

To this day, it is difficult to characterise the quality of Magnetic Resonance Imaging (MRI) images as there is no agreed or reference quality measure, as every MRI protocol shows its own technical limitations and requirements. The current method of obtaining information about image data is visual rating. This is very time-consuming and not always objective by different reviewers. In the case of a study, visual assessment cannot be done for every data set, but only on a random basis. Since large studies, such as the NAKO Gesundheitsstudie (2013-2023), currently collect more than 30,000 whole body MRI data. These cannot all be manually reviewed in terms of their quality.

For this purpose automated pipelines are a good way to get an overview of data. Therefore, a pipeline was created that can be used everywhere without the images having to leave the location where they were taken. Therefore a use in a clinical environment would be a possible field of application.

The development of such a pipeline which is independent, performant and flexible was the focus of the project.

2   Material and Methods

2.1   MeVisLab

MeVisLab [1] is a development environment for medical imaging. The software is developed by MeVis Medical Solutions AG. It is a powerful framework for medical image processing and useful for prototyping using visual program-

ming. The software has a C++ and Python backend for development.

2.2   Python

Python is a programming language which was designed in the 1980s. Since then it has become more and more popular and is currently the third most used programming language. The easy extensibility and the many possibilities this language offers makes it an ideal language for development. Therefore the language was used for the development of the pipeline as there are already many libraries that were needed. These include Bokeh for the visualization of the final results as well as libraries developed at Fraunhofer MEVIS such as *liqa* [2] which was used for the calculation of various tests.

2.3   Elastic Search

Elastic Search [3] is a non relational database with a REST API. REST stands for REpresentational State Transfer, API for Application Programming Interface. This refers to a programming interface that is oriented towards the paradigms and behaviour of the World Wide Web (WWW) and describes an approach for communication between client and server in networks.

The interfaces for queries and results are seperated from the database. This allows better scaling and faster access to the data. For the data request, it needs a query string which has a format like a JSON file structure and can be submitted through the API to get a result. A tool for monitoring the

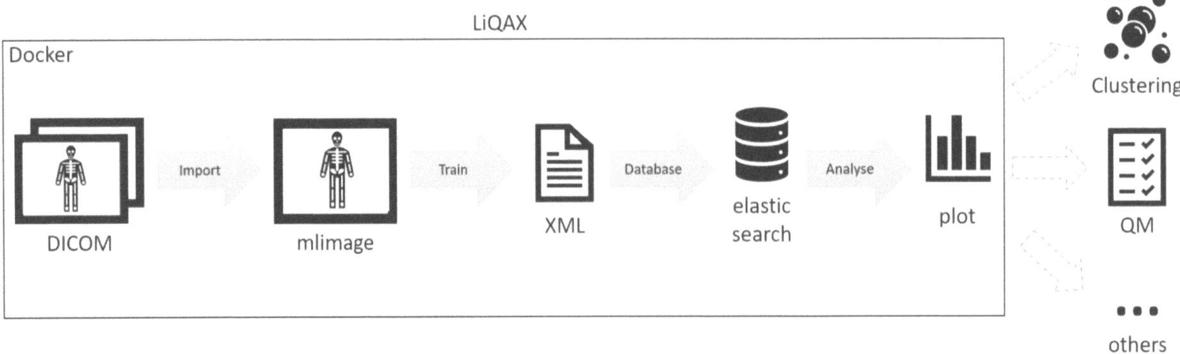

Figure 1: Schematic overview of the pipeline with its inputs and outputs and possible use cases for the data. The system is embedded in a docker image.

database and its entries is Kibana which was also developed by elastic. Kibana is useful to visualize the work with the database and to read out entries quickly.

2.4 Docker

Docker [4] is a virtualisation software for applications and services. The docker software packages are called containers. The containers are isolated virtual machines with their own operating system and software. All started containers will run in a single kernel of the host OS. The containers can be connected to each other in internal networks for communication or data transfer. Docker images are called software packages which will run in a container. This has the advantage that the software can be run on every system with docker without any special settings.

2.5 Data

The implementation was based on two types of data. Firstly, MRI DICOM image data, which still have to be preprocessed and secondly, already calculated quality assurance parameters, which are available in an Elastic Search database. The existing data served the purpose of testing the software in terms of stability, performance and usability. Evaluations were made on the basis of this data only in the context of usability and visualization.

2.6 Quality Assurance

Quality Assurance (QA) is important for medical imaging. It is necessary to calculate QA measurements for medical imaging systems from phantoms. For the measurement of the QA protocols, a phantom will be used from the American College of Radiology (ACR). The important assessments for the MR images are: geometric accuracy, high-contrast resolution, slice thickness accuracy, slice position accuracy, image intensity uniformity, percent of signal ghosting, and low-contrast object detectability. The results for signal-to-noise ratio, sharpness and quality index are used as well for the quality [2], [5], [6], [7].

The quality index is an index, which contains the characteristics of noise and sharpness. It is calculated as absolute average of the Laplace convolution over the image. The mean value of the image and the mean value of the edges in x and y direction. These are created by subtracting the convolution around a 3x3x1 kernel [8].

3 Results

A web-based pipeline was developed to standardize the evaluation of the quality of MRI data. This pipeline is called LiQAX based on an existing Python library called liqa. LiQAX was developed in the development environment MeVisLab, but is mainly based on Python and for the web service Node.js. During development it was packaged in a docker image to take advantage of its stand-alone and flexibility.

3.1 Pipeline LiQAX

The LiQAX pipeline has a modular and independent design for flexible and easy integration into new applications. The pipeline is divided into four sections: the import of MRI DICOM data, the evaluation of QA characteristics of MRI images, the communication with the database and the display and statistical evaluation of data from the database. Each section in the pipeline requires certain input data and generates output data. These input data do not necessarily have to come from the previous section but can also be generated by other applications.

A schematic overview of the pipeline with its inputs and outputs is shown in Figure 1.

Import of MRI DICOM images

The pipeline starts with the import of MRI DICOM slice images. These are loaded recursively from folders by the system and by means of settings at the frontend. The layers are merged to volume images. These are then saved as *mlimages* (a MeVisLab proprietary format) at a specified location.

Evaluation of images

The evaluation section loads *mlimage* images from a selected folder structure. There are different test sets which are responsible for the calculation of specific QA measures. This calculation is performed by scripts from the liqa library which dynamically creates networks in MeVisLab to calculate the values. Beside there the images are displayed as slice images in the web frontend.

The calculation can be performed on a single image or on multiple images. The results are written to an XML file for each image. This file contains the computed results and general information about the image extracted from the DICOM tags. The XML file will be expanded by each test set that is calculated for the file.

Communication with the Database

The following section focuses on communication with the Elastic Search database. The previously created XML files are read in, visualized and stored in the database. For the visualization, the XML is parsed and adapted for the web frontend. The information extracted from the XML can be transferred to an Elastic Search database. For this purpose, IDs are automatically generated for each image by using relevant data. The ID is used to transfer the data from the XML to the database and store it. Further settings that determine the storage location can be set additionally. After successful entry, the IDs are displayed as a link to the database and can be called up.

Statistic evaluation

The pipeline has as a last step an evaluation where existing databases can be accessed. All quality parameters contained in the database are available as evaluation criteria. One or more of them can be selected to be visualized. In order to limit the data and to select appropriate MRI data for comparisons some filter options are available. After the data have been selected, they are retrieved from the database using a query and then evaluated and visualized. For each QA parameter a separate plot is generated in which all data are plotted as data points. In addition, a box plot is generated that contains statistical information. To make a comparison it is possible to select different data which are displayed in a plot side by side or separated in time. Figure 2 shows an example for the comparison of two data series.

The data can also be forwarded to further processing or statistical evaluations. As an example, a clustering algorithm was applied to group QA values due to specific technical characteristics (e.g. acquisition failures or image artifacts). An example would be finding comparable data by clustering the QA values. An example of how this can be done with DBSCAN [9] is shown in Figure 3. Here it is tested if clusters can be found by using QA values outside the norm.

Configuration

The configuration of the LiQAX settings and the default values will be set by a configuration file or environment variables. These variables can also be set in the docker compose file. The settings file is a JSON document with some defined settings for the pipeline e.g. default test set, URLs or paths.

With the configuration, it is possible to change the data structure for the image loading. This has the advantage to calculate the QA values for *mlimages* from other import pipelines.

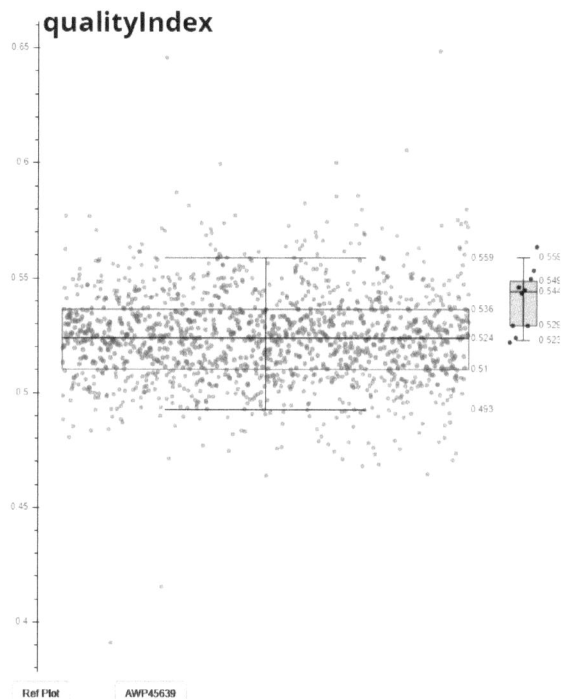

Figure 2: Example output of a quality parameter with statistical information. The larger left reference group contains about 1000 data points and the right comparison group contains about 10 data points.

4 Discussion

We successfully developed a pipeline as a package that meets the requirements of modularity, independence and autonomy. Additionally, it is flexible and easy to integrate into other evaluation processes (e.g. deep learning algorithms). Due to its modularity, the pipeline can also be combined with other applications that do pre- or post-processing. The independence and autonomy have the advantage that the Docker image with the pipeline can also be operated in closed clinical setups where strict data protection is requested.

5 Conclusion

In conclusion, a functional pipeline has been created that can be used in a variety of ways. Due to the modularity of the individual sections it is possible to integrate them easily into existing processes. The autonomy and indepen-

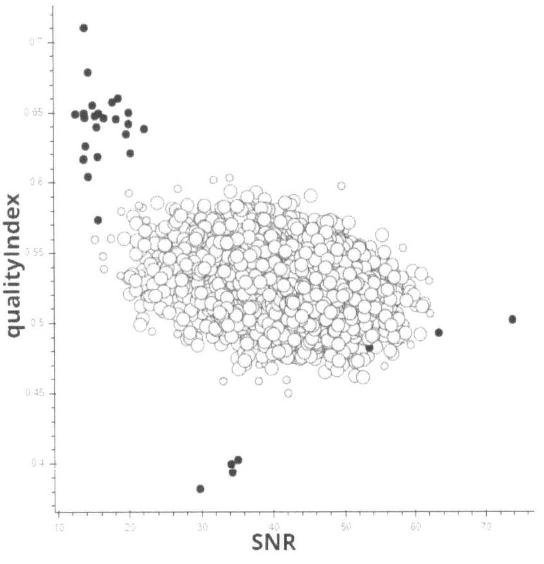

(a) Clustering of 3 QA parameters to find the outliers

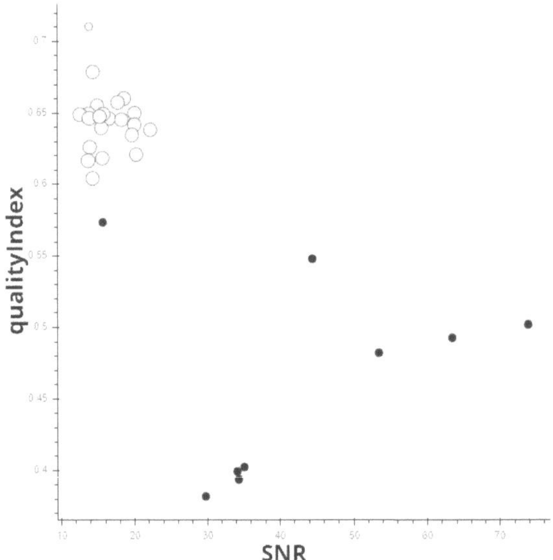

(b) Remove the norm cluster to find clusters outside the norm

Figure 3: Example of clustering data and finding clusters outside the norm. About 5,000 data were used to create the clusters.

dence also allows the use everywhere and no additional infrastructure is necessary. This also allows the use in closed systems such as everyday hospital life with high data protection standards. The whole application has been built into a docker image to be able to use it across different systems. The pipeline simplifies the workflow and offers an objective determination of quality parameters which can then be used for various evaluations. For this purpose, an analysis can be carried out in the software, which compares the selected data with each other and with other data. Outliers in the data series can be determined for various evaluations. The web interface allows a simple and intuitive operation which is an advantage in both research and routine data management environments.

Acknowledgement

The work has been carried out at Fraunhofer MEVIS, Bremen and supervised by the Institute of Medical Informatics, Universität zu Lübeck.

6 References

[1] MeVis Medical Solutions AG, `https://www.mevislab.de/`

[2] J.G. Hirsch, A. Köhn, D.C. Hoinkiss, J. Peter, A. Thomsen, M. Günther and the German National Cohort Study Investigators (2016). *Fully Automated Data Management and Quality Assurance in Very Large Prospective Cohort MR Imaging Studies – the MR Imaging Study within the German National Cohort*; Proc. 24th Scientific Meeting of ISMRM. Singapore, 2016.

[3] Elasticsearch, `https://www.elastic.co/de/elasticsearch`

[4] Docker, `https://www.docker.com`

[5] L. J. Erasmus, D. Hurter, M. Naude, H. G. Kritzinger and S. Acho, *A short overview of MRI artefacts : review article*, SA Journal of Radiology, 2004

[6] F. Schick, *Grundlagen der Magnetresonanztomographie (MRT)*, Der Radiologe, 2005

[7] J.G. Hirsch, A. Köhn, D.C. Hoinkiss, J. Singe, E. E.-B. Öztürk, M. Günther, and the NAKO MRI Study Investigators (2017). *Quality Assessment in the Multicenter MR Imaging Study of the German National Cohort (NAKO)*; Proc. 25th Scientific Meeting of ISMRM. USA, HI, Honolulu, 2017

[8] Zhou Wang and A. C. Bovik, *A universal image quality index*, IEEE SignalProcessing Letters, vol. 9, no. 3, pp. 81–84, Mar. 2002

[9] Martin Ester, Hans-Peter Kriegel, Jörg Sander and Xiaowei Xu *A Density-Based Algorithm for Discovering Clusters in Large Spatial Databases with Noise*, KDD'96: Proceedings of the Second International Conference on Knowledge Discovery and Data Mining, 1996

Efficient Self-Supervised Context Learning on MRI data

Christoph Großbröhmer [1], Max Blendowski [2] and Mattias P. Heinrich [2]

[1] Medical Engineering Science, Universität zu Lübeck, christoph.grossbroehmer@student.uni-luebeck.de
[2] Institute of Medical Informatics, Universität zu Lübeck, {blendowski, heinrich}@imi.uni-luebeck.de

Abstract

Self-supervised learning has experienced a growing research interest in medical data processing in recent years, as it promises possibilities to leverage with large, unlabeled data sets. First steps in this direction have shown that from 3D image data meaningful descriptors can be extracted, proving the possibilities to recognise and learn data-inherent patterns. However, these techniques still offer potential regarding efficiency and acceleration. In this paper, we present a novel approach to efficiently extract features by processing a whole 3D image at once using Convoluional Neural Networks. We assess its performance both in computational efficiency and quality of extracted descriptors indirectly.

1   Introduction

The use of Convolutional Neural Networks (CNNs) has been proven to solve many Computer Vision tasks efficiently. However, in general, CNNs require large amounts of training data to produce meaningful results. In tasks like multilabel segmentation, this training data consists of labeled organs, muscles and other structures, as the training calls for strong supervision. While in other fields related to Computer Vision annotations can be easily produced by laymen, interpretation and labeling of medical images usually requires expert knowledge. As producing these labels is cost and time-intensive, most available medical study data is not associated with labels or annotations. In an effort to overcome this problem, weakly and self-supervised learning techniques can be used.

Self-Supervised Learning, a subclass of unsupervised learning, uses auxiliary tasks to acquire knowledge of the applied data. Possible tasks for gaining an intrinsic understanding of images include the prediction of the spatial relation of image patches [1] or the colorization of grayscale images [2]. The auxiliary task used in this paper is closely connected to the proposed techniques of Blendowski et al. [3] and Doersch et al. [1], who could demonstrate the ability of CNNs to derive an intrinsic understanding of image content by predicting spatial relations of image patches.

In [3], this general technique has been applied to three dimensional CT data. The authors were able to show that in this case the use of 2.5D almost planar subvolumes resulted in significantly better representations compared to a naive 3D Doersch classification approach. Thus, the lower spatial resolution of medical image volumes in comparison to 2D images with a high amount of details used in Computer Vision, can be compensated by assigning fine-granular mapping tasks, effectively using a regression task instead of a rigid classification of 6 classes.

Feature representations may serve versatile downstream tasks, such as automatic registration or segmentation of medical images, enabling the idea of transfer learning known from the Computer Vision Community. In these scenarios, extracting features is just one early step of the whole process, therefore resulting in an urge for efficiency. While [3] shows good results concerning the quality of representations, its network structure could be improved regarding efficiency.

2   Material and Methods

In this paper, the proposed approach of Blendowski et al. [3] is used on MRI data published by the NAKO Gesundheitsstudie [4]. To improve its computational efficiency during runtime the technique has been altered, i.e. that a whole 3D image is passed to the CNNs, resulting in its feature representation. The auxiliary task, determining the spatial relation of elements drawn from this tensor can now be assigned for many voxel positions at once.

2.1   Feature Representation Learning

As in [3], we train an individual Convolutional Neural Network for every spatial direction of the 3D MRI image volume. The lower part of Fig. 1 shows the basic training principle for one of our three descriptor networks: In every epoch a whole 3D image is passed to the CNN yielding a tensor containing features of the size [Height/4, Width/4, Depth/4] and 64 channels as the effect of using e.g. stride operations (see Fig. 2 for details).

Hereinafter, for the sake of clear notation, the spatial axes of the feature tensor are assumed to be normalized to $[-1, 1]$ with a length of 2. The inplane axes are denoted as x and y, whereas the third axis is referenced as z. The main idea of

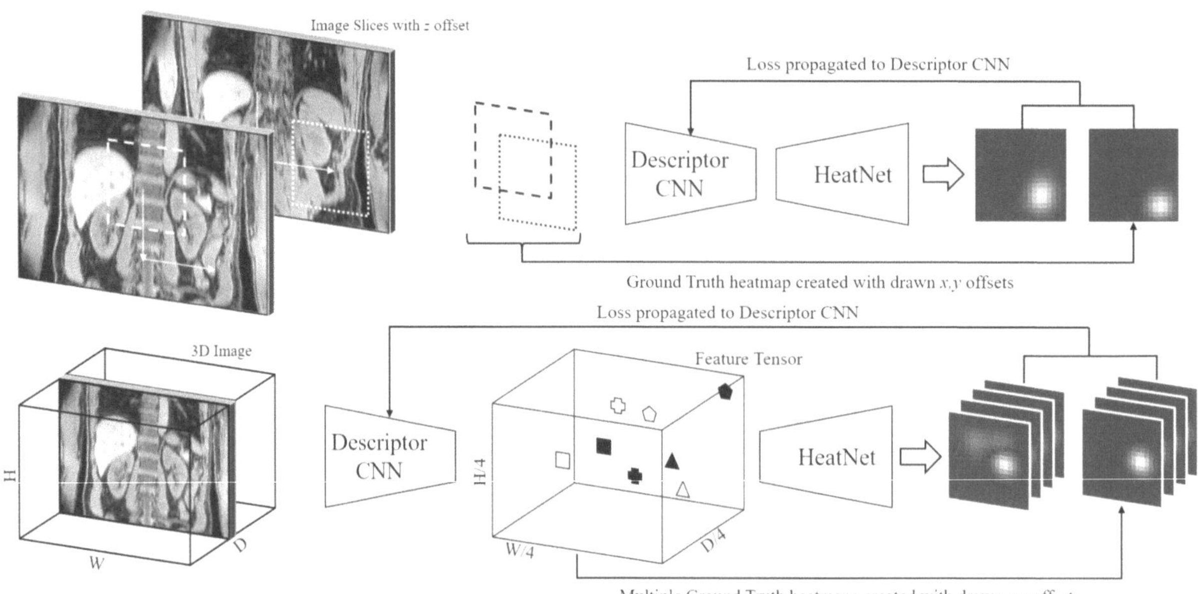

Figure 1: Top: Proposed architecture of Blendowski et al. [3]. Almost planar patches are drawn from a 3D volume, which are subject to a certain offset in z-direction. The patches are passed to a descriptor network, producing features which are used in the HeatNet to estimate a heatmap of the offsets in x and y direction. From these, a loss for the descriptor network is determined by comparison with the ground truth heatmap. Bottom: Our approach to improve efficency. We pass the entire 3D volume to the descriptor network, which then produces a feature tensor with downsampled spatial dimensions, from which n pairs of feature vectors are drawn at different spatial positions. As in [3], the HeatNet generates heatmaps from these features, propagating their mean squared error back to the Descriptor CNN. By processing several auxiliary tasks simultaneously, the Descriptor CNN can be trained more efficiently and finally, process whole images at once during inference.

this network is to construct a tensor, from which randomly drawn pairs of feature vectors are presented to the auxiliary HeatNet network, as it is detailed in [3]. The anchor element of each pair is drawn uniformly from $[-0.3, 0.3]^3$, resulting in a rather central position. To ensure that near-planar subvolumes represented by the feature vectors do not share characteristics like continuing lines, an offset in z-direction of at least 0.125 and up to 0.25 is enforced. Using these offsets counters the 'body border' problem [3], which would lead to oversimplified tasks regarding the heatmap prediction, resulting in inefficient learning. To accelerate initial training, random inplane offsets range from $[0.25, 0.3]$ in the beginning to $[0, 0.7]$ in later stages. The 64-dimensional feature vectors at both positions are handed to the HeatNet to predict a 17×17 heatmap regarding spatial displacements, as it offers a better gradient flow than predicting the sole numerical offset values [5]. Using the given shifts, a ground truth heatmap is created using

$$heat(i, j, \Delta_1, \Delta_2) = 10 \cdot e^{-15(i/9-\Delta_1)^2 + (j/9-\Delta_2)^2} \quad (1)$$

with $(i, j) \in \{-9, -8, ..., +8, +9\}$. The mean squared error between these heatmaps is used as a penalty term to change the descriptor net's weights. To obtain a final feature vector for each spatial element, all three feature vectors are concatenated, resulting in a 192-dimensional representation at every fourth voxel.

2.2 Details on the Descriptor Network

The structure of the descriptor nets is illustrated in Fig. 2. In order to obtain similar results to [3], the architecture was designed to rely on a similar receptive field. While Blendowski et al. used patches of size $42 \times 42 \times 3$, the convolutionial layers and striding operations selected yield a receptive field of $47 \times 47 \times 3$ in our network. Since a separate descriptor was trained in each spatial direction, it is possible to adapt the training hyperparameters to the characteristics of individual spatial directions in the data, such as voxel spacing. We train the networks processing the frontal and sagittal MRI images as inplane input with an initial learning rate of $5 \cdot 10^{-3}$, while in the transverse direction an initial learning rate of $5 \cdot 10^{-5}$ was applied. In each iteration, we extract 64 different pairs of feature vectors and compare their heatmaps. After every 50 iterations, we draw a new input image. Each descriptor network was trained with 3000 iterations, resulting in over 800.000 random displacements used for training. In total each descriptor network contains 221.289 trainable parameters.

2.3 Evaluation

Goal of this work was to provide a CNN design producing meaningful feature descriptors in a rapid fashion. To evaluate these goals, three experiments have been carried out. 1.) Since we imagine the descriptor network to be

Figure 2: Schematic representation of the architecture of the descriptor network. The operations inside the layers are abbreviated as followed: 1.) RepPad3d $\equiv$ ReplicationPad3d, 2.) Conv3d $(c_{in}, c_{out}, kernel, dilation) \equiv$ 3D-Convolution, 3.) GN $\equiv$ GroupNorm, 4.) LR $\equiv$ LeakyReLU. Please note that the first convolution in the descriptor network (denoted as Conv3d*) features a custom kernel, effectively slicing in axial direction.

used as an early step in different tasks like segmentation or registration, we validate its computational time against the patch-based method introduced in [3]. 2.) To verify if our architecture is able to produce meaningful features on MRI data, we perform a k-nearest-neighbour (kNN) based atlas label transfer segmentation task using an increasing number of labeled datasets. 3.) In order to evaluate further potential of acceleration, we investigate possibilities to reduce the CNN parameter space by statistical measures.

As a comparison for the indirect kNN segmentation-based evaluation of feature expressiveness, two other methods were implemented: One is a model of our network, which was initialized using the Xavier method but did not receive further training. A second comparison method is given by the BRIEF algorithm [6], which allows extracting features based on gray value differences of a random, normally distributed 3D pattern. To produce comparable results, the field of view of this method was adjusted to contain roughly the same as the CNN's receptive field.

2.4 Data

All experiments are performed on thoracoabdominal Magnetic Resonance Images published by the NAKO Gesundheitsstudie [4]. The imageset used consists of 91 T1-weighted 3D volumetric interpolated breath-hold examination sets acquired at three different MR imaging centers, recorded with a two-point Dixon method. Of these 91 water-only images, 80 were used to train the network. No further preprocessing was applied. Each voxel represents a volume $1.4mm \times 1.4mm$ inplane and $3mm$ in axial orientation with image sizes of $320 \times 260 \times 96$ (AP-LR-SI). In 11 images, liver and kidneys have been segmented and labeled to serve evaluation purposes.

3 Results and Discussion

3.1 Execution Time

To evaluate the acceleration gained by restructuring the descriptor net approach, we measured the time while extracting features for every fourth voxel in 50 image sets using a Intel Xeon E5-2620 v4 processor and a Geforce 1080 Ti GPU. Since we do not evaluate the quality of feature descriptors generated by the patch-based approach of

Blendowski et al. (further denoted as patch2D), we use untrained, but initialized models. The results presented in Table 1 show the mean of these measurements. As expected, we are able to outperform the patch2D approach, even with a batch size of 64 patches, by a considerable margin, both on CPU and with GPU support.

Table 1: Mean computational times in seconds.

	CPU	GPU
Our Work	16.13	0.21
patch2D	189.74	7.70

3.2 Feature Quality

We generate a feature vector for every fourth voxel on the labeled evaluation dataset with all 3 methods, resulting in 124.800 feature vectors per method and image. Prior to a 5NN search on labeled atlas data to achieve a segmentation, a principal component analysis (PCA) is performed on the features to accelerate evaluation, preserving 32 of 192 dimensions. Additionally, we perform a five-fold cross validation with random splits. In order to measure the quality of segmentation, we compute Dice scores of the predicted labels. Please note that we did not put any effort into fine-tuning these results for the segmentation task to purely assess the descriptive power of our self-supervised trained descriptors. Table 2 presents the mean dice scores for each method over 5 folds with 7 labeled datasets. Clearly, our trained network outperforms both the Xavier net and the BRIEF descriptors, resulting in the highest Dice scores in mean and on individual labels. Considering that the Xavier net has not undergone any training, its results are remarkable. This indicates that our network architecture is generally capable of capturing anatomical context in medical volumetric data. It is also noteworthy that training could increase the average dice score for the left kidney from about

Table 2: Mean Dice scores in % over 5 folds with 7 labeled evaluation images.

	Liver	L.Kidney	R.Kidney	Mean
Trained Net	84.75	45.32	53.12	61.07
Xavier	75.93	4.69	28.24	36.29
BRIEF	43.91	20.65	13.90	26.15

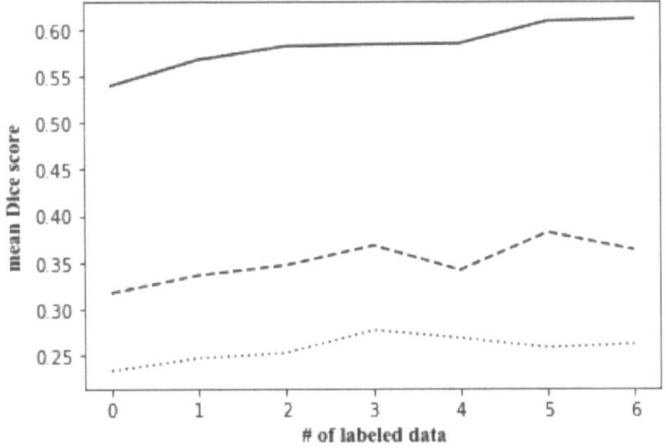

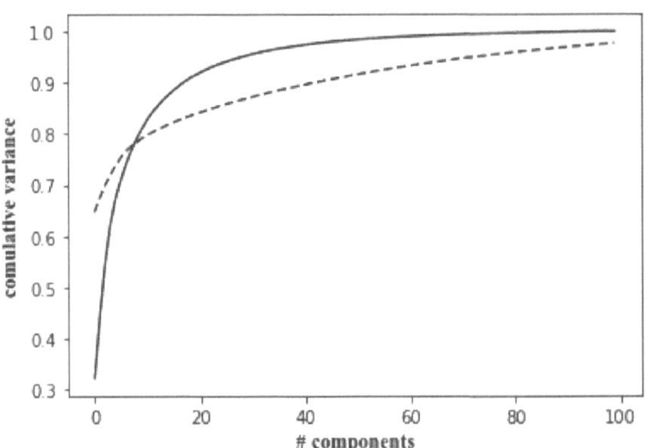

Figure 3: Mean dice scores for segmentations generated with our method (continuous line), the Xavier initialised net (dashed line) and BRIEF features (dotted line).

Figure 4: Percentages of cumulative variance over components for our trained (continous line) and untrained but initialized net (dashed line).

5% to over 45%. We analysed further how the number of labeled datasets influences the dice scores. The results can be seen in Fig. 3. As expected, the number of atlas records in combination with the 5NN search increases the average dice score in both the trained and the Xavier network.

3.3 Dimension Analysis

Since the previous results show that the features still contain meaningful information about the images even after dimensional reduction, we also evaluated the relative variances of components after performing a PCA on this data. Percentages of the cumulative variance of the total variance in respect to the first 100 components regarding both our trained net and the Xavier net are shown in Fig. 4. For our method, 29 components are sufficient to cover over 95% of the total variance. To reach this value with the Xavier initialized net 79 components are needed.

This leads to the conclusion that 32 was a good choice for dimensional reduction in the kNN-task, since it drastically shortened computational time, but still retained much information. The fact that the network is able to preserve a high level of information in fewer components after the training, prompts interest in research regarding dimensional reduction. However, these results already indicate that the architecture of the descriptor networks can be streamlined further.

4　Conclusion

We have developed a more efficient procedure for the self-supervised efficient extraction of significant features from volumetric magnetic imaging data sets. The method based on [3] uses 3 separate descriptor networks for all spatial directions that can successfully understand and describe the anatomical context. During training, heatmaps are generated from spatial differences on the feature tensors, output by the descriptor network. Therefore the speed of extracting descriptors could be increased by at least a factor of 10. With an evaluation by segmentation using a 5NN search on

unseen image data, average dice scores of over 60% could be achieved. Furthermore, the relevance of feature dimensionality was investigated. Beside dimensionality reductions, further work may also consider the effects of different network architectures and hyperparameters as well as application areas such as image registration.

Acknowledgement

The work has been carried out and supervised by the Institute of Medical Informatics, Universität zu Lübeck.

5　References

[1] C. Doersch, A. Gupta, and A. A. Efros, "Unsupervised visual representation learning by context prediction," *ICCV*, 2015.

[2] R. Zhang, P. Isola, and A. A. Efros, "Colorful image colorization," *ECCV*, 2016.

[3] M. Blendowski, H. Nickisch, and M. P. Heinrich, "How to learn from unlabeled volume data: Self-supervised 3d context feature learning," in *Lecture Notes in Computer Science*. Springer International Publishing, 2019, pp. 649–657.

[4] F. Bamberg *et al.*, "Whole-body MR imaging in the german national cohort: Rationale, design, and technical background," *Radiology*, vol. 277, no. 1, pp. 206–220, oct 2015.

[5] C. Payer, D. Štern, H. Bischof, and M. Urschler, "Regressing heatmaps for multiple landmark localization using CNNs," in *Medical Image Computing and Computer-Assisted Intervention – MICCAI 2016*. Springer International Publishing, 2016, pp. 230–238.

[6] M. Calonder, V. Lepetit, C. Strecha, and P. Fua, "BRIEF: Binary robust independent elementary features," in *Computer Vision – ECCV 2010*. Springer Berlin Heidelberg, 2010, pp. 778–792.

Development of a Deep Learning based classification of respirator masks for a mobile application

Luisa Bartram[1], Bartholomäus Dedersen[2], Mattias Heinrich[3]

[1] Medical Informatics, Universität zu Lübeck, luisa.bartram@student.uni-luebeck.de
[2] Drägerwerk AG & Co. KGaA in Lübeck, Bartholomaeus.Dedersen@draeger.com
[3] Institute of Medical Informatics, Universität zu Lübeck, heinrich@imi.uni-luebeck.de

Abstract

This paper describes an application that makes it possible to detect dust masks from competing companies and find the equivalent Dräger product. The application is integrated into an app. With the help of the app, Dräger sales staff will be able to present Dräger products to customers faster and better. The aim is to increase the market share of dust masks and increase competitiveness. The application is based on neural networks and cloud computing algorithms. A total of three different networks were trained and tested. The network with the best results was then used for the application. The algorithm classifies an image and then returns the three most probable masks with the corresponding probabilities. Afterwards the Dräger product equivalent can be displayed. This application is intended to make the work of the sales staff easier and more efficient.

1 Introduction

Drägerwerk AG & Co. KGaA is one of the leading companies in medical and safety technology. In the field of safety technology, the company primarily develops respiratory protective equipment, gas measuring systems, diving technology and alcohol and drug measuring devices.
The goal was to develop an application that enables the detection of dust masks from competing companies and to find the equivalent Dräger product.
Dust masks can be divided into three different protection classes - FFP1, FFP2 and FFP3. In order to increase competitiveness, Dräger is now offering its sales staff in the Safety Supplies and Consumables segment a digital product catalog in the form of an app, the SSC (Safety Supplies Consumables) App. The SSC App is a product catalogue app that displays the extensive non-digital catalogue. It contains gas masks, gas detectors, test gases, filters and many other products. Among other things, the very time-consuming updating of the analogue catalogue is no longer necessary. Furthermore, the sales people save on the transport of the analogue catalogue.
The developed classification application must be integrated into the SSC App. For this application different neural networks should be compared and evaluated. The network with the best results should then be integrated into the existing app. For this task, it has to be evaluated whether the network is located locally on the mobile device or whether it is connected via an interface to the cloud.

2 Material and Methods

2.1 Data

Supervised learning methods require enough annotated data to generalize unseen data correctly. It must be taken into account, that the data for the training are as different as possible or have a high variance in order to reflect the true density distribution of the data. Since no data was available yet, the number was limited to 17 different masks and a total of around 6500 images were taken. Different backgrounds, perspectives and mobile phones were used to generate as many different photos as possible.

Respirator masks Figure 1 shows two different respirators from the manufacturer MSA. Both masks have a built-in valves. The left mask Figure 1(a) belongs to protection class FFP1 and the right mask Figure 1(b) belongs to protection class FFP2. Further it has a built-in odour filter. In addition to MSA also masks of the manufacturer 3M were used. Optically the masks differ on the one hand in the form and on the other hand in the colours. The manufacturer 3M uses different rubber band colours depending on the protection class. In addition, the manufacturer's name and serial number are printed directly on the mask. Beside 3M and MSA there are many other manufacturers. However, 3M followed by MSA are the market leader in dust masks. For this reason, the two manufacturers were used in the work.

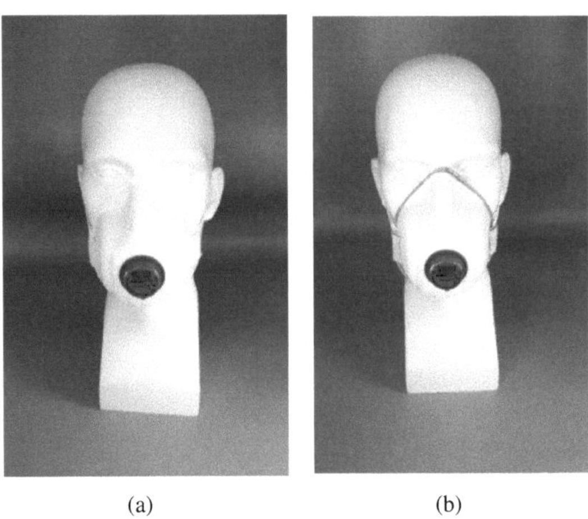

(a)   (b)

Figure 1: Examples of the training data. Shown are two masks of the provider MSA with different protection classes. The left mask belongs to protection class FFP1 and the right one belongs to protection class FFP2.

2.2   MobileNet

MobileNet is a pre-trained network available in Keras and can be used for transfer learning. Keras is a so-called high-level neural network API that runs on top of Tensorfow. It is based on the programming language Python. With the release of Tensorfow 1.4 Keras became a fixed part of the Tensorflow Core API [6].

MobileNet has been specially designed in 2017 for mobile applications due to its low storage capacity and low latency time. In MobileNet, the principle of separated convolution is used. There are two types of convolution. One is spatial separable convolution and the other is depthwise separable convolution. The architecture of the network is based on depthwise separable convolution, which explains the low storage capacity. The convolution consists of a depthwise convolution and a pointwise convolution. During the depthwise convolution the depth is not changed. For each chanel a different filter is used.The convolution is illustrated in Figure 2. A 12x12x3 image is folded with three separate 5x5x1 kernels. A filter folds one channel at a time. The resulting 8x8x1 matrices are merged to an 8x8x3 large image. Then follows the pointwise convolution with a 1x1 large kernel. The depth of the kernel corresponds to the output of the first convolution so in this example it would be 3. If you iterate with the 1x1x3 kernel through the 8x8x3 image, you get an output of size 8x8x1. You can repeat this process as often as you want and then combine the output to get the desired depth. This type of factorization has the effect of saving an extremely large amount of computing time and also reducing the model size [1].

For the transfer learning it has to be evaluated during the training, how many layers should be retrained to achieve the best results.

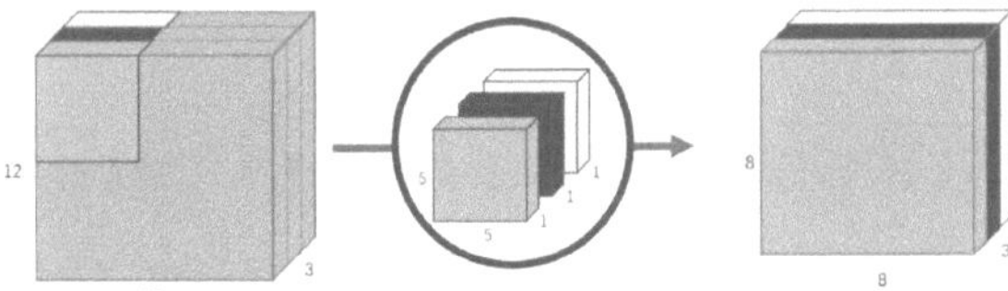

Figure 2: Depthwise Separable convolution with 3 kernels for transformation of an 12x12x3 image into an 8x8x3 feature map [2].

2.3   Azure

Azure is a cloud computing platform from Microsoft that has been officially available since early 2010. This platform offers a variety of different resources, which can be hosted for a fee. Resources include virtual machines. These exist in many different forms with different memories and operating systems. Among other things, there are also virtual machines that provide a graphics card (GPU). This also includes the DataScientist Virtual machine, that offers an NVIDIA Tesla K80 graphics card. The virtual machine was used to train the MobileNet.

In addition to virtual machines, other resources like an App Service and Azure Functions were used. WebApps can be hosted in an Azure App Service. The WebApps can be scaled as needed to deliver the desired performance. A WebApp was used to predict the masks based on the trained MobileNet.

Azure Functions can be used to develop event-driven, serverless compute services that execute code on demand. Thus, one can execute a section of code in response to various events. Such events include HTTP triggers and timers. A function is called as an endpoint and can be addressed via it. This has the advantage that resources are only used when they are actually needed. Thus, the application does not run permanently and costs can be reduced.

Moreover, Blob Storages were in use most of the time. A blob of storage is a scalable cloud storage for any type of data. This was used to store for the pictures and the trained neural networks [3].

3   Results and Discussion

The results of the work are presented in the following section. First the result of the neural network is explained and then the integration into the App.

3.1   Training results

The MobileNet was trained with nearly 4500 images and additionally validated with 1200 images. The quality was then determined with 787 test images. Moreover, Data augmentaion was used to extend the data set and reduce overfitting. The existing images are changed in different ways. This can be different mathematical operations, like flipping or rotating an image or adding noise. This should make the network more robust and better generalizable.

The best result that was achieved with MobileNet is a test accuracy of 0.98 and a loss of 0.079. The loss was computed with the categorical crossentropy function. In total, the network on the 787 test images only determined 11 masks incorrectly. The faults of the network are shown in a Confusion Matrix in Figure 3. The Matrix shows which mask has been classified incorrectly and which has been classified correctly. On the x-axis, you can see the masks determined by the mesh. For this purpose, the y-axis depicts the correct masks, i.e. the ground truth. This means that on the main diagonals the correctly classified masks can be seen. All fields next to of the main diagonals are errors. In the illustration you can see that 11 masks are misclassified.

During the training it turned out that the best results could be achieved by retraining the last 32 layers. All other weights have not been changed. For this reason, the training time can be reduced.

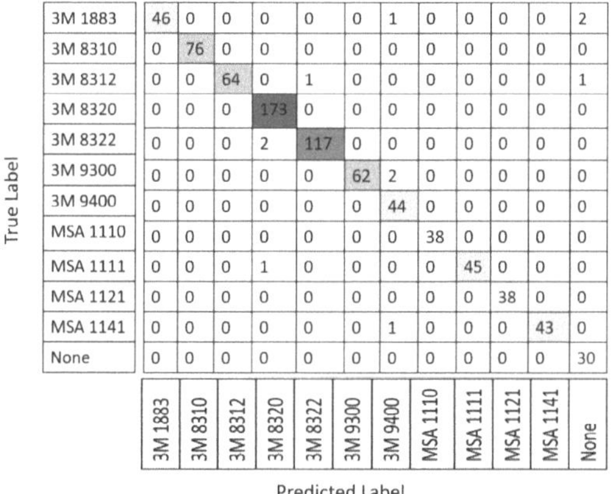

True Label \ Predicted Label	3M 1883	3M 8310	3M 8312	3M 8320	3M 8322	3M 9300	3M 9400	MSA 1110	MSA 1111	MSA 1121	MSA 1141	None
3M 1883	46	0	0	0	0	0	1	0	0	0	0	2
3M 8310	0	76	0	0	0	0	0	0	0	0	0	0
3M 8312	0	0	64	0	1	0	0	0	0	0	0	1
3M 8320	0	0	0	173	0	0	0	0	0	0	0	0
3M 8322	0	0	0	2	117	0	0	0	0	0	0	0
3M 9300	0	0	0	0	0	62	2	0	0	0	0	0
3M 9400	0	0	0	0	0	0	44	0	0	0	0	0
MSA 1110	0	0	0	0	0	0	0	38	0	0	0	0
MSA 1111	0	0	0	1	0	0	0	0	45	0	0	0
MSA 1121	0	0	0	0	0	0	0	0	0	38	0	0
MSA 1141	0	0	0	0	0	0	1	0	0	0	43	0
None	0	0	0	0	0	0	0	0	0	0	0	30

Figure 3: Confusion Matrix shows the result of test images with the trained MobileNet. The x-axis shows the masks predicted by the net and the y-axis shows the ground truth.

3.2　WebApp

In order to be able to use the network from remote and make predictions about an image, a web service was created using Flask, which is used for communication between frontend and backend [5]. The backend contains the Python code, which loads the model and sends an image through the network to get a prediction for the image. The frontend is a HTML website that uses jquery to send a post to the backend. The Flask service was containerized using Docker and then deployed to an Azure App Service [4]. Docker is free software used to isolate applications using container virtualization. A Docker container contains the application and all the required resources from the host system. It is a lightweight, standalone software package that contains the code, system libraries, and all the settings needed to run the software.

Deploying the application to Azure makes it accessible from anywhere and runs permanently. In addition, the application can be scaled horizontally and vertically. The complete technology stack is shown in Figure 4.

In the WebApp, it is possible to upload an image and then classify it. The selected image is base64 encoded and then sent to the Flask server. The addressed method decodes the image again. The previously trained model is already loaded from the Azure Blob Storage. The results are then displayed on the user interface.

Figure 4: Technology stack for creating the Azure WebApp [3]-[7].

3.3　Integration into the SSC App

The app is based on Ionic/Cordova, which makes it possible, to use only one code base for different systems, i.e. Android and IOS. The app is implemented in TypeScript and the interfaces in HTML.

Two basic approaches are available to integrate the fully trained neural network into the app. It is possible to load the network directly onto the mobile phone and thus use the application online, or the network is located on a server and addressed via an interface. This means, however, that access would only be possible with an existing internet connection. Since the network is loaded directly onto the mobile phone, this is only possible if the network is not too large in means of memory. MobileNet is well suited for this purpose. However, the integration directly into the device varies depending on the operating system. Since the app has only one code base for Android and IOS, the local integration would be very complex and time consuming and it would result in having two different apps at the end. Thus, the network was addressed by interface and is not stored on the device.

The functionality for the classification of a mask was implemented in the Azure WebApp and must now be addressed from the app via Representational State Transfer (REST) API. The captured image is base64 encoded and then sent via HTTP Post to the Azure App Service, which expects an image and processes it through the trained network. The three highest probabilities are returned to the app and presented to the user. After receiving the labels, the user must confirm whether the classification is correct or incorrect. If the classification is correct, it is possible to search for the

Dräger equivalent to the classified competitor product. However, if the user confirms that the mask was incorrectly classified, a search for a Dräger product is not possible. Instead, a pop-up window appears, giving the user tips on how to photograph the mask in proper way. The user is also informed that the misclassified image will be loaded into an Azure Blob Storage for later retraining the network. In the background, a HTTP Post is sent to an Azure Function that is triggered by an HTTP trigger and uploads the image to a blob. The collected misclassified images in the blob are intended for later training and should be used for that purpose.

The whole process described above is shown at following diagram.

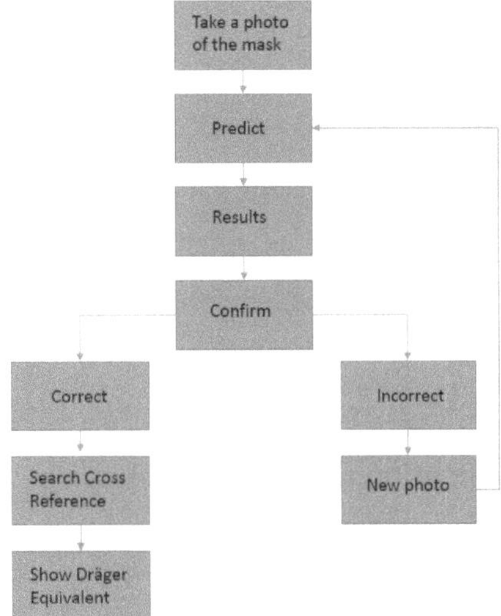

Figure 5: Flow chart of the classification process inside the SSC App.

3.4 Discussion

In the beginning there were only a very small number of dust masks available. MobileNet achieved the best results with 17 masks. However, the application will be expanded in the future. In total there are about 100 dust masks to be classified. But first the competing products have to be purchased in order to collect enough pictures.

Since the recognition of masks is still limited to a few products, the recognition task in the SSC App is hidden. Thus, it is only available for selected users, who are trained where to find the application. In the future, the detection will be designed for more masks, the application is obviously integrated so that every user can test it.

For this work various resources were used by Azure. In total, the costs cumulated to almost 400 euros. The biggest cost factor is the virtual machine with about 245 euros. The machine was used to train the neural network. Another cost factor with only 65 euros are the rented storage facilities. All other resources have only a small influence on the total costs. In general, the use of Azure has proven to be extremely helpful. The large number of different resources could be easily rented and used. Furthermore, the expenses are often much lower than hardware resources.

4   Conclusion

The result is an integrated function for the classification of dust masks in an existing app. A total of 17 different dust masks can currently be classified. All the user has to do is take a photo of the mask in the app and then click on the button to predict the mask. The user is always shown the three most likely masks. If the classification is correct, the user can display the equivalent Dräger product. The images of incorrectly classified masks are stored in a cloud memory for later training purposes. The classification task in the app is easy to use because of a simple and understandable design. Furthermore, the function is flexibly integrated and can therefore also be adapted according to requirements.

In the future the classification for more masks is to be extended.

Acknowledgement

The work has been carried out at Drägerwerk AG & Co. KGaA and supervised by the Institute of Medical Informatics, Universität zu Lübeck.

5   References

[1] Howard, A. G., Zhu, M., Chen, B., Kalenichenko, D., Wang, W., Weyand, T., et al. (2017). *Mobilenets: Efficient convolutional neural networks for mobile vision applications*. arXiv preprint arXiv:1704.04861.

[2] Chi-Feng Wang. *A basic introduction to separable convolutions*. Available: https://towardsdatascience.com/a-basic-introduction-to-separable-convolutions-b99ec3102728 [last accessed on 2019-12-05].

[3] Microsoft Azure. *Azure. Erfinden mit dem Ziel im Blick*. Available: https://azure.microsoft.com/de-de/ [last accessed on 2019-12-05].

[4] Docker. *The Docker Enterprise Difference*. Available: https://www.docker.com/ [last accessed on 2019-12-05].

[5] Kushal Das. *Introduction to Flask*. Available: https://pymbook.readthedocs.io/en/latest/flask.html [last accessed on 2019-12-05]

[6] Keras io. *Keras Documentation*. Available: https://keras.io/ [last accessed on 2019-12-05]

[7] Python.org *Python* Available: https://www.python.org/ [last accessed on 2019-12-05]

Automatic multi-object organ detection and segmentation in abdominal CT data

Oliver Mietzner [1] and Andre Mastmeyer [2]

[1] Medical Informatics, Universität zu Lübeck, oliver.mietzner@student.uni-luebeck.de
[2] Digital Health Management, Hochschule Aalen, andre.mastmeyer@hs-aalen.de

Abstract

The ability to generate 3D patient models in a fast and reliable way, is of great importance, e.g. for the simulation of liver punctures in a virtual reality simulation. The aim is to automatically detect and segment abdominal structures in CT scans. In particular among the selected organ group, the pancreas poses a challenge. We use a combination of random regression forests and U-Nets to detect bounding boxes and generate segmentation masks for five abdominal organs (liver, kidneys, spleen, pancreas). Training and testing is carried out on 50 CT scans from various public sources. The results show Dice coefficients of up to 0.71. The proposed method can theoretically be used for any anatomical structure, as long as sufficient training data is available.

1 Introduction

Virtual reality (VR) based simulation of interventions for training and planning is slowly gaining importance in clinical teaching and routine. VR methods can be used for various tasks, ranging from training scenarios for the medical student and staff, to individual patient related simulations of planned operations [1], [2], [3], [4]. The necessary individual patient models should be available fast and be accurate to guarantee a plausible simulation.

The first step in producing patient models is to acquire patient image data. This data will be used for the following steps in the modelling process. It is desirable to use high-quality image data, because inaccuracies will be carried over to the resulting model. The next step is the coarse localisation of the organs inside organ-specific volumes of interest (VOI) to simplify their subsequent segmentation.

The aim of our study, is to automatically detect bounding boxes for organs in abdominal CT data by a learning-from-example method. The detected bounding boxes are then used for segmentation. Detection of the abdominal organs can be a challenge, because of the variety of shapes these organs can have. The intensity based features of these organs are also challenging, because the intensity range of neighbouring organs often overlaps. We detected five abdominal organs (liver, right kidney, left kidney, pancreas, spleen), that are commonly used in simulations.

We used a random-forest method, that is developed by Criminisi et al. [5] to automatically detect the bounding boxes of abdominal organs. The method uses random-regression-forests to predict the location of organ bounding boxes in CT data. Furthermore, we used the U-Net, proposed by Ronneberger et al. [6], to automatically segment organs. The training and testing is carried out on a database of 50 abdominal CT scans, that had to be segmented beforehand.

2 Material and Methods

The database, that is used for training and testing, consisted of 50 CT scans from different sources. The data sets contained only abdominal CT data. The sets also varied in quality and field of view, to capture the variability of organs and scans. The scans are not only different in the number of slices (64-861), thickness of slices (1-5 mm) and field-of-view, but also in image noise. This variability in the datasets is important, to ensure robustness vs. typical inference factors of clinical image data during training and application. All five target organs (liver, pancreas, left kidney, right kidney, spleen) are included in the scans.

Only a small fraction of the segmentation maps contained all five target organs, thus manual structure segmentation was necessary frequently.

2.1 Definition of ground truth bounding boxes

The segmentation maps associated to the CT scans are the basis for the extraction of ground truth bounding boxes to be learned. A bounding box can be described as a cubical polyhedron that completely encloses an object.

The corresponding organ bounding boxes of each of the scans can be created by scanning the segmentation maps for coordinate direction extremes of label occurrence. To create a three-dimensional bounding box, for each coordinate direction (x, y, z), we iterate slicewise through the segmentation map and save the extreme limits with an object label.

2.2 Training of the models

Our method is composed of two different machine learning approaches, random regression forests (RRF) and U-Nets. The RRF's are used to detect organ bounding boxes in CT data. The U-Nets make use of the detected bounding boxes and segment the organ contained in them.

2.2.1 Training random regression forests for bounding box detection

The decision trees used in RRFs split by minimising variance, then each leaf node outputs the mean of all label values in the node. We use RRFs to determine the location and extent of abdominal organs [5]. The main difference between random classification forests and RRFs, is the type of output that is predicted. While classification forests try to categorize objects, regression forests predict continuous values. Regression forests partition the data into manageable chunks to predict average values. As seen in Fig. 1, the method expects CT scans and ground truth bounding boxes as input.

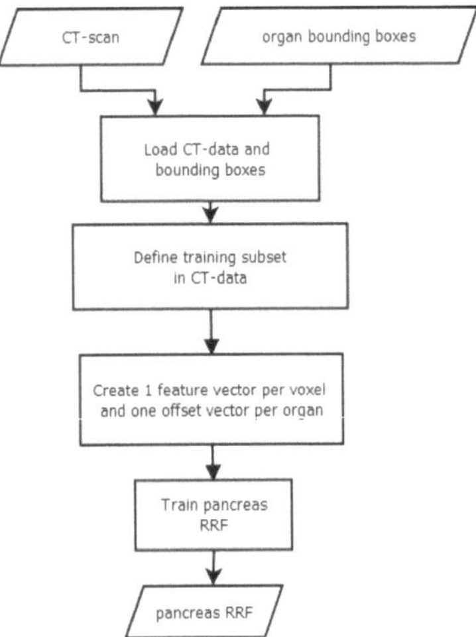

Figure 1: Training of a RRF: The inputs for the training process are CT scans and ground truth bounding boxes of the targeted organs. We create one feature vector and one offset vector for each voxel that is part of a predefined medial cylinder subset in the scan [5]. The trained RRF is able to predict the offset between a voxel and an organ bounding box.

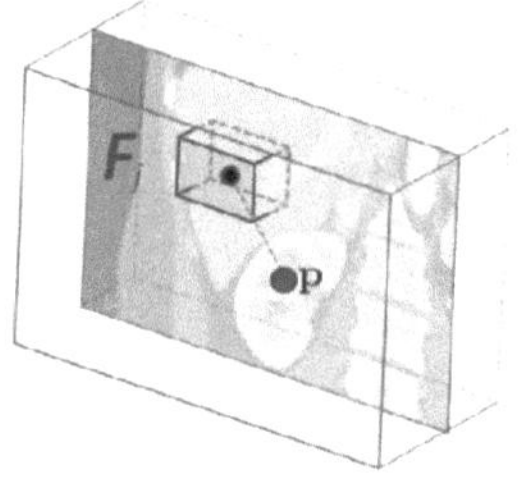

Figure 2: Example feature box: The feature box F_j is generated in correlation to the current voxel and calculated the mean value of a 3D image section [7].

A three-dimensional bounding box b_c of an organ c can be described by using a six-dimensional vector $b_c = (b_c^{Left}, b_c^{Right}, b_c^{Anterior}, b_c^{Posterior}, b_c^{Head}, b_c^{Foot})$ with coordinates in mm [5]. We run over all voxels $p = (x_p, y_p, z_p)$, which are within a specified radial distance (r = 5 cm) from the scan medial axis. The distance d between such a voxel and each of the bounding box walls, can be calculated by using $d(p) = (x_p - x^{Left}, x_p - x^{Right}, y_p - y^{Anterior}, y_p - y^{Posterior}, z_p - z^{Head}, z_p - z^{Foot})$ and is saved as the offset-vector to be learned. In contrast to Criminisi et al. [5], we use only 50 feature boxes, that are evenly distributed on three spheres (r = 5 cm, 2.5 cm, 1.25 cm) to generate the input feature vector. The feature boxes F_j are intended to capture the spatial and intensity context of the current voxel. For this purpose, the mean intensities of the feature boxes are calculated and saved in the feature vector. An example feature box is shown in Fig. 2. While training, the RRF learns the distance vector (later: output) to the reference bounding box using the feature boxes.

2.2.2 Training of a U-Net for semantic segmentation

The training data for our U-Net consists of the expert segmentations and ground truth bounding boxes, as schematised in Fig. 3. The bounding boxes (VOIs) are then used to locally extract the intensity and label data from the CT scans and their corresponding label maps. As input, the U-Net receives a VOI from the intensity data, while the same region within the label data is connected to the output. We use the U-Net architecture proposed by Ronneberger et al. [6], which consists of nine layers with four down- and up-scale steps. The network was trained using batches of size 15 over 50 epochs. In addition, Adam optimization and a cross entropy loss function were used. We trained one U-Net for each organ, using a ReLU activation function.

2.3 Application of the models

In the first step, the organ specific RRFs predict the organ bounding box candidates. Then, a distance vector is selected by majority voting and converted into a six-dimensional vector to describe the final organ-specific bounding box. Now, the U-Net model gets the corresponding bounding box as input. Since the U-Net only accepts

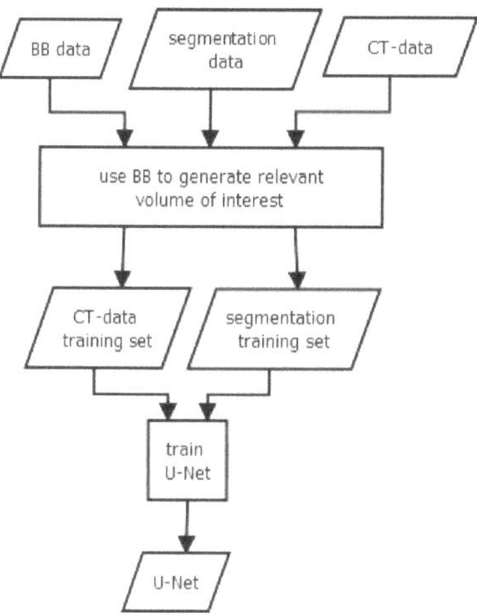

Figure 3: Training of a U-Net: The inputs for the training process are ground truth bounding boxes (BB), CT scans and their corresponding segmentation maps. The box is applied to the CT- and segmentation data to extract the relevant image region. Inside the organ VOIs, the segmentation is learned. The process results in an organ-wise U-Net, that can segment image regions.

a static number of voxels, the BB is resized to a fixed size. The U-Net uses the data contained inside the given bounding box, to segment the corresponding organ. The output is a segmentation map of the full target organ.

2.4 Evaluation

We used a five-fold Monte Carlo cross-validation based on a 30:20 (train:test) data split. The target bounding boxes extend along the three axis x, y, z. Based on the proposition by [5], the quality, of the prediction can be determined, by measuring the distance, between the position of the detected bounding boxes and the position of the ground truth bounding boxes. The distance is equal to the absolute value of the difference E between a ground truth coordinate GT and a predicted coordinate $Pred$, given in (1).

$$E = |GT - Pred|. \tag{1}$$

The resulting segmentations of the combined application are compared against the reference segmentations by using the Dice coefficient [8].

3 Results and Discussion

Table 1 shows the bounding box localisation error in mm, for our version of the method proposed in [5]. We compared our results to the results in [5], except the pancreas, which was not examined by Criminisi et al. [5].

All five organs produced similar results. The left kidney achieves the best value with only 17.3 mm difference between predicted and ground truth bounding box, while the liver produced the highest results with 23 mm. The

overall detection accuracy could not reach the results of [5], with results that deviated with up to 6.3 mm. Nevertheless the standard deviation (SD) was lower than in [5]. The results could also be calculated for each individual axis. This showed that the z-axis produces the biggest errors.

The boxes we estimated are comparable to the results shown in [5]. The loss in accuracy can be accounted to a fewer number of training samples. The use of a wider variety of scans may help to detect organs with varying forms. While the extent of the kidneys and spleen were similar in different patients, the extent of the liver and pancreas may vary, resulting in a bigger localisation error. Especially the prediction along the z-axis is not accurate enough and resulted in boxes that are displaced. This can be attributed to deviations in the resolution of the scans. Due to misalignment, some segmentation masks are cut off, leading to parts of some organs missing. The standard deviations are too high, indicating highly varying results.

Table 1: Comparison of bounding box localisation error (LE) in mm, for our version of the method proposed in [5].

Organ	LE ± SD	LE ± SD ([5])
liver	23.0±8.2	15.7±14.5
kidney (right)	19.1±9.1	16.1±15.5
kidney (left)	17.3±8.2	13.6±12.5
spleen	20.6±7.6	15.5±14.7
pancreas	22.2±7.7	-

Table 2: Comparison of [3] mean - over all per organ results - Dice coefficients and standard deviation (SD) of our general method vs. other one organ-focused methods.

Organ	Dice ± SD	Other Dice Coefficients
liver	0.71±0.31	0.97 [9]
kidney (right)	0.55±0.34	0.91 [10]
kidney (left)	0.67±0.32	0.91 [10]
spleen	0.71±0.26	0.94 [11]
pancreas	0.32±0.28	0.74 [11]

Table 2 shows the Dice coefficients achieved by our method for all five target organs. We compared our values to studies, that tried to segment the same organs with automatic methods.

The best results show up for the liver and spleen. Both organs achieve a dice coefficient of 0.71. The segmentation of the pancreas achieved the lowest dice coefficient with only 0.32. The standard deviation is similar for all organs ranging from 0.26 to 0.34. Varying results can be attributed to different target organs. Especially the pancreas is a challenge, because of its shape and intensity similarity to surrounding tissue. The right kidney is slightly worse vs. left. This could be attributed to the also larger localisation error from Table 1. The position errors of the right kidney BBs seem to be affected by the vicinity of the liver. An example segmentation is presented in Fig. 4.

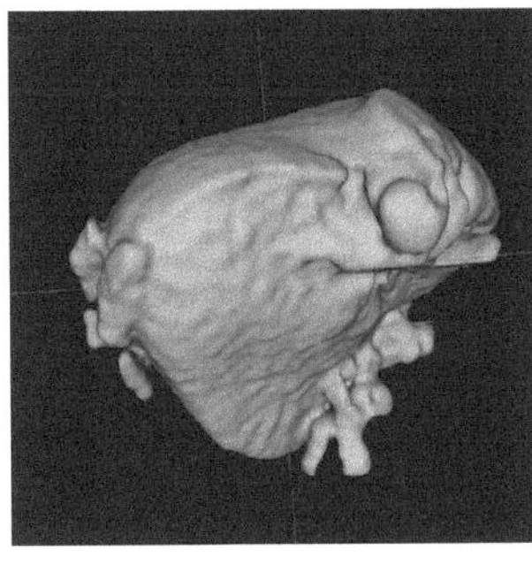

(a)

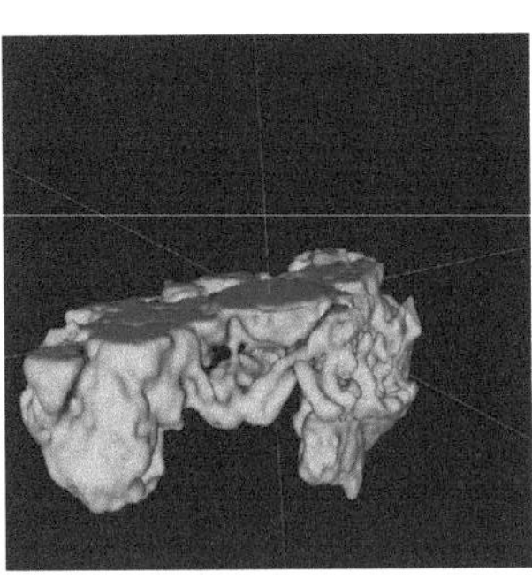

(b)

Figure 4: Example segmentations: (a) A good segmentation of the liver, with some leakage into surrounding structures. (b) A poor segmentation of the pancreas. The bounding box was shifted along the z-axis. This cut off the upper half of the pancreas, resulting in an incomplete segmentation. The segmentation also included parts of surrounding structures

Segmentation of unseen patient image data scan takes between 30 seconds and one minute depending on the size of the data, i.e. on an Intel-i7 processor with a NVIDIA GTX 1050 GPU.

4 Conclusion and Future Work

We were able to detect organ bounding boxes automatically. Nonetheless, the overall positioning and extension of the predicted bounding boxes have to be improved to ensure a satisfying segmentation.

We used predicted bounding boxes for a completely automatic method. Future works should focus on the accuracy and robustness of the BB detection.

Acknowledgements

This work was carried out at the Institute for Medical Informatics at the Universität zu Lübeck. DFG-MA 6791/1-1, Nvidia GPU grant, EXPLOR program by Stiftung Kessler + Co. für Bildung und Kultur.

5 References

[1] A. Mastmeyer, D. Fortmeier, and H. Handels, "Random forest classification of large volume structures for visuo-haptic rendering in CT images." p. 97842H, 2016.

[2] A. Mastmeyer, M. Wilms, D. Fortmeier, J. Schröder, and H. Handels, "Real-time ultrasound simulation for training of US-guided needle insertion in breathing virtual patients," in *Studies in health technology and informatics*, vol. 220. IOS Press, 2016, p. 219.

[3] A. Mastmeyer, M. Wilms, and H. Handels, "Interpatient respiratory motion model transfer for virtual reality simulations of liver punctures," *Journal of World Society of Computer Graphics - WSCG*, vol. 25, no. 1, pp. 1–10, 2017.

[4] ——, "Population-based respiratory 4D motion atlas construction and its application for VR simulations of liver punctures," in *SPIE Medical Imaging 2018: Image Processing*, vol. 10574. International Society for Optics and Photonics, 2018, p. 1057417.

[5] A. Criminisi, D. Robertson, E. Konukoglu, J. Shotton, S. Pathak, S. White, and K. Siddiqui, "Regression forests for efficient anatomy detection and localization in computed tomography scans," *Medical Image Analysis*, vol. 17, no. 8, pp. 1293 – 1303, 2013. [Online]. Available: http://www.sciencedirect.com/science/article/pii/S1361841513000029

[6] O. Ronneberger, P. Fischer, and T. Brox, "U-Net: Convolutional networks for biomedical image segmentation," *Med Image Comput Comput Assist Interv – MICCAI 2015*, p. 234–241, 2015.

[7] A. Criminisi, D. Robertson, O. Pauly, B. Glocker, E. Konukoglu, J. Shotton, D. Mateus, A. Martinez Möller, S. G. Nekolla, and N. Navab, *Anatomy Detection and Localization in 3D Medical Images*. London: Springer London, 2013, p. 198. [Online]. Available: https://doi.org/10.1007/978-1-4471-4929-3_14

[8] A. A. Taha and A. Hanbury, "Metrics for evaluating 3D medical image segmentation: analysis, selection, and tool," *BMC med imaging*, vol. 15, no. 1, p. 29, 2015.

[9] H. Meine, G. Chlebus, M. Ghafoorian, I. Endo, and A. Schenk, "Comparison of u-Net-based convolutional neural networks for liver segmentation in CT," *CoRR*, vol. abs/1810.04017, 2018. [Online]. Available: http://arxiv.org/abs/1810.04017

[10] F. Isensee and K. H. Maier-Hein, "An attempt at beating the 3D U-Net," 2019. [Online]. Available: https://arxiv.org/abs/1908.02182

[11] E. Gibson, F. Giganti, Y. Hu, E. Bonmati, S. Bandula, K. Gurusamy, B. Davidson, S. P. Pereira, M. J. Clarkson, and D. C. Barratt, "Automatic multi-organ segmentation on abdominal CT with dense V-Networks," *IEEE Transactions on Medical Imaging*, vol. 37, no. 8, pp. 1822–1834, Aug 2018.

Fast Pulmonary Fissure Detection in CT Scans using Deep Learning on Point Clouds

Marvin Lee Gillner [1], Lasse Hansen [2], Mattias P. Heinrich [2]

[1] Medical Engineering Science, Universität zu Lübeck, m.gillner@student.uni-luebeck.de
[2] Institute of Medical Informatics, Universität zu Lübeck, {hansen,heinrich}@imi.uni-luebeck.de

Abstract

This work proposes a fast approach for automatic pulmonary fissure detection by using point clouds instead of regular grid data. Determining the fissures' structure and position is an important element of precise automatic pulmonary lobe segmentation. The concept is based on three steps: First, a keypoint algorithm is applied to 3D chest CT data in order to extract significant points. Second, a semantic segmentation is applied to the resulting point cloud using the PointNet architecture. The last step contains a post processing method that fits a smooth and continuous fissure surface to the classified points. This procedure achieves a mean surface error of 5.1 mm over a 7-fold cross validation. The average runtime of the presented method is 1.93 s on CPU and 41.1 ms on GPU. Comparison values of the U-Net are 2.1 mm surface distance, 40.25 s on CPU and 281.1 ms on GPU.

1 Introduction

The lung consists of the right and left lung whereas each of them is parted in pulmonary lobes. The separation lines are called pulmonary fissures. There are five lobar compartments and thus three pulmonary fissures. The left lung is divided by one major fissure named left oblique (major) fissure (LOF). The right lung is composed of two pulmonary fissures, one major and one minor. The right oblique (major) fissure (ROF) separates the lower from the middle part and the right horizontal (minor) fissure (RHF) differentiates the middle from the upper lobe. The main imaging modality for the human lung is Computed Tomography (CT). This is reasoned in the fast application, scanning the whole chest in less than one second, and an isotropic spatial resolution in submillimeter range [1]. Detecting pulmonary fissures is an essential step in order to segment the pulmonary lobes precisely. The assignment of a lesion to the corresponding pulmonary lobe, for example, is a valuable information for the radiologist [1]. Fasten this process could lead either to real-time applications, e.g. during operation, or to usages on slower devices such as tablets.

In this paper, we introduce a method which predicts pulmonary fissures faster than current approaches based on the euclidean domain. This is achieved by using point clouds instead of regular grid data. The point clouds consist of keypoints extracted from a 3D chest CT scan. Other approaches such as FissureNet [2] use Convolutional Neural Networks (CNN) which learn feature extraction and classification simultaneously. But the whole CT data set must be fed to the network and thus it needs a lot of computational operations for calculating pulmonary fissures. By using the PointNet [5], which works on geometric data, only coordinates of the keypoints are passed to the network which attains a significant reduction of calculations and therefore a faster application. However, due to the information loss of the intensity distribution, we expect a decrease in accuracy. Geometric data lie in a non-euclidean domain which includes graphs, point clouds or manifolds and excludes regular grid data such as CT images [3]. The PointNet extracts local and global features and combines them for a label prediction of every keypoint individually. Because the data set has only expert segmentation of the major fissures, the middle and upper lobe of the right lung are combined. So the network is trained with segmentations of four lobes and two major fissures and thus has six labels to predict. Finally, the keypoints which belong to the LOF and ROF are extracted from the output and post processed to a continuous prediction of both pulmonary major fissures. In order to evaluate this method, a comparison with the U-Net [4], a well established CNN for medical image segmentation tasks, in respect of accuracy and inference time is performed.

2 Material and Methods

The presented concept consists of three major steps: Step one deals with the extraction of keypoints from a 3D Chest CT data set. Step 2 is based on the classification of keypoints belonging to the pulmonary fissures predicted by the PointNet. Step three is composed of fitting a smooth surface to the labeled keypoints. A graphical flowchart is presented in Fig. 2.

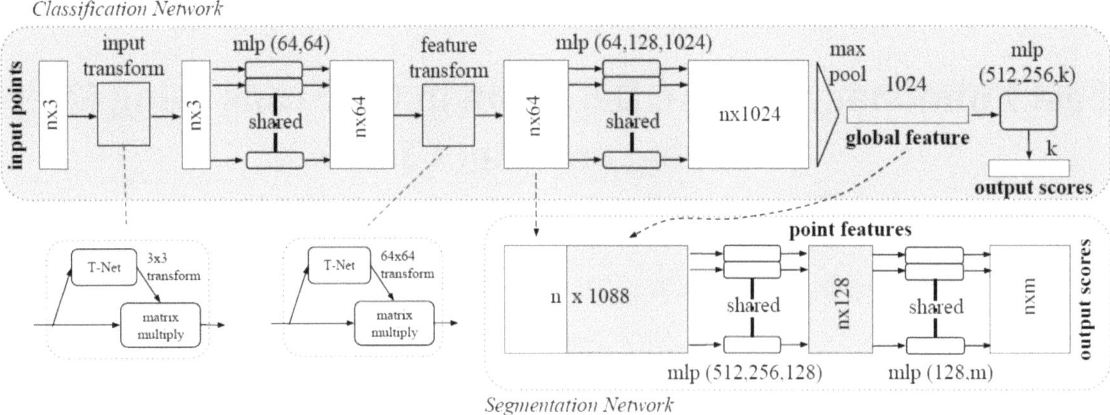

Figure 1: This figure illustrates the architecture of the PointNet which gets 3D point clouds as input and generates either class labels for the entire cloud or for each point independently. This figure is taken from the original paper [5].

2.1   Dataset

The data set for training the PointNet and U-Net consists of 14 3D chest CT scans which were originally used for pulmonary inspiration to expiration registration [6]. Keypoints were extracted with an algorithm based on the Foerstner operator which is further specified in the original paper [7]. The ground truth is based on expert segmentations of the whole lung as well as the major pulmonary fissures ROF and LOF. For the PointNet, lobe segmentations are generated from these expert segmentations. Therefore six different labels exist for training the network. Based on this, every keypoint was assigned to a label which builds the ground truth.

2.2   PointNet

The PointNet [5] describes a neural network developed by Charles R. Qi and Hao Su for mainly object classification, part and semantic segmentation. It works on 3D non-euclidean geometric data which includes for instance point clouds and meshes. The architecture is shown in Fig. 1. The input consists of n points each with three coordinates of x, y and z. First, the input coordinates are aligned with a 3×3 transformation matrix predicted from a mini transformation network called T-Net. Afterward, a shared multi-layer perceptron (MLP) is applied which increases the number of features from 3 to 64. The following T-Net applies a 64×64 transformation matrix to the features to align them in feature space. This T-Net is equal to the T-Net used for input alignment before. A Frobenius norm is applied to the T-Net in order to get the transformation matrix as orthogonal as possible. The current 64 features are not only stored as local features but also further processed to global features as well. The latter are generated through a shared MLP which raises the number of features from 64 to 1024 (64,128,1024) in combination with the following max pooling across points. In this paper, the segmentation part of the PointNet is needed and thus will be further explained. Because segmenting each point needs a combination of local and global context, both recently calculated features were concate-

nated for each point of the point cloud. Afterward, a shared MLP reduces the number of features from 1088 to 128 (512,256,128) and again from 128 to the number of different labels m. Finally, every point of the cloud gets a score for each label which leads to the final prediction.

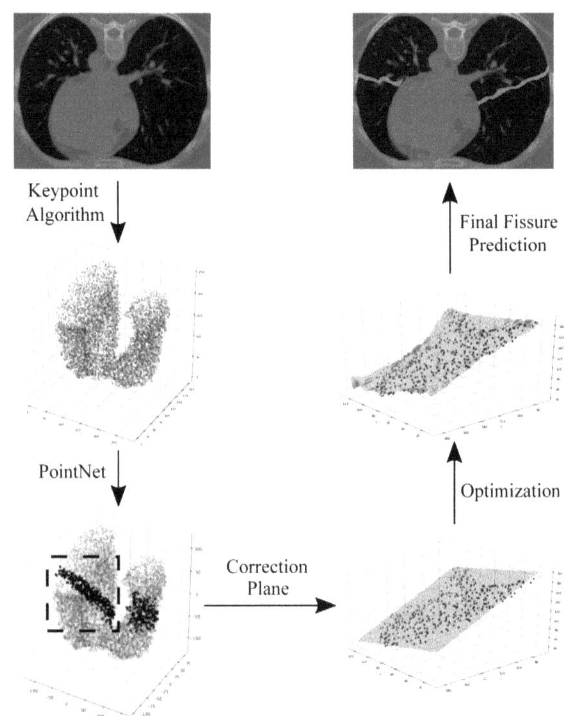

Figure 2: Flowchart visualizing the processing of CT data to obtain pulmonary fissure segmentations.

2.3   Training/Testing Details

During training, one pair of inhalation to exhalation CT scans was left out and used for evaluating the network. Consequently, all available data sets were split into 12 sets for training and 2 for testing at each cross validation step. To find the best combination of hyperparameters, a parameter search has been done over a 7-fold cross validation. As

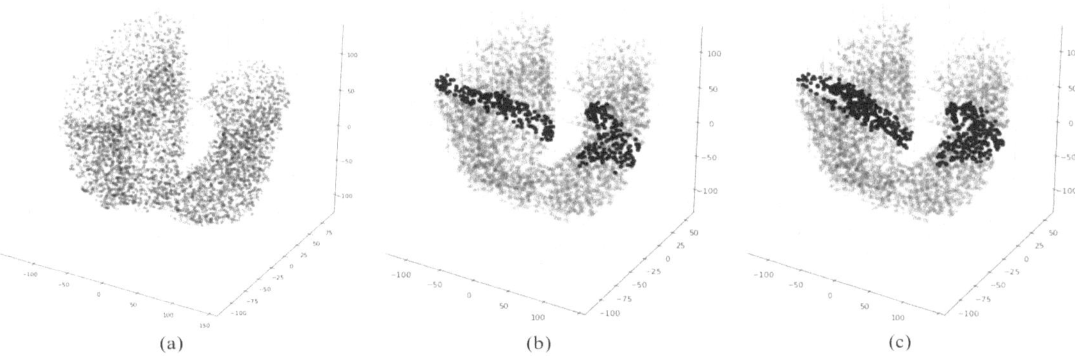

Figure 3: This figure shows the input to the PointNet (a), the output as label predictions (c) and the ground truth (b). In the output and ground truth, the pulmonary fissures are highlighted darker.

a result, the PointNet was trained for 110 epochs with a batch size of 4 randomly chosen point clouds. We used an Adam optimizer with a weight decay of 0.006. The learning rate was set to 0.003 with an exponential scheduler of factor 0.98. Moreover, the Frobenius norm, which is applied to the transformation matrix of the T-Net, was added to the loss with a weighting factor of 0.001. In order to give higher weight to the underrepresented labels belonging to the pulmonary fissures, two measures were implemented. First, we used a label weighting developed especially for point clouds. The class weights w_c were calculated with the formula $w_c = 1/log(1.2 + p)$ where p describes the probability of occurrence [8]. As a second step, we triplicate the number of labeled pixels belonging to the original fissure segmentations.

The hyperparameters of the U-Net are as follows: 190 epochs, a batch size of 2, an Adam optimizer with the learning rate of 0.002 and a weight decay of 0.002 as well as an exponential scheduler of factor 0.99. The U-Net was trained with the CT data as input and the fissure expert segmentations as ground truth.

All tests have been conducted on the following hardware:

- CPU: Intel Xeon(R) CPU W3520 @ 2.67GHz

- GPU: GeForce GTX 1080 Ti

2.4 Post Processing from Keypoints to Pulmonary Fissure

The PointNet's output includes label predictions for each keypoint of the point cloud. By extracting keypoints corresponding to the ROF and LOF separately, a correction plane for each pulmonary major fissure can be calculated. Both correction planes lie within the xy-dimension and reflect global position and orientation of the predicted keypoints. The correction planes are set up as a xy-grid with z-coordinates as a function value at each grid position. An Adam optimizer is used to update z-coordinates at each optimization step. We define a loss L consisting of both, L1 norm in z-direction of the current grid position z_i and the nearest predicted keypoint z_{kpt} as well as 2D bending en-

ergy regularization E_b [9]:

$$L = \sum_{i=1}^{n} \frac{\mid z_i - z_{kpt} \mid}{n} + \lambda E_b. \qquad (1)$$

With i as the current grid position index, λ as a weighting factor and the regularization term:

$$E_b = (\frac{\partial z_{xy}}{\partial x \partial x})^2 + (\frac{\partial z_{xy}}{\partial y \partial y})^2 + 2(\frac{\partial z_{xy}}{\partial x \partial y})^2. \qquad (2)$$

The L1 norm pushes the grid function value in direction of the nearest keypoint. The regularization contains prior knowledge about the fissure. The second derivatives in the regularization term of (2) ensure that the predictions proceed smoothly with a moderate curvature.

The Adam optimizer was initialized with a learning rate of 1 and applied for 50 epochs. The regularization term was weighted with 0.34.

3 Results and Discussion

One point cloud example of 3D CT based keypoints of the human lung which was given to the PointNet as input is shown in Fig. 3 (a). Furthermore, Fig. 3 illustrates the resulting assignment by the PointNet (c) as well as the corresponding ground truth (b). The labels belonging to the pulmonary fissures are emphasized darker. The average Dice score for the predicted labels over a 7-fold cross validation achieved 76.6 %. For the ROF a Dice score of 51.2 % was calculated and the LOF reached a score of 39.9 %.

The lower Dice scores of the pulmonary major fissures are reasoned in their thin structure and the ensuing underrepresentation of the labels. Despite we implemented measures against this issue, it is still difficult to learn. Nevertheless, the result presented in Fig. 3 shows that the distribution of the labeled keypoints belonging to the ROF and LOF fits well to the ground truth.

The calculation of the correction plane in combination with further optimization led to the final result shown in Fig. 4 (b). For evaluation and comparison with the results of the U-Net, a surface error from the estimated fissure to the

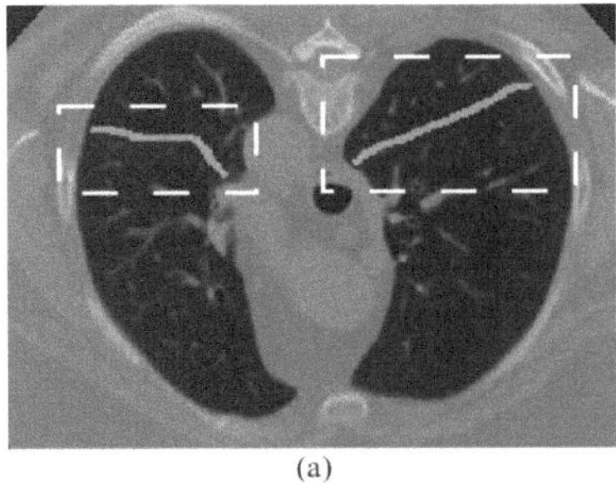 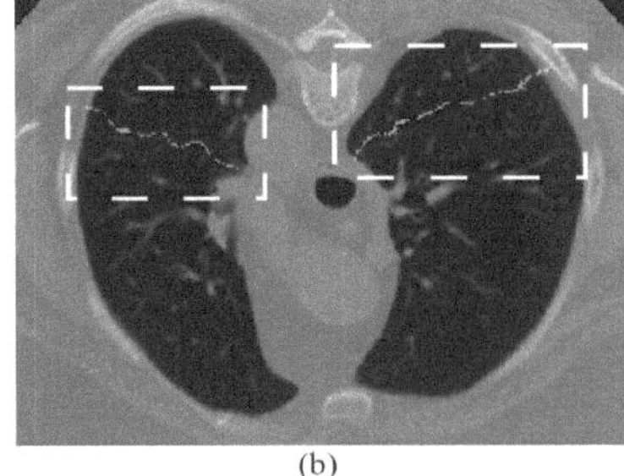 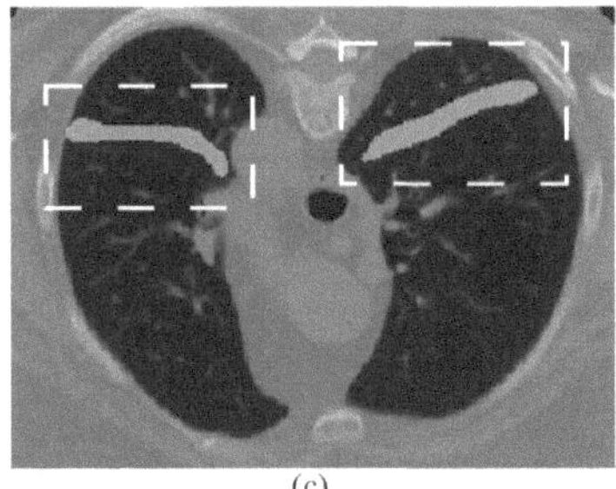

(a)   (b)   (c)

Figure 4: This figure illustrates the comparison of the results from the PointNet including subsequent post processing (b) and the U-Net (c) with the ground truth (a).

ground truth was calculated. Therefore the euclidean distance from every point of the estimated fissure to the nearest point of the ground truth was computed. The average value resulted in 5.1 mm, whereas the U-Net performs a mean surface distance of 2.1 mm. The output of the U-Net is shown in Fig. 4 (c).

Fig. 4 illustrates that neither the PointNet nor the U-Net predicts the fissure exactly. However, this is due to the low quantity of 14 data sets. But it should be emphasized that the position and orientation of the predicted fissures fit very well with the ground truth. So we can conclude, that the presented method is working but needs improvement regarding accuracy. The fact that the U-Net's output has a lower surface error was expected because we lost the information of the intensity distribution.

The runtime of the PointNet with following post processing is 1.93 s on CPU and 40.1 ms on GPU. Reference values of the U-Net are 40.25 s on CPU and 281.1 ms on GPU. Analyzing the results, the PointNet is about 21 times faster on CPU and 7 times faster on GPU.

4 Conclusion

This method was developed for applications which need to be fast. In summary, the presented results allow concluding that this objective is achieved while taking further steps in order to improve accuracy. One of these is the expansion of the data set for training and testing. More 3D CT data sets with corresponding expert segmentations have to be created. Further improvement could be achieved by training an additional network which learns to extract keypoints from the CT data. It would replace the Foerstner operator that provides general keypoints and specify this task by learning to select points according to the pulmonary fissures. This would not only generate more keypoints for preventing underrepresentation but also improve the post processing part by pushing the optimization parameters into a more reliable direction.

Acknowledgement

The work has been carried out and supervised by the Institute of Medical Informatics, Universität zu Lübeck.

5 References

[1] E. M. van Rikxoort, B. van Ginneken, M. Klik, and M. Prokop, "Supervised enhancement filters: Application to fissure detection in chest CT scans," *IEEE Transactions on Medical Imaging*, vol. 27, no. 1, pp. 1–10, 2008.

[2] S. E. Gerard, T. J. Patton, G. E. Christensen, J. E. Bayouth, and J. M. Reinhardt, "FissureNet: A deep learning approach for pulmonary fissure detection in CT images," *IEEE Transactions on Medical Imaging*, vol. 38, no. 1, pp. 156–166, 2019.

[3] M. M. Bronstein, J. Bruna, Y. LeCun, A. Szlam, and P. Vandergheynst, "Geometric deep learning: going beyond euclidean data," *IEEE Signal Processing Magazine*, vol. 34, no. 4, pp. 18–42, 2016.

[4] O. Ronneberger, P. Fischer, and T. Brox, "U-net: Convolutional networks for biomedical image segmentation," *Medical Image Computing and Computer-Assisted Intervention*, vol. 9351, pp. 234–241, 2015.

[5] C. R. Qi, H. Su, K. Mo, and L. J. Guibas, "Pointnet: Deep learning on point sets for 3d classification and segmentation," *Proceedings of the IEEE Conference on Computer Vision and Pattern Recognition*, 2017.

[6] R. Castillo *et al.*, "A reference dataset for deformable image registration spatial accuracy evaluation using the COPDgene study archive," *Physics in Medicine and Biology*, vol. 58, no. 9, pp. 2861–2877, apr 2013.

[7] M. P. Heinrich, H. Handels, and I. J. A. Simpson, "Estimating large lung motion in COPD patients by symmetric regularised correspondence fields," in *Lecture Notes in Computer Science*.   Springer International Publishing, 2015, pp. 338–345.

[8] A. Dai *et al.*, "Scannet: Richly-annotated 3d reconstructions of indoor scenes," *Proceedings of the IEEE Conference on Computer Vision and Pattern Recognition*, 2017.

[9] D. Rueckert *et al.*, "Nonrigid registration using freeform deformations: Application to breast mr images," *IEEE Transactions on Medical Imaging, vol. 18, no. 8*, 1999.

7

Medical Electronics

Analysis of current regulation circuits for solenoid valves in a bus communication architecture for a dialysis machine

David Alejandro Briceño Peniche [1], Armin Rieß [2], Christian Schleicher [2], Tim Jürgens [3]

[1] Biomedical Engineering, Luebeck University of Applied Sciences, david.alejandro.peniche@stud.th-luebeck.de

[2] R&D Extracorporeal Blood Treatment , B. Braun Avitum AG,{armin.riess; christian.schleicher}@bbraun.com

[3] Institute of Acoustics, Luebeck University of Applied Sciences, tim.juergens@th-luebeck.de

Abstract

As part of an architectural modification towards a fully modular structure of a dialysis machine, this study evaluates changes to the electronics control. A bus node communication is intended to be implemented in a dialysis machine. The creation of the nodes implies modifications in some control circuits, including the valve control. Solenoid valves are widely used in the machine and one goal of the architectural modification is to limit their power dissipation. An analysis of the current valve control and regulation system characteristics was accomplished and a new design was developed. The work principle of both, the current and the new design is described. A software simulation is performed for the new design and a comparison of solenoid currents of the two approaches is done. The novel proposal limits the current through the solenoid and requires 36% less space in the circuit board.

1   Introduction

A dialysis machine is intended to apply and monitor treatments that help to maintain the homeostasis for patients with established renal failure. The most common applied therapy is hemodialysis, in which the blood from the patient is extracted and pumped through a dialyzer, which acts like a filter and separates the metabolic waste products from the blood. The machine adds electrolyte and bicarbonate to the dialysis fluid, which removes substances from the blood. The removal of substances from the blood takes place by diffusion and convection processes, while other substances are added to the blood at the same time. The metabolic products are transported to the dialysate outlet and the treated blood is returned to the patient [1].

In order to reduce the harming risk to the patient, several control and monitoring systems are included in the machine. As part of the control mechanisms, solenoid valves, driven by the main control system, are used to regulate the flow of the different fluids that intervene in the therapy. Because of production purposes, the machine consists of different racks.

It is intended to implement a bus communication between the racks in order to reduce the amount of wiring connections and the length of the wires, introduce the possibility of a modular architecture for the machine and reduce costs. The bus communication requires the creation of nodes that will abstract the individual physical connections by logical bus gateways that shall transfer their information to the main application controller(s). The present architecture of the machine includes a centralized assembly that contains all the electronics components that participate in the control, regulation and connection of the different elements that constitute the machine, including the main controller and the system for the valve control and regulation.

Due to production purposes, every rack will represent a node. The number of valves in every node varies due to the particular function of the rack. The present design is capable to control all the valves included in the machine but, due to the change in the architecture, it is necessary to analyze if the present regulation system allows space optimization in the printed circuit board (PCB) of the node.

The solenoid valves are components that have a power consumption related to the characteristics of the valve, such as the size, the current required for the opening and if the valve is normally open or closed. A solenoid consists of a coil that carries the current, an iron shell and a movable pole that acts as a working element [2]. The valve is activated by a continuous voltage that causes a current flow through the coil. During a constant voltage application, the temperature of the coil increases due to power consumption and dissipation in form of heat.Eliminating unnecessary power consumption in solenoid coils saves money, improves valve performance, and extends valve operating life [3].

The current regulation through the solenoid coil gives design benefits and for the particular case of a hemodialysis machine, where a considerable number of valves are present, the implementation of the valve current regulation is a must. One of the most spread techniques to avoid the excessive power dissipation in the coil is the use of a pulse width modulation (PWM) signal. As an initial phase, the coil is supplied with a constant voltage, in order to overcome the inertia and open the valve. After the constant voltage, a PWM signal is supplied in order to decrease the current through the coil until the holding current necessary to maintain the valve opened is reached. Fig. 1 shows the ideal waveform for the PWM current regulation where the start time corresponds to the opening phase and the hold time illustrates the PWM supply to hold the valve opened.

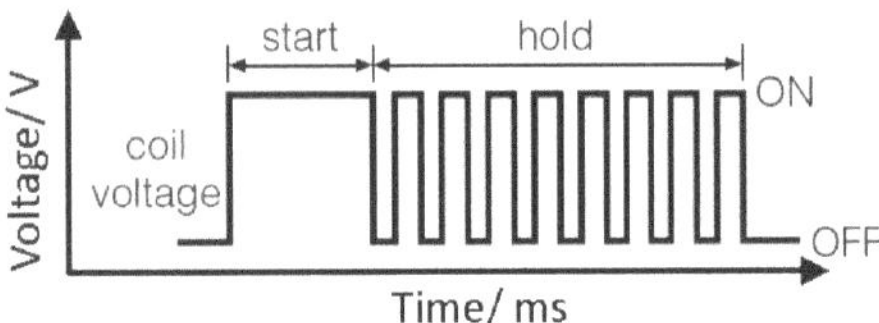

Figure 1: Valve voltage supply [4].

The PWM should have a high frequency, in the range of kHz, in order to avoid the closing of the valve during the off time of the cycle. The current supply to the coil is determined by the duty cycle of the PWM and can be calculated for the specific holding current of the valve. The electrical characteristics of the valve, such as the minimum opening time and voltage, or the holding current can be found in the datasheet provided by the valve manufacturer.

Fig.2 illustrates the change in the coil current in accordance with the change in the voltage supply. The peak of current is reached during the opening time, then the current decreases by the action of the PWM. The ripple observed during the hold time is caused by the change in the PWM.

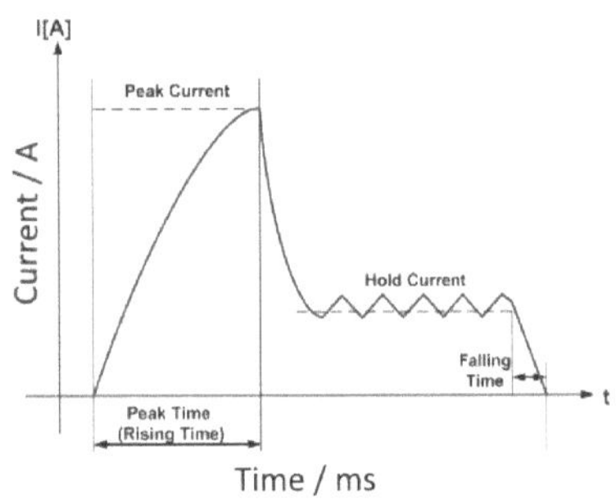

Figure 2: Current through the coil [5].

The aim of the current regulation is to supply the solenoid with a minimal initial current of $255\ mA$ for a minimum time of $30\ ms$ and then decrease the current to $220\ mA$ applying a PWM to maintain the solenoid position. The nominal electrical power of the valve is $9.6\ W$. The values were taken from the datasheet of the valve.

2 Material and Methods

2.1 Present-day design

Fig. 3 shows the circuit diagram of the valve control in its present form. The solenoid is represented by an inductor and a resistance in series (R_V) corresponding to the coil resistance. The switching of the valve is done by a metal oxide semiconductor field effect transistor (MOSFET) that grounds one of the terminals of the solenoid, while the other terminal is connected to the power supply, causing that

when the MOSFET gate is supplied with a positive voltage the valve is energized. When the solenoid is de-energized after being activated and because of the inductive nature of the coil, it acts as a power source that produces a current flow in the opposite direction. When this occurs, the coil supplies a voltage that could be several times higher than the initial voltage supplied to the coil. A protection diode is used, in order to avoid damage to the controller circuit.

The present architecture of the machine centralises the valve control in a module composed by a complex programmable logic Device (CPLD) which is responsible for activating and regulating the valve during the opening time and for supplying the PWM signal during the hold phase. The CPLD switches every valve according to the indications coming from the responsible main controller. In order to send the status of the valve to the CPLD, the MOSFET switches the valve to ground. An additional shunt resistor is employed for the current measurement. The amplifier compares the shunt voltage drop with a set threshold voltage, providing feedback to the CPLD, which generates the PWM, if the current exceeds certain limit. The switching signal comes from the CPLD and the output valve_comp is used to regulate the switching.

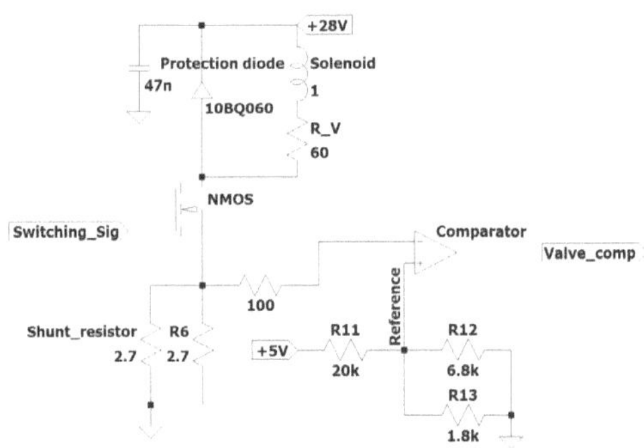

Figure 3: Circuit diagram of the current design for the valve control made using LT-Spice.

2.2 Oscillator approach

2.2.1 Proposed design

The voltage waveform in Fig.1 can be seen as the combination of two independent signals: a constant pulse with a duration that corresponds to the opening time, and a square wave signal with a duty cycle and frequency according to the coil current requirements. Based on the idea that the waveform is obtained by a combination of a single pulse at the beginning and a frequency signal that starts after the pulse, an oscillator and a monostable multivibrator can be used and combined logically, to create a system which produces an output resembling the desired waveform but without the CPLD. This *oscillator approach* is shown in Fig.4. The oscillator used to produce the square signal is built with an amplifier. This oscillator is also implemented in other control systems in the machine. The monostable multivibrator was implemented with an integrated circuit 74HCT123PW, which is a retriggerable monostable multivibrator with output pulse width control by an external resistor and an external capacitor [6].

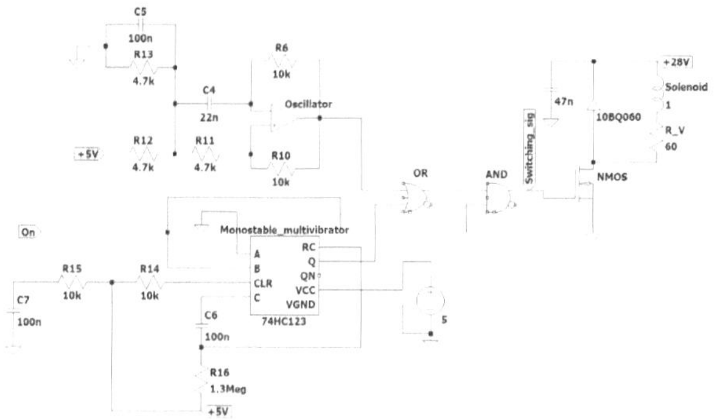

Figure 4: Circuit diagram of the controller using the external oscillator made using LT-Spice.

The pulse and the square signal are combined by an OR logic. The main controller sends a high level signal to activate the valve. This *On*-signal is used as a trigger for the multivibrator and in junction with the combined signal it is also used to produce the switching signal via AND logic. The switching waveform can (according to [7]) therefore be expressed by

$$Switching\_Sig = (On) \wedge (Pulse \vee PWM). \quad (1)$$

2.2.2 Duty cycle calculation

The voltage across the solenoid is given by

$$u = L\frac{dI}{dt} = L\frac{\Delta I}{\Delta t}. \quad (2)$$

Due to PWM, when the solenoid is grounded, meaning that the valve is on, (2) can be expressed as

$$\Delta I = \frac{U_S - (R_V + R_{ON})I_V}{L}\Delta t_{on}. \quad (3)$$

Where U_s is voltage supply, R_V is solenoid resistance, R_{ON} stands for drain-source on resistance of the MOFSET and I_V is the current through the solenoid. The fraction of the period where the valve is energized is represented by Δt_{on}. Equation (4) expresses Δt_{on} in terms of the duty cycle D and the signal frequency for the total *On*-period T and the pulse active time PW.

$$\Delta t_{on} = PW = D \cdot T = \frac{D}{f} \quad (4)$$

When the valve is off, the variation in the current ΔI is defined by (5), where U_D is the voltage drop in the diode and U_V is the voltage in the valve. Δt_{off} represents fraction of the period where the valve is de-energized which is defined by (6).

$$\Delta I = \frac{U_D + U_V}{L}\Delta t \quad (5)$$

$$\Delta t_{off} = T - PW = T - D \cdot T = T(1-D) = \frac{1-D}{f} \quad (6)$$

Solving the system for (3) and (5) by incorporating (4) and (6), the duty cycle is given by

$$D = \frac{U_D + R_V I_V}{U_V - R_{ON}I_V + U_D}. \quad (7)$$

The duty cycle therefore depends on the solenoid characteristics, concretely in the holding current, the coil resistance and the voltage supply. The calculated duty cycle is approximately 50% for a current of 220 mA, a resistance of 60 Ω and a supply of 28 V. The values for U_D and R_{ON} are given by the chosen components. For this application, the MOSFET SUD15N06-90L and the diode 10BQ060 models were used. Calculation of the duty cycle is shown in (8).

$$D = \frac{0.42V + (60\Omega \cdot 220mA)}{28V - (0.065\Omega \cdot 220mA) + 0.42V} \quad (8)$$
$$= 0.4794 \approx 50\%.$$

Because of the duty cycle $\approx 50\%$, a relaxation oscillator was used to realize the square waveform with equal on- and off-times in the implementation. The period duration of the pulse was increased to 60 ms, to take into account the coil tolerance. For multiple valves the intention is to use one oscillator for all the valves, but every coil should have a particular opening constant signal.

The nominal working voltage for the valve is 27 V, but due to tolerances and the use of a backup battery, the voltage supply varies between 21 and 29 V. In (8), the value of 28 V is used, due to that is the average voltage provided by the power supply.

3 Results and Discussion

Fig. 5a shows the voltage switching signal obtained in the simulation of the diagram in Fig. 4. Due to the high frequency of the PWM, the square signal can not be distinguished without making a close up to the trace. Fig.5b shows the resulting current through the coil that reaches a peak of around 450 mA and with the application of the PWM, drops approximately until 250 mA.

In comparison with the present approach, the oscillator circuit requires more elements, but those elements use less space in the PCB. With the available offer in the market for programmable logic, other devices like a field programmable gate array (FPGA), or even a micro-controller, can provide the switching signal. The use of a more complex component could be justified by the amount of valves to control and by the use of the same components for different proposes.

For the case of controlling multiple valves, the oscillator design could be implemented one time and the same signal could be used as the PWM for all the valves with the same holding current. In contrast, the initial pulse has to be generated for every valve. If the activation of the valves occurs in a sequential way, meaning that two valves are not activated at the same time, the same monostable generator could be used for a group of valves with the same characteristics. If

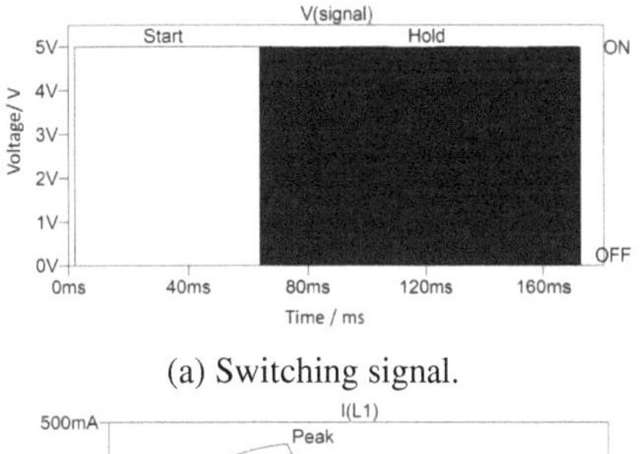

(a) Switching signal.

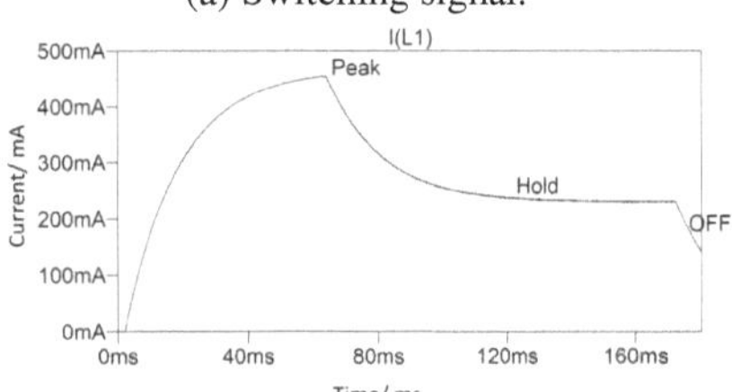

(b) Current measured through the coil.

Figure 5: Simulation results.

the activation frequency is not high enough to re-trigger the initial pulse for an extended period, the *On*-signals for the different valves could trigger the same monostable generator and supply the initial pulse to the solenoids connected to the system.

For this particular project, the oscillator that was built with the amplifier gives a grade of standardisation in the electronics. The oscillator is used for other signal acquisitions in the machine, but if the calculated duty cycle is different than 50%, a new circuit configuration needs to be applied. Due to that the oscillator circuit does not measure the current in the solenoid, the calculation of the duty cycle value shall take in consideration the tolerances of the coil and the components that constitute the oscillator and the multivibrator, in order to provide the opening voltage and to maintain the holding current above the minimal value. The measurement of the voltage supply could solve the problem that the calculation of the duty cycle changes due to the variation of the voltage supply. Towards the goal of standardisation, the monostable multivibrator can be built using NOR gates; the combination of the signal can be achieved with NOR logic, that in conjunction with the NOR multivibrator, will give a similar output. This NOR approach uses one integrated circuit per valve, in addition to the general oscillator. The *On*-signal is a low level signal (proposed in [7]), according to the logic in (9).

$$Switching\_Sig = \overline{\overline{(On)} \vee \overline{(Pulse \vee PWM)}} \quad (9)$$

4 Conclusion

The space-intensive, cost-intensive CPLD in the present electronic design was replaced by a logical combination of monostable multivibrator and oscillator. This produces the same desired switching signal for the valves, but uses 36% less space in the node for a single valve application. The power dissipation is achieved due to the current limitation. Because of the bus communication, every node will include a FPGA that will transmit information to the main controller. It would be beneficial if the target integrated circuit has enough free I/O, such to that the hardware configuration of the FPGA could be used to generate the switching signal, in combination with the circuit diagram presented in Fig. 3. The use of FPGAs in the proposed design reduces the amount of space in the PCB dedicated for the controlling of the valves and maintains the current measurement, but at the same time, requires to use software resources additional to the necessary routines for the bus communication. Due to the different number of valves in every node and the limitation of the oscillator approach for multiple valves, the most suitable option is the use of the FPGA in combination with the present measurement circuit.

Acknowledgement

The work has been carried out at B.Braun Avitum AG, Melsungen in the Research and Development department.

References

[1] *Dialog iQ dialysis machine. instructions for use SW 1.03xx EN*, B.Braun Avitum AG, 2018.

[2] *Solenoids. glossary. resources for engineers*, Johnson Electric, 2020. [Online]. Available: `https://www.johnsonelectric.com/-/media/files/resources-for-engineers/solenoids/solenoids-glossary.ashx?v=2005cd535a1c49678f48fcefd1b7c6f9`, [last accessed on 2020-01-21)].

[3] S. Glaudel, *Optimizing power management in solenoid valves*, ASCO Valve. [Online]. Available: `https://www.emerson.com/documents/automation/white-paper-asco-optimizing-power-management-en-us-134560.pdf`, [last accessed on 2020-02-04)].

[4] *Low consumption solenoid valves*, SMC Energy Saving Products, 2020. [Online]. Available: `https://www.smc.eu/portal_ssl/WebContent/corporative/content/energy_saving09/documentation/ES-Low_pamphlet_en.pdf`, [last accessed on 2020-01-21)].

[5] J. Lee, J.-H. Lee, and M.-S. Kim, "Analysis of hydraulic characteristics of two solenoid-driven injectors for CRDi system", Korea Institute of Science and Technology Information, 2011.

[6] *Dual retriggerable monostable multivibrator with reset, models:74hc123 and 74hct123 . Product data sheet*. Nexperia, 2015.

[7] M. Knaust, "Entwicklung eines busfähigen Gateways zur Ansteuerung von Dialyse Aktorik und Sensorik", B.S. thesis, Universität Kassel, 2015.

Development of a GUI-based Interface for the Optical Determination of Blood Flow Velocity

Anup Dixit [1], Reza Behroozian [2], Stefan Müller [3]
[1] Biomedical Engineering, Lübeck University of Applied Science, anup.dixit@stud.th-luebeck.de
[2] Graduate School in Biomedical Engineering, Lübeck University of Applied Science, reza.behroozian@th-luebeck.de
[3] Medical Sensors and Device Laboratory, Lübeck University of Applied Science, stefan.mueller@th-luebeck.de

Abstract

Blood flow velocity is an important medical parameter particularly during major operations such as open-heart surgery or when a patient is under anesthesia. There are various methods to measure blood flow velocity. It can be measured using ultrasound sensors, electromagnetic flow probes and optical methods[7]. The aim of this project was to create an interface (GUI) that will help in making fast and accurate measurements to check the function of an optical catheter for the determination of the blood flow velocity while logging the measurement data securely. This allows for easier post-processing of the respective data. The GUI initiates handshaking between the test setup and the PC while also allowing the user to interact and control the connected devices.

1   Introduction

Measuring the Blood flow speed can be a difficult task as many steps are required to conduct a measurement. The test setup as seen in Fig. 1 shows the overview of the interfaced devices. It mainly consists of a *spectrometer*, a *centrifugal pump* and *solenoid valves* to form a *artificial blood circulating system*. This is a difficult aspect, as the blood should be introduced into the system *void* of air bubbles. The test setup should be purged of the blood, once a test cycle has been conducted. This means that there are several parameters that need to be considered while controlling the connected devices and also carefully introducing the blood into the system, concurrently conducting the measurements. This will not only make the process slow but prone to human errors. Due to these difficulties there was a need to be able to control the whole test setup through a central command centre. A GUI would allow us to interact with the respective devices through a microcontroller which is the central unit. As depicted in Fig. 1, The blocks outside the dotted line represent the main instruments and devices of the test setup itself while blocks inside the dotted line are the components that have been developed during the course of this project.

This system would control each device in the test setup while governing the conduction of the measurements to check the reliability and function of an optical catheter for determination of the blood flow velocity. By creating a GUI, it will allow a single point where the user can manage the tasks of the devices and communicate with them simply through a click of a button. This will not only speed up the process of conducting the measurements but also create an environment where they are done fast and accurately.

2   Material and Methods

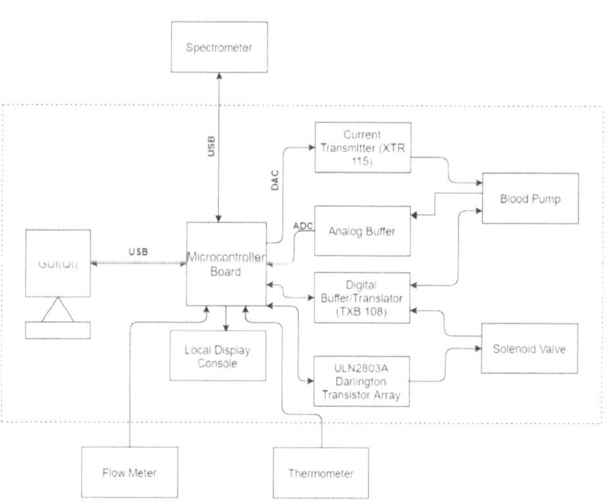

Figure 1: The major devices and components interfaced in the test setup

As shown in Fig. 1, the devices are to be interfaced with the GUI via a microcontroller board. The microcontroller being the core of the project.

Within the framework of this project, the required hardware was developed to provide the device connections to the PC via microcontroller The software that was needed to be developed for the GUI was programmed in Qt [6]. It was one of the first items developed along with the microcontroller programming that was coded in *Arduino C*. The required hardware such as a current transmitter were also designed for interfacing between the microcontroller and the setup devices like circulation pump and solenoid valves.

2.1 Graphical User Interface using Qt

The GUI was developed on an open source environment software called Qt version *5.13.2* [6]. Qt is a widget toolkit that is used to create a GUI as well as cross platform applications. The software is highly robust, and has a long list of widgets and third-party libraries that can be easily implemented [6]. The software is based on the *C++* programming language. This software was chosen to develop the GUI because it has immense documentation [6], and third party support from many developers. It has a healthy online forum and many well written textbooks. The software interacts with the microcontroller through a USB serial connection.

2.2 Microcontroller

The microcontroller is the heart of this project. It serves as a bridge between the developed GUI and the devices of the test setup such as centrifugal pump and the spectrometer. The microcontroller, currently being used in the project is an *Arduino M0 pro*. The development board is based on the Atmel's *SAMD21 MCU*, featuring a *32-bit ARM Cortex M0 core*. It has a operating voltage of *3.3*V, an output current of *7*mA, and a clock speed of *48*Hz [1]. The microcontroller program is developed in *Arduino C*, and run in *Arduino IDE version 1.8.9*. It is connected to the PC using a USB serial connection. As seen in Fig. 1, many devices included in the test setup are connected to the microcontroller allowing it to interface and control them, while conveying measurement data to the PC.

The microcontroller has its own *digital to analog converter* (DAC). The on-board DAC converts a digital signal to an analog signal needed to control the speed of the pump. The microcontroller has an *analog to digital converter* (ADC), used to read the the signals from the pump, which shows the adjusted speed of the blood flow.

2.3 Current Transmitter (XTR 115)

The XTR 115 IC, is a precision current output converter designed to transmit analog *4-20*mA signals. It provides accurate current scaling and output current limit functions. It can be seen in Fig. 2 shows the structure of this IC and a simple current transmitter based on that.

The XTR 115 was selected because it provides the user with transformation of a voltage range to a current range or another voltage range with the desired change in slope and offset. The centrifugal pump requires *4-20*mA to adjust its rpm, where *4*mA corresponds to *0rpm*, and *20*mA corresponds to *6000rpm* [2]. The output of the microcontroller provides the voltage input, and can be seen in the Fig. 2 as V_{in}. Circuit is connected as shown while the resistors R_{in} and R_o adjust the current output of the circuit with desired slope and offset. The output current can be measured from pins 4, and 7. The value of R_{in} can be calculated by the following equation.

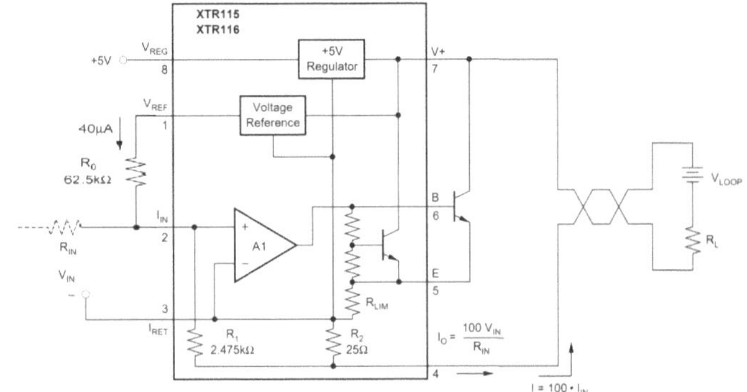

Figure 2: Circuit diagram of the developed Current Transmitter. This uses the XTR 115 IC, to adapt the DAC output of microcontroller to the input of the circulation pump [3]

$$R_{in} = \frac{100 * V_{in}}{I_o} \quad (1)$$

where,
R_{in} is the resistor which determines the slope and the offset, V_{in} is the output voltage of the microcontroller and I_o is the output current of the IC.

2.4 ULN2803A Darlington Transistor Array

As can be observed in Fig. 1, there are solenoid valves in the test setup. The respective valves are elctromechanical, and direct the flow of the blood. These valves require a large input current for their normal functioning. The microcontroller has a comparatively small current output of 7mA [1]. The ULN2803A consists of darlington connected NPN transistors [4]. It has a very high output current drive with a very-low input current. The connected diode is used to suppress the kick-back voltage from an inductive load.

This circuit allows the microcontroller to control the solenoid valves using its on-board pins. The circuit diagram is shown in Fig. 3, where V_{in} is the input from the microcontroller.

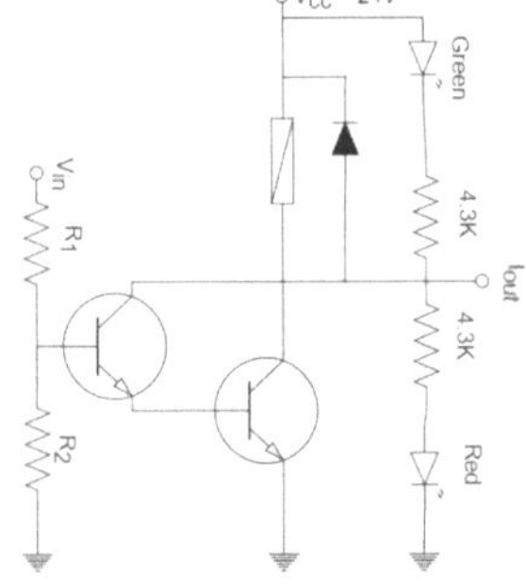

Figure 3: Circuit diagram of the ULN2803A darlington array

2.5   Analog Buffer

The analog buffer is connected between the centrifugal pump and the microcontroller. The main role of this buffer is to act as a protection barrier between the two. If and when a voltage/current surge were to occur it would be able to protect the microcontroller and connected devices from any damages that might occur.

2.6   Local Display Console

Local display console is an LED display, that allows the status of the connected devices to be easily seen. It displays whether the devices are ON, OFF, disconnected or if a fault is to be detected. This is helpful to observe whether the devices are in normal operation and working safely.

3   Results and Discussion

During the course of the project, there were several items that had to be developed such as the GUI, the Arduino C code, the current transmitter, the analog buffer, the digital buffer and the local display console. The results and respective discussion will be explained in detail in the following sections.

3.1   Developed Graphical User Interface

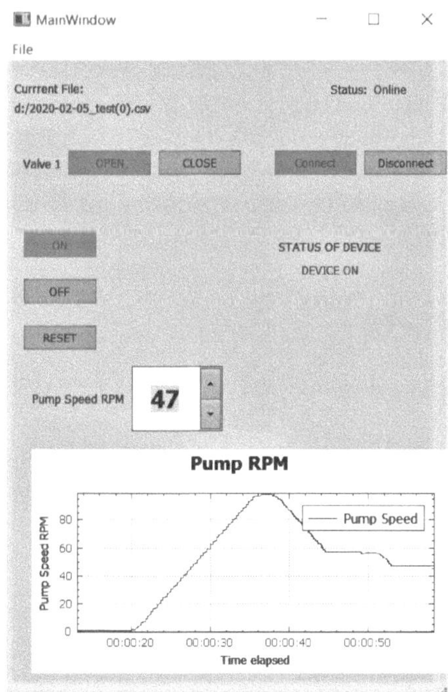

Figure 4: The developed GUI for the Circulation pump

In Fig. 4, the developed GUI can be observed. For the first phase of the project, the functionality is being tested on a development board to simulate the working of the circulation pump. The pump speed test is conducted by varying the brightness of a LED. This is possible since both an LED, and a motor work on the same principle i.e., varying

voltages to increase or decrease the brightness, similarly it will increase or decrease the speed of the motor. The major point to be noted is that the developed current transmitter will convert this voltage to a current and hence allowing it to be interfaced with the circulation pump.

In the top right corner there is the status display of the pump. This will indicate whether the pump is truly online or offline.

This is achieved by initiating the communication between the arduino and the software through the serial connection. The arduino sends a confirmation to the software to indicate successful communication.

It is read by the software and the displays the respective status on the GUI. If by any chance, the serial connection was to be interrupted, then a prompt will appear on the screen to indicate the same. Below the status, there is the *connect* and *disconnect* button. These two buttons will initiate and end the serial communication between the PC and the microcontroller which are connected through the USB serial port. To the left of the GUI, the location and the file name of the *CSV* file to which the measurement data will be stored is located. *CSV* or *comma separated variable*, format was chosen as a file type to store data because it is very trivial to implement while being small in size and easy to parse. An example of the way the data is stored can be seen in Fig. 5. By clicking *file* in the top left corner, a *drop-down menu* will display *new*. By clicking *new*, the data being recorded can be stored in new file.

Val TRNSMTTD by QT	Val RCVD from ARDUINO	Pump Speed RPM	Date & Time:
connect		0	2019.12.04-13:42:43
connect	off	0	2019.12.04-13:42:43
high	off	0	2019.12.04-13:42:46
high	not overload	0	2019.12.04-13:42:46
high	on	0	2019.12.04-13:42:46
high	on	1	2019.12.04-13:42:50
high	on	2	2019.12.04-13:42:50
high	on	3	2019.12.04-13:42:50

Figure 5: Example of the data stored in the CSV file

Below the filename are the buttons that control the solenoid valves. As depicted, it turns the solenoid valves on or off. On the bottom left are the control functionality. As the name suggests the *ON* and *OFF* are the buttons to turn on and off the respective centrifugal pump. The *reset* button below, is there to acknowledge the system that an *overload* has occurred. The status of the the device is displayed on the right of it.

Below the status of the device, a widget known as *spinbox* [6] can be seen. It has small arrows pointing up and down which increase or decrease the respective pump speed.

Below the *spinbox* widget, the real time graph can be observed. This displays the feedback from the circulation pump. The real time plotting in the GUI has been implemented using *Qcustomplot* library [5].

3.2 Developed Electrical Circuit using (XTR 115)

The XTR 115 Integrated Chip plays an important role in the development in the interface between the PC, the microcontroller and the centrifugal pump. As stated in subsection 2.3, the pump needs an operating current of $4\text{-}20$mA[2] to be able to adjust the speed of the centrifugal pump. In table 1 are the comparisons between the theoretical and the measured current output. The tests were conducted by connecting a voltage source to the IC as to simulate the microcontroller output and measure the output current using a multimeter. After testing the circuit it was seen that result was as expected. The measured output values can be seen in Table 1. Similarly, Fig. 6 depicts the linearity between the microcontroller output voltage and the theoretical and real output current that drives the pump.

Table 1: Comparison of Theoretical vs Measured output current of the XTR115

Input Voltage V	Theoretical Current mA	Measured Current mA
0.0000	4.000	4.000
0.4125	6.000	6.002
0.8250	8.000	8.007
1.2375	10.000	10.004
1.6500	12.000	12.003
2.0625	14.000	14.001
2.4750	16.000	16.001
2.8875	18.000	18.000
3.3000	20.000	20.000

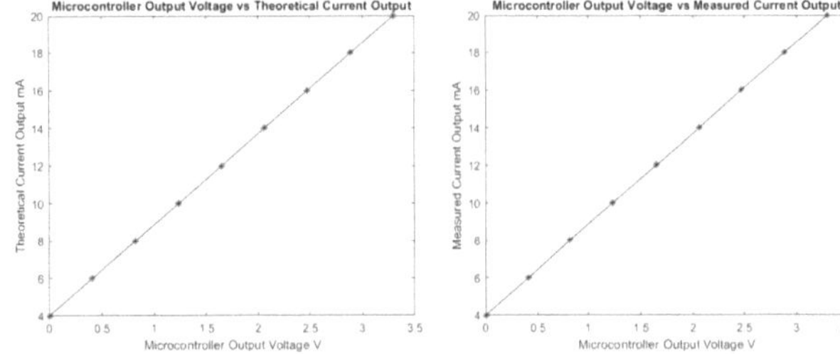

Figure 6: Comparison between the Theoretical and Measured current output

4 Conclusion

Currently the GUI-based interface has been created for the centrifugal pump and the solenoid valves. The functioning of the GUI has turned out as planned at the beginning of the project. The spectrometer is still to be integrated in the GUI. There is still room for improvement. Currently the location where the data file is saved is default. Future improvements will enable the file to be stored in user specified folders or drives. This will allow for flexibility in the way the data is being handled. Currently The number of connected valves is one. There are plans on interfacing more valves using the same developed ULN2803A IC.

The next phase is to interface the spectrometer to the microcontroller and the GUI. This will include logging the measured data in a format that can be read for post-processing, and also displaying the measured light spectrum on the GUI.

In Fig. 7 the developed hardware is seen along with the arduino and the solenoid valve.

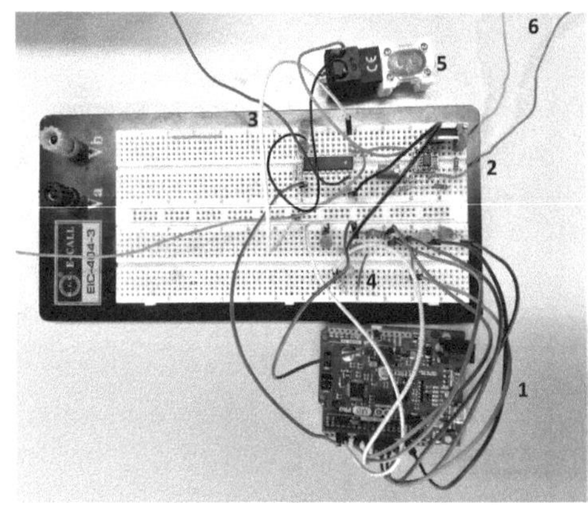

Figure 7: Developed Hardware. 1: Arduino M0 Pro, 2: Current Transmitter, 3: ULN2803A IC, 4: Circulation Pump LED Display, 5: Solenoid valve, 6: Supply Wire for the Circulation Pump

Acknowledgement

The work has been carried out at the Medical Sensors and Devices Laboratory, Lübeck University of Applied Science. I would like to convey my sincere gratitude to Reza Behroozian for his immense support and suggestions during the internship.

References

[1] Arduino. *Arduino M0 Pro*. URL: `https://store.arduino.cc/m0-pro`.

[2] Cole-Parmer GmbH. *ISMATEC*. URL: `http://www.ismatec.de/int_e/pumps/g_gearpumps/bvp_z.htm`.

[3] Texas Instruments. *Texas Instruments*. URL: `https://www.ti.com/lit/ds/symlink/xtr115.pdf`.

[4] Texas Instruments. *ULN2803A Darlington Transistor Arrays*. URL: `http://www.ti.com/lit/ds/symlink/uln2803a.pdf`.

[5] *Qcustomplot*. URL: `https://www.qcustomplot.com/`.

[6] *Qt Documentation*. URL: `https://doc.qt.io/`.

[7] Michael K Pugsley Reza Tabrizchi. "Methods of blood flow measurement in the arterial circulatory system". In: (2000).

Design of a Temperature Control System for the Three-Dimensional Magnetic Particle Spectrometer

Xi Zhang [1], Xin Chen [2], Thorsten M. Buzug [2]

[1] Medical Engineering Science, Universität zu Lübeck, xi.zhang@student.uni-luebeck.de
[2] Institute of Medical Engineering, Universität zu Lübeck, {chen, buzug}@imt.uni-luebeck.de

Abstract

Magnetic particle imaging (MPI) is a new imaging technique that can be used to determine the local distribution of superparamagnetic iron oxide nanoparticles (SPIONs) [1]. Meanwhile, the magnetic characteristics of SPIONs can be measured by using a magnetic particle spectrometer (MPS) [2]. Temperature affects the sensitivity of the magnetic material due to the significant dependence that magnetic characteristics of magnetic material have on temperature [3]. In this project, a temperature control system is designed and assembled to adapt to the 3D MPS. The corresponding software is developed to control the system automatically. The results show that the design has reached its requirements of automatic temperature control and could be used for the magnetic particle spectrometer. There are also recommendations regards of the defects in the experiments following after results analysis and discussion.

1 Introduction

In 2005, Bernhard Gleich and Jürgen Weiznicker presented magnetic particle imaging (MPI) as a new method for medical imaging [1]. MPI can determine the distribution of the magnetic tracer, which typically consists of SPIONs with a diameter of tens of nanometers. MPI takes advantage of the non-linear magnetization characteristic of SPIONs. If an oscillating magnetic field is applied, the nanoparticles exhibit a time-varying magnetization, which will induce a signal with higher harmonics in the receive coils [4]. Temperature affects SPIONs in magnetic fields, which means the higher the temperature, the faster the magnetic moment changes [3]. Briefly said, temperature control is the main significance of this experimental project because SPIONs are sensitive to the change of temperature.

A temperature control unit in the magnetic particle spectrometer (MPS), as shown in Fig. 1, is a cylindrical setup connected to a temperature controller [5]. This unit consists of two concentric cylinders: one cylinder has an outer diameter of 25 mm and a thickness of 1 mm, and the other has an outer diameter of 20 mm and a thickness of 1 mm. The inner tube, where the SPIONs sample are placed, is 18 mm in diameter. There is a 1.5 mm gap—the water flow channel—between these two cylinders. In addition, there are two annular devices with outlets fixed on both sides of the cylinder, through which water circulates. By controlling the temperature of water, the internal temperature of the cylinder can be controlled within a temperature range.

In previous study, in order to manipulate the sample temperature, a prototype of the temperature control system has been implemented to guarantee the system is capable of

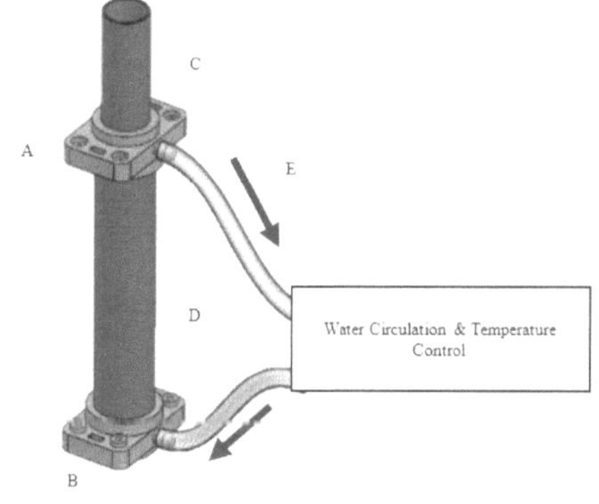

Figure 1: The temperature control unit in MPS. A. Fixture with outlet; B. Fixture with inlet; C. Inner pipe; D. Outer pipe; E. Water tube; The arrows show the direction of the water flow.

changing the thermal equilibrium state inside the sample chamber. However, the setup is not practical [5]. It also failed to control the temperature automatically.

In this project, a setup for automatic temperature control will be designed. First, a hardware design will be determined, which can implement functions such as temperature monitoring, heating, cooling, and circulating water. Second, there will be written software programs to make sure that the temperature can be displayed in real time and controlled at the PC interface. Programs can automatically reach and maintain the required temperature. Later, an separate test of the setup and a test in MPS are performed. The purpose of both experiments was to test the performance of the temperature control system. After a required temperature was set, the system could reach the temperature quickly and remain stable within a target range. This temperature range represents not exceeding 1 °C above or below the target temperature. The test results are displayed graphically after gathering data.

2   Material and Methods

The temperature control system requires rapid heating or cooling process to bring the sample to setting temperature. The target temperature has a working range from 10 °C to 60 °C. This control system includes both hardware and software parts.

2.1   Main Hardware

There are some of the main hardware required in this design: temperature sensor, Peltier Element, microcomputer control chip – Arduino, and relay. In addition, hardware such as water pumps, check valves, water pipes, and wires are needed to form the entire hardware system.

Temperature sensor DS18B20 is a commonly used digital temperature sensor with digital output signal. This sensor has many advantages as small size, low cost, strong anti-jamming capability, and high precision. Importantly, DS18B20 communicates over a 1-Wire bus which requires only one data line (and ground) with a central microprocessor, so only one interface is needed to read temperature parameters [6].

Peltier Element A Peltier element is an electrothermal transducer. When the current flows through a Peltier element, the element produces a temperature difference. Normally, a chiller always needs a refrigerant and most often a compressor, within the Peltier element, only a 12V DC needed to guarantee the performance of the chiller. Moreover, the Peltier element makes it easier to switch in-between cooling and heating sides by simply changing the direction of current. The Peltier Elements used in this project has small size (40 mm * 40 mm), low weight (20 g), and can also easily be set up. The current project uses four Peltier elements for the hardware design, including two for cooling and two for heating.

Arduino Arduino is a flexible and easy-to-use open-source hardware product based on an Atmel AVR microcontroller. The Arduino programming language is simple and intuitive. The Arduino hardware is reliable and durable and can be switched on and off through the control interface to control the system circuit precisely. Here, Arduino detects the temperature via the sensors, and controls the motors to circulate the water and relays to switch on/off the Peltier elements. The current project uses Arduino MEGA 2560.

Relay Relay is an electronic control unit and is normally used in automatic control circuits. It is also an automatic switch that uses a small current to control a large current. It plays a vital role in the safety protection and conversion circuitry in the circuit. The relay has three interfaces: NC (Normally Closed)—denotes a closed contact; NO (Normally Open); and COM (Common). This project uses three 4-channel relays to connect the Peltier elements and DC source.

2.2   Software

The program of this design includes heating program, cooling program, and water mixing program. The room temperature is defined as 23 °C artificially and stored as parameter of the program. The heating program starts operating when the setting temperature is higher than 23 °C; the cooling program operates when the setting temperature is below 23 °C; and when the setting temperature is in between the temperature of cold water and hot water, the water mixing program is started.

Temperature control program mainly uses two logic systems: positive feedback logic system and temperature hysteresis. Positive feedback logic simply relies on the temperature sensor to monitor the temperature and uploads the data to the Arduino. The hardware operates based on a setting temperature. Once it has reached the setting temperature, the heating program automatic stops functioning; once the temperature drops below the setting point, the heating program restarts operating. However, this will cause the relay to switch on and off according to the water temperature fluctuation, which will possibly damage the relay. Thus, a temperature hysteresis is recommended during the process of temperature control. It is the same that when water temperature has reached the setting point, the heating elements stop working automatically. However, with the temperature hysteresis, the heating program only restarts when the water temperature drops 1°C below the setting point, which will maintain the water temperature within a specified temperature range of -1°C to +1°C, and also reduce the frequent relay switches and potential damage.

2.3   The Structure of the Design

The hardware design is shown in the Fig. 2 and the real objects of the hardware in the Fig. 3 below: the left side is the heating unit, the right side is the cooling unit, and the water mixing unit is in the middle.

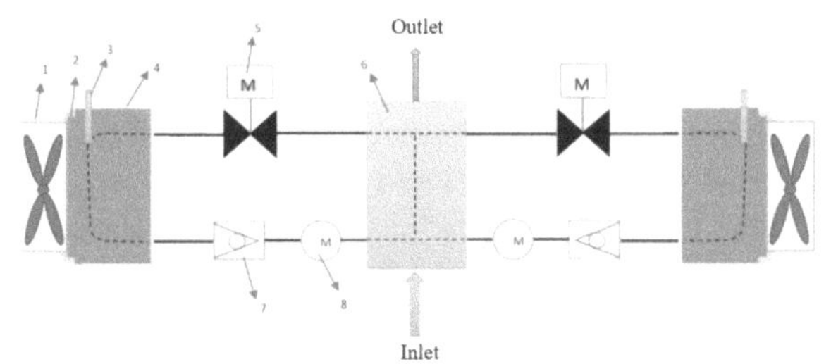

Figure 2: Structure of the hardware. 1. Fan; 2. Peltier Element; 3. Temperature Sensor; 4. Water box; 5. Magnetic valve; 6. Water box for mixing; 7. One-way valve; 8. Water pump

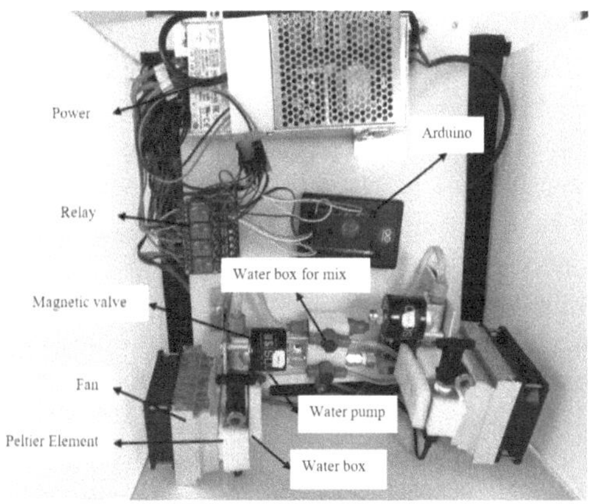

Figure 3: View of the hardware

To allow the sample to reach and maintain the desired temperature, the system circulates water to heat or cool the sample. Whether operate the heating or cooling function is determined by the room temperature. When the setting temperature is higher than the room temperature, the heating unit operates. As shown in the Fig. 2 und Fig. 3, the solenoid valve and water pump on the left side are turned on, and the solenoid valve and water pump on the right side are turned off, which let the water heat up the sample while it flows through. If the setting temperature is below the room temperature, then this system works in the opposite way. If the setting temperature is in between the temperature of the cooling and heating unit, the water-mixing program will start operating. Water with different temperature will be mixed in the intermediate tank to reach the specified temperature quickly.

Two sets of experiments will be performed due to the difficulty to connect the entire hardware to the MPS, which are separate experiments and experiments with MPS. After the data from separate tests shown that the setup meets the requirements of automatic temperature control, the entire hardware system will be connected to the MPS for further testing. Both tests were performed several times, and the program parameters were debugged, and the hardware was optimized to achieve better test results. Due to space limitations, each experiment in the next chapter only shows the results of one of the experiments. In both tests of the hardware design, the thermometer measured the internal temperature of the water box. The setting temperature started at 10 °C and raised 5 °C each time until it reached 60 °C. Each temperature point remained for five minutes to observe temperature fluctuations. In the meantime, the temperature sensor measured the temperature of the water and recorded it.

3 Results and Discussion

3.1 Separate Test of Hardware

As shown in Fig. 4, the solid line represents the temperature in the cold-water section, and the dashed line represents the temperature in the hot-water section. The program doesn't operate when the setting temperature equals to room temperature. When the setting point is above room temperature, heating program will start heating up on the hot-water section. Meanwhile, the valve will only open to allow sample flows through hot-water section to be warmed up. And the cold-water program starts operating when the setting temperature is below room temperature. In the separate test, the minimum requires temperature that the system can reach is 15 °C.

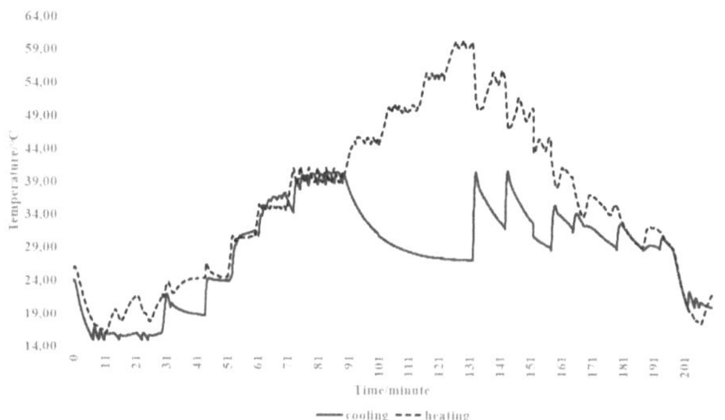

Figure 4: The temperature curve in the separate test

It can be observed from Fig. 4 that the actual temperature fluctuated slightly around the setting temperature points during the process of temperature increases. However, during the process of temperature decreases, the temperature fluctuated greatly around the setting temperature, which being said, the water temperature was unstable. The reason of the instability is the wrong parameter setting of the temperature hysteresis, which caused a dramatic change in temperature and a significant fluctuation around the setting temperature. Therefore, the temperature hysteresis parameter was reset during the temperature decreases. Once the temperature dropped to 50 °C, the process of mixing water was stopped. The heating unit was operated later and afterwards the hot water was delivered.

3.2 Hardware Test with the MPS

After completing the separate test, the entire temperature control device was connected to the MPS. The sample and the high sensitivity thermometer were both placed in center tube. Water flows through the interlayer between the outer and inner tubes to heat or cool the sample in the central tube. As can be seen from the Fig. 4, it was difficult to adjust the setting temperature precisely, when it should be close to the room temperature. Thus, choosing a temperature much higher than room temperature for the experiment is a suitable approach.

A fiber-optic probe (PRB-100, Osensa Innovations, Canada) was inserted into the sample to monitor the temperature. The program parameters were adjusted, and the starting temperature was set from room temperature 20 °C. Because the water path is relatively long, there is an absolute difference between the temperature of the water box and the sample temperature. To allow a quick reach of the set temperature, a manual adjustment has been made. For

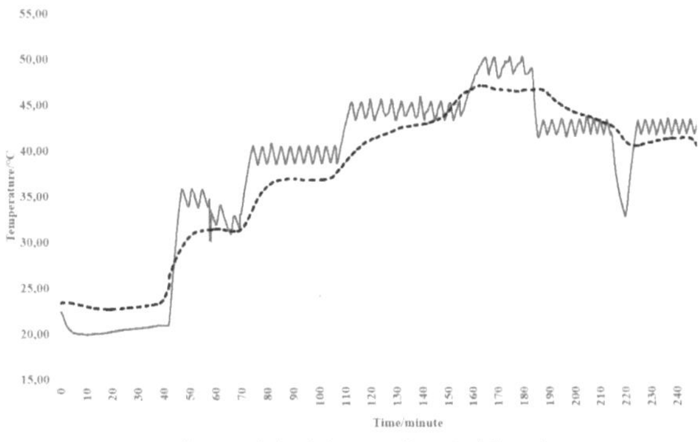

Figure 5: The temperature in water box

example, when the desired temperature was 33 °C, in order to speed up the test process, the temperature of the water box was artificially set to 35 °C. When the temperature change of the sample was visible, the temperature was reset to 33 °C. When the sample need to cool down rapidly from 50 °C to 40 °C, the setting temperature must be artificially set to 30°C. That is the reason why there was a short temperature fluctuation at a period of 60 to 70 minute and 215 to 225 minute (see Fig. 5). Therefore, the temperature of the water box was adjusted to different temperature (i.e., 40 °C, 45 °C, 50 °C), while the changes of sample temperature were tested.

The whole experiment, in which two temperature curves were obtained, lasted about four hours. The analysis of the two temperature curves shows that the temperature control module has reached its original design goal. However, the cooling effect of this design was not ideal, and it did not successfully set the temperature below room temperature. There are two main reasons: Firstly, the water path was too long, thus, the dissipation was severe and the cooling rate became smaller; Secondly, the cooling efficiency was low, because the heat dissipation from the hot side is not fast enough.

4   Conclusion

Temperature has a significant influence on the behavior of SPIONs, which will then influence the quality of imaging. Therefore, temperature controller is one of the necessary components of the MPS. In this project, a hardware solution is designed. This solution combined the heating and cooling functions. The setup has advantages like small size and compact structure. The graphic and data analysis has shown that the temperature control system has reached every target temperature stably when the setting temperature is above the room temperature. Which being said, that the heating program functioned well and matched its original goal—to control the temperature automatically and reach the setting temperature successfully.

However, this design has a major disadvantage: the cooling problem cannot be solved. In the separate experiment, the minimum required temperature of the water box was 15 °C.

Meanwhile, in the experiment with MPS, the minimum required temperature of the water box was 20 °C, which did not meet the required temperature range from 10°C. There were several reasons that would cause the failure of reaching the minimum required temperature: cooling Peltier elements were inadequate and their cooling capacity were low, the waterways were too long.

The following recommendations are given to optimize experiments in the future. First, higher efficient cooling Peltier elements should be purchased and used in the experiments. Second, the design should be improved by using a flat metal water box instead of the original water box. As a result, there will be more cooling elements on both sides which could increase the system cooling efficiency. Third, the software should be written by a PID control algorithm, such as proportional-integral-derivative control algorithm. Because the algorithm can not only automatically calculate the difference between the target and actual temperature, but also automatically adjust the output of the heating or cooling, which could help the system reach the setting temperature more quickly.

Acknowledgement

The work has been carried out at Institute of Medical Engineering at the University of Lübeck, Germany.

5   References

[1] B. Gleich and J. Weizenecker. Tomographic imaging using the nonlinear response of magnetic particles. *Nature*,435(7046):1214-1217,2015. doi: 10,1038/nature03808.

[2] S. Biederer, T. Knopp, T. F. Sattel, K. Lüdtke-Buzug, B. Gleich, J. Weizenecker, J. Borgert, and T. M. Buzug. Magnetization response spectroscopy of superparamagnetic nanoparticle imaging. *Journal of Physics D: Applied Physics*, 42(20): 205007, 2009. doi: 10.1088/0022-3727/42/20/205007.

[3] S. Draack, T. Viereck, C. Kuhlmann, M. Schilling, and F. Ludwig. Temperature-dependent MPS measurements. *International Journal on Magnetic Particle Imaging*. 3(1), 2017. doi: 10.18416/ijmpi.2017.1703018.

[4] T. M. Buzug. *Magnet-Partikel-Spektrometer*. Springer Vieweg, Universität zu Lübeck, 2012. doi: 10.1007/978-3-8348-2407-3.

[5] X. Chen, A. Behrends, A. Neumann, And T. M. Buzug. Sample Temperature Control in a Three-Dimensional Magnetic Particle Spectrometer. *International Workshop on Magnetic Particle Imaging*. 9th IWMPI: 17-19 March 2019.

[6] DS18B20 Programmable Resolution 1-Wire Digital Thermometer. Maxim Integrated 160 Rio Robles, San Jose, CA 95134 USA 1-408-601-1000. 2018.

Experimental investigation of the sEMG-force relationship of the biceps brachii

Christian Kaußler [1], Eike Petersen [2], Julia Sauer [2], and Philipp Rostalski [2]

[1] Medical Engineering Science, Universität zu Lübeck, christian.kaussler@student.uni-luebeck.de

[2] Institute for Electrical Engineering in Medicine, Universität zu Lübeck, {eike.petersen, j.sauer, philipp.rostalski}@uni-luebeck.de

Abstract

Lifting various weights is an everyday task, with complex force transmissions involving muscles and tendons. How the muscle activity can be measured while using the electromyography (EMG) to gain a better understanding of the relationships will be examined in this article. A series of measurements will be conducted to investigate the generated force and the corresponding electromyographic measurements of the musculus biceps brachii. Using a custom-built measuring station, force measurements were taken during isometric movement in relation to different elbow angles and contractions of varying strength. Using a biomechanical model, the biceps force can be calculated from the measured force. As part of this experimental setup, a study with 10 subjects will be executed. The data of the study will be used to describe a model that allows an empirical investigation of the dynamic transfer function from EMG to force values.

1 Introduction

Electromyography (EMG) is the measurement of electrical fields that arise during muscle contraction. The electromyogram is closely related to the mechanical force generated by the muscle [1]. In surface electromyography (sEMG), which will be used in this research, electrodes are attached to the skin surface above the target muscle. The electrodes measure the signal generated by the superimposition of action potentials, which spreads on the surface of the active muscle fibers, in relation to a reference electrode in an area with few muscles. These action potentials are representations of the surrounding muscle activity. However, the exact relationship between the EMG signal and the generated muscle power depends on a number of physiological and technical factors, such as the current muscle length, the relative degree of muscle activation and the impedance of the measuring electrodes [2]. Simple heuristics (e.g. EMG envelopes) are often used in practical applications to infer a force signal from the EMG signal. The relationship between electrical activity and the resulting force development during isometric movement has already been investigated in many studies [3], [4], but none of the published studies examined the dynamic transfer function empirically. If advanced knowledge of the dynamic relation between EMG and force signal depending on muscle length and degree of activation were known, it would be advantageous for many medical applications, such as assisted mechanical ventilation [5].

2 Material and Methods

The following section describes the measuring equipment and the different measuring methods. In order to record the sEMG and force signals, an sEMG amplifier (BN-TX) and a corresponding analysis device (MP160) from Biopac Systems, Inc. were used. This system makes it possible to transfer the recorded EMG data wirelessly to the analyzer. A goniometer (Twin-axis Goniometer 150, TSD130B, Biopac Systems, Inc.) is also used to check the arm angle position continuously. The force detection is realized by an S-shaped load cell (Load-Cell 3138, Phidgets, Inc.), which converts a force acting on the sensor into small electrical signals by means of strain gauges. The analysis software AcqKnowledge (Biopac Systems, Inc.) enables synchronous recording of all data files.

To achieve a well controllable data acquisition regarding the sEMG and force signals, a specially developed measuring station was developed at the Institute of Electrical Engineering in Medicine. Fig. 1 shows a structural sketch of the experimental setup. The loop of a rope is lifted by a controlled upward movement and pulled against a fixed resistance. The rope is guided by two pulleys and is connected to the S-load cell. The different arm angles are implemented by means of various rope lenghts. The sEMG signal is recorded by four electrodes on the upper arm in the biceps brachii area according to the European recommendations for surface electromyography [6] (seniam guidelines) and a reference electrode on the elbow. A pair of electrodes is applied to both the long and the short biceps head in order to further differentiate the activities of both muscle groups.

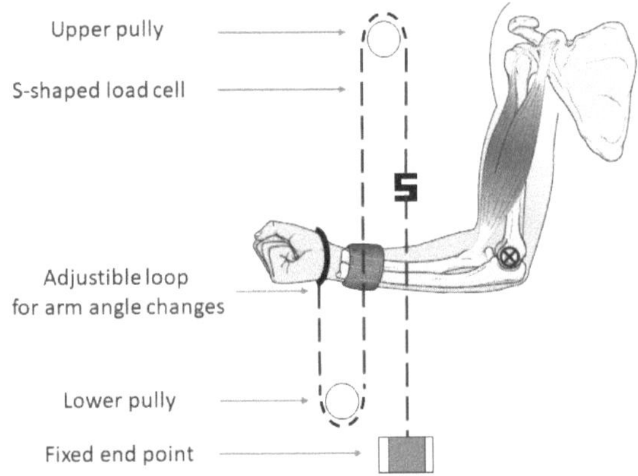

Figure 1: Structure of the measuring station: The traction cable is fixed to the lower rail and is first guided over the upper pulley. Then the cable is connected to the load cell and is guided around a lower pulley. The loop at the end of the rope can be adjusted in length according to the desired elbow angle. Modified image from [1].

The data recording methods vary in measured values with maximum contraction in different arm angle positions and fixed angle of $100°$ with 20%, 30%, 40%, 50%, 60% and 70% of the previously measured maximum force. However, the force values determined in these two measuring procedures do not yet refer to the pure strenght that must be applied in the biceps, but rather represent a required force effort at the transition point to the measuring station around the wrist. Yamaguchi [7] presented a model that can distinguish between the force generated by the biceps muscle and the force that acts via the elbow joint and is applied to the forearm and hand. Torques are generated because the forces act transverse to the axis of rotation at the elbow joint. With regard to the forces that arise during a contraction, shown in Fig. 2, the following three torques are of decisive importance. The torque determined by the generated biceps force $\vec{M}_{BIC}$, the torque consisting of the weight of the forearm in conjunction with the gravitational force g called $\vec{M}_{FA}$ and the torque generated by the point of application of the forearm attachment $\vec{M}_{APP}$. The force generated by the biceps resulting in the equation for the torsional moment

$$\vec{M}_{BIC} = \vec{M}_{FA} + \vec{M}_{APP}. \tag{1}$$

The torque $\vec{M}_{BIC}$ is the Cartesian product of the lever $\vec{r}_{BIC}$ and the biceps force $\vec{F}_{BIC}$. From the vector notation of the torque

$$\vec{M}_{BIC} = \vec{r}_{BIC} \times \vec{F}_{BIC} \tag{2}$$

we can extract a scalar expression

$$M_{BIC} = r_{BIC} \cdot F_{BIC} \cdot \sin\phi \tag{3}$$

where ϕ is the angle between an extended line of the force vector and the lever [8]. Pigeon et al. [2] expressed the distance from the center of rotation in the elbow to the point of insertion of the biceps brachii at the ulna, which is described

by $r_{BIC} \cdot \sin\phi$, with an elbow angle dependent function

$$r_{BIC(\theta)} = 14.660 + (4.5322 \cdot 10^{-1})\,\theta + (1.8047 \cdot 10^{-3})\,\theta^2$$
$$+ (-2.9883 \cdot 10^{-5})\,\theta^3 \tag{4}$$

The elbow angle is defined by θ, used in degrees and is illustrated in Fig. 2. The parameter $r_{BIC(\theta)}$ is calculated in millimeters. A further torque in which the elbow serves as the center of rotation is described by

$$\vec{M}_{FA} = \vec{r}_{FA} \times \vec{F}_{GF} \tag{5}$$

and results from the gravitational force of the forearm $\vec{F}_{GF}$ and the distance from the center of rotation to the center of the forearm mass $\vec{r}_{FA}$. The scalar notation of the torque can be written as

$$M_{FA} = m_{FA} \cdot g \cdot r_{FA} \cdot \sin\theta \tag{6}$$

and shows a direct dependence on the elbow angle θ. Parameter m_{FA} describes the weight of the forearm. The force F_{APP} acts close to the wrist as shown in Fig. 1. The distance from the center of rotation to the application point forms the lever r_{APP}. This results in the torque

$$M_{APP} = r_{APP} \cdot F_{APP} \cdot \sin\theta \tag{7}$$

which also depends on the elbow angle θ. Using the scalar functions of the forces and levers involved in (1) and solve it to the biceps force F_{BIC}, the corresponding equation

$$F_{BIC} = \frac{m_{FA} \cdot g \cdot r_{FA} \cdot \sin\theta + r_{APP} \cdot F_{APP} \cdot \sin\theta}{r_{BIC(\theta)}} \tag{8}$$

is obtained.

Fig. 2 shows an overview of all the forces and levers that are significantly involved. To calculate the mass of the forearm m_{FA} which is required for the torque M_{FA} the arm is placed in a loop right below the load cell in a right-angled elbow position. The weight of the forearm is measured via two mass points in

$$m_{FA} = m_1 + m_2 \tag{9}$$

which are located at the elbow joint with m_1 and at the wrist reference point with m_2. A typical model for calculating a center of mass with two mass points is described by Katz [9] with

$$r_{FA} = \frac{m_1 \cdot x_1 + m_2 \cdot x_2}{m_1 + m_2}. \tag{10}$$

As the mass m_1 is suspended at the elbow joint and the distance measurement starts at this joint, x_1 is set to zero and the equation is simplified to

$$r_{FA} = \frac{m_2 \cdot x_2}{m_1 + m_2}. \tag{11}$$

The distance between the center of elbow rotation and the reference point of the rope suspension at the wrist is described by x_2 and represents the parameter r_{APP}. Patla et al. [10] were able to show that the center of the forearm

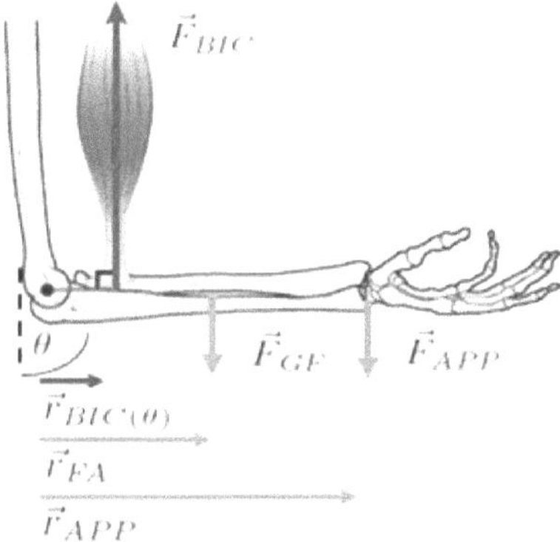

Figure 2: Overall representation of all decisive forces and levers that occur during isometric contraction of the arm using the measuring station. $\vec{F}_{BIC}$ represents the force in the biceps, $\vec{F}_{GF}$ the gravitational force that affects the forearm and $\vec{F}_{APP}$ the transmitted force to the traction rope. The distances from the center of rotation of the elbow to the point of insertion of the biceps-brachii at the ulna is represented by $\vec{r}_{BIC(\theta)}$, to the center of the forearm by $\vec{r}_{FA}$ and to the point of application of the traction rope at the wrist by $\vec{r}_{APP}$. Modified image from [1].

mass is located at 43% of the forearm length and thus the equasion

$$r_{FA} = x_2 \cdot 0.43 \qquad (12)$$

applies. Following (9) the result

$$r_{FA} = \frac{m_2 \cdot x_2}{m_{FA}} = x_2 \cdot 0.43 \Rightarrow m_{FA} = \frac{m_2}{0.43} \qquad (13)$$

is a simplified expression for calculating the weight of the forearm. The force transmitted to the rope during measurement and determined by the load cell is defined as F_{APP}. After discussing all necessary steps to calculate the actual biceps force F_{BIC}, the evaluation of the two measuring methods and the corresponding force calculation according to (8) will be shown. The sEMG values in the following figures were plotted using the root mean square method (RMS) to allow direct comparisons.

3   Results and Discussion

An extract of the maximum force series at an elbow angle of 100° is presented in Fig. 3. In direct comparison, Fig. 4 shows an example of data recording of sEMG signals at a force generation of 20% of the maximum force and an arm angle of 100°. The proband follows a visual specification of the percentage force on a live monitor. The clearly recognizable peak value after the beginning of the contraction is caused by the fact that it is difficult to regulate the force precisely at the beginning. It is also noticeable that the sEMG values fall back into a slightly lower, constant progression after the initial peak value. As expected, these sEMG values have a significantly smaller amplitude than in the example in Fig. 3. The following comparison in Fig. 5 shows a significantly higher calculated force at the biceps compared to

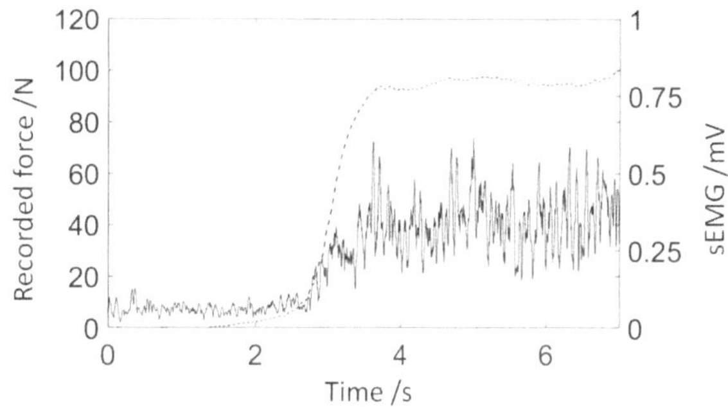

Figure 3: Recording the sEMG values (RMS, solid line) and the corresponding force generation (dashed line) at maximum contraction during an elbow angle of 100°.

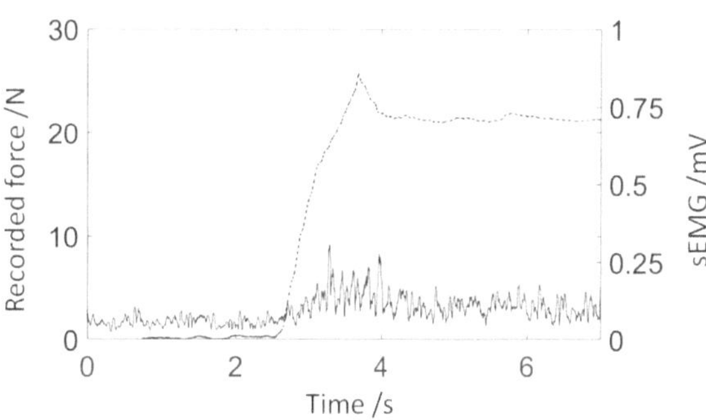

Figure 4: Recording of the sEMG values (RMS, solid line) at 20% of the maximum force (dashed line) during an elbow angle of 100°.

the measured force at the reference point of the measuring station. This behaviour clearly shows the strong influence of the lever forces on the biceps muscles. Especially the force dependent peak value at an arm angle of 100° can be clearly recognized. This is due to the length of the arm moment $r_{BIC(\theta)}$ which, according to (4), reaches its maximum value at an angle of 94° and thus enables maximum power transmission. As mentioned before, the sEMG sig-

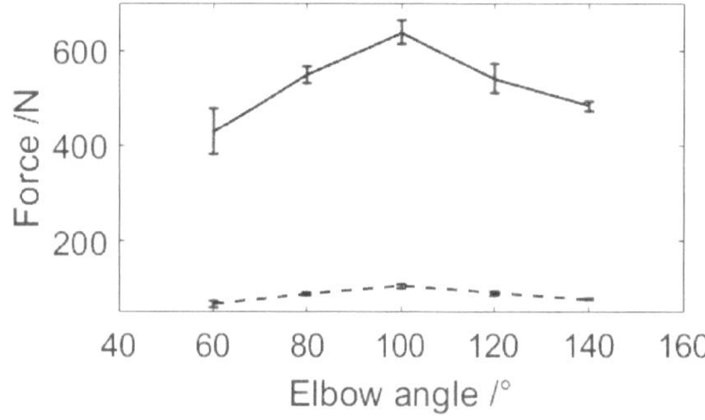

Figure 5: Comparison of recorded force value F_{APP} (dashed line) by the load cell and calculated force value in the biceps F_{BIC} (solid line) in relation to different arm angle positions θ.

nals were recorded at the short and the long biceps head in order to study the strain on both muscle groups with respect to force generation during isometric movement. A direct comparison of the sEMG values of short and long biceps head shows a clear trend towards higher muscle activity in the area of the short biceps head. A further trend behavior represents the maximum signal strength of both muscle groups in the area of an arm angle of $100°$ and depends on the optimal length of the arm moment, as described for the maximum force values. A similar trend behavior becomes

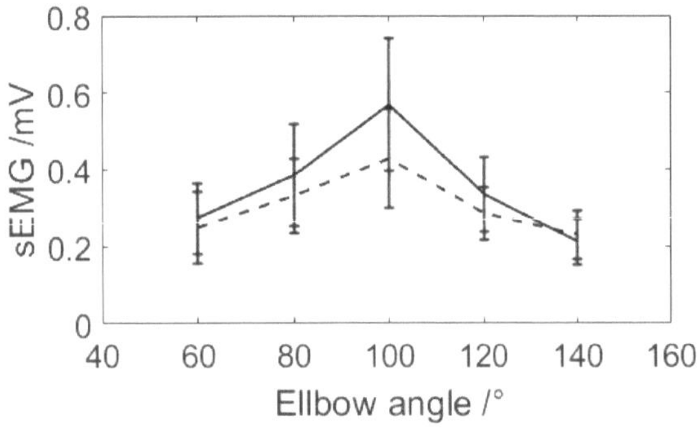

Figure 6: SEMG data representation (RMS) of short (solid line) and long (dashed line) biceps head in relation to different arm angle positions θ at maximum contraction force.

visible when comparing the percentage sEMG data evaluation. As expected, a strong increase in signal strength can be seen in Fig. 7 with a simultaneous increase in the percentage of maximum force generation.

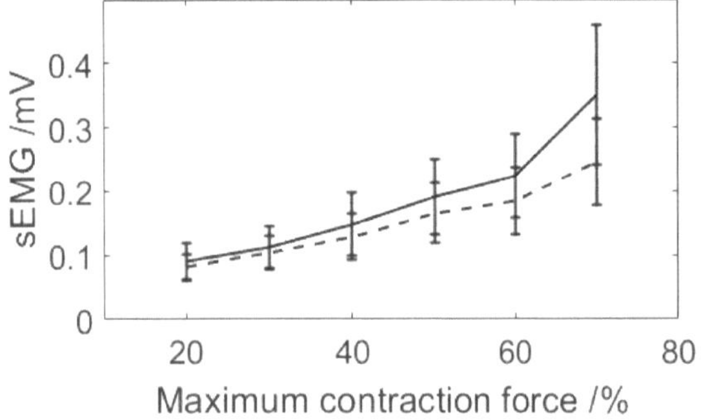

Figure 7: Recording of the sEMG data (RMS) of short (solid line) and long (dashed line) biceps head in relation to the percentage maximum force values.

4   Conclusion

The preparation of this report was based exclusively on trial data of the measuring station. A low-risk study with 10 healthy subjects was approved by the local ethics comittee and will be performed during the next few weeks. Due to large standard deviations, the trends must be verified by a multitude of measurement data. However, the evaluation to date and the knowledge about trend developments suggest that the structure and execution of the test series offer a promising method of data recording. Furthermore, the

study results will be used to describe a model that allows an empirical investigation of the dynamic transfer function from the EMG and force values. The evaluation of the study and the progress of the model description will be presented at the student conference in March 2020.

5   Acknowledgement

The work has been carried out and supervised by the Institute for Electrical Engineering in Medicine, Universität zu Lübeck.

6   References

[1] P. A. Houghlum and D. B. Bertoti, *Brunnstorm's Clinical Kinesiology*. F.A. Davis, Philadelphia, vol. 6, pp. 10, 65 and 126, 2012.

[2] P. Pigeon, L. Yahia, and A. G. Feldman, *Moment arms and lengths of human upper limb muscles as functions of joint angles*. Journal of Biomechanics, vol. 29, pp. 1365–1370, 1996.

[3] J. Leedham and J. Dowling, *Force-length, torque-angle and emg-joint angle relationships of the human in vivo biceps brachii*. Journal of Applied Physiology, vol. 70, pp. 421–426, 1993.

[4] M. Schwartz, *EMG Methods for Evaluating Muscle and Nerve Function*. InTechOpen, Rijeka, ch. 3, pp. 31-54, 2012.

[5] G. Bellani et al., *Measurement of Diaphragmatic Electrical Activity by Surface Electromyography in Intubated Subjects and Its Relationship With Inspiratory Effort*. Respiratory Care, vol. 63, pp. 1341-1349, 2018.

[6] D. Stegeman and H. Hermens, *Standards for surface electromyography: The European project Surface EMG for noninvasive assessment of muscles*. Available: http://www.seniam.org [last accessed on 2020-01-21].

[7] G. T. Yamaguchi, *Dynamic Modeling of Musculoskeletal Motion: A vectorized Approach for Biomechanical Analysis in Three Dimensions*. Springer US, vol. 1, pp. 51, 103-110, 2001.

[8] K. Lüders and G. von Oppen, *Lehrbuch der Experimentalphysik*. vol. 12, Walter de Gruyter GmbH & Co. KG, Berlin, p. 219, 2008.

[9] D. M. Katz, *Physics for Scientists and Engineers: Foundations and Connections*. vol. 1, Cengage Learning, 2015.

[10] A. E. Patla, M. G. Ishac, and D. A. Winter, *Anticipatory control of center of mass and joint stability during voluntary arm movement from a standing posture*. Exp Brain Res, pp. 318-327, 2002.

Modelling of a PEEP-Valve for an Anaesthesia Workstation

Christian Ehlers [1], Georg Männel [2]

[1] Medical Engineering Science, Universität zu Lübeck, christian.ehlers@student.uni-luebeck.de
[2] Institute for Electrical Engineering in Medicine, Universität zu Lübeck, ge.maennel@uni-luebeck.de

Abstract

In modern system design and development models are commonly used for early virtual development and optimisation of the system performance. For instance, advanced control approaches, such as robust control, require a model of the dynamics of the underlying system. Within this paper a model of the system dynamics for the PEEP-valve (*PEEP: Positive-End-Expiratory Pressure*) used in medical ventilators is developed with the focus on controller design. The model is designed to be easily adaptable to different valve design types, using the inlet pressure and the control voltage as model inputs and the flow through the valve as model output. The model parameters are estimated using non-linear system identification. The received model has a good fit to the measured system data. Though a revision of the model is recommended.

1   Introduction

The increasing complexity of systems lead to new approaches in system engineering and development, by using models describing the system under development. A field were the usage of models emerged early on, was in classic controller design, using a model of the systems dynamics to optimize the closed loop performance [1]. This further lead to advanced control approaches, such as robust control or model predictive control, where the controller is designed using optimizations techniques, for an optimal performance [2]. In order to find a good model fit of the system dynamics, two main different modelling approaches are commonly taken into consideration. For one whitebox models where the system structure and parameters are derived using first principle methods. The second approach is blackbox modelling were the model is derived from data, with no prior knowledge about the physics of the system. The combination of both approaches, for example a model structure derived from first principle methods and from parameter identified by data, are so called grey box models [3].

Mechanical ventilation as well as anaesthesia are important elements in todays medicine. The anaesthesia workstation is therefore an essential medical device in every operating theatre and intensive care unit.

Within this paper a model for the dynamics of the PEEP-Valve (*PEEP: Positive-End-Expiratory Pressure*) of an anaesthesia workstation is derived. The PEEP-valve commonly resembles one of the two actuators required in mechanical ventilators and anaesthesia workstations to controll the pressure of at the patients airways. A PEEP-valve is necessary during the mechanical ventilation of patients with critical health conditions or during surgery. It is used to maintain a certain pressure level in the lungs of the patient in order to prevent the alveoli from collapsing. Since the gas exchange takes place within the alveoli and alveolar capillaries, their collapse would reduce the oxygenation of the blood. This is a reversible reaction but the reopening of the alveoli can lead to a ventilator induced lung injury due to the large pressure applied to the patients lungs during the process.

In this paper an approach is taken to model a PEEP-valve by using the measurements of the expiratory air pressure and the expiratory air flow. Those measurements are retrieved by the designated built-in sensors of the ventilation system of an anaesthesia workstation, as seen in figure 1. The model is therefore independent of the general system structure as long as measurements of the expiratory pressure and flow, as well as for the actuator input for the valve, are available. Known physical characteristics and dynamics of the moving parts of the valve were taken into account in the design of the underlying structure of the model. The parameters of the designed grey box model were estimated using MatLab with a non-linear least-squares solver [7].

2   Material and Methods

2.1   System Description

A schematic of the ventilator system of an anaesthesia workstation is depicted in figure 1. The patient is connected via an y-piece to two hoses, one for the inspiration and one for the expiration. During the respiration cycle the air-anaesthesia mixture flows through one of the hoses depending of the respiration phase, this is guaranteed using two check valves, one for the inspiratory path (I) and one for the expiratory path (E). In the inspiratory part the reused air from the reservoir bag is sucked through the ab-

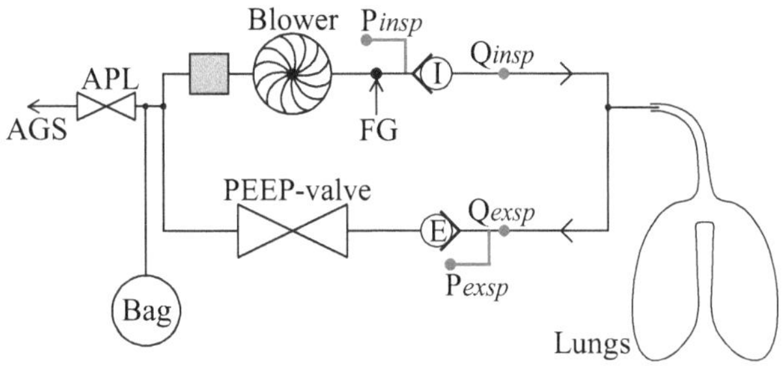

Figure 1: Simplified schematic structure of a ventilator system with a PEEP-valve with the absorber represented as the grey box.

sorber (grey box in Figure 1) by the blower. The absorber is purifying the air by removing the CO_2. The blower is responsible for the circulation of the air-anaesthesia mixture in the ventilation system. To compensate for the oxygen uptake of the patient it is required that fresh gas (FG) is constantly added into the semi-closed system. Additionally an adjustable pressure limitation (APL) is required as a safety measure for manually ventilating a patient e.g. in case of a long-lasting power outage. The scavenging system (AGS) is responsible for collecting the waste gas after it gets exhaled by the patient. During the expiration the blower reduces the inspiratory pressure P_{insp} and air can flow out of the lungs due to the opening of the PEEP-valve. The PEEP-valve closes when a preset positive pressure level in the lungs is reached so the alveoli will not collapse. The expiration is mainly done by the restoring forces within the lungs.

For the experiments during this work a research demonstrator of an anaesthesia workstation is used. The ventilation system, e.g. PEEP-valve, blower, etc. is controlled by two microcontrollers, one Infineon XMC-4800 and one Infineon XMC-4300. Both controllers communicate using an EtherCAT-bus with a sampling rate of 1kHz. The XMC-4300 is used to read and pre-process the sensor data. The XMC-4800 is connected to the actuators in the system and the control architecture is implemented there due to the larger internal memory.

2.2 Model of the PEEP-Valve

A PEEP-valve is a pressure controlling valve, where a tappet presses onto the membrane and keeps the valve closed. When the force from the pressure building up in the area above the valve exceeds the force applied to the PEEP-valve by the tappet the valve will open and gas can escape. The dynamics of the moving membrane, as it is a vibrating system, inside the PEEP-valve can be abstracted as a classical spring-mass-damper-system [4], which is seen in figure 2. The goal is to model the system with a grey box approach, using knowledge about the physical system to derive the model structure [3]. The model parameters are then estimated from input and output data of the model.

Applying Newton's second law, the acceleration of the

membrane can be expressed by the sum of all acting forces. The motion of the membrane inside the valve acts in direction of x. With the location of the membrane being x_0, changing over time. The spring force F_k with spring constant k

$$F_k = k \cdot x_0 \qquad (1)$$

and the force

$$F_d = d \cdot x_1 \qquad (2)$$

applied by the damper with damping coefficient d. According to Newton's second law, the system can be described by the following second-order ordinary differential equation.

$$m \cdot \dot{x}_1 + d \cdot x_1 + k \cdot x_0 = 0 \qquad (3)$$

The variable m represents the mass of the moving parts of the valve. The movement of the membrane is described by the acceleration $\dot{x}_1$ and the velocity x_1. The model is extended by the force F_{act} applied by the actuators with an input PWM-signal (*PWM: Pulse-Width-Modulation*). It is assumed that the force of the actuator is proportional to its input, and that it is so fast that the effects of its dynamics can be neglected. The PWM-signal is the voltage driving the actuators underneath the membrane. The actuators are acting against the restoring forces of the spring and the damper. In addition the forces working on the membrane by the expiratory pressure $F_{P_{exsp}}$ and the expiratory flow $F_{Q_{exsp}}$ had to be taken into account. This leads to

$$m \cdot \dot{x}_1 + d \cdot x_1 + k \cdot x_0 + F_{P_{exsp}} + F_{Q_{exsp}} = \kappa \cdot F_{act} \qquad (4)$$

with κ being the input gain for the PWM-signal and $F_{P_{exsp}}$ representing the force applied by the expiratory pressure P_{exsp} on the membrane. This force depends on the area k_1 the pressure is applied to [5]

$$F_{P_{exsp}} = k_1 \cdot P_{exsp}. \qquad (5)$$

In this case k_1 represents the area of the membrane. $F_{Q_{exsp}}$ describes the lift force applied by the expiratory flow, lift occurs due to the moving air being diverted by the membrane of the PEEP-valve. Therefore the lift force acts perpendicular to the movement of the membrane [6]. With the lift stated as

$$F_{Q_{exsp}} = c_w \cdot k_1 \cdot \rho \cdot v^2 \qquad (6)$$

c_w standing for the lift coefficient, k_1 for the area being exposed to the lift, ρ for the density of the moving fluid and v for the relative velocity between the fluid and the membrane [5]. Equation 6 will be simplified by summarising the coefficients c_w, k_1 and ρ as k_2. And the velocity v can be presumed as being proportional to Q_{exsp} which leads to

$$F_{Q_{exsp}} = k_2 \cdot Q_{exsp}^2. \qquad (7)$$

With k_2 representing the input gain for the expiratory flow Q_{exsp}. The inputs into the model are therefore the PWM-signal, the expiratory pressure P_{exsp} and the expiratory flow Q_{exsp}. As the output of the model a combination of the expiratory pressure P_{exsp} and the location x_0 of the membrane was used. Additionally the spatial boundary, the maximum way of movement of the membrane, was taken into

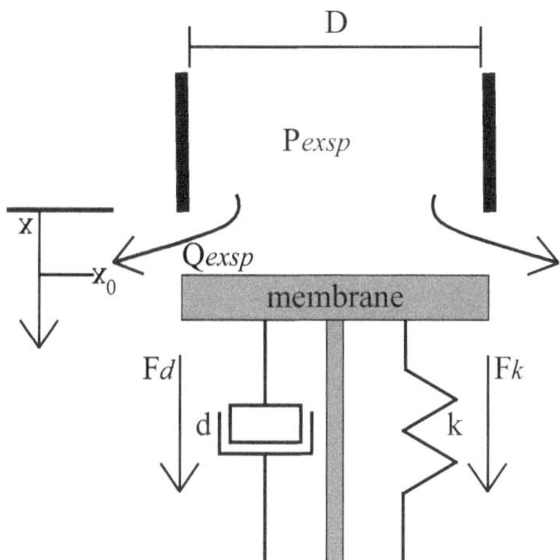

Figure 2: Various forces work on the membrane of a peep-valve, the forces can be abstracted by a spring (spring constant: k) and a damper (damper constant: d). In this case the spring and damper work against the force applied by the actuators which receive the PWM-signal. The model shown here works when the peep-valve is not exposed to an airflow. Otherwise the forces applied $F_{P_{exsp}}$ and $F_{Q_{exsp}}$ had to be accounted for.

account. In our case the maximum way of movement of the membrane is described as w, otherwise the membrane could be damaged. The expiratory flow Q_{exsp} should be equal to

$$Q_{exsp} = -\zeta \cdot (w - x_0) \cdot (P_{exsp})^2. \qquad (8)$$

With ζ representing the output gain of our model.
The model can be transformed into state space form which is then used for the parameter estimation using MatLab.

$$\begin{bmatrix} \dot{x_0} \\ \dot{x_1} \end{bmatrix} = \begin{bmatrix} 0 & 1 \\ -\frac{k}{m} & -\frac{d}{m} \end{bmatrix} \begin{bmatrix} x_0 \\ x_1 \end{bmatrix} + \begin{bmatrix} 0 & 0 & 0 \\ \kappa & k_1 & k_2 \end{bmatrix} \begin{bmatrix} \text{PWM} \\ P_{exsp} \\ Q_{exsp}^2 \end{bmatrix} \qquad (9)$$

The model parameter were estimated using MatLab system identification toolbox [7].

2.3 Data processing

The data for the expiratory pressure P_{exsp} and the expiratory flow Q_{exsp} was acquired using the respective system sensors built into the ventilation tubing, also seen in figure 1. The PWM-signal, which is the input signal into the actuators of the PEEP-valve, was recorded. Those signals were sampled with a sampling rate of 1kHz. The PWM was adjusted every second alternating between a minimum value of one and the previous high adding 4 % each step. The blower was continuously driven with a PWM of 40 % during the whole experiment.
The data for the three signals was then imported into Mat-Lab. In Matlab the System Identification Toolbox [7] for the estimation of the parameters seen in table 1 was used. No pre-processing was applied to the imported data in Mat-Lab, as there was no significant noise on the datasets. The

data was then formatted to the *iddata* format [8] to be used with the MatLab System Identification Toolbox [7].

3 Results and Discussion

As initial states for the system, the states were set to zero. Initial values which have been normalised for the estimated parameters, a description of the estimated parameters can be found in table 1. To estimate our non-linear greybox model the MatLab function *nlgreyest* [9] was used. The estimator chose to not change the initial values for the damping coefficient d, the spring constant k, the input gain for the PWM κ nor the input gain k_2 for the flow. As it only adjusted the output gain, it is very likely that there are problems in the model structure or in the normalisation of the different model parameters. In figure 3 the measured Q_{exsp} is compared to the output of the generated model. The fit between the model and the system is at 85.3 %. This is mainly due to the fact that the system was operated in saturation, in which there are no dynamics present. For the first 25 seconds the output of the model has a lower amplitude than the amplitude of the measured system output. This might be due to an offset error. If one would adapt two curves to this section the curve for the model would have a steeper slope than the slope of the curve adapted to the system. In the next section until around 40 seconds the models accounts too much for the overshoots in the system measurements. The model has a higher amplitude than the measured system output. The negative overshoots in the systems output can be explained by the vibrating of the membrane and therefore not closing completely. From around 20 seconds until the end of the measured system positive peaks of the measured system output are visible, those might occur due to the valve not opening correctly. For the last 20 seconds of the experiment the systems output measurements match the model output to a high degree, here the system is operated in saturation. The noise in the plateaus of the model output is significant due to the throughput of the input signal Q_{exsp} to the output. For the next model the passthrough of the input Q_{exsp} should be eliminated. This could be achieved by introducing another state into the model, which decouples the passthrough. Also the use of pre-processing, e.g. with a moving average filter, could help to receive a better fitting model after the estimation procedure.

4 Conclusion

In this paper a model of a PEEP-valve was proposed and a parameter estimation for the model has been performed. The underlying model structure should be a good abstraction of the dynamics of the PEEP-valve, but only when the system is operated in saturation, in which case there are no dynamics. The parameter estimation process lead to no change in four of five parameters, which might be due to the non-existing dynamics. It is recommended to revise the model structure. First of so the model is less influenced by the overshoots and the system noise. Secondly, a more ap-

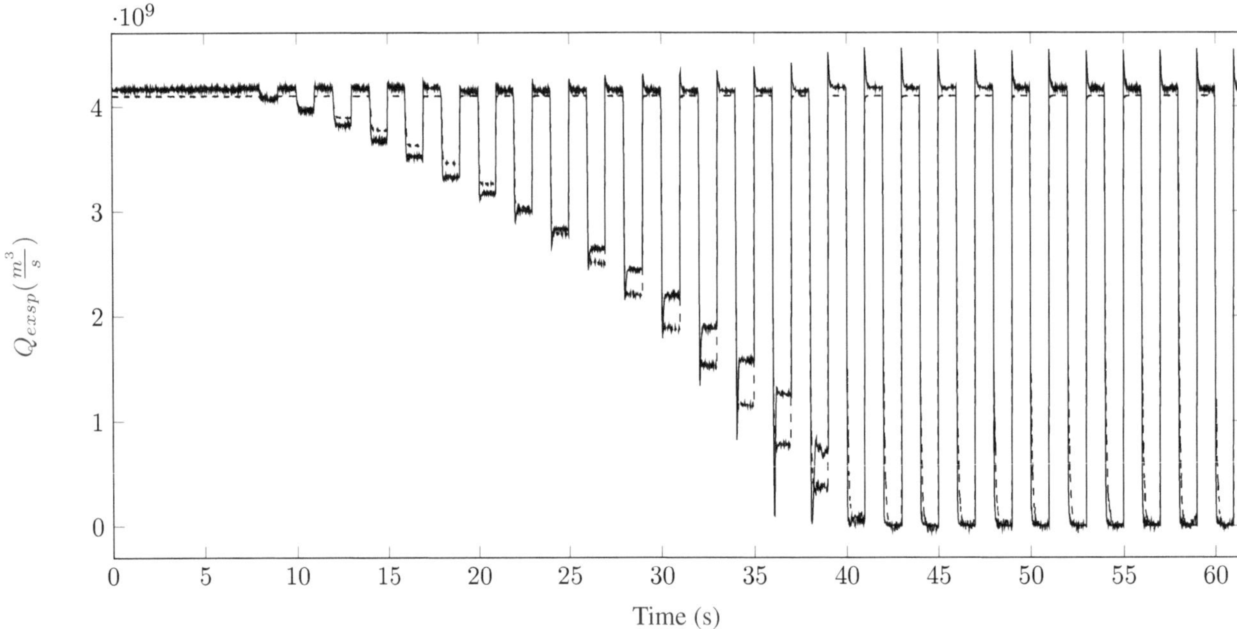

Figure 3: The measured output-signal Q_{exsp} is shown in black. The dashed line represents the output data of our model. The fit of the estimated model to the measured data is at 85.3 %.

propriate normalisation for the estimated parameters should be chosen, to account for the the problem in the parameter estimation process.

Parameter	Description
d	damping coefficient
k	spring constant
κ	input gain PWM
k_2	input gain Q_{exsp}
ζ	output gain

Table 1: The parameters of the model estimated by MatLab using the non-linear-least-squares solver *lsqnonlin*.

Acknowledgement

The work has been carried out at Institute of Electrical Engineering in Medicine, Universität zu Lübeck. It has been supervised by Prof. Dr. Philipp Rostalski, with the help of Georg Männel, both from the Institute of Electrical Engineering in Medicine.

5   References

[1] Daniel E. Rivera, Manfred Morari and Sigurd Skogestad, *Internal model control: PID controller design* , Ind. Eng. Chem. Process Des. Dev., Vol. 25, No. 1, pp. 252-265, 1986.

[2] Carlos E. Garcia, David M. Prett and Manfred Morari, *Model Predictive Control: Theory and Practice - a Survey*, Automatica, Vol. 25, No.3, pp. 335-348, 1989.

[3] Bill Whiten, *Model completion and validation using inversion of grey box models*, ANZIAM Journal 54 (CTAC2012), pp. C188-C189, 2013.

[4] Xue-Guan Song, Young-Chul Park, Joon-Hong Park, *Blowdown prediction of a conventional pressure relief valve with a simplified dynamic model*, Mathematical and Computer Modelling 57, pp. 279-288, 2013.

[5] Horst Kuchling, *Taschenbuch der Physik*, Fachbuchverlag Leipzig im Carl-Hanser-Verlag, 2014.

[6] National Aeronautics and Space Administration, *What is Lift?*. Available: https://www.grc.nasa.gov/WWW/K-12/airplane/lift1.html, last accessed 19.01.2020.

[7] Mathworks, *System Identification Toolbox (R2019a)*. Available: https://de.mathworks.com/help/ident/index.html, last accessed 19.01.2020.

[8] Mathworks, *iddata (R2019a)*. Available: https://de.mathworks.com/help/ident/ref/iddata.html, last accessed 19.01.2020.

[9] Mathworks, *nlgreyest (R2019a)*. Available: https://de.mathworks.com/help/ident/ref/nlgreyest.html, last accessed 19.01.2020.

8

Medical Imaging

Segmentation of fluorescently labelled axons in a model of brain hemorrhage-induced axonal degeneration using convolutional neural networks

Narayan Kumar Menon[1,2,3], Alex Palumbo[2,3,4], Philipp Grüning[5], Svenja Kim Landt[2,3], Lara Heckmann[2,3], Luisa Bartram[2,3], Amir Madany Mamlouk[5] and Marietta Zille[2,3,4*]

[1] Biomedical Engineering, Lübeck University of Applied Sciences, Lübeck, Germany, narayan.kumar.menon@stud.th-luebeck.de

[2] Fraunhofer Research Institution for Marine Biotechnology and Cell Technology, Lübeck, Germany

[3] Institute for Medical and Marine Biotechnology, University of Lübeck, Lübeck, Germany

[4] Institute for Experimental and Clinical Pharmacology and Toxicology, University of Lübeck, Lübeck, Germany, m.zille@uni-luebeck.de

[5] Institute for Neuro- and Bioinformatics, University of Lübeck, Lübeck, Germany

Abstract

The main goal of this study was to investigate the vitality of axons based on morphological substructures in an *in vitro* model of hemorrhagic stroke. Therefore, murine primary cortical neurons exposed to different hemin concentrations were recorded using a fluorescence microscope. Since traditional computational tools have limitations when it comes to automatically detecting axons, we here used a convolutional neural network (CNN) that was trained to segment different features of axonal degeneration. Our data suggests that the CNN-based image segmentation allows efficient and automatic quantification of fluorescence microscopy data. This will promote the investigation of axonal degeneration using fluorescent markers.

1 Introduction

Current computational tools for image segmentation are limited when it comes to the quantitative analysis of axonal degeneration. In addition, computing image segmentation tasks on fluorescence microscopy data is challenging because of the overlapping of axons, low-contrast datasets or background intensity discontinuities [1].

Some of the most frequently used methods for quantifying changes in axonal morphology include [2]: a) Manual thresholding to recognize discontinuous image components by detecting local intensity maxima; b) ImageJ-based methods that involve manually defining the region of interest; c) Filament tracking that detects the continuity of axons based on manual definition. The main drawback of these methods, however, is that they require manual annotation and are thus only semi-automatic. In addition, they only allow the quantification of axons in low density areas. Deep convolutional neural networks (CNNs) have become a widely used machine learning method since they provide state-of-the-art results in tasks such as classification and segmentation [3]. Recently, our group trained a CNN model to segment features of axonal degeneration on phase-contrast microscopic images (unpublished data). The aim of this study was to train a CNN model for the automatic detection of axonal degeneration for fluorescence microscopy data.

2 Material and methods

2.1 Convolutional neural networks

CNNs are networks employing convolutional layers inspired by the human nervous system. In composition with non-linearities and pooling layers, long sequences of convolutional layers (i.e. deep models) can detect increasingly complex and abstract features, such as eyes, cells, wheels etc. [3]. Deep learning can be performed in a supervised or unsupervised manner. In this study, a supervised form of machine learning was implemented. During supervised learning, the model algorithm is provided with a dataset, whose desired output is already known. Hence, during training, the values generated by the CNN (the predictions) are compared to the targeted output labels [4]. The model learns by adjusting the weights of the neurons in the network after every time it has passed through a collection of image samples (batch) [5].

We separated our deep learning approach into three distinct processes: Training, validation and testing (Fig. 1). If the validation results are insufficient, the model has to be trained additionally [4].

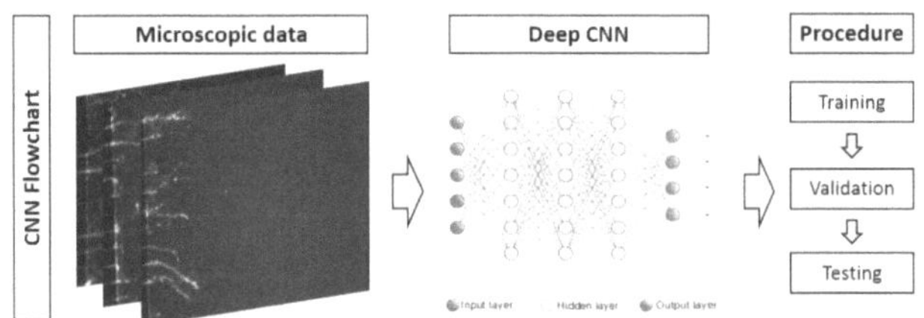

Figure 1: **Flowchart of the CNN to analyze data from fluorescence microscopy for the quantification of axonal degeneration.** Different, independent microscopic datasets are used for training, validation and testing.

2.2 Fluorescence dataset

Primary cortical neurons were isolated from E14 embryos from Crl:CD1 (ICR) Swiss outbred mice (Charles River) (prospective contingent animal license number 2017-07-06 Zille, approved by the Schleswig-Holstein Ministry for Energy Transition, Agriculture, Environment, Nature and Digitalization). Cells were grown in axon isolation devices as previously described by Zhang and colleagues [6] for 7 days and axons were exposed to 0, 50, 100 or 200 μM of hemin to model hemorrhagic stroke *in vitro*. After 24 hours, the axons were stained with Calcein AM to visualize living cells as previously described by Miles and colleagues [7] and images were acquired with an Olympus IX81 microscope.

2.3 Training of the CNN

For the CNN to learn to detect axonal degeneration in fluorescence images, i.e. to perform correct segmentation, 20 fluorescence images of different hemin concentrations were labelled using GIMP v.2.10.14. The different classes were assigned different colors on separate layers, as shown in Table 1 and Fig. 2. We used a variant of the U-net architecture [5] using the VGG16 network as an encoder [8]. We trained the network for 120 epochs using the adam optimizer, a batch size of 2 and a learning rate of 0.01.

Table 1: **Class segregation**

Class	Labelled structures	Color
Axon	Only healthy axons	Blue
Axonal spheroid	Large, bright spherical structures within the axons	Green
Axonal fragment	Degenerated or separated structures originating from the axons	Red
Background	All pixels not included in the other classes remain un-labelled	Black

2.4 Validation of the CNN

To objectively validate our trained CNN, we labelled six additional images that were unknown to the CNN. During

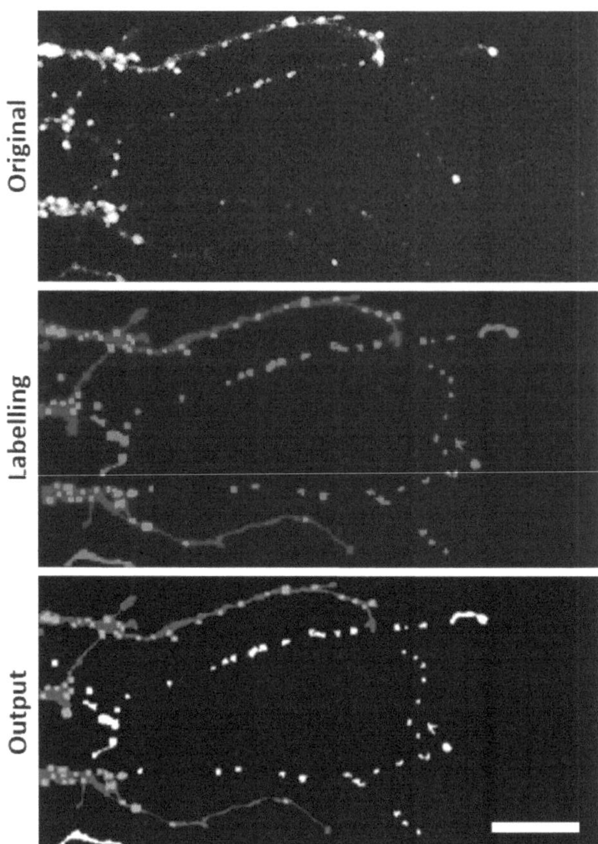

Figure 2: **Manual labelling of the fluorescence datasets.** Representative image of axons stained with Calcein AM that was manually labelled according to the class segregation in Table 1 and segmented by the CNN. Scale bar = 50 μm.

validation, the prediction of the CNN model was compared to the ground truth (labels) and the number of true positive (TP), false positive (FP), true negative (TN) and false negative (FN) pixels were counted. For each class, the accuracy, precision, recall and F1 scores were calculated as follows [9]:

$$Accuracy = \frac{TP + TN}{TP + FP + FN + TN} \qquad (1)$$

$$Precision = \frac{TP}{TP + FP} \qquad (2)$$

$$Recall = \frac{TP}{TP + FN} \qquad (3)$$

$$F1\ Score = \frac{2 * (Recall * Precision)}{(Recall + Precision)} \qquad (4)$$

The F1 score is the harmonic mean of precision and recall, which includes both false positives and false negatives. This is useful for uneven class distribution, as is the case in the present segmentation task.

2.5 Testing

We used the CNN on a new, unlabelled dataset of the hemin concentration response with eight different positions per concentration to analyze axonal degeneration in the model of hemorrhagic stroke. The area covered by each class was calculated relative to the image size. The axonal degeneration index was calculated as the ratio between the relative area of axonal fragments and the relative area of the sum of axons, axonal spheroids and axonal fragments [10].

3 Results and discussion

The goal of this study was to implement a CNN-based image segmentation analysis of axonal degeneration to automatically evaluate fluorescence microscopy data.

3.1 Validation results

A CNN model was trained to segregate the fluorescence microscopy data into different classes (background, axons, axonal spheroids, axonal fragments). Table 2 shows the performance values of the validation dataset for the trained CNN. The accuracy for all the classes was high (> 0.92). Accuracy is a good measure to depict the performance of a model, but only for symmetric datasets that have an even class distribution. For uneven class distributions, the F1 score is preferable as it disregards TNs, which can be misleading. In our case, we do not consider accuracy as an appropriate performance measure for our CNN as the background pixels cover most parts of the fluorescence images [11].

Table 2: **Performance results of the CNN trained with fluorescence microscopy datasets**

Class	Accuracy	Precision	Recall	F1 Score
Background	0.922	0.999	0.919	0.958
Axon	0.924	0.293	0.915	0.444
Axonal spheroid	0.996	0.289	0.649	0.400
Axonal fragment	0.993	0.220	0.434	0.292

The precision, recall and F1 score were high for the class background. This is not surprising as background dominates the area covered in the fluorescent images and is comparably easy to identify. As for the class axon, the precision and F1 score were low because the FP rate was high, meaning that some of the pixels labelled for the class axon belonged to other classes. In contrast, the recall score was high (0.915) due to the low FN rate, meaning that only few pixels that were labelled for other classes belonged to the class axon.

The performance values in the classes axonal spheroid and axonal fragment were comparably lower, probably because they occur at a lower frequency on the images and thus, the

CNN has less information to be trained with. Another possibility is that the labelling is subjective and likely harder for these classes, which can lead to more uncertainty. Further training data will be needed to improve the fluorescence CNN model to recognize the different classes.

With increasing number of training samples provided to the CNN model, the degree of generalization also increases and so does the ability of the model to automatically recognize different axonal features. In contrast, other frequently used methods for quantifying morphological changes in axons are semi-automatic [2].

One of the main drawbacks of using CNNs is the increase in the computational time with increasing the dataset size to train the system. Additionally, since a supervised form of learning was implemented here, perfect labelling of the training datasets is of great importance. Hence, this requires additional manual labor. Although significant progress was achieved in terms of the detection of the various features of axonal degeneration, the model can be further improved. This can be predominantly achieved with increasing the number of training samples.

3.2 Testing results

The testing was done using 32 images of different concentrations of hemin (0, 50, 100, 200 μM, eight images per concentration). The relative area of the axons decreased with increasing hemin concentration, while the axonal fragments and the axonal degeneration index increased. The axonal spheroid area, however, was similar between the different hemin concentrations (Fig. 3). Our results are in accordance with the fluorescence images (Fig. 4) and demonstrate that the CNN model can accurately detect changes in axonal degeneration on fluorescence images.

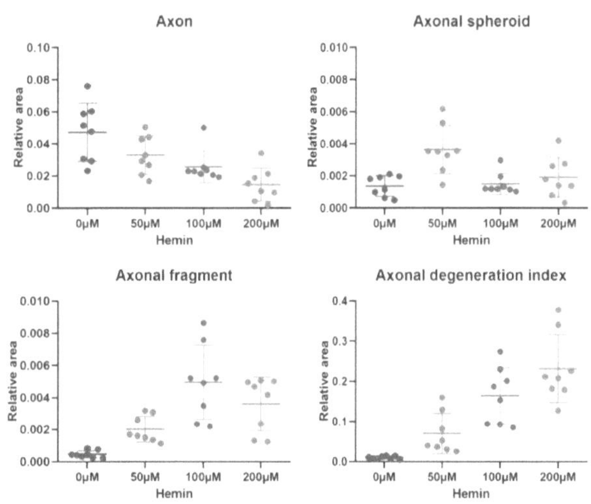

Figure 3: **Testing results showing the different features of axonal degeneration for the different hemin concentrations.** Eight images per hemin concentration were analyzed and the individual data points are shown with mean +/- standard deviation.

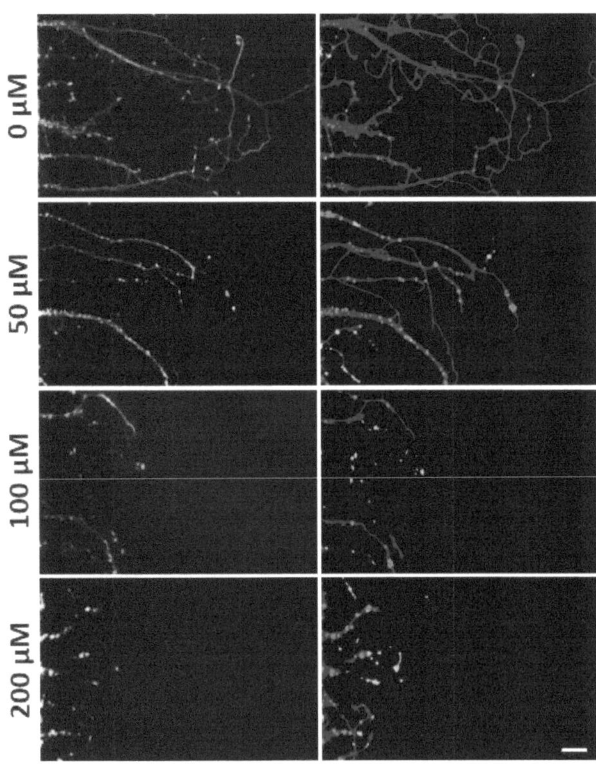

Figure 4: **Calcein AM images (left) and corresponding output masks (right) of the different hemin concentrations.** Scale bar = 50 μm.

4 Conclusion

We successfully trained a CNN model using fluorescence microscopy datasets for the analysis of axonal degeneration. In the future, we plan to investigate axonal degeneration using the newly trained CNN in the *in vitro* model of hemorrhagic stroke.

5 Contribution of authors

M.Z. conceived the project. A.P. performed the cell culture experiments and provided the microscopic datasets. P.G. developed the deep learning tool and L.H. and L.B. developed the algorithms to retrieve the output. N.K.M. and S.K.L. labeled the datasets. N.K.M., A.P., and M.Z. discussed and interpreted the data. N.K.M. and M.Z wrote the manuscript. All authors discussed and commented on the final version of the manuscript.

Acknowledgements

This work has been carried out at the Fraunhofer Research Institution for Marine Biotechnology and Cell Technology and was supported by a Fraunhofer MEF grant (Project number 600199) to M.Z.

6 References

[1] E. Meijering, M. Jacob, J.-C.F. Sarria, P. Steiner, P. Hirling and M. Unser, Design and *Validation of a Tool for Neurite Tracing and Analysis in Fluorescence Microscopy Images*, Cytometry Part A, vol 58A, 167–176, 2003.

[2] L. Yang, M. Yang, H. Zhuo, C. Xiaoping, T.M. Maloney, Z. Li, et al., *AxonQuant: A Microfluidic Chamber Culture-Coupled Algorithm That Allows High-Throughput Quantification of Axonal Damage*, Neurosignals, vol 22, 14-29, 2014.

[3] E. Moen, D. Bannon, T. Kudo, W. Graf, M. Covert and D.V. Valen, *Deep learning for cellular image analysis*, Nature methods, vol 16, 1233–1246, 2019.

[4] T. Hastie, R. Tibshirani, and J.H. Friedman, *The elements of statistical learning: data mining, Inference and Prediction*, Springer series in statistics, vol 2, 390-395, 2016.

[5] O. Ronneberger, P. Fischer, T. Brox, *U-net: Convolutional networks for biomedical image segmentation. In International Conference on Medical image computing and computer-assisted intervention* , MICCAI 2015 Lecture Notes in Computer Science-Springer, vol 9351, 234-241, 2015.

[6] K. Zhang, Y. Osakada, M. Vrljic, L. Chen, H.V. Mudrakola, and B. Cui, *Single-molecule imaging of NGF axonal transport in microfluidic devices*, Lab on a chip, vol 10, 2566-2573, 2010.

[7] F.L. Miles, J.E. Lynch, and R.A. Sikes, *Cell-based assays using calcein acetoxymethyl ester show variation in fluorescence with treatment conditions*, Journal of Biological Methods, vol 2(3), e29, 2016.

[8] K. Simonyan and A. Zisserman, *Very deep convolutional networks for large-scale image recognition*, Published as a conference paper at ICLR 2015, 2014.

[9] D.M.W. Powers, *Evaluation: from Precision, Recall and F-measure to ROC, Informedness, Markedness and Correlation*, Journal of Machine Learning Technologies, vol 2(1), 37-63, 2011.

[10] Y. Sasaki, B.P.S. Vohra, F.E. Lund and J. Milbrandt, *Nicotinamide Mononucleotide Adenylyl Transferase-Mediated Axonal Protection Requires Enzymatic Activity But Not Increased Levels of Neuronal Nicotinamide Adenine Dinucleotide*, The Journal of Neuroscience, vol 29(17), 5525–5535, 2009.

[11] A. Moses, *Statistical modeling and machine learning for molecular biology*, CRC Press Taylor Francis Group, vol 1, 3-26, 2017.

A low-cost, open-source device for characterizing super-paramagnetic nanoparticles

Katrin Brandt [1], Eli Mattingly [2,3], Erica E. Mason [2,3], Monika Śliwiak [2], Lawrence L. Wald [2,4], and Thorsten M. Buzug [5]

[1] Medical Engineering Science, Universität zu Lübeck, katrin.brandt@student.uni-luebeck.de

[2] MGH/HST A.A. Martinos Center for Biomedical Imaging, Dept. of Radiology, Massachusetts General Hospital, Boston, MA, USA, msliwiak@mgh.harvard.edu, wald@nmr.mgh.harvard.edu

[3] Harvard-MIT Health Sciences & Technology, Cambridge, MA, USA, {ematting, ericamas}@mit.edu

[4] Harvard Medical School, Boston, MA, USA

[5] Institute of Medical Engineering, Universität zu Lübeck, buzug@imt.uni-luebeck.de

Abstract

Magnetic Particle Imaging (MPI) is an emerging imaging modality that utilizes superparamagnetic iron oxide nanoparticles as a tracer agent. MPI boasts high sensitivity and resolution; these characteristics fundamentally depend on the magnetic properties of the particles. Consequently, it is essential to understand these properties. Magnetic Particle Spectroscopy (MPS) is a type of magnetometry with many similarities to MPI and therefore offers an ideal method for characterizing MPI agents. Currently, existing spectrometers can be prohibitively expensive restricting the number of new investigators looking to employ this technique. To fill this need for an accessible platform, we developed a low-cost, open-source MPS device capable of providing detailed magnetization curves. The system allows one to examine and compare different nanoparticles under various excitation parameters. By providing the design information on a public GitHub page, new MPI researchers can more easily enter the field and benefit from an affordable particle analysis.

1   Introduction

Magnetic Particle Imaging (MPI) was introduced as a new imaging technology in 2005 by B. Gleich and J. Weizenecker [1]. It functions by mapping the distribution of super-paramagnetic iron oxide nanoparticles (SPIOs). MPI uses the non-linearity of the SPIO's magnetization curve combined with an oscillating external field to localize the particles as seen in Fig. 1. The magnetization increases with increasing magnetic field strength up to a certain point and then saturates, as described by the Langevin theory of super-paramagnetism. A sharp Langevin curve is desired since it corresponds to a narrow point spread function and thus to high resolution. This can be achieved by a large magnetic moment and rapid saturation [2].

A time-dependent, sinusoidal magnetic field with the excitation or "drive" frequency f_1 is applied for signal generation (Fig. 1 left). The time-dependent magnetization of the particles is detected as an induced voltage in the surrounding solenoidal receive coil via faraday detection. Since the Langevin curve is non-linear, the spectrum of the response also contains higher order harmonics [1]. A selection field with a central region of net-zero flux (field-free region, FFR) is superimposed to obtain spatial encoding (Fig. 1 right). This leads to saturation of the particles outside the FFR and therefore eliminates the particle signal because this part of the Langevin curve has zero slope [1].

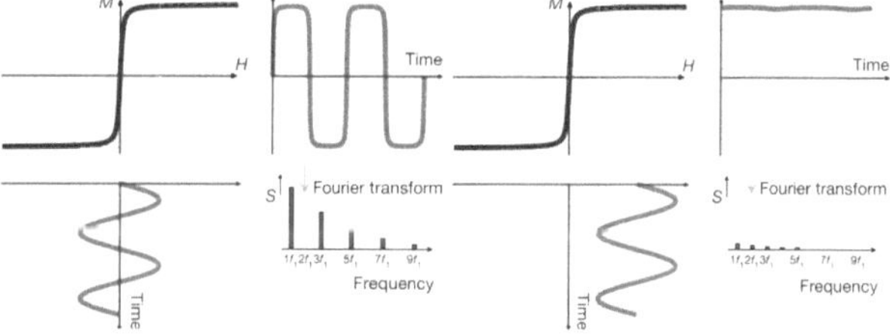

Figure 1: Signal generation and spatial encoding in MPI. Figure reproduced from: [1].

Since its introduction in 2005, there has been substantial research to improve MPI quality. The image quality not only depends on the imaging system itself (drive field strength, selection field gradient, etc.), but also on the magnetic nanoparticles. To determine their properties, it is possible to use Magnetic Particle Spectroscopy (MPS) [3]. The configuration of an MPS system is the same as an MPI scanner, but without a selection field. Its purpose is to determine the properties of the nanoparticles, such as the magnetization curve, Brownian and Néel relaxation effects or the frequency-dependent hysteresis, all which contribute to the MPI signal. Together, these are used to define a particle model for some MPI reconstruction algorithms [2], [3].

Traditionally, MPS instruments are limited to mono-frequency excitation by a bandpass transmission filter. Furthermore, the particle signal at f_1 is often eliminated by a bandstop receive filter to attenuate the transmission feedthrough interference. However, several methods have already been presented that overcome these limitations. References [2] and [3] present the use of multiple drive frequencies to more completely analyze particle properties. This allows different drive field frequencies to be explored for MPI. For multiple drive frequencies, it is necessary to filter the transmit signal and cancel out the feedthrough without relying on fixed filters. We employ a gradiometer receive coil, which provides an alternative and broadband method for eliminating transmit feedthrough. The gradiometer utilises a second coil, the "cancellation coil", with opposite winding direction in series to the main receive coil. Both coils of the gradiometer should be located symmetrically in the drive coil to receive the same transmitting signal but with 180° phase shift. This attenuates the induced voltages from the drive coil allowing the particle signal to better cover the dynamic input range of the analog-to-digital converter [4].

In its simplest mode of operation, the MPS instrument simply measures the frequency spectra of the SPIOs excited by the drive field (high amplitude AC field). By adding a near-DC biasing field, relaxometry and magnetometry can be performed. In relaxometry, a large near-DC field is added to the AC field using a separate coil. This emulates the effect of the selection field [5]. The magnetometry feature attempts to measure the magnetization curve directly by reducing the drive field to a small amplitude (creating only small, approximately linear, perturbations along the magnetization curve) and modulating the operating point with a high amplitude near-DC field. This is similar to the Superparamagnetic Quantifier [6].

The presented MPS project is part of the open-source project "OS-MPI", which is accessible on a GitHub page: `https://github.com/OS-MPI`. The goal of this project is to lower the bar for new groups to enter into the field of MPI. Providing system designs and thoughts from the design process in a repository shall give them an easier start. Since the SPIOs are important for the MPI system's function and especially characterizing their properties enables more informed reconstructions, one part of the project is the development of a cheap, simple 1D MPS system.

2   Material and Methods

The basic configuration of the developed MPS device is illustrated in Fig. 2. The data acquisition system (DAQ, USB-6211, National Instruments) is interfaced to a computer and generates all the applied waveforms and records the data. Custom software controls the transmission sequences for the biasing and drive (Tx) coil and also postprocesses the acquired data from the receive (Rx) coil. The 20 Hz sinusoidal biasing field is created by the biasing coil which is driven by an audio amplifier (XTi2002, Crown International). The biasing coil consists of magnet wire (20 AWG, 6 layers, 100 turns per layer) and is wound around a copper tube, which is represented by a dashed line in Fig. 2. The drive field which excites the SPIOs is created

by the Tx coil driven in the kHz-range by a high-current op-amp (Apex Microtechnology model PA12A). To have more flexibility to analyze different particle properties, one of the three different frequency ranges (10 kHz, 25 kHz, and 40 kHz) can be applied after manually switching the variable Tx filter. The drive coil is a Litz wire solenoid coil (20 AWG, 2 layers, 70 turns per layer) to reduce losses at high frequencies. The Rx coil is inside the Tx coil and both are inside the copper tube. The receive coil is a gradiometer out of magnet wire (34 AWG, 450 turns per coil). The samples are inside 3 mm glass bulbs (18 μl) and can be moved in and out of the coils (5 mm bore) with a stepper motor. A low noise instrument amplifier (INA217, Texas Instruments Inc.) amplifies the detected signal to resolve the particle signal.

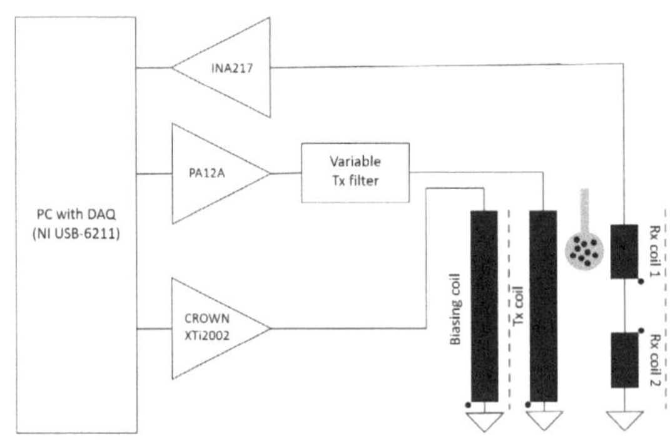

Figure 2: Overview of the system. The PC with the DAQ generates the biasing and the excitation signal, which are subsequently amplified. The gradiometer cancels the transmission feedthrough and detects the particle signal, which is amplified before the following data analysis.

Most of the system components are designed in the CAD-Software Autodesk Inventor (Autodesk Inc.). A half-section view of the main parts can be seen in Fig. 3. The middle part shows the coil set-up. The Rx and as well the Tx coil holders are 3D-printed. The Tx coil is mounted inside the copper tube and is movable in axial direction to decouple the Tx-Rx-assembly from the biasing coil, which is wound around the copper tube. The Rx coil is fixed to a threaded rod and with the use of a custom bearing, the gradiometer is adjustable in position to reach the maximal transmit feedthrough cancellation. With this mechanism, we reached an attenuation of 71 dB. The sample needs precise positioning to achieve a reproducible measuring process. The left part of Fig. 3 shows the sample holder which moves the sample in and out of the system and position it accurate and repeatable with the use of a stepper motor.

The purpose of the Tx filter is to reduce the inductive impedance of the Tx coil at the drive frequency and simultaneously attenuate spurious harmonics which can mimic the signal. To achieve this, a fourth-order bandpass filter was simulated in LTspice (Analog Devices Inc.). The given output impedance of the PA12A and the measured values of the drive coil have been considered. Fig. 4 shows the filter schematic in which the two parallel capacitors have to be exchanged with the drive frequency. Table 1 lists the

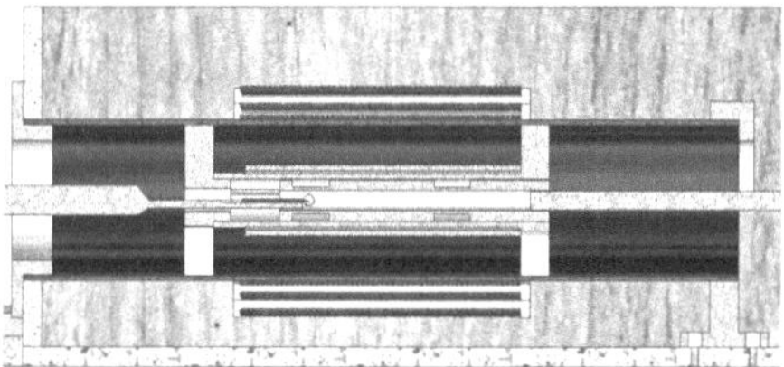

Figure 3: Half-section view in Autodesk Inventor. The left part shows the sampleholder. The center illustrates the concentrically arranged coils.

capacitor values for the different excitation frequencies and the corresponding simulated attenuations of the respective third harmonics. The filter responses are presented in Fig. 5.

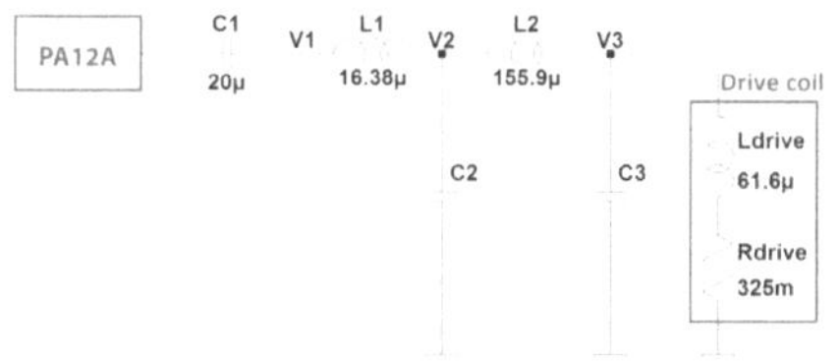

Figure 4: Schematic of Tx filter design. To change the filter response and thereby the drive frequency the capacitors C2 and C3 are varied.

Table 1: Values for different filter layouts.

Frequency	C2	C3	f_3 attenuation
10 kHz	11 μF	5 μF	60 dB
25 kHz	5 μF	0.94 μF	76 dB
40 kHz	0.56 μF	0.33 μF	72 dB

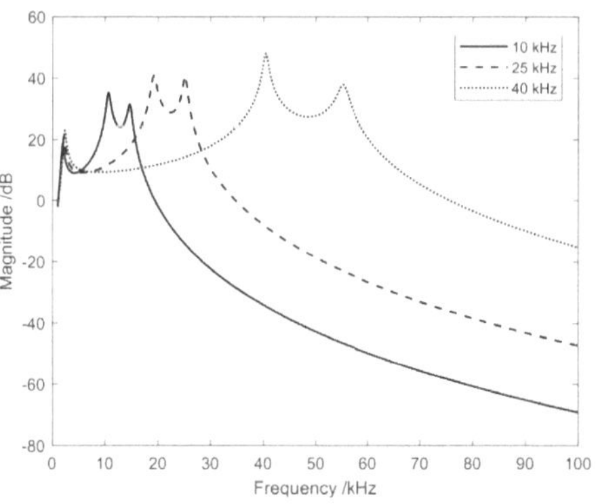

Figure 5: Simulated filter responses.

The presented filter is implemented inside of an aluminum box for shielding the outer parts of the system from the high-power transmitting chain. Air core toroidal inductors wound with Litz wire are chosen to limit flux leakage which might couple into the Rx chain. Capacitive feedthroughs reduce the coupling of higher frequency interference into the Tx and Rx chain.

Fig. 6 shows a picture of the designed MPS device including the filter inside the aluminum box. The PA12A is mounted

to this box. The coil set-up is fixed in a wooden box and the sample holder assembly is mounted to it. A zoom shows the sample in the holder in front of the bore. The smaller box contains the low noise amplifier in the receive chain. Not shown is the XTi2002.

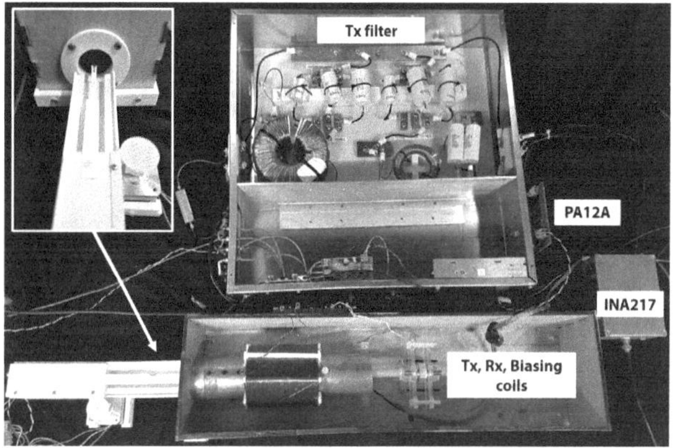

Figure 6: Picture of designed system. The lower part shows the main body with the three coils (Rx and Tx inside the copper tube) and the gradiometer adjusting mechanism inside a wooden box. On the left side is the sample positioning construction and also a detailed view of it.

All software is written in MATLAB (The MathWorks Inc.). The software controls the sending of the drive and biasing signal, the movement of the stepper motor via an Arduino and consequently the positioning of the sample. Within associated functions, the detected signal is processed. The measuring process consists of two primary steps. First, the positioning of the sample: The sample is moved into the bore and at every step of the motor (0.3 mm) a 0.2 s excitation with the drive field is applied and the received signal is recorded. The position generating the highest particle signal is then used in the main measurement. During the next step, the particles are moved to their position and drive and biasing fields are applied. The detected voltage in the Rx coil and the current in the biasing coil (measured through a hall effect current sensor) are acquired simultaneously during transmission. In the subsequent algorithm, the received signals are averaged over time. We finally generate the magnetization curve of the particles and can fit the Langevin function to the data. Thereby, different particles are comparable. For the experiments two different particles are used: Ocean NanoTech (SPP-25-25, 5 mg/ml concentration, Ocean NanoTech) and VivoTrax (5.5 mg/ml concentration, Magnetic Insight Inc.).

3 Results and Discussion

The designed MPS system with a receive chain bandwidth of 125 kHz is capable of performing the introduced three modes of operation: spectroscopy, relaxometry and magnetometry. The magnetometry mode produces magnetization curves of SPIOs in ~12 s. As a result, it is possible to compare different particles and their properties for imaging. Currently, the detection limit is 500 ng VivoTrax particles

(50 mT$_p$ biasing field at 20 Hz and 5 mT$_p$ drive field at 24.3 kHz).

Fig. 7 shows an example of the acquired signal from undiluted Ocean NanoTech particles. The used settings in magnetometry mode were: 50 mT$_p$ biasing field at 20 Hz, 1.25 mT$_p$ drive field at 10.7 kHz, 4 s measurement time, 3 repetitions. The plot shows the measured biasing current and the induced voltage in the receive coil averaged over one biasing period and normalized. These results show that the induced voltage in the Rx coil decreases with increasing biasing field. The maximum signal is detected when no biasing field is present. This indicates that the SPIOs go into saturation due to the applied field and therefore the magnetization change becomes small or vanishes and thus the induction decreases. The remaining signal can be caused by noise or transmission feedthrough. To see the signal in more detail, the figure also illustrates a zoom into the induced receive coil voltage. It demonstrates that the particle signal consists of a high-frequency oscillation. The particles are excited by the drive field and respond with the excitation frequency and higher order harmonics.

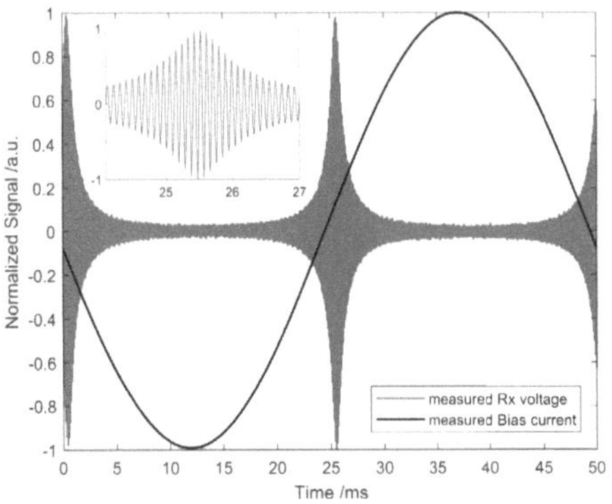

Figure 7: Averaged and normalized data from a measurement of Ocean NanoTech particles. A zoom into the Rx coil signal shows a high frequency oscillation.

The magnetization curve calculated from this data is presented in Fig. 8. In addition, the curve of the VivoTrax particles, which was recorded under the same measurement conditions, is overlaid. The Ocean NanoTech particles exhibit a steeper magnetization curve compared to the VivoTrax ones. This corresponds to a narrower point-spreadfunction, which can be an advantage in imaging. The Ocean NanoTech particles show saturation at $\sim$15 mT, the VivoTrax ones at $\sim$10 mT, important information for the selection field in MPI.

4   Conclusion

The presented system is capable of examining and comparing SPIOs under different experimental conditions and determining their magnetization curves. Looking forward, the device will be used to provide particle analysis for informing reconstruction algorithms in MPI systems.

Future work will also include replacing the XTi2002 with the TDA7391 (STMicroelectronics N.V.). This substitution

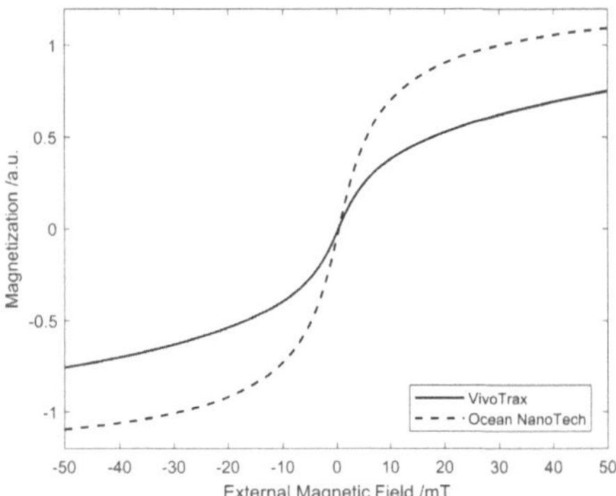

Figure 8: Calculated magnetization curves for VivoTrax and Ocean NanoTech particles in comparison (Langevin fits not shown for clarity).

will reduce the cost of the overall system to about 500 USD excluding the PC and the DAQ. Because of the low manufacturing costs and the dissemination of the design information on GitHub, the presented system provides new groups in the field of MPI the equipment to evaluate their SPIOs and thus obtain important information for imaging.

Acknowledgement

The work has been carried out at the A.A. Martinos Center for Biomedical Imaging, and supervised by the Institute of Medical Engineering, Universität zu Lübeck.

5   References

[1] B. Gleich and J. Weizenecker, "Tomographic imaging using the nonlinear response of magnetic particles," *Nature*, 2005.

[2] Z. W. Tay et al., "A high-throughput, arbitrary-waveform, MPI spectrometer and relaxometer for comprehensive magnetic particle optimization and characterization," *Scientific Reports 6*, 2016.

[3] A. Behrends, M. Graeser, and T. M. Buzug, "Introducing a frequency-tunable magnetic particle spectrometer," *Current Directions in Biomedical Engineering*, 2015.

[4] M. Graeser, T. Knopp, M. Gruettner, T. F. Sattel, and T. M. Buzug, "Analog receive signal processing for magnetic particle imaging," *Medical Physics*, 2013.

[5] N. Garraud et al., "Benchtop magnetic particle relaxometer for detection, characterization and analysis of magnetic nanoparticles," *Physics in Medicine & Biology*, 2018.

[6] M. van de Loosdrecht et al., "A novel characterization technique for superparamagnetic iron oxide nanoparticles: The superparamagnetic quantifier, compared with magnetic particle spectroscopy," *Review of Scientific Instruments*, 2019.

The semantic congruence effect is not impaired in older subjects but associated with changes in neural processing

Ricardo J. Alejandro [1], Pau A. Packard [2], Nico Bunzeck [2]

[1] Biomedical Engineering, Lübeck University of Applied Sciences, ricardo.alejandro.benavides@stud.th-luebeck.de
[2] Institut für Psychologie I, Universität zu Lübeck, {pau.packard, nico.bunzeck}@uni-luebeck.de

Abstract

Long-term memory can be promoted when incoming information is congruent with pre-existent semantic information. This so-called congruence effect has been widely shown in younger adults but age-related changes and underlying neural mechanisms remain unclear. In this study, semantic congruence improved recognition memory in younger and older adults. Electroencephalography (EEG) data showed that congruence during encoding led to differences in event-related potentials (ERPs) during retrieval. However, a later ERP component (400-600 ms) revealed the congruence effect only in the younger subjects. Together, with overall enhanced ERP amplitudes in the older subjects, our findings suggest that congruence drives recognition memory across the life span but there are age-related changes at the neural level.

1 Introduction

The encoding of novel information into long-term memory can be promoted by presenting it within a known context. This well-described behavioral effect is known as the semantic congruence effect (i.e. 'congruence effect') and has been demonstrated in humans with a priori semantic information followed by congruent (vs. incongruent) material [1], [2]. Input information is supposed to be linked to pre-existent representations in memory, finding correspondences and associations to form predictions and facilitate cognition. Therefore, pre-existent general knowledge plays a key role for selecting inputs according to goals [3].

Since most studies have focused on younger adults (i.e. 18 – 35 years), the potential age-related changes and neural mechanisms involving the congruence effect remain unclear. Memory deficits are expected in older adults, as a normal well-described aspect of healthy aging, semantic memory (i.e. long-term memory for facts independent of time and date) is often preserved until old age [4].

Electroencephalography (EEG) provides high-temporal resolution to observe the probable neural correlates of these cognitive processes. We used EEG to investigate the neural mechanisms involved in the age-related differences of the congruence effect. To this end, we implemented an adapted word list paradigm from a previous study [2], which found a strong memory enhancement for congruent stimuli. The experiment comprised a learning phase (i.e. encoding phase) and a retrieval phase (i.e. test phase). Results form the encoding phase have already been published online [5]; this work summarizes the findings from the test phase. We expected a weaker congruence effect in older adults as well as specific age-related effects in post-stimulus ERPs.

2 Material and Methods

2.1 Participants

Twenty-three young (ages $18 - 28$ years, mean 22.25 ± 2.82 years, 9 males) and twenty-five older participants (ages 52 $- 79$ years, mean 65.33 ± 5.12 years, 8 males) took part in the experiment. As described in [5], all participants were healthy, right-handed, had normal or corrected-to-normal vision (including color vision) and reported no history of neurological or psychiatric disorders, or current medical problems (excluding blood pressure). This study and the protocol were carried out in accordance with the recommendations and approval of the local ethics committee (University of Lübeck). Each participant gave written informed consent. Cognitive abilities were assessed by using the Montreal Cognitive Assessment version 7 [6].

2.2 Materials

As explained in [5], experimental stimuli consisted of 66 categorical 6 word lists [2] translated into German. Each list consisted of the 6 most typical instances (e.g., Mercury, Mars, Neptune, Earth, Venus, Saturn) of a semantic category (e.g., planet). All of the 396 typical instances, belonging to the 66 semantic categories, were presented in separate encoding trials. For the test phase, we used another set of semantically unrelated words as control.

2.3 Behavioral procedures

We used a modified version of a previous EEG experiment [2] itself adapted from other paradigms [7]. The paradigm (as described in detail by [5]) for the encoding and test phase is shown in Fig. 1, a) and b), respectively. The encoding phase consisted of 396 separate one word trials, presented in random order. A fixation cross was shown on the screen for a random duration of 2000-3000 ms at the beginning of each trial. Subsequently, a category name (semantic cue) appeared for 1500 ms, followed by a fixation cross for 2000 ms. Participants were then sequentially shown the subsequent word for 1000 ms. In the congruent condition, the subsequent word belonged to the semantic category [1], for example, 'music' followed by 'jazz'. In the incongruent condition, the subsequent word did not correspond to the presented category, for example 'fruits' followed by 'bed'. While the second word was shown, participants pressed one of two buttons indicating whether the word was congruent or incongruent with the semantic category.

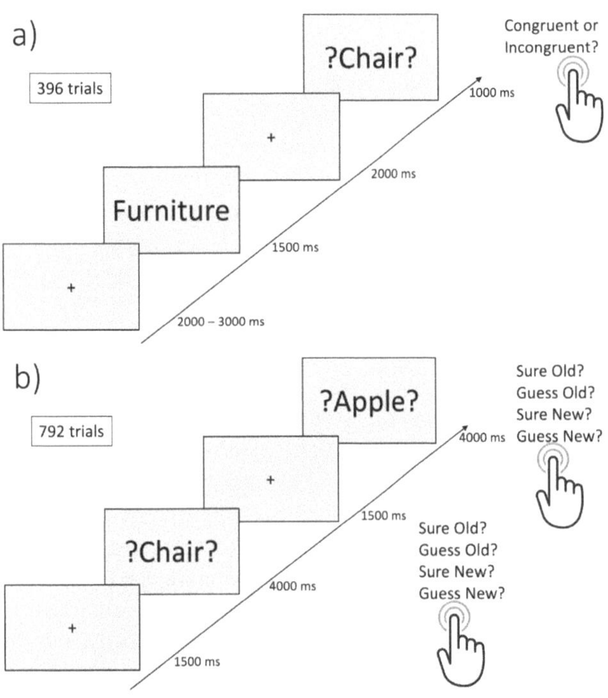

Figure 1: Experimental Paradigm. a) Encoding phase. Each word is shown after a semantic category. b) Test phase. All the words from the encoding phase, in addition to distractors (new words), are sequentially shown.

Each category had an equal amount of congruent and incongruent word pairs during encoding. The retention interval between the encoding and test phases was 10 min, following previous work [2] and identical to [5], allowing the paradigm to capture encoding and retrieval components and not the consolidation-dependent processes underlying the congruence effect, which require longer intervals.

The test phase contained 396 Old-word (all encoding-items) and 396 New-word trials. The trials were presented pseudo-randomly for each participant. Each of the 792 trials started with a fixation cross on the screen (1500 ms). Then a word was displayed for 4000 ms. Next, participants had to deem said word as "sure old," "guess old," "guess new," or "sure new," by pressing the corresponding key, within 4000 ms.

2.4 Statistical analyses of memory results

Analysis of Variance (ANOVA), with encoding condition (two levels: Congruent vs Incongruent) as a within-participant factor, and age group (two levels: Young and Older participants) as a between-participant factor, was performed on the response rates, using Jamovi Version 1.0.8.0. For all analyses, α (Type I error rate) was set at 0.05. To estimate effect sizes, we used η_p^2 (Partial Eta Squared). Given that older participants typically show age-related impairments for high-confidence responses [8], we focused on high-confidence responses. Corrected Hit Rates (CHR) were calculated by subtracting the proportion of false alarms (FA, 'old' responses to 'new' items) from the proportion of hits (correct 'old' responses).

Post-hoc t-tests were used to break down significant interactions detected in the ANOVAs, with Bonferroni correction by lowering the significance level according to the amount of post-hoc tests run for each ANOVA respectively. Tests which were not significant according to the lower Bonferroni adjusted significance levels are stated as such.

2.5 EEG Analysis

As detailed by [5], EEG activity was acquired with an Easy Cap system by BrainProducts with 32 standard active electrodes. For detecting vertical and horizontal eye movement (VEOG / HEOG), 4 electrodes were used. Impedances were maintained under 20 kΩ. Electrodes FCz and AFz served as reference and ground, respectively. The sampling rate was 500 Hz with online high-pass (0.1 Hz) and low-pass (240 Hz) filters. EEGLAB (version 2019; [9]) and customized MATLAB version 2019b (The MathWorks) tools were used for pre-processing the EEG data offline.

2.6 ERP Analysis

For the ERP analysis, we followed the procedures from the encoding data analysis [5]; the data were filtered offline with a low-pass filter (Hamming window, cut-off frequency at 40 Hz, filter order at 166), and no additional high-pass filtering. All trials were epoched and down-sampled to 125 Hz. ERPs were studied by extracting event-locked EEG epochs of 900 ms, ending 800 ms after stimulus onset, with 100 ms prior to stimulus onset used for the baseline. Trials with an amplitude exceeding 100 μV were rejected. Major artifacts, trials with amplifier saturation, and bad channels were visually identified and removed (maximum 4 channels, mean = 0.76). Afterwards, blinks, muscle and eye movement artifacts were removed with independent component analysis (ICA; [9]). Finally, bad channels were interpolated. Electrode Oz was selected to re-reference the data, as re-referencing to average can mask the effects of EEG differences with a broad distribution across the scalp [10].

For the memory analysis, trials correctly classified during encoding were divided into 'Remembered', and 'Forgotten' (including both congruent and incongruent trials). Only high-confidence correct 'old' responses were included in the 'Remembered' condition. Four young and five older participants were excluded from the analysis due to excessively noisy data or exceedingly low performance (compared to their age group, identified with Jamovi using a step of 1.5 x Interquartile Range). Fieldtrip [11] and customized MATLAB scripts were used for statistical data analysis. To detect reliable differences between Congruent vs. Incongruent, the conditions were contrasted using Fieldtrip via a two-tailed non-parametric cluster-based permutation test [12]. The test included all time points between 0 and 800 ms at 27 scalp electrodes (reference and eye electrodes are not considered). As implemented in [5], for all contrasts, a t-test was performed for every sample ((channel, time)-pair). All t scores corresponding to uncorrected p-values of 0.05 were formed into clusters. The sum of the t scores in each cluster was calculated and the maximum sum in each of the sets of tests was recorded and used to estimate the distribution of the null hypothesis. The Monte Carlo estimate was calculated by running 1000 random permutations of the condition labels and comparing the cluster statistics from the real data with the random data. The p-value is thus obtained with the proportion of cluster statistics in the random data exceeding that in the real data. Clusters were formed from significant samples ($p < 0.05$), considering only effects with at least three significant neighboring channels.

3   Results and Discussion

3.1   Behavioral findings

3.1.1   Main effect of congruence for CHR

The proportions of high-confidence 'Sure' responses during the test phase were analysed in a 2x2 ANOVA. This analysis showed a significant main effect for congruence ($F_{(1,46)} = 264.09$, $p < 0.001$, $\eta_p^2 = 0.85$) with higher CHR for congruent words (mean 0.57, Standard Error of the Mean (SEM) 0.02), than for incongruent words (mean 0.33, SEM 0.02). There was no main effect of age ($F_{(1,46)} = 0.123$, $p = 0.727$, $\eta_p^2 = 0.003$), and no congruence by age interaction effect ($F_{(1,46)} = 0.468$, $p = 0.497$, $\eta_p^2 = 0.010$).

3.1.2   Main effect of congruence and of age for Reaction Times (RTs)

For participants' RTs, there was a main effect of congruence ($F_{(1,46)} = 22.38$, $p < 0.001$, $\eta_p^2 = 0.33$), incongruent words took longer times to be identified (see Table 1). Age also had a significant effect ($F_{(1,46)} = 5.72$, $p = 0.02$, $\eta_p^2 = 0.11$). There was no significant congruence by age interaction ($F_{(1,46)} = 0.302$, $p = 0.585$, $\eta_p^2 = 0.007$).

Table 1: Mean (SEM) of test Reaction Times in ms.

	Congruent	Incongruent
Younger adults	862.32 (50.46)	902.28 (52.91)
Older adults	1001.35 (36.99)	1045.16 (40.29)

3.2   EEG findings

3.2.1   ERP Cluster Analysis

A Monte Carlo cluster-based permutation test was performed on the 'Sure' responses for correctly 'Remembered' congruent vs. incongruent words, of young and older participants grouped together, from 0 ms to 800 ms after word onset. A negative cluster ($p = 0.004$) due to congruence was detected, from approximately 472 to 528 ms, with a mainly frontal topography but also including central and parietal electrodes (Electrodes: F3 Fz F4 FC5 FC1 FC2 FC6 T7 C3 Cz C4 T8 CP5 CP1 CP2 CP6 P3 Pz P4, as depicted in Fig. 2). The effect of congruence between young and older participants was contrasted in the same way (congruent minus incongruent in young participants vs congruent minus incongruent in older participants), with no significant results.

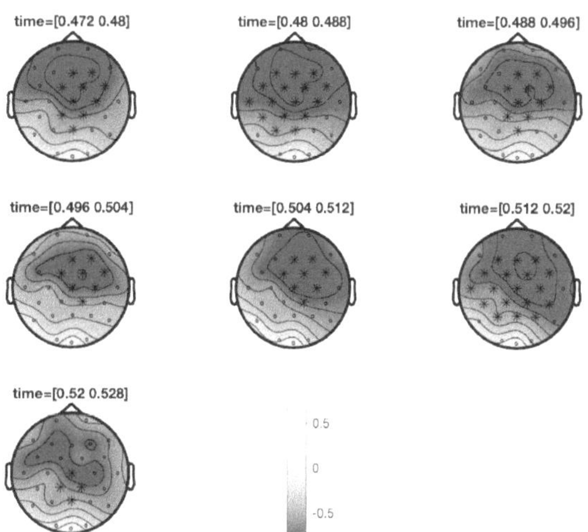

Figure 2: Negative cluster (472 - 528 ms). Units are μV.

3.2.2   ERP Analysis for Fz

In order to more thoroughly characterize the ERP effects, we studied the activity from the electrode Fz in the identified cluster. We included a time window from 100 ms before to 800 ms after stimulus, and, after visual inspection, focused the analysis on an early (100 – 250 ms) and a late (400 – 600 ms) time window. On a descriptive level, a clear difference between congruent and incongruent conditions can be seen in the obtained ERPs shown in Fig. 3, especially for younger participants. Additionally, the older group had a higher electric potential in both time windows. The data from each time window were analyzed using a 2x2 ANOVA. The early time window exhibited a main effect of congruence ($F_{(1,46)} = 3.97$, $p = 0.05$, $\eta_p^2 = 0.08$), but there was no effect of age ($p = 0.11$) and no congruence by age interaction ($F_{(1,46)} = 3.82$, $p = 0.057$, $\eta_p^2 = 0.077$). The late time window revealed a main effect of congruence ($F_{(1,46)} = 10.02$, $p = 0.003$, $\eta_p^2 = 0.179$), and a main effect of age ($F_{(1,46)} = 8.91$, $p = 0.005$, $\eta_p^2 = 0.16$), as well as a congruence by age interaction effect ($F_{(1,46)} = 4.39$, $p = 0.04$, $\eta_p^2 = 0.087$) . Subsequent post-hoc t-tests showed a significant

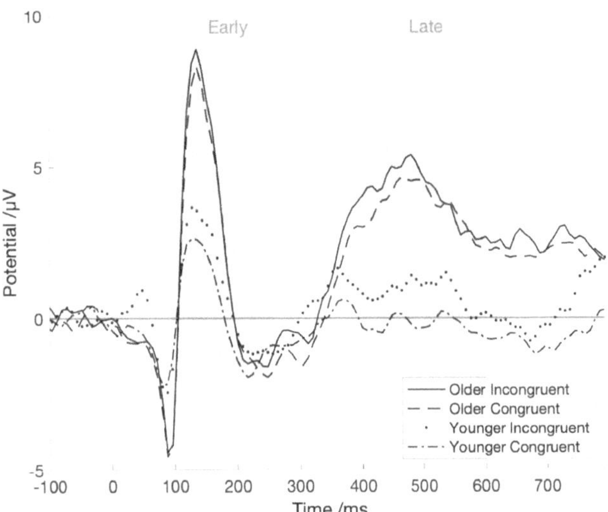

Figure 3: Contrasted ERPs of both conditions (Congruent vs. Incongruent) and both groups (Young vs. Old).

difference in congruence for younger participants ($t_{(46)}$ = -3.65, p = 0.004), which was absent in older participants ($t_{(46)}$ = -0.77, p = 1.00).

4   Conclusion

We used EEG to investigate the neural mechanisms associated with the congruence effect and possible age-related impairments. Congruence improved long-term recognition memory in young and older adults with no significant differences between age groups. At the neural level, an early component was associated with semantic congruence in both age groups suggesting that congruence during encoding drives the neural representation and early memory retrieval. In a later time-window, a congruence effect could only be observed in young participants. Although our older subjects did not differ in memory performance, this effect could be linked to age-related neural degeneration, which might be compensated by overall higher ERP amplitudes. Together, these findings give new insights into the behavioral and neural processes associated with congruence-dependent memory enhancements across the life span.

Acknowledgement

The work has been carried out at the Institut für Psychologie I, Universität zu Lübeck, supervised by Prof. Nico Bunzeck and supported by the German Research Foundation (Deutsche Forschungsgemeinschaft, Grant BU 2670/7-1 to N.B.). Special thanks to the entire Life-Span group for their continuos guidance and friendly support.

5   References

[1] F. I. M. Craik and E. Tulving, "Depth of processing and the retention of words in episodic memory," *Journal of Experimental Psychology: General*, vol. 104, no. 3, pp. 268–294, 1975.

[2] P. A. Packard, A. Rodríguez-Fornells, N. Bunzeck, B. Nicolás, R. de Diego-Balaguer, and L. Fuentemilla, "Semantic Congruence Accelerates the Onset of the Neural Signals of Successful Memory Encoding," *The Journal of Neuroscience*, vol. 37, no. 2, pp. 291–301, 2017.

[3] M. Bar, "The proactive brain: memory for predictions," *Philosophical Transactions of the Royal Society B: Biological Sciences*, vol. 364, no. 1521, pp. 1235–1243, 2009.

[4] T. Hedden and J. D. E. Gabrieli, "Insights into the ageing mind: a view from cognitive neuroscience," *Nature Reviews Neuroscience*, vol. 5, no. 2, pp. 87–96, 2004.

[5] P. A. Packard, T. K. Steiger, L. Fuentemilla, and N. Bunzeck, "The congruence effect is impaired in healthy aging: a relationship to event-related potentials and theta-alpha oscillations," Neuroscience, preprint, 2018.

[6] Z. S. Nasreddine, N. A. Phillips, V. Bédirian, S. Charbonneau, V. Whitehead, I. Collin, J. L. Cummings, and H. Chertkow, "The Montreal Cognitive Assessment, MoCA: A Brief Screening Tool For Mild Cognitive Impairment," *Journal of the American Geriatrics Society*, vol. 53, no. 4, pp. 695–699, 2005.

[7] H. Kim and R. Cabeza, "Trusting Our Memories: Dissociating the Neural Correlates of Confidence in Veridical versus Illusory Memories," *Journal of Neuroscience*, vol. 27, no. 45, pp. 12 190–12 197, 2007.

[8] Y. L. Shing, M. Werkle-Bergner, S.-C. Li, and U. Lindenberger, "Committing memory errors with high confidence: Older adults do but children don't," *Memory*, vol. 17, no. 2, pp. 169–179, 2009.

[9] A. Delorme and S. Makeig, "EEGLAB: an open source toolbox for analysis of single-trial EEG dynamics including independent component analysis," *Journal of Neuroscience Methods*, vol. 134, no. 1, pp. 9–21, 2004.

[10] S. J. Luck, *An introduction to the event-related potential technique*, ser. Cognitive neuroscience. Cambridge, Mass: MIT Press, 2005, oCLC: ocm57574045.

[11] R. Oostenveld, P. Fries, E. Maris, and J.-M. Schoffelen, "FieldTrip: Open Source Software for Advanced Analysis of MEG, EEG, and Invasive Electrophysiological Data," *Computational Intelligence and Neuroscience*, vol. 2011, pp. 1–9, 2011.

[12] E. Maris and R. Oostenveld, "Nonparametric statistical testing of EEG- and MEG-data," *Journal of Neuroscience Methods*, vol. 164, no. 1, pp. 177–190, 2007.

Development of a Monitoring System for the Team Foundation Server-Build and Test Infrastructure

Jan Magonov [1], Chethan-Kumar Mahadevaswamy [2], Marco Schett [2], and Thorsten M. Buzug [3]

[1] Medical Engineering Science, Universität zu Lübeck, jan.magonov@student.uni-luebeck.de
[2] Siemens Healthcare GmbH, Forchheim
[3] Institute of Medical Engineering, Universität zu Lübeck, buzug@imt.uni-luebeck.de

Abstract

The Build and Test management within the development of software for computed tomography requires a large IT infrastructure. In order to use the IT resources as efficient as possible, most components are virtualized. Since there are usually several hundreds of virtual machines in the infrastructure, it is essential to monitor the entire environment to ensure its availability. The system administrator must have an overview of the infrastructure and must be able to quickly detect anomalies that may occur. As a result of this publication a compact monitoring system is introduced, showing the most important metrics representing the infrastructure healthiness within two dashboards. Therefore state-of-the-art open-source software was used. Additional details about the systems architecture were integrated into the monitoring. Furthermore a framework for anomaly detection was tested and a script for the automated generation of network diagrams was developed.

1 Introduction

Computed tomography (CT) has become indispensable in today's clinical diagnostics [1]. To cope with the increasing complexity in software development, companies use the concept of Product Lifecycle Management (PLM). PLM is a process that manages the entire lifecycle of a product from the idea through design and manufacturing to service and retirement [2]. One subdiscipline within the PLM chain is the Build and Test management. Therefore a large Build and Test infrastructure is required. It is based on the Microsoft Team Foundation Server (TFS), a platform that provides source code management and executes all requests, triggered by the whole development department, to ensure the product quality. The availability of this infrastructure is one of the key factors for the successful software development. Consequently a monitoring of the infrastructure is essential.

One distinctive feature of the infrastructure is that it is mostly virtualized by using software like VMware vSphere. This minimizes the environment to a few physical high-performance servers, which contain virtualization layers, called ESXi-Hosts as shown in Fig. 1. A virtualization layer allows multiple virtual machines (VMs) with various operating systems and applications to run isolated side-by-side on the same physical machine [4]. ESXi-Hosts and VMs are also known as Hypervisors and Clients. Virtualization provides greater IT resource utilization, efficiency and flexibility [10, 11]. Resources like CPU or

RAM capacity can be managed like a shared utility and can be dynamically deployed to different VMs [5].

The infrastructure that needs to be monitored within this project consists at present of about 250 VMs running on about 20 ESXi-Hosts. Performance metrics of VMs and ESXi-Hosts can only be monitored directly with the vSphere Client. Although vSphere collects all important metrics, only a few of them can be visualized at the same time. This makes it difficult to detect the cause of an anomaly within the performance data.

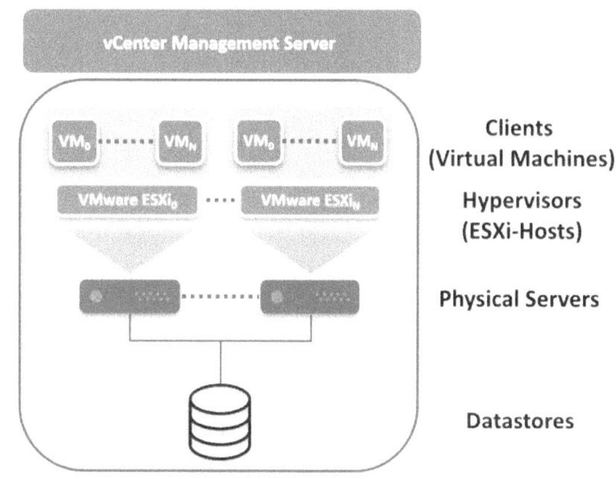

Figure 1: Overview of the VMware architecture showing the ESXi-Hosts running on physical servers providing VMs. The total infrastructure is supervised by the VMware vCenter.

Target of this project is the development of a monitoring platform that represents the key parameters showing the infrastructure healthiness. Observed metrics should be picked from both client and hypervisor layer. The monitoring system should be easy to use and display all the important information about the infrastructure in a compact way.

2 Material and Methods

2.1 Architecture

The setup of the monitoring system is based on an open-source platform called the TICK Stack provided by Influx-Data. TICK Stack is an acronym and describes the included components:

- **T**elegraf: collects time-series metrics from a source

- **I**nfluxDB: high performance time-series database

- **C**hronograf: real-time visualization engine

- **K**apacitor: native data processing engine

Each of these components can be used separately, but together they provide a scalable, integrated open-source system for processing time-series data. Time-series data consists of measurements that were tracked, monitored and aggregated over time. Simply put, the measurements get time-stamped. Fig. 2 shows the workflow of the monitoring setup using the TICK Stack. All parts of this platform are entirely implemented in Go, a compiled programming language designed at Google. For this monitoring setup all components were installed and configured on a Windows system.

to collect performance metrics from the entire infrastructure.

Telegraf writes the collected metrics directly into the InfluxDB, a time-series database designed to handle high write and query loads [7]. One of the key features of this database is the expressive SQL-like query language, wich allows to easily query aggregated data. Tags within the data provide series to be indexed for fast and efficient queries. By using retention policies the database is also capable to auto-expire stale data. This prevents an overloading of the database.

Kapacitor can be used to pre- or postprocess the aggregated data. Depending on the purpose, data is collected either directly from Telegraf or is pulled from InfluxDB, as shown in Fig. 2. Kapacitor uses a domain specific language called TICKscript to define tasks involving the transformation and loading of data and especially the tracking of arbitrary changes within the data [8]. Alerts can be reported to the user for instance by a logfile or via e-mail. In this project Kapacitor was used to integrate user-defined functions (UDFs) for an anomaly detection and alerting. This approach is further described in subsection 2.3.

The last and most important part of this monitoring system is the graphical visualization engine. It is used to query series of data from the database and represent the series in a proper way. Instead of Chronograf another open-source visualization engine called Grafana provided by Grafana-Labs was used. In comparison to Chronograf, it offers a wide range of official and community built panels to visualize the data. In general Grafana is structured like a sandbox system. Various panels can be easily created and configurated within the dashboard. Since it runs as a web application, Grafana can be accessed from anywhere within the infrastructure.

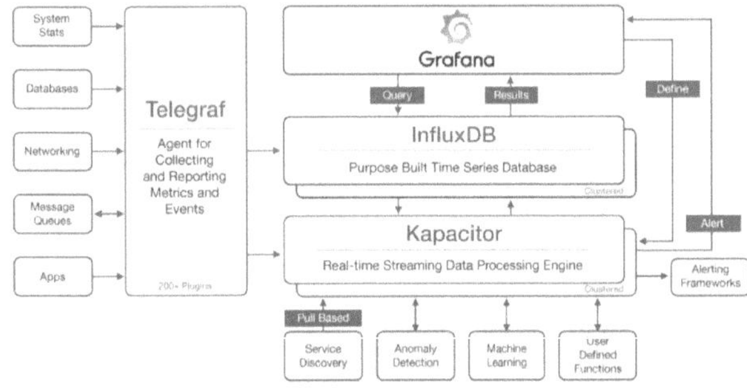

Figure 2: Workflow of the TICK Stack showing the main components. In this case Grafana was used instead of the basic visualization engine, Chronograf.

The first component of the TICK Stack, called Telegraf, is a plugin-driven server agent for collecting and reporting metrics [6]. Telegraf provides plugins to collect metrics directly from the system it is running on or gather metrics from third party application programming interfaces (APIs). In this setup the vSphere plugin was used to collect performance data directly from the vSphere API. The advantage of this approach is that only one active Telegraf service is required

2.2 Metrics

At the beginning of this project a research had to be done which metrics are the most important and need to be included within the monitoring. Due to the virtualized infrastructure of the observated environment, these metrics should be picked out from all layers, especially the client and hypervisor layer. About 20 different metrics were chosen, covering the operating system uptime, CPU performance, RAM utilization, disk and virtual disk usage, network usage and datastore capacity. Telegraf collects this data entirely from the vSphere API.

In addition to these performance metrics some supplementary details concerning the structure of the observed system should be integrated. The aim is to offer a direct overview about Controller and Build Agents running on a specific VM. An infrastructure report is generated daily providing this information as an excel spreadsheet file. A PowerShell script was created to transform this excel file to a comma-separated values (CSV) file and import the information into the database via Telegraf. The data is subsequently reported in a table panel within the Grafana dashboard.

2.3   Anomaly Detection

Grafana provides the definition of customized static thresholds within the panels. However, for some metrics it is not reasonable to set static thresholds. When observing the free disk space of a system, for example, it is relatively easy to define a critical value from which on the admin should be alerted. But when observing metrics like CPU usage, the definition of a static threshold will not perform well, since the data is very spiky and a short utilization of $100\,\%$ should not be immediately interpreted as an anomaly. The idea is to provide an anomaly detection that observes a metric automatically and creates an alert when detecting atypical behavior.

For this purpose a framework for flexible anomaly detection was integrated into Kapacitor. It basically maintains a dictionary of a metrics normal behavior and compares new windows of data to this dictionary [9]. If new datapoints do not fit with the dictionary, they are considered anomalous. To compare the distribution of data windows a non-parametric algorithm called Kolmogorov-Smirnov test (KS test) is used, which is not assuming a normal distribution like for example standard deviation does [9]. In addition an algorithm called Lossy Counting Algorithm (LCA) is used to maintain the dictionary of normal behavior [3]. To conserve space it drops less frequent items. There are two parameters in this algorithm, error tolerance and minimum support, defining the minimum percentage of frequency at which items will be dropped and the minimum percentage of frequency at which an item is considered frequent [3]. The metric that should be observed by the anomaly detection is determined via TICKscript in Kapacitor. Once an anomaly is detected, an alert is sent to the logfile.

2.4   Infrastructure-Diagram

Besides monitoring the system healthiness, it is desirable to have a network diagram that represents the current structure of the observed infrastructure at any time. To this end a PowerShell script was created that automatically builds a network diagram within Microsoft Visio upon execution. The VMware Power-Command-Line-Interface (Power-CLI) was used to provide a connection between PowerShell and the vSphere API. An example for the resulting network diagram is presented in section 3.2.

3   Results and Discussion

As a result of this project two Grafana dashboards containing the performance metrics form the client and hypervisor layer and an automated infrastructure network diagram were presented. In addition some investigation has been made with the framework for the automated anomaly detection.

3.1   Monitoring Dashboards

One Dashboard was created for both the ESXi-Hosts and the VMs each, as shown in Fig. 3 and Fig. 4. These dashboards contain all selected performance metrics and detailed information about the infrastructure. Several variable drop-down fields have been placed in the upper row of the dashboards to allow users to select which VM or ESXi-Host should be visualized.

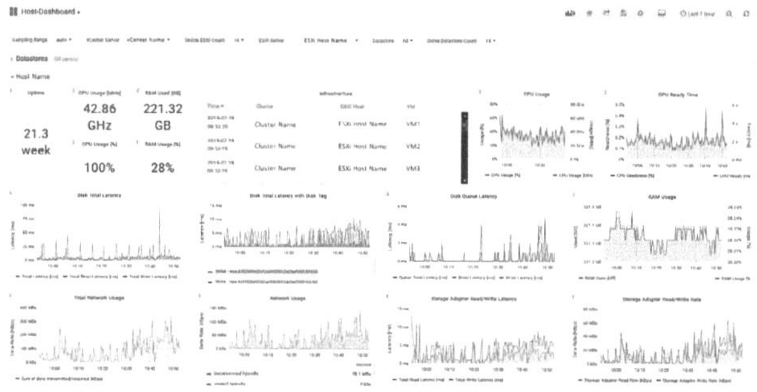

Figure 3: Grafana Host-Dashboard

Additionally, a variable has been added that shows the exact number of online ESXi-Hosts and VMs. Grafana creates a new dashboard instance for each ESXi-Host or VM selected by the user and inserts it below the previous one. This allows the user to decide exactly which objects to view and permits a direct comparison between the selected objects.

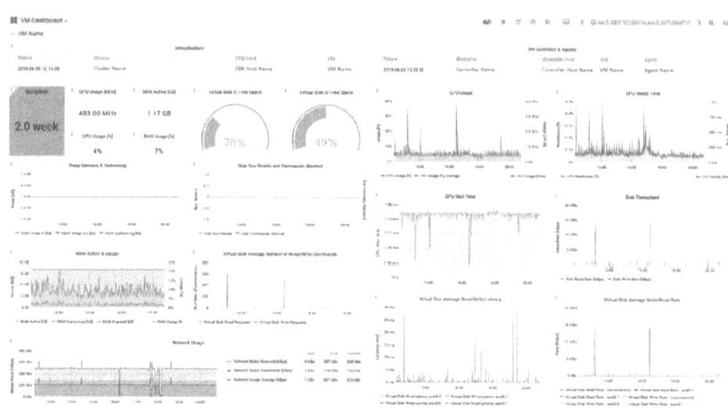

Figure 4: Grafana VM-Dashboard

Six tiles in the upper left part of the dashboard show the current uptime and the CPU and RAM utilization of the selected object. Subsequently, the performance metrics are visualized, mostly represented as graphs. The detailed information about the infrastructure is presented in tables. Each panel within these dashboards is labeled with a description of the metric displayed in that specific panel. Several descriptions also contain hints regarding the normal behavior of the selected metric. Another function is that the user can precisely select the period of observation in the upper right corner of the dashboard. Depending on the retention policy of the database a specific period or a predefined interval like "yesterday" or "last week" can be selected. Both dashboards were merged with an existing Grafana platform that is focused on observing more specific parameter concerning the Build and Test Agent execution.

The major advantage of this monitoring approach in relation to the monitoring by using the vSphere Client is that all key metrics are easily and compact accessible for each VM and ESXi-Host.

The automated anomaly detection was evaluated based on the CPU usage metric of a computer located beyond the Build and Test infrastructure. During the evaluation the CPU was heavily loaded in irregular intervals. It could be shown that most of the deviations were successfully detected. One problem is that it costs a lot of resources to perform the anomaly detection for all metrics observed in the monitoring system. Even if the amount of metrics investigated by the anomaly detection would be reduced to a minimum, it must be taken into account that the infrastructure consists of about 250 VMs and 20 ESXi-Hosts that need to be observed. One possible approach would be to use the anomaly detection for a limited number of fragile VMs or ESXi-Hosts and only for a specific amount of different metrics.

3.2 Infrastructure-Diagram

An example for the developed network diagram is shown in Fig. 5. The diagram captures all the VMs, ESXi-Hosts, clusters and datastores included in the infrastructure and presents their relation in a distinct way. Several different views were built within the Visio file, representing either the entire environment or just individual clusters.

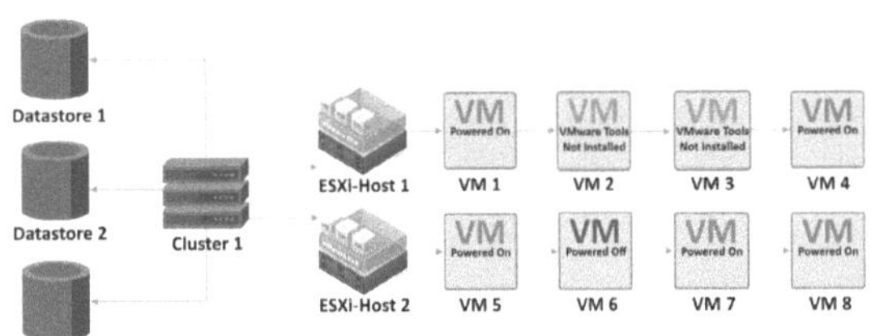

Figure 5: An exemplary infrastructure network diagram showing the ESXi-Hosts and VMs running on a specific cluster and the datastores that are connected to this cluster. The diagram also reveals which VMs are powered off or do not have the latest VMware tools installed.

Each ESXi-Host and VM is connected to a hyperlink that points to its specific Grafana dashboard. A great advantage of this network diagram is that it maintains itself always up-to-date since a periodic automated execution of the Power-Shell script can be enabled.

4 Conclusion

Altogether a compact monitoring system is introduced which sensibly complements the existing monitoring platform and offers the user a simple way to keep track of all metrics showing the system healthiness. The monitoring is very flexible and can be easily complemented if certain data needs to be added. In addition to performance metrics, important infrastructure information has been included into the dashboard. The network diagram provides a better insight into the whole system. Further steps would be the evaluation of the monitoring system in daily use. Further improvements are also needed in terms of anomaly detection and the associated alerting.

Acknowledgement

The work has been carried out at Siemens Healthcare GmbH, Forchheim and was supervised by the Institute of Medical Engineering, Universität zu Lübeck.

5 References

[1] T. M. Buzug, *Computed Tomography*. Springer-Verlag, Berlin Heidelberg, 2008.

[2] Siemens PLM Software, *Warum ALM und PLM zusammen eingesetzt werden sollten*. Whitepaper, intern document

[3] G. S. Manku, R. Motwani, *Approximate Frequency Counts over Data Streams*. Proceedings of the 28th International Conference on Very Large Databases, Hong Kong, 2002.

[4] VMware Inc., *VMware Infrastructure Architecture Overview*. Whitepaper, VMware Inc., 2006.

[5] VMware Inc., *Einführung in VMware vSphere*. VMware Inc., 2009.

[6] InfluxData Inc., *Telegraf documentation*. Available: https://docs.influxdata.com/telegraf/v1.11/ [last accessed on 2019-07-24].

[7] InfluxData Inc., *InfluxDB documentation*. Available: https://docs.influxdata.com/influxdb/v1.7/ [last accessed on 2019-07-24].

[8] InfluxData Inc., *Kapacitor documentation*. Available: https://docs.influxdata.com/kapacitor/v1.5/ [last accessed on 2019-07-24].

[9] N. Cook, *Morgoth documentation*. Available: http://docs.morgoth.io/docs/ [last accessed on 2019-07-24].

[10] W. Ding, B. Ghansah and Y. Wu, *Research on the Virtualization Technology in Cloud Computing Environment*. In: Student Conference on Medical Engineering Science 2012, Grin Publishing, München, pp. 59–62, 2012.

[11] R. Scroggins, *Emerging Virtualization Technology*. Global Journal of Computer Science and Technology, vol. 17, issue 3, version 1.0, 2017.

Monte-Carlo simulation of positron emission tomography combined with Compton-camera imaging

Andreas Bolke [1,2], Milan Zvolský [2], Magdalena Rafecas [2]

[1] Medical Engineering Science, Universität zu Lübeck, andreas.bolke@student.uni-luebeck.de
[2] Institute of Medical Engineering, Universität zu Lübeck, [zvolsky;rafecas]@imt.uni-luebeck.de

Abstract

Combining positron emission tomography (PET) with Compton-camera (CC) imaging offers a variety of new concepts in nuclear imaging. This work investigates a novel approach for PET/CC in silico. The design of the simulated setup is inspired by the combination of two existing prototypes and allows a first investigation of the radionuclides ^{89}Zr and ^{124}I. Since the simulation software is not able to describe such a bi-modal system, a post-processing tool was developed. Images of two point sources were reconstructed from PET data and their profiles were compared. The results include a first classification of PET and CC data, separately, for both radionuclides.

1 Introduction

In this work, we investigate the combination of two imaging modalities, Compton-camera (CC) imaging and positron emission tomography (PET) to measure radioactive isotopes in silico. While PET is a well established functional imaging tool, the medical application for near-field CC are still being researched. Our long-term goal is to investigate if a CC can be utilized to improve the information of images acquired by the small animal PET device of the MER-MAID project (*Multi-Emission Radioisotopes - Marine Animal Imaging Device*) [1]. Acquiring and analysing datasets of the two modalities is the first step required before developing novel algorithms for joint tomographic reconstruction. PET is based on the quasi-simultaneous (*coincidence*) detection of the two annihilation photons which arise from positron annihilation. To reconstruct the activity distribution, CCs use the Compton scattering effect and the detection in coincidence of the initial and the scattered photon. This work focuses on long-lived positron (β^+) emitters, in particular ^{89}Zr and ^{124}I, which in addition to positrons, also emit single photons. In this scenario, the combination of PET with a CC could be advantageous for improved image reconstruction, compared to PET alone. Non-pure isotopes can produce more false coincidences if the prompt gamma (PG) lies near the energy window. Various PG correction methods for non-pure isotopes exist [2]; however, using the additional PG information could be beneficial.

2 Material and Methods

A combined PET/CC system has been constructed in silico using the Monte-Carlo simulation software GATE [3], version Gate 8.2.

2.1 Setup

The PET system consists of 18 detector modules arranged in a ring (Fig. 1) of radius 28 mm. Each module is composed of a 8 x 8 crystal matrix with the dimension of $1.2 \times 1.2 \times 15.0\,\text{mm}^3$ for each crystal. The crystal's material is Lutetium-Yttrium oxyorthosilicate (LYSO). The centre of the PET-detector-system is the origin of the coordinate system. The CC consists of three equally sized detectors which are made of Lanthanum(III) Bromide (LaBr$_3$). The general geometry of the CC is based on an existing prototype [4]. Length and height of the CC detectors measure 28.5 mm each, while the thickness is 5 mm. Similar to the PET detectors, the distance from the coordinate system's origin to the first edge of the first detector is 28 mm and the edge-to-edge distance between the detectors is 23 mm. Contrary to the PET detectors, the detectors of the CC are not pixelated, but consist of monolithic crystal blocks.

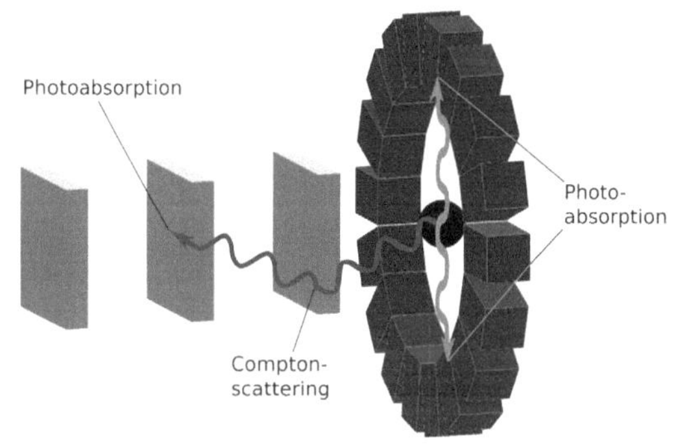

Figure 1: Setup of the simulation. PET (dark), CC (light) and water phantom (darkest). The wavelike lines display gamma photons as a potential outcome of the decaying nuclide inside the phantom.

2.2 Phantom and Radionuclides

We simulated a spherical water phantom (radius: 5 mm). Within the phantom, a radioactive point source was placed. ^{89}Zr (half-life $T_{1/2}$: 78.42 h) decays 76.2% by electron capture (EC) and 22.8% by positron decay to ^{89m}Y [5]. For PET imaging only the latter is relevant. With a half-life of 15.84 s, ^{89m}Y then undergoes an isomeric transition to the ground state of ^{89}Y by emitting a 909 keV gamma photon. The positron mean range is 3.8 mm. In this work, emitted gamma photons under 1% emission rate are not of interest for the CC. While ^{89}Zr has a simple decay scheme, ^{124}I has a rather complex one [6]. ^{124}I ($T_{1/2}$: 100.2 h) decays 73.3% by EC and 22.7% by positron decay into the stable ^{124}Te. The positron decay consist of the pure β_2^+ decay (10.7%), the positron emission of β_1^+ (11.7 %) and β_3^+ (0.3 %). Alongside β_2^+, a PG γ_1 of 602 keV and with β_3^+, a PG of 723 keV in cascade with γ_1 is emitted. The mean range of β_1^+, β_2^+ and β_3^+ is 2.8 mm, 4.4 mm and 1.1 mm, respectively. Among the non-prompt gamma transitions, the energies 1,325 keV, 1,376 keV, 1,509 keV and 1,691 keV are of interest for the CC.

2.3 Simulation and Postprocessing

To implement the CC, we used a dedicated GATE module, which is still under development [7]. Data acquisitions corresponding to 300 s were simulated for an activity of 1 MBq for both radionuclides. The output is written to two separate files for each modality. In GATE, each interaction of a particle with matter is referred as a *hit*. Multiple hits in the same volume constitute a *single*. In PET, two singles that happen in a predefined time and energy window as well as in two separate detector modules are defined as a coincidence. In CC imaging, multiple singles that occurred in different detectors and in a specified time window are defined as coincidence. Unfortunately, GATE is not able yet to output data from two different detector systems in one simulation. Therefore, we had to develop a dedicated C++-based post-processing pipeline to mimic the finite energy and time resolutions and then identify coincidences from the blurred singles. Post-processing allows to create multiple datasets with different resolutions from an ideal simulated dataset. Energy and time resolution were both implemented by random sampling of Gaussian distributions. In order to mimic the energy-dependent energy resolution of a scintillation detector, we use the relationship

$$R = R_0 \cdot \frac{\sqrt{E_0}}{\sqrt{E_{dep}}}, \tag{1}$$

where R represents the full width at half maximum (FWHM) at the deposited energy E_{dep}. E_0 and R_0 represent the energy of reference and the resolution at that energy, respectively, and are user-defined. In this work, $R_0 = 10\,\%$ for $E_0 = 511$ keV and the time resolution was 1 ns FWHM for both modalities. To build coincidences from the single events, two sorting algorithms were implemented: for the CC a single event opens a time window of 2 ns, this together with all other singles registered within this time interval in the other detectors are stored as a coincidence. The algorithm for the PET system works similarly, excepting that the singles within a coincidence are constrained to an energy window ($\Delta E = E_0 \pm 2 \cdot$ FWHM) and a coincidence of a detector module with itself is forbidden.

The coincidences are then classified in *true*, *random* and *scattered* coincidences. Scattered events are defined as coincidences, where one of the photons undergoes a Compton scattering inside the water phantom. Random events are those coincidences which originate from two different nuclear decays. Everything else is considered a true coincidence. For the Compton-data dataset, GATE does not provide a direct information if a photon was scattered in the water phantom or not. Therefore, CC coincidences could only be classified as true, random and rejected. Here, the deposited energy of all singles per coincidence is summed up. If that sum lies within an energy range of interest and all singles originate from the same nuclear decay, the coincidence is classified as true. If that sum lies within the energy windows but the underlying singles originate from different nuclear decays, the coincidence is classified as random. Everything else is classified as a rejected coincidence. For both modalities true mainly means that the detected singles are related to the same nuclear decay. Thus it must be noted that, for example, a true PET coincidence could be the result of a 909 keV gamma, which only deposits 511 keV in the detector and then escapes, and the full absorption of an annihilation photon, provided that both originated from the same decay. This problem results i.a. from outputting only GATE singles. From the imaging perspective, this coincidence should be considered as a random; therefore, we call this kind of events *pseudo-true*.

Using only PET data, images were reconstructed with the maximum likelihood expectation maximisation algorithm [8]. As voxel size $0.4 \times 0.4 \times 0.4$ mm^3 was chosen.

3 Results and Discussion

In the ideal singles spectrum of the PET data for ^{89}Zr (Fig. 2), the photopeaks at 511 keV and 909 keV as well as the corresponding Compton edges can be clearly identified. The peak in the first bin is due to characteristic X-ray of ^{89}Zr followed by a peak due to characteristic X-ray of ^{177}Lu. The histogram also shows two smaller X-ray escape peaks to the left of the two photopeaks (511 keV, 909 keV). X-ray escape peaks occur in small detectors where a gamma photon is absorbed by an inner-shell electron which leaves a vacancy to be filled by an outer-shell electron. During the transition the electron releases energy as a characteristic X-ray. If the X-ray then leaves the current crystal, the deposited energy in the crystal is decreased by the amount of the X-rays energy. If the X-ray gets absorbed by another crystal, a peak can be seen characteristic to the scintillation crystal's material which is ^{177}Lu in our case. The ideal singles spectrum for the CC does not show X-ray escape peaks.

Fig. 3a and 3b show the summed energy per coincidence of

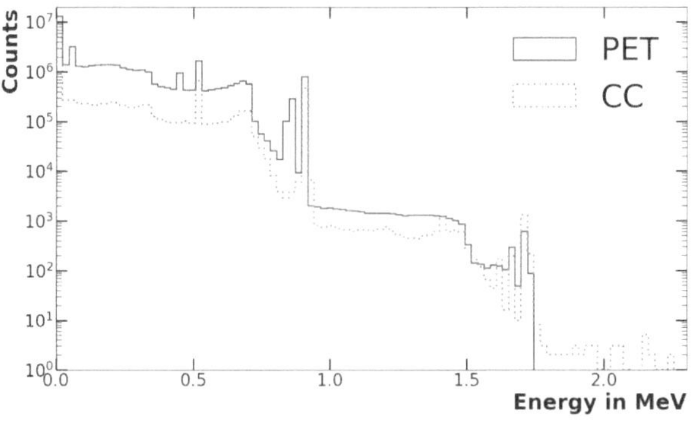

Figure 2: Ideal singles spectrum of ^{89}Zr measured with PET and CC.

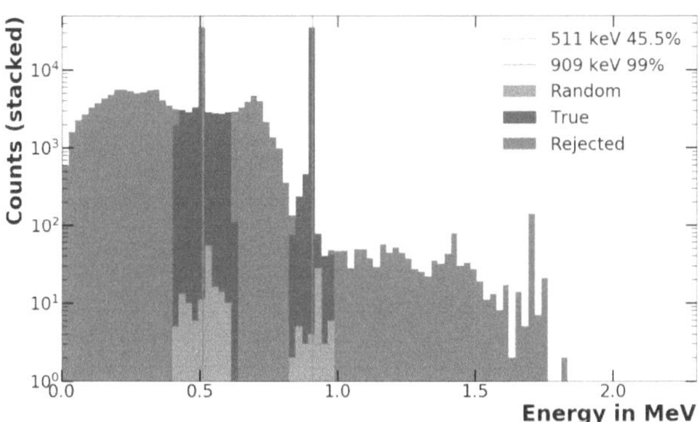

(a) Perfect resolution.

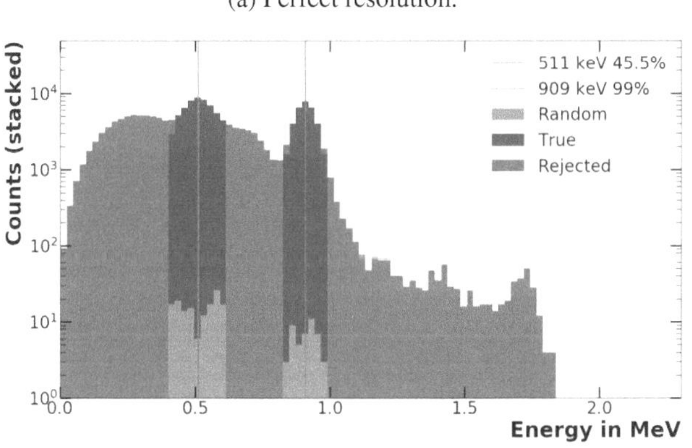

(b) $R_0 = 10\,\%$ @511 keV, time resolution 1 ns FWHM.

Figure 3: Energy summed over each coincidence of the CC for ^{89}Zr (The histogram is stacked). Vertical lines depict the center of an energy window.

the CC data with and without finite energy and time resolution in a stacked histogram for ^{89}Zr. In Fig. 3b, the photopeaks of 511 keV and 909 keV are clearly visible. Similar to the PET data, the low number of random coincidences is probably due to the relatively low activity combined with a small coincidence window of 2 ns and the relatively good time resolution of 1 ns FWHM.

In the ideal singles spectrum of the PET data for ^{124}I (Fig. 4), the photopeaks as well as the Compton edges are more difficult to distinguish. Given that the 511 keV,

602 keV and the 723 keV photopeaks are relatively close to each other, the corresponding X-ray escape peaks fall close to the 511 keV photopeak. The first two peaks are characteristic X-rays of ^{124}I mixed with ^{177}Lu. In general, even with perfect energy resolution the different peaks are already difficult to distinguish, and they might merge into each other for a finite energy resolution.

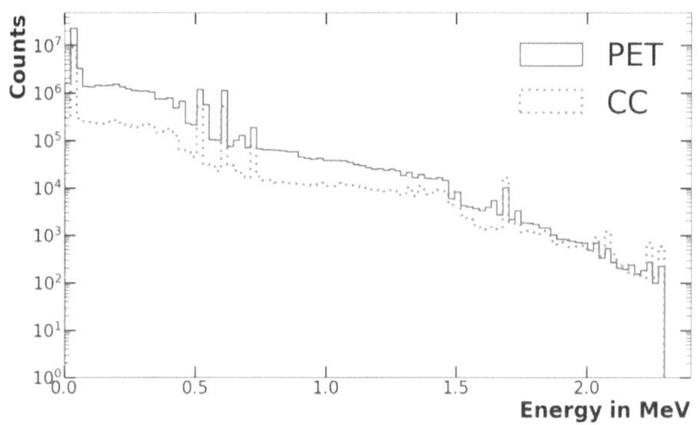

Figure 4: Ideal singles spectrum of ^{124}I measured with PET and CC.

The PET spectrum of ^{124}I in Fig. 5 shows that with an energy resolution of 10 % at 511 keV, the photopeak of 511 keV can still be distinguished from the PG peak at 602 keV. We also observe that X-ray escape peak, which lies approximately at 550 keV, is not visible and most likely merged with the photopeak at 511 keV. Therefore, a considerable amount of random and pseudo-true coincidences might later result in a noisier background, loss of resolution and a reduction of the signal to noise ratio in the reconstructed image.

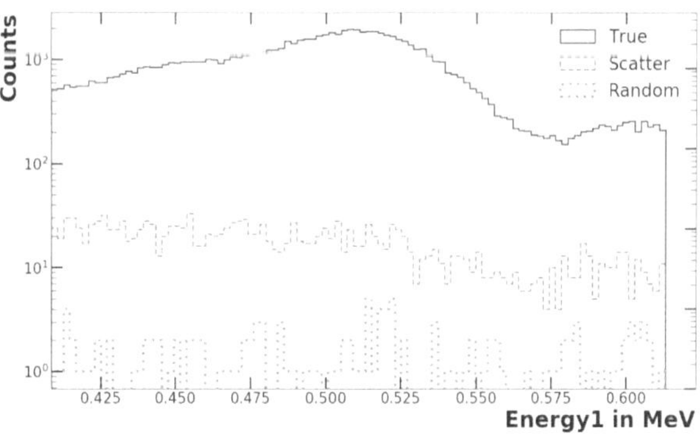

Figure 5: ^{124}I spectrum of one of the two deposited energies per coincidence measured with PET.

In Fig. 6a and 6b the summed energy over each coincidence in the CC for ^{124}I is displayed. We can clearly see that the different energy windows lie so close to each other that they merge. At 511 keV, 602 keV and 723 keV the peaks have merged together as well as the peaks 1,325 keV, 1,376 keV and the 1,509 keV. The 1691 keV peak is clearly visible. Fig. 7 displays the profiles of the reconstructed point sources ^{89}Zr and ^{124}I after 15 iterations. At this stage, no CC information was used for reconstructing these images.

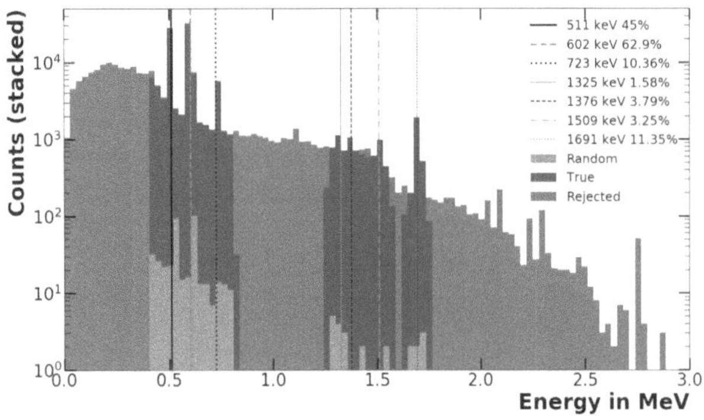

(a) Perfect resolution.

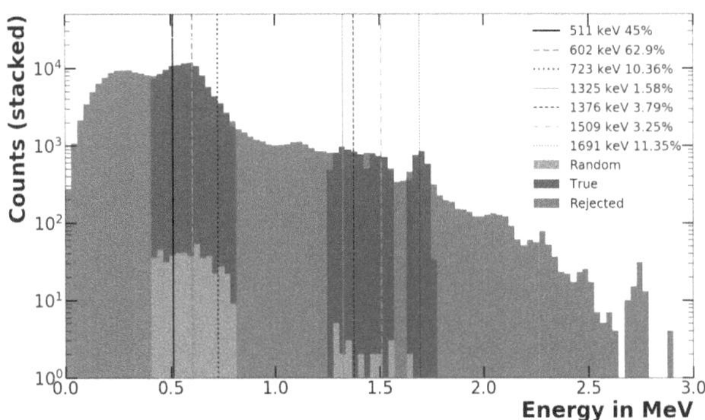

(b) $R_0 = 10\% \ @511\,\text{keV}$, time resolution $1\,\text{ns}$ FWHM.

Figure 6: Energy summed over each coincidence of the CC for ^{124}I (The histogram is stacked). Vertical lines depict the center of an energy window.

The profile of ^{124}I is broader, probably due to the large positron range of ^{124}I compared to ^{89}Zr. The amount of random/pseudo-true coincidences in ^{124}I might be responsible for the non-zero background at the tails of the profile.

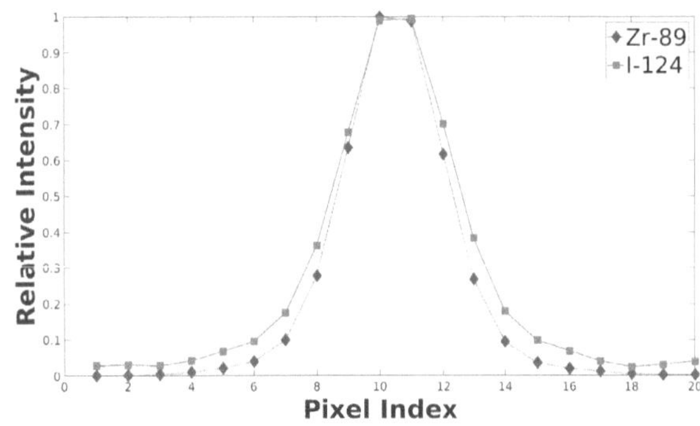

Figure 7: Transverse intensity profiles of reconstructed point sources ^{89}Zr and ^{124}I from the PET data. Voxel size is $0.4 \times 0.4 \times 0.4 \ \text{mm}^3$.

4 Conclusion

A novel concept for nuclear imaging has been simulated: the combination of PET and CC imaging in one simulation.

The spectra of two radiotracers and two imaging modalities were thoroughly analysed and possible challenges for the future image reconstruction of both modalities have been identified. Random coincidences which originate from the same decay (pseudo-true) are still considered true in our classification and should be correctly identified in the future. The reconstructed images are an indication that the coincidence sorting was done correctly. Long positron range and random coincidences of ^{124}I need a dedicated modeling and correction in the reconstruction. Reducing the size of the energy windows for the CC offers a variety of possibilities to improve the amount of true coincidences. Furthermore, co-registrating the images of both modalities after separate reconstructions could be a future step before developing a joint reconstruction. Further PG nuclides will be investigated.

Acknowledgement

The work has been carried out at the Institute of Medical Engineering at the University of Lübeck, Germany.

5 References

[1] M. Zvolský et al., "Nukleare in-vivo Bildgebung von Zebrafischen," in 50. Jahrestagung der Deutschen Gesellschaft für Medizinische Physik (DGMP) - Abstractband, C. Gromoll and N. Wegner, Eds., 2019, p. 35.

[2] M. Conti and L. Eriksson, "Physics of pure and non-pure positron emitters for pet: a review and a discussion," EJNMMI physics, vol. 3, no. 1, pp. 8–8, Dec. 2016.

[3] S. Jan et al., "GATE v6: a major enhancement of the GATE simulation platform enabling modelling of CT and radiotherapy," Physics in Medicine and Biology, vol. 56, no. 4, pp. 881–901, jan 2011.

[4] J. Roser et al., "Evaluation and Validation of a Sensitivity Model for a Three-layer LaBr3 Compton Telescope," in 2018 IEEE Nuclear Science Symposium and Medical Imaging Conference Proceedings (NSS/MIC), Nov 2018, pp. 1–5.

[5] Laboratoire National Henri Becquerel. Accessed 12 Jan 2020. [Online]. Available: http://www.lnhb.fr/en/

[6] The National Nuclear Data Center. Accesses 12. Jan 2020. [Online]. Available: https://www.nndc.bnl.gov/

[7] A. Etxebeste et al., "CCMod: a GATE module for Compton Camera imaging simulation," Physics in Medicine & Biology, 2019.

[8] L. A. Shepp and Y. Vardi, "Maximum likelihood reconstruction for emission tomography," IEEE Transactions on Medical Imaging, vol. 1, no. 2, pp. 113–122, Oct 1982.

9

Regulatory Affairs

Development of a digital tool for Risk Management of Medical Devices

Shristi Shrestha

Biomedical Engineering, Luebeck University of Applied Sciences, shristi.shrestha@stud.th-luebeck.de

Abstract

For market access in the European Union (EU), medical devices have to go through a complex process called conformity assessment. With a constantly changing technological landscape, there is a need of higher safety and higher regulatory compliance. In order to demonstrate this compliance, manufacturers need to prepare a technical documentation for each device. A central part of this documentation is related to Risk Management (RM)- with a focus on post-market activities is a new regulatory requirement. Within RM, there is a high chance of data inconsistency, because manufacturers often use spreadsheets or an isolated software solution that leads to inefficient manual work. This research aims to develop a concept of a collaborative tool for RM by identifying the regulatory requirements of manufacturers, developing an integrated concept of RM and identifying interfaces and redundancies with other regulatory requirements that enables manufacturers to improve and maintain their RM File.

1 Introduction

Before a medical device is placed in the EU market, manufacturers have to follow Regulation (EU) 2017/745 known as Medical Device Regulation (MDR). The new regulation came into force on 25th May 2017 and will be fully applicable on 26th May 2020. MDR will replace the existing directives, 90/385/EEC on Active Implantable Medical Devices (AIMDD) and 93/42/EEC on Medical Devices (MDD) [1]. Under the new regulation a device has to undergo a conformity assessment procedure, which demonstrates whether the legislative requirements have been fulfilled. To show compliance of a medical device with MDR, manufacturers have to prepare a technical documentation for each device, which is a cumbersome task. A part of it is to perform Risk Management (RM). RM process helps in analyzing, evaluating, controlling and monitoring risk of a medical device. Risks presented in a medical device have to be reduced as far as possible maintaining a high level of protection of health and safety taking into account the state of the art [2]. RM is regarded as a lifecycle process and has to be updated regularly because once device is placed on the market and comes into operation, it may bring previously unknown risks to the light. This may be detected by post-production monitoring under MDR. The standard 'EN ISO 14971: Application of risk management to medical devices' provides more details on the incorporation of production and post-production information into the product specific RM file. The information from the production and post-production phase can relate to new hazards, hazardous situations or harms, and can affect the risk estimations or the balance between benefit and overall residual risk [3]. The continuous

and systematic incorporation of new information into the RM file in order to keep it up-to-date may be a tedious task. Especially when isolated documentation, or even worse, paper based documentation is used, this may be a challenging duty. In the worst case, it can even lead to manufacturers losing track of documentation creating document inconsistencies, which may be detected during Notified Body reviews leading to justified delays in conformity assessment. For patients, this might lead to delays in medical device availability and treatment. For manufacturers, the workload increases leading to additional effort and expenses. In order to harmonize processes and avoid data redundancy there is a need for a digital solution which would allow manufacturers to collect and document risk data in a convenient way whenever information becomes available. The main aim of this research is to develop a concept of a collaborative tool for a risk management by identifying the regulatory requirements of manufacturers, establishing an integrated concept of RM, identifying an interface with other related regulatory requirement, determining redundancies with interface processes and creating a set of user requirements and template for a potential software. An Integrated Content Management System (ICMS) can address the key weaknesses of today's process enabling manufacturers to improve their RM files with respect to documentation readability, quality and avoidance of inconsistencies throughout the technical documentation of a medical device.

2 Material and Methods

A systematic approach was carried out to analyze the regulatory requirements and develop software requirements for a digital tool. The concept of analysis of the requirement was on three different levels: Document, Process and Content.

- In the beginning, it was crucial to have an overview of manufacturers requirements as it provided information on what task needs to be performed. So, as a first step, the legislative requirements with regard to risk management for manufacturers of medical devices were identified according to MDR and ISO 14971. The necessary documents were determined and Gap Analysis was also performed between MDD and MDR, to have an overview on changes and additional requirements for RM.

- In the second step, an integrated concept of a risk management procedure, in a flowchart form, was developed according to MDR and standard ISO 14971. A document-based approach – as applied in general in industry was used in this project to better understand the complex lifecycle process of risk management in detail and ensure required elements of the process. As risk management is considered as a continuous lifecycle process of a device, so an interface of it with other regulatory requirements was established.

- In the third step, contents relevant to other processes were listed in order to identify the redundancies. The content was broken down into content items to have a clear view on data repetition. A list containing all the unique content items was created and lastly screening was done in a way that displayed only the relevant content of risk management.

- To understand what a tool should do to ensure that the end-result meets the user expectations, a list of requirement from user perspective was analyzed based on the requirements, process and contents. And described from the point of view of a risk management responsible that works on the tasks of risk management, e.g. creating a risk matrix. The user requirement would be 'I want to create a risk matrix combining severity of harm and probability of harm to define acceptability levels for all risks'.

- Finally, out of the contents, set of templates were created for each identified document of RM.

3 Results and Discussion

3.1 Gap Analysis of Risk Management in MDD and MDR

Although performing risk management is an essential requirement in MDD, it has not been explicitly defined. In contrast to that, MDR clearly states in Article 10 - 'General obligation of manufacturers' point (2) that "Manufacturers shall establish, document, implement and maintain a system for risk management as described in Section 3 of Annex I" [4]. Annex I - 'General safety and performance requirements (GSPR)', Chap I (3) describes clearly and in detail, the information on risk management requirements. According to GSPR (3), manufacturers have to establish and document a Risk management plan for each device, identify hazards, estimate the risk, control the risk and evaluate the impact of information from production and Post-Market Surveillance (PMS) on hazards, overall risk, benefit-risk ratio and risk acceptability and if necessary apply risk control measures. GSPR (4) gives some more insights on the risk control measures. In addition, more emphasis is given to RM as a lifecycle process that needs to be systematically updated regularly. Some new rules are added to MDR in comparison to MDD. A new class Ir - reusable surgical instrument, has been introduced in the new regulation. Depending on the class of medical device the conformity assessment procedure also varies. The manufactures of class I devices can declare conformity of their device without involvement of notified body. The class I devices that are placed on the market in sterile condition (Is), have a measuring function (Im) or are reusable surgical instruments (Ir) need involvement of notified body. However, the involvement of the notified body is limited. For higher classes, such as IIa, IIb and III, involvement of notified body is required.

3.2 Role of EN ISO 14971 in MDR

Like in the MDD, the new regulation allows the use of harmonized standards to show compliance to GSPRs of the MDR. According to Article 8 in MDR, when a device is in compliance with a harmonized standard or relevant part of this standard then they are presumed to be in compliance with the requirements in MDR. Only the references that are published in the Official Journal of the European Union are considered as a harmonized standard. With regard to this, EN ISO 14971:2012- Medical devices - Application of risk management to medical devices is considered as harmonized standard for the implementation of a risk management system under MDD. Currently, no standard is harmonized under MDR, but the new EN ISO 14971:2019 has been published and reflects the current state of the art. Therefore, manufacturers should consider its application.

3.3 Risk Management process

The RM process defined according to ISO 14971 provides a framework to systematically manage the risk associated with each individual device. A lifecycle process of RM is shown in Fig.1. The starting point of RM process is preparing a RM Plan, which provides a roadmap to RM activities that are carried out during the entire product lifecycle. The RM Plan is a living document that is updated regularly whenever new information is available [3]. Afterwards risk analysis is performed. The intended use of a medical device is the starting point of risk analysis, which helps in determining the correct use and reasonably foreseeable misuse of a device [3]. With these considerations the hazards related to a device that could affect safety, can be identified. For

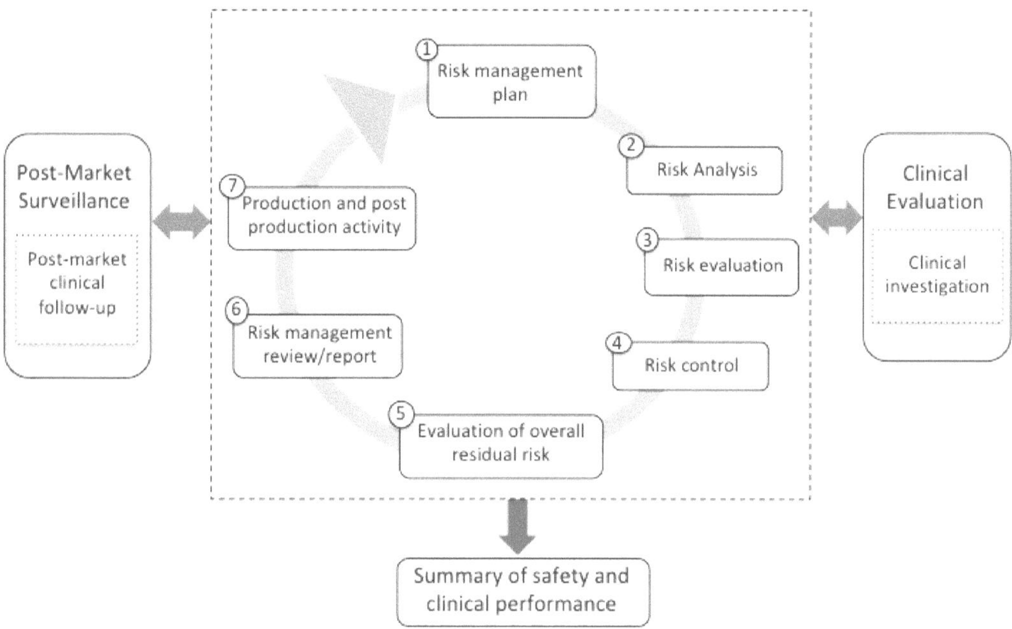

Figure 1: A schematic diagram of lifecycle process of Risk Management and its interface with different systems such as Post-Market Surveillance, Clinical Evaluation and Summary of Safety and Clinical Performance.

each identified hazard a sequence of events is determined which leads to a hazardous situation. For each hazardous situation the associated harm is determined. The risk acceptability criteria documented in the RM Plan help to determine the acceptance level of a risk. The acceptance level depends on the severity level of harm and probability of occurrence of harm. Those risks that are not acceptable needs to be reduced as low as possible using an appropriate risk control measures. Implementation and effectiveness of risk control measures has to be verified and documented. Thereafter, a residual risk is estimated using the same criteria as before. If a risk is still not judged acceptable, a benefit-risk analysis is performed. Benefit-risk determines whether the expected benefit of intended use of the medical device outweighs the residual risk. Then, using the method and the criteria for acceptability of the overall residual risk defined in the RM Plan, the overall residual risk is evaluated. The manufacturer needs to disclose the residual risks to the users. Out of these, a RM Report is established that gives a summary of the RM activities. And finally, the information from production and post-production are collected and reviewed and an appropriate action is taken. Considering the overall RM process there are five essential documents i.e. RM Plan, Risk Analysis, List of Hazards, List of Risk Control and RM Report, which a manufacturer needs to prepare to meet the requirements of MDR. All of the documents are recorded in a RM File - the most integral part of the risk management process, which is either organized for one type of medical device or for a medical device family and is regularly updated. The RM File provides traceability for every risk detected. Compliance is checked by inspection of RM File [3].

3.4 Interface of Risk Management with other systems

Clinical Evaluation (CE) and PMS are interlinked with RM. For the post-market phase, manufacturers need to actively collect and review the information of medical devices. That is performed according to the PMS Plan. When the collected information is related to safety several action is taken to the medical device or risk management process. For eg. If new hazards or hazardous situation is present then associated risk needs to be assessed and controlled. Or if a risk is no longer acceptable then it needs to be assessed again. Or sometimes information can affect the balance between benefit and overall residual risk [5]. Information from Post-Market Clinical Follow-Up (PMCF), which is a part of PMS, also enhances the clinical evidence for safety and performance once the device is placed on the market but there could be a case where it is justifiable not to perform a PMCF. CE summarizes the benefit-risk profile and weighs it against the benefit-risk profile of the state of the art. The assumptions made for benefit-risk acceptance in RM File are justified and confirmed by CE. This and the RM review/report are fed into Summary of Safety and Clinical Performance (SSCP). Thus, this shows that various systems and processes complete the lifecycle of RM.

3.5 Integrated Content Management System (ICMS) for RM

As discussed previously, RM File provides traceability for every risks detected. This works only in a perfect world where the RM File is well-structured and updated continuously. But in real life it is a complex process and often fails due to isolated processes and inefficient tools. This is where an ICMS can address the problem and help manufacturers to work efficiently. An ICMS have capability to link the documents, processes and contents. It provides a solution to problems like data redundancy, document inconsistencies that often occurs in a RM File. For an example some contents like intended use, intended population, intended operator, scope, benefit-risk profile and status of device are common in all the documents i.e. RMR, PMSR and CER as shown in Table 1. With an ICMS connected to database system, all of the unique content item can be created and stored and extracted to the required document whenever needed. This will solve the problem of copy-pasting. Additionally, an ICMS have a potential to monitor data inconsistencies by displaying an alert message. Similarly, any changes, updates, addition or deletion can also be tracked saving a lot of time.

Table 1: Example of some common contents in RM Report (RMR), PMS Report (PMSR), CE Report (CER)

Content	RMR	PMSR	CER
Intended use	X	X	X
Intended population	X	X	X
Intended operator	X	X	X
Scope	X	X	X
Benefit-Risk Profile	X	X	X
Status of device	X	X	X

4 Conclusion

The study was carried out keeping in view of medical device manufacturers difficulties in preparing an integrated RM process, which defines the interfaces with other regulatory systems providing a comprehensive RM File to show compliance with conformity assessment. Some of the problems in RM are unidentified hazards, unspecified risk control measures, document updating, data redundancy and inconsistencies. A concept of a collaborative tool established in this research will solve problems of the document, process and content level maintaining document consistency and keeping track on inconsistencies. The gap analysis showed that the RM requirements have been explicitly defined in MDR unlike MDD and more emphasis is given to RM as a lifecycle process with regular updating.

The development of an ICMS will certainly provide an organization an opportunity to work in a single harmonized environment. It can help to reduce time to perform a certain activity without interruption and increase productivity. The set of templates which are created for each document will be adequate to build a prototype of an ICMS. In future, research on incorporation of different techniques that support risk analysis in ICMS can be conducted.

Acknowledgement

The work was carried out by PRO-LIANCE GLOBAL SOLUTION GmbH, Münster at Galenos Solution GmbH, Fulda. I would like to express my sincere gratitude to my supervisor, Mr. Florian Tolkmitt, for his guidance throughout the project. I would like to thank Mr. Michael Kania for providing workspace. My special thanks goes to Prof. Dr. Folker Spitzenberger, Centre for Regulatory Affairs in Biomedical Sciences (CARBS), Technische Hochschule Luebeck for supervising and guiding my project.

5 References

[1] European Comission, *Internal Market, Industry, Entrepreneurship and SMEs, Medical Devices*. Available: https://ec.europa.eu/growth/sectors/medical-devices [last accessed on 2020-01-15].

[2] J. j. Vroonhoven, *Risk management for medical devices and the new ISO 14971*. Available: https://www.bsigroup.com/en-GB/medical-devices/resources/whitepapers/].

[3] International Organization for Standardization *ISO/FDIS 14971:2019(E): Medical devices — Application of risk management to medical devices*. ISO, 2019

[4] European Union, *Regulation (EU) 2017/745 of the European Parliament and of the Council of 5 April 2017 on medical devices, amending Directive 2001/83/EC, Regulation (EC) No 178/2002 and Regulation (EC) No 1223/2009 and repealing Council Directives 90/385/EEC and 93/42/EEC*. Official Journal of the European Union, 2017.

[5] International Organization for Standardization *ISO/DTR 24971:2019(E): Medical devices — Guidance on the application of ISO 14971*. ISO, 2019

The benefit and specifications of a digital tool to optimize Clinical Evaluation of Medical Devices

Umesh Satyal [1],
[1] Biomedical Engineering, University of Applied Sciences Lübeck, umesh.satyal@stud.th-luebeck.de

Abstract

One of the focused areas of reinforcement in the new Medical Device Regulation (MDR) is the clinical evaluation of medical devices. With higher demands for clinical data, medical device manufacturers are toiling with planning, preparation, generation, submission, and update of clinical evaluation documentation. This research aims to evaluate the need, potential benefits, and specifications of a digital tool that would facilitate manufacturers through all these processes. To achieve this, all requirements derived from regulations and guidance documents were collected. A gap analysis between the Council Directives 93/42/EEC concerning medical devices (MDD) and the MDR was performed. Based on the requirements, the redundancies in the content items were identified. Each sub-process was established as a stand-alone document and templates were created. The study suggests that the documentation effort for clinical evaluation is significantly higher when comparing MDR requirements against MDD requirements. To conclude, there is a clear need for a digital tool to handle clinical data, maintain documentation quality and consistency in the clinical evaluation process.

1　Introduction

The regulatory landscape in the medical industry is changing rapidly. The primary motive of this change is to ensure the effective and safe use of medical devices to the user. With the motive of making the European Union (EU) regulatory framework more modernized and robust, the European Parliament came up with two new European Regulations i.e. Medical Device Regulation (MDR) and the In-Vitro Diagnostic Device Regulation (IVDR) [1]. These regulations will be replacing three existing directives. These regulations entered into force on 25th May 2017 and become enforceable law in all member states supra-nationally. They aim for rigorous affirmation of patient safety, increase transparency, traceability, and foster overall quality in the medical/healthcare industry with reinforcement of various current systems [2].

Before a medical device is put on the market, every device has to go through a convoluted procedure of conformity assessment for market access in the EU [3]. One of the pivotal processes is clinical evaluation. It is one of the major areas of the reformation in MDR and has been specified as one of the general obligations of the manufactures regardless of medical device class [3]. "Clinical evaluation is a set of ongoing activities that use scientifically sound methods for the assessment and analysis of clinical data to verify the safety, clinical performance, and/or effectiveness of the medical device when used as intended by the manufacturer" [4]. Thus, clinical evaluation is a process that should be realized as a life- cycle approach from the early development of a medical device to the post-market phase until the end of life span. Clinical data lies central to clinical evaluation and the process of evaluation is required to be critical [5]. Therefore it needs to identify, appraise and analyze both favorable and unfavorable data. In addition, depending on the stage in the life-cycle of the product, the plan and the report should cover different aspects. In this regard, clinical evaluation requires rigorous tasks. While failure to establish, implement and manage reports of the clinical evaluation process often may lead to the delay in conformity assessment or suspension in the market access (CE mark) of medical devices in case of marketed product [3]. The consequences are therefore higher efforts and costs for the manufacturers, delayed and much effort to review by a notified body. Subsequently, this is leading to the deferral in the availability of medical treatments and may increase the healthcare cost.

The current approaches for clinical evaluation include manual search in electronic databases, bidirectional handling of multiple information from Risk Management(RM), Post-Market Surveillance (PMS) and its manual update and multiple redundant works. Thus, maintaining documentation quality, readability and consistency throughout the technical documentation, document traceability and regulatory submission in the entire life-cycle of medical devices is crucial to manufacturers.

This research aims to address the key weaknesses of current practice, to develop and to evaluate the feasibility of a digital tool that works as an integrated management tool for clinical evaluation of medical devices.

2   Materials and Methods

The goal of the digital tool pertaining to this research is to help plan, create, maintain, update and generate the documents for clinical evaluation that are compliant to MDR. Thus, MDR was chosen as the primary source of the requirements. In addition, to identify the changes and gaps between MDR and MDD all the requirements from MDD were also collected. Besides, there are specific guidance documents for clinical evaluation released by the European Commission and International Medical Device Regulatory Forum (IMDRF). These guidance documents were considered to list out the contents in depth for clinical evaluation. Finally, to consider the realistic practice of manufactures, some of the real-world experience and operating procedure was considered. After the identification of all the materials used, the methods deployed follow the following order.

- The collection of requirements from Annex XIV Part A: clinical evaluation, Chapter VI Article 61: clinical evaluation from the new Medical Device Regulation (EU) 2017/745 (MDR) of the European Parliament and of the Council on medical devices[3].

- The collection of requirements from the Council Directives 93/42/EEC concerning medical devices (MDD[6].

- The collection of requirements from guidance documents MEDDEV 2.7/1 Rev 4 for clinical evaluation (MEDDEV), IMDRF MDCE WG/N56 FINAL:2019, guidance documents related to sufficient clinical data, and clinical evidence [4] [5] .

- Gap Analysis between MDD and MDR and comparison between MEDDEV and MDR.

- Requirements collection from regulatory/clinical professionals and real-world experience like standard operating procedure (SOP) and comparison of the requirements with the real world functioning procedures.

- Grouping of relevant requirements into the same content items.

- Transforming the content items into objective flow process and graphical representation.

- Breakdown of the life cycle approach of clinical evaluation into different sub-processes such as Clinical Evaluation Plan (CEP), state of the art analysis, clinical data identification, clinical data appraisal, assessment of equivalence, clinical data analysis, and Clinical Evaluation Report (CER).

- Identifying the documents generated from different sub-processes of clinical evaluation.

- Establishing the identified documents as a stand-alone document with the possibility of combining with different other processes like Risk Management (RM), Post-Market Surveillance (PMS).

- Identifying the redundancy in contents in comparison to RM, PMS, technical documentation, and within the different documents of clinical evaluation.

- Development of final templates for different documents of the clinical evaluation process like CEP, state of the art analysis, clinical data identification, clinical data appraisal, assessment of equivalence and CER for implementation into software.

3   Results and Discussion

3.1   GAP ANALYSIS

Clinical Evaluation Definition: "Clinical evaluation means a systematic and planned process to *continuously* generate, collect, analyze and assess the clinical data pertaining to a device in order to verify the safety and performance, *including clinical benefits*, of the device when used as intended by the manufacturer." (Article 2(44)) [3]. MDR defines clinical evaluation as a newer approach than MDD, taking into consideration the intended clinical benefits of the device. This indicates that focusing on compliance with general requirements of safety and performance is not sufficient for manufacturers. They have to identify the clinical benefits and verify it throughout the life-cycle of the device. Also, "clinical benefits" is used in the definition of "clinical performance" (Article 2(51)), and "clinical evidence" (Article 2(52)). This is made more evident with the description of intended clinical benefits to patients as the basic requirements of the CEP. Thus manufacturers should identify the clinical benefits of the device prior to elaborating of risk-benefit profile.

Clinical data: MDR identifies the clinically relevant information coming from PMS, in particular, Post-Market Clinical Follow-Up (PMCF) as additional sources of clinical data [3]. This clinical data has possibility to trigger the update of clinical evaluation process and update on the manufacturers claim of clinical benefits, and potentially Instructions For Use (IFU). Thus manufacturers should establish PMS and continuously generate clinical data to verify the safety and performance and continued acceptability of identified risks on the basis of factual evidence.

MDD requires that clinical evaluation be based on data in scientific literature, clinical investigations, or both whereas MDR (Article 61(3)) mentions explicitly that "clinical evaluation shall follow a defined and methodologically sound procedure based on the following 1) a critical evaluation of the relevant scientific literature currently available relating to the safety, performance, design characteristics and intended purpose of the device ... 2) a critical evaluation of the results of all available clinical investigations... 3) a consideration of currently available alternative treatment options for that purpose if any"[3]. The requirement (3) is new in MDR in contrast to MDD. So manufacturers have to consider the available medical options of treatment and state of the art analysis of the device should be properly docu-

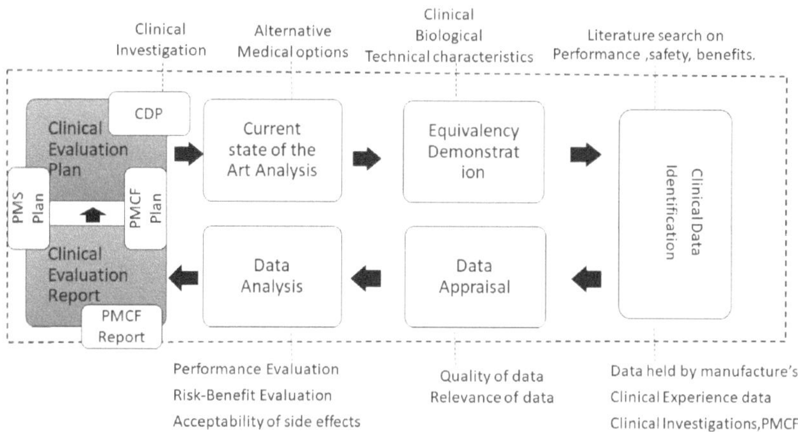

Figure 1: Representation of the clinical evaluation process with all-important sub-processes

mented.

Equivalence: MDR has stronger limitations on the use of clinical data from other devices. In order to claim equivalence to another device, manufacturers must consider clinical, technical, and biological characteristics and have scientific evidence of the comparability. In cases of differences identified, justification for acceptability of difference that there is no concern regarding the safety and performance is required. Thus, equivalence concept is strengthened in MDR than MDD in terms of requirements. It is required to have access to more clinical and non-clinical data relating to devices with which equivalence is claimed in order to justify it.

However, the MDR concepts seems to be more flexible than MEDDEV with the change of word "same" to "similar" for some characteristics. An example is a technical characteristic comparison where it has mentioned "the device shall have similar condition of use" in contrast to "the same condition of use" of MEDDEV.

CEP: MDR also explicitly mentions the need to establish and conduct the CEP, with enlisted basic contents of CEP. Such list of contents is missing in MEDDEV and MDD. Thus manufacturers should have a CEP stating all the parameters used for quantitative and qualitative evaluations of clinical safety including the benefit-risk profile of each indication and intended purpose (Annex XIV (part A-1)).

Summary of Safety and Clinical Performance (SSCP): MDR under Article 32, states that the manufacturer must summarize the main safety and performance aspects of the device and outcome of clinical evaluation and make publicly available through the European database on Medical device (EUDAMED). However, this is limited to implantable and Class III medical devices. Thus, higher risk medical device manufactures should establish such system for transparency of clinical performance and safety.

Scrutiny procedure: The requirements of the notified body regarding the clinical evaluation is also reinforced with the provision of clinical evaluation consultation Procedure.

MDR (Article 54(1)) states that when performing conformity assessment process the notified body is obliged to request that the expert panel scrutinizes the Clinical Evaluation Assessment Report (CEAR) along with manufacturer's clinical evaluation documentation (CER, CEP and PMCF plan) for class III implantable devices and IIb active devices intended to administer and/ or remove medicinal product .

These reinforcements, in general, indicate that the higher is the risk class, the higher the requirements for clinical evaluation and evidence, including higher regulatory demands like review of the assessment of CER within the clinical evaluation consultation procedure. There is also a higher focus on external control and involvement of the notified body and authority responsible for notified bodies.

3.2 Requirements Grouping and Identification of Redundancies

Table 1: Example of content items of Clinical Evaluation and its redundancies in the clinical evaluation documents as well as other documents such as RMR: Risk Management Report, PMSR: Post-Market Surveillance Report

Content Items	CEP	CER	RMR	PMSR
Device-Identification	✓	✓	✓	✓
Device description	✓	✓	✓	✓
Intended purpose	✓	✓	✓	✓
Benefit-Risk profile	✓	✓	✓	✓
State-of the Art Analysis	✓	✓	-	✓
Literature Appraisal	-	✓	-	✓

Clinical evaluation can be realized in a process approach as depicted in Fig.1. It can be observed that clinical evaluation is a continuous process starting from the CEP. Different sub-processes follow it and finally, a CER is prepared which gets its input from sub-processes in-between. It is observed that these sub-processes generate documents that can be established as stand-alone documents. The implication of establishing stand-alone documents is that the requirements can be grouped into the content items, which are used multiple times inside the clinical evaluation process and others. Table 1 presents the example of a few content items used in clinical evaluation which are redundant in other different documents. An example is "device intended purpose" which is a content item in clinical evaluation that contains requirements like device indications, contraindications, special claims and warning pertaining to the device. This content item has multiple redundancies that includes but is not limited to documents like CEP, CER, RM report and PMS Plan. This method enlisted the major redundant information of clinical evaluation which can be handled with a digital tool.

The requirements collection from regulatory/clinical professionals and real-world experience as mentioned in methodology also suggested that consistency is another bigger issue in regulatory document planning, preparation, generation, submission, and update. For example, clinical experience data will improve the manufacturers understanding of clinical benefits for their device which can potentially change the claims of intended clinical benefits in subsequent clinical evaluation. This triggers an update in a number of regulatory documents like CEP, CER, PMS report simultaneously. In this regard, the digital tool that could keep track of the changes, notify changes and ask for verification simultaneously between different departments within the company, would be beneficial and more efficient than the current practice of working in consecutive order. Ultimately, providing consistent information tracking between the departments could help to make work easier for regulatory professionals.

4 Conclusion

Compared to the MDD, the clinical evaluation requirements are highly reinforced in MDR. The gap analysis shows that the quality and quantity of clinical data required for the demonstration of sufficient clinical evidence has been raised. Furthermore, the equivalency route used for clinical evaluation will be rigorously examined by the notified body. Also, the regulations indicate that clinical data and CER, in particular, will face different levels of scrutiny especially in the case of high-risk medical devices because the notified body, expert panels and authority responsible for notified bodies will be involved for subsequent assessment. The notified body will be much stricter under MDR and will check the consistency in documents submitted by manufacturers due to more detailed requirements for them and scrutiny by bodies observing them. In general, the regulatory/clinical department are overloaded with document creation and updates. Furthermore, for the transition to MDR, it was observed that a digital tool that can assemble all requirements, manage the clinical data, and establish sub-processes of the clinical evaluation can help the regulatory professionals to spend more time on content and less time on handling redundant work. Which in run, can plan,

create, update, record and track the clinical evaluation process and its document flow in a systematic and planned way avoiding the inconsistencies.

However, future research should be conducted in more realistic settings for the proper interfacing with other processes for regulatory submission and establishing the digital tool as an integrated management tool to ensure increased work efficiency.

Acknowledgement

This research project was carried at PRO-LIANCE GLOBAL SOLUTIONS GmbH. My sincere gratitude to Mr. Florian Tolkmitt for guidance in the entire research period. Also my sincere thanks to Galenos Solutions GmbH for the access to their software. Finally, I would like to thank my professor Dr. Folker Spitzenberger for guiding and supervising my research work.

5 References

[1] European Union, *Internal Market, Industry, Entrepreneurship and SMEs medical devices.* Available: https://ec.europa.eu/growth/sectors/medical-devices [last accessed on 2020-2-3].

[2] Martelli, Nicolas, Déborah, Carole, judith, Patrice et al. *New European Regulation for Medical Devices: What Is Changing?.*CardioVascular and Interventional Radiology, vol. 42, no. 9, pp. 1272–1278, 2019.

[3] European Union,*REGULATION (EU) 2017/745 OF THE EUROPEAN PARLIAMENT AND OF THE COUNCIL of 5 April 2017 on medical devices, amending Directive 2001/83/EC, Regulation (EC) No 178/2002 and Regulation (EC) No 1223/2009 and repealing Council Directives 90/385/EEC and 93/42/EEC.*Offical Journal of the European Union, vol. 117, 2017.

[4] IMDRF,*International Medical Device Regulatory Forum.*Available: https://www.imdrf.org[last accessed on 2020-1-14].

[5] European Comission, *MEDDEV 2.7/1 revision 4.Clinical evaluation:a guide for manufacturers and notified bodies under directives 93/42/EEC and 90/385/EEC.* Available:https://ec.europa.eu/docsroom/documents/17522/ attachments/1/translations.[last accessed on 2020-2-1].

[6] European Union,*Council Directive 93/42/EEC of 14 June 1993 concerning med- ical devices.*Offical Journal of the European Union,2017.Available:https://eur-lex.europa.eu/legal-content/EN/TXT/?uri=CELEX[last accessed on 2020-1-15].

Development of a digital tool for Post-Market Surveillance of Medical Devices

Reeshav Pradhan [1]
[1] Biomedical Engineering, Lübeck University of Applied Sciences, reeshav.pradhan@stud.th-luebeck.de

Abstract

After passing through a complex conformity assessment procedure for CE marking, manufacturers of medical devices in the EU are required to proactively monitor their devices on the market and to assure quick and adequate response in case of any serious issues. This process is called Post-Market Surveillance (PMS) and the new European Union Medical Device Regulation (EU) 2017/45 (MDR) requires a corresponding and complex documentation to be complied for each medical device. Although PMS is linked to several activities such as risk management and clinical evaluation and should therefore follow an integrative approach, the methods used within PMS are often unconnected from other activities that manufacturers perform. To overcome the resulting challenges like delays and gaps in the documentation, this project aims at a detailed analysis of the regulatory requirements for PMS and at the development of an integrated approach supported by a digital tool for PMS for medical devices.

1   Introduction

The current European Union (EU) Medical Device Directive 93/42/EEC (MDD) and the Active Implantable Medical Device Directive 90/385/EEC (AIMDD) will be replaced by Regulation (EU) 2017/745 known as Medical Device Regulation (EU MDR). The date of application for manufacturers to comply with the EU MDR is on 26th May 2020. The new Regulation is expected to create a transparent and sustainable regulatory framework by introducing new requirements for medical devices, their manufacturers and other stakeholders of the medical device regulatory system. A special focus is now laid on the requirements related to the whole device life cycle including the so-called post-market phase. According to the MDR, "post-market surveillance system shall be suited to actively and systematically gathering, recording and analyzing relevant data on the quality, performance and safety of a device throughout its entire lifetime, and to drawing the necessary conclusions and to determining, implementing and monitoring any preventive and corrective actions" [1].

Manufacturers should therefore establish a comprehensive Post Market Surveillance (PMS) system set up under their quality management system and based on a PMS plan [2]. As new element, the MDR even comprises of a separate Annex III for the technical documentation on PMS. An integrated digital system that allows inter-process data handling may be a helpful solution for manufacturers to run the PMS system smoothly. The main aim was to understand overall PMS requirements, understand the content redundancies within PMS and Post-Market Clinical Follow-Up (PMCF) and redundancies with the interface processes and lastly capture them all in order to be able to derive user require-

ments for the development of Integrated Content Management System (ICMS). The ICMS can enable manufacturers to improve their post-market surveillance activities with respect to documentation quality, readability and avoidance of inconsistencies throughout the technical documentation of the medical device and especially to improve reaction time and seamless integration of complaint handling with preventive and corrective action.

2   Material and Methods

- The most important requirements from a comprehensive list of legislation, standards and guidance documents were identified and analyzed. Requirements for manufacturer were the focus of the analysis. For this purpose and as first step, identification of the regulatory requirement for PMS of medical devices was performed. Starting from MDR, Article 83 till Article 86 were focused on Post Market Surveillance. Nevertheless, other requirements like those out of Article 15 (Person Responsible for Regulatory Compliance) were also considered. Documents such as Post-Market surveillance for manufacturers ISO TR 20416 draft and guidance documents on a medical devices such as vigilance system MEDDEV 2.12-1 rev.8, Post Market Clinical Follow-up studies MEDDEV 2.12/2 rev.2 were considered for understanding the overall PMS process and its contents.

- Another objective was identification of interfaces between PMS and related other processes. This was achieved by creating a flowchart which provides an insight about documents essential during PMS process

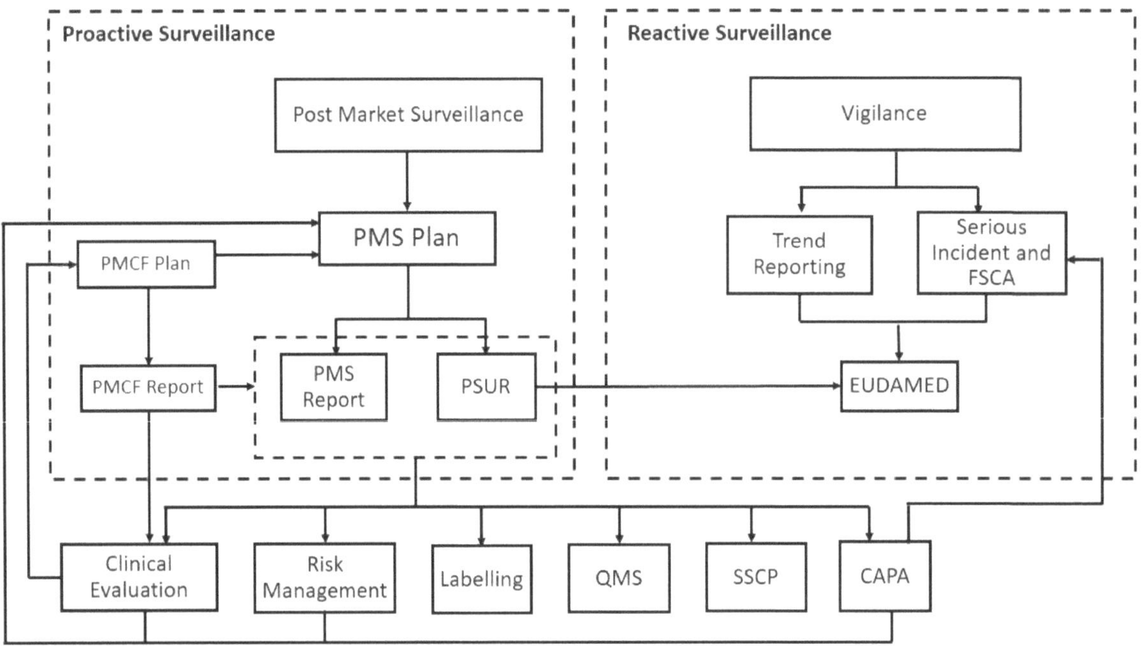

Figure 1: Illustration of the PMS and vigilance documentation according to MDR requirements and its interaction with other processes and systems.

and interaction of PMS with other processes. (see Fig. 1).

- A systematic process (with sub-processes) is required to compile the necessary content leading to the creation of the required documentation for PMS starting with a PMS Plan and leading to the Post-Market Surveillance Report (PMSR) or Periodic Safety Update Report (PSUR), respectively. The required content was listed for each of the sub-processes and these were mapped against other processes (i.e. clinical evaluation and risk management). Many of the content items were redundant between these processes such as description of the device, intended purpose and others. Screening and identifying redundancies was performed to eliminate them.

- After elimination of the redundant contents, for the remaining content items, a user requirement for a potential software was written. Finally, a set of generalized and predefined templates were created which includes PMS Plan, PMS Report (PMSR), Periodic Safety Update Report (PSUR), PMCF Plan, PMCF Report, Field Safety Corrective Action (FSCA) and many more which will be incorporated in the ICMS.

3   Results and Discussion

3.1   Interaction of PMS with other processes

Fig. 1 reflects a complete set of PMS documents and its interaction with other systems and processes that manufactur-

ers need to establish to be in compliance with the MDR. The major PMS documents that manufacturers need to develop are the PMS Plan followed by a PMSR or PSUR respectively. The PMSR is intended for low risk devices (Class I) whereas PSUR is and extension of PMSR containing information about moderate and high risk devices (Class IIa, IIb, III, implantable). PMCF being a part of PMS should also be carried out throughout the life cycle of the medical device. PMCF corresponds to two major documents namely the PMCF Plan and the PMCF Report [5]. The PMCF Plan should be part of the PMS Plan and the main findings of the PMCF Report must be included into the PSUR.[1] The output from PMS acts as input to other required processes such as Risk Management, Clinical Evaluation, CAPA, QMS and others. Similarly, the output from other processes like Risk Management, Clinical Evaluation, QMS acts as an input to PMS as shown in Fig. 1. This shows the dependency and importance of each process in order for the whole system to function smoothly.

According to Fig. 2 it can be observed that the PMS process starts with the planning reflected through the PMS Plan which should be ready before the device is placed on the market. The PMS Plan defines the overall PMS activities and responsibilities for the device including data collection methodologies, person responsible for each task such as gathering, analysis, summarization of data and the time frame to conduct a certain task such as the compilation of the PMS Report/ PSUR. These steps are followed to carry out the further process where the collected data are analyzed using the predefined templates which were created. Finally

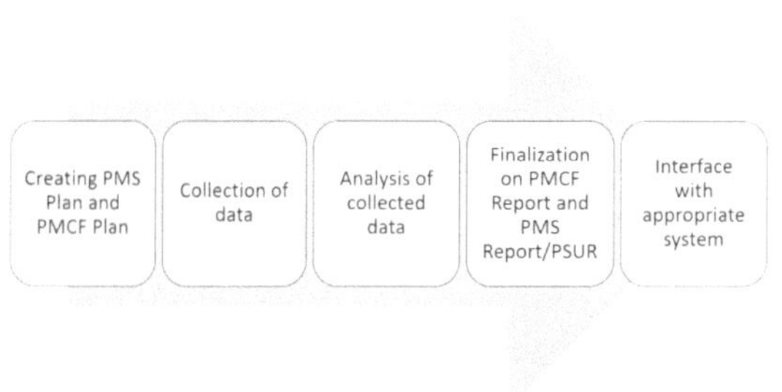

Figure 2: Illustration of the PMS process according to MDR requirements

manufacturers have to create a summary of the analysis as PMSR/PSUR which also acts as input to the other processes identified.

3.2 Gap Analysis for PMS requirement between MDD and MDR

The new EU MDR has additional requirements for PMS when compared to the current MDD. According to Annex IV/3 and Annex VII/4, the MDD requires PMS but fails to provide detailed requirements.
MDD:
PMS according to the MDD is the source of clinical data which is used to update the clinical evaluation report. In order to be in compliance, the manufacturer needs to keep up-to-date a systematic procedure and when necessary needs to apply necessary corrective action [3]. MDD fails to provide sufficient information about the overall process of PMS.
MDR:
In the MDR requirements for the PMS are explained in more detail in Articles 83 to 86. A detail about PMS plan is mentioned in Annex III of technical documentation. For each device regardless of the class of device, manufacturers are obliged to conduct PMS and have to plan, establish, document and update the respective documents. The MDR requires PMS to be an integrated part of the manufacturer's Quality Management System. The MDR has also introduced EUDAMED for submitting PSUR [1]. Currently, guidance documents under the MDR (endorsed by Medical Device Coordination Group) are not provided in support of PMS yet. The PMS plan is however mentioned and described according to ISO TR 20416 which is ahead of publication and which would provide guidance for manufacturers to carry out the PMS process.

3.3 Gap Analysis between MEDDEV 2.12/1 rev. 8 and MDR for elements of the vigilance system

According to the MEDDEV 2.12/1, manufacturers must report serious incidents and near incidents. The reporting timelines defined in MEDDEV 2.12/1 are[4]:

- event of a serious public health threat: Immediately, and not later than 2 days

- event of death or unanticipated serious deterioration in state of health: Immediately, and not later than 10 days

- other reportable incidents: Immediately, and not later than 30 days

Where as according to the MDR, manufacturers must report only serious incidents. The reporting timelines defined in MDR are:

- event of a serious public health threat: Immediately, and not later than 2 days

- event of death or unanticipated serious deterioration in a person's state of health: Immediately, and not later than 10 days

- other serious incidents: Immediately, and not later than 15 days

According to MEDDEV, manufacturers need to report any serious incidents and Field Safety Corrective Action (FSCA) to the corresponding National Competent Authority. According to the MDR, the report must be submitted to the new centralized electronic system EUDAMED.
There is no description about PSUR in MEDDEV compared to the detailed explanation about the content to be included into the PSUR according to the MDR.

3.4 Effect of risk class on PMS documentation

The risk class of a device has a major effect on the medical device. The risk class of the medical device determines the PMS documentation of the devices which is outlined in Chapter VII, Article 85 and Article 86 of the MDR. According to Article 85, manufacturers need to generate PMSR for a low risk Class I devices. The PMSR summarizes the results and conclusion of the collected PMS data along with a rationale and description of any corrective actions taken. According to Article 86, manufacturers need to generate the PSUR for moderate and high-risk devices (Class IIa, IIb, III, Implantables). The PSUR can be seen as an extension of the PMSR which consist of additional information to that of the PMSR.

Table 1: PMS reporting frequency according to risk class

MDR Classification (report type)	Update
Class I (PMSR)	atleast every 2 years
Class IIa (PSUR)	atleast every 2 years
Class IIb (PSUR)	atleast every year
Class III (PSUR)	every year

Report as PMS report: For class I devices, manufacturers have to prepare a PMSR and that should be updated at least once in 2 years as shown in Table 1. They shall make it available to the competent authority upon request. The major content that the PMSR must include:

- Information concerning serious incidents including FSCA

- Information from trend reporting

- Relevant technical literature, databases and/or registers

- Information, including feedback and complaints, provided by users, distributors and importers

- Publicly available information

Report as PSUR: For class IIa, class IIb and class III devices, manufacturers have to prepare a PSUR that should be updated once in 2 years for class IIa or yearly for high risk devices such as class IIb and class III (See Table 1). The PSURs for high risk devices have to be submitted via the electronic system "EUDAMED". The PSUR must include the content covered by PMSR and in addition, also needs to indicate the following:

- The conclusion of the benefit-risk determination

- The main findings of PMCF

- The volume of sales of the device and an estimate evaluation of the size and other characteristics of the population using the device and, where practicable, the usage frequency of the device.

3.5 Integrated Content Management System for PMS

PMS has a close relation with other processes as shown in Fig. 1. Manufacturers have a demanding task for keeping all required documents in order and up-to-date.

Data redundancy is one of the major problem for manufacturers. Manufacturers have to compile many reports to prepare their technical documentation. Among these reports many redundant contents exists in different documents. Some of the major redundant contents commonly found in documents are intended purpose, device description and others. An ICMS connected to a database system provides manufacturers a platform to store all the content as a unique content item. Whenever necessary, the content can be accessed by the user to incorporate it in all the required documents ultimately reducing the data redundancy. In additions, an ICMS also has the potential to improve efficiency by providing consistent documents. Any inconsistent document may be flagged, so the user may ultimately save time avoiding revision of the whole documentation again to determine the lacking documents. Documentation tracking can also be a major challenge for manufacturers which can be overcome by use of an ICMS.

4 Conclusion

In the study, we demonstrated that the EU MDR significantly increases the requirements for PMS activities to be undertaken by manufacturers of medical devices. A well-structured PMS system including adequate planning and reporting documents plays a central role for the proactive safety evaluation of medical devices. An ICMS can be one of the solutions for manufacturers problem of data inconsistency and data redundancy. The developed set of PMS requirements along with predefined templates is sufficient to build the first prototype of an ICMS for PMS which can provide a platform for manufacturers to avoid redundancy in the collection and archiving of relevant information and to achieve a higher degree of consistency. However, further improvements can be considered and implemented such as platform for root cause analysis, an interface for reporting the documents directly to the electronic system EUDAMED and others.

Acknowledgement

I sincerely would like to thank Mr. Florian Tolkmitt, PROLIANCE GLOBAL SOLUTION GmbH, Münster, Germany for conducting and supervising the internship and Mr. Michael Kania, Galenos Solution GmbH, Fulda, Germany for providing a work place. I would also like to thank Prof. Dr. Folker Spitzenberger, Center for Regulatory Affairs in Biomedical Sciences (CRABS), Technische Hochschule Lübeck, Lübeck for his continous guidance and supervision during the project.

5 References

[1] European Union, *Regulation (EU) 2017/745 of the European Parliament and of the Council of 5 April 2017 on medical devices, amending Directive 2001/83/EC, Regulation (EC) No 178/2002 and Regulation (EC) No 1223/2009 and repealing Council Directives 90/385/EEC and 93/42/EEC*. Official Journal of the European Union, 2017.

[2] International Organization for Standardization. Technical Committee Quality management and corresponding general aspects for medical devices, *ISO 13485: Medical Devices-Quality Management Systems-Requirements for Regulatory Purposes*. ISO, 2003

[3] Directive, Council, *93/42/EEC of 14 June 1993 concerning medical devices*. Official Journal of the European Communities, 1993.

[4] European Union, *Guidelines on a Medical Devices Vigilance System (2013) (MEDDEV 2.12/1 rev.8)*. Available: https://ec.europa.eu/health/home, Accessed on January 10, 2020

[5] European Union, *Guidelines on a Medical Device Post Market Clinical Follow-Up Studies (2012) (MEDDEV 2.12/2 rev.2)*. Availabale: https://ec.europa.eu/health/home, Accessed on January 10, 2020

Approval of safety valves and globe valves in South Korea

Sandra Henn [1], Christoph Kunath [2], Marc Zaubitzer [2] and Thorsten Cordes [2]

[1] Medical Engineering Science, Universität zu Lübeck, sandra.henn@student.uni-luebeck.de
[2] HEROSE GMBH, Bad Oldesloe, {christoph.kunath; marc.zaubitzer; thorsten.cordes}@HEROSE.com

Abstract

This paper contains the procedures for the approval of the company HEROSE GMBH for the South Korean market. HEROSE GMBH is a company which manufactures valves such as safety valves and globe valves. These valves are used under extreme conditions (high pressure up to 550 bar, temperatures from -270 °C to 400 °C) using partly toxic or explosive media. It must be ensured that systems operated at high pressure pose no danger to people or the environment. For this reason, high safety and quality requirements are placed on the valves all around the world. In Korea there are different types of approval. Only the approval by the Korean Gas Safety Corporation (KGS) is relevant for HEROSE GMBH. KGS has its own regulations (KGS code) prescribing the requirements for the products/production process of a company. The approval was successfully carried out on the basis of the specifications issued by the Korean authority.

1   Introduction

HEROSE is a globally operating company. The basis for the distribution of goods in other countries are the corresponding approvals for the products. HEROSE GMBH currently has six different approvals from classification societies and nine country-specific approvals.

Approvals are an official permit for access to a market. The requirements depend on the laws of the individual countries. Within the framework of the project, the regulations necessary for South Korean market access will be determined and the requirements set out in these regulations analysed in order to derive the necessary measures. The approval is carried out in cooperation with all involved parts of the company (design, development, marketing, etc.).

2   Material and Methods

There are various approvals in South Korea, of which only the KGS approval is relevant for the products [5]. KGS stands for Korean Gas Safety Corporation, which is a governmental organization for testing, inspection and education. It reports to the Ministry of Trade, Industry and Energy (MOTIE) and ensures uniform quality of gas products in Korea. The legal basis is formed by the High-Pressure Gas Safety Control Act, the High-Pressure Gas Safety Control Enforcement Decree and the High-Pressure Gas Safety Control Enforcement Rule [1, 6, 7]. The requirements derived from the laws are published in a KGS Code for various product groups by the KGS. The KGS Code is a detailed description of the requirements and instructions for action and thus represents the practical aspects in contrast to the generally formulated laws. The code applies to equipment, technologies, inspections and manufacturing and is comparable to the principle of European standards.

The market analysis carried out by product management has shown that there is a need for safety valves and globe valves in South Korea. Safety valves are pressure relief devices that protect e.g. containers and pipes from excessive pressure. They respond as soon as the pressure exceeds the maximum permissible operating pressure. After blowing the gas into the surrounding area, the valve closes again. Globe valves can be used to control the flow of liquids and gases in pipes. The globe valves have a handwheel with which the opening can be adjusted. The KGS Code AA319 provides the instructions for the safety valves and for the globe valves it is the KGS code AA335. It follows from these documents that the procedure for KGS approval includes a technical review and a factory audit [1]. There is no experience with this procedure for these products.

2.1   Technical Review

	Document	Responsibility
1	Company profile/ history	Marketing
2	Existing certificates	Quality Management
3	List of production articles	Product Management
4	List of subcontractors	Quality Management
5	List of production equipment	Production
6	List of measuring instruments	Quality Assurance
7	Quality manual	Quality Management
8	Product drawings	Design
9	Strength calculations	Development

Table 1: Required documents [1]

The technical review is only a document verification. The documents 1-6 were requested from the respective departments (see table 1).

In the course of digitization and modernization the company HEROSE has transferred the quality manual into the a WIKI. Every employee can access the WIKI from anywhere at any time. However, the WIKI is still under construction and for this reason there is no English version available which is required for international applications. For this reason, a quality manual in English is maintained in parallel, which is used for approvals. So the existing quality management manual was compared in detail with the Korean requirements and then revised accordingly [2].

2.1.1 Product Drawings

Product drawings illustrate all components of a product and are additionally provided with a parts list containing the corresponding material information of the individual parts. In addition, these drawing sheets contain all relevant information about the product (e.g. size variants, operating conditions). For the design of drawings the CAD program Creo Parametric 3D modelling software from PTC is used. The drawings can be edited in the workspace. First, the desired section is selected from the 3D model of the product. This is then transferred to a 2D view. The 2D worksheet must be configured to meet the requirements for Korea. Since creating a drawing is very time-consuming, it is advisable not to create a separate drawing for each product number, but to combine variants on one drawing sheet. Once the drawing is finished, it must be uploaded via Windchill and an application for approval must be made. Approval is given by the head of the design department. All drawings for the KGS approval are summarized in a file and provided with a cover sheet. The cover sheet contains a list of all relevant information that is important for understanding the documents.

2.1.2 Strength Calculations

The strength in relation to the applied pressure is an important measure for the safe use of valves. Therefore, the strength calculations must be carried out for all pressure-bearing components. Pressure-bearing components in the valves are the seat inserts and the body. Various methods can be used to determine the strength of a product.
In the past it was usual to carry out a burst test. In a burst test, the corresponding product is tested on a burst test bench. The data collected are all recorded in a protocol. The pressure is increased until the required pressure is reached. In the protocol, the result is noted whether the body of the product is damaged or shows no cracks or deformations. Since HEROSE GMBH has been selling many products for a long time, there is evidence of this for many valves. During the examination of the documents, however, it was noticed that some proofs are missing or incorrect. For this reason, new burst tests had to be carried out in some cases. The strength could be confirmed. The burst test method is not common for high-pressure valves. For this reason, calculations were carried out to prove the strength. The corresponding documents with the already existing calculations were checked and various problems were found in the calculations. For this reason, the strength had to be recalculated.

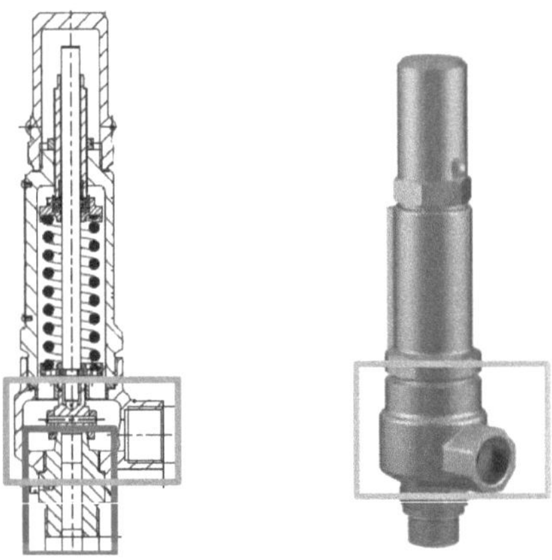

Figure 1: 2D and 3D model of a safety valve (body bright box, seat insert dark box)

Fig. 1 shows an example of a valve of the products to be approved. The dark box encloses the seat insert and the bright box contains the body.

Seat inserts
The calculation refers to DIN EN 12516-2 [3].

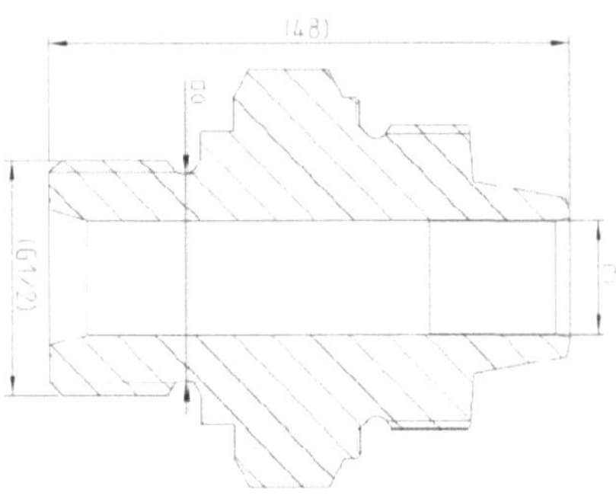

Figure 2: Seat insert

For the seat inserts, the following parameters are relevant for the calculation:

- outer diameter d_0 (Fig. 2) [4]
- inside diameter d_i (Fig. 2)
- calculation Strength f
- operating temperature t_d
- operating pressure p
- strength parameter Rp1.0
- safety factor SF
- welding factor k_c
- manufacturing tolerances c_1
- corrosion tolerances c_2
- effective wall thickness e_a

The condition for strength is [3]:

$$e_a > e_c + c_1 + c_2 \tag{1}$$

e_c is the calculated wall thickness. The formula for the wall thickness depends on the ratio of outer to inner diameter. For

$$d_0/d_i > 1,7 \tag{2}$$

the following formula is taken from DIN EN 12516-2 [3]:

$$e_c = \frac{d_i}{2}\left(\sqrt{\frac{f \cdot k_c + p}{f \cdot k_c - p}} - 1\right) \tag{3}$$

For

$$d_0/d_i < 1,7 \tag{4}$$

the following formula is taken from DIN EN 12516-2 [3]:

$$e_c = \frac{d_0 \cdot p}{(2 \cdot f - p) \cdot k_c + 2 \cdot p} \tag{5}$$

with

$$f = \frac{R_{p1.0}}{SF} \tag{6}$$

The strength could be verified for all seat inserts.

Body

The minimum wall thickness of the outlet chamber and minimum wall thickness of the outlet are calculated according to the same principle like the strength calculation for the seat inserts.

The strength could be verified for [3]

$$e_{a(chamber)} > e_c + c_1 + c_2 \tag{7}$$

and

$$e_{a(outlet)} > e_c + c_1 + c_2 \tag{8}$$

To verify the strength of the whole body, the acting forces per mm^2 must be determined. The surface comparison method is used for this purpose. First, the load-bearing areas must be determined on the basis of rules and regulations. These are marked in the 3D model (Fig. 3) and the surfaces for the surface comparison method are determined from them.

The areas are shown in Fig. 3. p is the operating pressure, k_c is the welding factor and f is the calculation strength (6) [3].

$$p \cdot \left[\frac{A_{pI}}{A_{fI} \cdot k_c} + \frac{1}{2}\right] \leq f \tag{9}$$

$$p \cdot \left[\frac{A_{pII}}{A_{fII} \cdot k_c} + \frac{1}{2}\right] \leq f \tag{10}$$

$$p \cdot \left[\frac{A_{pIII}}{A_{fIII} \cdot k_c} + \frac{1}{2}\right] \leq f \tag{11}$$

The strength could be verified for all valves.

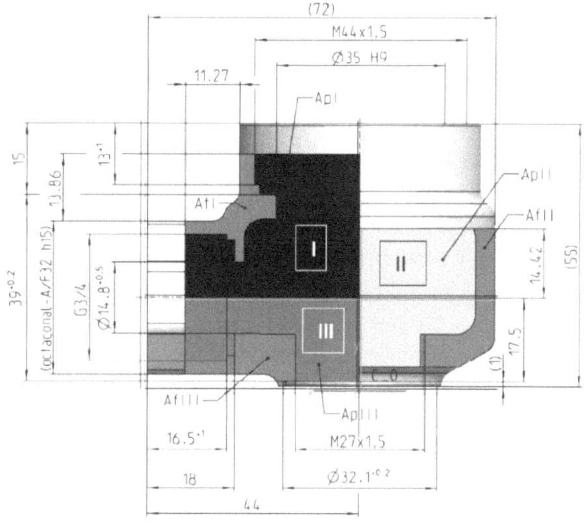

Figure 3: Areas for surface comparison method

2.2 Factory Audit

The KGS has communicated a detailed audit plan in advance. The audit takes place for three days at the factory in Bad Oldesloe. It begins with an opening meeting and a tour of the company. Documents are then viewed in detail. The third day ends with a closing meeting and the result of the audit. The audit team consists of three persons from KGS. One auditor is specialized in checking the code and standards. Another auditor is taking over the inspection of the manufactory and the test equipment. The third auditor is responsible for controlling the quality management system.

2.2.1 Tour along the material flow

The main part of the audit is the tour along the materal flow. It is a test procedure of the auditors in which all processes of the company can be examined. With this, a continuous quality in all areas of the company can be determined. In this test procedure, however, not only the various processes are checked, but also the products themselves are tested. At first the quality of the incoming goods is checked with measuring and testing equipment (microscope, gauges, contour measuring device). Also the spring rate is inspected. In addition, the chemical composition of metals is determined by X-ray fluorine essence analysis (XRF). The goods that meet the requirements are marked for release. If the delivered goods have significant defects, they are blocked and may not be used. For the auditors, it is important to ensure that no defective components get into production.

The components are then uniquely marked with an automatic marking machine. Marking and traceability is an essential aspect of auditing. Another important aspect of the audit is ensuring the quality of the semi-finished products. The quality is determined by means of a 100% hydrostatic pressure test and a leak test.

In the welding shop, the components are first cleaned (degreased) and then brazed or welded. For the audit team, the approval of the welding company and the welding supervisor are relevant. All documents and certificates are examined in detail.

For the auditors, the test benches play an important role in assembly. The test benches must be certified and regularly maintained. Corresponding seals and markings are located on each test bench. Further documents are also stored in the Wiki. The Wiki also contains the work instructions which specify how the valves must be mounted, tested and marked. It is of interest to the auditors that all employees have access to the necessary information and that work instructions are executed correctly. Various test benches are located in the test area. These test benches are used to carry out sample tests. An important point of the KGS audit is the performance test of the valves. It is important in this test that it can be shown that the valves are safe and functional. The functionality and safety of the valves can be demonstrated.

2.2.2 Document analysis

After the tour through the company, the documents are discussed more in detail with the quality management. Individual areas of the QM manual are examined in more detail. All documents collected during the tour of the company are checked for conformity with the KGS code.

3 Results and Discussion

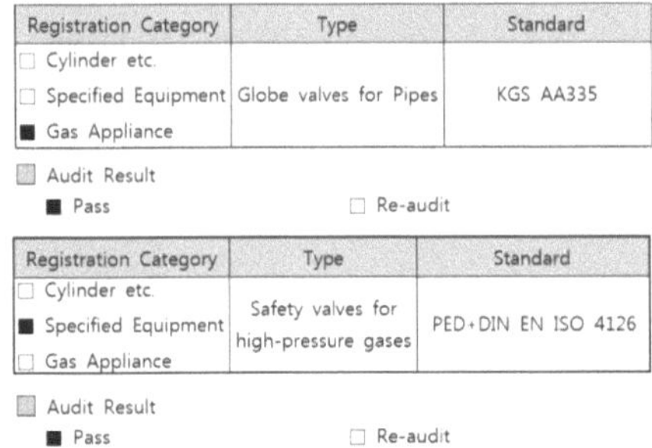

Figure 4: Result of the document review and the factory audit

The auditors catalogue their results in 4 categories:

1. Conform with the KGS Code
2. Suggestion for improvement: These tips have no influence on the first approval as they are not safety relevant. However, the suggestions are expected to be implemented by the next audit.
3. Minor non-conformities: If a non-conformity is found either to the KGS regulations or to the manufacturer's specification, but these do not matter centrally.
4. Major non-conformities: If the relevant KGS regulations or manufacturing instructions are not complied with. If a major fault is found. If several minor defects are found which impair the production quality.

No minor or major non-conformities were found so that the result is *PASS* (Fig.4).

This result was to be expected because HEROSE GMBH has a comprehensive quality management system which has already been tested and approved by various approval bodies and classification societies. However, suggestions for improvement were made. In addition to the calibration by the TÜV (every 5 years), the plate of the spring measuring instruments should be checked internally every year. In addition, KGS requires that every product exported to Korea must be tested by an independent body (TÜV). Previously, this was only done at the request of the customer.

4 Conclusion

Because all products and processes were comprehensively checked for conformity with the requirements in preparation for the audit, no non-conformities were found during the audit. This approval is valid for three years. An application for renewal must be submitted approximately 6 months before the expiry date. The certificate will then be renewed after a positive control audit. A control audit checks whether there have been changes in the company processes (e.g. new managers or new test benches) and whether suggestions for improvement have been implemented. In addition, it is checked whether the production quality is continuous and meets the same standards as in the first factory audit. It can be assumed that HEROSE GMBH will also pass this next examination.

Acknowledgement

The work has been carried out at HEROSE GMBH, Bad Oldesloe and supervised by Dr. Young-Hwa Song, Institute of Physics, Universität zu Lübeck. I would like to thank you for your support.

5 References

[1] Korea Gas Safety Corporation, *Factory Registration Guide for Foreign Manufacturers of Gas Product.* South Korea, 2018

[2] Korea Gas Safety Corporation, *Quality Control System Requirements for manufacture registration of foreign products.* South Korea, 2011

[3] DIN EN 12516-2:2014 *Industriearmaturen - Gehäusefestigkeit - Teil 2: Berechnungsverfahren für drucktragende Gehäuse von Armaturen aus Stahl*

[4] DIN 76-1:2016-08 *Gewindeausläufe und Gewindefreistiche - Teil 1: Für Metrisches ISO-Gewinde*

[5] KOSHA, *Safety Certification Guide.* South Korea, 2014

[6] KLRI *High-Pressure Gas Safety Control Act.* South Korea, 2016

[7] KLRI *Safety Control and Buisiness of Liquefied Petroleum Gas Act.* South Korea 2015

Regulatory requirements for laboratory developed tests (LDTs) – International comparison and conclusions for best-practice requirements for safety and performance of "in-house" IVD medical devices

Abdulrahim Safi,

Biomedical Engineering, University of Applied Sciences Lübeck, rahim.safi1985@gmail.com

Abstract

While regulatory requirements for medical devices including in vitro diagnostic medical devices (IVD) are constantly increasing all over the world, there is a controversial regulatory debate with regard to so-called "in-house" IVD or laboratory developed tests (LDTs). Although, - especially in the case of rare and emerging diseases -, LDT have a central impact on patient diagnosis and treatment, there is currently no common understanding or internationally harmonised approach for minimum or essential requirements for LDTs. With the aim to elaborate recommendations for a best-practice regulatory framework for LDT, an international comparison and critical analysis of requirements for LDT, as established in major regulatory systems, have been carried out. As a result, a list of essential regulatory elements for LDT has been identified that represents the core of a regulatory best–practice approach for ensuring quality, safety and performance of LDT.

1   Introduction

Laboratory developed tests (LDTs) or "in-house IVDs" are broadly used in medical laboratories. LDTs are in vitro diagnostic testing methods that are performed by using an In Vitro Diagnostic Medical Device (IVD) which are developed, manufactured and used within a single health institution and its equivalent laboratory. Normally, they are not available on an industrial scale.

The medical relevance of LDTs is highly significant though, because LDTs are used for patient diagnosis and monitoring in a wide scope of rare and/or emerging diseases in almost all fields of medical laboratory testing [1]. Consequently, the setting of quality, safety and performance requirements within a regulatory framework for LDT is relevant and is expected to have a significant impact on both individual patient care and on public health and safety.

Currently, however, the regulatory situation for LDT in different countries inside and outside the EU is versatile and sometimes controversial.

In Europe, for example, a new regulation was published in May 2017 which includes for the first-time harmonized requirements for IVD that are manufactured and used in the same health institution. These requirements are set out in Article 5 paragraph 5 of the Regulation (EU) 2017/746 ("IVDR") [2]. Germany, Austria, the United Kingdom and Switzerland are currently using their national regulatory requirements while preparing for compliance with the new requirements as set by the IVDR.

Australia, a traditionally well-regulated country, has a legislation for LDTs under the Therapeutic Goods (Medical Devices) Regulations 2002 which is actively in force since several years.

Many countries or regions from emerging markets like Brazil, Taiwan, China and India have no concrete requirements for LDTs yet.

Regulatory elements and aspects such as, for example, requirements for quality, safety and performance of the test systems prior use, the implementation of a quality management system as well as of a risk management system etc. are discussed controversially between different countries and regions. The aim of the study is to perform an international comparison and analysis of the so far existing different approaches including their regulatory elements leading to recommendations for a best-practice approach for quality, safety and performance of LDTs.

2   Material and Methods

2.1   Applicable legislation, standards and guidance documents for LDTs

According to hierarchy of regulation documents, guidance documents which are also called "soft law" and they are not legally binding but might be helpful to fulfill the legal requirements. The applicable legislation and normative documents for the countries and regions like U.S., Australia, EU (Germany, Austria and Switzerland) were analysed [1]-[9].

Also further regions and countries were elaborated which did not have any LDT requirements. Furthermore, GHTF and IMDRF documents were also analyzed [10] which currently set criteria for safety and performance MD/IVD, but they do not offer specific guidance for LDTs.

2.2 Gap Analysis

A gap analysis is a method which plays a central role for the comparison procedure of the actual performance and the desired performance of regulatory procedures and concepts. Gap analysis helps to determine the gaps which are currently existing in procedures related to legal and normative requirements. Performing a gap analysis may disclose some parts or elements (gaps) that are essential to establish or to improve an effective regulatory framework for LDT [11].

The gap analysis was done in two different steps. The first step was to review and analyse all applicable legislation and normative documents for LDTs that were identified. The second step was to develop a list of criteria from various regulatory systems which helps to perform an international comparison.

3   Results and Discussion

3.1   Common regulatory elements as essential criteria for LDT

As the analysis of the regulatory framework for LDT among major markets and regulatory systems revealed, a harmonized approach for LDTs at the international level is currently not established or available. Although IMDRF and its preceding organisation, the GHTF, offer guidance on essential requirements for safety and performance of medical devices and IVD, there is no specific recommendation for LDT available.

It is also evident that regulatory requirements for LDT are so far mostly omitted in countries of less regulated and/or emerging medical device markets. Therefore, countries like China, India, Brazil and others do not provide regulatory requirements or guidance with regard to LDT yet.

Several well-established regulatory systems like those in the EU, Australia, U.S.A. and Canada do however cover LDT by either their primary and secondary legislation or by national/regional standards.

The gap analysis between the existing legislation and standards revealed that a set of essential and common criteria for LDTs can be concluded as most significant regulatory elements for ensuring quality, safety and performance of LDTs (see Table 1). Although discussed contentiously, these elements are part of all the regulatory systems evaluated within this study.

Out of these common elements, a best practice regulatory approach for LDTs can be developed. The analysis and international comparison also revealed differences between the regulatory systems. The rationales for differences are discussed controversially among the regulators.

For example, in Australia, the laboratories which are manufacturing LDTs are obliged to be accredited by the National Association of Testing Authorities (NATA) according to ISO 15189 (Medical laboratories - Requirements for quality and competence).

In the EU, the standard ISO 15189 is for the first time acknowledged in the legal text of the IVDR, but its implementation remains optional according to the national provisions of the different Member States. For example, accreditation according to ISO 15189 is voluntary in Germany, while it is obligatory in France for all medical laboratories without consideration of whether LDT are used or not.

In the U.S., all laboratories including laboratories which are manufacturing LDTs are accredited according to the Clinical Laboratory Improvement Amendment (CLIA). CLIA acknowledges the criteria of ISO 15189. However, the CLIA accreditation scheme does neither include the assessment of the validation of tests nor the assessment of the clinical validity for LDTs. This is one of the concerns that FDA considers with regard to LDTs approved under CLIA. To demonstrate the currently existing regulatory situation for LDT, two common elements are further elucidated within the following sections.

3.2   Requirements for quality, safety and performance prior use

One of the pivotal elements in the field of medical device regulatory system is the conformity assessment procedure. The conformity assessment procedure will ensure the safety and performance of a medical device throughout its life cycle. The risk classification of a medical device take part as an important aspect in conformity assessment procedure. Because the higher the risk class of medical devices the more rigorous the conformity assessment procedure will be in comparison to low risk class medical devices.

There are some specific rationales for LDTs regard to requirements for quality, safety and performance prior use. These requirements exist in the new regulation (EU) 2017/746 on in vitro diagnostic medical devices according to Article 5 (5). The IVDR describes that regardless of the classes (A, B,C,D), the manufacturer of in – house testing or (LDTs) must fulfil the relevant requirements of general safety and performance according to Annex (I) of new regulation (EU) 2017/746. The rest of the regulation is not applied to in-house manufacturing devices (LDTs). Also, compared to commercially available IVDs, there is no CE – mark and there is no notified body for LDTs, unless the laboratory must fulfil some rationales before requirements which are described in Article 5(5) index (a) – (i) of the new regulation.

In Australia the situation is diverse. They have different kind of conformity assessment procedure for in-house IVDs such as, class 1-3 in-house IVDs and class 4 in -house IVDs. The conformity assessment procedure for class 4 in-house IVDs is described according to part 1 or part 6B, schedule 3, of Therapeutic Goods (Medical Devices) Regulations 2002, and describes that the laboratories that are manufac-

Table 1: List of major regulatory elements identified by international comparison of legislation and standards for LDT

No	Regulatory elements
1	Regulatory oversight mechanism
2	Risk -based approach towards regulatory control of LDTs (risk-based classification)
3	Requirements for quality, safety and performance prior use.
4	Quality management system requirements
5	Risk management system requirements
6	Post market surveillance and vigilance activities (CAPA, complaints handling)

turing class 4 in-house IVDs must register them in Australian Register of Therapeutic Goods (ARTG). To do this the laboratories have two approaches. The first approach is, to obtain Therapeutic Goods Administration (TGA) conformity assessment certificate prior to register in ARTG. The second approach is, to use their current NATA accreditation which is complied with ISO 15189. It should be noted that there is no registration for class 1-3 in-house IVDs in ARTG. The conformity assessment procedures for laboratories manufacturing class 1-3 in-house IVDs must fulfill the requirements which are described in part 6A, Schedule 3 of Therapeutic Goods (Medical Devices) Regulations 2002.

In the U.S, the LDTs will go under FDA Premarket review, when they publish the final draft guidance. According to this draft, FDA will review all high-risk class (III) subject to Pre-market Approval (PMA). The most moderate risk class (II) will go under review of accrediting third party which is requiring the submission of a premarket notification (510K). It should be noted that quality system registration (QSR) requirements will be generally performed along with premarket review requirements.

There are some additional categories for LDTs which are not covered by EU and Australia. These additional categories of LDTs such as LDTs for rare diseases, traditional LDTs and LDTs for unmet needs will generally remain under the current enforcement discretion (An appropriate policy of FDA, to take enforcement action when necessary to protect the public health).

3.3 Quality management system

Quality management system is one of the other core elements for LDTs that is discussed very contentiously between different regulatory systems. Medical laboratories are mostly complying with EN ISO 15189 which is an international standard for quality and competence of the medical laboratories. This situation is not always same. In Europe, according to the new regulation framework, Article 5 (5) index (c) defines that the laboratory of a health institution is compliant with EN ISO 15189 or where appropriate national requirements, including national provisions regard to accreditation. This statement indicates that the application of EN ISO 15189 is optional. For more precision, for example, in Germany the application of EN ISO 15189 is not obligatory, instead they use RiliBÄK - Labor which is abbreviated form in German or " Guidelines of the German Medical Organization for quality assurance in laboratory testing" [12].

One of the differences between ISO 15189 and RiliBÄK is the application. EN ISO 15189 is an international standard used only for medical laboratories and it is very comprehensive for quality assurance process. On the other side, RiliBÄK is obligatory in Germany and has minimal requirements for quality and assurance process for Laboratory testing. RiliBÄK is not only used for medical laboratories but also used for doctor's office and even the Pharmacies that are selling commercially IVDs, etc.

In UK the accreditation process for medical laboratory is through UKAS (United Kingdom Accreditation Service). They are using ISO 15189 as appropriate standard for accreditation of medical laboratories.

In Australia they need a full quality assurance procedure for class 4 in-house IVDs and class 1-3 in-house IVDs must comply with ISO 15189.

In U.S the manufacture of medical device must implement minimal quality system requirements (21 CFR part 820) to ensure the safety and effectiveness of medical devices. FDA will apply the current enforcement discretion policy with respect to quality system registration requirement (QSR), until the manufacturer of LDTs submit a PMA or (510k) clearance. During the time, the medical laboratory manufacturing or using the LDTs will be responsible to have quality system in place and comply with 21 CFR part 820. As mentioned before, at this time CLIA governs the accreditation of all laboratories compliant with ISO 15189.

4 Conclusion

Nowadays the innovation of medical devices is growing fast and even it has numerous consequences on laboratory developed tests. Modern laboratory developed tests are very complex. Therefore, LDTs need harmonized regulatory requirements to assure the care and safety of the patient. Unfortunately, there are no minimum harmonized regulatory requirements available for LDTs so far. The existing harmonized regulatory requirements are very certain. In order to fill this gap, an international comparison of regulatory requirements of different countries and regions has been done. During an international comparison the most significant elements that were so far identified, will be used to develop a best – practice approach for safety and performance of LDTs in the future, for example within a standardization project in ISO level. A few of those core elements have been already elaborated. A list of these elements as shown in Table 1.

In the future, the existence of minimum harmonized requirements for LDTs is highly crucial. By having the harmonized regulatory requirements, there would be plenty opportunities for medical laboratories, manufacturers, patient, and conformity assessment bodies.

Acknowledgement

This work has been carried out at the University of Applied Sciences of Lübeck, Germany and supervised by Prof.Dr. Folker Spitzenberger, Centre for Regulatory Affairs for Biomedical Science (CRABS), University of Applied Science of Lübeck.

5 References

[1] Draft Guidance for Industry, Food and Drug Administration Staff, and Clinical Laboratories 2014, Framework for Regulatory Oversight of Laboratory Developed Tests (LDTs). Available: https://www.fda.gov/regulatory-information/search-fda-guidance-documents/framework-regulatory-oversight-laboratory-developed-tests-ldts [last accessed on 2020-01-16].

[2] Regulation (EU) 2017/745 of the European Parliament and of the Council of 5 April 2017 on in vitro diagnostic medical devices. Available : https://eur-lex.europa.eu/legal-content/EN/TXT/?uri=CELEX:32017R0746 [last accessed on 2019-12-20].

[3] Federal Register of Legislation 2019, Therapeutic Goods (Medical Devices) Regulations 2002, Therapeutic Goods Administration, Canberra. Available : https://www.legislation.gov.au/Details/F2019C00603[last accessed on 2020-01-15].

[4] TGA: Regulatory requirements for in–house IVDs, version 2.2, 2018. Available: https://www.tga.gov.au/publication/regulatory-requirements-house-ivds [last accessed 2020-01-16].

[5] UK.gov: Draft guidance on the health institution exemption (HIE). Available : https://www.gov.uk/government/consultations/health-institution-exemption-for-ivdrmdr[Last accessed on 2020-01-17].

[6] International Standards Organization (ISO) : Medical Laboratories – requirements for quality and competence : ISO 15189:2012 : chapter 4.2 : Quality management system

[7] The Swiss Federal Council, Medical device ordinance 2019. Available: https://www.admin.ch/opc/en/classified-compilation/19995459/index.html [Last accessed on 2020-0-17].

[8] FDA: discussion paper on the laboratory developed tests 2017. Available : https://www.fda.gov/media/102367/download [last accessed on 18.01.2020].

[9] BfArM : Gesetz über Medizinprodukte (Medizinproduktegesetz - MPG). Available: http://www.gesetze-im-internet.de/mpg/MPG.pdf [last accessed on 2020-01-18].

[10] IMDRF (International Medical Device Regulatory Forum).Available: http://www.imdrf.org[last accessed on 2020-01-16].

[11] Gap analysis: Definition and advantages. Available: https://en.wikipedia.org/wiki/Gap_analysis [last accessed on 2020-01-13].

[12] DE GYRUTER: Revision of the "Guideline of the German Medical Association on Quality Assurance in Medical Laboratory Examinations – Rili-BAEK" (unauthorized translation). Available : https://www.degruyter.com/view/j/labm.2015.39.issue-1/labmed-2014-0046/labmed-2014-0046.xml [last accessed on 2020-01-20].

10

Safety and Quality

Design and implementation of a quality awareness process to reduce the cost of non-quality

Benedikt Abel[1], Gerhard Buntrock[2]

[1] Medical Engineering Science, Universität zu Lübeck, benedikt.abel@student.uni-luebeck.de
[2] Institute for Software Engineering and Programming Languages, buntrock@isp.uni-luebeck.de

Abstract

To reduce the cost of non-quality, the error rate in the production and thus improve the quality of their products, a company in Northern Germany decided to design and implement a Quality Awareness process. The individual phases of the business process lifecycle were applied as well as various methods from Lean Management. The result is a two-phase process to sensitize the employees from the production to the topic of quality. The first initial phase provides employees with basic knowledge and understanding of the topic in a workshop. The second phase serves as a reminder and consolidation of the learned points. The process has already been started as a pilot process in one area of production (14 employees) and is now in the second process phase. The next step is to evaluate this pilot process, make necessary improvements and see whether an increased understanding of quality fulfils the pursued objectives.

1   Introduction

Cost of non-quality, i.e. the costs that arise when a medical device does not meet the quality requirements at the end of the production chain, must be kept as low as possible. Since quality standards must be maintained, various methods are used to detect or prevent errors in production steps at an early stage. One of these methods is the training of employees. Training employees at the manufacturing level is an inexpensive and effective method, as employees work directly on the product and can easily identify any damage or visual defects on the product. A big company for medical devices in northern Germany decided, that a specific process for quality awareness in the production is needed. This process should ensure, that not just a few people in the check-up areas are aware of damaged items. Furthermore, all people that are involved in the production line will watch out and are aware of the consequences on the cost of non-quality.

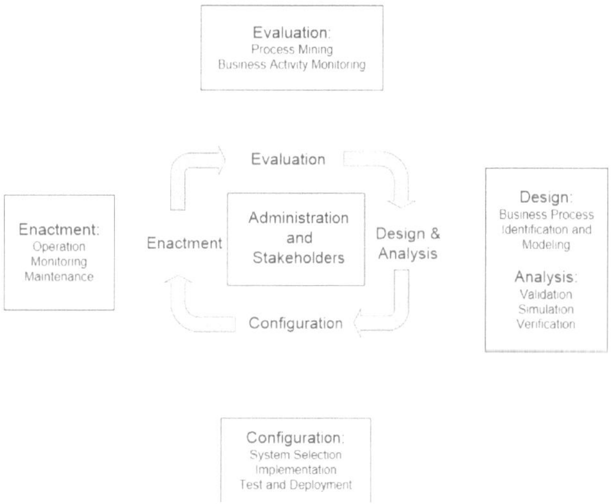

Figure 1: Four phases of the Business Process Lifecycle [1].

2   Material and Methods

2.1   Business Process Lifecycle

Business Process Management aims at the sustainability of business processes. On the one hand, this involves the quality, traceability and transparency of processes, as well as the identification of weaknesses. To realize, evaluate and improve processes, they pass through several phases: Design and analysis, configuration, implementation and evaluation. This leads to an improvement of business processes, which is called Business Process Lifecycle (BPL) [1]. The sequence of the phases is shown in Fig. 1.

2.1.1   Design and Analysis

The BPL begins with the design and analysis step. The first action here is to analyse the enviroment in wich the planned process takes place. For this purpose existing processes and their organizational and technical environment are analysed. Based on the findings of this survey possible gaps and problems in the existing process flows are identified by using various techniques of process modelling, validation, simulation and verification. For a successful start, all parties, subsequently involved in the process, must participate in the design and analysis step.

2.1.2 Configuration

Once an initial model is designed and analysed, a detailed process must be configured. It must be decided which methods can be used to integrate the objectives into the existing environment and how they can be achieved there. No superfluous steps should be inserted. Once a process with exact methods has been found, it is simulated and tested in the existing system. At this point the testing allows to identify any problems and give an early opportunity to eliminate them.

2.1.3 Enactment

In the enactment phase the process comes into effect. This phase is the actual time in which the process runs. The enactment phase is often started by a specific event. During this phase, process management must continuously check whether the process is being carried out correctly, planned steps are being executed and if all involved parties are fulfilling their role. The previously set goals should be fulfilled in this phase.

2.1.4 Evaluation

The last phase of the BPL is used to evaluate and improve the new process. For this purpose, data that have been documented and obtained during the process, are analysed. There are various methods for this purpose with the aim of evaluating the implementation, the added values and the quality of a process. The last phase ensures that no resources are wasted during the process, process errors are eliminated and gaps in the process chain are closed. Thus, the existing processes are continuously improved.

2.2 Lean management

Lean Management is a method of continuous process optimization with the goal of effectively designing the entire value chain. With the help of various tools and principles, it pursues the approach of coordinating processes in such a way that the waste of resources along the value chain is avoided, thus creating a "lean" production system. The central aspects of Lean Management include both customer orientation and cost reduction. The basis for the successful implementation of the methodology is the training of employees in the Lean Management philosophy and the creation of motivation to achieve a common goal. The following four tools from Lean Management were used to design and implement the Quality Awareness Process [2],[3],[5].

2.2.1 5S

The 5S method is a tool that improves the working environment and thus the production process. It is one of the basic introductions to Lean Management and is composed of the following five Japanese words

- Seiri - sorting, selection

- Seiton - systematics

- Seiso - cleaning

- Seiketsu - standardize

- Shitsuke - discipline

The challenge here is to design the working methods at both the production and management level fundamentally and in the long term for the 5S components. If the 5S method is applied consistently, it helps to strengthen the identification of the employees with the Lean Management philosophy and thus the company.

2.2.2 Kaizen

Kaizen describes the basic attitude of an employee towards his own work, the workplace and the quality of processes and products. Those who "live" Kaizen are firmly convinced that there is always something to improve, simplify or optimise. Therefore Kaizen is described as a principle of continuous improvement. Kaizen is neither a method nor a tool, but a way of thinking that all employees should internalize and observe in their activities. The term Kaizen is often used in the context of an event, to describe that this event should be used to improve existing conditions.

2.2.3 Gemba

The basic assumption of the Gemba methodology is that all problems can be observed directly on site. Information on these problems must therefore be obtained from the responsible managers where the actions and processes take place, in order to immediately identify both value-adding activities and waste. These self-made experiences on site replace theoretical assumptions, serve for decision making, can offer starting points for optimization measures and reveal the causes of problems. Especially in quality management Gemba is therefore an important component. The exchange with those involved on site is crucial, as it is possible to profit directly from their knowledge of the work processes. In addition, not only individual persons are involved in finding the problem, but also employees from production are motivated to contribute suggestions to the optimisation.

2.2.4 Kamishibai

A visual control for performing improvements within a process by making normal and abnormal conditions clearly and quickly visuable. The Kamishibai board makes the lean management concept part of the standardized work of management. It gives managers a schedule of when to visit a process and what they can improve. The board indicates whether or not the required improvements have taken place, the results of the audits and, if necessary, notes on anomalies and countermeasures. The aim is to take immediate action in case of anomalies.

2.3 Start up workshop

A self-developed start up workshop was planned to gather the opinions of employees from the various departments, that are part of the production chain. This workshop has two objectives. On the one hand, it is intended to create a first picture of how employees see the topic of quality. On the other hand, the existing quality policy is to be analysed with these employees and action goals are to be developed from this. Both goals are to be fulfilled by the employees reacting to scenarios and questions. The reactions are then discussed in the group.

3 Results and Discussion

3.1 Design and Analysis

3.1.1 Current status and regulatory request

As a first step of the BPL, existing onboarding processes on the topic of quality were examined. Although the existing ones always emphasise the importance of sensitising employees to the topic of quality, there are only isolated sections on the topic of quality awareness. A further process on the topic of awareness is already in place, as it is required by regulatory requirements. The requirements according to ISO 13485 are as follows:

- EN ISO 13485, 6.2.d Human resources
  d) ensure that its personnel are aware of the relevance and importance of their activities and how they contribute to the achievement of the quality objectives; [4]

In the existing process, the focus is on training employees for regulatory requirements. Direct references or possibilities to act quality conscious are not mentioned. A new quality awareness process must therefore start at exactly this point and give employees goals that are directly related to their work and mention possibilities to achieve them.

3.1.2 Start up workshop - goals

The next analysis step was the start up workshop with employees from the divisions: Manufacturing, Factory Engineering, Operations Quality, Quality and Regulatory and Process Engineering. In this meeting four key points, that are in line with the company's quality policy, were developed and discussed. These four points are summarized in Fig 2.

Subsequently, three possible guidelines, that each member of the production team can adopt, were derived from the points described above:

1. living quality in every small process step
   -> How would the customer see the current product? Is he satisfied?

2. behaviour where quality is the focus of attention
   ->Motivate others to do the same.

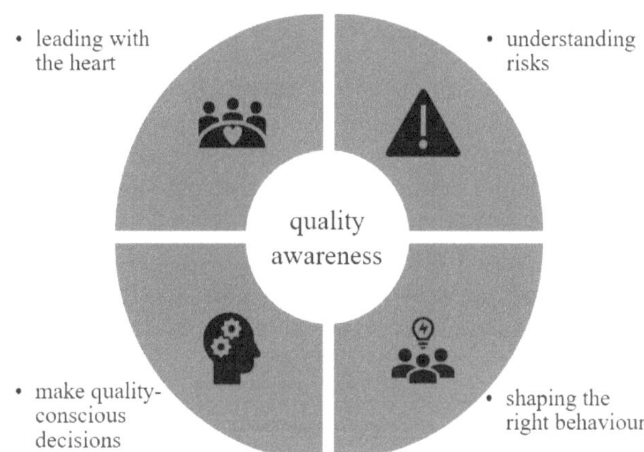

Figure 2: Four key points that sum up the company's quality policy.

3. take action
   -> If you see a problem that could affect the quality, report it and inform the responsible contact person. The earlier the better

3.2 Configuration

After the key points that the process should achieve have been identified and discussed, the next step is to configurate the process. Since various lean methods are already being used in the company, tools from Lean Management are also used to plan and implement this process. According to the Kaizen principle, the awareness process can never be completed entirely, as the understanding of the employees towards quality must be constantly improved. The process can therefore be divided into two phases. An initial phase, where a basic understanding of quality is created, and a long-term phase where the basic understanding is consolidated and expanded. The two phases are shown in Fig. 3.

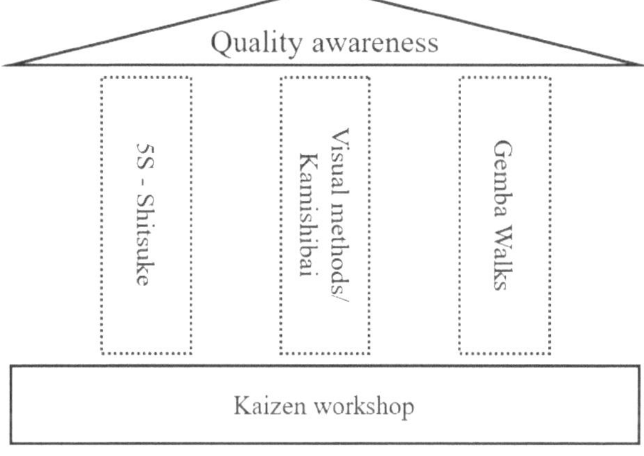

Figure 3: Two phases of the Quality Awareness Process. The first initial phase, consisting of the Kaizen workshop, and the second long-term phase, consisting of 5S - Shitsuke, Visual methods/Kamishibai and Gemba walks.

3.2.1 Initial Phase

In order to create a basic understanding of quality awareness and to bring all employees to the same level, a Kaizen workshop takes place. The workshop consists of three chapters: The first chapter shows what can happen if the desired quality is not achieved. This is done on the one hand by negative examples, as well as by the incurred costs. It is made clear to the employees that even small defects in medical technology can have a strong impact on patients and customers. In the second chapter the quality policy of the company and the aim of this policy is explained. The policy is gone through step by step and employees are given positive examples to explain why the company focuses on qsuality. The examples cover the company's medical products and how they help patients.
In the last chapter direct possibilities for action on the workplace and for behaviour in production are given.

3.2.2 Long term Phase

In the long-term phase, three tools are used to deepen and expand the acquired knowledge. First and foremost, employees are reminded by various visual methods, such as the Kamishibai board or posters at the workplace, of the learned points and possible actions. Secondly, the already applied 5S method appeals to the self-discipline (shitsuke) of the employees to live what they have learned. And thirdly, management and quality staff must carry out regular Gemba walks. These serve to find further errors and possible actions with the employees, which can further improve the quality. In the long-term phase, therefore, the management, the quality staff and the production staff have a role to play. Success can only be achieved if everyone fulfils their roles and thus always looks for possibilities of improvement (Kaizen).

3.3 Enactment

In order to test the process in a suitable framework, it was introduced in a single assembly area. A total of fourteen employees work in this assembly area, twelve of them took part in the Kaizen workshop. Posters with the quality policy and the possibilities for action were hung up at all workplaces and in the entrance area. In addition, a Kamishibai board, with the subject of quality, was introduced in the daily group meeting. In intervals of three weeks Gemba walks have been performed and will continue to be held. Three months after the introduction of the process a meeting is planned, where the employees will be asked questions about the process and the topic of quality. The employees involved in the process can also give feedback there.

4 Conclusion

In the Quality Awareness Process presented here, the first two phases of the BPL were successfully completed, and the third phase was started as a pilot process for one specific assembly area. With an initial workshop, which makes it clear to the employees why quality in production is crucial for the product, the customers and the entire company, the foundation for an awareness process has been laid. With the tools from Lean Management described above, it should now be possible to ensure that this knowledge of quality is maintained. After a sufficient period of time in the "enactment" phase, the fourth phase "evaluation" must be started. In this phase the success of the process must be presented and analysed. It has to be examined if the workshop has conveyed the necessary sensitivity and whether this sensitivity is maintained continuously. In case that the process is successful, the goal should be met and the error rate and cost of non-quality in this area reduced. Both must be found out through data analysis. In addition, the acquired awareness of the employees must also be checked. This should be done by questionnaires and direct employee interviews. At this point the first round in BPL will be successfully completed. In the second round, the process can then be improved and extended to the entire production area.

Acknowledgement

The work has been carried out at a company for medical devices in northern Germany and supervised by the Institute Institute for Software Engineering and Programming Languages, Universität zu Lübeck. I sincerely thank the involved people from this company for providing workspace, material and knowledge.

5 References

[1] M. Weske, *Business Process Management - Third Edition*, Springer, Potsdam, 2019.

[2] S. Gao and S. Pheng Low, *Lean Construction Management - The Toyota Way*, Springer, Singapore, 2014.

[3] W. Jakoby, *Qualitätsmanagement für Ingenieure*, Springer Vieweg, Trier, 2019.

[4] Iso 13485:2016 *Medical devices - Quality management systems - Requirements for regulatory purposes*.

[5] J. Pinto, J. Matias, C. Pimentel, S. Azevedo and K. Govindan, *Just in Time Factory - Implementation Through Lean Management Tools*, Springer, Boston, 2018

Connecting Old Devices Safely To The Cloud

Niclas Kath [1], Arne Seidel [2] and Josef Ingenerf [3]

[1] Medical Informatics, Universität zu Lübeck, niclas.kath@student.uni-luebeck.de
[2] seca, Hamburg, arne.seidel@seca.com
[3] Institute of Medical Informatics, Universität zu Lübeck, josef.ingenerf@uni-luebeck.de

Abstract

Security is a crucial topic in medical environments. More and more software solutions are migrating to the cloud. This opens a closed system inside a hospital to a system with connection to the internet. Medical devices often have a rather long life cycle. Connecting them to the cloud can provide several problems as their product age makes them not compatible with newest secure communication protocol. In this work we evaluate several options, including implementing TLS or the usage of Raspberry Pi's as communication gateways, for medical devices which shall be connected to the cloud but do not provide secure communication. The options are evaluated under their provided security, interference with existing administration's admissions and the needed work.

1 Introduction

Any data breach can cause financial losses and can leak potential highly sensitive data to the public. Data breaches in medical context are highly dangerous as very sensitive data can be leaked. Lately more and more medical data is stored to support healthcare which leads to more devastating data breaches if anyone gets unauthorized entrance to the systems. Data breaches in medical environments are the most expensive ones with up to an average of 6.54 million USD per breach. In comparison to an average of 3.92 million USD per data breach in non-medical setups as shown in Fig. 1. Each stolen medical record equals a financial damage of approximately 380 USD while the average cost for a single record is about 141 USD [1].

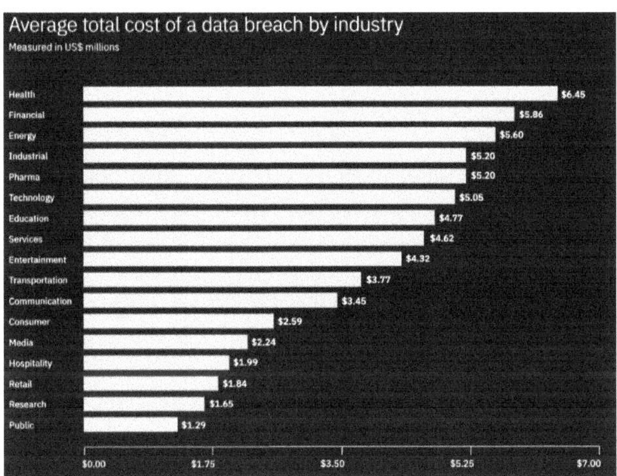

Figure 1: Overview on avergage total costs of a data breach. Data breaches in health industry are by far the most expensive ones [2].

An even more concerning result of security breaches in medical environments is the potential influence on patient health. New researches show that security breaches in clinics can lead to an increased myocardial infarction mortality [3].

More and more software solutions are moving to the cloud. Even though clinical systems are used to be a closed system in order to maximize security, opening them up for cloud computing provides more chances than risks [4]. Cloud based solutions offer many new opportunities, not only for patient health like new medical findings due to data mining but also for clinics as cloud computing is often cheaper than buying expensive hardware.

Adding a connection to the internet also simplifies the update process for medical devices and software. Furthermore, an internet connection fulfils the required postmarked surveillance of U.S. Food and Drug Administration (FDA) and Medical Device Regulation.

Adding an internet connection opens up a normally closed system which normally require a physical connection to the closed clinic system to perform an attack. There are numerous options on how to secure internet communication. The most commonly known option is the Transport Security Layer (TLS) [6]. Medical devices often have a long life cycle because replacing them is an expensive task for a hospital. Therefore many old and often outdated devices can be found in a hospital which do not support TLS from the start. Long lifetime is also based on the required local market admission. Getting an admission is an expensive task. For example the FDA admission fee for one device can cost over 340,000 USD [5]. The man-hours the manufacturer has to spend in order to prepare the FDA admission are excluded. The topic of this paper is to explore options on how to establish safe and secure connections to the internet even if the device does not provide TLS from the start.

2 Material and Methods

In this section several options are discussed. Three major option will be evaluated in the scope of this work:

- Implementing TLS

- Mini-PC which handles the communication to the outer world

- Proxy-service in a local network

Implementing TLS will focus on the effort to implement a TLS version on the system.

The Mini-PC will explore the options on adding an extra device to the medical measurement system. This device can be any kind of a Mini-PC, however most likely it will either be a Raspberry Pi or an Arduino.

The proxy-service will evaluate the option of a lightweight software which receives data in a local network from the device. The proxy-service will work encrypt the datastream and establish a secure connection the to cloud. The proxy-service works as a safe connection the to cloud.

Evaluation will focus on:

- Security

- Implementation work

- Interference with existing FDA admissions

We will evaluate each option with related work if possible. The evaluation will focus on devices with Windows CE 6.0 running on .NET Compact Framework 3.5.

3 Results and Discussion

3.1 TLS

The most straight forward option is to implement a safe version of TLS on a given system. However, Windows CE 6.0 does not include any newer TLS versions higher than TLS 1.0. TLS 1.0 is heavily outdated and possesses strong vulnerabilities which some do not require huge amount of time and computing power to exploit [7]. So TLS 1.0 is not a viable option. Even TLS 1.1 or 1.2 are prone to several attacks but those attacks require several right circumstances to be successful [8]. For example the German Federal Office for Security in Information Technology recommends the use of at least TLS 1.2 [9]. Even though TLS 1.3 is wanted for better security, given the set-up, TLS 1.2 should and can be considered a good option. Furthermore, TLS 1.2 introduced a better implementation of Elliptic Curve Cryptography (ECC). Compared to the the standard Rivest–Shamir–Adleman (RSA) encryption ECC provides better security while using smaller keys [10]. This is especially important for systems with relatively low computing

power as using long keys can lead to bad user-experience. Since Windows CE 6.0 does not offer native support for TLS 1.2 an own TLS 1.2 implementation is needed. Even though there are some references on how to implement TLS 1.2, it is not recommend to do so as the chance of implementing vulnerabilities is quite high. There are several verified open-source libraries out there like mbedTLS or OpenSSL which should be used instead. On the other hand adding TLS 1.2 natively to an existing system with an admission in place will not result in a needed readmission. Therefore no additional man-hours have to be invested into readmission.

However, even though these libraries exist, they cannot be implemented directly on a running Windows CE 6.0 system. Both libraries require certain C-header files which Windows CE 6.0 does not provide. It might be possible to compile the libraries without the missing header files but there is no official guideline in doing so and the body of work ends mostly in a try-and-error approach. It is not possible to estimate this amount of work in a proper way.

3.2 Mini-PC

Due to the recent emergence of mini PCs, such as a Raspberry Pi or Arduino, they have received some attention in health systems. Most presented work has set its focus on monitoring health parameters mostly for third-world countries since those systems only cost a small part of what industrial monitoring systems cost. In contrast the work of Jassas et. al [11] has shown that it is possible to use a Raspberry Pi as a gateway for medical sensors into cloud systems.

From a security point of view this approach does not offer any security breaches which can happen in real-world usage. There is no vulnerability in TLS 1.3 that does not require several special circumstances. The only possible security problem can be the communication between device and Mini-PC. Most likely they will be connect via USB, therefore any sniffing of the data requires some kind of physical intervention with the system which is not suitable in real-world usage. However, there are approaches to close this possible sniffing attacks [12].

From a development point of view using a Mini-PC is quite intriguing. These systems have a really good documentation and several work has shown that those Mini-PCs can be used in many possible ways. Since most of them are based on a Linux distribution which already have TLS implemented. This should make the needed work of implementation reasonable. Most medical measurement devices have some kind of data transfer already implemented which has to be connected to the Mini-PC. The Mini-PC only works as a gateway to the expected API of the internet application. Building this connection does not require much effort as most of the needed functions are already implemented on the device.

Furthermore, most of these systems are really cheap to buy which makes them interesting from a manufacture's point of view.

The biggest problem with this approach is the admission. Most health application running on a Raspberry Pi are designed for countries with low health care standards. However getting an admission for the big markets especially an FDA admission on a Raspberry Pi can be difficult. Even if the device, which shall be extended, has an admission already in place, adding additional hardware will most likely require a readmission. There is no Raspberry Pi system at date of publish which has received a FDA certification. This makes the Mini-PC approach from a manufacture's point of view quite risky as the whole admission process has to be done. This requires a huge amount of man-hours.

3.3 Proxy-Service

This approach is in some way quite similar to the approach with the Mini-PC. Instead of adding a Mini-PC, a proxy-service on a PC in the local network will be created. This requires a established local network which will already be in place in almost any real-world scenario. The proxy-service will handle then the communication just like the outlaid approach in 3.2.

The only security concern is the communication between the device and the proxy-service. As in 3.2 pointed out, a USB connection can be considered safe. An other approach is to use an existing local network in the clinical environment. This opens up possible security problems. However, this can be solved quite easily. Firstly, a possible attacker needs entrance into the local network. This should not be possible in the first place but if this happens an attacker should not be able to sniff the data of any device communication. Therefore, the communication between device and proxy-service has to be encrypted. Even quite outdated environments like Windows CE 6.0 and .NET Compact Framework 3.5 provide encryption options which should be enough to secure the data communication. Device and proxy-service have to know the key. Symmetric encryption can be used, however this results in hard-coding the encryption key for the device and proxy-service. Under the pre-condition that the local network is protected against unauthorized usage, symmetric encryption can be considered safe. Asymmetric encryption is more safe but requires more computing power and implementation work. The proxy-service only has to receive the data, decrypt it, encrypt it according to TLS and send it to the cloud. The proxy-service also can handle the translation process if device and cloud have different data formats. This simplifies the implementation and guarantees that existing communication ways can still work since the device remains the same output format.

From a developer's point of view the required amount of work for this approach should not be a problem. Developing a proxy-service with TLS in modern frameworks is not a problem and in most scenarios the device already has some kind of communication implemented. Changing the target for the communication is not a problem.

Admission-wise there are no physical changes or admission-breaking changes to the code which means the admission will persist. When it comes to the proxy-service however this can require an admission if not any existing software with an admission in place is used. This proxy-service is a lightweight software which does not require huge amount of work for the admission. In a best-case scenario the device already communicates with some kind of software which only has to be added with the proxy-service.

3.4 Discussion

The results of the examined options are summarized in Table 1.

Table 1: Overview on positive and negative factors of each approach.

Approach	Positive	Negative
TLS	- safe communication - no additional software and hardware - admission	- implementation work - costs
Mini-PC	- safe communication to internet - implementation work	- additional hardware - security communication to Mini-PC - costs of new admission
Proxy-Service	- admission - implementation work	- possible need for new software - security communication to proxy-service

The TLS approach is intriguing. It offers the best standpoint from an admission's point of view. No physical changes or additional software are needed. If it is possible to implement TLS on the device this is the approach which can be recommended the most. Third-party-libraries should be used which possess a validation of their correctness. However, if the device cannot use any third-party-library, as given in the scope of this work, since its based on Windows CE 6.0 and .NET Compact Framework 3.5, this approach can require huge amount of work.

The Mini-PC approach is the least usable of the three. Even though the positive arguments are quite good, the negative side overwhelms. The uncertainty arising from the needed admission for the Mini-PC is really concerning. Since there are no practical examples of someone who has gotten an FDA admission for a Raspberry Pi or other commercial Mini-PCs makes this approach highly risky. Obviously it is possible to design an own Mini-PC but that requires a huge amount of work. If that road is taken, you can redesign the device itself with better results. This approach is interesting for markets which do no have high standards of certification.

The proxy-service approach somewhat combines the other two approaches. It shares advantages and disadvantages with both. If there is any communication in place to some kind of data representing software this is the go-to approach since there only has to be added more communication to the outside.

However, before looking for options on how to establish secure internet connection on outdated systems it should be asked, if this is really needed or if a substitution of the whole system does make more sense. This is time-consuming and cost-intensive but with each year passed, more and more vulnerabilities will be found in old systems. Fixing them will become harder and harder. For each device this has to considered before making a decision.

4   Conclusion

This work has shown that there are several options on how to implement and establish a secure internet connection on devices when the software running on the device does not provide the needed requirements for them itself. Furthermore, each presented approach can work but comes with some advantages and disadvantages. This work has evaluated them and given an advice which of them is the best one under the given circumstances of this work. With different circumstances some evaluated points will differ but the bottom-line stays the same. Implementing TLS directly is the most straight forward approach which is the best option in most scenarios. There are circumstances, like the scope of this work, in which implementing TLS directly is not the best option.

Acknowledgement

The work has been carried out at seca, Hamburg and supervised by the Institute of Medical Informatics, Universität zu Lübeck.

5   References

[1] Ponemon Institute, *2017 Cost of Data Breach Study: Global Overview*. Available: https://www.ibm.com/downloads/cas/ZYKLN2E3 [last accessed on 2020-01-04].

[2] Ponemon Institute, *2019 Cost of a Data Breach Report*. Available: https://www.ibm.com/security/data-breach [last accessed on 2020-01-04].

[3] S.J. Choi, M.E. Johnson and C.U. Lehmann, *Data breach remediation efforts and their implications for hospital quality*. Health services research, vol. 54, no. 5, pp. 971–980, 2019.

[4] E. AbuKhousa, N. Mohamed and J. Al-Jaroodi, *e-Health cloud: opportunities and challenges*. IN: Future internet, Molecular Diversity Preservation International, vol. 4, no. 3, pp. 621-645, 2012.

[5] U.S. Food and Drug Administration, *Medical Device User Fee Rates for Fiscal Year 2020* Available: https://www.federalregister.gov/documents/2019/07/31/2019-16270/medical-device-user-fee-rates-for-fiscal-year-2020 [last accessed on 2020-01-04]

[6] L.C. Paulson, *Inductive analysis of the Internet protocol TLS*. In: TISSEC, vol. 2, no. 3, pp. 332–351, 1999.

[7] A. Fardan, J. Nadhem and K.G. Paterson, *Plaintext-recovery attacks against datagram TLS*. NDSS 2012, 2012.

[8] A. Fardan, J. Nadhem and K.G. Paterson, *Lucky thirteen: Breaking the TLS and DTLS record protocols*. In: 2013 IEEE Symposium on Security and Privacy, pp. 526–540, 2013.

[9] Bundesamt für Sicherheit in der Informationstechnik, *Technische Richtlinie TR-02102-2 - Kryptographische Verfahren: Empfehlungen und Schlüssellängen*. Avalaible: https://www.bsi.bund.de/SharedDocs/Downloads/DE/BSI/Publikationen/TechnischeRichtlinien/TR02102/BSI-TR-02102-2.pdf [last accessed on 2020-01-11]

[10] D. Hankerson and A. Menezes, *Elliptic curve cryptography*. Springer, 2011

[11] M.S. Jassas, A.A. Qasem and Q.H. Mahmoud, *A smart system connecting e-health sensors and the cloud*. In: CCECE, pp. 712–716, 2015.

[12] M. Neugschwandtner, A. Beitler and A. Kurmus, *A transparent defense against USB eavesdropping attacks* In: 12th EuroSec, pp. 6, 2016.

Medical Packaging Recyclability and Environmental Impact Analysis

Gonzalo Chavez Gomez [1], Martje Timmermann [2]
[1] Biomedical Engineering, Luebeck University of Applied Sciences, gonzalo.chavez.gomez@stud.th-luebeck.de
[1] Next Generation Packaging, Stryker Trauma GmbH, martje.timmermann@stryker.com

Abstract

The main function of Medical Packaging is to maintain the sterility of the femoral nailing systems until their use on the operating room, but one important aspect that needs to be looked into is the ecological impact it has and how to improve it. For this an assessment of the packaging systems was required to know if the materials used were recyclable and how to dispose of them correctly. Company documentation, design drawings, industry best practices and online resources were researched in order to get data regarding the recyclability and environmental impact of the packaging. With this, insight was gained over the materials, waste generated, ease of recycling and procedures needed. The information gathered was documented in order to serve as reference for which aspects need to be improved during packaging development.

1 Introduction

It's important that the medical packaging developed is successful on keeping the sterile barrier and the medical devices intact, but usually not much attention is put during the design process to what will happen with these systems after they are used and disposed of, how much waste they generate, how it's managed and how it's recycled.

That's why there is an ever-increasing demand from customers and industry for medical packaging to be environmentally-friendly, easier to recycle and less wasteful. In order to achieve this, it is important that during the design of the packaging it is planned how resources are used and the environmental impact it will have [1].

One of the objectives of this assessment is to present and collect information of the packaging systems regarding the recyclability of the materials that compose them as well as alternatives for recyclability improvements.

This is needed so that the engineers have at hand all the ecological impact information, so that packaging can be created and modified to improve its use of resources, waste, recyclability and sustainability. Is also important that the users in the operating room are informed on the improvements being made and how to manage the waste generated by the packaging.

2 Material and Methods

In order to address this, it was needed to assess the current recyclability and ecological impact. The focus was on two of the trauma and extremities medical devices packaging from Stryker, one of which is an older blister packaging system and a newer one with a pouch packaging, these systems are referred to as BL-1 and PP-1 and PP-2 respectively on this document. These packaging systems can be seen on figure 1, figure 2 and figure 3 respectively.

Packaging Systems drawings and specifications were analyzed to obtain the information required regarding the components, internal and external Voice of Customer sessions were carried, ISO and EU regulations regarding packaging and medical packaging were revised as well as guides and resources from organizations in the EU regarding reciclability analysis and implementing environmental considerations to reduce the packaging impact [1] [2] [3] [4] [8] [9] [10].

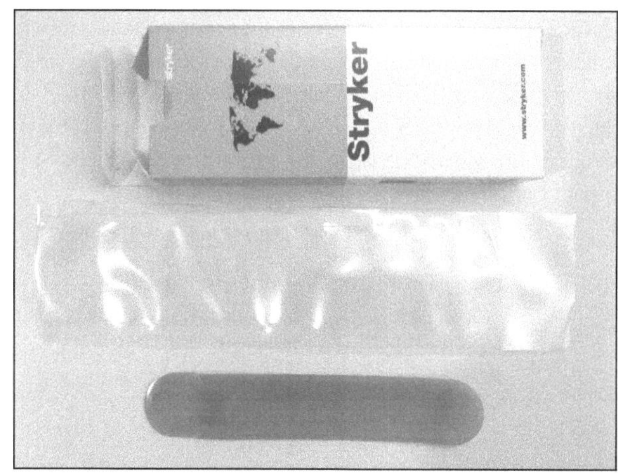

Figure 1: Packaging System PP-1 configuration example

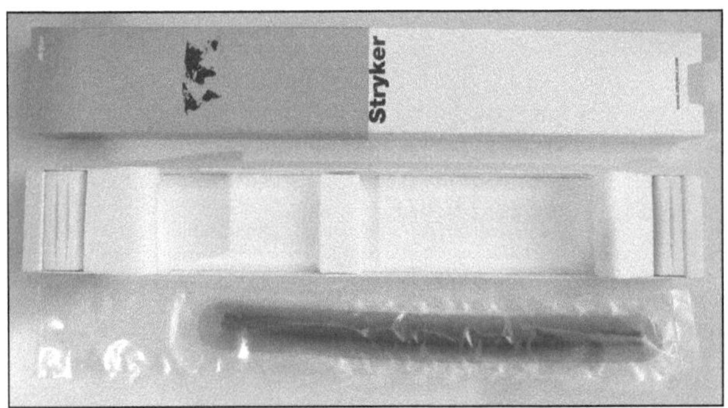

Figure 2: Packaging System PP-2 configuration example

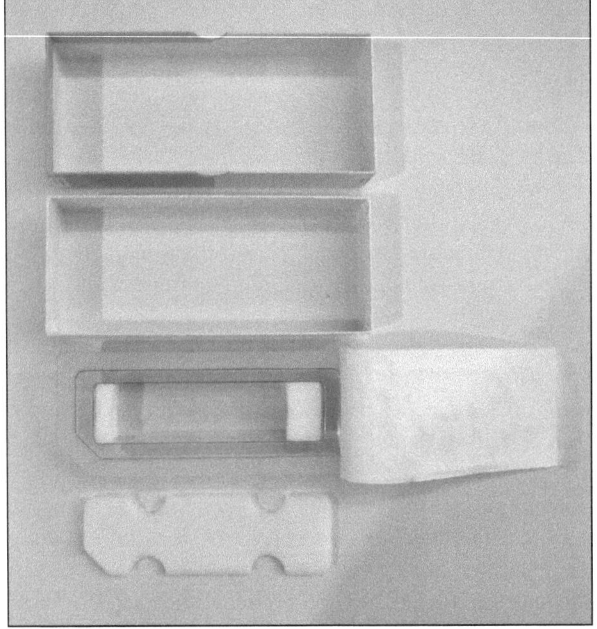

Figure 3: Packaging System BL-1 configuration example

3 Results and Discussion

3.1 Waste

Medical packaging systems, being formed by various components in order to keep the medical devices and their sterile barrier intact, can end up generating between 31.87 to 971.9 cm^3 of waste per packaging system, (as shown on Table 1), which if not disposed of in the correct way, can have an impact on the environment greater than it should.

Due to the fact that Operating Rooms "produce more than 30 percent of a facility's waste and two-thirds of its regulated medical waste" [2], it is important to know how much waste the packaging systems generate and how much space it takes once it's disposed of. This to ensure that packaging waste is recycled correctly and the materials can be recovered for further use as a different product, as mentioned in the Directive 2008/98/EC [3].

Correct waste management of Stryker packaging system components in the hospitals can become a problem, due to the lack of recycling information regarding the packaging material and its composition to correctly sort it. That has to be provided to the hospital personal alongside the packages and the fact that, unlike cardboard and carton, "medical plastics are not similar in shape, size, volume or even sometimes plastic type" [2]. Part of the difficulty to correctly sort and recycle the Stryker packaging lies also in the lack of ecological inputs during their design process, which has contributed on making the packaging end up being composed of a variety of materials with different recycling characteristics and capacities, in the aim of trying to achieve the best packaging possible from a technical point of view.

As mentioned on the directive (EU) 2018/852 regarding packaging waste the design of packaging should protect, preserve and improve the quality of the environment, making sure that the natural resources are being used in an efficient and mindful way [4]. That's why producers of medical packaging have the responsibility to develop with plans to use responsible natural resources for packaging designs, help reduce the impact on the environment of the packaging and improve management of the waste generated by making the material sorting and recycling easier.

According to the waste hierarchy, one of the best ways to prevent waste and its impact on the environment is to avoid generating unneeded waste and by helping reduce the existent [5].

To reduce paper waste the company has replaced the physical paper IFUs (instructions for use), that were added on the packaging systems, for an electronic IFU which is indicated on a label on the package. This way the information can be accessed still without being printed and generating waste. From November 2018 to November 2019, 590.800 IFUs were removed from Stryker packaging resulting in saving around 6,91 t of paper, which can be translated to roughly 82 trees saved [6].

Table 1: Packaging Volumes Information

Packaging System	Volume Storage (mm^3)	Volume Discard (mm^3)	Volume Percentage compacted
PP-1	180,375.00	31,874.29	18%
	284,375.00	87,605.79	31%
PP-2	1,160,250.00	848,094.35	73%
	1,755,250.00	971,900.15	31%
Bl-1	316,000.00	187,895.00	59%
	1,293,824.00	881,200.88	68%

Table 2: Components and Unused Volume

Packaging System	Number Components	Unused Volume (mm^3)
PP-1	3	173,926.25
	3	278,846.00
PP-2	6	552,874.80
	6	1,134,244.00
BL-1	6	225,147.98
	11	820,151.25

Continuing with these actions, the change of packaging system from the old blister system to the newer pouch system reduced the number of packaging components to around half, from 6 to 3 components on the smallest packaging and from 11 to 6 components on the biggest ones, as seen on table 2.

This change also had an improvement on the volume the packaging components waste occupy once discarded, the biggest pouch system takes only 73 percentage of the original packaging volume while the blister system takes 68 percentage. On the smallest packages the difference is more obvious. The blister takes 18 percentage of the original packaging volume and the blister 68 percentage, as shown on Table 2. This is mainly due to the nature of the rigid blister material that ends up being a bulky waste while the pouch components are almost flat and flexible.

Waste does not only comes from sources like an excessive number of components, but also empty/wasted space on the packaging systems. As shown on table 1 and 2, the unused volume compared against the volume at storage values are too close, which means that there is an amount of empty space inside each package being sent that is grater than the volume occupied by the medical devices in it.

This wasted empty space inside the packaging boxes exemplifies an inefficient use of resources for this components. It would be preferable to prevent this to help reduce its environmental impact [5]. To get a better idea of how much empty space is being left unused on the packaging systems, for the smallest packaging of the pouch system PP-1 we can see on table 2 it occupies a volume of 180,375.00 mm3, if we took a 20 foot metallic shipping container, (with an internal volume of 33.1 m^3) [7], and filled it up with the small PP-1 packages we could fit around 183,481 of them, which would have an internal unused volume that would be equivalent to approximately 176,921 more units of these packaging system.

Table 3: Components Recycling and Waste type

Packaging System	Component Description	Waste Type	Recycling Path
PP-1	Folding box	Recyclable	Paper, cardboard
	Pouch	Recyclable	Mixed plastics (flexible)
	Sterile Pouch	Biohazard	Incineration
PP-2	Telescopic box	Recyclable	Paper, cardboard
	Pouch	Recyclable	Mixed plastics (flexible)
	Sterile Pouch	Biohazard	Incineration
	Foam inlay	Recyclable	Mixed plastics (flexible)
	Foam buffer	Recyclable	Mixed plastics (flexible)
BL-1	Slip box	Recyclable	Paper, cardboard
	Tyvek Lid	Recyclable	PE and PP
	Blister	Biohazard	Incineration
	Foam inlay A	Biohazard	Incineration
	Foam inlay B	Biohazard	Incineration
	Foam block B	Biohazard	Incineration
	Silicone Stopper	Biohazard	Incineration
	Silicone Shell	Biohazard	Incineration
	Foam Blister	Biohazard	Incineration

3.2 Recyclability

An important point to consider when recycling the different materials of the packaging components, is that, on the operating room there is a non-sterile and sterile area, and the materials that are handled on the sterile area can get contaminated becoming biohazard waste, which cannot be recycled and must be treated in another way, most commonly being incinerated [9], as it can be seen on table 3.

The blister packaging system has more elements at risk of becoming contaminated during operation, due to the high number of foam elements inside the blister holding the medical device. While it is possible that not all of these internal components get contaminated, the risk is high and for that reason they were classified as such. Meanwhile the number of components on the pouch packaging system is much smaller and the only element at risk of getting contaminated is one pouch.

At this point the importance of being able to identify the different material types of the medical packagings is more visible, to prevent that they are classified and thrown away incorrectly. To ensure materials are being recycled and re-

utilized, as well as preventing that material is incorrectly disposed of as biohazard waste, which would be incinerated causing an unnecessary impact to the environment.

The recycling paths mentioned on table 3 for the material waste of the packaging system, are the ones established for the common recovery of waste in Germany [8], with the exception of the management of biohazard waste [9]. There might be other options available for waste management, where the plastics are recycled in a way that impacts less the environment, recovers the constituents of it or transforms them into energy, but this is dependent on how each institution manages their waste.

3.3 Sustainability

Currently after revising the materials that form each the packaging system components, shown in table 4, the carton boxes and part of their plastic components can be recycled, but some of them don't have an specific recycling path on the waste management system, table 3, which poses a problem for their sorting and reuse.

Table 4: Components and Materials

Packaging System	Component Description	Material
PP-1	Folding box	Carton
	Pouch	OPA-PE
	Sterile Pouch	TPU
PP-2	Telescopic box	Carton
	Pouch	OPA-PE
	Sterile Pouch	TPU
	Foam inlay	Polyethylene foam
	Foam buffer	Polyethylene foam
BL-1	Slip box	Carton
	Tyvek Lid	Tyvek
	Blister	PET-G
	Foam inlay A	Polyethylene foam
	Foam inlay B	Polyethylene foam
	Foam block	Polyethylene foam
	Silicone Stopper	Silicone rubber
	Silicone Shell	Silicone rubber
	Foam Blister	Polyethylene foam

In addition, some of the plastic materials used can be a problem at the moment of being sorted and recycled. For example, the polyamide (PA) that is part of the pouch laminate materials can drastically reduce the quality of the recycled material that it gets recycled with, also the PETg from the blisters can be problematic in this same way [10]. If it gets mixed with low melting temperature plastics on the recycling stream, due to its high melting temperature, it would create lumps in the recyclate [10]. This would be the case for the pouches used in the packaging systems, because the laminates that conform it also contain polyethylene (PE), which has a low melting temperature.

Because the carton of the packaging systems have plastic labels glued on, the quality of the material obtained after the recycling process will be affected in some degree, diminishing the quality of the recyclate obtained [8].

4   Conclusion

The new pouch packaging systems can be recycled by 50% and the quantity of biohazard material produced in the operating room was reduced to 25%, in comparison with the old blister packaging systems with 31% recyclability and 69% of biohazard material generated.

After this initial assessment of the recyclability of the packaging systems, it can be seen that some improvements are still needed to be made regarding the wasted space inside the packaging boxes and materials used in the components. Something else that should be looked into would be to search for material alternatives easier to recycle to the ones mentioned on section 3.3.

At the end of this analysis the information gathered was documented and presented for future reference at the moment of designing new packaging systems.

Acknowledgement

The work has been carried out at Stryker Trauma GmbH and supervised by Prof. Dr.-Ing. Stephan Klein of the Medical Sensors and Devices Lab, Luebeck University of Applied Sciences.

5   References

[1] ISO 11607-1 Packaging for terminally sterilized medical devices — Part 1: Requirements for materials, sterile barrier systems and packaging systems *GREENING THE OR*. Available: https://www.iso.org/standard/70799.html [Last accesed on 18-01-2019]

[2] Practice Preenhealth, *GREENING THE OR*. Available: http://bit.ly/30x40wZ [Last accesed on 18-01-2019]

[3] DIRECTIVE 2008/98/EC. Available: http://bit.ly/2v53KcZ [Last accesed on 18-01-2019]

[4] DIRECTIVE (EU) 2018/852. Available: http://bit.ly/2G1NYll [Last accesed on 18-01-2019]

[5] EPA, *the-waste-hierarchy*. Available: http://bit.ly/2sCKJ0x [Last accesed on 18-01-2019]

[6] Conservatree, How much paper can be made from a tree? Available: http://bit.ly/3axsZVw [Last accesed on 18-01-2019]

[7] Intermodal container, Specifications. Available: http://bit.ly/2G42o4c [Last accesed on 18-01-2019]

[8] Verification and examination of recyclability, Requirements and assessment catalogue of the Institute cyclos-HTP for EU-wide certification Available: http://bit.ly/2uamEhW [Last accesed on 18-01-2019]

[9] IJIC, Healthcare waste management in Germany. Available: http://bit.ly/2ugnk5e

[10] Mepex Consult, Basic Facts Report on Design for Plastic Packaging Recyclability. Available: http://bit.ly/3ake5BH

Trajectory Control with Lane Boundary Constraints for Autonomous Vehicles Using Model Predictive Control

Robin Kensbock [1], Philipp Huß [1] and Georg Schildbach [2]

[1] Medical Engineering Science, Universität zu Lübeck, {robin.kensbock, philipp.huss}@student.uni-luebeck.de

[2] Institute for Electrical Engineering in Medicine, Universität zu Lübeck, georg.schildbach@uni-luebeck.de

Abstract

Recent works have shown that model predictive control (MPC) is a viable solution for vehicle control in autonomous driving. This paper investigates the usage of MPC, which benefits from using constraints in comparison with classic controllers, for trajectory following for autonomous vehicles. The customized MPC is applied on a one-tenth scaled car to follow the optimal racing line of a specified driving track. In this experimental setup, the addition of constraints for lane borders and collision avoidance is tested. The presented possibility to use these constraints to solve our task has led to the desired driving behavior. The results demonstrate the advantage of MPC over classical controller design.

1   Introduction

Due to more computational power and more capable algorithms, modern control methods, like *model predictive control* (MPC), have become practical to use. MPC, a technique for optimal control, is attractive for its good performance and the possibility of using constraints on inputs and states [1]. This makes it flexible for a variety of today's control applications. For example, the possibility of using MPC either in medical robotics [2] or drug infusion control [3] has been shown. As autonomous driving is developing into a broad field of research, the examination of MPC for tasks such as trajectory control for vehicles has become a potential application [4]. In comparison to more conventional techniques for path tracking in autonomous driving, such as pure pursuit control, MPC offers the possibility of explicitly considering constraints on the system states and control inputs. This paper investigates the use of MPC for trajectory control, for a one-tenth scaled electric car. All relevant information about the environment, such as the ideal trajectory and the lane boundaries of the track, which must not be crossed, are assumed to be available to the MPC. For simplicity, the localization of the vehicle is done by a motion capturing camera system. The goal of this paper is to examine the performance in terms of lane border consideration and obstacle avoidance, in the presence of forbidden zones on the driving track.

2   Material and Methods

2.1   System Dynamics

The basic principle of MPC is to solve an optimization problem over a finite time horizon, regarding the given con-

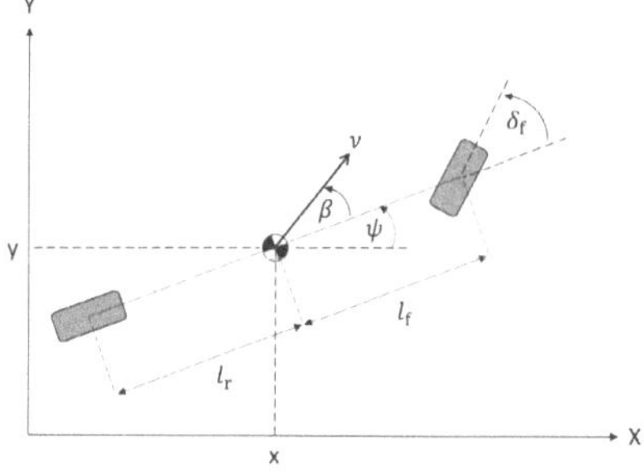

Figure 1: A kinematic bicycle model.

straints, to get the, in terms of a cost function, optimal control signals to reach the desired trajectory. Therefore, it predicts the behavior of the system for the calculated inputs by means of a prediction model. For this reason, a mathematical model of the system is needed. Here a kinematic bicycle vehicle model (Fig. 1) is used to describe the behavior of the car. The main idea is that the two front wheels and the two rear wheels of the vehicle are lumped into one front and one rear wheel, respectively. In the related equations of motion [5]

$$\dot{x} = v \cos(\psi + \beta) \tag{1a}$$

$$\dot{y} = v \sin(\psi + \beta) \tag{1b}$$

$$\dot{\psi} = \frac{v}{l_\mathrm{r}} \sin(\beta) \tag{1c}$$

$$\dot{v} = a \tag{1d}$$

$$\beta = \tan^{-1}\left(\frac{l_\mathrm{r}}{l_\mathrm{f} + l_\mathrm{r}} \tan(\delta_\mathrm{f})\right) \tag{1e}$$

$\dot{x}$ and $\dot{y}$ describe the velocities in x and y direction with regard to the center of mass. ψ is the angle from the x-axis

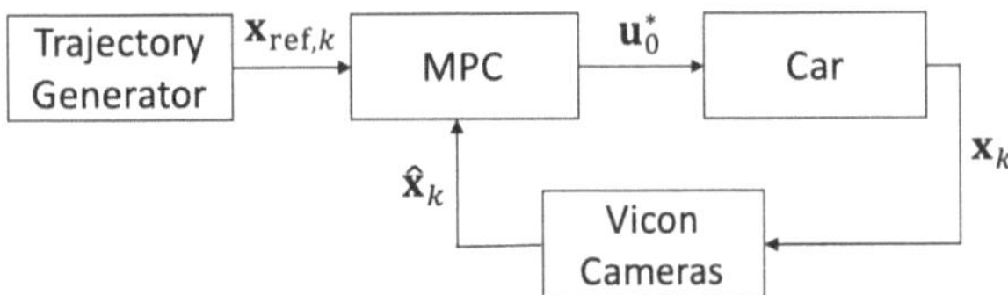

Figure 2: A graphical illustration of the control loop. In every time step, the MPC provides the car with the optimal control input $\mathbf{u}_0^*$ by solving the optimization problem with the measured states $\hat{\mathbf{x}}_k$ and the reference trajectory $\mathbf{x}_{\mathrm{ref},k}$. Because of the sub-millimeter accuracy of the motion capturing system one can assume $\hat{\mathbf{x}}_k \approx \mathbf{x}_k$.

to the axis of the vehicle. l_f and l_r are the lengths of the vehicle from the center to the front and rear end. v equals the velocity. The control inputs are the acceleration a and the front steering angle δ_f, which corresponds to angle β of the direction of a at the center of mass.

2.2 Control Loop

The MPC controller is realized using ACADO [6], an open-source toolkit for optimal control. ACADO uses the qpOASES solver. The differential equations are given to ACADO as well as the constraints and the properties of the MPC algorithm. The constraints include the limitations of the hardware, like the maximum steering angle of the front wheels or the maximum acceleration and are therefore called input constraints. In every time step, the MPC solves a new optimization problem over the shifted horizon and the first calculated control input is passed to the car. The optimal control problem is formulated as a constrained least squares minimization [7]:

$$\min_{\mathbb{X},\mathbb{U}} \sum_{k=0}^{N-1} \|h(\mathbf{x}_k,\mathbf{u}_k) - \mathbf{x}_{\mathrm{ref},k}\|_{\mathbf{Q}}^2 + \|h_N(\mathbf{x}_N) - \mathbf{x}_{\mathrm{ref},N}\|_{\mathbf{P}}^2 \tag{2a}$$

subject to:

$$\mathbf{x}_0 = \mathbf{x}(t_0) \tag{2b}$$

$$\mathbf{x}_{k+1} = F(\mathbf{x}_k,\mathbf{u}_k),\ k=0,\ldots,N-1 \tag{2c}$$

$$\mathbf{x}_k^{\min} \le \mathbf{x}_k \le \mathbf{x}_k^{\max},\ k=0,\ldots,N \tag{2d}$$

$$\mathbf{u}_k^{\min} \le \mathbf{u}_k \le \mathbf{u}_k^{\max},\ k=0,\ldots,N-1 \tag{2e}$$

$$\mathbf{x}_k \in \mathbb{X} \tag{2f}$$

$$\mathbf{u}_k \in \mathbb{U}. \tag{2g}$$

In (2a), $\mathbf{Q} \in \mathbb{R}_+^{6\times6}$ and $\mathbf{P} \in \mathbb{R}_+^{4\times4}$ denote positive-semidefinite weighting matrices. N is the length of the prediction horizon. The state vector $\mathbf{x}_k$ is formulated as $\mathbf{x}_k = \begin{bmatrix} x & y & v & \psi \end{bmatrix}^\mathrm{T}$ and $\mathbf{u}_k$ contains the controls $\mathbf{u}_k = \begin{bmatrix} \delta_\mathrm{f} & a \end{bmatrix}^\mathrm{T}$. Both can be bounded by maximum and minimum values. $\mathbf{x}_0$ denotes the current state and $\mathbf{x}_{\mathrm{ref},k}$ the reference trajectory. To formulate the least squares problem, the states and inputs are given to the reference function h. The MPC runs with a frequency of 50 Hz and is configured with a prediction horizon length of 3.0 seconds, which

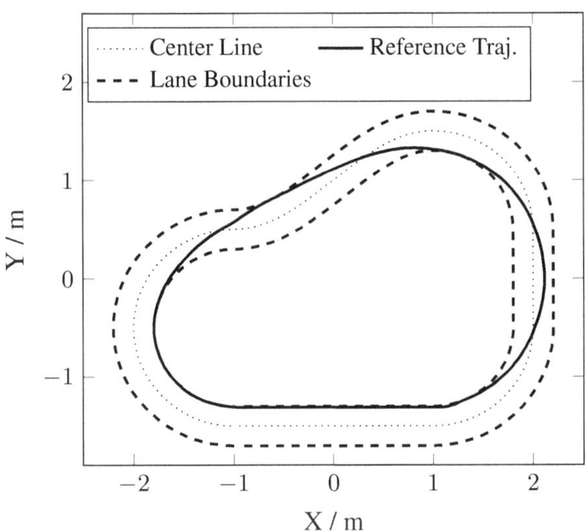

Figure 3: A visualization of the driving track including the center line, lane boundaries and the optimal racing line.

results in $N = 30$ for a discretization time of 0.1 seconds. To simulate objects on the track, state constraints, representing forbidden positions of the vehicle, can be included in the optimization problem (2). In all experiments, the car is set to drive with a constant velocity and hence only the steering angle δ_f is varied by the controller. Nonetheless, in addition to the bounds on the steering angle, constraints on the velocity and acceleration are set as

$$a = 0.0\ \mathrm{m/s}^2 \tag{3a}$$

$$v = 0.5\ \mathrm{m/s} \tag{3b}$$

$$-30^\circ \le \delta_\mathrm{f} \le 30^\circ. \tag{3c}$$

From these properties, ACADO generates C code, which is integrated using Robot Operating System (ROS) on an one-tenth scaled electric car. Further information about an MPC, similar to the one used in this paper, can be found in Philipp Huß's master thesis [8].

2.3 Trajectory Generation

To test the capabilities of the MPC, the racing track shown in Fig. 3 is used. The discretization of the trajectory points is 0.05 meters. The optimal racing line is calculated using an optimization tool [9], based on a total lane width of 0.4 meters and is passed as a reference trajectory to the MPC in all experiments. This width sets the area, where the center of mass of the car is allowed to be. Because the car has a width itself, the total width of the real driving lane would be wider in aspect to the width of the vehicle.

2.4 Hardware

The F1tenth hardware kit consists of multiple sensors in combination of an NVIDIA Jetson TX2 board with a quad-core ARM processor, on top of a one-tenth scale chassis of a Ford Fiesta Traxxas model ($l_\mathrm{r} = 0.16$ m and $l_\mathrm{f} = 0.17$ m). The controls are calculated online on the car's hardware

Figure 4: A F1tenth car with three markers for motion capturing.

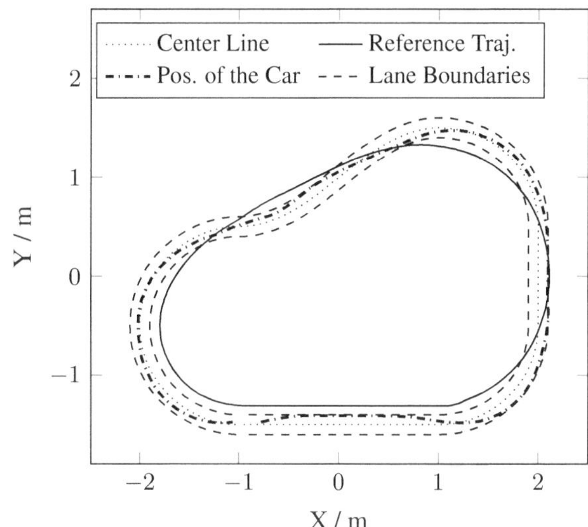

Figure 5: A measurement of the car's position, while following the computed optimal racing line.

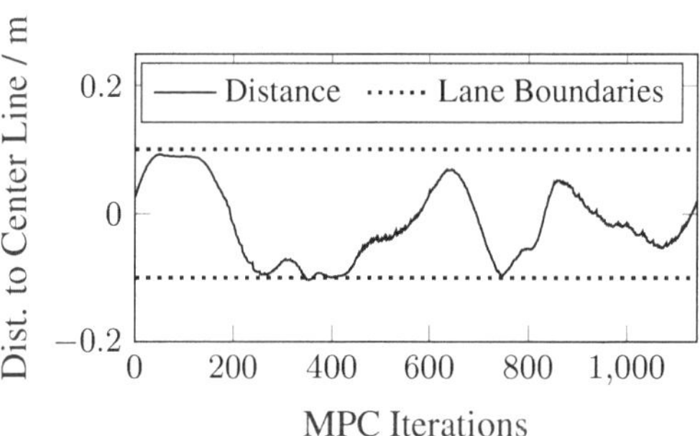

Figure 6: The distance (see Fig.5) between the car and the center line.

with the transmitted position, yaw angle and velocity data received from a Vicon motion capturing system [10]. Further, the motion capturing system is used to examine the performance of the MPC, in terms of accuracy of the trajectory following and avoidance of obstacles on the test track.

3 Experiments and Results

In every of the subsequent experimental setups, the car will drive in the counterclockwise direction, starting at the coordinates (-1.0 m, -1.5 m) and follow the optimal racing line. At the same time, the position of its center of mass is measured by the motion capturing system.

3.1 MPC with Lane Constraints

In reality, driving lanes are limited on each side of the road. To achieve this behavior during the experiment, the constraints of the optimization problem are customized as

$$\sqrt{(x - x_{cl})^2 + (y - y_{cl})^2} \leq l_{width}/2. \tag{4}$$

This constraint directs the MPC to steer the vehicle inside a corridor with a width of l_{width} around the center line coordinates (x_{cl}, y_{cl}). To demonstrate that this method works properly, the car has to follow the reference line. By narrowing the MPC's internal lane boundary constraints (4) down to a total width of $l_{width} = 0.2$ meters, the car will not be able to follow the reference over the whole track. Fig. 5 shows a measurement of the vehicle positions in this case. It is clearly seen, that the car stays inside the lane boundaries while following the reference. In Fig. 6 this behavior is demonstrated by plotting the distance between the center of mass and the center line of the track over the MPC iterations. Observe, that the vehicle successfully keeps the constraint set in (4) at all times.

3.2 Obstacle Avoidance

In this part, the performance of the MPC is examined, when obstacles are added to the track. Because this setup does not rely on camera data, the objects are realized by creating restricted areas throughout optimization constraints in the manner of

$$\sqrt{(x - x_{obs})^2 + (y - y_{obs})^2} \geq r_{obs}. \tag{5}$$

Those obstacles are shaped like circles, where x_{obs} and y_{obs} are the center coordinates and r_{obs} is the radius. In this experiment, the total width of the driving lane is defined as $l_{width} = 0.4$ meters, so the car is not prevented from following the reference by boundaries. Three obstacles are added to the track (see Table 1). Every object is placed in a way

Table 1: Position and radius of the obstacles.

Obstacle	X / m	Y / m	Radius / m
1	1.0	-1.3	0.2
2	-0.32	1.0	0.2
3	-1.8	-0.5	0.2

that it overlaps over one half of the lane at some point. To demonstrate the effect of the obstacles two paths of the vehicle are shown in Fig. 7. The first one is recorded with setting the objects on the track and the second one without. It is seen, that the car maneuvers differently in the areas

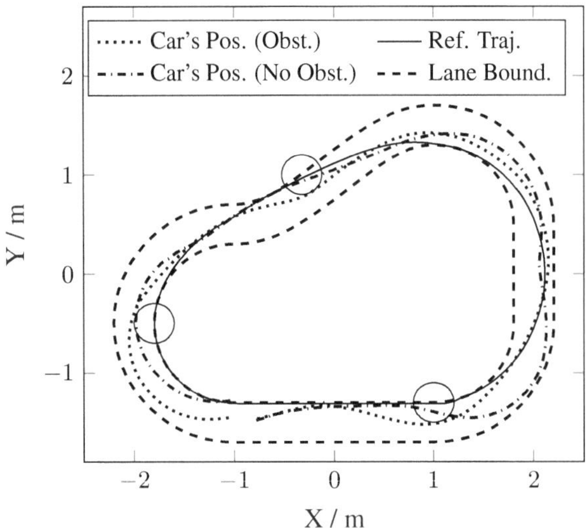

Figure 7: Recordings of the car's position, while following the reference trajectory with and without considering the obstacles (represented by the circles) on the track.

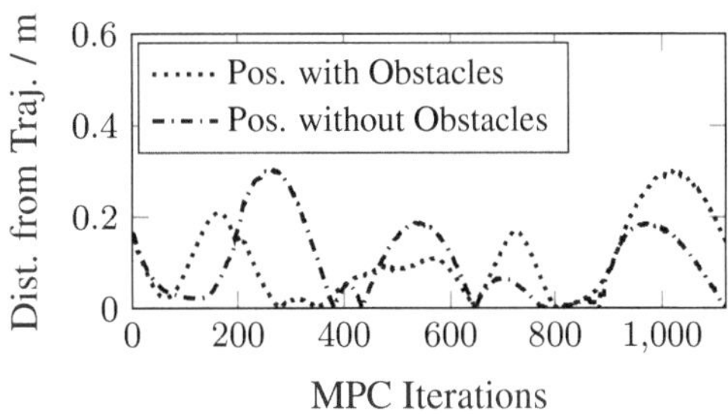

Figure 8: The distances (see Fig. 7) between the car and the reference trajectory.

around the obstacles to avoid a violation of the constraint. For further comparison Fig. 8 visualizes the absolute distance between the car's center of mass and the reference. In terms of the first object, it is clearly seen, that the car starts earlier to move away from the optimal racing line to evade.

4   Conclusion

This paper presented a design for trajectory following for autonomous driving, including additions for considering lane boundary constraints. Further, the simultaneous avoidance of obstacles was examined. The results have shown that this goal was achievable throughout the use of MPC, due to the ability to consider constraints during the control process. The controller is capable of steering around the obstacles while keeping the car inside of the driving lane. Our presented method is able to run in real-time on the car's hardware. A future task would be to compare our approach to simpler controllers, like a pure pursuit controller, and to evaluate the benefits of using explicit constraints.

Acknowledgement

The work has been carried out at Laboratory for Autonomous Driving, at the Institute of Electrical Engineering in Medicine, Universität zu Lübeck. We gratefully acknowledge the donation of a Jetson TX2 development kit by NVIDIA Corporation.

5   References

[1] F. Borrelli, A. Bemporad and M. Morari, *Predictive control for linear and hybrid systems*, Lecture Notes, University of California Berkeley, 2014. Available: http://www.mpc.berkeley.edu/mpc-course-material [last accessed on 2020-01-14].

[2] J. Gangloff, R. Ginhoux, M. de Mathelin, L. Soler and J. Marescaux, "Model predictive control for compensation of cyclic organ motions in teleoperated laparoscopic surgery", in: *IEEE Transactions on Control Systems Technology*, vol. 14, no. 2, pp. 235-246, 2006. doi: 10.1109/TCST.2005.863650. Available: http://ieeexplore.ieee.org/stamp/stamp.jsp?tp=&arnumber=1597194&isnumber=33591

[3] J. Cordingley, D. Vlasselaers, N. Dormand, et al., "Intensive insulin therapy: enhanced Model Predictive Control algorithm versus standard care", in: *Intensive Care Medicine*, vol. 35, p. 123, 2009. https://doi.org/10.1007/s00134-008-1236-z [last accessed on 2020-01-14].

[4] J. Kong, M. Pfeiffer, G. Schildbach and F. Borrelli, "Autonomous Driving using Model Predictive Control and a Kinematic Bicycle Vehicle Model", in: *American Control Conference*, 2015.

[5] R. Rajamani, *Vehicle Dynamics and Control.* Springer Science and Business Media, 2012.

[6] ACADO-Toolkit. Available: http://acado.github.io [last accessed on 2020-01-14].

[7] D. Ariens, M. Diehl, H. Ferreau, B. Houska, F. Logist, R. Quirynen and M. Vukov, *ACADO Toolkit Users Manual*, p. 103, 2014. Available: http://acado.sourceforge.net/doc/pdf/acado_manual.pdf [last accessed on 2020-01-14].

[8] P. Huß, *Trajectory optimization with collision avoidance for autonomous model vehicles using model predictive control*, Master's thesis, Universität zu Lübeck, 2020.

[9] L. Flegel, *Concept of a numerical optimization method for the calculation of optimal lines for autonomous model vehicles*, Bachelor's thesis, Universität zu Lübeck, 2019.

[10] Vicon Vero Motion Tracking System. Available: https://www.vicon.com/hardware/cameras/vero/ [last accessed on 2020-01-14].

Modeling of a Failure Mode and Effects Analysis in SysML on a respiratory alcohol detector

Jonah Mateo Goldyn [1], Georg Männel [2], Christian Brendle [3]

[1] Medical Engineering Science, Universität zu Lübeck, jonah.goldyn@student.uni-luebeck.de
[2] Institute for Electrical Engineering in Medicine, Universität zu Lübeck, ge.maennel@uni-luebeck.de
[3] Drägerwerk AG & Co. KGaA, Lübeck, christian.brendle@draeger.com

Abstract

An increase in complexity of technical systems poses the challenge to ensure efficient communication between the multiple teams involved in the development process. Model-Based Systems Engineering leads the path to a model-based description of the system in development with the goal of gathering all necessary information regarding the system in a single model instead of a distributed document-based description. As of now, the risk assessment is not commonly integrated directly into the system model. Therefore, different approaches regarding an integration of a Failure Mode and Effects Analysis (FMEA) into the modeling environment SysML were analysed. The focus was on traceability and usability. An approach was selected and exemplified on a simple system model of a respiratory alcohol detector. Attaching items from the FMEA directly to elements in the system model enables traceability of interrelations across structures.

1   Introduction

According to the ISO / IEC 15288 [1] Systems Engineering (SE) is the interdisciplinary process to develop and realise technical systems with the focus on the whole product life cycle. SE is commonly done on a document basis were all information is stored in distributed documents with different focuses on the same system. With an increasing system complexity it becomes hard to maintain consistency over all documents, thus more efficient means of communication and data exchange are required. Modeling techniques aim at capturing connections between different engineering domains and therefore recognising system wide effects of modifications at an early stage of development. Model-Based Systems Engineering (MBSE) is the formalised application of modeling to support activities regarding system requirements, architecture, analysis, verification and validation during the whole system life cycle.

One aspect of SE is risk assessment to identify risks and failures in the systems' functions and components. This is especially important regarding medical products due to the severe risk it might pose to a patient's health. However, the integration of the risk assessment into the system model is not common. A commonly used method for risk assessment is the Failure Mode and Effects Analysis (FMEA). When fully integrated into the model, it could be carried out partially automatic as first approaches, e.g. [8], show. This should ensure an automated chain and lead to improved traceability of interrelations.

The aim of this paper is to review different existing approaches to integrate the risk assessment into the system model - also regarding the usage of a single model. One of the reviewed methods is exemplarily implemented on a respiratory alcohol detector model in SysML with the aim to evaluate the approach regarding usability and traceability in the system model. The focus is not on the correct construction of the FMEA table, but on the application of an FMEA profile. Therefore, all results regarding a concrete FMEA given in section 3 are just examples. The FMEA of the system was not carried out by a safety expert

2   Material and Methods

With the goal of evaluating an approach in integrating an FMEA into a system model, requirements for the chosen methodology will be derived based on the current state of the art regarding FMEA and system modeling.

As an example, a simple system of a respiratory alcohol detector based on a former project carried out at the Institute for Electrical Engineering in Medicine, Universität zu Lübeck, is used. The system contains the sensor, an amplifier, a microcontroller, a display and three LEDs as well as a voltage supply. The test person breaths into the sensor and the output voltage correlates with the concentration of the gas. This voltage is amplified, measured and used to calculate the alcohol concentration which is then displayed and the respective LED lights up.

2.1   System modeling with SysML

For system modeling, specifically developed languages such as the semi-formal Systems Modeling Language

(SysML, [2]) - an extension of the Unified Modeling Language (UML, [2]) - can be used. UML as well as SysML offer the expansion mechanism of stereotypes which expand the general vocabulary and make it applicable to specific domains. For this work the mainly used diagram for modeling systems in SysML is the internal block diagram (ibd). It provides the white box or internal view of a the respiratory alcohol detector, the main function *measure_alcohol* and the interactions of the composite blocks/functions (instantiated as parts) through connectors and ports.

The system model is divided into two models: The logical architecture (Figure 1) and the functional architecture (Figure 2). "Logical" in this context refers to the technical concepts and principles of the system and contains abstract components of the system. For modeling the system in SysML the tool MagicDraw developed by No Magic, Inc. is used [3].

Requirements for the FMEA integration are a direct linkage between components/functions and failure modes/other FMEA items. Furthermore, all information should be kept in a single model, so there should be no export to another model or tool.

2.2 Failure Mode and Effects Analysis

According to [5] the Failure Mode and Effects Analysis is a bottom-up, inductive analytical method in risk assessment and is particularly used in the development stage of new products. The goal of the FMEA is the identification of possible failures, their causes and their effects. The effects get assigned a value to rate the severity (S) and the causes get assigned a value to rate the probability of occurrence (O). Afterwards, prevention and detection controls get assessed and the detection controls get assigned a value to rate the detectability (D). At Dräger the value range for all three items is set between 1-10 but can be customised for different projects. To evaluate the highest risks and/or the priority of addressing failures, the risk priority number (RPN) is calculated by multiplying:

$$RPN = S \cdot O \cdot D. \tag{1}$$

Following the calculation, mitigation actions can be taken to prevent failures and the RPN should be reduced.

The FMEA can be distinguished into different kinds: the system-FMEA focusing on the system as a whole, the design-FMEA focusing on the system design and the process-FMEA focusing on the production.

In [4] the construction of an FMEA is divided into five steps: the structural analysis that gives physical and functional structures, the functional analysis which maps functions to elements, the failure analysis that assignes failures to functions, the action analysis which assesses the current state of the mitigation and the optimisation that adds more mitigation and assesses the modified state.

For the construction of an FMEA, Dräger works with Plato SCIO^{TM}(-FMEA) [12], which allows the creation of trees (structure and functions) and therefore contains all system components and functions as well as the hierarchy of the components. The FMEA table contains the following elements: The function that is affected, the failure mode, the effect of the failure, the severity of the effect, the cause of the failure, the occurrence of the cause, current detection and prevention controls, the detectability of the failure, the RPN, recommended actions, responsibilities and results of the recommended action. However, it is not possible to show component interrelations, ports or interfaces in one product.

The integrated FMEA should produce a table similar to the one in SCIO regarding the contained elements as well as the possibility to assign more than one element to a failure mode. It should also give the possibility to create Fault Trees or fault propagation chains.

3 Results and Discussion

There are few approaches concerning an integration of the FMEA as well as other methods of safety and reliability analysis into SysML and MBSE in general. Many of the proposed approaches create extra stereotypes and/or profiles, others use exports of the SysML model.

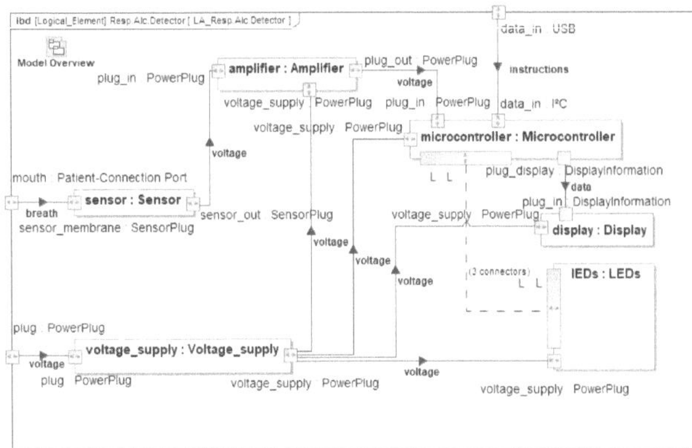

Figure 1: Logical architecture (ibd) of the respiratory alcohol detector.

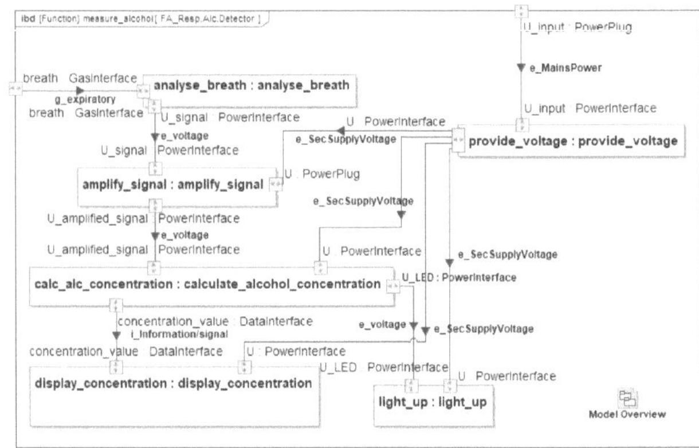

Figure 2: Functional architecture (ibd) of the respiratory alcohol detector.

3.1   Review of SysML modeling for FMEA

The Cameo Safety and Reliability Analyzer Plugin [6] supports the failure mode, effects and criticality analysis (FMECA) - an extended version of the FMEA - and hazard analysis according to medical standards. It provides an FMEA profile, a profile for Traceability of Safety And Reliability Analysis and one for Medical Risk as well as a template for reliability analysis. It is not possible to create Fault Trees yet.

The proposed OMG (Object Management Group) standard SafeML (Safety and Reliability specification) by Biggs et al. [7] contains a SysML profile as well as a model library. For the FMEA process it is recommended to establish a simulation context which then includes an FMEA analysis presented by FMEAItems. These items are connected to requirements via stereotyped dependency relations and in case of violation of requirements result in new requirements. The items also contain «*Situation*» and «*FailureMode*» blocks that show relations to other blocks, e.g. its relevance to another element. Furthermore, the elements Cause, FailureMode and Effect can be specialised (e.g. intermediate or final effect). Causes, failure modes and effects can also be chained for fault propagation. This specification seems to satisfy all of the derived requirements, but is not yet available for use.

Improved SafeSysE proposed by Baklouti et al. [8] automatically generates a preliminary FMEA that is completed by a safety expert. However, the FMEA is not directly integrated into the system model but in an external table. After the FMEA analysis architecture modifications are made by using redundancy i.e. the critical component is doubled. A Redundancy Profile is used in order to generate a Dynamic Fault Trees (DFT) from SysML diagrams. Those DFTs are analysed and the minimal cut sequence is used to create a state machine diagram. Using redundancy is often not an optimal solution because of additional weight and cost.

MéDISIS proposed by Cressent et al. [9] also starts with automatically generating a preliminary FMEA that is completed by a safety expert. The second step is the mapping between the SysML model and the AltaRica language model in which dysfunctional models are constructed. This approach is not an optimal solution because there is no linkage between elements directly in the system model.

The "Failure" Stereotype proposed by Kunnen et al. [10] does not integrate a defined risk assessment method such as FMEA or Fault Tree Analysis (FTA), but it uses an ontology that offers the possibility to link block diagrams across all structures. The stereotype also provides attributes such as *Action*, *ID*, *IsFailureOf*, *Risk*, *Risk Type* and *Scenario*, but does not seem to offer a possibility to create a table.

The UAV Dependability Profile proposed by Steurer et al. [11] also does not integrate a defined risk management tool but uses a Dual-graph Error Propagation Model (DEPM) which takes control flow and data flow graphs from the system model and uses those as well as probabilities of control flow transitions and properties of system elements (like fault activation and error propagation) to produce discrete Markov chains. It also provides an algorithm to transform SysML models to the DEPM. Because of the transformation into another model this approach does not satisfy all requirements.

3.2   Application of an approach

The Cameo Safety and Reliability Analyzer Plugin is chosen for the application because of its availability in contrast to the other approaches.

The FMEA profile included in the Cameo Safety and Reliability Analyzer Plugin allows the engineer to create failure modes and connect them to system elements (actions, blocks, part properties, requirements, operations and activities). From the elements with assigned failure modes the FMEAItems for the FMEA table can be generated. The FMEA table for the respiratory alcohol detector is shown in Figure 3.

The plugin also allows the creation of causes, effects (local and final) and design controls (prevention and detection controls). These items can be put into the FMEA table (via drag and drop or the selection window) and numbers for severity, occurrence and detection can be assigned. The RPN is automatically calculated (multiplication by default). Recommended actions, mitigation and responsibility as well as custom columns (e.g. for comments) can also be added in the table. It is also possible to create an FMEA configuration adjusted to the project e.g. regarding the maximal value of severity or the calculation of the RPN.

Based on the RPN a priority order can be determined to first address the most critical components - in the respiratory alcohol detector those are the sensor and the detector in general. The main failure in the sensor is "not enough breath". A possible mitigation for that failure is the change of component i.e. to use a different sensor that is more robust against smaller breath volumes or has a more effective mouth piece. A new requirement can be created and added to the mitigation column in the FMEA table.

Because the FMEA is divided into different types (see section 2.2) that are connected to each other, i.e. the cause in the system-FMEA is the failure in the design-FMEA and the effect in the process-FMEA, and because many components have the same failures, causes, effects and/or controls it is important to create a database for reusability. The plugin allows the creation of a database - it is also possible to create a profile or library. However, intuitively it is only possible to use elements as one type e.g. a cause of failure cannot be used as an effect of failure. This leads to reduced traceability between elements and fault propagation or the creating of Fault Trees, which give a hierarchy of failures, is therefore not possible.

Although it is possible to add several elements into one row e.g. causes to a failure, it seems to not be possible to assign more than one number (severity, occurrence, detectability each) per row. This is currently avoided by cloning rows which is not an optimal solution regarding clarity. It is also not intuitively possible to get to the right row by clicking the failure mode of an element.

It is not possible to automatically create a preliminary

#	Id	Item	Failure Mode	Local Effect Of Failure	SEV	Cause Of Failure	Localisation	OCC	Prevention Control	Detection Control	DET	RPN
1	F-1	measure_alcohol	Alcohol concentration not measured	Alcohol concentration not measured	8	Alcohol concentration not correctly measured	Resp.Alc.Detector	2	Choice of component	Prototype test	3	48.0
2	F-1a	measure_alcohol	Alcohol concentration not measured	Alcohol concentration not measured	8	Not fluid-resistant	Resp.Alc.Detector	2	Choice of material	Leak test / Prototype test	2	32.0
3	F-2	measure_alcohol	Alcohol concentration not correctly measured	Alcohol concentration not correctly displayed	8	Not fluid-resistant	Resp.Alc.Detector	2	Choice of material	Leak test / Prototype test	2	32.0
4	F-2a	measure_alcohol	Alcohol concentration not correctly measured	Alcohol concentration not correctly displayed	8	Not enough breath	Sensor	3	Fix sensor position / Choice of component	FAI / Prototype test	2	48.0
5	F-3	analyse_breath	Breath not analysed	No calculation / Alcohol concentration not measured / Alcohol concentration not displayed / No LED lights up	8	No voltage provided	Voltage_supply	2	Choice of material - plug connections / Constructive prevention: Poka Yoke	HALT / Visual inspection	2	32.0
6	F-3a	analyse_breath	Breath not analysed	No calculation / Alcohol concentration not measured / Alcohol concentration not displayed / No LED lights up	8	Not enough breath	Sensor	3	Fix sensor position / Choice of component	FAI / Prototype test	2	48.0

Figure 3: Part of the FMEA table of the respiratory alcohol detector produced with the Cameo Safety and Reliability Analyzer Plugin in MagicDraw.

FMEA table per element as it is done in some approaches, e.g. [8], so it has to be done manually. For larger systems or systems with more components this is time inefficient.
All listed problems were reported to NoMagic, but have not yet been addressed by the time of publishing this paper.

4 Conclusion

Integrating risk assessment methods such as an FMEA into SysML can improve traceability of interrelation between elements and their possible failure modes if the profile / application allows it.
There are few approaches that integrate risk management into MBSE/SysML but they either work with exports from SysML, i.e. not directly in the model, or do not fully support the FMEA or the FTA.
Although the Cameo Safety and Reliability Analyzer Plugin satisfies the requirements to some extent - all information is kept in a single model, traceability between functions and failure modes is possible (but not intuitively) and the table is similar to the one in SCIO. The traceability between system elements and the FMEA tables seems to be impossible. An automated generation of a preliminary FMEA and the FTA support are yet to be introduced.

Acknowledgement

The work has been carried out at Drägerwerk AG & Co. KGaA, Lübeck and supervised by Prof. Dr. Philipp Rostalski, Institute for Electrical Engineering in Medicine, Universität zu Lübeck.

5 References

[1] ISO (International Organization for Standardization) / IEC (International Electrotechnical Commission) 15288: *Systems engineering – System life cycle processes*.

[2] Object Management Group. *Systems Modeling Language*. [Online]. http://www.omgsysml.org/; *Unified Modeling Language*. [Online]. https://www.omg.org/spec/UML/About-UML/. [Accessed Dec. 10, 2019].

[3] No Magic, Inc. [Online]. https://www.nomagic.com/. [Accessed Dec. 17, 2019].

[4] -, *Bosch Schriftenreihe, Heft 14: Fehler-Möglichkeits- und Einfluss-Analyse*, technical report.

[5] K. Wälder and O. Wälder, *Methoden zur Risikomodellierung und des Risikomangements*. Springer Vieweg, Springer Fachmedien Wiesbaden GmbH, 2017.

[6] No Magic, Inc. *Cameo Safety and Reliability Analyzer Plugin*. [Online]. https://www.nomagic.com/product-addons/no-cost-add-ons/cameo-safety-and-reliability-analyzer-plugin. [Accessed Dec. 10, 2019].

[7] G. Biggs, A. Armonas, T. Juknevicius, K. Post, N. Yakymets and A. Berres, *OMG standard for integrating safety and reliability analysis into MBSE: Concepts and applications*. 29th Annual INCOSE international, 2019.

[8] A. Baklouti, N. Nguyen, F. Mhenni, J.-Y. Choley and A. Mlika, *Improved Safety Analysis Integration in a Systems Engineering Approach*. Appl. Sci. 2019, 9, 1246.

[9] R. Cressent, V. Idasiak, F. Kratz and P. David, *Mastering Safety and Reliability in a Model Based Process*. 2011 Proceedings - Annual Reliability and Maintainability Symposium, Jan 2011, Lake Buena Vista, FL, United States. 6 p.

[10] S. G. Kunnen, D. Adamenko, R. Pluhnau, A. Nagarajah, *An Approach to Integrate Risk Management in Cross-structure SysML-models*, in Proceedings of the 22nd International Conference on Engineering Design (ICED19), Delft, The Netherlands, 5-8 August 2019.

[11] M. Steurer, A. Morozov, K. Janschek and K.-P. Neitzke, *SysML-based Profile for Dependable UAV Design*. 2018, IFAC (International Federation of Automatic Control) Hosting by Elsevier Ltd. All rights reserved.

[12] Plato SCIO™. [Online]. Available: https://www.plato.de/produkte/scio-produktfamilie/. [Accessed Dec. 10, 2019].

Evaluation of Measurement Methods for Determining the Humidification Output of a Humidifier System

Lena Luers [1], Timo Matthews [2]
[1] Medical Engineereing, Universität zu Lübeck, l.luers@student.uni-luebeck.de
[2] Drägerwerk AG & Co. KGaA, Lübeck, timo.matthews@draeger.com

Abstract

During artificial respiration, the patient must be provided with humid breathing air. For the humidification output of an active humidifier, minimum requirements are given in standards but no satisfying way to measure it. This project evaluates different methods to measure the output. Using a humidity sensor turns out to be insufficiently accurate. More accurate is a gravimetrically method. For this purpose, the mass change of the humidifier system and the volume of air, conducted through the system are measured. This method provides reliable results with an accuracy of $\pm$ 0.9 mg/l, which meets the standard. However, the method is impracticable and affects the performance of the humidifier. A promising alternative is to use a dehumidifier like silica gel, which absorbs moisture from the air. Initial measurements show, that this method facilitates the mass determination and delivers accurate results without affecting the performance of the humidifier.

1   Introduction

Under normal circumstances, the air inhaled by a healthy patient is naturally conditioned by the upper airways. A physiological humidity level prevents depletion of moisture from the mucous membranes and transport system of the airways and lungs. The upper airway provides 75 % of the heat and moisture supplied to the alveoli. The optimal moisture level below the carina is 44 mg/l which corresponds to 100 % relative humidity at 37 °C [1]. Where absolute humidity is defined as the total mass of water vapor per given volume of gas and is given in mg/l. Relative humidity on the other hand is a percentage and defined as the ratio of water vapor in a gas compared to the capacity of the gas for water vapor. Whereby the capacity of water vapor in a gas increases with temperature [2], [3]. Patients whose upper airways have been bypassed by artificial respiration have to be supplied by humidified breathing air to prevent the secretions of the artificial respiratory tract from drying out. The breathing air can be heated and humidified with an active humidifier [1]. To preserve the patient's safety the required humidity level at the patient port opening is specified in the corresponding standard DIN EN ISO 8185 and is also referred to as humidification output. To verify that the required humidity is reached, the humidification output of the humidifier system needs to be measured. There are different approaches to measure the humidity produced by the system, those will be examined in the following. Furthermore, the standard also specifies that a measurement error of $\pm$ 1 mg/l must not be exceeded [4].

2   Material and Methods

In this section, the setup of three different methods to measure the humidification output of a humidifier system and the implementation of the measurements are presented. First the humidifier system used for all measurements will be described and the test conditions will be defined.

2.1   Humidifier System

The used active humidifier system is based on the most common method of humidifying respiratory gases, the pass-over humidification. As shown in Fig. 1 the humidifier system consists of a humidifier which is driven by an external energy source. The device heats an exchangeable water chamber which is pushed on the heating plate of the humidifier and is automatically filled with water to the optimal level through a water bag (called auto feed function). The respiratory gas is conducted over the heated water surface in the chamber. Depending on the surface temperature of the water and the gas flow velocity, there is an interchange of heat and humidity between water and gas. The humidified air is then conducted to the patient by a heated breathing tube [5], [6]. For the measurement of the humidification output only the inspiratory breathing tube is of interest. The used neonatal breathing tube system is designed for low flow therapy on infants. Therefore, the measurements are performed with 12 l/min.

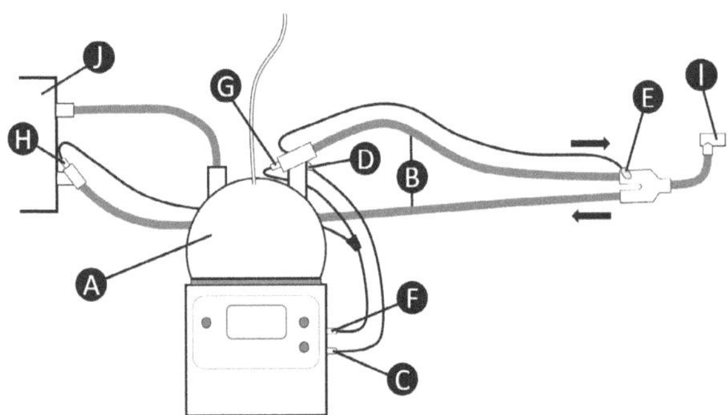

Figure 1: Humidifier system with auto feed water chamber through a water bag (not shown). A: humidification chamber can be slided onto humidifier base. B: breathing circuit. C: temperature probe plug on humidifier base. D: temperature probe at chamber exit. E: temperature probe at inspiratory airway exit. F: heater wire adaptor plug on humidifier base. G: heater wire adaptor at inspiratory breathing circuit socket. H: heater wire adaptor at expiratory breathing circuit socket. I: patient connection. J: ventilator

2.2 Measuring Devices

To determine the measurement accuracy, the precision of the individual measuring devices must be known. The used devices and their accuracy are listed in Table 1.

Table 1: Accuracy of the used measuring devices.

Device	Accuracy
weight scale	0.01 g, max. 1210 g
flow sensor	2 % of reading or 0.1 l/min whichever is greater
temp. sensor	0.05 °C
temp./humidity sensor MH8D46 (AHLBORN):	
- rel. humidity	± 2.0 % RH (range: 10 to 90 % RH)
	± 4.0 % RH (range: 5 to < 98 % RH)
- temperature	typical ± 0.2 K at 5 … 60 °C
	max. ± 0.4 K at 5 … 60 °C

2.3 Test Conditions

To make different measurements comparable, the measured gas volume must be converted to standard conditions. The volume V_{ATPS} (ATPS: Ambient Temperature Pressure Saturated) measured at ambient conditions must be converted into the equivalent volume under standard condition of the lung V_{BTPS} (BTPS: Body Temperature Pressure Saturated with water vapor) as shown in (1). For the calculation of the desired volume the temperature of the measured gas T_1 [K] and the ambient pressure p_{amb} [hPa] need to be known. Furthermore, the saturation vapor pressure p_s [hPa] and T_K [K] at body temperature (37 °C = 310 K) is required [4].

$$V_{BTPS} = V_{ATPS} \cdot \frac{T_K}{T_1} \cdot \frac{p_{amb}}{p_{amb} - p_s} \qquad (1)$$

p_s can be calculated as shown in (2) with T_K being the temperature in Kelvin (here 310 K) [4].

$$p_s(T_K) = 10^{30.6 - 8.2 lg(T_K) + 2.5 \cdot T_K - \left(\frac{3142.31}{T_K}\right)} \ [kPa] \qquad (2)$$

2.4 Method A - Gravimetric Measurement of the Humidifier System

This method is presented in the standard DIN EN ISO 8185, the setup is shown in Fig. 2. The mass of the moisture which reaches the patient is determined by this setup. First the initial mass m_0 of the system is measured after an appropriate warm-up phase of the humidifier system. The system is then started for an appropriate runtime t_r. After t_r has passed the mass m_1 of the system is determined. The mass difference Δm [mg] represents the moisture that has left the system. To maintain the required accuracy, the humidifier itself and the connecting cable are not measured. The total mass can be reduced this way and a more accurate weight scale can be used. The water chamber needs to be disconnected from the humidifier for this measurement. To calculate the humidification output H_{out} as shown in (3), the mass difference Δm and the volume of the air passing through the humidifier system to the patient V_{BTPS} is required [1].

$$H_{out} = \frac{\Delta m}{V_{BTPS}} \ [\frac{mg}{l}] \qquad (3)$$

V_{BTPS} is the product of the gas flow [l/min] and t_r [min] converted according to (1). Where T_1 is the temperature of the supplied dry gas whose flow is controlled by a pressure reducer and a flow sensor.

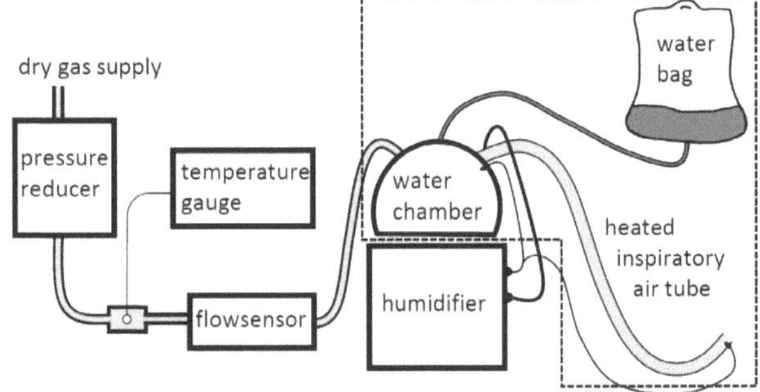

Figure 2: Measurement setup method A - gravimetric measurement of the humidifier system. The mass of the items inside the dotted box is measured without the cables before and after the system runtime t_r to determine Δm.

2.5 Method B - Measurement with Humidity Sensor

Fig. 3 shows the setup of method B, where the humidity sensor MH8D46 from the company AHLBORN is used to determine the humidification output. The humidity sensor is positioned in a T-piece at the end of the inspiratory breathing tube close to the patient.

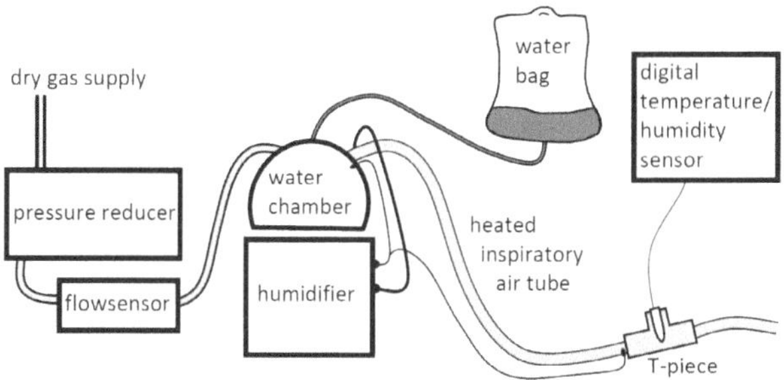

Figure 3: Measurement setup method B - measurement with humidity sensor. The humidity sensor is positioned in a T-piece at the end of the inspiratory breathing tube.

The measured temperature T_K [K] and relative humidity RH [%] values are used to calculate the humidification output. Since H_{out} can be written as the molecular weight of water m_W (18.016 g/mol) times the amount of substance n [mol] per volume V [l], it can be calculated using the ideal gas equation with the water vapor pressure p_{rel} [hPa] and the universal gas constant R (8.3145 J/(mol · K)) as shown in (4).

$$H_{out} = \frac{n \cdot m_W}{V} = \frac{m_W}{R} \cdot \frac{p_{rel}}{T_K} \cdot 100 \left[\frac{g}{m^3}\right] \qquad (4)$$

Equation (5) shows how to determine p_{rel} from the saturation vapor pressure p_s and the measured RH [3].

$$p_{rel} = \frac{RH \cdot p_s}{100\,\%} \, [hPa] \qquad (5)$$

2.6 Methode C - Gravimetric Measurement of a Dehumidifier

The setup of this method can be seen in Fig. 4. The heated and humidified gas is conducted through a container filled with silica gel as shown in Fig. 5. Silica gel consists of amorphous silicon dioxide, it is strongly water-attractive and serves as a dehumidifier in this measuring method.

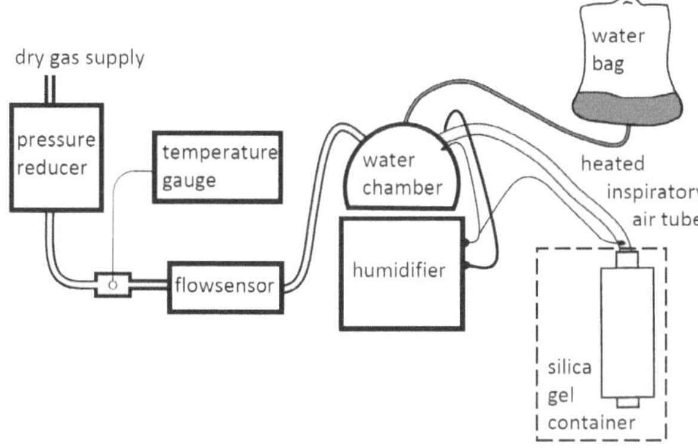

Figure 4: Measurement setup method C - gravimetric measurement of a dehumidifier. The mass of the silica gel container (highlighted with dotted box) is measured before and after the system runtime t_r to determine Δm.

Figure 5: Silica gel container. Left: empty container with schematic drawn airflow represented by arrows. Right: container filled with silica gel.

The material extracts the moisture from the air, which is conducted by the humidifier system to the patient. The mass of the dehumidifier is determined before and after the runtime. The mass difference Δm is used to calculate H_{out} as shown in (3).

2.7 Realization of Measurement

The measurements according to methods B and C are carried out simultaneously in one setup with the measurement according to method A to obtain a direct comparison. Furthermore, a humidity and temperature sensor will be placed at the exit of the silica gel container to obtain more information about the discharged air.

The measurements are performed with a flow of 12 l/min, a t_r of 20 min and the invasive mode of the humidifier. In this mode the temperature in the water chamber is controlled to 37 °C and at the breathing tube end to 40 °C.

3 Results and Discussion

As listed in Table 1, the measurement results of methods A and C show similar values with an accuracy within the limits given by the standard. The results of method B are clearly below the measured values of method A and C. Furthermore, the accuracy of method B is not within the limits of the specification.

With method A the humidification output can be calculated very accurately. The disadvantage of method A is, that the water chamber must be separated from the humidifier for the mass determination of the system. The water chamber and the heated inspiration tube thus cool down during the

Table 2: Measurement results of the different measuring methods at a flow of 12 l/min, a measurement runtime t_r of 20 min and invasive mode of the humidifier.

Measurement	Method	H_{out}	Accuracy
01	A	42.28 mg/l	$\pm$ 0.9 mg/l
	B	39.47 mg/l	$\pm$ 2.4 mg/l
02	A	43.72 mg/l	$\pm$ 0.9 mg/l
	C	42.43 mg/l	$\pm$ 0.9 mg/l

measurement of the mass and must first be reheated during the runtime. The performance of the humidifier system is therefore affected by the measurement.

Method B, where the humidity is measured with a humidity sensor, does not affect the performance and is simple to perform. It is, however, inaccurate because the humidity sensors cannot provide enough accuracy. Humidity sensors are generally not designed for such high temperatures and humidity. Besides, it is difficult to measure directly in the tube with the sensor. Therefore, the sensor is positioned in a T-piece, but the air cools of quickly at the cold surface and the moist air condenses. This causes the humidity to decrease before the air reaches the sensor. Which leads to the observation, that with method B the measured humidification output is lower than it actual is at the end of the inspiratory breathing tube.

Method C on the other hand delivers promising first results for the humidification output. Furthermore, the accuracy complies with the specification of the standard. The biggest advantage of method C is however, that the measurement does not interrupt the performance of the humidifier system. The measurement is simple to perform by plugging the container with the silica gel in and out of the inspiratory breathing tube.

The measurements at the exit of the silica gel container with the humidity sensor are listed in Table 2. The results confirm that hardly any humidity leaves the container. However, a significant heating of the material during the measurement can be observed, which can affect the water absorption of the silica gel in the long run. The disadvantage of this method is that after being saturated with moisture, the material must be dried in a climate chamber at over 100°C for a few hours. After drying, the silica gel can be used as before.

Table 3: Measurement results of the humidity sensor, positioned at the exit of the silica gel container. Measured during the runtime t_r of the humidifier system to examine the discharged air. With t_x being the time of measurement after the runtime started.

t_x	Relative Humidity	Temperature	Absolute Humidity
3 min	0.9 %	22.21 °C	0.18 mg/l
11 min	0.7 %	23.06 °C	0.14 mg/l
19 min	0.6 %	24.28 °C	0.13 mg/l

4   Conclusion

Although the measurement according to method A has major disadvantages in terms of practicability and the influence on the performance of the humidifier system, it provides reliable results for the humidification output during the runtime. Method A should therefore be preferred to method B for measurements requiring high accuracy. Furthermore, it could be established that method B with the humidity sensor significantly underestimates the humidification output. Depending on the objective of the measurement, this method may be sufficient if the accuracy has a subordinate role or the deviation is included in the evaluation of the results.

Although the measuring method C with the silica gel as dehumidifier is promising, further measurements are needed to clarify whether the method works well under other circumstances. On one hand it must be investigated whether the silica gel can extract moisture quickly enough from the air at a higher flow rate. On the other hand, it has to be evaluated how many measurements can be made until the silica gel is saturated. Furthermore, the container should be cooled to conduct the heat away that is absorbed by the silica gel and thus maintain the performance of the material over a longer time.

Acknowledgement

This work has been carried out at the Drägerwerk AG & Co. KGaA, Lübeck and supervised by the Institute of Physics, University of Lübeck.

5   References

[1] International Standard ISO 80601-2-74:2017-05, *Medical electrical equipment – Part 2-74: Particular requirements for basic safety and essential performance of respiratory humidifying equipment*. First Edition, 2017.

[2] M. Schertenleib and H. Egli-Broz, *Globale Klimatologie: Meteorologie, Wetterinformation und Klimatologie*. Compendio Bildungsmedien AG, 2003.

[3] G. Wiegleb, *Gasmesstechnik in Theorie und Praxis: Messgeräte, Sensoren, Anwendungen*. Springer-Verlag, Wiesbaden, 2016.

[4] DIN EN ISO 8185:2007-09, *Anfeuchter für Respirationsluft für medizinische Zwecke – Besondere Anforderungen an Anfeuchtersysteme für Respirationslusft*. Deutsche Fassung, 2007.

[5] Fisher & Paykel, *Instruction Sheet - MR850 Respiratory Humidifier*. 2015.

[6] ResMed, *Clinical Guide - HumiCare D900 System*. 2015.

Literature analysis of the competitive landscape of minimally invasive treatments for benign prostate hyperplasia

Afshan Momin
Biomedical Engineering, Lübeck University of Applied Sciences, afshan.momin@stud.th-luebeck.de

Abstract

The aim of the study is to evaluate the safety and performance of the new iTIND (Temporary Implantable Nitinol Device) in the competitive environment of minimally invasive techniques for the treatment of BPH (Benign Prostate Hyperplasia), with the focus on two devices (Urolift and Rezum) having commercial relevance and acceptance in the medical societies of EU and US. The study is performed at three levels starting with systematic literature search, identification of relevant outcome parameters and their analysis. The results from this study will be used in two ways, as an evidence-base for the proof of treatment efficacy as well as for future clinical study programs resulting from identified evidence-gaps.

1 Introduction

Benign prostate hyperplasia refers to a non-cancerous increase in the size of the prostate gland [1]. It usually develops after 40 years of age and is common among most of the men by the age of 85 [2]. Obstructive enlarged prostate results in LUTS (lower urinary tract symptoms).It includes either obstructive urinary problems, irritative symptoms or combination of both [1].

At initial stage with mild LUT symptoms, medications are preferred as a first-line treatment with watchful waiting [2]. However, 'TURP' (Transurethral resection of the prostate) remained the gold standard for moderate and severe LUT symptoms in last century. TURP is a surgical procedure, very effective in symptom relief, but results in long and short-term complications, ejaculatory and erectile dysfunction. In addition, it requires longer hospital stays under general or spinal anesthesia. After the emerging concept of minimally invasive procedures, many MIT (minimally invasive treatments) were developed as an alternate to TURP, with lower side effects [5]. This research is based on three minimally invasive treatments including prostatic urethral lift (PUL), water vapor thermal therapy (Rezum) and Nitinol stents.
PUL (Prostatic urethral lift) or Urolift is a permanent implant that retracts the obstructed lateral lobes of prostate by placing hooks on both sides and opens urinary stream without cutting a tissue [4]. Whereas, Rezum uses convective water vapor energy transferred to the transition zone causing cell necrosis in the prostatic urethra [3]. TIND is a temporary implantable stent therapy associated with the ischemic necrosis of the tissue. The stent is left in position for 5 days and then removed. It reshapes the urethra due to the high pressure applied by stent, blocking the blood supply to tissue and causing tissue necrosis

as shown in Fig.1 [6]. iTIND is the second generation of TIND with an improved design. It has three double intertwined struts instead of four single-layer struts which creates large incisions at three positions due to the exerted struts pressure [5]. For the assessment of new technology

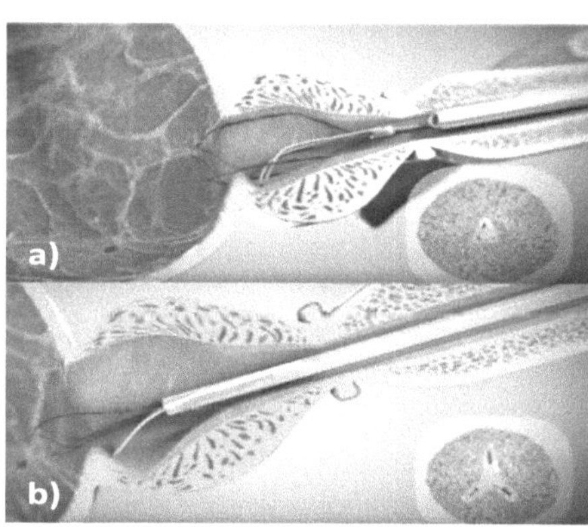

Figure 1: iTIND retrieval and reshaping of prostate

a) Device placement in the prostatic urethra.
b) Device retrieval through a catheter and change in the shape of urethral channel after removal.

as iTIND, which has limited research in the field, evidence mapping will help in collecting evidence on clinical performance and safety of new temporary implantable nitinol device in comparison to Urolift and Rezum. The identified evidence will support further claims for marketing of a product to make it stand-out among competitors products. The identification of potential evidence gaps steers future research for conducting high quality studies and to collect in-depth data, which demonstrates efficacy of medical device with large treatment effect.

2   Material and Methods

2.1   Questions Development

Before starting the literature research, some questions on performance parameters were defined, by reviewing marketing claims of the selected devices available on their official web pages [7]-[9]-[8]. American and European Guidelines were studied to understand about the treatment process and the tools that are used to see the effectiveness of the therapy. Some questions are as follow:

- How are the procedures performing for rapid symptom relief ? (Follow-up: 2 weeks to 6 months).

- How are the procedures performing for long term symptom relief ? (Follow-up: 1 to 3 years).

- How safe are these procedures? (Adverse events rate).

- Do they preserve sexual function?   (Erectile function (IIEF-5) and Ejaculatory function (MSHQ-EjD) scores).

- How ideal is the treatment for out-patient settings? (Time to discharge, operative time and without general anesthesia)

- Does patients require re-treatment within one year after the procedure? (re-treatment rate)

2.2   Determining inclusion criteria for publications

Using PICO Tool ('Problem/Patient', 'Intervention', 'Comparator' and 'Outcome') foundation for literature analysis was defined.

Patient/Problem: Benign prostate hyperplasia among middle and elderly age men (50-85 years)

Intervention: iTIND

Comparators: Urolift and Rezum.

Outcomes: Diagnostic tool – IPSS (International prostate symptom score) or AUASI (American urological Association symptom Index) measured at short and mid term (2 weeks to 6 months) and long term (1 to 3 years), Qmax (urinary flow), QoL (quality of life), PVR (Post void residual volume), sexual function assesment tools including IIEF (International index of erectile function) and MSHQ-EjD (Male sexual health questionnaire for assessing ejaculatory dysfunction), adverse events (hematuria, dysuria, pelvic pain/discomfort, UTI (urinary tract infection), urge incontinence, micturition/urinary urgency and urinary retention) and retreatment rates after procedure.

The long term results were limited to 3 years, because this time was defined by clinicians as expected duration of treatment effect. For analysis of short-term outcomes of urolift and iTIND, studies with 2 weeks to 6 months results were used.

Following study types were defined to be included in the search:

- RCT (Randomized controlled trials) - blinded or non-blinded.
- Cohort studies - retrospective or prospective and comparative or non-comparative.
- Systematic reviews.

2.3   Literature research

PubMed database was majorly used for all literature research on clinical trials and RCT and Cochrane for getting systematic reviews. Specific search strategy was used to get the most relevant publications. For the initial search in the PubMed database the following terms were used:

Minimally Invasive surgical procedure [MeSH terms-Medical subject heading] OR "Minimally Invasive" AND (Rezum OR ((convective OR "convective radiofrequency") AND "water vapor" AND ("thermal therapy" OR "thermal ablation")) OR ("water vapor" AND (therapy OR "energy treatment")) OR ("UroLift" OR "Prostatic urethral lift" OR PUL OR L.I.F.T) OR ("Temporary implantable nitinol device " OR TIND OR iTIND)).

2.4   Literature review

Systematic literature selection criteria was used, analogous to the PRISMA tool to further filter the identified studies. Starting with the title and abstract screening, studies with non-relevant topics, study design, review articles and with non-specified treatment groups were excluded. Then, the most relevant publications were procured for full-text screening and organized in an Endnote reference-managing library. At this stage, some papers were eliminated due to the small patient population (less than 28) resulting in less accurate mean values, as well as studies with same population and study design but with minimum follow-up results. All metadata extracted from these publications was organized in excel database. The following data was extracted: sample size, study design, inclusion criteria (age, prostate volume, obstructive lobes), exclusion criteria (Post void residual volume, obstructive lobes), treatment failure rate, sexual dysfunction, adverse events and diagnostic outcomes (short to long-term). The key results were further organized in tabular charts for easier analysis and comparison of the outcomes.

All the publications with 1 to 3 years follow-up on outcome parameters were used to get the values of IPSS, Qmax, Qol, PVR, IIEF and MSHQ-EjD. The baseline score and change in outcome at each stage of follow-up were taken from the studies and the percentage change was calculated from these values at each time point. The outcomes in randomized controlled trials, were based on two analysis intent-to-treat and per-protocol analysis. Per-protocol analysis outcomes were considered which included only those patients who completed the treatment. The main focus of the analysis is on changes from the baseline with-in each of the treatment group, because no studies directly compared TIND/iTIND with Urolift or Rezum.

Further, we collected the rate of recorded postoperative complications within 3 months among patients from all publications to get an overview of safety of the procedure. Moreover, re-treatment rates were also included to understand the long term efficacy of the outcomes and patient satisfaction. Additionally, parameters that will help in understanding the benefits and limitations of the treatment, were extracted from the studies. To describe the performance of Urolift and Rezum compared to sham procedures in randomized controlled trials, statistical significance of differences between groups was reported.

3   Results & Discussion

After getting 96 search results total 60 abstracts were reviewed, further 31 review studies were excluded and 29 full text articles were assessed for eligibility. After full screening of papers, 14 studies were further left-out. Overall, 15 clinical studies were included for detail analysis of Urolift, Rezum and TIND/iTIND. The study design of these publications can be seen in Table 1. The 5 years L.I.F.T study by Roehrborn (1) is a comparative study between sham and PUL group whereas, the study by Rukstalis is a continuation of L.I.F.T study in which some patients from the sham group were enrolled for crossover PUL treatment. On the other hand, Gratzke compared PUL to TURP treatment in his publication. For Rezum, Mc Vary conducted a comparative study between sham and REZUM treatment. Whereas, Roehrborn (2) study is a continuation of the this study with the outcomes of sham to crossover Rezum treatment. No study was found with direct comparison of TIND/iTIND to Urolift or Rezum treatment. All the above studies are multicenter studies except Kim JH, Mollengarden and TIND study that were conducted in single center.

Table 1: Study Design and patient population

Authors	Study Design	Patients
	Urolift	
Eure G et al.	Single-arm retrospective	1248
Sievert et al.	Single-arm prospective	86
Kim JH.	Single-arm retrospective	32
Neal Shore et al.	Single-arm prospective	51
McNicholas et al.	Single-arm retrospective	102
Roehrborn et al.(1)	Randomized controlled trial	137
Gratzke et al.	Randomized controlled trial	44
Rukstalis et al.	Prospective crossover study	53
	Rezum	
Mc Vary et al.	Randomized controlled trial	135
Roehrborn et al.(2)	Prospective crossover study	53
Darson M et al.	Single-arm retrospective	131
Mollengarden et al.	Single-arm retrospective	129
Dixon et al.	Single-arm prospective	65
	TIND/iTIND	
Propiglia.(TIND)	Single-arm prospective	32
Propiglia.(iTIND)	Single-arm prospective	81

All of the publications included IPSS, Qmax and QoL parameters, which are the primary outcomes to evaluate clinical effectiveness. Change calculated comparing outcomes to the baseline in 1-year follow-up is shown in Table 2.

Decrease in IPSS and QoL symptoms were selected as an indicator of LUT's relief. Whereas, increase in Qmax indicated better recovery. These results cannot be compared directly from one publication to another because of different study criteria and design, but are comparable within a same study population to see the change in symptoms by the time. Fig.2 shows the change in IPSS symptom score in 5 years among Urolift, and iTIND studies. TIND study results are not included in the comparison of outcomes because TIND device is off-market and currently second generation device outcomes are more of significance. iTIND one year study shows significant IPSS drop and seems to be effective as Urolift. But the iTIND study did not yet provided the long term data for detail analysis on long term effect. The primary outcomes analysed from Rezum studies showed nearly 50% decrease in IPSS symptoms in 3 years with highest effectiveness among all. Some of the comparative results and analysis on Qmax and QoL were also presented for these treatments as above but are not presented in the paper.

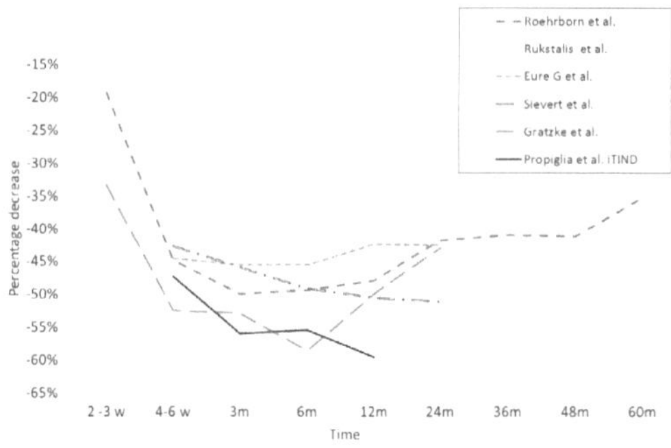

Figure 2: IPSS comparison of Urolift and iTIND

The secondary endpoints including IIEF and MSHQ-EjD for sexual function assessment after treatment were not found in iTIND feasibility studies, as primary outcomes were major endpoint of these studies [6].

Most of the publications mentioned mild and transient complications within the first 3 months, with no major related serious adverse events. Adverse events rate collected from the studies included hematuria, dysuria, pelvic pain/discomfort, urinary tract infection, urge incontinence, micturation/ urinary urgency and retention. Urolift showed a trend of increased adverse events compared to Rezum and iTIND within 30 days after treatment. Some studies also mentioned the complication rates in 1 year. The limitation of iTIND study is that it requires each complication rate and its occurrence in certain time, also with long term follow-up study at-least in one year.

The re-treatment rate is the important tool to determine the success rate of the procedure among the study population. According to the analysis, urolift re-treatment rates increased over the time from 1-5 years with 13.6% retreatments in 5 years. Whereas, re-treatment rates in Rezum goes upto 4.4% in 4 years study. Follow-up for iTIND study is limited to 1 year, with the re-treatment rate of 5%, which was nearly equivalent to urolift re-treatment rates within 1 year.

iTIND studies shows that this treatment is a potential office solution with the lowest operative time of 5.8 mins, the procedure can be performed in an outpatient settings with local anesthesia which is common in all three treatments.

Table 2: One year outcomes in comparison to the baseline.

Authors	Follow-up (Months)	IPSS $\Delta\%$	Qmax (ml/s) Δ	QoL $\Delta\%$
Urolift				
Eure G et al.	24	-42%	-0.4	-43%
Sievert et al.	24	-51%	2.9	-47%
Kim JH.	12	-42%	3.2	-61%
Neal Shore et al.	1	NA	NA	NA
McNicholas et al.	12	-51%	4.1	-52%
Roehrborn et al. (1)	60	-48%	4.0	-51%
Gratzke et al.	24	-50%	4.0	-60%
Rukstalis et al.	24	-40%	4.0	-43%
Rezum				
Mc Vary et al.	48	-53%	5.5	-52%
Roehrborn et al. (2)	12	-56%	5.9	-55%
Darson M et al.	12	-48%	1.5 *	-47%
Mollengarden et al.	6	NA	NA	NA
Dixon et al.	24	-58%	4.6	-61%
TIND/iTIND				
Propiglia et al. (TIND)	36	-45% **	5.1 **	-67% **
Propiglia et al. (iTIND)	12	-60%	7.3	-75%

Note: All $p < 0.05$ (significant p-Values in treatment groups).
**p-Value (Non-significant).*
***Significance level not mentioned.*

Through this study, some evidence voids were determined, where we need further research and collection of data. To evaluate preserved continence ISI (Incontinence Severity Index Score) should be measured. The downtime of a patient also effects on quality of life, so data should be collected on downtime of patients or their return to post-operative activity. Patients should also be evaluated through VAS (Visual analogue scale) for QoR (Quality of recovery) and for pain. All these data should be clearly presented in relation to the baseline scores and at different stages after treatment. Moreover, future studies with evidence on cost-effectiveness can also help in determining better options of BPH treatment.

In addition to the currently published iTIND study there are three other studies which are currently being carried out for iTIND treatment. A randomized control trial conducted in USA and Canada, which is an IDE (Investigational device exemption) study for US FDA approval and two prospective multi-center single-arm studies. These studies include primary as well as secondary endpoints [6]. The results presented through this work will help in designing future studies.

The identified evidence of outcome parameters will help to develop marketing claims that addresses physician's and patient's needs. From marketing perspectives, this will help patients to choose most favourable treatment with least complications. It will also be valuable for physicians in terms of providing best patient care in cost-sensitive environment.

4 Conclusion

Performance of iTIND seems to be comparable to Urolift and Rezum. However, we require more data from high quality randomised studies to demonstrate clinical benefits with efficacy of treatment in long-term. This analysis will help to determine the future clinical study program of iTIND to collect the required evidence. The performance and safety data evaluated is valuable in marketing as well as to fill the evidence gaps, which will be covered by future research.

Acknowledgement

The work has been carried out at Olympus Europa SE Co. KG (OEKG) and supervised by Prof. Dagmar Lühmann, the Department of Applied Sciences, Technische Hochschüle, Lübeck.

5 References

[1] R. C. Langan, *Benign Prostatic Hyperplasia*,The New England Journal of Medicine, pp. 248-57, 2012.

[2] AUA Practice Guidelines Committee.*"AUA guideline on management of benign prostatic hyperplasia (2003). Chapter 1: Diagnosis and treatment recommendations."* The Journal of urology vol. 170, 2003

[3] Williams, D. et al., *Minimally-Invasive Treatments for Lower Urinary Tract Symptoms in People with Benign Prostatic Hyperplasia: A Review of Clinical Effectiveness.* Canadian Agency for Drugs and Technologies in Health, 2019

[4] Magistro, G. et al., *Mini-Review: What Is New in Urolift?*, European urology focus, 36–39, 2018.

[5] Porpiglia, F. et al., *Second-generation of temporary implantable nitinol device for the relief of lower urinary tract symptoms due to benign prostatic hyperplasia: results of a prospective, multicentre study at 1 year of follow-up.* BJU international, 1061–1069, 2019.

[6] Amparore, D. et al., *First-and Second-Generation Temporary Implantable Nitinol Devices As Minimally Invasive Treatments for BPH-Related LUTS: Systematic Review of the Literature.* Current urology reports, 47, 2019.

[7] Urolift, Teleflex, 2020. Available:www.urolift.com

[8] Water therapy with Rezum, Boston Scientific, Aug 2019. Available: www.treatmybph.com.

[9] Treatments for IPB, Medi-Tate. Available: itind.it.

Systematic review on comparative effectiveness of high cut-off membrane haemodialysis to haemodialysis with other types of membrane

Wiktor Łuczak [1]
[1] Biomedical Engineering, Lübeck University of Applied Sciences, wiktor.mikolaj.luczak@stud.th-luebeck.de

Abstract

Objective: The aim of this work is to evaluate available evidence on the effectiveness of high cut-off (HCO) membrane in patients with multiple myeloma. *Methods*: A systematic literature search yielded 2611 unique records, from which the pool of the 5 articles were only eligible according to created protocol. A qualitative analysis of those articles was conducted. *Results*: Reports on the survival of patients are unclear. Renal recovery and protein reduction are stated to be better in case of the HCO than other membranes. *Conclusion*: The HCO membrane is effective and can be used for the elimination of the Free Light Chains (FLCs) from the patients. Use of that membrane also results in bigger number of patients recovering the kidney functions in patients with multiple myeloma. However, further proof in form of the good-quality Randomized Controlled Trials (RCTs) is needed to further confirm such assessment.

1 Introduction

Multiple myeloma affects terminally differentiated anti-body secreting B cells, known as plasma cells [1]. Common complications of the disease are bone destruction, hypercalcemia, anemia, renal damage, renal impairment and increased susceptibility to infection. In the majority of patients with the multiple myeloma elevated levels of the light chains, which are produced by plasma cells are observed [2]. The levels of monoclonal light chains are important in determining the prognosis and monitoring the disease [3][4]. The free light chains can deposit in the cells and be the source of amyloidosis [5]. That can lead to kidney disease and kidney injury. Therefore, it is necessary to find an effective way to remove FLCs from the patient. Renal failure is the inability of the kidneys to perform its' extraction functions, which in turn results in the nitrogenous compounds (including the proteins – free light chains) being kept in the blood flow [6]. In multiple myeloma, the light chains levels are higher than the capacity of the kidneys to catabolize them, resulting in creation of the casts. As a result, the tubules are obstructed and impairs further kidneys functions [7]. The high cut-off (HCO) membrane should have the specifics of the cut-off of molecules/compounds between 40 and 60 kDa, which according to the Shum et al. classifies as an HCO membrane [8]. That specification is supposed to allow better clearance of the molecules with bigger molecular weight, like free light chains Kappa and Lambda (22.5 kDa and 45 kDa respectively) [9].

Recently, a new category of membranes that can fit in the HCO category was developed - the medium cut-off membranes. They were introduced, with the cut-off of 45 kDa and are intended for the routine use in haemodialysis. In turn, the HCO membrane is supposed to be used for the acute applications [10]. The aim of the analysis was to assess the effectiveness of the HCO membrane in comparison to the other used membranes. Research questions of the work were as follows: 1. Does the treatment with the high cut-off membrane during dialysis increase the patient's survival? Does it lower the morbidity of the patients? 2. Does the treatment with the high cut-off membrane during dialysis increase the levels (amount) of removed light chains in the patients with multiple myeloma?

2 Material and Methods

2.1 Screening process and inclusion criteria

The Department of the Health Care Management team in the spring 2019 conducted a literature search of the following databases in search of the records: The Cochrane Central Register of Controlled Trials (CENTRAL), Cochrane Library, PUBMED, EMBASE, MEDLINE, clinicaltrials.org, WHO registries. From that search, all published Randomized Controlled Trials (RCTs) plus prospective and retrospective, comparative studies, that compared the haemodialysis of patients with multiple myeloma, with the aim of removing the light chains from the patient's body, using high cut-off and other membranes (i.e. high-flux) were identified. Complete search strategy will be provided upon request.

Then, the abstract and title screening of the literature was conducted. The included studies had to fulfil the inclusion

Table 1: Studies details

Study	Design	No of patients in group		Outcome reporting time		
Group	-	HCO	control	Survival	Morbidity	Protein reduction
Bridoux	RCT	50	48	12 months	3/6/12 months	unclear
Hutchison	Prospective	3	12	-	-	unclear
Buus	Retrospective	10	10	9 months	-	unclear
Gerth	Retrospective	42	17	12 months	3 months	-
Peters	Retrospective	5	5	390 days	390 days	1 and 2 weeks

Table 2: Studies details cont.

Publication	Intervention membrane	Control membrane	Cointerventions	Time of dialysis (intervention)
Bridoux	Theralite 2100	HF	Albumin + anticoagulation	5h
Hutchison	HCO 1100	SF and HF	Few of patients received albumin	up to 12 h
Buus	Theralite 2100	LF	Albumin + anticoagulation	8h
Gerth	HCO 1100	HF	Anticoagulation	4-6h
Peters	HCO 1100	unspecified	Albumin + anticoagulation	5h

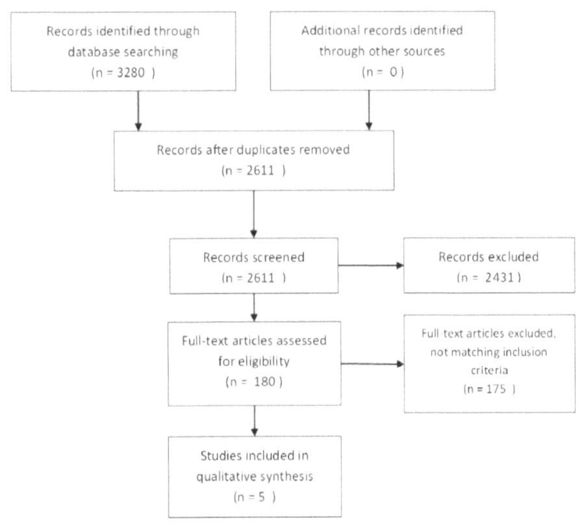

Figure 1: Short screening scheme.

2.2 Analysis description

Three types of outcomes were taken into consideration, extracted and analysed from each of the study, if possible. Said outcomes were: survival/mortality, morbidity and protein levels. The survival/ mortality was defined as number of patients who died at different time points. The morbidity was defined as independence of the dialysis/renal recovery of patients within different time points. Finally, the protein level was a percentage reduction of the serum free light-chain proteins after the treatment, calculated in respect to the baseline. Data from each study was extracted into forms adjusted from the Cochrane Collaboration data extraction forms for clear analysis. Due to the small number of control patients and poor reporting, the prospective study could be used only for assessment of the protein levels.

The meta-analysis of the results was not possible, therefore the analysis was conducted in form of qualitative analysis of a narrative systematic review only.

criteria developed in form of the Protocol. In short, studies designed for and conducted on patients with the diagnosis of the multiple myeloma and requiring haemodialysis (due to kidney impairment) were included. Patients only with the age of 18 years or older were considered, as multiple myeloma mostly affects that population [11]. There were no sex or race restrictions. The comparator could be a haemodialysis with any other type of the membrane (i.e. high-flux, low-flux, etc.). The use of the two dialysers in series was a reason of exclusion of a study. Special care was taken when considering membranes used in intervention group, excluding the studies with medium cut-off -classifying them as different type than the HCO. The process is shortly summarised on Fig. 1. The bias in the studies was assessed with help of adjusted tools developed by the Cochrane Collaboration. For RCTs the RoB 2 tool was used, while for the rest the ROBINS-I.

3 Results and Discussion

3.1 Description of studies

The literature search resulted in 3280 records. After duplicate removal, 2611 unique documents were identified. Abstract screening resulted in 180 entries treating with the high cut-off membrane. After full-text screening only five of the studies were meeting the criteria defined in protocol (patients suffering due to multitple myeloma; kidney impairment; adults; treated with haemodialysis with high cut-off membrane; comparative studies). Among them, only one RCT was identified [12], with the rest being prospective or retrospective comparative studies [13][14][15][16]. Details of the analysed studies can be found in Tables 1 and 2.

Table 3: Outcomes of the studies

Study	No of dead patients		No of patients with recovered renal functions		Reduction in the protein levels in comparison to baseline, in %	
Group	HCO	control	HCO	control	HCO	control
Bridoux	9	10	19(3m)/26(6m)/28(12m)	16(3m)/17(6m)/18(12m)	89	71
Hutchison	-	-	-	-	49.62	14.12
Buus	2	7	N/A	N/A	61	N/A
Gerth	13	9	27	5	N/A	N/A
Peters	1	4	3	0	74.6(1w)/80.8(2w)	N/A

3.2 Extracted data on outcomes

The extracted data in shortened form is presented in Table 3 (with no. of patients who died, achieved renal recovery and the mean reduction of the protein concentration (in %) since the beginning of the treatment). Unfortunately, the articles do not report on the protein baseline concentration separately for intervention and control (when there is any information on that). However, the baseline varied greatly, i.e. Buus 2015 reports that the baseline concentration of the sFLC in patients at the start of the treatment was between 1557 mg/L and 31691 mg/L.

3.3 Results

The analysed studies give quite a contradicting data on survival - from almost no difference in case of the Bridoux 2017 (9 deaths vs 10 between intervention and control), to better survival in Buss 2015 (2 deaths vs 7), Peters 2011 (1 death vs 4) and Gerth 2016 (31% vs 53% deceased - presented data is in percentage as the groups were vastly unequal in term of number of patients). Keeping in mind the limitations of the study, it can be concluded that there is no difference in survival between the intervention and control groups. The morbidity and protein level reduction is better in the intervention groups than the controls. The morbidity data shows an advantage of HCO in recovering the kidney function within three studies (mean percentage of recovered patients was 61.72% in intervention group vs 22.17% in control). The reduction of protein levels in comparison to the reported controls (71% in Bridoux 2017 and 14.12% in Hutchison 2007) in case of the HCO is reported superior by Bridoux and Peters (2 weeks), similar by Peters (1 week) and inferior by Buus and Hutchison in comparison to the control group from the Bridoux 2017. On the other hand, in comparison to the Hutchison control group, all of the intervention groups resulted in improved reduction of FLCs.

3.4 Discussion

The analysis shows that the HCO membranes have better protein clearance (reduction rate) than the comparators - answering the second research question. Also, the kidney recovery is achieved in bigger number of patients in treatment with HCO membrane in all of the analysed studies with available data. The survival, however, is not that clear, where the analysed studies cannot confirm whether intervention or the control is undoubtedly advantageous. As a result, there is no clear answer to the first research question of the paper. Therefore, without further studies conducted to clarify that issue, the HCO membranes should not replace the current treatment. Concerning the quality of used data, Bridoux 2017 presents the most balanced design, with the least bias. Gerth 2016 brings the second biggest number of patients, however, due to imbalances in the intervention and control group and lack of randomisation can bring a bias. Finally, the three other studies had small number of participants (no more than 20). The Hutchison reports on the control of only three patients who were treated with 16 treatments in total. Therefore, the comparison with the Bridoux is more vital and is not overestimating the effectiveness of the HCO membrane.

In all cases, the treatment with the haemodialysis was not the only intervention. In each of the studies the patients were also treated with chemotherapy as co-intervention. Also, the time of the dialysis treatment differed between studies. The dialysate flow for the retrospective studies was set at 500 ml/min. Bridoux 2017 reports on the dialysate flow of at least 500 ml/min and the Hutchison 2007 reports only on 7 patients with flow between 300 and 500 ml/min. The reporting on the methods and results of the studies is poor in most cases, with exclusion of the Bridoux 2017. The quantitative analysis with the meta-analysis was impossible due to differences in the protocols of the studies. Not only co-interventions and the time of the dialysis varied between the studies, but also the reporting and analysis of the data was conducted with different approach. For instance, only survival for two studies was reported for a fixed time point - 12 months (plus a study presenting the survival through the four years allowing the analysis). The morbidity was reported in three studies, for each study different time point was reported: 90 days, 3,6, 12 months and around 390 days. The worst presentation of the results was concerning the protein levels, which were reported in unclear way (only one study - Peters 2011 reports on the exact time of the measurement). The articles mostly focus on description of the intervention, while the control was only mentioned in passing. Therefore, only separate analysis of the studies was possible and the qualitative analysis was conducted.

As mentioned before, due to the imbalance in the number of patients in the analysed group in Gerth 2016, the data should be taken with reserve. Similarly, small number of patients analysed in other two retrospective studies makes the results doubtful and less then convincing. Therefore, only the RCT is of quality good enough to yield meaningful results. It is clear that due to the difference in the protocols and reporting, there is no clear answer to the question of whether the HCO membrane is more effective than others.

The studies differed in use of HCO membranes (HCO 1100 and Theralite 2100). The comparators varies greatly (low-flux, high-flux and super-flux). Moreover, the comparators were only named in one case (Hutchison 2007, where Toray BK-F 2.1 and Braun Hi-PeS 18 were used). That additionally complicates and prevents effective comparison between the membranes. That shows the need for better and more precise reporting on the methods and results of the studies, as well as more uniform protocol for conducting studies. For the more refined and informative analysis, the studies need to be of better quality and with smaller bias.

4 Conclusion

On basis of the analysed studies, the haemodialysis with high cut-off membrane is an effective way of eliminating the FLCs from the patient as the other membranes. Such treatment also yields faster renal recovery. Nonetheless, due to small number of comparative data, it is not conclusive whether the use of HCO membranes can be superior in such treatments to the conventional membranes. Therefore, more studies and trials proving the capabilities of the HCO and comparing them to other membranes need to carried out. Moreover, the need of more specific and strict protocol in those studies is needed to give them enough quality to be considered and accepted by the scientific community.

Acknowledgement

The work has been carried out at the Department of Health Care Management at Technical University of Berlin under supervision of Dr D. Panteli and M.Sc. H. Eckhardt. The supervisor responsible within the Univeristy of Lübeck was Dr med. D. Lühmann.

5 References

[1] B.G. Barwick et al., *Multiple myeloma immunoglobulin lambda translocations portend poor prognosis*. Nature Communications, vol. 10, article no. 1911, 2019

[2] C.T. Hansen, P.T. Pedersen, L.C. Nielsen, N. Abildgaard, *Evaluation of the serum free light chain (sFLC) analysis in prediction of response in symptomatic multiple myeloma patients: Rapid profound reduction in involved FLC predicts achievement of VGPR*. Eur J Haematol., vol 93, no. 5, p. 407–413, 2014

[3] N.M. Shahzad, C. Huang, *Significance of Serum Free Light Chains in Prognosis and Monitoring of Multiple Myeloma: A Review*. Journal of Blood Disorders and Transfusion, vol. 10, issue 2, no. 426, 2019.

[4] A. Rafae, M.N. Malik, M. Abu Zar, S. Durer, C. Durer, *An Overview of Light Chain Multiple Myeloma: Clinical Characteristics and Rarities, Management Strategies, and Disease Monitoring*. Cureus, vol 10, no. 8, e3148, 2018

[5] I.B. Bayer-Garner, B.R. Smoller, *The spectrum of cutaneous disease in multiple myeloma*. J. Am. Acad. Dermatol, vol. 48, pp. 497–507, 2003

[6] S. Bindroo, H.J. Challa, *Renal Failure*. In: StatPearls [Internet]. Treasure Island (FL): StatPearls Publishing; 2019. Available from: https://www.ncbi.nlm.nih.gov/books/NBK519012/ [Updated 2019 Mar 5].

[7] M. Dimopoulos, E. Kastritis, L. Rosinol, J. Bladé, and H. Ludwig, *Pathogenesis and treatment of renal failure in multiple myeloma*. Leukemia, vol. 22, no. 8, pp.1485–1493, 2008

[8] H.P. Shum, K.C. Chan, W.W. Yan, T.M. Chan, *Treatment of Acute Kidney Injury Complicating Septic Shock with EMiC2 High-cutoff Hemofilter: Case Series*. Kidney International, vol. 21, no. 11, pp. 751–757, 2017

[9] P. Yadav, N. Leung, P.W. Sanders, P. Cockwell, *The use of immunoglobulin light chain assays in the diagnosis of paraprotein-related kidney disease*. Kidney International, vol. 87, no. 4, pp. 692–697, 2015

[10] M. Hulko et al., *Pyrogen retention: Comparison of the novel medium cut-off (MCO) membrane with other dialyser membranes*. Scientific Reports, vol. 9, no. 6791, 2019

[11] D. Alexander, P. Mink, H. Adami, P. Cole, J. Mandel, M. Oken and D. Trichopoulos, *Multiple myeloma: A review of the epidemiologic literature*. International Journal of Cancer, vol. 120, no.S12, pp.40–61, 2007

[12] F. Bridoux et al. *Effect of High-Cutoff Hemodialysis vs Conventional Hemodialysis on Hemodialysis Independence among Patients with Myeloma Cast Nephropathy: A Randomized Clinical Trial*. JAMA, vol. 318, no. 21, p. 2099–2110, 2017

[13] C.A. Hutchison et al., *Efficient removal of immunoglobulin free light chains by hemodialysis for multiple myeloma: In vitro and in vivo studies*. JASN, vol. 18, no. 3, p. 886–895, 2007

[14] N.H. Buus, J.M. Rantanen, S.P. Krag, N.F. Andersen and J.D. Jensen, *Hemodialysis using high cut off filters in light chain cast nephropathy*. Blood Purification, vol. 40, no. 3, p. 223–231, 2015

[15] H.U. Gerth et al., *Impact of high-cut-off dialysis on renal recovery in dialysis-dependent multiple myeloma patients: Results from a case-control study*. PLoS ONE, vol. 11, no. 5, p. e0154993, 2016

[16] N.O. Peters, E. Laurain, J. Cridlig, C. Hulin, T. Cao-Huu and L. Frimat, *Impact of free light chain hemodialysis in myeloma cast nephropathy: A case-control study*. Hemodialysis International, vol. 15, no. 4, p. 538–545, 2011

Selection and implementation of a data pipeline to display the test results of automated tests in a user-defined dashboard

Hendrik Schönmuth [1] and Lutz Aulmann [2]

[1] Medical Engineering Science, Universität zu Lübeck, hendrik.schoenmuth@student.uni-luebeck.de
[2] R&D Anesthesia Development, Drägerwerk AG Co. KGaA Lübeck, lutz.aulmann@draeger.com

Abstract

Keeping an overview of a large amount of data, e.g. from software tests, is a big challenge. Developers in particular are very interested in receiving timely feedback on test results and the necessary information to perform error analysis. This should at best be possible in live time and for structured and unstructured data. Within the present work, the selection and implementation of a suitable framework for the representation of test data within automated tests and secondly the creation of a user-defined dashboard, for individual analysis of the remote test framework (RTF) test results used at Dräger will be presented. Through both tasks a comprehensive concept was developed, which offers the individual solution for analysis and presentation of RTF test data and was implemented within the automated tests into the company network.

1   Introduction

When developing large software projects, the software is commonly developed in distributed teams, which only work on parts of the larger project. Therefore the interaction of the system components should be tested regularly and at an early stage of the development cycle. In this way the correct functioning of the software is provided. A high degree of test automation is required to work economically. This is called *Continuous Integration* (CI). Essential for the success of the CI is that there is timely feedback on the status of the tests such that in the event of errors it is possible to react quickly. With the large number of tests that are performed on a daily basis, it is difficult to keep track of the CI status. It is therefore important that the test results are presented in a suitable form to quickly identify the causes of errors and to be able to correct them promptly. In order to do so, the large amount of test data hast to be processed and stored in such a way, that it can be accessed fast and selectively.

In this paper an end-to-end solution called *ELK-Stack* of the company *elastic* [3] is implemented for the procedure of data processing and adapted to the data format used by *Dräger*. Fig. 1 shows the principle of operation. The *ELK-Stack* is defined here as an end-to-end solution, since it offers a concept that ranges from data collection to user-defined presentation of the information it contains. The *ELK-Stack* (or *Elastic-Stack*) consists of three applications (*Elasticsearch, Logstash, Kibana*), which are described in more detail below. The *Filebeat* application, also provided by *elastic*, was used to forward and centralize the test results. The *Drägerwerk AG Co. KGaA* specific data format is defined by the *Remote Test Framework* (RTF) which is a self-developed framework for test automation. The RTF tests run on *Dräger's* end devices and allow testing on var-

ious hardware configurations.

The goal of this paper is to create a clear understanding of the *Elastic-Stack*. As an example, the implementation of the stack is shown using the problem of data processing and later visualization given here.

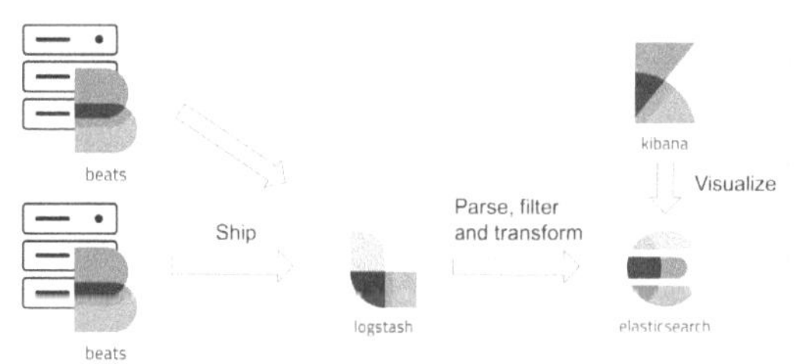

Figure 1: The *Elastic-Stack* consists of four components and represents the data pipeline used here.[8]

2   Material and Methods

The data visualization approach described in this paper uses the Elasticstools *Filebeat, Logstash, Elasticsearch* and *Kibana* for the implementation of a user-defined data pipeline to display the RTF test results. *Filebeat* collects the data and sends it to *Logstash. Logstash* parses, filters and transforms the incoming data and sends it to *Elasticsearch. Elasticsearch* acts as a database. The data is indexed and stored. *Kibana* accesses the stored data and can display certain selected content using predefined fields and indexes.

The tools of the *ELK-Stack* were integrated into the *Dräger* network. Fundamental changes had to be made to the configuration and *YML* files and adapted to the network.

Markup languages such as *YAML* (or *YML*) are human-readable data serialization languages. *YAML* is often used for configuration files in which data is stored or transferred. In this paper *YAML* is used for the configuration of data transfer between the individual Elastictools.

The tools *Logstash* and *Filebeat* needed more complex adjustments, as these are essential for data analysis and later display of RTF data. The individual adaptations include filtering of structured data, using the Grok- and XML-filter and filtering of unstructured data, which are analyzed and filtered by *regular expressions* (regex). Grok- and XML-filters are *Logstash's* own filters for processing structured data. Regex are used to describe single or multiple elements within a string, which enables regex to act as filter criteria for text searches. Each of the individual filters has the goal of making the data clearer for the user. More details about the individual filters are described in the respective sub chapters.

2.1   Filebeats

Filebeat is a lightweight "Data-Shipper", which monitors log data or even storage locations. It collects text-based log events and passes them either to *Logstash* for processing or to *Elasticsearch* for indexing [5].

Filebeat collects the test results of the RTF tests, which are stored as XML format. Before the XML files are sent to *Logstash*, they must be configured using the multiline setting. For the configuration of the multiline XML messages regex are defined for the pattern. The multiline pattern specifies which beginnings of lines belong together and connects the corresponding sections. The pattern of the XML file can be solved using relatively simple regular expressions:

multiline pattern: $\underbrace{^\wedge \backslash <}_{(a)} \ | \ \underbrace{^\wedge <\backslash}_{(b)} \ | \ \underbrace{^\wedge <?}_{(c)}$

Within the XML files, the individual lines begin with a start tag (a), an end tag (b) or a tag for XML decleration (c). Accordingly, the multiline pattern searches for line starters beginning with either (a), (b), or (c) and connect them together. With "$^\wedge$" the beginning of a line is indicated and with "|" another possible beginning of line can be added. According to the multiline pattern, *Filebeat* sends the XML file as a uniform message to *Logstash* for further filtering and analysis. If the *ELK-Stack* is used in combination with *Filebeat* (or *Beats*), it is called *BELK-Stack*.

2.2   Logstash

Logstash is a Java application used for simultaneous processing and normalization of data streams from different data sources such as HTTP, messaging queues or various logging frameworks. An incoming message is converted into a JSON-like structure consisting of key-value pairs [1]. Additionally, each incoming event can be enriched with information and its content can be changed. For exam-

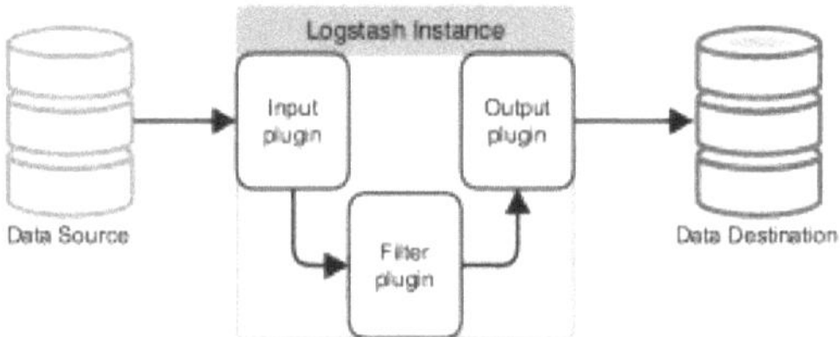

Figure 2: *Logstash* allows simultaneous processing and normalization of data streams. *Logstash* offers a number of plugins (input, filter, output plugin) to process the received data stream.[1]

ple, information such as a simple timestamp can be added. *Logstash* also offers further possibilities to process the incoming messages[1]. The data can be enriched, transformed and sent via a wide range of input, filter and output plugins.

In this paper XML files were analyzed and filtered by *Logstash*. The *Logstash* configuration is fully described in JSON. The input, filter, and output plug-ins are configured application-specifically by the plug-in structure shown in Fig. 2.

Input and Output-Plugin

The input and output plugin contain largely standard configurations in which only the port numbers and host addresses of the respective input (*Filebeat*) or output instance (*Elasticsearch*) are passed.

Filter-Plugin

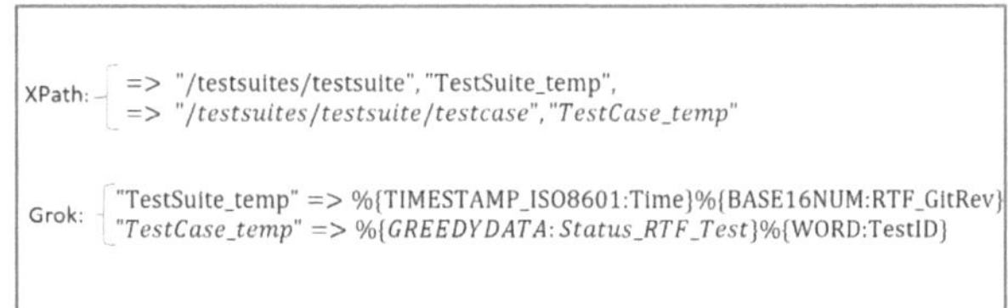

Figure 3: Within the filter, mainly the two filter plugins Grok and XML are used, which are provided by *Logstash*. The XML plugin is used to access the nodes of the XML file and write them to user-defined fields, and the Grok filter is used to parse the user-defined fields that have been extracted by the XML plugin.

The filter plugin is the most important part of the overall data analysis. It processes the incoming information from the XML files sent by *Filebeat* and stores it in user-defined fields for later analysis.

Logstash offers an XML filter plugin for filtering XML files. Via *XPath*, that filter plugin accesses important nodes of the XML files by using path expressions. The fields filtered out with *XPath* are stored and then filtered a second time with a Grok filter, as shown in Fig. 3. The fields filtered out by *XPath* correspond to long multinline messages, which can be precisely analyzed by the Grok filter. The Grok filter applies a previously defined Grok-Pattern to the messages. *Logstash* simplifies the handling of the Grok-Pattern and provides predefined Grok-Pattern expressions with nearly 100 executable files. The Grok-Pattern-Expressions are predefined regex expressions. A list of these files can be viewed via *GitHub* [7]. The difficulty is to find a Grok-Pattern that matches the message. When the Grok-Pattern and the message match, the filter returns the found value for each pattern in a structured way and writes the found value into a user-defined field. With the help of the user-defined fields it is possible to create meaningful diagrams and thus design a dashboard that is specific to the application.

Regular expressions

```
\\\\(\w+)\\\\([0-9]+)\\\\
```

TEST STRING

```
C:\\Temp\\testresults\\A350RandomExecutor_Ravel\\215\\Ravel_TestR
un_191219-1047\\TestResult_CI_Ravel_TestRun.xml
```

Figure 4: Within the file path there is important information that is of interest in the later evaluation of the data. The information within the file path was filtered out with *regular expressions*.

Fields such as the file path ([log][file][path]), are analyzed with regular expressions. Fig. 4 shows the regex pattern used for this and explains it in detail in the following. The file path is designed in such a way that it contains important information and must be taken into account for filtering. On the one hand, the used regex pattern matching gives information about the build number and the specific test name of the RTF tests. On the other hand, it has been specially adapted to the pattern of the file path, as this could change in its structure. In this case there are no predefined Grok-Pattern-Expressions so a self defined regex pattern was defined. In this way, the unstructured data was also matched and evaluated.

- $\backslash\backslash$: matched all backslash characters ($\backslash$).

- $\backslash w+$: matched all word characters. This expression is equivalent to $[a - zA - Z0 - 9\_]$. This is important because the underscore ($\_$) in the test name must also be matched.

- $[0-9]$: matched all single characters in the range from 0 to 9.

- $+$: indicates in this context that the matched character can occur between one and infinite times.

- Full Match:

$$\backslash\backslash \underbrace{A350RandomExecutor\_Ravel}_{Test\ name} \backslash\backslash \underbrace{215}_{Build\ number} \backslash\backslash$$

2.3 Elasticsearch

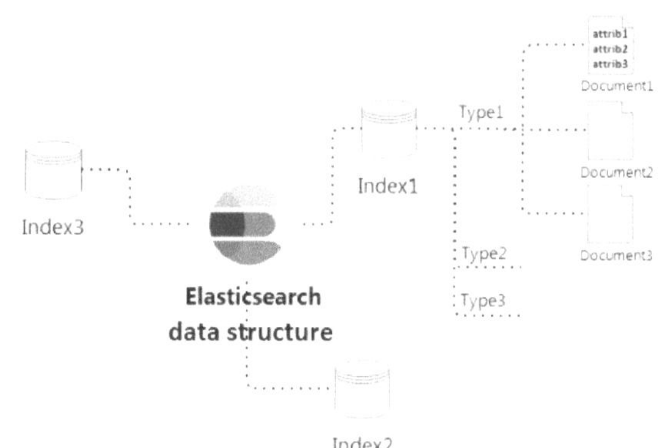

Figure 5: The data is stored as a JSON document. The data structure used by *Elasticsearch* stores the data in indexes, which in turn are divided into types and documents. The actual data is located within the documents and is assigned to attributes there.[6]

Elasticsearch is a database developed in Java, which is based on *Apache Lucene* (full text search engine). The big advantage compared to other database types (e.g. a SQL database) is that *Elasticsearch* enables the search and analysis of large amounts of data so efficiently that it can be displayed almost instantly [1].
The data is stored as a JSON document and made available through a *Representational State Transfer* (REST) API [2]. The data structure used by *Elasticsearch* stores the data in indexes, which in turn are divided into types and documents, as can be seen from the structure shown in Fig. 5.

- **Index:** A kind of database in which the data can be stored and searched. An index can be considered as a collection of documents with similar properties and is divided into several types [6] [1].

- **Types:** A type is a logical partition of an index, which allows a more structured classification of an index document [6] [1].

- **Document:** A document is the smallest information unit within *Elasticsearch* and is formed by a single JSON object [1]. The documents contain the actual data and are provided with "attributes" (atrribut1, attribut2, attribut3, ...)[6].

2.4 Kibana

Kibana is the web-based visualization tool of the ELK stack. *Kibana* allows you to access, search and interact with the data in the *Elasticsearch* indices. Based on this, custom graphics and dashboards can be created.

3 Results and Discussion

The result of this work is the data pipeline shown in Fig. 6, which is implemented in the *Dräger* network and displays

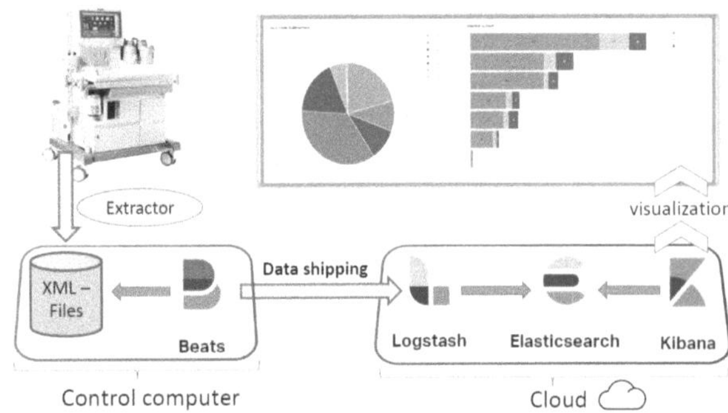

Figure 6: The Elastictools are installed on different network locations. The RTF-Tests are running on the anaesthesia devices and stored the test results on a control computer via an extractor. *Filebeat* is installed on the control computer as well and monitors the storage location of the test results. New test results sending to a cloud system on which the *ELK-Stack* is running. *Kibana* displays the filtered test results on a web-based interface.

the RTF test results on a dashboard. Within the project, a local test environment was first implemented, which works with *Elastic* versions 7.4.2 and was used for the results presented here.

For experimental setup the Elastictools were installed on different network environments. The RTF tests are currently running on the anaesthesia devices. The test results of the RTF tests (XML files) are stored on a control computer by means of an extractor. *Filebeat* is installed on the control computer and monitors the storage location of the test results. As soon as new test results are stored, *Filebeat* sends them to the cloud system on which the *ELK-Stack* is running. After the test results are filtered, parsed and indexed, they are displayed via *Kibana*. The created dashboard allows an early analysis of the RTF tests and ensures a permanent feedback of the automated tests running on the end devices.

4 Conclusion

A concept was developed that displays the RTF test results within automated tests via a dashboard. With the *ELK-Stack* the company *Elastic* offers a very efficient end-to-end solution that can be used for a variety of data analysis variants. Also in this paper the *ELK-* or rather the *BELK-Stack* is used, because it allows a live time analysis and can be used mostly free of charge. By integrating the *BELK-Stack* within the *Dräger* network it is possible to collect, filter and present the RTF test results in a suitable form. This is done by using the filter plugins provided by *Logstash*. By using the XML and Grok filter, the important information for this work could be returned in a structured form and stored in user-defined fields. Unstructured data, such as the file path, is filtered using the regex pattern and also stored in user-defined fields. The user-defined fields help to process the user-specific requirements and display them in the form of graphics or diagrams. The result is a meaningful dashboard tailored to the needs of the developers.

Acknowledgement

The work has been carried out at *Drägerwerk AG Co. KGaA* R&D Anesthesia Development, Lübeck and supervised by the Institute for Electrical Engineering in Medicine. Special thanks to G. Männel and Prof. Dr. P. Rostalski for being my supervisors.

5 References

[1] P. Kleindienst, *Building a real-world logging infrastructure with Logstash, Elasticsearch and Kibana*. Media Science, Stuttgart Media University, 2016.

[2] F. Beck, *Simulation und KPI-basierte Analyse von Geschäftsprozessen in Anwendungslandschaften*. Master thesis, HAW Hamburg, 2019.

[3] Elastic 2019, *Elastic Stack and product documentation*. Available: https://www.elastic.co/guide/index.html [last accessed on 2020-01-09].

[4] S.R. Ramchandra, S.S. Santosh, S.V. Krischna, P. Adiga *Data Analytics with Elk Stack and Custom Dashboard*. In: International Journal of Research in Engineering, Science and Management, IJRSEM, Mijar, India, pp. 763–765, 2019.

[5] J. Hamilton, B. Schofield, M.G. Berges, J.-C. Tournier *SCADA STATISTICS MONITORING USING THE Elastic Stack (Elasticsearch, Logstash, Kibana)*. In: 16th Int. Conf. on Accelerator and Large Experimental Control Systems, JACoW Publishing, Barcelona, Spain, pp. 451–455, 2019. DOI:10.18429/JACoW-ICALEPCS2017-TUPHA034.

[6] Abdullah Zahir 2018, *Elasticsearch als Suche im E-Commerce*. Available: https://www.tudock.de/blog/e-commerce/artikel/elasticsearch-als-suche-im-e-commerce/ [last accessed on 2020-01-09].

[7] GitHub, *Grok Pattern Executable Files*. Available: https://github.com/elastic/logstash/blob/v1.4.2/patterns/grok-patterns [last accessed on 2020-01-10].

[8] bmc 2018, *Using the Elastic Stack with Remedy Logs - Part 1*. Available: https://communities.bmc.com/community/bmcdn/-bmc_remedy_ar_system/blog/2018/07/31/using-the-elastic-stack-with-remedy-logs-part-1 [last accessed on 2020-01-10].

11

Signal Processing

Acoustic scene classification with neural networks

Markus Kuttner [1], Jürgen Tchorz [2]

[1] Auditory Technology , Universität zu Lübeck, markus.kuttner@student.uni-luebeck.de

[2] Institute for Acoustics, Luebeck University of Applied Science, juergen.tchorz@th-luebeck.de

Abstract

Through machine learning, complex problems can be solved with the help of large datasets without human intervention. A subfield of machine learning is deep learning, which can use virtual neurons to build up a multi-layer neural network. This independently finds an optimal solution to a complex problem by repeated learning. A competition that deals with machine learning and the solution of different acoustic classification problems is the DCASE Challenge. In one task, Mel Frequency Cepstral Coefficients (MFCCs) were used to extract features from audio files that served as data for a neural network. This master internship deals with the optimization of the DCASE neural network with residual probabilities of incorrectly decided classifications. The accuracy of the rebuilt DCASE baseline model was 61.7 %. After optimization of the neural network the optimized model could not achieve an improvement with the same average accuracy. For future experiments, a modified optimizer or activation function could produce better accuracy.

1 Introduction

Artificial intelligence, or AI, simplifies a variety of technical processes in the modern world. The basic idea of AI is to imitate human thinking with computers and enables them to solve complex problems independently, without external interaction. A optimization method of artificial intelligence is Deep Learning. This optimization method tries to build up a neural network with the help of artificial neurons and obtain the optimal solution by repeated practice. With the neural network, the computer learns from observation data and finds its own solution to the problem at hand. A neural network can roughly be divided into three types of layers. The first layer type is also called input layer. On this layer are the input neurons, which are responsible for the mapping of the input data. The second layer type is also called hidden layer. The term "hidden" refers to the property of the layer. These properties cannot be observed in their work, so they work hidden. The number of hidden layers is not fixed, and can be freely defined. The third layer type is the output layer. This layer has a predefined number of output neurons with defined properties. A neural network consists, as the name suggests, of a combination of many neurons in different layers [1]. With this resulting model, unknown data can then be analysed and classified. Currently the biggest advantage of Deep Learning is the complex and automatic processing of very large amounts of data. A competition, which takes place annually with different tasks in the field of detection and classification, for example by deep learning, is the DCASE Challenge (challenge for detection and classification of acoustic scenes and events). Every year, this competition manages to produce new, innovative solutions to acoustic classification problems by the participants.

Five tasks were advertised for the DCASE Challenge 2019. These tasks were, acoustic scene classification, audio tagging with noisy labeling and minimal supervision, sound localization and detection, sound detection in domestic environments and urban sound tagging [2]. In my master internship I focused on the problem of classifying different acoustic scenes. The goal of this classification is to correctly assign recorded test data to one of the predefined acoustic scenes. The data set for this task consists of recordings of different acoustic scenes. These were recorded in 12 major European cities, providing five to six minutes of audio material for each scene class. The original recordings were divided into segments of 10 seconds and stored separately as dataset. Those 12 European cities were Amsterdam, Barcelona, Helsinki, Lisbon, London, Lyon, Madrid, Milan, Prague, Paris, Stockholm and Vienna. The acoustic scene classes to be classified were airport, indoor shopping mall, metro station, pedestrian street, public square, street with medium level of traffic, travelling by tram, travelling by bus, travelling by underground metro and urban park. A total of over 40 h of audio data with a sampling rate of 48 kHz was available. Through data processing and segmentation, 14400 audio files (120 per class, per city) are generated [3]. The basis for the recognition of scene classes are spectral features of the audio data. In the case of the DCASE Challenge, Mel Frequency Cepstral Coefficients (MFCC) are used for feature extraction. This method has been known in speech recognition for a long time. Already in 1994 Young described the advantages of MFCCs using the example of a HTK toolkit of the Hidden Markov Model [4]. To obtain the MFCC features, the audio files are usually divided into frames and multiply with a window function to avoid edge effects. After determining the frequency spectrum with a discrete Fourier transform, the logarithmic amplitudes are calculated based on the audiological level perception of the auditory system. Also the frequencies have different importance in perception. The audio data were reduced by using frequency bins for individual frequency bands instead of the entire spectral components. The distribution of the frequency bands below 1kHz is linear and above 1kHz logarithmic. This type of arrangement is also known as Mel scaling [5]. The goal of this internship is an improved classification with a modified network structure.

2 Material and Methods

The first step of acoustic scene classification using neural networks was based on the baseline program of the DCASE Challenge 2019 Task 1a. Therefore a Linux-based operating system with Keras, a front end of Tensorflow, was chosen. For a faster processing of the neural network a GeForce GTX 970 graphics card and the Tensorflow GPU v.1.9.0 driver were used.

2.1 Baselineprogram

The entire baseline program could be divided into several sub-processes. The first part provided for the download and read in all audio data. The next step was the extraction of the features from the acoustic data. The audio signal was first split into 10 second segments and spectral features were extracted using MFCCs. The audio frame length was 40 ms with an overlap of 50% and 40 Mel frequency bins were fixed. After this the maschine learning process starts with a two-layer Convolutional Neural Network (CNN). The structure of both CNN layers was:

- CNN layer #1

  - 2D Convolutional layer (filters: 32, kernel size: 7) + Batch normalization + ReLu activation
  - 2D max pooling (pool size: (5, 5)) + Dropout (rate: 30 %)

- CNN layer #2

  - 2D Convolutional layer (filters: 64, kernel size: 7) + Batch normalization + ReLu activation
  - 2D max pooling (pool size: (4, 100)) + Dropout (rate: 30 %)

For the training network it was necessary to convert the entire data matrix into a data vector. For this, the baseline program had an additional flatten layer which contained the acoustic data matrix line by line. For training the neural network a dense layer with 100 neurons and a Rectified Linear Unit activation function (ReLu) were given. In order to avoid overfitting, the dropout was also set to 30 % during training. The output layer of the training network worked with a Softmax activation function. The learning process was defined with an audio segment count of 9185, an epoch count of 200, a batch size of 16 and an Adam optimizer. Between the epochs, the data was remixed. This allowed the

audio data to be trained to the given acoustic scenes. After the training of the network followed the testing. For this purpose 30 %, 4185 audio files, were taken from the entire acoustic database. Attention was paid that no audio file of the training data was included in the test data. The test data also contained acoustic audio files from all given acoustic scenes and European countries of the DCASE-Challenge 2019. The outputs of the test network were probabilities between 0 and 1, which were buffered in a matrix. This matrix represented the probability distribution of the audio file for all of the given acoustic scene classes. After the testing network, the results were validated to assess the quality of the programmed classifier. The validation used the highest probability of the probability distribution for the decision of the classifier. Because the real acoustic scene classes were stored for the validation of the acoustic test data of the DCASE Challenge, a comparison of the predicted acoustic class of the programmed classifier with the actual acoustic class could be made. The output of the validation showed the result of the comparison, i.e. the correctness of the detection of the individual acoustic scene classes with the baseline neural network [6].

2.2 Method for improvement of the scene classification

The goal of the master internship was to optimize a neural network for acoustic scene classification by building a 2 step classification using the baseline system of the DCASE-Challenge 2019. The basic idea of the optimization was that the remaining probabilities of a classification, i.e. the remaining percentage points of incorrectly assigned acoustic scene classes, were significant for the real acoustic scene class. If there are correlations between the residual percentages of the neural network and the given acoustic scene class, the accuracy of a classifier could be increased by using a new neural network or adding another network layer. For testing this hypothesis, the first step was to generate the probability distribution of each audio file of the training data over all acoustic scene classes. The test process of the baseline program was used for this. By inserting the training data into the test process, the probability distribution of all audio files over the scene classes could be generated and stored separately before validation. The real class assignment of the training data was known to the neural network, so the probability distributions for each acoustic class could be stored separately in folders. To illustrate the data used for the optimized neural network, Table 1 shows the distribution of the probabilities for the first 10 audio files of the class "Airport". For a clearer representation, the classes "Bus", "Park", "Public Square", "Street, trafic" and "Tram" are not shown, because the prediction probability was 0 for all audio files.

From these data files, it was possible to generate a confusion matrix for visualization the individual probabilities distribution of an actual acoustic scene class versus the predicted scene class of the programmed classifier. A confusion ma-

Table 1: Probability distribution of audiofile examples of the class "Airport". For a better overview the scene classes with 0% probability were removed.

Airport	Metro	Metro Station	Shopping Mall	Street, pedestrian
0,979	0,000	0,000	0,021	0,000
0,390	0,000	0,023	0,586	0,000
0,991	0,000	0,000	0,009	0,000
0,677	0,000	0,038	0,286	0,000
0,983	0,000	0,000	0,013	0,003
0,676	0,000	0,173	0,099	0,050
0,312	0,260	0,253	0,173	0,001
0,920	0,000	0,012	0,043	0,011
0,832	0,000	0,000	0,168	0,000
0,434	0,000	0,006	0,560	0,000

trix allows the evaluation of the accuracy of a classifier. In the resulting matrix, the percentage of the correctly predicted class can be read in the main diagonal. The other values are the confusion of the target class with the output class. In this report the values were generated with a function of scikit learn [7]. A confusion matrix of the tested training data with the baseline network is shown in Fig. 1.

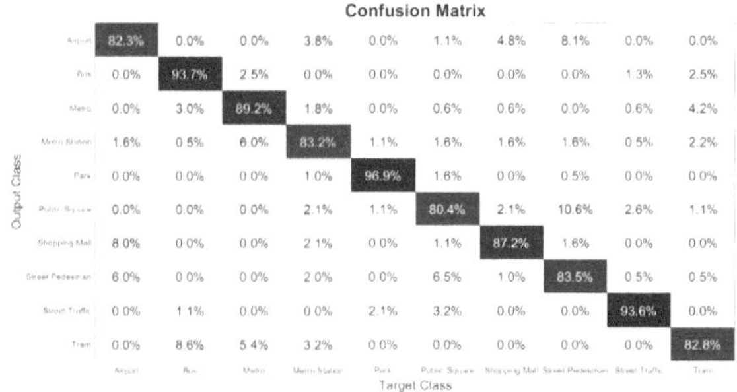

Figure 1: The confusion matrix of the tested baseline training data.

2.3 Neural network due to mispredicted class probabilities

To achieve the goal of an optimized acoustic scene classifier, a new neural network was trained with the probability data of the class prediction from the baseline network shown in Figure 1. The acoustic scene classes and the distribution of the probabilities over the scene classes were retained. For the training of the neural network a Dense Layer with 100 neurons, a ReLu activation function and an Adam optimizer was used. The training process was repeated with 50 epochs and a batch size of 256. The output with a Softmax activation function was set to 10 neurons. To verify the new classification method, the test audio segments of the baseline program were first tested with the baseline network. Then the resulting probability distributions were tested with the second neural network and the resulting probability distributions of this network were

saved. The maximum probability of each test audio segment for an acoustic scene was stored as the prediction of this scene class. For a validation of the optimized neural network, the predicted scene classes were compared with the actual scene classes of the test audio segments. The stored acoustic scenes of the test data were used for this purpose. A confusion matrix is presented to show the accuracy of the optimized neural network (Fig. 2).

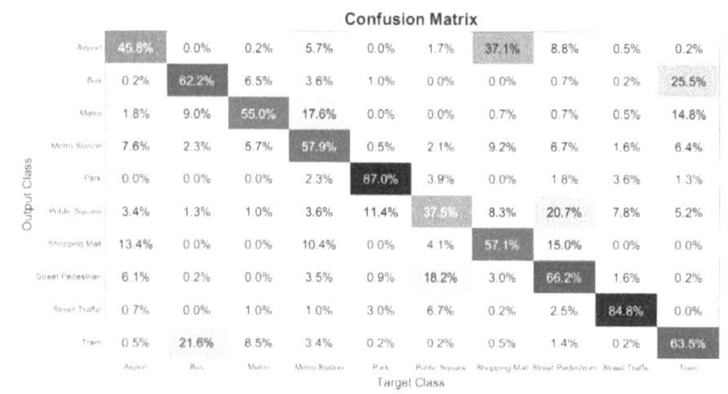

Figure 2: The confusion matrix of the optimized neural network.

3　Results and Discussion

The probabilities in Table 1 show that the training audio data is often assigned to the correct scene class. However, it can be seen that some audio files are also classified in other classes, such as "Shopping Mall". The accuracies of the confusion matrix, resulting from all audio files, shown in Fig. 1 are very high and even exceed 90 % in 3 scene classes. This high accuracy rate is due to the use of the training data as a test of the neural network. The classifier was trained with the training data and can therefore easily assign them. But some scene classes had residual probabilities of up to 10% for an incorrectly predicted class. For example, the acoustic scene class "Street Pedestrian" was often confused with "Public Square" and "Airport". During the validation run of the baseline network an average recognition rate of 61.7% (Table 2) was achieved. This value is a few percentage points lower than the given average value of 62.5% in the DCASE own publications (Table 2). In addition to the small percent difference in the mean value, there is a difference of up to 10% in the example of the scene class "Metro".

The generally low recognition rate of the acoustic scene classes is due to the very similar audio files. Through acoustic perception of the audio files, only few differences between the individual scene classes can be found. The achieved deviations could be caused by the random initialization of the weights of the neuron layers. Also possible version differences in the individual Python packages could be a possible reason. The validation of the optimized neural network resulted in an average detection rate of 61.7% (Table 2). Contrary to expectations, this value is exactly the same as the average detectability of the baseline program by own validation. But there are differences in accuracy in the respective acoustic scene classes. The accuracy of the class "Metro Station" could be increased by 4%, and the class "Public Square" could gain 2%. However, the class "Metro", "Shopping Mall" and "Street Traffic" lost 2% accuracy compared to the non-optimized neural network. These results were obtained through a single test run. For more significant results, the tests would have to be repeated several times to obtain an average value and to prevent large fluctuations. Furthermore, changing different properties of the neural network can lead to different accuracies. First experiments with increasing the epochs and doubling the number of neurons brought no improvement and was therefore not used. For further optimization attempts the optimized neural network can be tested with different activation functions, a significantly larger number of neurons or changing optimizers. Finding optimal properties of a neural network usually takes a lot of time, because every change has to be tested.

Table 2: Accuracy of the probability distribution of the individual scene classes for the DCASE baseline validation, own baseline validation and the optimized neuronal network (NN).

Scene	Accuracy DCASE baseline	Accuracy Baseline	Accuracy optimized NN
Airport	48,40%	45,60%	45,80%
Bus	62,30%	62,20%	62,20%
Metro	65,10%	57,50%	55,00%
Metro Station	54,50%	54,30%	57,90%
Park	83,10%	86,50%	87,00%
Public square	40,70%	35,40%	37,50%
Shopping mall	59,40%	59,90%	57,10%
Street, pedestrian	60,90%	67,40%	66,20%
Street, traffic	86,70%	86,10%	84,80%
Tram	64,00%	62,60%	63,50%
Average	62,50%	61,70%	61,70%

4   Conclusion

The report presents a possibility of optimizing an acoustic scene classifier. For this purpose, the baseline program of the DCASE-Challenge 2019, a competition for, among other things, the classification of acoustic data using neural networks, was used. With the help of data of faulty classification and another neural network an improvement of the recognition of acoustic scene classifiers should be achieved. It was shown that the first attempt of the optimized network did not lead to an improvement of the average recognizability. Only some classes were recognized better and others worse. These results do not mean that the idea must be rejected. Working with neural networks always requires a lot of time. By repeated testing with differently defined attributes of the neural network the results can be improved. Future work could be concerned with finding the optimal attributes, like optimizer, number of neurons or layers.

Acknowledgement

The work has been carried out at the Luebeck University of Applied Science and supervised by Prof. Dr. rer. nat. Jürgen Tchorz.
I thank Dr. Tchorz for the provision of a suitable PC for working with neural networks and for comments and hints that greatly improved the internship.

5   References

[1] M. Nielsen, *Neural Networks and Deep Learning*. Determination Press, 2015.

[2] *DCASE2019 Challenge.*
Available: http://dcase.community/challenge2019/index [last accessed on 2019-10-30].

[3] *Acoustic scene classification-Task description.*
Available: http://dcase.community/challenge2019/task-acousticscene-classification [last accessed on 2019-10-30].

[4] S.J. Young, *The HTK Hidden Markov Model Toolkit.* Cambridge University Engineering Department, Cambridge, 1994.

[5] B. Logan, *Mel Frequency Cepstral Coefficients for Music Modeling.* Cambridge Research Laboratory, Cambridge, 2000.

[6] *DCASE2018 Baseline.*
Available: http://dcase.community/challenge2018/task-acoustic-scene-classification#baseline-system [last accessed on 2020-01-21].

[7] *sklearn.metrics.confusion_matrix.*
Available: https://scikit−learn.org/stable/modules/generated/sklearn.metrics.confusion_matrix.html [last accessed on 2020-01-21].

Measurement of Room Impulse Responses and Comparison of Different Interpolation Methods

Glenn Röwekamp [1], Fabrice Katzberg [2], and Alfred Mertins [2]

[1] Medical Engineering Science, Universität zu Lübeck, glenn.roewekamp@student.uni-luebeck.de
[2] Institute for Signal Processing, Universität zu Lübeck, {katzberg, mertins}@isip.uni-luebeck.de

Abstract

Basic bandlimited interpolation methods (e.g. linear interpolation) for spatio-temporal room impulse responses (RIRs) fail to adapt the characteristics of sound propagation. For the reconstruction of signals with high frequencies, it is necessary to have sampling points of short distances, which results in complex hardware or time-consuming measurements. To compare different interpolation methods, measurements of RIRs inside a volume of interest (VOI) have been gained with a microphone mounted on a hexapod robot. Based on these measurements, a comparison of different interpolation methods has been achieved. Basic bandlimited interpolation methods are tested as well as a totally different, physically inspired approach using spherical harmonic functions.

1 Introduction

Room impulse responses (RIRs) are a significant indicator for information about the local acoustic properties inside a closed room. Various audio applications are based on the knowledge of these RIRs e.g. listening-room compensation [1], [2] or room acoustic reproduction [3]. To measure various RIRs inside a volume of interest (VOI), there are mainly two ways to do so. It is possible to use an array of static microphones, which results in a more complex and expensive measurement setup, or the use of a moving microphone [4], [5]. In this work the data is measured in a static setup. Therefore a single microphone was used, successionally positioned by a hexapod robot, enabling a precise positioning in six degrees of freedom. To limit the number of measurements for a given VOI while preserving relevant information of the RIRs for the room in between the sampled points, it is feasible to use interpolation methods. To measure RIRs between a loudspeaker and a microphone at a fixed location, common techniques such as perfect sequences [6] or sine sweeps [7] can be used.

The measurement of a set of RIRs using a moving microphone is possible in different ways. The dynamic method described by Ajdler, Sbaiz and Vetterli [4] uses a microphone moving with constant speed along a defined trajectory and leads to the reconstruction of RIRs. Therefore it is necessary to determine the location for a given time and consider the influence of Doppler effect in measurements. The applied approach is moving the microphone to given coordinates. When the microphone has reached its desired static position, the measurement takes place and the microphone moves to the next position. Considering a bandlimited signal with cutoff frequency f_c, the duration of the measuring process depends on the desired bandwith of the interpolated RIRs and the size of the VOI. An obvious approach for setting the measuring coordinates is to define a three-dimensional volume with equidistant sampling points inside.

For the interpolation of RIRs at low frequencies, it is possible to make use of compressed sensing, that leads to a reduction of sampling points using less measurements than required by the sampling theorem [8], [9]. High frequency interpolation yet fails to signifcantly reduce the number of measurements. Common interpolation approaches as trilinear or Lagrange interpolation require information from surrounded sampling points at the desired interpolation position. In a conventional sound-field sampling with equidistant sampling points, spatial intervals of

$$\Delta < \frac{c_0}{2f_c} \qquad (1)$$

are required for aliasing-free reconstruction [10], with c_0 as the speed of sound and f_c as the temporal cutoff frequency. Trilinear and Lagrange interpolation only preserve simple mathematical assumptions, while the method with spherical harmonics described by Rafaely [11] introduces a robust physical model of sound propagation with interpolation based on Bessel functions. The three different interpolation methods are being applied on the measurements and compared with each other.

2 Material and Methods

Considering a linear time-invariant (LTI) system consisting of a emitter-receiver pair, the temporal connection between an emitted sound signal $s(t)$ and the received signal $x(t)$ in a closed-room scenario is given by

$$x(t) = \int_{-\infty}^{\infty} h(\tau)s(t-\tau)\mathrm{d}\tau. \tag{2}$$

The received signal $x(t)$ is a filtered version of $s(t)$ with the room impulse response $h(t)$. The RIR contains information about the acoustic properties of a room and sound propagation in it. The direct sound that reaches the microphone first, early reflections of the walls and diffuse late-field reverberation are building up the RIR. This can be considered by looking at Figure 1 showing a measured RIR. As sound pressure is described by sound fields for a single sound source at a fixed position at time t and receiver position $\boldsymbol{x} = [x, y, z]^T$, the sound pressure field is

$$p(\boldsymbol{x}, t) = \int_{-\infty}^{\infty} h(\boldsymbol{x}, \tau)s(t-\tau)\mathrm{d}\tau. \tag{3}$$

Whereby $h(\boldsymbol{x}, t)$ is the spatially varying RIR from the source location to the point $\boldsymbol{x}$.

2.1 Measurement

To measure RIRs, it is possible to use perfect sequences or sine sweeps. In this work, exponential sine sweeps in the form

$$s(t) = \sin[\theta(t)] = \sin\left[K \cdot \left(e^{\frac{t}{L}} - 1\right)\right] \tag{4}$$

have been used, with

$$K = \frac{T \cdot \omega_1}{ln(\frac{\omega_2}{\omega_1})}, L = \frac{T}{ln(\frac{\omega_2}{\omega_1})}, \tag{5}$$

T being the duration in seconds, ω_1 and ω_2 as the lower and higher bounds of the angular frequency range of the measurement. Considering (2) in the form $x(t) = s(t) * h(t)$ and with the use of the so-called inverse signal $\tilde{s}(t)$, the room impulse response can be calculated in a deconvolution process as

$$h(t) = x(t) * \tilde{s}(t). \tag{6}$$

We can exemplarily see a RIR in Figure 1 at a given position $\boldsymbol{x}$ for a speaker-microphone pair. If the sound propagation shall be analysed in a VOI, it is necessary to have a set of measured RIRs with accurate positioning.

2.2 Interpolation methods

When having sampled a VOI with equidistant sampling points that satisfy the spatial intervals provided by (1), bandlimited interpolation methods can be applied to provide information at any point inside the VOI. A simple approach is trilinear interpolation given by

$$p = \boldsymbol{k}^T \boldsymbol{q} \tag{7}$$

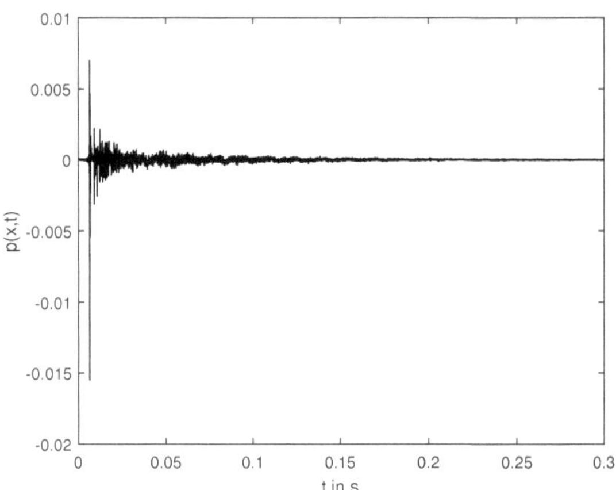

Figure 1: A room impulse response $h(\boldsymbol{x}, t)$ at a given position $\boldsymbol{x}$.

with $\boldsymbol{k}$ being the coefficient vector and $\boldsymbol{q}$ the distance vector. This method interpolates linearly at the desired position based on the direct three-dimensional neighbouring sampling points.

Lagrange interpolation in one dimension can be expressed by

$$P(x) = \sum_{i=1}^{n} f_i l_i(x) \tag{8}$$

where f_i are the supporting points and l_i the Lagrange basis polynomials. Depending on the degree of Lagrange polynomials, this interpolation method can be time consuming, because of the calculation of a high number of polynomials for different supporting points from the neighbourhood.

2.2.1 Interpolation with spherical harmonics

An alternative method for interpolation is based on spherical harmonics and has been described by Rafaely [11]. Due to the characteristics of sound fields, it is possible to measure these at the surface of a sphere and reconstruct functions at positions inside the encircled volume. Proceeding from the homogeneous acoustic wave equation

$$\nabla_r^2 p(\boldsymbol{r}, t) - \frac{1}{c^2}\frac{\delta^2}{\delta t^2}p(\boldsymbol{r}, t) = 0, \tag{9}$$

with $c \approx 343\,\mathrm{m\,s^{-1}}$ as the speed of sound in air, ∇_r^2 denoting the Laplacian in spherical coordinates and $\boldsymbol{r} = [r, \theta, \phi]^T$ as spherical coordinates with the radius r, inclination θ, and azimuth angle ϕ, a function

$$p(k, r', \theta', \phi') \approx \sum_{n=0}^{N} \sum_{m=-n}^{n} \frac{j_n(kr')}{j_n(kr)} p_{nm}(k, r) Y_n^m(\theta', \phi') \tag{10}$$

can be deduced. This function calculates the sound pressure at any position (r', θ', ϕ') inside of a sphere with the knowledge of the sound pressure at the surface of the sphere $p_{nm}(k, r)$, where $k = \frac{\omega}{c}$. The spherical harmonic function of order n and mode m is given by Y_n^m and the Bessel function of first kind is

$$j_n(x) = (-1)^n x^n \left(\frac{1}{x} \frac{d}{dx} \right)^n \frac{\sin(x)}{x}. \qquad (11)$$

As we will see in the results in Section 3.1, the interpolation method with spherical harmonics calculates inaccurate values for specific frequencies which is caused by ill-conditioned problems. Those frequencies can be calculated before interpolation and the ill-conditioned problems occur for very small values of the Bessel functions.

3   Results and Discussion

The setup for the measurement of RIRs inside a VOI consists of a speaker, a microphone mounted on a moveable roboter, a static reference microphone, and a audio interface *RME Fireface 800* that is recording the emitted signal. The measurements took place in a room of the dimensions $6.91\,\mathrm{m} \times 5.23\,\mathrm{m} \times 2.61\,\mathrm{m}$, where the microphone is located at the position $1.08\,\mathrm{m}, 2.50\,\mathrm{m}, 1.5\,\mathrm{m}$ mounted on a rod on top of a *PI H-820 6-Axis Hexapod Microrobot* with a microphone-speaker distance of $2.30\,\mathrm{m}$. Because of the limited travel range of the *H-820 Hexapod* in x- and y-direction of $\pm 50\,\mathrm{mm}$ and $\pm 25\,\mathrm{mm}$ in z-direction, only a small volume could be covered. To achieve meaningful results, a spatial sampling with high density took place. For a larger VOI, the distance between sampling points would need to be reduced, if the total amount of measurements is restricted.

Based on (1) the distance between sampling points needs to be $\Delta < 10.7\,\mathrm{mm}$ with a cutoff frequency $f_c = 16\,\mathrm{kHz}$. First, a cube with edge length $x = y = z = 3.6\,\mathrm{cm}$ and equidistant sampling intervals of $d = 4.0\,\mathrm{mm}$ has been measured. This results in $n = 1000$ sampling points with a covered VOI of $V_c = 46.6\,\mathrm{cm}^3$. To compare the interpolated RIRs with the real RIRs at a given position, measurements at the points where the RIRs are interpolated took place. The interpolation positions are centered in between their threedimensional neighbours, meaning a distance of $\Delta x = \Delta y = \Delta z = 2\,\mathrm{mm}$.

3.1   Comparison of the interpolation methods

To compare the different interpolation methods, the mean energy spectral density of the error (ME-ESD) is defined and calculated as

$$\text{ME-ESD}(f) = \frac{1}{N} \sum_{u=1}^{N} |H_u(f) - \hat{H}_u(f)|^2 \qquad (12)$$

with $H_u(f)$ and $\hat{H}_u(f)$ as the Fourier transforms of the true RIR and the corresponding reconstructed RIR, respectively [5]. Figure 2 shows the associated frequency error of the linear and the Lagrange interpolation method. The Lagrange polynomials are of second order and the number of interpolated RIRs is $N = 9^3 = 729$. It can be seen that for the high resolution sampling both the linear and Lagrange

interpolation methods result in small ME-ESD in the range of $\text{ME-ESD}(f) \leq 2.6 \times 10^{-7}$.

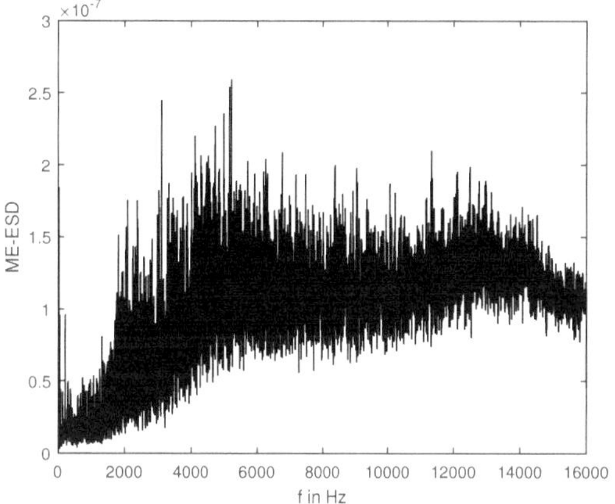

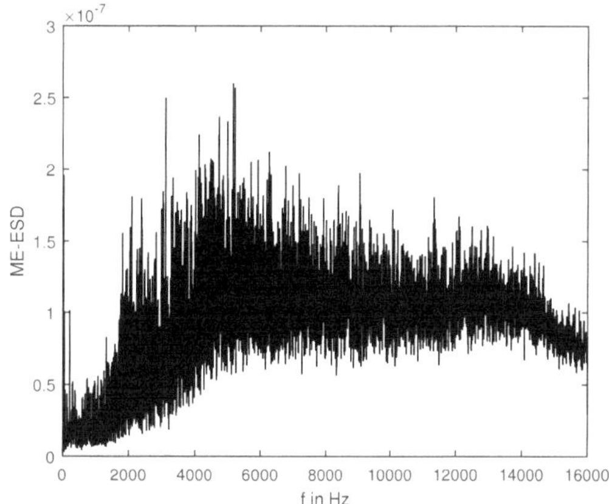

Figure 2: The ME-ESD from linear (top) and Lagrange (bottom) interpolation of $N = 729$ interpolated RIRs.

To apply the method of interpolation with spherical harmonics, a different spatial alignment for the measurements has been applied. Due to the limited travel range, measurements took place at the surface of a sphere with radius $r = 2.5\,\mathrm{mm}$ and an encircled VOI $V_s = 65.45\,\mathrm{cm}^3$ covered, with evenly distributed sampling positions and with the same amount of sampling points $n = 1000$ as for the other interpolation methods. The number of interpolated RIRs is $N = 8^3 = 512$ because the sampling positions near the edges of a cube of the size $3.6\,\mathrm{cm}^3$ are not covered by the sphere. The corresponding frequency error can be seen in Figure 3 (top). It is recognizable, that only specific frequencies have a high ME-ESD. As mentioned in the context of (10), those inaccuracies are well-known, caused by ill-conditioned problems and can be partially compensated by limiting the maximum Bessel order. This has been done for the major ME-ESD at $f = 15.14\,\mathrm{kHz}$ and the result can be seen in Figure 3 (bottom).

Comparing the three interpolation methods based on the ME-ESD to each other, it can be seen that the interpolation method with spherical harmonics shows the least error over all frequencies, except for the specific frequencies where $\text{ME-ESD}(f) \gg 0$ the spectrum of interpolated RIRs are

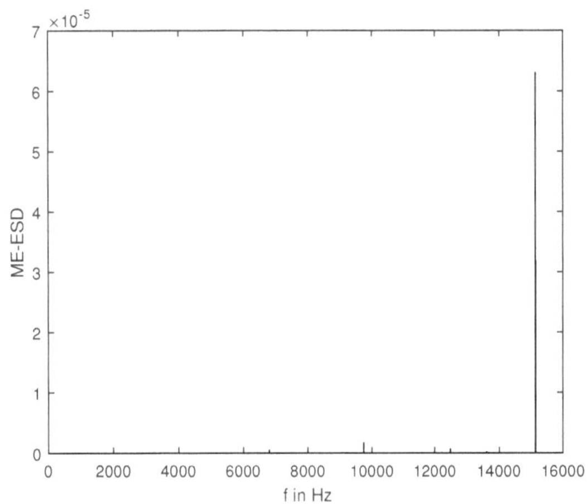

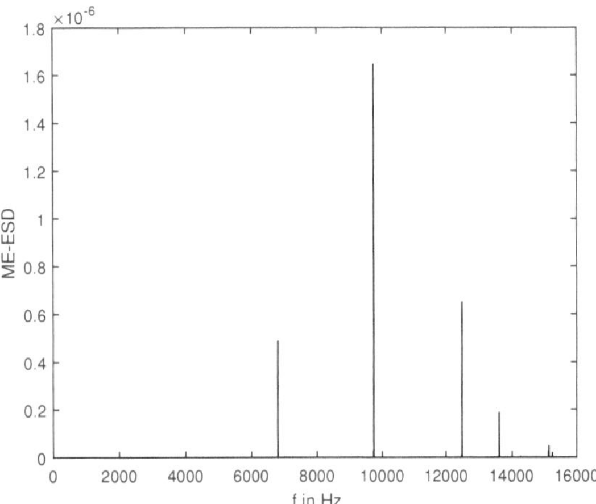

Figure 3: The ME-ESD from interpolation with spherical harmonics (top) and corrected maximum order (bottom) of $N = 512$ interpolated RIRs.

equal to those of the measured RIRs. The trilinear interpolation method shows the highest frequency error for higher frequencies, which was to be expected because it is the simplest method. Both the trilinear and Lagrange interpolation have good interpolation results for lower frequencies what can be explained with the spatial density of sampling points.

4   Conclusion

In this work a setup for the measurement and reconstruction of RIRs inside a VOI has been developed. Within measurement series, well-known interpolation methods such as trilinear and Lagrange interpolation have been applied and a new interpolation approach using spherical harmonics is performed.

The comparison of these interpolation methods has shown that the basic interpolation methods for RIRs like trilinear and Lagrange interpolation have good results at lower frequencies, providing a dense spatial sampling. By the use of interpolation with spherical harmonics at the same number of measurements, a larger VOI can be sampled while achieving more precise results for most of the frequencies.

Acknowledgement

The work has been carried out at the Institute for Signal Processing (ISIP), Universität zu Lübeck.

5   References

[1] M. Miyoshi and Y. Kaneda, *Inverse filtering of room acoustics*. IEEE Trans. Audio, Speech, Lang. Process, vol. 36, no 2, pp. 145-152, Feb. 1988.

[2] S. Tervo, J. Pätynen, A. Kuusinen, and T. Lokki, *Spatial decomposition method for room impulse responses*. J. Audio Eng. Soc., Vol. 61, No. 1/2 Jan./Feb. 2013.

[3] D. de Vries and E. M. Hulsebos, *Auralization of room acoustics by wave field synthesis based on array measurements of impulse responses*. 12th European Signal Processing Conference., Sept. 2004.

[4] T. Ajdler, L. Sbaiz, and M. Vetterli, *Dynamic measurement of room impulse responses using a moving microphone*. J. Acoust. Soc. Am. 122(3) pp. 1636-1645, Sept. 2007.

[5] F. Katzberg, R. Mazur, M. Maass, P. Koch, and A. Mertins, *Sound-field measurement with moving microphones*. J. Acoust. Soc. Am. 141(5) pp. 3220-3235, May 2017.

[6] H.-D. Lüke, *Sequences and arrays with perfect periodic correlation*. IEEE Trans. Aerosp. Electron. Syst., vol.24(3) pp. 287-294, May 1988.

[7] Q. Meng, D. Sen, S. Wang, and L. Hayes, *Impulse response measurement with sine sweeps and amplitude modulation schemes*. 2nd International Conference on Signal Processing and Communication Systems, pp. 1-5, Dec. 2008.

[8] R. Mignot, G. Chardon, and L. Daudet, *Low frequencies interpolation of room impulse responses using compressed sensing*. IEEE/ACM Transactions on Audio, Speech and Language Processing (TASLP), vol, 22(1) pp. 205-216, Jan. 2014.

[9] F. Katzberg, R. Mazur, M. Maass, M. Böhme, and A. Mertins, *Spatial interpolation of room impulse responses using sompressed sensing*. Proc. International Workshop on Acoustic Signal Enhancement (IWAENC), Tokyo, Japan, Sept. 2018.

[10] T. Ajdler, L. Sbaiz, and M. Vetterli, *The plenacoustic function and its sampling*. IEEE Trans. Signal Process., vol.54, no. 10, pp. 3790-3804, Sept. 2006.

[11] B. Rafaely, *Fundamentals of Spherical Array Processing*. Springer Topics in Signal Processing, Vol. 16, Springer, 2019.

Single-Channel EEG - Detection of the Event-Related Potential (ERP) P300 Using an Acoustic Stimuli Synchronization

Katrin Mrotzeck [1], Hugo Gamboa [2], and Alfred Mertins [3]

[1] Biomedical Engineering, Luebeck University of Applied Sciences, katrin.mrotzeck@stud.th-luebeck.de

[2] PLUX wireless biosignals, S.A. , Lisbon Portugal, hgamboa@plux.info

[3] Institute for Signal Processing, Universität zu Lübeck, mertins@isip.uni-luebeck.de

Abstract

We present an approach on how to detect the P300 event-related potential (ERP) by an acoustic stimuli using single-channel Electroencephalography (EEG) with the goal of providing the source code as a tool to get started for easier data acquisition and signal processing for a specific EEG-system. The stimuli were generated according to the oddball paradigm to trigger the event in two test subjects with a total amount of 90 stimuli for each acquisition. For determination of the event, signal processing of the EEG data such as filtering and averaging over time windows, starting at the target stimuli onset, were applied. The synchronization was performed with an accuracy of $\pm 2\,\mathrm{ms}$ and the extraction of the P300 from both subjects was achieved. This enables for example the use of the P300 event with the given device and tools for the use as a feature extraction in brain computer interfaces (BCIs).

1   Introduction

The brain frequencies of an awake adult range from about $4\,\mathrm{Hz}$ to $100\,\mathrm{Hz}$ with theta (4-$7\,\mathrm{Hz}$), alpha (8-$12\,\mathrm{Hz}$), beta (13-$30\,\mathrm{Hz}$) and gamma (30-$100\,\mathrm{Hz}$) waves. Lower frequency waves such as delta (0.5-$3\,\mathrm{Hz}$) appear in sleeping stages. The brain activity stages of an adult are divided into further parts of beta and gamma waves for mental activity such as mathematical calculations (beta) and cognitive and motor functions. Usually electroencephalography (EEG) data is filtered using a bandpass filter with cutoff-frequencies in the range of $0.1\,\mathrm{Hz}$ to $70\,\mathrm{Hz}$ depending on the frequencies of interest as well as a notch filter of $50\,\mathrm{Hz}$ or $60\,\mathrm{Hz}$ conditional on the mains frequency of the location [1],[2],[3]. Time windows, which divide the signal into several segments, are applied over the signal and averaging over several signal segments is done to reduce noise and extract re-occurring signal information such as event-related potentials (ERP) [4].

ERPs can be triggered by either auditory, visual or somatosensory stimuli and can be detected in the EEG signal. One example is the P300, which is a positive deflection at roughly $300\,\mathrm{ms}$ (between $200\,\mathrm{ms}$ and $700\,\mathrm{ms}$) after a target stimuli event [5]. It is elicited by a target stimuli occurring amongst other more frequent stimuli. The oddball paradigm is the method to trigger the P300 as well as other ERPs. In order to properly detect the ERPs, a synchronisation of the EEG and the precise moment of the stimuli are essential. This is needed because some potentials are evoked a few hundred milliseconds after the onset of the stimuli [6]. The detection of event-related potentials such

as the P300 are used as input signals for Brain Computer Interfaces (BCI) for patients with spinal cord injury, amyotrophic lateral sclerosis (ALS) and other neuromuscular disease. It is applied as an assistive technology to provide communication when the level of illness impaires the ability of speaking, as well as for studies of cognitive processes for instance attention and functional memory [6],[4],[7]. Several BCIs have been described in previous studies, for example in [5], where a specific brain region for electrode positioning was found to trigger the highest peak amplitude in the central regions (Cz,Pz), see Fig. 1, of the international 10-20 system for EEG electrode positioning where visual or auditory stimuli were implemented. In addition, studies have shown that the classification accuracy could be maintained by reducing the amount of electrodes [8]. Previous works showed that BCI classifiers such as wavelet based Independent Component Analysis (ICA) can distinguish relatively well between P300 and non-P300 using single trial and single-channel recordings [9].

In this paper the EEG response to an acoustic stimuli is measured using a single-channel EEG sensor on a specific location on the scalp to improve further investigations on BCIs providing a more affordable, faster and easier experimental set-up. The source-code for data acquisition and signal processing will be made available as an open-source tutorial (www.plux.info/57-notebooks).

2   Materials and Methods

For conventional EEG data acquisitions electrodes are placed according to the international 10-20 system to measure the electric potentials from the skin. The potential differences are measured between each electrode and a reference electrode whereas in bipolar recording, such as the given single-channel EEG sensors, the potential difference is measured in between adjacent electrodes [1].

Usually, averaging over a huge amount of trials as well as many electrode channels is crucial in order to detect the ERP from the signals with entailing a high spatial resolution. However, using many electrodes results in adverse effects. The response time of the system is diminished and the large amount of electrodes result in a complex user interface that would render the system impractical for most BCI applications. Thus, reducing the amount of electrode channels to remove redundant features as well as preventing overfitting should be considered for most BCI applications [8],[10]. Hence single-channel EEG is applied in this study to ascertain the possibility of P300 detection with the smallest possible and most user friendly device. Due to noise, the small changes in EEG amplitude can not be seen and vanish in the signal. Averaging over each time window, starting after each known stimuli onset, enables the detection of ERP because the EEG noise and the ERP do not correlate to each other [6].

2.1 Sensors and Electrodes

The single-channel EEG sensors (PLUX biosignalsplux EEG Sensor, www.biosignalsplux.com) with reference SENSPRO-EEG, a gain of 40000, and a bandwidth of 0.8-49 Hz, consist of two measurement sensors as well as one reference sensor of which all are connected to pre-gelled electrodes. The pre-gelled electrodes are then placed on the scalp to measure the potential difference in one specific scalp region with respect to a reference which is placed in a region with low muscular activity.

2.2 Generation of Acoustic Stimuli

The acoustic stimuli was generated according to the oddball paradigm with two different sine waves with frequencies of 330 Hz for the regular and 440 Hz for the target stimuli. A sampling frequency of 1000 Hz was applied for optimal acquisition and synchronization using OpenSignals Software. A total amount of 90 acoustic stimuli was generated with 77 regular and 13 target stimuli for each acquisition.

2.3 Experimental Setup

Both electrodes of the two bipolar sensors were positioned vertically on the center line at Cz/Pz position according to the 10-20 system as illustrated in Fig. 1, to achieve the maximum potential occurrence [9]. The electrode of the ground sensor was positioned behind the ear to eliminate the possibility of artefacts. The electrodes were fixed with a headband to assure an optimal alignment on the scalp. The test subjects were both healthy male with ages of 23 and 46 years. Experimental runs were performed

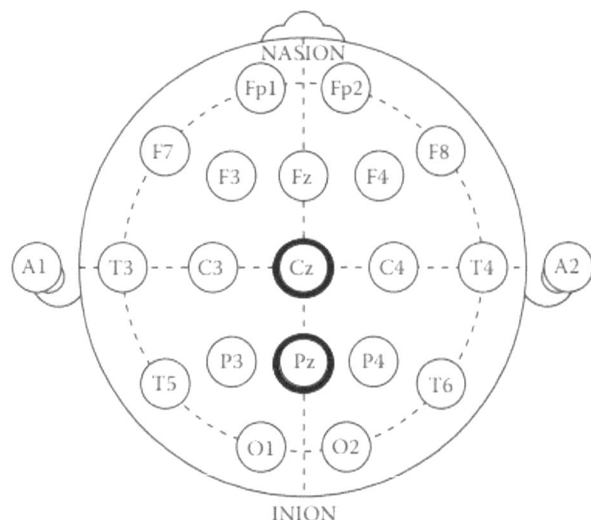

Figure 1: Electrode configuration for ERP detection, acquisition by a single-channel EEG with two electrodes positioned at the central positions of Cz and Pz (framed) located according to the 10-20 system. Image from [3].

where a single trial of both subjects with highest signal to noise ratio (SNR) was used for data evaluation. For synchronisation of EEG and acoustic stimuli, both signals were connected to the device hub using a connecting cable for both sound output and headphone input and recorded throughout the experiment using OpenSignals software (www.biosignalsplux.com/index.php/software). The acoustic stimuli was played through the headphones once the test subject was in a relaxed state with eyes closed to prevent movement artefacts.

2.4 Signal Processing of the Acquired Data

2.4.1 Filtering Methods

The acquired sound stimuli signal was filtered using a low-pass filter with a cutoff frequency of 440 Hz. Then the shift from the baseline was removed and the absolute value of the filtered signal applied. A stimuli vector was created to store the precise onset of the stimuli with a precision of ± 2 ms (visual inspection). The acquired EEG signal was filtered with a bandpass filter with a lower cutoff frequency of 0.1 Hz (f1) and an upper cutoff frequency of 40 Hz (f2) to reduce noise, the shift from the baseline was removed and the absolute value was applied. Windowing was applied with a size of 1000 ms after each onset of stimuli and the average of time windows extracted to detect the ERP P300 and remove unwanted noise in the signal.

2.4.2 Synchronisation of EEG and Stimuli

The synchronisation of both the acquired EEG signal and the acoustic signal was performed using a threshold method to detect the onset of each stimuli window containing sine waves of different frequencies. The index of either target or non-target stimuli onset was stored in a stimuli vector and the given time indices applied to the EEG signal for windowing.

2.4.3   Coherent Averaging

For P300 ERP detection, the bandpass filtered EEG data was further processed using specific time windows. The mean value was applied over all time windows to eliminate noise from the signal and detect the evoked potential, as in (1). The average signal response A is described by

$$\forall i \in [\,1,k\,]\,, A_i = \frac{1}{n}\sum_{j=1}^{n} a_i^j.\qquad(1)$$

Where a denotes the element of the given time window, i and n the number of time windows and k the number of elements in a window.

3   Results and Discussion

3.1   Synchronisation of EEG and Stimuli

The synchronisation was successfully generated using specific thresholds to detect the first index of either target or non-target stimuli and generating a stimuli vector to store the index values. A precision of $\pm 2\,$ms for the detection of stimuli onset was observed by visual inspection. This was achieved by overlapping the stimuli vector with the acquired stimuli signal. The stimuli vector was then used for ERP P300 detection as window indices for averaging.

3.2   EEG Response for Target and Non-Target Stimuli

Fig. 2 illustrates the stimulus response (EEG signal) in micro volt (μV) over time in milliseconds (ms) after the stimuli onset at a time of $0\,$ms for subject 1 (upper) and subject 2 (lower). The solid line represents the response to the target stimuli (odd stimuli) and the dashed line represents the response to the non-target stimuli (regular stimuli). The vertical dotted line represents the index of peak amplitude and peak latency after stimuli onset for both test subject in the time window of $200\,$ms to $700\,$ms. As shown in Fig. 2, a different EEG response appears after target and non-target stimuli. In the time window of $200\,$ms to $700\,$ms, after the target stimuli onset, the peak amplitude for subject 1 reaches a value of $4.97\,\mu$V and for subject 2 a value of $2.95\,\mu$V, as presented in Table 1. The peak latency of subject 1 and subject 2 reaches a value of $302\,$ms and $295\,$ms, respectively. Both non-target EEG responses indicate values of $-1.04\,\mu$V and $1.14\,\mu$V at time indices of peak amplitude of the target response, see Table 1.

The results illustrate that the ERP P300 was detected, providing that the given open-source tools for data acquisition and signal processing can be applied for further investigations on P300 detection. Hence for evaluation of data acquisitions when applying the given tools, a larger amount of subjects must be considered.

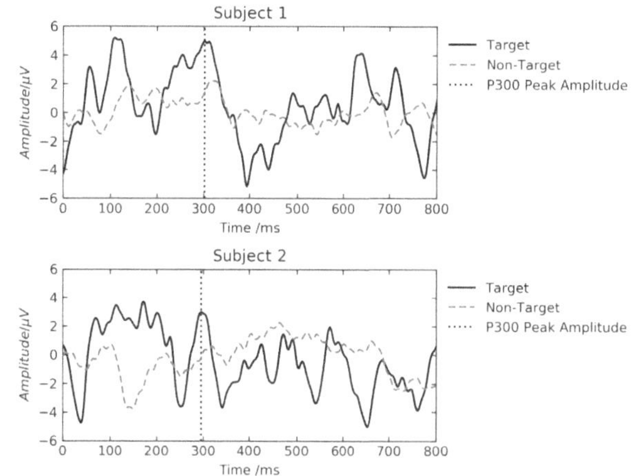

Figure 2: Averaged EEG Response Signal in μV over time in ms of Target and Non-Target Acoustic Stimuli after Stimuli onset at time $0\,$ms for Subject 1 (upper) and Subject 2 (lower) and Peak Amplitude of Target Response of the time window from $200\,$ms to $700\,$ms.

For sharper P300 detection, the following aspects should be considered. An impact on the averaged peak size may be caused by an existent delay jitter due to physiological causes, inducing a different phase of the ERPs and consequently a smaller average. Splitting the signal into its frequency bands could overcome this problem as described in [11]. Also the synchronization accuracy of $\pm 2\,$ms influences the precision for P300 detection because it affects the overlapping of time windows at the correct onset of stimuli that has to be very precise for the correct averaging of the signal. Moreover, sharper P300 peaks should be detected by applying a larger quantity of stimuli for averaging over a greater number of time windows. The signal-to-noise ratio (SNR) is the measure for this and is described as follows. The SNR decreases as a function of the square root of the amount of trials performed with $R/\sqrt{N}$, where R and N denote the amount of noise in a single trial and the number of trials, respectively [3]. Hence, making the feature extraction more time intensive, where one needs to decide between accuracy and speed. Compared to multi-channel EEG system, single-channel EEG ERP detection is more feasible when it comes to usability and cost. Though for real time BCIs, the coherent averaging of multiple events that can be extracted from a number of electrodes in multi-channel EEG systems and in single-channel EEG when increasing the number of stimuli, decreases the speed in the latter case but increases the spatial resolution.

Studies have been performed where P300 was detected based on EEG Shape Features such as in [6], which should be evaluated in further studies. As other papers show, the P300 detection is much preciser when using a minimum amount of electrodes of 8 [8],[10]. Where precision does not change much after adding more electrodes, with an error rate of around 0.6-0.7 for single-channel and below 0.2

Table 1: Amplitude and Latency of Averaged Response at Peak Amplitude/Latency of target stimuli response for time window 200 ms to 700 ms after stimuli onset.

Subject	Stimuli	Amplitude [μV]	Latency [ms]	Position
1	Target	4.97	302	Cz/Pz
	No-Target	-1.04	302	Cz/Pz
2	Target	2.95	295	Cz/Pz
	No-Target	1.14	295	Cz/Pz

for more than 8 channels [8]. Therefore for one-channel electrodes a high error rate must be considered when these are used in a BCI. In this case, it must be taken into account the extend of error in real-time application such as health and risk for patients.

4 Conclusion

As shown in Fig. 2, it was possible to trigger and detect the P300 ERP potential using an acoustic stimuli and single-channel EEG sensors. The synchronization of the acoustic stimuli with the EEG signal was successfully executed to define the correct indices with an accuracy of ± 2 ms, using specific thresholds and a stimuli vector. The results show peak amplitudes and peak latencies of both test subjects in the time window of 200 ms to 700 ms after the stimuli onset with values in the range of about 3-5 μV and approximately 300 ms when applying the sensors in the positions of Pz/Oz of the 10-20 system for EEG acquisitions.

An open-source tutorial has been developed providing the tools to get started for data acquisition and signal processing for ERP detection when using the specific EEG system. Due to the low amount of test subjects, this study only contributes to the creation of the open-source tutorial. When utilizing the P300 detection results for further BCI studies, the experiments must be repeated with a larger amount of test subjects. Further investigations must be implemented to make the system compatible for precise and fast communication BCI applications, such as averaging over a larger amount of data to improve precision or applying shape features to improve speed. Hence, it needs to be decided whether precision or speed is more important in specific cases. With the goal of a real-time BCI for patients suffering from communication loss, a fast alternative must be found and considered in future studies.

Acknowledgement

The work has been carried out at PLUX wireless biosignals S.A. in Lisbon, Portugal and supervised by the Institute for Signal Processing, University of Lübeck.

5 References

[1] M. Libenson, *Practical Approach to Electroencephalography E-Book*. Elsevier Health Sciences, 2012.

[2] A. Hassanien and A. Azar, *Brain-Computer Interfaces: Current Trends and Applications*. Intelligent Systems Reference Library, Springer International Publishing, 2014.

[3] K. Nidal and A. S. Malik, eds., *EEG/ERP Analysis*. Taylor & Francis Ltd., 2014.

[4] E. Donchin, K. M. Spencer, and R. Wijesinghe, "The mental prosthesis: assessing the speed of a p300-based brain-computer interface," *IEEE transactions on rehabilitation engineering*, vol. 8, no. 2, pp. 174–179, 2000.

[5] A. Furdea, S. Halder, D. Krusienski, D. Bross, F. Nijboer, N. Birbaumer, and A. Kübler, "An auditory oddball (p300) spelling system for brain-computer interfaces," *Psychophysiology*, vol. 46, no. 3, pp. 617–625, 2009.

[6] M. Alvarado-González, E. Garduño, E. Bribiesca, O. Yáñez-Suárez, and V. Medina-Bañuelos, "P300 detection based on eeg shape features," *Computational and mathematical methods in medicine*, vol. 2016, 2016.

[7] J. N. Mak, D. J. McFarland, T. M. Vaughan, L. M. McCane, P. Z. Tsui, D. J. Zeitlin, E. W. Sellers, and J. R. Wolpaw, "Eeg correlates of p300-based brain–computer interface (bci) performance in people with amyotrophic lateral sclerosis," *Journal of neural engineering*, vol. 9, no. 2, p. 026014, 2012.

[8] C.-Y. Kee, S. Ponnambalam, and C.-K. Loo, "Multiobjective genetic algorithm as channel selection method for p300 and motor imagery data set," *Neurocomputing*, vol. 161, pp. 120–131, 2015.

[9] N. Haghighatpanah, R. Amirfattahi, V. Abootalebi, and B. Nazari, "A single channel-single trial p300 detection algorithm," in *2013 21st Iranian Conference on Electrical Engineering (ICEE)*, pp. 1–5, IEEE, 2013.

[10] H. Cecotti, B. Rivet, M. Congedo, C. Jutten, O. Bertrand, E. Maby, and J. Mattout, "A robust sensor-selection method for p300 brain–computer interfaces," *Journal of neural engineering*, vol. 8, no. 1, p. 016001, 2011.

[11] W. Giroldini, L. Pederzoli, M. Bilucaglia, S. Melloni, and P. Tressoldi, "A new method to detect event-related potentials based on pearson's correlation," *EURASIP Journal on Bioinformatics and Systems Biology*, vol. 2016, no. 1, p. 11, 2016.

Removal of ECG artifacts caused by mechanical CPR using externally recorded compression markers

Lukas Boudnik [1], Johannes Hochreiter [2] Gregor Müllegger [2] and Mattias Heinrich [3]

[1] Medical Informatics, Universität zu Lübeck, lukas.boudnik@student.uni-luebeck.de
[2] GS Elektromedizinische Geräte G. Stemple GmbH, Kaufering, {johannes.hochreiter, gregor.muellegger}@corpuls.com
[3] Institute of Medical Informatics, Universität zu Lübeck, heinrich@imi.uni-luebeck.de

Abstract

Cardiopulmonary resuscitation (CPR) produces interfering artifacts in ECG signals. Shock advisory algorithms used in automated external defibrillators are unable to determine a shock decision with chest compressions performed in parallel. The CPR must be stopped to examine the patient's heart rhythm. But CPR interruptions are negative correlated with the patient's outcome. Artifact reduction filters promise to minimize these interruptions. In this paper, artifacts caused by mechanical thorax compressions are investigated. An artificial dataset was generated using ECGs recorded on swine. A Least-Mean-Square (LMS) was tested in its ability to distinguish ventricular fibrillation from artificially added artifacts. The results indicate that the LMS is able to improve the signal quality. The average signal-to-noise ratio (SNR) improved over 10 dB in all tests. Yet the dataset is limited in its clinical value and needs further improvement.

1 Introduction

Resuscitation of patients with a sudden cardiac arrest (SCA) are challenging situations for emergency medical services. The German Resuscitation Registry (GRR) reported that 69 out-of-hospital cardiac arrests (OHCA) per 100.000 citizens and year were treated in 2018 [1].

Two of the most important factors affecting the patient outcome are the cardiopulmonary resuscitation (CPR) and defibrillation. To increase the chances of a return of spontaneous circulation (ROSC) the European Resuscitation Council emphasized the impact of early defibrillation and high-quality CPR in 2015 [2]. The guidelines demand a chest compression depth of at least 5cm at a rate of 100-120 compressions per minute. The "hands-off" time should be reduced to a minimum. Patients with ventricular fibrillation (VF) should receive an early defibrillation, but an artifact free ECG interval is needed to analyze the rhythm with a shock advisory algorithm. Normally the chest compressions are paused for a few seconds while the automated external defibrillator (AED) examines the ECG and charges for the following shock.

Aramendi et al. [3] showed that CPR artifacts can be removed from an ECG. Effective removal of artifacts would allow an ECG analysis with ongoing chest compressions, resulting in reduced "hands-off" time [4]. The reduction of pre-shock-pauses would increase the chances of survival [5] and patients with a nonshockable rhythm would benefit from an increased chest compression fraction [6].

The American Heart Association (AHA) recommended minimal performance requirements, arrhythmia analysis algorithms should achieve [7]. The sensitivity must be 90% or higher for shockable rhythms and the specificity must be above 95% for nonshockable rhythms. A shock advisory algorithm in combination with a CPR artifact filter should therefore fulfill these requirements to provide a benefit for patients. Especially a lower specificity is critical, as it leads to erroneous defibrillations.

In the last twenty years different artifact filtering methods were presented. In [8] and [9] the existing methods are compared and reviewed. Independently of each other, both authors came to the conclusion that current methods are promising. Since the requirements of the AHA are not met, the solutions are not yet suitable for use on patients. In [10] ECG signals were first filtered and then a shock decision was made by clinical experts. Also, in this study the specificity was below the recommended 95%.

2 Material and Methods

In order to gain a better understanding of the challenges and problems of CPR artifact filtering, an ECG dataset was created for this paper. As a first approach to separate compression artifacts from the ECG signal, a Least-Mean-Square (LMS) filter is described.

2.1 Dataset

A controlled setting, in which the distinction between artifacts and ECG is clear, is generated by using an artificial mixture approach. Therefore, VF signals and CPR artifacts were recorded independently and added at a desired signal-to-noise ratio (SNR). This model assumes that the CPR artifact is an additive noise [11].

The VF signals s_{VF} were collected from 4 swine with a corpuls3. The sampling rate is 400 Hz and the defibrillator pads were used to record the ECG. The signals were filtered with an order-four Butterworth bandpass filter (0.3–35 Hz) to remove baseline wander and high frequency noise. Overall, 286 VF samples with a length of 15 s were collected.

The CPR signals s_{CPR} were recorded from 21 swine. For all animals, the underlying heart rhythm was asystole. The same recording infrastructure and preprocessing settings were used. After filtering, each segment was normalized to unit variance. The CPR was performed with a corpuls cpr, a mechanical thorax compression device. The compression rate was constant at 100 min^{-1} and the compression depth was set to 6 cm. Manual registration of the ECG and the corpuls CPR signal was needed to match the compression markers with artifacts in the ECG. The compression markers were used to generate the reference signal for the LMS, see chapter 2.2. To increase the sample size and test different starting points, the segments were extracted with an overlap of 7.5 seconds. In total the dataset contains 224 compression artifact segments.

Following the description in [11], the input signal s_{in} for the LMS is computed as

$$s_{in} = s_{VF} + C * s_{CPR}. \tag{1}$$

The constant C is a scaling factor to ensure a fixed SNR in dB over the whole dataset and is determined as

$$C = \sqrt{\frac{\sigma^2_{s_{VF}}}{10^{\frac{SNR}{10}}}}. \tag{2}$$

The factor C must be calculated for each test sample s_{in} individually.

2.2 Method

ECG artifacts caused by a mechanical thorax compression device have a consistant frequency. In Fig. 1 the power spectral density (PSD) of CPR artifacts recorded in an animal with asystole is shown. The harmonics of the fundamental frequency (1.666 Hz) are prominent in the frequency domain. CPR performed by humans show a wider spread in the main frequency and are more difficult to model [3].

In [12], an often used LMS artifacts removal filter is described. The underlying idea is to model the artifacts as a periodic signal through its Fourier series representation.

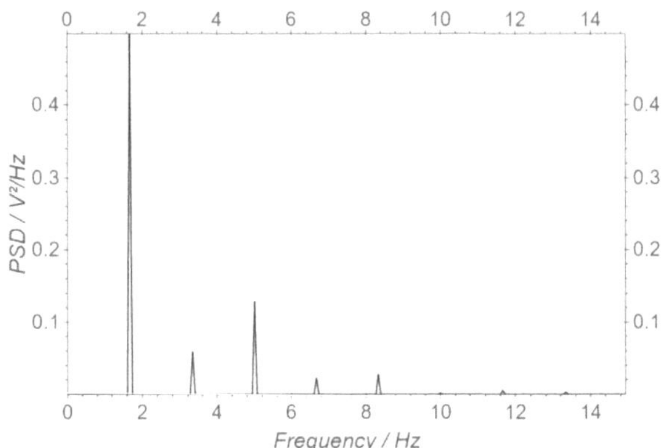

Figure 1: PSD of CPR artifacts with mechanical compression. The segment was recorded during asystole. The main spectral components of the compressions are present at the compression rate and its harmonics. The compression rate is constant at 1.666 Hz during the segment.

The recorded compression instances i are used to compute the phase of the compression at every time point n

$$\phi(n) = \frac{2\pi}{\Delta n_i}(n - n_i) + i2\pi, \quad n_i \leq n < n_{i+1} \tag{3}$$

with a distance between two following compressions of

$$\Delta n_i = n_{i+1} - n_i. \tag{4}$$

The estimated CPR artifact $\hat{s}_{CPR}$ is modeled as a Fourier series representation, using N harmonics

$$\hat{s}_{CPR}(n) = \sum_{k=1}^{N} a_k(n)cos(k\phi(n)) + b_k(n)sin(k\phi(n)). \tag{5}$$

The filter weights of the time varying in-phase, $a_k(n)$, and quadrature, $b_k(n)$, amplitudes are initalized to zero and updated by the LMS filter. The estimated ECG $\hat{s}_{ECG}$ is calculated by

$$\hat{s}_{ECG}(n) = s_{in}(n) - \hat{s}_{CPR}(n) \tag{6}$$

and is used as the error signal in the update process of the coefficients. Every harmonic k has its own update step size μ_k. To reduce the free parameter, the step sizes are chosen as in [12], with

$$\mu_k = \frac{1}{k} * \mu_0, \quad k = 1, \dots, N. \tag{7}$$

The only parameters which must be set manually by the user are μ_0 and N.

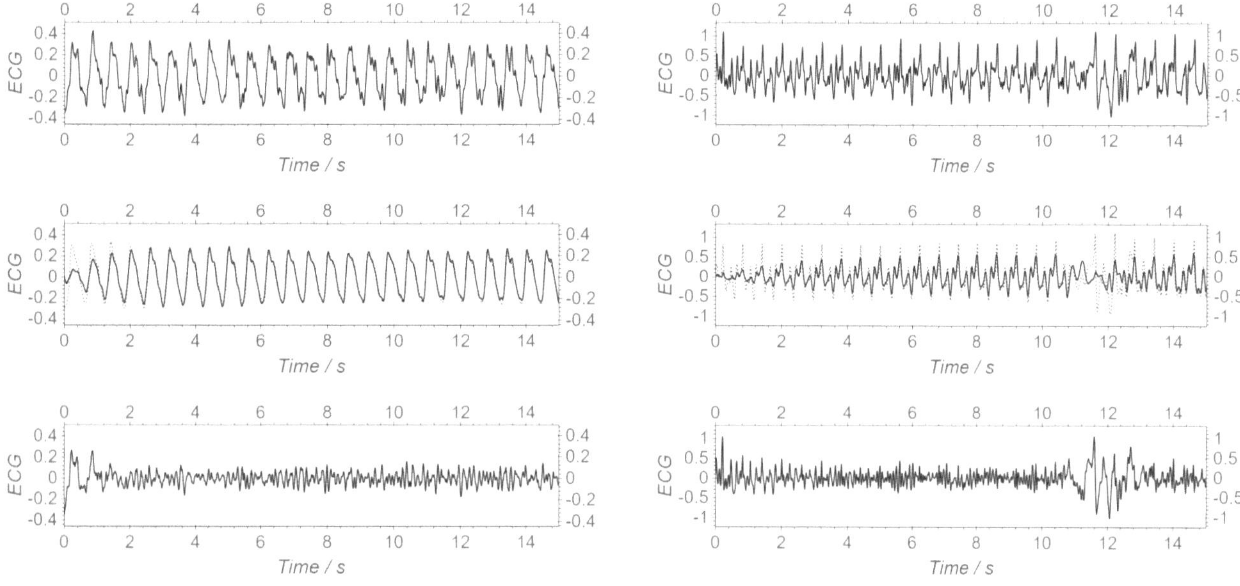

Figure 2: Two examples of filtering with $N = 4$ and $\mu_0 = 0.0025$. In each top row the input signal s_{in} is shown. In the center the estimated artifact $\hat{s}_{cpr}$ (solid curve) and the input artifact s_{cpr} (dotted curve) are shown. The estimated ECG $\hat{s}_{ecg}$ is in the bottom row. The adaption phase is present in the positive example (left). After two seconds the LMS method can effectively filter the artifacts. On the right side a negatively example is displayed. The LMS cannot model the artifact correctly. After 11 seconds one compression is skipped, resulting in a large artifact.

3 Results and Discussion

Two experiments were designed to test the LMS filter on the dataset described in chapter 2.1. First, the optimal parameter for a fixed SNR was searched. In the second experiment the most promising LMS is applied on data samples with variable SNR values. SNR improvement is the metric for both tests.

3.1 Parameter optimisation

To estimate optimal parameter N and μ_0, all 224 artifact signals were added with a random VF signal using the formulas (1) and (2). The SNR was set to 0 dB. For N, values between 3 and 5 were tested, because a larger number would increase the computational complexity.

In Fig. 3 the average improvements are displayed. The best result (17 dB) is obtained when N is 4 and $\mu_0 = 0.0025$. When only three harmonics are used, the best average restored SNR is 1 dB lower. Using more than four harmonics does not improve the results. The best working range of the LMS filter as of values of μ_0 is between 0.002 and 0.003. A positive and a negative filtering example are displayed in Fig. 2. The LMS needs less than 2 s to converge, but when one compression is skipped the LMS produces a large artifact. These pauses are produced after 100 hundred compressions to reevaluate the position of the mechanical compression device.

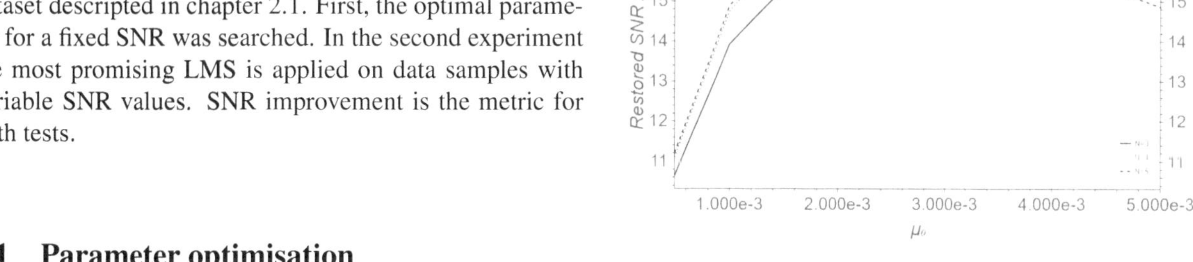

Figure 3: Performance of the filter removal in terms of the adjustable perameters N and μ_0. The highest average restored SNR in dB is achieved with $N = 4$ and $\mu_0 = 0.0025$. While there is no significant difference between four and five harmonics, the LMS with only three harmonics is not able to achieve the same performance.

3.2 Artifact variability

In the second experiment, the ability to model artifacts with different amplitudes was examined. The highest scoring LMS from experiment one was applied on input signals with SNRs between -10 and +10 dB. In Fig. 4 the results are displayed. Regardless of the SNR in the input signal, a noise reduction of at least 10 dB is achieved. The more present the artifacts are, the worse the filtering is. The peak of the filter performance is at +5 dB. At +10 dB the filter performance decreases. But those signals do not need strong filtering, because the artifact amplitudes are small.

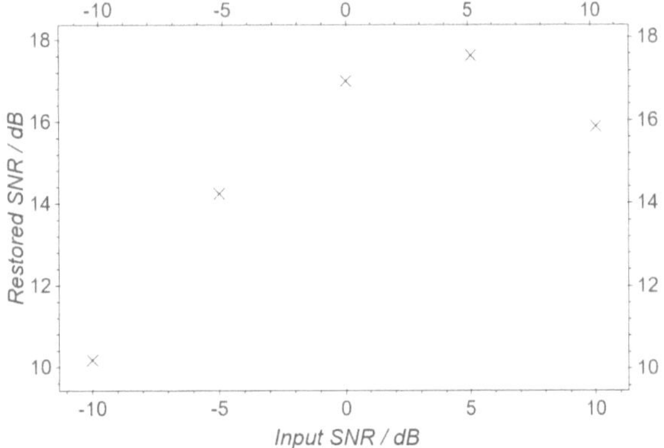

Figure 4: Performance of the LMS with parameters $N = 4$ and $\mu_0 = 0.0025$ for different level of SNRs. The lowest noise reduction is achieved for signals with large artifacts. The best artifact removal is achieved at input $SNR = 5$ dB. If the artifact amplitude is too small the performance decreases.

4 Conclusion

It was shown that only four harmonics are needed to model CPR artifacts caused by a mechanical thorax compression device. Especially for mobile devices like an AED, the computing power is limited. A filtering solution may only demand little computing power. The LMS fulfills this requirement and was able to improve the signal quality on all tested signals. Using the recorded compression markers by the compression device reduces complexity, because no additional compression detection algorithm is needed. Readjustment pauses could be overcome with communication between the CPR machine and the patient monitor.

The work is currently being continued. The focus will be on building a database of real CPR missions and modeling compression pauses. An improved database is needed to better assess the clinical significance of CPR filtering.

Acknowledgement

The work has been carried out at GS Elektromedizinische Geräte G. Stemple GmbH, Kaufering and supervised by the Institute of Medical Informatics, Universität zu Lübeck.

5 References

[1] J. Wnent et al., "Jahresbericht des deutschen reanimationsregisters außerklinische reanimation 2018," *Anästh Intensivmed*, vol. 60, pp. V91–V93, 2019.

[2] K. G. Monsieurs et al., "European resuscitation council guidelines for resuscitation 2015: Section 1. executive summary," *Resuscitation*, vol. 95, pp. 1–80, 2015.

[3] E. Aramendi, U. Irusta, U. Ayala, H. Naas, J. Kramer-Johansen, and T. Eftestøl, "Filtering mechanical chest compression artefacts from out-of-hospital cardiac arrest data," *Resuscitation*, vol. 98, pp. 41–47, 2016.

[4] R. Partridge, Q. Tan, A. Silver, M. Riley, F. Geheb, and R. Raymond, "Rhythm analysis and charging during chest compressions reduces compression pause time," *Resuscitation*, vol. 90, pp. 133–137, 2015.

[5] S. Cheskes et al., "The impact of peri-shock pause on survival from out-of-hospital shockable cardiac arrest during the resuscitation outcomes consortium primed trial," *Resuscitation*, vol. 85, no. 3, pp. 336–342, 2014.

[6] C. Vaillancourt et al., "The impact of increased chest compression fraction on return of spontaneous circulation for out-of-hospital cardiac arrest patients not in ventricular fibrillation," *Resuscitation*, vol. 82, no. 12, pp. 1501–1507, 2011.

[7] R. E. Kerber et al., "Automatic external defibrillators for public access defibrillation: Recommendations for specifying and reporting arrhythmia analysis algorithm performance, incorporating new waveforms, and enhancing safety. a statement for health professionals from the american heart association task force on automatic external defibrillation, subcommittee on aed safety and efficacy," *Circulation*, vol. 95, no. 6, pp. 1677–1682, 1997.

[8] S. Ruiz de Gauna, U. Irusta, J. Ruiz, U. Ayala, E. Aramendi, and T. Eftestøl, "Rhythm analysis during cardiopulmonary resuscitation: Past, present, and future," *BioMed research international*, vol. 2014, p. 386010, 2014.

[9] R. Affatato, Y. Li, and G. Ristagno, "See through ecg technology during cardiopulmonary resuscitation to analyze rhythm and predict defibrillation outcome," *Current opinion in critical care*, vol. 22, no. 3, pp. 199–205, 2016.

[10] E. Alonso et al., "Evaluation of chest compression artefact removal based on rhythm assessments made by clinicians," *Resuscitation*, vol. 125, pp. 104–110, 2018.

[11] S. O. Aase, T. Eftestøl, J. H. Husøy, K. Sunde, and P. A. Steen, "Cpr artifact removal from human ecg using optimal multichannel filtering," *IEEE transactions on bio-medical engineering*, vol. 47, no. 11, pp. 1440–1449, 2000.

[12] U. Irusta, J. Ruiz, S. R. de Gauna, T. Eftestøl, and J. Kramer-Johansen, "A least mean-square filter for the estimation of the cardiopulmonary resuscitation artifact based on the frequency of the compressions," *IEEE transactions on bio-medical engineering*, vol. 56, no. 4, pp. 1052–1062, 2009.

Interpolation of individual HRTF datasets using principal component analysis

Andreas Biel [1], Fabrice Katzberg [2] and Alfred Mertins [2]

[1] Auditory Technology, Universität zu Lübeck, andreas.biel@student.uni-luebeck.de
[2] Institute for Signal Processing, Universität zu Lübeck, {katzberg, mertins}@isip.uni-luebeck.de

Abstract

Binaural synthesis of audio signals can create a realistic 3D audio environment for headphone applications. For this an accurate dataset of head related transfer functions (HRTFs) is needed, which needs to be measured with a great expense of equipment and time. This paper focuses on a method to reduce the necessary work by interpolating partially measured HRTFs using principal component analysis (PCA). This statistical tool aims at the reduction of the dimensionality while maintaining the characteristics of the dataset. The interpolation of partially measured HRTF datasets is computed by estimating weight vectors and using the orthogonal bases of the PCA to calculate the missing directions of the dataset. The accuracy of this method has been determined and an amount of principal components for data interpolation is recommended. Phase information was reconstructed using minimum phase of the data and a model for interaural time difference calculation. A brief listening test indicated plausibility of data.

1 Introduction

In recent years the accurate reproduction of 3D-sound has gained more prominence in both professional and non-professional applications. Especially in virtual reality scenarios with a stereoscopic head-mounted display a realistic soundscape is of great importance to support the general impression of an authentic environment.

Since these devices are most commonly paired with headphones for sound reproduction, binaural signals are used. These can either be directly recorded by dummy heads or artificially created by convoluting the desired signal with suitable head-related transfer functions (HRTFs).

A HRTF dataset consists of a large number of transfer functions that describe the acoustical transmission path from a point source to the ear canal [1].

The measurement is usually carried out as pairwise impulse response measurement from multiple directions at a fixed distance, since the head-related impulse responses (HRIR) for both ears are required to synthesize a binaural signal for one combination of azimuth (horizontal angle), elevation (vertical angle) and distance. To be able to generate virtual 3D-sound from nearly all directions, a fine grid of sampled angles is needed, which leads to a complicated measurement setup and a lot of precise work for every individual HRTF dataset measurement.

The measurement setup is usually placed in an anechoic chamber and consists of an arc of small equalized loudspeakers that play back a certain calibration sound like a sinusoidal sweep. Through deconvolution of the calibration sound and the recorded signals from microphones placed inside the ear canals, the desired impulse responses can be calculated. Additional to this elaborate equipment, the HRTF measurement for one subject can take multiple hours and requires the subject to be stationary, since even small head movements can disturb the measurement by altering the incident angle of the sound.

This paper focusses on the possibility to simplify the process of HRTF measurement through only measuring a few directions of head-related transfer functions and interpolating the missing data with principal component analysis of existing HRTF databases.

2 Material and Methods

2.1 HRTF Database

The used database was taken from the Acoustics Research Institute (ARI) of the Austrian Academy of Sciences (ÖAW), whose HRTF measurements are part of a large HRTF repository of worldwide collected and standardized measurements. The data are available in the SOFA file format (Spatially Oriented Format for Acoustics), that has been provided by the Audio Engineering Society (AES).

A total amount of 97 individual datasets was used, each of which contains pairwise HRIR data of 1550 measured source positions. Since this paper only focusses on interpolation of data from the horizontal plane (elevation $0°$), only the 90 measurements of different azimuth angles in this plane were considered.

Each impulse response has a length of 256 points at a sampling frequency of 48000 Hz.

2.2   Principal Component Analysis

"Principal components [sic] analysis is a statistical procedure that attempts to provide an efficient representation of a set of correlated measures. The central idea of PCA is to reduce the dimensionality of a data set in which there are a large number of interrelated measures, while retaining as much as possible of the variation present in the data." [2]

In this process an orthogonal transformation is utilized to create a new n-dimensional Cartesian coordinate system whose axes are called principal components (PC). The first PC explains the greatest part of variance of the database, all following PCs are chosen as the linear independent vector from the previous components that explains the greatest portion of remaining variance.

Inside these new axes every value of the original data corresponds to a score in the new orthogonal basis.

The database was preprocessed similarly as in [3].

To apply the PCA to the HRTF database, the logarithmic magnitude responses of all IRs were calculated. Since the magnitude response contains redundant data, all 256 point magnitude responses were reduced to 129 points to retain all information but the phase. Phase information will be reconstructed after interpolation. 80 datasets were used for analysis, the remaining 17 randomly selected sets were used for evaluation. As every dataset contains data for two ears, the angles of the right ear were reversed and considered as a new set, resulting in a total of 160 datasets for PCA.

The datasets are stored in column vectors $\mathbf{h}_i$, that are defined as

$$\mathbf{h}_i = [\mathbf{h}_{i1}\mathbf{h}_{i2}...\mathbf{h}_{i90}]^T \qquad (1)$$

with $\mathbf{h}_{ij}$ being the log magnitude response of the jth direction of the ith dataset.

All vectors $\mathbf{h}_i$ are merged into the matrix

$$\mathbf{H} = [\mathbf{h}_1\mathbf{h}_2...\mathbf{h}_{160}]. \qquad (2)$$

Furthermore, $\mathbf{h}_i$ can also be written as

$$\mathbf{h}_i = \mathbf{C}\omega_i + \bar{\mathbf{h}}, \qquad (3)$$

where $\mathbf{C}$ is the principal component matrix, ω_i is the score or weight vector and $\bar{\mathbf{h}}$ is the mean of all 160 dataset vectors $\mathbf{h}_i$. While $\mathbf{C}$ can be directly calculated from the covariance matrix of $\mathbf{H}$ as it consists of its eigenvectors, the weight vector ω_i can be obtained by solving (3).

2.3   Interpolation of partially measured HRTF Datasets

Assuming that $\mathbf{C}$ and $\bar{\mathbf{h}}$ are valid also for similarly measured HRTFs, the measured HRTF $\mathbf{g}_k$ can be described as

$$\mathbf{g}_k = \mathbf{C}\omega_k + \bar{\mathbf{h}}. \qquad (4)$$

This means that if $\mathbf{g}_k$ is only partially measured, it can be fully recreated, if a good estimation of ω_k can be calculated. If $\tilde{\mathbf{g}}_k$ is a partially measured dataset with measurements out of the n measured directions $[d_1, d_2, ...d_n]$, $\tilde{\mathbf{g}}_k$ can be written as

$$\tilde{\mathbf{g}}_k = [\mathbf{g}_{kd_1}\mathbf{g}_{kd_2}\cdots\mathbf{g}_{kd_n}]^T. \qquad (5)$$

Similarly an incomplete matrix $\tilde{\mathbf{C}}$, only consisting out of the columns correlating to the measured directions, is defined as

$$\tilde{\mathbf{C}} = [\mathbf{C}_{d_1}\mathbf{C}_{d_2}...\mathbf{C}_{d_n}]^T. \qquad (6)$$

A good approximation for ω_k can be computed in the least-squares sense from (4), resulting in

$$\omega_k = \tilde{\mathbf{C}}^+(\tilde{\mathbf{g}}_k - \tilde{\mathbf{h}}) \qquad (7)$$

with $\tilde{\mathbf{C}}^+$ being the pseudo inverse matrix of $\tilde{\mathbf{C}}$ and $\tilde{\mathbf{h}}$ being the mean vector of the measured directions $[d_1, d_2, ...d_n]$ only. With this estimation of ω_k and (4) the log magnitude responses of the partially measured HRTF dataset can be interpolated.

If a specific amount of m PCs should be used for the interpolation, only the first m columns of $\mathbf{C}$ or $\tilde{\mathbf{C}}$ and only the first m rows of ω_k are considered in the equations.

2.4   Phase Reconstruction

During the PCA and interpolation all phase information is lost. But since it can be shown that HRTFs may be considered as minimum phase system [4], phase information can be reconstructed by splitting the calculated transfer functions into minimum-phase and all-pass systems, retaining the minimum phase and delaying it according to a model for interaural time difference (ITD) [2].

A simple way to model the ITD is the spherical head model by Woodworth [5], which can be discribed with the formula

$$\text{ITD} = \frac{a}{c}(\theta + \sin\theta), \qquad (8)$$

where a is the head radius, c is the speed of sound and θ is the azimuth angle.

More complex head models like an ellipsoidal head model with three adjustable head dimensions and ear position offset have shown more accuracy [6], but since this is not the focus for this paper and those detailed biometrical head data are missing in the used database, the spherical model is considered in the following.

3   Results and Discussion

3.1   Optimal Parameters for Interpolation

The main parameters of PCA data interpolation that can be changed are the amount of measured directions and the number of principal components used for reconstruction. To evaluate which combination of these parameters yields the best results in relation to the necessary work load, the average standard deviation of difference of real and calculated HRTF dataset vectors $\mathbf{h}_i$ and $\mathbf{g}_i$ was computed as measure of accuracy. This process was performed for the 17 randomly selected datasets as evaluation group as well as for the 80 datasets that were used for PCA. Additionally the measure of accuracy of a mean HRTF was calculated as a reference point.

The amount of measured directions was varied from 1, 2, 3, 4, 6, 8, 9, 12, 18, 24, 36 up to 72. The frontal direction ($0°$) was always included, the remaining directions were divided equally.

As expected, the average standard deviation (SD) of the recreated PCA data is highly dependent on the amount of PCs used for recreation (Fig. 1). On the other hand the amount of measured directions for interpolation plays a minor role, since the average SD is not significantly lowered for more than two measured directions.

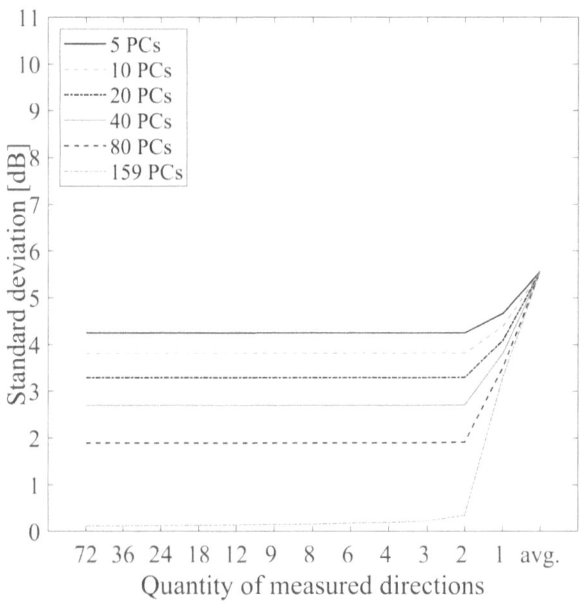

Figure 1: Average accuracy of the calculated datasets for different combinations of measured directions and utilized principal components. Displayed are the analysis data that were used for the PCA.

When analysing the interpolation of the datasets excluded from PCA, different effects can be observed. While the amount of PCs is beneficial for the dataset recreation when many directions were measured, this relation is inverted for few measured direction (Fig. 2). Using many PCs for interpolation with few measured directions leads to a big distortion and worse results than using an average HRTF dataset. This is an indicator that the orthogonal basis determined in the PCA is not true for the evaluation data. To improve the universal validity of this basis a significantly larger database is required.

But still the interpolation with at least three measured directions and around 10-20 used PCs lead to a better approximation than just using an average dataset. The author recommends to use 10 PCs when measuring three directions ($0°, 120°, 240°$) and 20 PCs when measuring six ($0°, 60°, 120°, 180°, 240°, 300°$) or eight directions ($0°, 45°, 90°, 135°, 180°, 225°, 270°, 315°$).

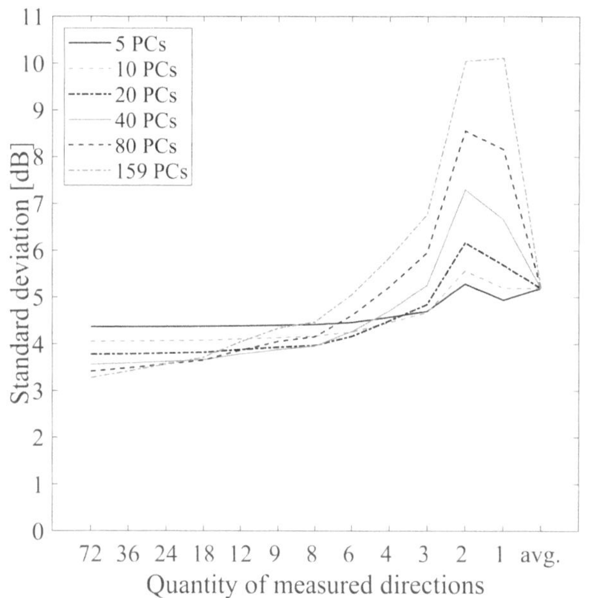

Figure 2: Average accuracy of the calculated datasets for different combinations of measured directions and utilized principal components. Displayed are the evaluation data that had not previously been used to perform the PCA.

3.2 Subjective Verification of Phase Reconstruction

To verify the acoustical plausibility of the interpolated datasets, the minimum phase of the inverse Fourier transform of the computed magnitude responses were considered and delayed by the ITD calculated with (8). An exemplary plot of the ITD for a head diameter of 18cm is displayed in Fig. 3.

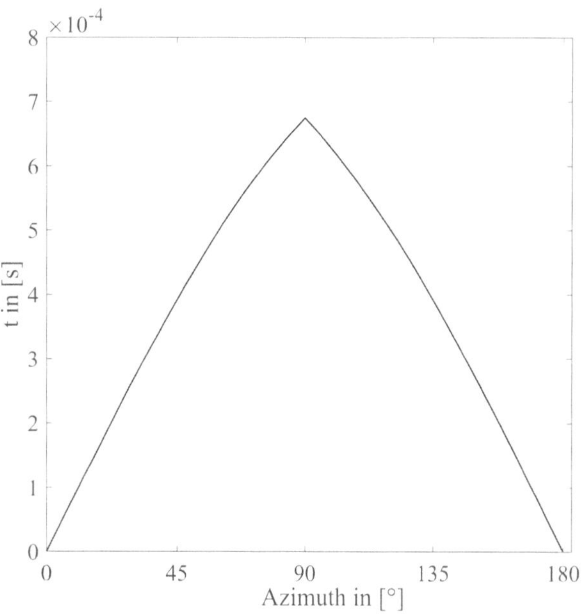

Figure 3: Interaural time difference according to a spherical head model with 18cm diameter for $0°$ elevation.

Four test subjects listened to multiple different combinations of measurement directions and used PCs and should tell their perceived direction of sound. Although the overall perception was good, front/back errors occured and three of the four subjects reported a perceived elevation, since none of the interpolated HRTFs matched the individual HRTFs of the test subjects. This was the case for all combinations. For future works an in-depth listening test with many different scenarios together with a more accurate ITD model is stongly advised. But due to a limited schedule the focus of this project and this paper was on the PCA.

4   Conclusion

A method for interpolation of HRTF datasets using PCA was tested and evaluated on a publicly available database. It was verified that this method leads to a better accuracy and personalization than using an average HRTF dataset. A certain amount of used PCs for interpolation was recommended depending on the amount of measured directions (three, six, eight). The phase information of the interpolated magnitude responses was reconstructed using the minimum phase and a time delay calculated by a simple spherical ITD model. Furthermore it has been hypothesised that the orthogonal bases from the PCA are not universal, but would benefit from a larger database.

Acknowledgement

The work has been carried out and supervised at Institute for Signal Processing, Universität zu Lübeck.

5   References

[1] C. I. Cheng and G. H. Wakefield, "Introduction to head-related transfer functions (hrtfs): Representations of hrtfs in time, frequency, and space," in *Audio Engineering Society Convention 107*.   Audio Engineering Society, 1999.

[2] D. J. Kistler and F. L. Wightman, "A model of head-related transfer functions based on principal components analysis and minimum-phase reconstruction," *The Journal of the Acoustical Society of America*, vol. 91, no. 3, pp. 1637–1647, 1992.

[3] K. Matsui and A. Ando, "Estimation of individualized head-related transfer function based on principal component analysis," *Acoustical science and technology*, vol. 30, no. 5, pp. 338–347, 2009.

[4] S. Mehrgardt and V. Mellert, "Transformation characteristics of the external human ear," *The Journal of the Acoustical Society of America*, vol. 61, no. 6, pp. 1567–1576, 1977.

[5] R. S. Woodworth, B. Barber, and H. Schlosberg, *Experimental psychology*.   Oxford and IBH Publishing, 1954.

[6] R. O. Duda, C. Avendano, and V. R. Algazi, "An adaptable ellipsoidal head model for the interaural time difference," in *1999 IEEE International Conference on Acoustics, Speech, and Signal Processing. Proceedings. ICASSP99 (Cat. No. 99CH36258)*, vol. 2.   IEEE, 1999, pp. 965–968.

Optimisation of support vector regression for the determination of haemoglobin derivatives in whole blood

Saidurga Karthikeyan[1,2,3], Benjamin Redmer[3,4], Bodo Nestler[3] Stefan Müller[3]
[1]Biomedical Engineering, Universität zu Lübeck, Germany, saidurga.karthikeyan@student.uni-luebeck.de
[2]Biomedical Engineering, Lübeck University of Applied Sciences, Germany, saidurga.karthikeyan@stud.th-luebeck.de
[3]Medical Sensors and Devices Laboratory, Department of Applied Natural Sciences, Lübeck University of Applied Sciences, Germany
[4]Graduate School for Computing in Medicine and Life Sciences, University of Lübeck, Germany

Abstract

The support vector machine is a powerful algorithm for medical applications. In this paper, it is used for regression analysis to determine haemoglobin derivatives in unaltered whole blood samples in the spectral range from 450 to 700 nm. Dysfunctional haemoglobin derivatives like carboxyhaemoglobin and methaemoglobin do not participate in vital oxygen transport and lead to life-threatening diseases. Optimisation is essential for accurate clinical diagnostics and makes the algorithm robust for implementing in small optical sensors. Various aspects of support vector regression were studied for optimisation and the largely influencing aspects - the hyperparameters and data preprocessing are taken into account for implementation. The python machine learning framework scikit-learn was used. The model performance obtained after hyperparameter optimisation is significantly better leading to a decrease in prediction error by a factor of 3 to 18. However, dimensionality reduction results in a relatively poor performance since the input feature space has a stable ratio. The model performance is evaluated using root mean square error, mean absolute error and coefficient of determination.

1 Introduction

Machine learning is a subset of artificial intelligence that deals with the study and implementation of algorithms and statistics. It enables computers to perform tasks without any explicit instructions. One such algorithm is the support vector machine (SVM). SVMs are used for classification and regression purposes out of which only the regression part is in the scope of this paper. The support vector regression (SVR) algorithm has been successfully used in automobile industries [1], face recognition [2], financial forecast [3] and web technologies [4] to name a few.

In this paper, it was applied to the field of medical technology to determine the haemoglobin derivatives in unaltered whole blood samples in the spectral range from 450 to 700 nm [5]. Haemoglobin is a protein molecule that is responsible for oxygen transportation in human body to various organs by effectively binding oxygen to its iron atoms. Unlike its physiological variations like reduced haemoglobin (HHb) and oxyhaemoglobin (O_2Hb), dysfunctional variations like methaemoglobin (MetHb) and carboxyhaemoglobin (COHb) can no longer participate in the vital oxygen transportation. This leads to life-threatening blood-related disorders. Hence, determination of these values is very important for clinical diagnostics and thereby for effective treatments. In addition to these values, the hematocrit is also an important parameter as it influences the optical behaviour of the blood sample. All

the above mentioned parameters are considered for determination by the machine learning model. Specific to our application, a powerful algorithm is needed to account to the complex scattering effects that occur during the optical analysis of whole blood [6]. In this case of whole blood analysis, there is no prior destruction of the red blood cells (RBCs) when compared to the existing techniques that usually perform haemolysis prior to the analysis.

The main objective of this paper is the optimisation of the SVR models to render a more accurate determination of the haemoglobin derivatives which is an important aspect to be considered for medical applications that have negligible tolerance. Technically, SVR optimisation aims at reducing the generalisation error which provides an equilibrium between the bias and variance of the model. This can be done in several ways. Various aspects of SVR that were considered to fine tune the model are loss function, kernels, quadratic programming (QP) solver, hyperparameters and data preprocessing. Out of these, hyperparameter tuning exhibited a greater influence on the model performance. Hyperparameters are the variables that determine the fit of the model. Methods like GridSearchCV were employed to highlight the right combination of the hyperparameters from a given practical set. A study on dimensionality reduction techniques was done and one of the techniques, called kernel principal component analysis (k-PCA) was applied to the feature space. The number of principal components was varied and its dependence on the model performance studied.

2 Material and Methods

Sample measurements were done in the Medical Sensors and Devices Laboratory (MSGT) at Lübeck University of Applied Sciences, Germany. The measurement set-up consists of a laser-driven light source that irradiates the blood sample. The blood sample flows continuously through a flow cell of $100\,\mu m$ optical path length which is placed between two integrating spheres. Three grating spectrometers were used to record the diffuse reflection, diffuse transmission, unscattered and total transmission spectra. In this paper, only the total transmission spectra was used. As it contains both unscattered and scattered components of transmission, no additional components were needed to separate them. This enhances its application in small optical sensors. Whole blood samples are taken from healthy blood donors. Samples of about 3 ml are taken in test tubes and centrifuged thrice. The plasma centrifuged was substituted by isotonic phosphate buffer solution (PBS) with a set osmotic concentration. By this, the hematocrit level and osmotic concentration was set by retaining some of the plasma solution and adding the PBS buffer. The pH of PBS is 7.4. After this, variation of MetHb concentration was done by adding sodium nitrate after retaining a specific volume of the buffer or plasma from the sample at regular intervals that accounts for the time-dependent oxidation of the central iron atom of haemoglobin. Subsequently, the fractional concentrations of O_2Hb, HHb and COHb were varied by flushing with O_2, CO_2 and CO in the tonometer respectively. The process was validated and monitored by a OSM 3 Haemoximeter (Radiometer, Denmark). These values were noted down along with the total transmission spectral values from the measurement set-up which were calculated by dividing the number of detected photons when the cell is filled with the sample to the number of detected photons when the cell is not filled with the sample. Detailed description of the sample preparation can be found in [5]. These total transmission spectral values were taken as the feature space for the model and the values from the oximeter as the output values to be predicted. This data was fed into the Python machine learning framework Scikit-learn 0.22.1 and fitted using the SVR estimator from the SVM library in the framework [7]. The program was written in Python 3.7.

3 Support Vector Regression

The data was organised in two matrices. The first matrix is the input matrix which carries the feature space that consists of the total transmission spectral values where the rows correspond to the samples and the columns correspond to the wavelength range from 450 to 700 nm. The second matrix is the target matrix which carries the corresponding parameter values which are determined by the OSM 3. So, there are 358 samples each with 426 features in the input matrix and 6 different parameter values in the target matrix. SVR poses to be a fitting estimator because of its kernel capability that maps non linear feature space to a high dimension where the regression is performed [8]. Radial basis function (RBF), also known as Gaussian kernel, is used in the

estimator with the three hyperparameters - ϵ, C and γ. ϵ directly impacts the fit of the model which acts like a tube to fit the support vectors. A small value of ϵ overfits the model as it fits not just the support vectors but almost all the points in the feature whereas a higher value underfits the model leaving out few support vectors. This affects the bias and variance of the model. C is the regularization parameter which provides a trade off between them. γ is a similarity measure between the sample points even when they are distributed far from each other [9].

The data is split into train (70%) and test (30%) sets. Data preprocessing is a method to scale and size the data prior to fitting. The RobustScaler method from the preprocessing library of scikit-learn is used for this purpose. The train data is fitted and transformed followed by transforming the test data before prediction. After this step, model fitting is done on train set and predictions are made on the test set. GridsearchCV method using five-fold cross validation with neg_mean_squared_error is employed to highlight the best parameter from a wide range of practical values of ϵ, C and γ. Each parameter is separately fit with its own best parameters and prediction is made on test set. Evaluating the model is the step to determine the performance of the model. The root mean squared error (RMSE), coefficient of determination (R^2) and mean absolute error (MAE) are used to scale the efficacy of the model. R^2 is a statistical measure to express how close the data are to the fitted regression line. It always has a value between $-\infty$ to 1. However, the model performance cannot be determined solely with this parameter because there could be cases where a good model can have a low R^2 and a poor model could have a high R^2 value. R^2 just expresses the adequacy of the model to perform the regression task here.

Mathematically, it is calculated as follows:

$$R^2(y, \hat{y}) = 1 - \frac{\sum_{i=1}^{n}(y_i - \hat{y}_i)^2}{\sum_{i=1}^{n}(y_i - \bar{y})^2} \tag{1}$$

where $\hat{y}_i$ is the predicted value of the i-th sample and y_i the corresponding reference value. $\bar{y}$ is the average of all reference values for total n samples.

MAE is defined as the average of all the errors over the test set. The error is calculated by taking the absolute difference between the predicted and the observed values. In case of MAE, all the errors are assigned equal weights. This is calculated as

$$MAE(y, \hat{y}) = \frac{1}{n} \sum_{i=0}^{n-1} |y_i - \hat{y}_i| \tag{2}$$

RMSE is similar to MAE. In the case of RMSE, the error is squared and the square root of the average is taken. Unlike MAE, RMSE calculation assigns different weights to its error vaues where a large weight is assigned to a large and undesirable error. Hence, this metric is used here for evaluation because large errors are undesirable. It is calculated as follows:

$$RMSE(y, \hat{y}) = \sqrt{\frac{1}{n} \sum_{i=0}^{n-1} (y_i - \hat{y}_i)^2}. \tag{3}$$

Both MAE and RMSE are negatively-oriented scores, meaning lower values are better.

4 Results and Discussion

The results after optimisation are illustrated in this section in the form of plots and tables. The first section deals with hyperparameter optimisation and the second section deals with dimensionality reduction.

4.1 Parameter Optimisation

In this section, the contour plots are shown that illustrates the effectiveness of hyperparameter optimisation. As we can see from the figures 1 to 6, the RMSE decreases towards the best parameter combination. A smaller value of RMSE means that the model is well optimised and performs the regression task effectively.

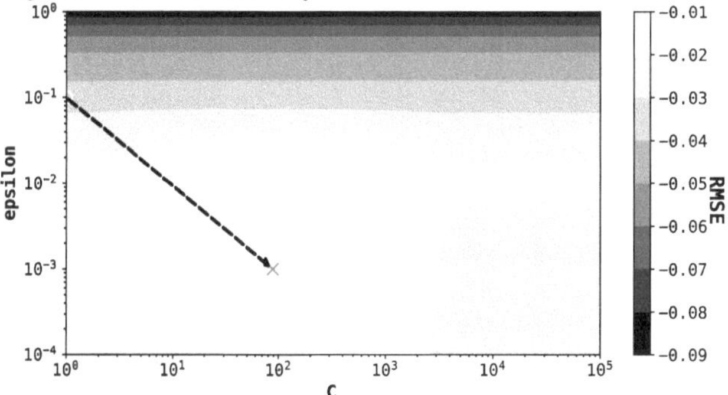

Figure 1: Hyperparameter optimisation for Haematocrit

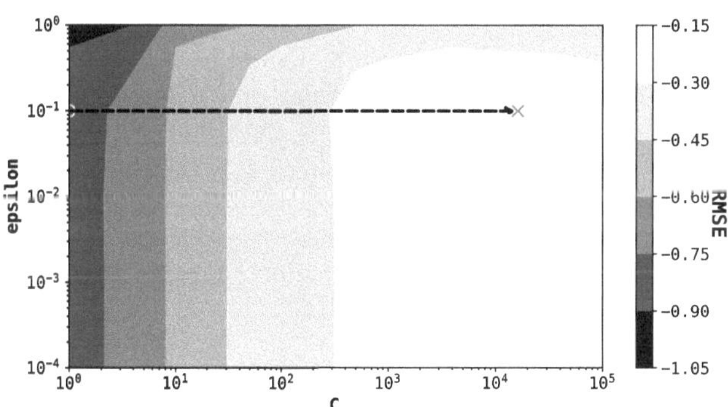

Figure 2: Hyperparameter optimisation for tHb

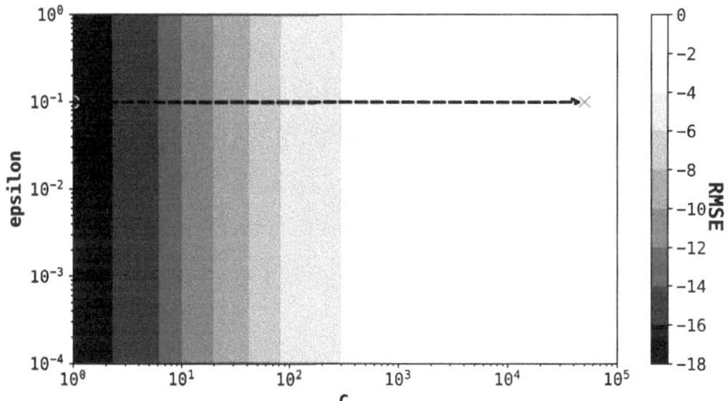

Figure 3: Hyperparameter optimisation for O_2Hb

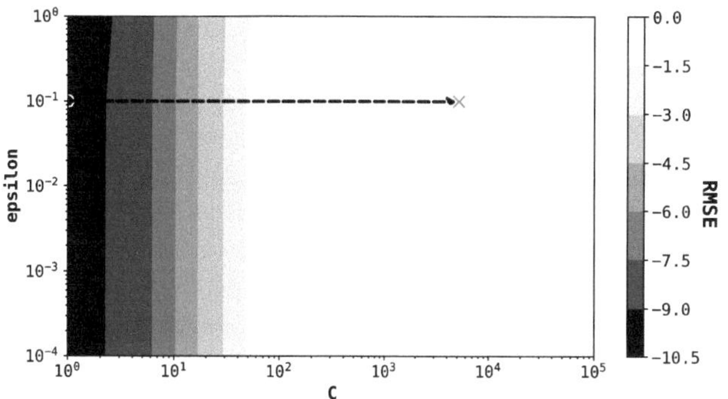

Figure 4: Hyperparameter optimisation for HHb

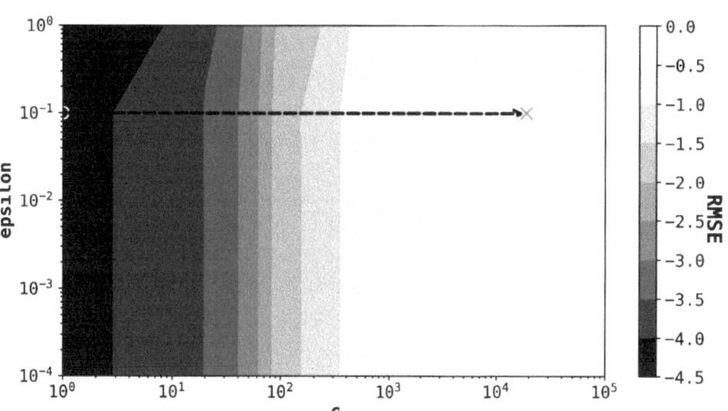

Figure 5: Hyperparameter optimisation for COHb

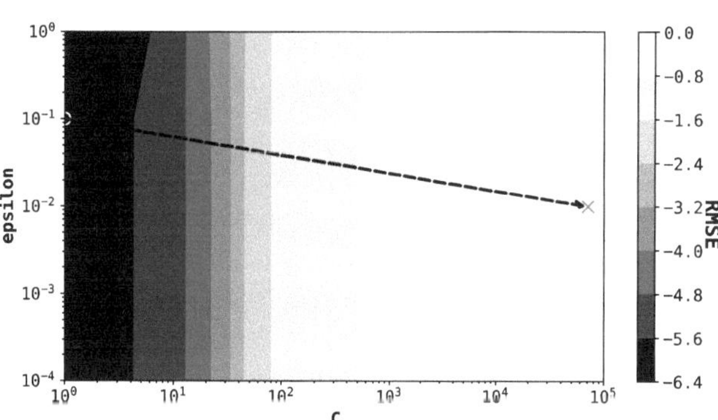

Figure 6: Hyperparameter optimisation for MetHb

In all the figures 1 to 6, the position of the gray circle represents the default values of C and ϵ as given in the SVR method which is 1 and 0.1 respectively. The cross represents the C and ϵ values that is suggested after optimisation by GridsearchCV. The decreasing shades of gray represents decreasing RMSE and the arrow depicts how optimisation reduces the RMSE in each of the cases. It has reduced the error values by a factor of 3 to 18.

4.2 Dimensionality Reduction

Dimensionality reduction is the process of selecting or extracting valuable features from a large set of features. There are different types of non-linear dimensionality reduction techniques out of which k-PCA is applied in the scope of this paper. The method KernelPCA from the library 'decomposition' is used for this purpose. This results in poor evaluation metrics when only five components are chosen but as the number of components is increased, the results get better. By default, when the number of components are not specified, the method transforms 426 features into 249 features. The table 1 illustrates the result for COHb. The case is the same for all the other parameters considered. Table 1 reflects the dependence of the model performance on the number of principal components used for the k-PCA technique. It also shows that the model performance without k-PCA is better. This could mean that the initial 358 x 426 feature space considered with the spectral range from 450 to 700 nm is already sufficient to produce the optimum result. In the table 1, the number of components are chosen at a reasonable interval in such a way that the improvement in the metrics can be seen legibly. When the n_components is not specified in the method, it keeps all the non-zero components in the feature space (here, 249).

Table 1: Comparison of metrics with k-PCA and its dependance on the number of components for COHb

Number of components	R^2	RMSE	MAE
5	-0.0465	7.7190	3.2427
10	0.9709	1.2864	0.9607
25	0.9845	0.9381	0.6369
50	0.9846	0.9351	0.6316
100	0.9855	0.9098	0.6172
249 (Default)	0.9854	0.9118	0.6157
426 (Without k-PCA)	0.9852	0.9177	0.4775

However, the performance remains at an acceptable range even when the feature space is reduced (eg., when number of components is equal to 50 or 100). Hence, it can be inferred from the above result that this model would remain effective even when the spectral resolution is low in case of small optical sensors. An important factor to be considered in k-PCA is to change the range of hyperparameters given to GridsearchCV.

5   Conclusion

On studying the various aspects of SVR, we derived that hyperparameters of SVR enhances the model performance and its implementation in the algorithm is feasible. Practically too, the results shown reflect the efficacy of using the right combination of the hyperparameters in SVR where the RMSE decreased by a factor of 3 to 18. Further, k-PCA which is one of the dimensionality reduction techniques was applied to transform the feature space. This type of data preprocessing is said to optimise the model [10] but in this case, it did not. The reason inferred was that the input feature space was already stabilized by shortening the spectral range from 450 to 700 nm. This could however be advantageous while installing this model in a small and cost-effective optical sensor used in blood gas analysers where the spectral ranges of the light source used could be low.

Acknowledgement

The work has been carried out at MSGT in Lübeck University of Applied Sciences, Germany. The blood bags were provided by Institut für Transfusionsmedizin at Universitätsklinikum Schleswig-Holstein (UKSH), Lübeck, Germany. This research work was keenly guided by Benjamin Redmer[3,4] who also performed the blood sample preparation along with Philip Schargus, student at MSGT.

6   References

[1] P. Zhu, F. Pan, W. Chen, and S. Zhang, "Use of support vector regression in structural optimization: Application to vehicle crashworthiness design," *Mathematics and Computers in Simulation*, vol. 86, pp. 21–31, dec 2012.

[2] P. J. Phillips, "Support vector machines applied to face recognition," *Advances in Neural Information Processing Systems*, vol. 11, Dec. 2001.

[3] D. Gupta, M. Pratama, Z. Ma, J. Li, and M. Prasad, "Financial time series forecasting using twin support vector regression," *PLOS ONE*, vol. 14, no. 3, p. e0211402, mar 2019.

[4] A. Corazza, S. D. Martino, F. Ferrucci, C. Gravino, and E. Mendes, "Applying support vector regression for web effort estimation using a cross-company dataset," oct 2009.

[5] B. Redmer, P. Schargus, S. Karthikeyan, B. Nestler, and S. Mueller, "Determination of hemoglobin derivatives in unaltered whole blood samples using support vector regression in the spectral range from 450 to 700 nm" in the SPIE proceedings 2020, to be published."

[6] A. Roggan, M. Friebel, K. Dorschel, A. Hahn, and G. Muller, "Optical properties of circulating human blood in the wavelength range 400–2500 nm," *Journal of Biomedical Optics*, vol. 4, no. 1, p. 36, 1999.

[7] F. Pedregosa et al., "Scikit-learn: Machine learning in Python," *Journal of Machine Learning Research*, vol. 12, pp. 2825–2830, 2011.

[8] A. J. Smola and B. Schölkopf, "A tutorial on support vector regression," *Statistics and Computing*, vol. 14, no. 3, pp. 199–222, aug 2004.

[9] V. N. Vapnik, *The Nature of Statistical Learning Theory*. Springer-Verlag New York, 2000. [Online]. Available: https://www.ebook.de/de/product/2082300/v_n_vapnik_the_nature_of_statistical_learning_theory.html

[10] B. Schölkopf, A. Smola, and K.-R. Müller, "Kernel principal component analysis," pp. 583–588, 1997.

Aortic Detection in Image Sequences of Electrical Impedance Tomography

Sarah Herrmann [1], Dominic Schneider [2], Julia Mrongowius [2] and Mattias Heinrich [3]

[1] Medical Informatics, Universität zu Lübeck, sarah.herrmann@student.uni-luebeck.de

[2] Innovation Center Computer Assisted Surgery, Universität Leipzig, {julia.mrongowius,dominic.schneider}@medizin.uni-leipzig.de

[3] Institute of Medical Informatics, Universität zu Lübeck, heinrich@imi.uni-luebeck.de

Abstract

Algorithms on Electrical Impedance Tomography (EIT) images are numerous, but a limited spatial resolution covers information, that has not been transformed into clinical practice for now. A reliable anatomical benchmark like the aorta is one step of improvement for the localization of unventilated parts of the lung in thoracic EIT images. We propose an aortic detection algorithm based on a method presented by Krukewitt et al. in 2019 [1] not relying on an aortic catheter or saline bolus injection. The algorithm considers timing and shape of an aortic pulse and was tested on data from fifteen pigs in apnea and under ventilation. Manual segmentations and a new saline bolus aortic detection algorithm using Pulse Contour Cardiac Output (PICCO) standards, are used for comparison. The paper presents a robust result of the aortic position at apnea and ventilation and a reliable tool for pursuing analysis of the cardiovascular system by EIT.

1   Introduction

EIT   Electrical Impedance Tomography (EIT) is an imaging method invented in the 1980s by Barber and Brown, being a promising new method for monitoring lung ventilation [2]. The big advantages of non-invasive, radiation-free imaging and the small and portable machine are leading to a cheap and fast diagnostic support. For an EIT measurement, electrodes are placed on the body surface around the chest in a transversal layer [3]. Small alternating currents are conducted between two electrodes while at all other electrodes the resulting voltages are measured. As different tissues have a different specific conductivity (reverse impedance) on alternating currents depending on the fluid content, an impedance image can be reconstructed from the voltages. In functional EIT, performed in this paper, only relative values are regarded, influenced by movements of air and blood during time. One of the problems with EIT is a limited spatial resolution. In horizontal as well as in vertical direction, where the electrodes are sensitive to approximately half the chest width [3]; so there always will be effects from surrounding layers in tomographical EIT images. This leads to non-intuitive interpretation for humans and to the need of improved algorithms to analyse the hidden information.

Medical Background   Monitoring of ventilation is a relevant application for patients with chronic lung disease or acute lung injury. It is also important for mechanical ventilation, because volumes and pressures not fitting the patient, may lead to ventilator-associated lung injury [3]. The dorsal parts of the lung are prone to potential impacts, such as atelectasis. In EIT images those injured and therefore unventilated dorsal parts can not be separated from tissue outside the lung border easily; they often stay unnoticed though they are important to interpret lung functionality. Hence, a constant landmark is needed to have a better spatial orientation. The aorta is located between the right and left lung, near the spine and little ventral from the dorsal lung border; a suitably fixed anatomical position. A main criterion to identify the aorta is the pulse arrival time (PAT), the time a pulse takes from its release in the heart until its arrival in a specific vessel. The pulse reaches the aorta before it reaches the lungs and right after the impedance in the heart had its minimum. A second anatomical fact we are using is the special shape of the aortic pulse. In systole the increasing blood pressure stretches the elastic vascular walls of the aortic arch and the big arteries. After the aortic valve has closed with a small decrease of pressure, in diastole the pressure decays exponentially because the vascular walls contract [4]. This phenomenon, known as Windkessel effect, leads to a double peak for the pressure curve and an slow, special shaped decrease in volume, that we try to detect.

Related Work   This paper relates to a publication by Krukewitt et al., introducing an solely EIT-based algorithm for aortic detection that does not rely on an aortic catheter or the injection of a saline bolus as prior publications did [1]. This method was modified and tested on EIT data from fifteen pigs in apnea and under ventilation. As a reference manual segmentations from computed tomography (CT) images were used as well as a new aortic detection algorithm for saline bolus injection using Pulse Contour Cardiac Output (PICCO) standards [5]. If methods like the one presented in this paper can be applied to a portable EIT system, they will enable a fast and broad monitoring of the patients cardiovascular system at the bedside and in emergency situations.

2 Material and Methods

The data used for this paper was collected in 2016 during studies on EIT based ventilation monitoring after experimental thorax trauma at the University of Leipzig. The study was approved by the governmental animal ethics committee (Regierungspräsidium Leipzig, 24-9168.11/14/38). Besides many respiratory measurements also one apnea series for CT scans was recorded by EIT as well as one apnea series for perfusion measurements. Both series were done in end-expiratory hold. The ventilation was done with the Draeger Evita XL at 8 ml tidal volume per kg body weight and 25 $^1/$min respiration rate. For the perfusion measurements 10 ml of a 10% NaCl solution were injected intravenously according to the standardized PICCO method. We only used the baseline measurements before the thorax trauma was induced. The Dräger Pulmovista 500 was used for the EIT measurements, having 16 electrodes and recording at a frame rate of 50 fps [6]. The raw EIT data was reconstructed using the GREIT algorithm, as proposed by Adler et al. [7]. The resulting data sets are sized 64 x 64 pixel on a timeline between 270 and 3670 frames. Preprocessing was performed, including the separation of ventilated sequences into ventilation and perfusion signal by a cascaded bandpass filter. Furthermore, a watershed segmentation was conducted on the absolute perfusion signal, summed up over all frames. The maximum pixel in each segmented region of the heart and the right and left lung is determined and used as representative pixel for following analysis. A priori knowledge in form of CT segmentations, is used for validation as well as for restriction. A mean mask for trunk, thorax, heart and aorta is calculated by summing over the whole measurement series. In addition, a mean pixel for the aorta for every measurement is calculated.

Aortic Detection   Krukewitt et al presented a method for contrast agent free aortic detection at the EIT conference 2019 [1]. They used prior knowledge to find a threshold criteria for EIT pixels. The three criteria were the pulse arrival time, the amplitude and the timing of the impedance minimum. We developed a slightly different algorithm based on a fuzzy logic instead of binary thresholds. Our calculations are done on an average cycle computed for every pixel as follows: A band pass filter designed by a high pass (order 40, cutoff frequency 0.1 Hz) and a low pass filter (order 10, cutoff frequency 5 Hz), is applied to remove noise. We also applied an anatomical mask to restrict the calculations

on the thorax. A heart region is defined size 5×5 pixels around the representative heart pixel from the segmentation. The cardiac cycle is defined as the time between two consecutive maxima of the first heart pixel. Now, all cycles are summed up after interpolating them on the average cycle time, leading to an average cycle of unitary length for every pixel. In addition to the PAT, we introduced a new main criteria for the aortic region based on the special shape of the aortic pulse due to the Windkessel effect. We compared every pixel to an anatomical expectation from one EIT data set, seen in Fig. 1(a), and mapped the similarity. The most reliable output had a combination of squared sum and cross-correlation with the fit function (Fig. 1 (b)).

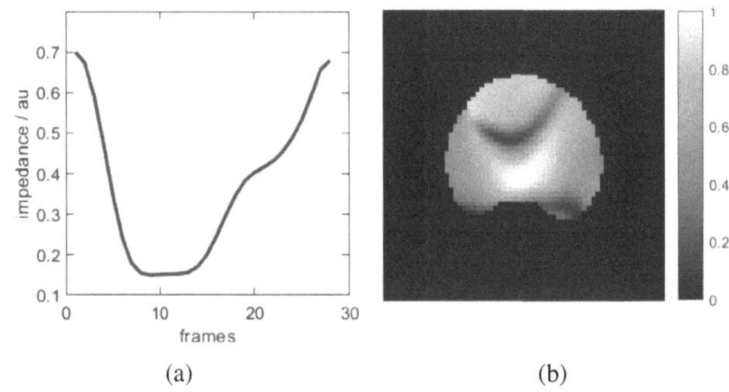

Figure 1: (a) Expected impedance curve of an aortic pixel used as a fit function for pixelwise comparison. (b) The normalized similarity to the fit function in range [0,1] for one EIT data set.

In iteration over all 25 heart pixels a fuzzy map (FMAP) for the location of the aorta is formed: As the current heartpixel reaches its minimum - at the maximal presence of blood - a reference point is set for the PATs; $t(minHeart) = 0$. The blood reaches the aorta very quickly, so that the smallest PAT values include the aortic pixels. An own FMAP is designed for every criterion, using a sigmoid function. We searched for the threshold separating the 20 % of the shortest PATs and defined it as the turning point of the sigmoid function with a gradient of -10. In this way, we excluded high PATs without setting a hard limit. Additionally by using 20 % of the data points - and not of the data values - outliers with very small or very huge PAT values do not impact the sigmoid function. After standardization we called it FMAP1. The second sigmoid function excludes all amplitudes < 70 % with a gradient of 10 and is applied on FMAP1. In the resulting FMAP2 margin tissue is excluded. The third criterion is also applied on FMAP1 and finds all pixels which reach their minimum value in the latest 10 % with a gradient of 10 resulting in FMAP3. This three criteria in relation to each other are also used by Krukewitt et al. [1]. The additional criterion of the similarity to the fit function is casted in FMAP0 by a sigmoid function of a gradient of 10 leaving the highest 10 % of the pixels. The four criteria are combined by pixelwise multiplication leading to the final fuzzy map as illustrated in Fig. 2. The resulting FMAPS of all 25 heart pixels are summed up and standardized. The

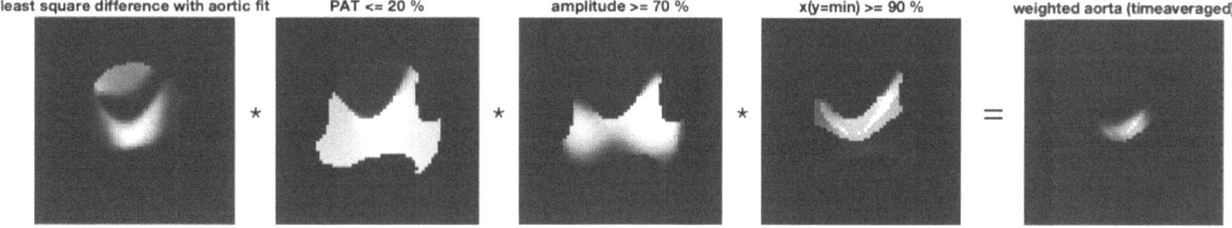

Figure 2: The four criteria for aortic detection (from left to right): similarity to physiological fit, pulse arrival time, amplitude, timing of the minimum. The pixelwise multiplication leads to the final fuzzy map.

final distribution is restricted in all directions by a mask designed from anatomical information and the maximum pixel of the region is returned as the middle aortic pixel.

Saline Bolus Aortic Detection   A popular validation method for the results of aortic detection algorithms is an analysis of the impedance image after a saline bolus injection as described by Wodack et al. [8]. This method also is consulted by Krukewitt et al.. A hypertonic saline bolus is injected into the ascending aorta. Conductivity rises sharply while the bolus passes a vessel and the impedance changes in general increase. The signal can be separated into perfusion and saline bolus impedance changes. Only the slow but wide change of the injection without the oscillating perfusion signal is used for the aortic detection. As mentioned before, our injection method differs from the data used by Wodack et al.. The saline bolus is injected into the central venous system according to PICCO standards and not into the aortic arch. Wodack et al. searched for a peak right after the injection, while the bolus in our case first passes the right ventricle, the lung and the left ventricle, before it reaches the aorta and finally the systemic circulation. The timing is less accurate and the intensity of the bolus already spread a little. For that reason a modified analysis method is needed. Because the bolus arrives in the aorta right after it starts leaving the left ventricle we focus on the left heart. In the impedance curve of a heart pixel, there are two local minima coming from the bolus passing the right and left ventricle. To find a distinct second peak, we moved our representative heart pixel from the segmentation 5 pixels down and right towards the main signal region of the left heart. If a peak can be found here, the x-value is chosen and a time frame of 50 frames is taken from that point. Examining video sequences we noticed, that the bolus passing the left lung significantly affects the impedance of pixels in the aortic region. Due to a fluent transition of impedance changes coming from the left lung and those from the aorta, an aortic peak can not be detected. Instead we searched for pixels with the smallest decrease in the defined time frame and still having a high amplitude. Anatomical bounds are applied, as with the bolus free aortic detection. The maximum pixel of the region is determined and returned. Because our comparison method is not validated, we additionally used aortic pixels from the CT segmentations, and transformed them depending on the electrode positions. Comparison is done by mean pixel distances.

3   Results and Discussion

The primary aim of this paper was to find an aortic detection algorithm with a broad applicability for generally noisy EIT data. The existing threshold algorithm [1] only had consistent results on our EIT images with adjusted thresholds for every data set. Besides few very noisy or incomplete measurements the new aortic detection algorithm worked on all data sets for apnea and ventilated sequences. Fig. 3 illustrates the generalizability of the fuzzy logic based algorithm, compared to binary thresholds. It is a demonstration and does not include the averaging process over different heart pixels for the binary detection (compare [1]), also the threshold values are individually chosen. On the left side, you can see the influence of the thresholds on the resulting fuzzy/binary map. The first row is an example for adjusted parameters c1 (1 % shortest PATs), c2 (30 % highest amplitudes) and c3 (10 % latest minima). The second row presents the parameters used for calculations on all data sets with the best global applicability. The third and fourth row

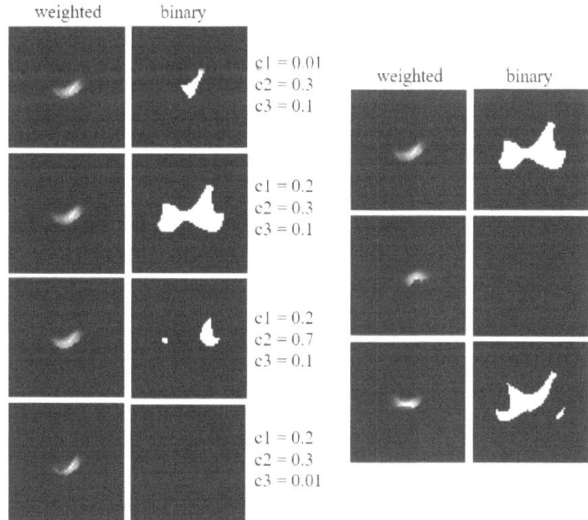

Figure 3: The influence of threshold parameters on the resulting fuzzy/binary maps (left). The first row shows adjusted parameters c1 (1 % shortest PATs), c2 (30 % highest amplitudes) and c3 (10 % latest minima). In the second row, the parameters used for calculations on all data sets are shown. The third and fourth row illustrate the effects by changing c2 and c3. On the right side, you can see the applicability on different data sets.

illustrate the effects by changing c2 and c3. On the right side, you can see the applicability on different data sets and the binary aortic region found with fixed thresholds. The impact of unadjusted parameters on the aortic region found by our detection algorithm are significantly smaller.

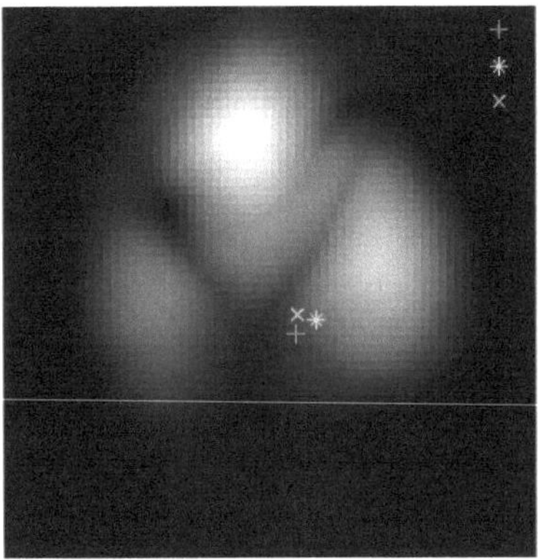

Figure 4: Results for the three algorithms showing the mean aortic pixels over all data sets. The EIT image again is represented by the absolute and summed up frames. The middle aortic pixel from the transformed CT segmentation, the saline bolus injection algorithm and the only EIT based algorithm are marked.

We evaluated the middle aortic pixels for the different methods and made a comparison by mean pixel distances. Over all measurements the mean value and standard deviation is 5.76 ± 2.27 pixels from the bolus free aortic detection to the segmented CT pixels. The apnea sequences had a better result with 4.65 ± 1.84 pixels. The ventilated sequences are at a distance of 6.73 ± 2.27 pixels. Comparing the saline bolus aortic detection to the segmented CT pixels results in a mean pixel distance of 4.26 ± 1.78. Between the aortic detection with and without saline bolus injection there is a mean distance of 4.53 ± 2.73 pixels. Fig. 4 connects the three results, showing their mean aortic pixels over all data sets. The EIT image again is represented by the absolute and summed up frames. The middle pixel from the transformed CT segmentation, the saline bolus injection algorithm and the only EIT based algorithm are marked. For a pig of trunk scope of 70 cm, which is the mean scope of all pigs of the study, the mean pixel distance of 4.53 ± 2.73 leads to an variability of approximately 1.68 ± 1.02 cm in horizontal direction. The mean pixel distance of Krukewitt et al. was 2.3 ± 1 compared to the bolus injection method by Wodack et al. ([1], [8]) which is a more precise result. However, we present a robust result, which we could not achieve by applying existing algorithms.

4 Conclusion

The algorithm was tested on data of fifteen pigs partly on apnea partly on ventilation and showed a robust result on almost all data. Further studies on existing data could test the generalizability and provide an assessment on the quality of the data used for this study. The new method pre-

sented for saline bolus injection based aortic detection also has shown consistently small distances to the CT pixels. As the PICCO standardized injection for advanced haemodynamic monitoring is a widely used method, studies are presumably conducted more frequent than studies including an injection into the aortic arch; this could give access to a bigger data pool for ongoing calculations, making big data based algorithms possible. Further studies could use the aortic position also for calculations on blood pressure and ventilation-perfusion-index.

Acknowledgement

The work has been carried out at Innovation Center Computer Assisted Surgery, Universität Leipzig and supervised by the Institute of Medical Informatics, Universität zu Lübeck.

5 References

[1] L. Krukewitt et al., *Entirely eit-based aortic detection.* Proceedings of the 20th International Conference on Biomedical Applications of Electrical Impedance Tomography, London, UK, p. 22, 2019.

[2] D. S. Holder, *Electrical Impedance Tomography - methods,history and applications.* Series in Medical Physics and Biomedical Engineering, vol. 29, 2015.

[3] I. Frerichs, *Chest electrical impedance tomography examination, data analysis, terminology, clinical use and recommendations: consensus statement of the translational eit development study group.* Thorax, vol. 72(1), pages 83-93, 2017.

[4] N. Westerhof, J. Lankhaar and B. E. Westerhof, *The arterial Windkessel.* Med Biol Eng Comput, vol. 47, pp. 131–141, 2009

[5] Pulsion Medical Systems SE, *PiCCO Technologie - Hämodynamisches Monitoring auf höchstem Niveau.* Available: https://www.getinge.com/siteassets/products-a-z/picco/de/picco-technology-brochure-de_r07-screen.pdf [last accessed on 2020-01-19], 2018.

[6] E. Teschner, M. Imhoff and S. Leonhardt, *Electrical Impedance Tomography: The realisation of regional ventilation monitoring.* Drägerwerk AG & Co. KGaA, vol. 2nd edition, 2015.

[7] A. Adler et al., *Greit: a unified approach to 2d linear eit reconstruction of lung images.* Physiological Measurement, vol. 30, pp. 35–55, 2009.

[8] K. H. Wodack et al., *Detection of thoracic vascular structures by electrical impedance tomography: a systematic assessment of prominence peak analysis of impedance changes.* Physiological Measurement, vol. 39, February 2018.

Model-based evaluation of magneto-mechanical oscillator signals for wireless sensing and localization

Justin Ackers [1], Bernhard Gleich [2], Jürgen Rahmer [2], Ingo Schmale [2] and Thorsten M. Buzug [3]

[1] Medical Engineering Science, Universität zu Lübeck, j.ackers@student.uni-luebeck.de
[2] Philips Research, Hamburg, Germany {bernhard.gleich, juergen.rahmer, ingo.schmale}@philips.com
[3] Institut für Medizintechnik, Universität zu Lübeck, buzug@imt.uni-luebeck.de

Abstract

Vital signs like temperature and blood pressure are essential parameters in medical diagnostics. To measure these quantities, a new, wireless, easily miniaturizable method based on a magneto-mechanical oscillator was proposed. Currently a simple Fourier approach is used to analyze the signal received from this oscillator and determine the frequency. However, this approach does not detect the natural frequency of the oscillation and shows a strong dependency on the excitation strength. This contribution describes a model-based method to fit the measured data to a physical model of the oscillator to overcome the shortcomings of the Fourier approach as well as to find additional parameters. It is shown that the model-based approach is able to compensate for varying excitation amplitudes in situations with large excitations. In addition the orientation of the oscillator is obtained, which is a first step towards full localization.

1   Introduction

The possibility to accurately and easily measure physiological parameters like body temperature and blood pressure has a large benefit for many medical procedures. The ideal sensor would need no wired connection and would be small enough to not disturb the region of interest.

While inductor-capacitor passive wireless sensors exist [1], they are challenging to build sufficiently small, because they suffer large losses in radiation efficiency when reducing the antenna size [2]. Therefore, a novel magnetic method was invented to overcome this limitation and make smaller sensors possible [3]. In this method a magneto-mechanical oscillator of around $1\,\mathrm{mm}$ in length is excited by an external pulse and the voltage induced by the free decay of the oscillation is recorded. A coil system is used for generation of excitation pulses and reception of the oscillator signals. The sensor is constructed in a way, that the natural frequency f_0 of this oscillation depends on the physical quantity (e.g. temperature or pressure) to be measured. Therefore, this frequency should be determined accurately.

Currently the signal of the decaying oscillation is analyzed using a Fourier approach, which takes a section of the signal and searches for a single peak in the frequency spectrum. Due to the non-linear nature of the mechanical process, the frequency changes during the decay of the oscillation. Therefore, the Fourier approach does not detect the natural frequency, but instead gives a result between the maximum and the natural frequency depending on the selected section and the maximum amplitude of the oscillation. The maximum amplitude depends strongly on the distance to the excitation coil, therefore the Fourier approach has an inherent distance dependency, which would have to be compensated. The approach proposed in this contribution addresses most of these issues by modeling both the oscillator and receive chain. The model parameters are determined by fitting the model simulation to the measured data. In addition to removing the dependency on the strength of the excitation, the proposed method promises to yield additional parameters, primarily the orientation of the oscillator in respect to the receive coil.

2   Material and Methods

The proposed model-based method consists of three main parts. First, a differential equation model of the oscillator is used to simulate a temporal sequence of rotational states for a set of physical parameters. After that a model of the receive chain is used to calculate a voltage which can then be compared to the actual measurement. As the third part, an optimization algorithm is used to find the optimal parameters for step 1 and 2 to produce the best fit to the measured signal.

2.1   Model of the oscillator

The structure of the magneto-mechanical oscillator is shown schematically in Fig. 1. It consists of two permanent magnet spheres inside a cylindrical casing acting as magnetic dipoles. One of these spheres is fixed to one side of the casing while the other is suspended on a string fixed to the other side.

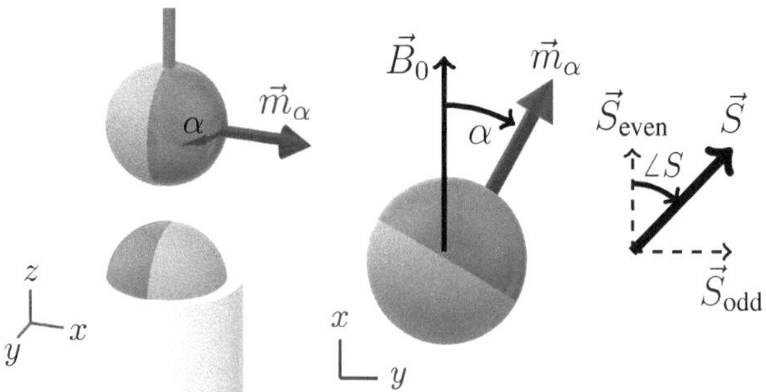

Figure 1: Structure of the magneto-mechanical oscillator. The upper sphere is rotated into the equilibrium position $\alpha = 0$ by the magnetic field $\vec{B}_0$ of the lower sphere. The sensitivity $\vec{S}$ of the receive coil can be split into even and odd parts, detecting different components of the oscillation.

The fixed sphere attracts the moving sphere, until the string retains it. The magnetic field of the fixed dipole applies a torque on the moving dipole to align it into an equilibrium position. Once the moving sphere is rotated out of this equilibrium, for example by a temporary external magnetic field, the restoring torque of the fixed sphere causes an oscillation with a natural frequency f_0 of a few kHz. This frequency depends on the strength of the torque and therefore in this case on the distance between the two spheres. The oscillation can be described by the following differential equation:

$$I_z \ddot{\alpha} = (\vec{m}_\alpha \times \vec{B}_0)_z + \dot{\alpha} D, \qquad (1)$$

where I_z is the moment of inertia of the moving sphere for the rotation axis, $\vec{m}_\alpha = (m \cos \alpha, m \sin \alpha, 0)^T$ is the magnetic dipole moment of the moving sphere rotated by the angle α and $\vec{B}_0$ is the magnetic field produced by the fixed sphere at the center of the moving sphere. Meanwhile $\dot{\alpha} D$ denotes a simple dampening term accounting for the quality factor of the oscillation.

Setting $\alpha = 0$ to the equilibrium and using the fact that $\vec{B}_0$ is parallel to the equilibrium orientation allows (1) to be simplified, resulting in:

$$I_z \ddot{\alpha} = -m \sin \alpha B_{0x} + \dot{\alpha} D. \qquad (2)$$

Using the small angle assumption $\sin \alpha = \alpha$ a rather simple analytic solution could be found. This would remove the non-linearity of the problem and would be equivalent to the Fourier approach with an added dampening term. Since the non-linearity should be included, the classic Runge-Kutta method was used to solve the original differential equation for different values of f_0 (and therefore B_0) and the initial angle α_{max} of the free oscillation at the end of the excitation. For this method a time-step of $10\,\mu s$ was used.

Quadratic dampening term

In addition to the friction dampening term $\dot{\alpha} D$, variable tension applied to the string adds a non-linear dampening mechanism to the free oscillation. This dampening effect is larger with bigger deflection angles.

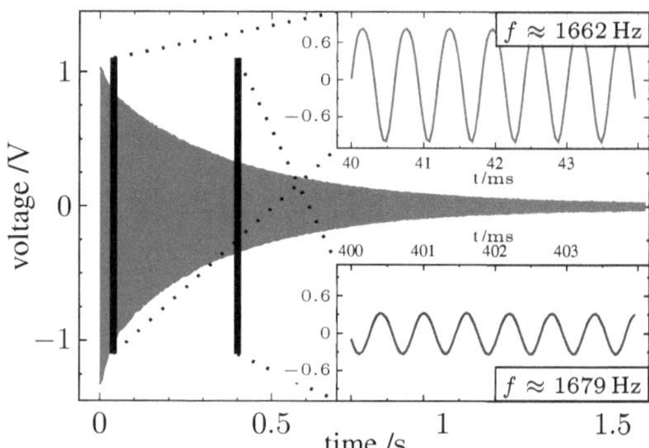

Figure 2: Typical signal received from the oscillator for $\alpha_{\text{max}} = 30°$. The frequency difference for different amplitudes results in wrong results for the Fourier approach.

To allow for an accurate modeling of the physical mechanisms the following quadratic dampening term was used:

$$I_z \ddot{\alpha} = -m \sin \alpha B_{0x} + \dot{\alpha}(D + D_q(\alpha^2)^\delta). \qquad (3)$$

The values for the parameters D, D_q and δ were chosen by fitting the shape of the simulated signal to a measured signal.

2.2 Model of the receive chain

To determine the voltage induced in the receive coil, the orientation of the oscillator in respect to the sensitivity of the receive coil at the location of the oscillator has to be considered.

If the detecting coil has a sensitivity component perpendicular to the equilibrium axis, the induced voltage includes base frequency and all odd harmonics. For a parallel sensitivity, the coil detects a signal that contains twice the base frequency and all even harmonics (see Fig. 1). For every other receive angle $\angle S$ the measured signal is a combination of these two components through the scalar product. The induced voltage can be written and simplified as:

$$U = \frac{\mathrm{d}}{\mathrm{d}t}(\vec{S} \cdot \vec{m}_\alpha) = S m \dot{\alpha} \sin(\angle S - \alpha). \qquad (4)$$

The receive angle $\angle S$ influences the spectral components of the signal by shifting the ratio of the even and odd harmonics of the base frequency. Therefore, this parameter has to be found as well. Since the higher harmonics have smaller amplitudes, in practice only the second harmonic and for larger deflections also the third harmonic contain relevant information.

Frequency response of the receive chain

The induced voltage is band-pass filtered before entering the ADC, in order to suppress various interferences. While attenuating the unwanted frequencies, these filters and the inductance of the receive coil also introduce different phase shifts and amplitude modulations to the different harmonics. To include this in the model, the transfer function of the receive chain was measured. To reproduce the frequency dependence of the acquisition system, the simulated signal was multiplied with the transfer function in the frequency domain.

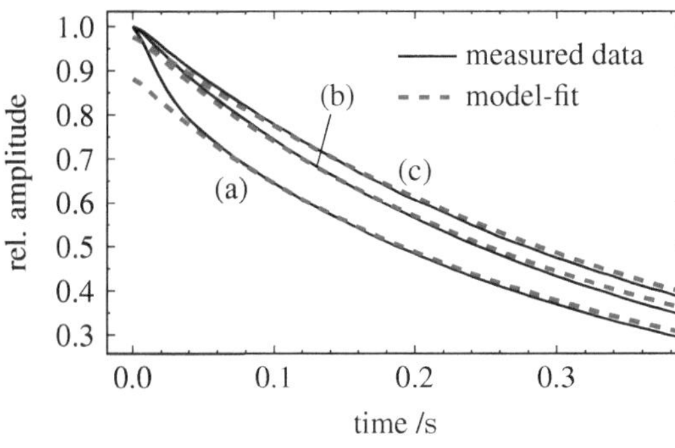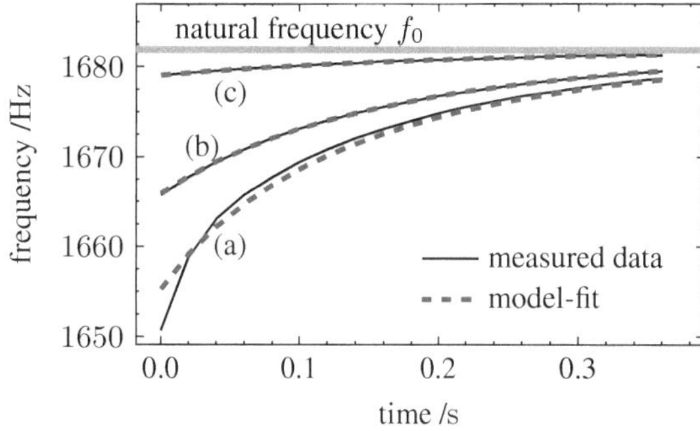

Figure 3: Change of amplitude and frequency of the oscillation over time for different excitation amplitudes of (a) $30°$, (b) $23°$ and (c) $10°$. The amplitude traces were scaled to their respective maximum amplitude. The model based method is generally able to reproduce well both amplitude and frequency traces, with increasing accuracy for lower amplitudes.

2.3 Fitting the model to the measurement

Fig. 2 shows a typical oscillator signal recorded by the receive system. It can be seen that the oscillation has a high quality factor. The slight asymmetry is caused by the phase shift between the first and second harmonic, which is only significant for larger deflection angles.

Using the model described in the previous sections it is possible to generate a time trace, which is approximately what an actual measurement of the oscillator would produce. The variable parameters are the natural frequency of the oscillation f_0, the maximum deflection angle α_{max} of the oscillation and the angle $\angle S$ under which the oscillator is oriented to the sensitivity of the receive coil.

An iterative Nelder-Mead algorithm [4] on the squared error is used to find the optimal parameters. For one specific oscillator the variation of f_0 due to changes of the measured quantity is small compared to the variation between different oscillators. Therefore, the initial searching point can be adjusted to be in the correct range for each oscillator.

The periodic nature of the receive angle and the fact that a deflection angle above a certain value would not allow a stable oscillation were not taken into account, because the local search never reached these regions, when given a reasonable starting point. For a future approach, this might be taken into account to find a more efficient minimization method. Since the starting phase of the oscillation and the magnitude of the sensitivity S are independent of the other parameters, they were not included into the main optimization, but instead added as an intermediate step.

2.4 Experiments

To evaluate the model-based approach two experiments were performed, one investigating the influence of distance, the other investigating the influence of orientation. All test data has been recorded using a single-channel system using separate excitation and receive coils and a temperature dependent oscillator. The signal was sampled with a temporal resolution of $40\,\mu s$.

For all measurements the temperature of the oscillator and therefore the frequency can be assumed as constant. For both experiments, an excitation of three pulses per second was used and for each excitation a voltage trace of 4096

samples was put into the Fourier evaluation as well as the model-based approach.

For the first experiment the oscillator was put in a fixed position and orientation and 60 excitations were recorded over $20\,s$. This was done for three positions in a distance of $5.5\,cm$, $7\,cm$ and $10\,cm$ to the send coil, which produced different maximum excitations α_{max}.

As the second experiment, the oscillator was fixed on a stand that allowed rotation without changing the distance to the coil setup. The initial angle was $125°$. During the measurement time of $60\,s$ the oscillator was rotated by hand 1.5 times counterclockwise with nearly constant velocity.

3 Results and Discussion

In Fig. 3 it can be seen, that the differential equation model with the quadratic dampening term is able to recreate the frequency and amplitude trace of actual measurement data. For larger angles, the model becomes less accurate, because additional dampening and oscillation mechanism become relevant. However the frequency traces are more accurate, since the dampening model was optimized to recreate accurate frequency traces.

In the next two sections the performance of the optimization algorithm to determine the frequency f_0 and detection angle $\angle S$ is evaluated.

3.1 Frequency

The results of the frequency test are shown in Fig. 4. It can be seen, that the model-based approach produces plausible results. Generally, it results in a higher frequency, since the actual natural frequency is detected. Over all three distances the model-based method gives a frequency of around $(1683.10 \pm 0.23)\,Hz$, while the Fourier method results in a range of $4.8\,Hz$. This shows that the distance dependency is decreased significantly. In (c) it can be seen, that the Fourier approach produces oscillating results, because at this distance the maximum excitation angle is also oscillating. This happens since the oscillator is not yet at rest, once the next excitation pulse occurs. The model-based method can compensate for these oscillations, reducing them by a factor of 7.2, but cannot remove them completely.

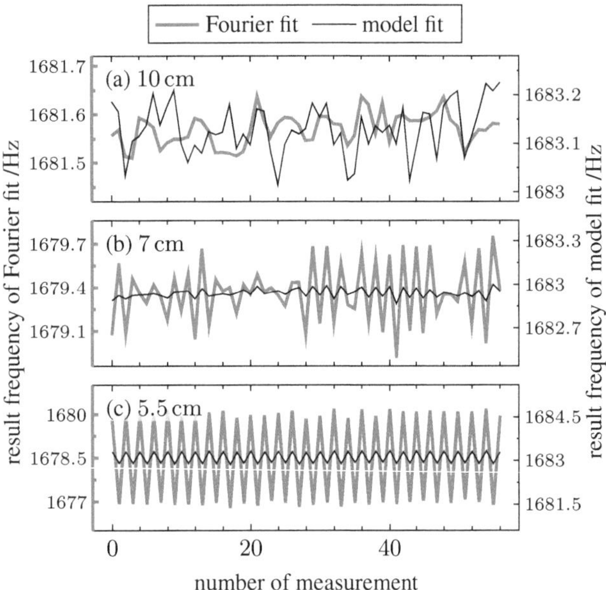

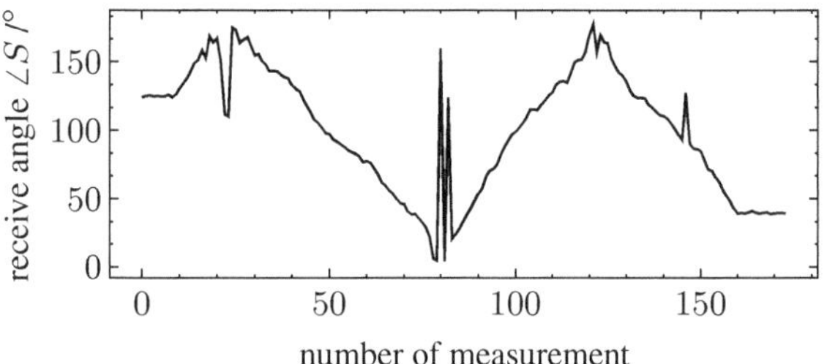

Figure 5: Results for $\angle S$ from the model-based method for the orientation experiment. The 1.5 counterclockwise rotations can be clearly seen.

Figure 4: Detected frequency for the Fourier approach and the model-based approach compared for different α_{max} of (a) 9°, (b) 13° and (c) 18°. For larger excitation amplitudes, the model-based approach to some extent compensates for the varying excitation and produces smaller frequency variations. For smaller angles the variation disappears and both approaches produce similar results.

With a more complex model and better identification of the dampening processes, the remaining dependency should be removable in a future approach.

For the larger distance to the send coil and therefore smaller excitation angles the Fourier fit and the model-based method produce similar results, with slightly larger deviations with the model fit. This is likely due to the fact, that the model-based method detects multiple variables and without additional regularization there is a trade-off between frequency performance and the other parameters.

It was difficult to find a good dampening model for different oscillators. The dampening model is essential to produce the correct frequency result for larger excitations, so this should be the first approach to improve the results.

3.2 Orientation

The trace for $\angle S$ detected by the model-based method during the rotation experiment is shown in Fig. 5. The slow counterclockwise rotation with start and end performed during the measurement is clearly visible. For the orientations parallel to the equilibrium (0° and 180°) there is no excitation and therefore no signal. Since we do not know the absolute phase of the oscillation, we cannot determine the sign of $\angle S$. Because the optimization starts with a positive angle, it return angles from 0° to 180° and an angle of $-160°$ is detected as 160°. The detected angle produces plausible results and varies less than 2° for a fixed position.

4 Conclusion

It has been shown, that the model-based approach produces results consistent with the Fourier approach. In addition, it is able to compensate for varying excitation amplitudes and is producing a frequency measurement with no distance dependency. Furthermore it has been shown, that the detection of the orientation is possible. With the use of multiple coils this may make the full localization possible in the future.

In future work it might be possible to improve the model by including additional physical effects and finding a better dampening term. When using a better model, the fit will also improve. In addition, including prior knowledge about the problem may help with regularization.

To conclude, the model-based approach proved to be a viable tool to combine with the simpler and faster Fourier method for quick and accurate frequency detection as well as finding additional parameters.

Acknowledgment

The work has been carried out at Philips Research, Hamburg and supervised by the Institute of Medical Engineering, Universität zu Lübeck.

5 References

[1] A.D. Dehennis and K.D. Wise, *A wireless microsystem for the remote sensing of pressure, temperature, and relative humidity*. Journal of Microelectromechanical Systems, 14:1, pp. 12–22, 2005.

[2] Q.A. Huang, L. Dong and L.F. Wang, *LC passive wireless sensors toward a wireless sensing platform: status, prospects, and challenges*. Journal of Microelectromechanical Systems, 25:5, pp. 822–841, 2016.

[3] B. Gleich and J. Rahmer, *Magnetic measurement device*. European patent EP3583890A2, 2019.

[4] F. Gao and L. Han, *Implementing the Nelder-Mead simplex algorithm with adaptive parameters*. Computational Optimization and Applications, 51:1, pp. 259–277, 2012.

Convolutional and recurrent neural networks for surface electromyography-based respiratory diagnostics and therapy

Nele Sophie Brügge [1], Marcus Eger [2], Thomas Handzsuj [2], Stephan Walterspacher [3,4] and Philipp Rostalski [5]

[1] Medical Engineering Science, Universität zu Lübeck, nele.bruegge@student.uni-luebeck.de
[2] Drägerwerk AG & Co. KGaA, marcus.eger, thomas.handzsuj@draeger.com
[3] II. Medizinische Klinik – Klinik für Pneumologie, Kardiologie und internistische Intensivmedizin, Klinikum Konstanz
[4] Lehrstuhl für Pneumologie – Lungenklinik Köln-Mehrheim, Fakultät für Gesundheit, Universität Witten/Herdecke
[5] Institute for Electrical Engineering in Medicine, Universität zu Lübeck, philipp.rostalski@uni-luebeck.de

Abstract

Patients unable to maintain adequate spontaneous breathing are ventilated mechanically. With artificial ventilation, additional physical work is provided when detecting a patient's respiratory effort. In order to prevent patient-ventilator asynchronies, recent studies have shown that it is beneficial to monitor the patient's respiratory activity using a NAVA catheter [1] instead of solely employing a conventional flow trigger. A noninvasive alternative is the electromyogram measured by skin surface electrodes (sEMG). In this work, a convolutional (CNN) and a recurrent neural network (RNN) model are proposed to predict active respiratory phases from sEMG signals for the purpose of ventilator triggering and cycling off. With a mean-squared-error (MSE) of 0.073 for this binary problem, the CNN achieves superior results compared to the RNN. Nevertheless, the RNN is advantageous regarding its inherent real-time capability and its ability to incorporate information from previous time steps, a feature which is to be exploited in future work.

1   Introduction

Artificial ventilation is used to support or replace insufficient or non-existent spontaneous breathing. Its life-sustaining function is a central component in anesthesia, emergency medicine and intensive care. A distinction is made between controlled and assisted ventilation. The latter pressure-assisted ventilation, which is the subject of this work, is intended to support the patient during spontaneous breathing. Therefore, the ventilator must detect the patient's inhalation and exhalation efforts using inspiration and expiration triggers such as the initial air flow which is generated by the patient. However, it can happen that despite great respiratory effort, breathing attempts are not detected for various reasons. These so-called missed efforts indicate poor patient ventilator interaction, which is a common problem with conventional assisted ventilation and can contribute to patient discomfort, higher use of sedation, ventilator-induced lung injury, prolonged mechanical ventilation, and ultimately lethality [1]. Furthermore, systems often react with a noticeable delay. To address this patient-ventilator asynchrony, in recent studies the electrical activity of the diaphragm (EA_{di}) is recorded with a nasogastric catheter [1]. This is intended to allow an estimation of the respiratory drive and the control of the ventilator. First studies have further proven an improved synchronisation between patient and ventilator as well as an efficient relief of the patient's respiratory muscles. The main disadvantages of the mentioned method are its invasiveness, the

therewith associated discomfort of the patient and the often challenging positioning of the catheter. For these reasons, it is desirable to exploit the information gathered from surface electromyography (sEMG) as a non-invasive alternative signal source for the control of a respirator. The sEMG is measured on the patient's thorax. When working with sEMG data it is particularly challenging that the signal to noise ratio is small, the signals are corrupted by electrocardiography (ECG) artefacts and subject to crosstalk from muscle activity not related to respiration. In addition, a correct placement of the electrodes is of high importance.

In this work, the problem of predicting the start and the end of inspiration efforts from raw, baseline filtered as well as ECG removed (ECGr) [2] sEMG data is addressed. The muscular pressure signal is used to create a reference signal. This signal quantifies the muscular effort of the patient and is calculated on the basis of the pneumatic data of the ventilator as well as the esophageal pressure measured by a balloon catheter. During the inspiratory phase, the muscular pressure (P_{mus}) values therefore are significantly increased, so that an adaptive thresholding can be applied to obtain a binary signal assessing the sequence of active and passive respiratory phases. For the purpose of predicting the mentioned binary signal, two different neural network architectures are proposed, namely a Convolutional Neural Network (CNN)- and a Recurrent Neural Network (RNN)-based approach comprising a Long Short-Term Memory (LSTM) cell.

2   Material and Methods

2.1   Data

The networks were trained using data recorded at Klinikum Konstanz (Germany), which stems from a total of 41 patients. For each patient several data recordings are available, corresponding to approximately 20 minutes of data. The recordings in turn are divided into different pressure support phases. For training the networks, only those phases which show intrinsic muscle activity in the sEMG signals are used. Each data set includes two raw sEMG channels, pneumatic ventilation data and the esophageal pressure signal which is required to calculate the reference signal. These signals are sampled at 1000 Hz and 100 Hz respectively. For sEMG data acquisition two electrodes are positioned bilaterally at the costal margin on the midclavicular line and two further electrodes are placed bilaterally in the second intercostal space. Exemplary signal segments are depicted in the upper left corner of Fig. 1.

Ideally, a reference signal should be used for the training, which indicates at any time whether breathing effort is present or not. Since the data is not annotated to this extent, the independently calculated P_{mus} signal [3] is used given the fact that this signal is increased during inspiration. Therefore, a target can be created by applying an adaptive thresholding to the signal according to Otsu's method [4], so that the resulting signal values become 1 during the active inhalation phase and 0 otherwise. To close gaps in the segmentation and to remove small false-positive outliers caused by local noise, a closing and a subsequent opening with respective kernel sizes of 30 and 6 samples was performed. This is exemplarily shown in the upper right corner of Fig. 1 where the light gray signal corresponds to the original P_{mus} and the black signal to the thresholded P_{mus}.

2.2   Proposed Network Architectures

With a convolutional and a recurrent neural network approach, two network architectures are proposed for the supervised learning classification task.

2.2.1   Convolutional Neural Network Model

The first mentioned model is a deep CNN containing 13 layers. It is illustrated in Fig. 1. Its structural characteristics are based on those of U-Nets [5] since it consists of a downsampling block followed by an upsampling block in which the equal-sized features of each block are linked by skip connections. In order to obtain a large receptive field and with regard to the high sampling rate, the kernel sizes of the first convolutional layers were set to a large value of 15. Downsampling to the factor of 64 was done by six strided convolution layers. Subsequent upsampling of the semantic features was achieved by simple linear upsampling operations each followed by a convolutional layer with a kernel size of 7. This leads to significantly better results in terms of accuracy than transposed convolutions according to recent studies [6] as a result of preventing checkerboard arte-

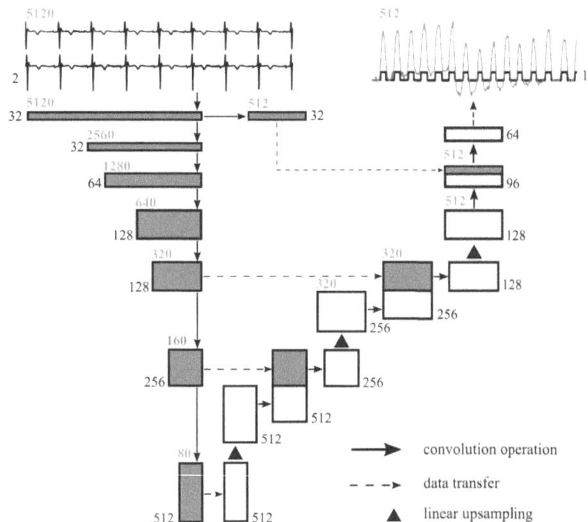

Figure 1: CNN architecture which is reminiscent of the U-Net from [5] due to the shortcuts from the downsampling to the upsampling block. Rectangles represent network features. Bold grey numbers above these rectangles indicate the size of the features, whereas the adjacent black numbers specify the channel size.

facts. With a value of 6.4, the upsampling factor is set to a smaller value compared to the downsampling factor due to the ten times lower sampling rate of the target signal generated from the P_{mus} signal. After the first and the second upsampling layer, the corresponding features are concatenated with the features of the same size from the downsampling block. After the concatenation of features from different levels of abstraction, a convolution layer is used to restore the original channel size. A final sample-wise classifier is used for the segmentation of the active respiratory phases on the basis of the calculated features. It consists of a further convolution layer with two output channels for the respective classes. With the exception of this last layer, batch normalization [7] is employed after convolution operations for faster training convergence and enhanced generalization and Rectified Linear Units (ReLUs) are used as nonlinear activation functions.

2.2.2   Recurrent Neural Network Model

Another proposed model is an LSTM-based framework [8]. This is an intuitive approach, since the signals are time sequences and LSTMs take into account the temporal dependencies of the signal. This is done by additionally considering the latest output data as input and specifying an internal state, which is propagated through all time steps. The internal state and the gates also ensure the prevention of so-called vanishing gradients, leading to an improved training process and enabling a long-term memory, which is where the name of the cell is derived from. In this work, a single LSTM cell with a state size of 128 is used as elementary network. As shown in Fig. 2, it is followed by a fully-connected layer with two output neurons. To create a class label the argmax is formed over this output.

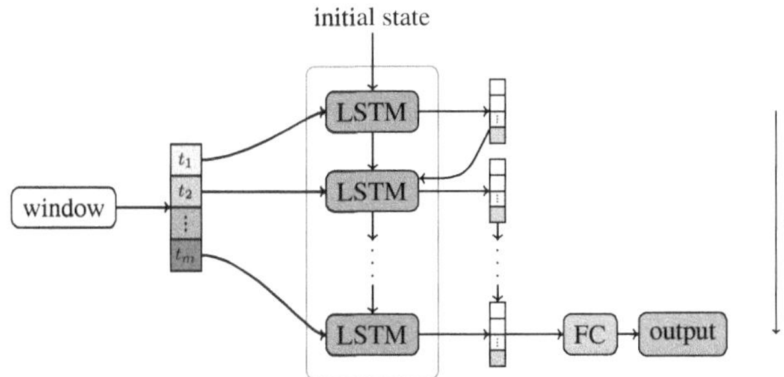

Figure 2: LSTM-based network architecture. One input time window is divided into m timesteps $t_1, t_2, ..., t_m$.

2.3 Network Training

During network training 26 out of 41 suitable recordings from 41 different patients were involved while the remaining were used for validation. To ensure the optimum comparability of the performances the training and test data sets of both networks were chosen identically. A widely used standardization procedure was applied for pre-processing the sEMG input data, which includes a channel-by-channel normalization of the signals in order to ease the training process. The mean value was subtracted and the signal was divided by its standard deviation to obtain mean-free data with unit standard deviation. As a classification task is given, a cross entropy loss is used both for the CNN and the RNN model. Learning rates were empirically determined.

2.3.1 Convolutional Neural Network

For training the deep CNN, overlapping and fixed-length raw signal patches of 5.12 s were cropped from the signals as network input and the corresponding thresholded P_{mus} signal patches as classification labels. These patches were randomly assembled into batches of size 32. The only pre-processing step applied to this data is a simple baseline filtering, which can be realized using a high pass filter. The network was trained for 10 epochs using Adam Optimization with an initial learning rate of 0.001. Gradient Clipping led to superior results and the possibility to use higher learning rates, hence its application in training. Since the network can only process signal components of the specified size, successive outputs, which are also fixed in size, are concatenated in test. This enables generating a complete signal of arbitrary length. It is thus a prerequisite that the signal parts do not overlap in this test case. For a signal part with a length of 60 s a processing time of approximately 240 ms on an INTEL Core i5-8265U CPU and 8 ms on an NVIDIA GeForce GPX 1080 GPU is required during inference.

2.3.2 Recurrent Neural Network

The same as the CNN, the RNN receives batches of randomly merged signal patches as input. A signal part length of only 1 s is chosen, since it determines the memory of the RNN. The learning rate is set to a value of 0.0004. Unlike

in the CNN, it is necessary that from the data the ECG has been removed to the most possible extent. Another difference is that not the entire signal is used as input at once, but each sampling point is passed to the RNN successively, as illustrated in Fig. 2. This is one reason why this network is capable of online use. Only the very last output value of the RNN is considered the final output and compared to the label via the loss function. In contrast to the CNN, signals of different lengths can be processed directly without any concatenation of successive output patches. This is made possible by the use of parameter sharing over two separate networks for training and testing. The sequential application of the network to the successive time steps eliminates the possibility of parallelisation as with the CNN. The inference time of a 60 s patch is nevertheless only about 3 s on an INTEL Core i5-8265U CPU.

3 Results and Discussion

Table 1 summarizes the performances of the two different nets on ECGr and raw data (RAW). For evaluation purposes, the mean squared error (MSE), true positive (TP) and true negative (TN) rates were calculated and listed respectively in Table 1 and Table 2. The latter are intended to serve as a measure of the percentage of activity correctly classified as activity and vice versa.

Table 1: Overall test results obtained with both network architectures on the hold-out set.

Net	Signals	MSE	TP	TN
CNN	ECGr	0.073 ± 0.030	95.1 %	88.9 %
	Raw	0.093 ± 0.039	86.8 %	92.9 %
RNN	ECGr	0.123 ± 0.068	67.3 %	99.2 %

Table 2: Test results obtained with the CNN architecture for 5 arbitrary patients on the hold-out set. Raw and ECG removed data was used and listed respectively.

Data	Patient	MSE	TP	TN
ECGr	15	0.055	91.0 %	96.8 %
	17	0.055	92.0 %	96.2 %
	24	0.060	88.1 %	97.3 %
	34	0.047	96.4 %	94.9 %
	41	0.053	94.6 %	94.8 %
RAW	15	0.057	93.4 %	94.9 %
	17	0.062	95.4 %	93.0 %
	24	0.081	86.8 %	94.8 %
	34	0.084	74.6 %	99.1 %
	41	0.057	95.3 %	93.7 %

The results clearly show that it is possible to detect active breathing phases from the sEMG data using a neural network approach. Since the reference does not necessarily segment these phases exactly, the MSE is rather a rough

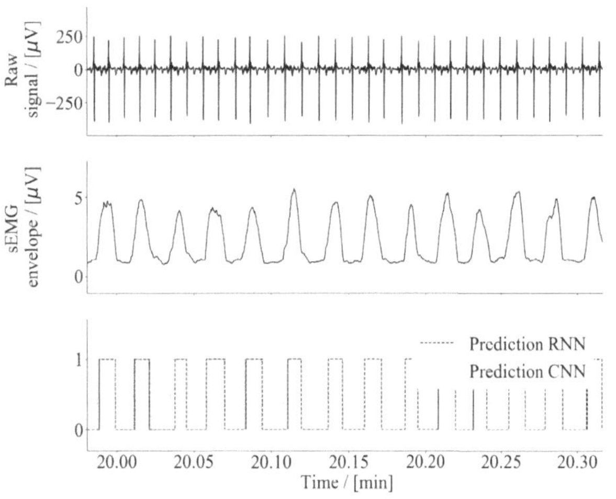

Figure 3: Raw data, sEMG envelope and an exemplary result on a signal part of the data from patient 34 for the RNN and the CNN. In order to get a continuous result of the CNN, which was trained on raw data, four output patches of approximately 5 s have been concatenated.

guide to the performance of the network and will not be strictly consistent with the predictions. Nevertheless, it is clearly visible and not surprising that the results obtained with the CNN are better if ECGr is used instead of raw data. As seen in the results of patient 34 (Table 2), for this patient the model performs considerably worse on raw data despite the fact that there is very strong activity in the data. It is conceivable that this is caused by the fact that the ECG of different patients can vary greatly and some patients' ECG may also show anomalies.

In terms of accuracy, the CNN is clearly superior to the RNN, as the latter misclassifies active inspiration phases more often. A major advantage of the RNN is that it is capable of operating in real time. Therefore, it would be feasible to use it as a ventilator trigger. On the contrary, the CNN can only process patches of fixed length and hence always implies a certain time delay. However, the RNN cannot be applied to raw data because of the dominant ECG spikes.

Fig. 3 depicts the results for an exemplary signal part. In the top panel a raw input signals is shown, revealing strong cardiac artefacts. By gating out these artefacts and applying a root-mean-square filter the sEMG envelope signal, displayed in the second panel, is obtained, which roughly indicates the relative muscle excitation. The last graph shows the binary trigger signals that were obtained with both network architectures. Compared to the CNN, which in contrast is provided with information from future and past time steps, the RNN model reacts with a slight delay, but still matches the slopes well in most cases.

4   Conclusion

This paper investigates two different neural network architectures and presents their advantages and disadvantages in predicting the time period of muscle activity during respiration from raw and ECGr sEMG signals. With an MSE of 0.0734 the CNN based approach outperformed the LSTM based approach (MSE of 0.123), even with use of raw data, but in contrast has only limited real-time capability. The time delay when using the RNN is negligible, which would make the model suitable as a potential trigger algorithm. Furthermore, it is very efficient in both training and interference due to the comparatively small number of parameters. However, a crucial shortcoming of this model is that it can only be used on the ECG removed data what limits its responsiveness in practical use. Thus, in further work the advantages of both models should be combined to build a hybrid framework for predicting inspiratory periods.

Acknowledgements

The work has been carried out at Drägerwerk AG & Co. KGaA in Lübeck and was supervised by the Institute for Electrical Engineering in Medicine, Universität zu Lübeck. We would like to thank Jan Grasshoff for his thoughtful comments and improvements to the paper.

5   References

[1] O. Moerer, J. Bargwing and M. Quintel, *Neurally adjusted ventilatory assist (NAVA)*. In: Anaesthesist, vol. 57, pp. 998–1005, 2008.

[2] L. A. van Eykern, *Apparatus for Detecting the Activity of the Respiratory Organs and the Heart of a Living Being*. United States patent US 4248240, Feb. 3, 1981.

[3] J. Grasshoff, E. Petersen, T. Becher and P. Rostalski, *Automatic Estimation of Respiratory Effort using Esophageal Pressure*. In: 41st Annual International Conference of the IEEE Engineering in Medicine and Biology Society (EMBC), pp. 4646–4649, 2019.

[4] N. Otsu. *A threshold selection method from gray level histograms*. In: IEEE Transactions on Systems, Man, and Cybernetics, vol. 9, no. 1, pp. 62–66, 1979.

[5] O. Ronneberger, P. Fischer and T. Brox, *U-Net: Convolutional Networks for Biomedical Image Segmentation*. In: Navab N., Hornegger J., Wells W., Frangi A. (eds) Medical Image Computing and Computer-Assisted Intervention – MICCAI 2015. MICCAI 2015.

[6] A. Odena, V. Dumoulin and C. Olah, *Deconvolution and Checkerboard Artifacts*. In: Distill vol. 1, 2016.

[7] S. Ioffe and C. Szegedy, *Batch normalization: Accelerating deep network training by reducing internal covariate shift*. In: Proceedings of the 32nd International Conference on Machine Learning 2015, vol. 37, pp. 448–456, 2015.

[8] S. Hochreiter and J. Schmidhuber, *Long Short-Term Memory*. In: Neural Computation, vol. 9, no. 8, pp. 1735–1780, 1997.